Construction Site Erosion and Sediment Controls

Planning, Design and Performance

Robert E. Pitt, Ph.D., P.E.
Shirley E. Clark, Ph.D., P.E.
Donald W. Lake, P.E., CPESC, CPSWQ

DEStech Publications, Inc.

HOW TO ORDER THIS BOOK

BY PHONE: 717-290-1660, 9AM–5PM Eastern Time

BY FAX: 717-509-6100

BY MAIL: Order Department
DEStech Publications, Inc.
439 North Duke Street
Lancaster, PA 17602, U.S.A.

BY CREDIT CARD: American Express, VISA, MasterCard

BY WWW SITE: http://www.destechpub.com

To the many past and future erosion and sediment control specialists, and to our families:

REP: Kathryn
SC: My parents
DL: Patricia

Construction Site Erosion and Sediment Controls

DEStech Publications, Inc.
439 North Duke Street
Lancaster, Pennsylvania 17602 U.S.A.

Printed in the United States of America
10 9 8 7 6 5 4 3

Main entry under title:
Construction Site Erosion and Sediment Controls: Planning, Design, and Performance

A DEStech Publications book
Bibliography: p.
Includes index p. 377

ISBN No. 978-1-932078-38-1

Table of Contents

Foreword

Construction Site Erosion and Sediment Controls: Planning, Design, and Performance, has been a work in progress for many years. Portions of the text have been used during advanced undergraduate and graduate classes on construction site erosion and sediment control at the University of Alabama at Birmingham, the University of Alabama, and at Penn State, Harrisburg, and at many short courses and workshops throughout the country. The final development of the text was made possible through partial support from the University Transportation Center of Alabama, and our respective departments. This support and the many helpful comments from students and colleagues are greatly acknowledged.

The purpose of this book is to supplement, certainly not replace, the numerous state, regional, and local guidelines available on the use of construction site erosion and sediment controls. The book provides background on regulations and problems caused by construction site runoff in Chapter 1. The U.S. Environmental Protection Agency's National Pollutant Discharge Elimination System (NPDES) stormwater regulations pertaining to construction site runoff are described in detail, stressing Phase II of this program, which affects all urbanized areas of the country and most construction sites. Basic approaches to the control of construction site erosion are outlined. Much of this material is summarized from official *Federal Register* announcements and relevant EPA publications. Chapter 2 goes into more detail on the selection of control practices and important site planning issues. The basic tenets of prevention (erosion control) are stressed, while supplemental treatment technologies (sediment control) are also addressed. Different checklists of needed components of an effective erosion control plan are presented, along with discussions of minimum standards for several states. Summaries of major components of most states' guidelines, along with Internet links to their erosion control manuals, are also given. The costs of construction site erosion and sediment controls are also presented in Chapter 2.

Local site conditions are the foundation of the factors affecting the selection, design, and performance of construction site erosion and sediment controls. Chapter 3 discusses the most important site factors: rainfall and resulting site hydrology conditions. Examples are given describing the appropriate selection of critical rains for the design of controls, considering varying levels of risk. Natural Resource Conservation Service (NRCS) site hydrology methods are described in detail, including many examples using the newest version of WinTR-55 to calculate the multiple variables of construction site hydrology needed for different steps in construction site erosion control planning and sediment device design. Chapter 4 presents detailed descriptions for the use of the Revised Universal Soil Loss Equation (RUSLE) to predict the benefits of the many erosion control options based on local soil and vegetation conditions. Example design guidelines for different vegetative controls are also given, and the new RUSLE2 modeling tools are described.

Quantifiable construction site erosion controls are discussed in Chapter 5, which covers channel and slope designs for stability. Much of the basic information developed by the US Army Corps of Engineers and the U.S. Department of Agriculture on these topics is summarized and modified to be applicable to the scale of construction sites in this chapter. Traditional, along with emerging, approaches for channel and slope stability are discussed, including summaries for the selection of appropriate grasses for different areas of the country, and the selection of turf mats and other flexible reinforcing materials that are used along with vegetation. Descriptions for a selection of these materials are provided, along with many examples on how channels and slopes can be designed in conjunction with these materials.

Chapter 6 analyzes the use of temporary sediment ponds and filter fences, and related sediment control devices, at construction sites. Pond design fundamentals are reviewed and many examples are provided showing how ponds can be sized for specific performance objectives. Emerging technologies, such as chemical-assisted settling, are also described in case studies. Filter fence design and performance expectations are presented.

Chapter 7 is a compilation of historical and new sources of information pertaining to the growing field of construction site erosion and sediment control.

This book is unique in that it covers the basic background

and history of problems associated with construction site runoff and the theoretical and practical aspects of control, stressing an engineering approach for the design of many types of planning and control options. Specifically, this book shows through many examples and case studies, how the different aspects of an effective construction site erosion and sediment control plan can be developed with cooperative elements that consider important site conditions and restraints.

Early versions of this book have been used as a text with students. All of the subject matter in these pages can be covered in a typical 3-unit course devoted to understanding and preventing site erosion. The authors typically assign a major project for the students. During the first week of class, the students are required to select a local and convenient construction site, which they must visit often during the term. (When the course is offered during an accelerated term, the use of the site-visit project can suffer, unless a nearby construction area undergoing rapid changes can be found.) A longer 12-to-14 week term usually works best, but a shorter term is certainly feasible, especially in communities experiencing growth. The students are to introduce themselves to the site foreman and describe the class project. They take care not to trespass or endanger themselves or the workers during the project. In addition, they are requested to ask for a copy of the site construction erosion and sediment control plan. The site contractor usually makes a copy for the students, or the students may need to obtain a copy from the local regulatory authority. The students are instructed to visit the site about once a week during the term and keep a diary describing the site conditions (construction phase, work activity, site-control status, needed or recent maintenance, and any erosion problems observed). They also try to visit the site soon after any major rains. Near the end of the term, the students are required to prepare an erosion control plan for the site. The report should reflect several different construction phases. The students make oral presentations and submit reports on their sites at the mid-term and during the final. This project has been an important part of the course and offers students the chance to work on real-world problems. Often, students in classes taught by the authors have been employed by local site-development firms. In such instances, they used one of their own company's sites for the class project, which adds a greater perspective to the assignment and benefits the whole class.

Introduction to Erosion and Sediment Control, Problems and Regulations

PROBLEMS ASSOCIATED WITH CONSTRUCTION SITE EROSION AND SEDIMENT LOSS

Observed Erosion Rates from Construction Sites

PROBLEMS associated with construction site runoff have been known for many years. More than 25 years ago, Willett (1980; Virginia 1980) estimated that approximately five billion tons of sediment reached U.S. surface waters annually, of which 30 percent was generated by natural processes and 70 percent by human activities. Half of this 70 percent is attributed to eroding croplands. Although urban construction accounted for only ten percent of this total, this amount equaled the combined contributions of forestry, mining, industrial, and commercial activities. While construction occurred on only about 0.007 percent of U.S. land in the 1970s, it accounted for approximately ten percent of the sediment load to all U.S. surface waters (Willett 1980), and the vast majority of the sediment load to urban streams. Increased development in many areas of the U.S. in recent years has only served to increase the need for construction site erosion controls. Developed in response to the increased awareness of these problems and to the public's demand that they be reduced, the EPA's Stormwater Permit Program includes regulations for the control of construction site erosion discharges. This chapter summarizes these emerging regulations and includes an appendix describing example regulation specifications for many areas in the country.

Construction accounts for a much greater proportion of the sediment load in urban areas than it does in the nation as a whole. Urban areas experience large sediment loads from construction site erosion because construction sites have extremely high erosion rates and because urban construction sites are efficiently drained by stormwater drainage systems installed early during the construction activities. Construction sites have measured erosion rates of approximately 20 to 200 tons per acre per year, a rate that is about 3 to 100 times that of croplands. Construction site erosion losses vary greatly throughout the nation, depending on local rain, soil, topographic, and management conditions. As an example, the Birmingham, Alabama, area may have some of the highest erosion rates in the U.S. because of its combination of very high energy rains, moderately to severely erosive soils, and steep slopes. The typically high erosion rates mean that even a small construction project may have a significant detrimental effect on local water bodies.

Extensive evaluations of urban construction site runoff problems have been conducted in Wisconsin for many years. Data from the highly urbanized Menomonee River watershed in southeastern Wisconsin illustrate the impact of construction site erosion on water quality. These data indicate that construction sites had much greater potentials for generating sediment and phosphorus than did areas in other land uses (Chesters, *et al.* 1979). For example, construction sites can generate approximately 8 times more sediment and 18 times more phosphorus than industrial sites, the land use that contributes the second highest amount of these pollutants, and 25 times more sediment and phosphorus than row crops. In fact, construction sites contributed more sediment and phosphorus to the Menomonee River than any other land use, although in 1979, construction comprised only 3.3 percent of the watershed's total land area. During this early study, construction sites were found to contribute about 50 percent of the suspended sediment and total phosphorus loading at the river mouth (Novotny, *et al.* 1979).

Similar conclusions were reported by the Southeastern Wisconsin Regional Planning Commission in a 1978 modeling study of the relative pollutant contributions of 17 categories of point and nonpoint pollution sources to 14 watersheds in the southeast Wisconsin regional planning area (SEWRPC 1978). This study revealed construction as the first or second largest contributor of sediment and phosphorus in 12 of the 14 watersheds. Although construction occupied only two percent of the region's total land area in 1978, it contributed approximately 36 percent of the sediment and 28 percent of the total phosphorus load to inland waters, making construction the region's second largest source of sediment and phosphorus. The largest source of sediment was estimated to be cropland; livestock operations were estimated to be the largest source of phosphorus. By comparison, cropland comprised 72 percent of the region's land area and contributed about 45 percent of the sediment and only 11 percent of the phosphorus to regional watersheds. This early study again pointed out the high pollution-generating ability of construction sites and the significant water quality impact a small amount of construction may have on a watershed.

Another early monitoring study of construction site runoff water quality in the Village of Germantown (Washington County, Wisconsin) yielded similar results (Madison, *et al.* 1979). Several large subdivisions under construction with single and multi-family residences were selected for runoff monitoring. All utility construction, including the storm drainage system and streets, was completed before monitoring began. Analysis of the monitoring data showed that sediment leaving the developing subdivisions averaged about 25 to 30 tons per acre per year (Madison, *et al.* 1979). Construction practices identified as contributing to these high yields included the following:

- Removing surface vegetation;
- Stripping and stockpiling topsoil;
- Placing large, highly erodible mounds of excavated soil on and near the streets;
- Pumping water from flooded basement excavations; and
- Tracking of mud in the streets by construction vehicles.

If the amount of sediment leaving the site during utility development had been added in, the total amount of eroded sediment leaving the site would have been substantially greater. Analysis of the Germantown data also showed that the amount of sediment leaving areas undergoing development was a function of the extent of development and was independent of the type of development. Almost all eroded sediment from the Germantown construction areas entered the receiving waters. The delivery of sediment to the receiving waters was nearly 100 percent when ten percent or more of the watershed was experiencing development. The smallest delivery value obtained during the Germantown monitoring was 50 percent, observed when only five percent of the watershed was undergoing development. These high delivery values occurred (even during periods with small amounts of development) because storm drainage systems, which efficiently transport water and its sediment load, had been installed during an early stage of development. When looking at the Milwaukee River as a whole, the highly-efficiency delivery system installed during urban land development ensures that construction is a major sediment contributor, even though the amount of land under active construction is very low (Figure 1.1).

A comparison of the contributions of agriculture and construction indicates that the low delivery of sediment from agricultural areas contrasted with very high deliveries from construction sites results in much greater unit area yields of sediment from construction areas (Sources of Sediment in Milwaukee River, WI DNR). For example, based on the calculations below assuming 4% delivery efficiency for agriculture and 100% efficiency for construction, the construction activities, while occupying far less land, generates more sediment.

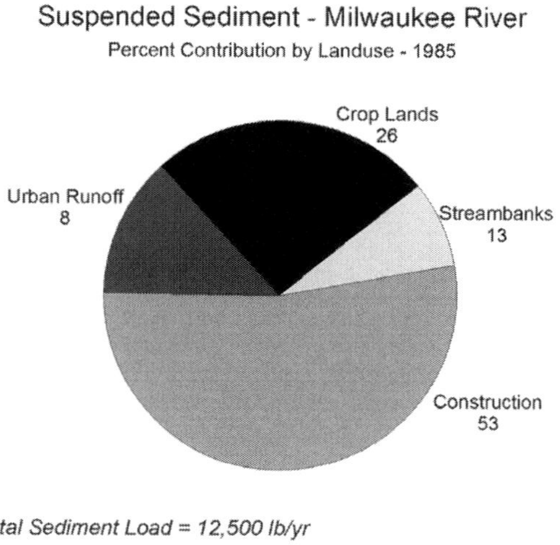

Figure 1.1. Soil Delivery to Streams.

Agricultural Field (assuming 4% efficiency of sediment delivery due to buffer zones, rough surfaces, flat surfaces with sedimentation depressions)

(10 tons/ac/yr) (4%) = 0.4 tons/ac/yr

Construction Site (assuming 100% efficiency of sediment delivery due to the direct connection between the construction area and the drainage system)

(20 tons/ac/yr) (100%) = 20 tons/ac/yr

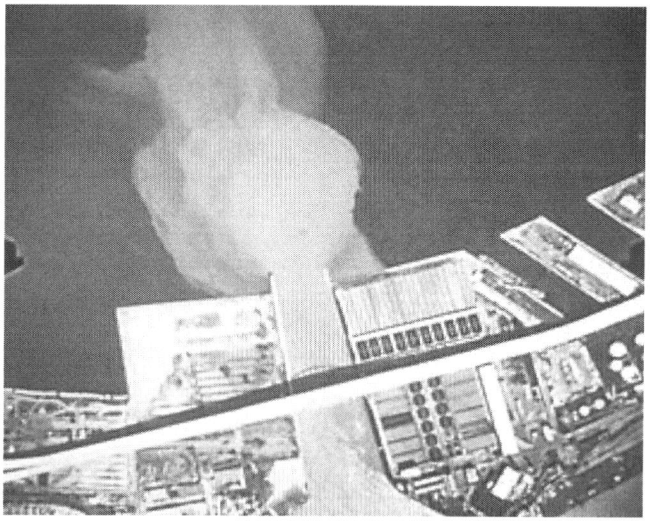

High sediment discharges from Inner to Outer Harbor in Milwaukee (WI) during heavy rains (WI DNR).

Why Construction Site Erosion Rates are Comparatively High in the Piedmont and Appalachian Plateaus of the Southeastern Region of the U.S.

Local Birmingham, AL, erosion rates from construction sites can be ten times the erosion rates from row crops and one hundred times the erosion rates from forests or pastures (Nelson 1996). The site specific factors affecting construction site erosion in the Birmingham, Alabama, area include the following:

- Rainfall Energy (Alabama has the highest in the nation)
- Soil Erodibility (northern part of the state has fine grained, highly erosive soils)
- Site Topography (northeastern part of the state has steep hills under development)
- Surface Cover (usually totally removed during initial site grading on hilly construction sites)

Rainfall energy is directly related to rainfall intensity, and the USLE (Universal Soil Loss Equation, described in Chapter 4) rainfall erosion index varies from 250 to 550+ for Alabama (most of the state is about 350), which is the highest in the U.S. Based on the rainfall energy distribution for Alabama, the months having the greatest erosion potential are February and March, while September through November have the lowest erosion potential in Alabama. Nelson (1996) monitored sediment quantity and particle size from 70 construction site runoff samples from the Birmingham area. He measured suspended solids concentrations ranging from 100 to more than 25,000 mg/L (overall median about 4,000 mg/L), while the turbidity values ranged from about 300 to >50,000 NTU (average of about 4,000 NTU). About 90% of the particles (by mass) were smaller than about 20 μm (0.02 mm) in diameter, with the median size being about 5 μm (0.005 mm). Local construction site erosion discharges were estimated to be approximately 100 tons/acre/year. Table 1.1 summarizes the measured suspended solids and median particle sizes as a function of rain intensity for this study. High intensity rains were found to have the most severe erosion discharges, as expected, with much higher suspended solids concentrations, compared to lower intensity rains. The typically small sizes of the erosion particulates make it very difficult to remove these particulates from the runoff water after they have been eroded from the site. The extreme turbidity values also cause very high in-stream turbidity conditions in local receiving waters for great distances downstream of eroding sites.

TABLE 1.1. Birmingham Construction Site Erosion Runoff Characteristics (Nelson 1996).

	Low Intensity Rains (<0.25 in/hr)	Moderate Intensity Rains (about 0.25 in/hr)	High Intensity Rains (>1 in/hr)
Suspended solids, mg/L	400	2,000	25,000
Particle size (median), μm	3.5	5	8.5

Receiving Water Impacts Associated with Construction Site Discharges

The following is a summary of a recent research project that investigated actual in-stream biological conditions downstream of construction sites having varying levels of erosion controls (none, the use of filter fences, and filter fences plus grass buffers) for comparison. The project title is: *Studies to Evaluate the Effectiveness of Current BMPs in Controlling Stormwater Discharges from Small Construction Sites* and was conducted for the Alabama Water Resources Research Institute, Project 2001AL4121B, by Drs. Robert Angus, Ken Marion and Melinda Lalor of the University of Alabama at Birmingham. The initial phase of the project, described below, was completed in 2002.

Research Objectives

This project examined the effectiveness of low-cost erosion controls, as well as the effects of the discharged silt on the receiving streams' biological communities. Since total control of fine sediment in construction site runoff is unlikely, this project attempted to determine the tolerable amount of fine sediment that can be discharged to a stream or river without causing serious detrimental conditions to the aquatic ecosystem. Currently available EPA-approved rapid bioassessment procedures were not derived specifically to measure the impacts of siltation on biological communities, nor have the sensitivities of the metrics to siltation-caused stress been evaluated. One of the project objectives was therefore to develop or refine metrics that are more sensitive for comparing the level of impairments between sites affected by construction site erosion. These improved metrics are expected to be extremely useful for evaluating the utility of alternative erosion controls.

Methods

This study was conducted in the upper Cahaba River watershed in north central Alabama, near Birmingham. The study areas had the following characteristics: (1) topography and soil types representative of the upland physiographic regions in the Southeast (i.e., southern Appalachian and foothill areas). Thus, findings from this study should be relevant to a large portion of the Southeast. (2) The rainfall amounts and intensities in this region are representative of many areas of the Southeast, and (3) the expanding suburbs of the metropolitan Birmingham area are rapidly encroaching upon the upper Cahaba River and its tributaries. The effectiveness of in-place erosion control devices (silt fences and grass buffers) were evaluated at small construction sites. Water passing through the filter fences was sampled during "intense" (≥1 inch/hr) rain events. The runoff samples were analyzed for turbidity (using a nephelometer), particle size distribution (using a Coulter Counter Multi-Sizer IIe), and total solids (dissolved solids plus suspended/non-filterable solids). Sampling was only carried out on sites with properly-installed and well-maintained silt fences, located immediately upgrade from areas with good vegetative cover. Stormwater runoff samples were collected from sheet flows above silt fences, and from points below the fence within the vegetated buffer.

Six tributary or upper mainstream sites were studied to investigate the effects of sedimentation from construction sites on both habitat quality and the biological "health" of the aquatic ecosystem (using benthic macroinvertebrates and fish). No other sediment sources, except for the construction areas, affected the study sites. Two of the sites generated heavy sediment loads, two were moderately impacted, and two (reference sites) had little, or no, sediment inputs. Each site was assessed in the spring to evaluate immediate effects of the sediment, and again during the following late summer or early fall to evaluate delayed effects. The EPA's "Revision to Rapid Bioassessment Protocols for Use in Streams and Rivers" was used to assess the habitat quality at the study sites.

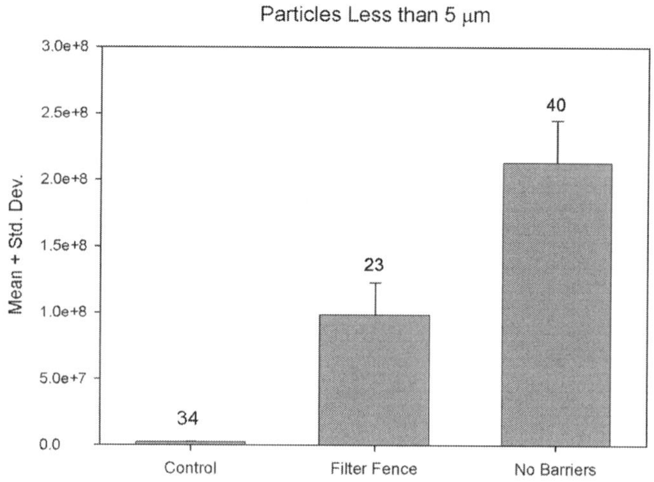

Figure 1.2. Comparison of numbers of small particles (<5 µm) from similar samples taken in undisturbed vegetated areas (control), below silt fences, and from areas with no runoff contr. *Note:* numbers of grab samples in each category are shown above each corresponding bar (Angus, et al. 2002).

Preliminary Results

Effectiveness of Silt Fences

Comparisons were made between samples collected immediately below silt fences and samples collected nearby but not below a silt fence (Figure 1.2 and Table 1.2). Silt fences were found to be better than no control measures at all, but not substantially. The mean count of small particles below the silt fences were about 50% less than that from areas with no erosion control measures, even though the fences appeared to be properly installed and in good order. However, the variabilities were large and the difference between the means was not statistically significant (Table 1.2). This level of control is similar to levels found during controlled laboratory tests. The silt fences obviously did not reduce particle counts to levels comparable to nearby undisturbed sites (Table 1.2). For every variable measured (turbidity, total solids, suspended solids, etc.), the mean values of samples taken below silt fences were significantly higher ($p < 0.001$) than samples collected from undisturbed vegetated control sites collected nearby and at the same time. These data therefore indicate that silt fences are only marginally effective at reducing soil particulates in runoff water. Surprisingly, the amount of silt in the runoff (as measured with the variables listed above) was not significantly correlated with slope of the site, or the amount or intensity of rainfall. This may reflect the fact that samples were only collected during "intense" (>1 inch/hour) rainfall events, the most erosive category.

TABLE 1.2. Mean values (± std. error) of particle counts in similar samples taken during >1″/hr rain events in unvegetated control sites, below silt fences, and in disturbed areas with no barrier (Angus, et al. 2002).

	No Barriers (n= 40)	Filter Fence (n =23)	Control (n = 34)
Total Particles	$2.18 \times 10^8 \pm 3.28 \times 10^7$	$1.01 \times 10^8 \pm 2.48 \times 10^7$	$2.45 \times 10^6 \pm 3.54 \times 10^5$
Small Particles	$2.13 \times 10^8 \pm 3.21 \times 10^7$	$9.82 \times 10^7 \pm 2.43 \times 10^7$	$2.36 \times 10^6 \pm 3.44 \times 10^5$
Large Particles	$4.37 \times 10^6 \pm 9.20 \times 10^5$	$2.91 \times 10^6 \pm 7.28 \times 10^5$	$8.56 \times 10^4 \pm 1.31 \times 10^4$

Note: In each row, the mean for the Control is significantly lower than for the other cells in the same row (ANOVA on log transformed data, $p << 0.001$). Means for the "No Barriers" and "Filter Fence" treatments were not significantly different for any particle size groups ($p > 0.05$), although the filter fence sites had apparently reduced particle counts.

Effectiveness of Filter Fences with Vegetated Buffers

Runoff samples were also collected immediately below filter fences, and below filter fences after flow over buffers having 5, 10, and 15 feet of dense (intact) vegetation. Again, only sites with filter fences which appeared to be properly installed and maintained were evaluated. Mean total solids in samples collected below silt fences and a 15 foot wide vegetated buffer zone were about 20% lower, on average, than those samples collected immediately below the silt fence. Preliminary analysis of the data indicate that the installation of filter fences above an intact, good vegetated buffer removes sediment from construction site runoff more effectively than with the use of filter fences alone. High variations in the effectiveness were observed, likely due to variations in site microenvironments. Longer buffer lengths (15 feet) generally resulted in greater removals of sediment than shorter buffer lengths (5 feet). An increase in the percent removal of sediment in the vegetated buffer strip appeared to correlate weakly with a decrease in the site slope.

Development of Biological Metrics Sensitive to Sedimentation Effects (Fish)

Analysis of the fish biota indicates that various metrics used to evaluate the biological integrity of the fish community also are affected by highly sedimented streams. As shown in Figure 1.3, the overall composition of the population, as quantified by the Index of Biotic Integrity (IBI) is lower, the proportion and biomass of darters, a disturbance-sensitive group, is lower; the proportion and biomass of sunfish is higher; the Shannon-Weiner diversity index is lower; and the number of disturbance-tolerant species higher.

Benthic Macroinvertebrates

A number of stream benthic macroinvertebrate community characteristics were also found to be sensitive to sedimentation. Metrics based on these characteristics differ greatly between sediment-impacted and control sites (Figure 1.4). Some of the metrics that appear to reflect sediment-associated stresses include the Hilsenhoff Biotic Index (HBI), a variation of the EPT index (%EPT minus Baetis), and the Sorensen Index of Similarity to a reference site. The HBI index is a weighted mean tolerance value; high HBI values indicate sites dominated by disturbance-tolerant macroinvertebrate taxa. The EPT% index is the percent of the collection represented by organisms in the generally disturbance-sensitive orders Ephemeroptera, Plecoptera, and Trichoptera. Specimens of the genus Baetis were not included in the index as they are relatively disturbance-tolerant. The HBI and the EPT indices also show positive correlations to several other measures of disturbance, such as percent of the watershed altered by development.

In collaboration with the Jefferson County Stormwater Management Authority, this project developed a method for predicting the soil erosion damage potential of a site to affect receiving water biological conditions. The cartographic model consists of selected data layers for the study area, including NRCS soils, multispectral satellite imagery, parcel level land use, and a digital elevation model. The derived layers are then combined in a Geographical Information System (GIS) to produce a Sedimentation Potential Index (SPI). This index is a measure of the "erodibility" of a site and an indication of the potential to produce excessive silt runoff if disturbed. Calculated SPI values for various subwatersheds were compared with measured biological characteristics (Figure 1.5). The calculated SPI scores strongly correlated with a number of metrics that reflected sedimentation impacts.

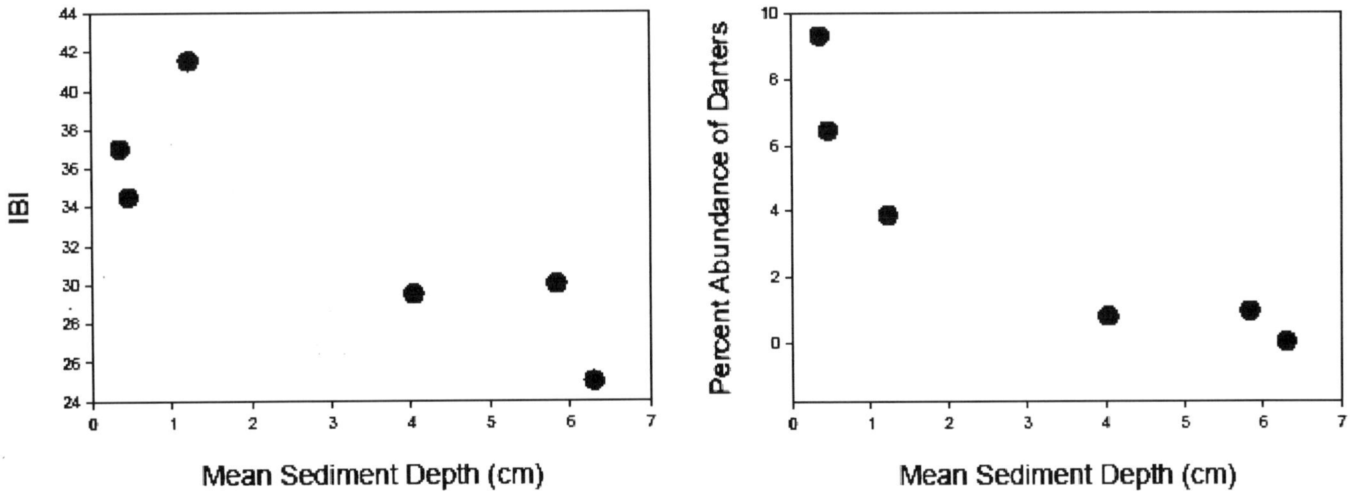

Figure 1.3. Association between two fish metrics and amount of stream sediment. *Note:* the IBI (Index of Biotic Integrity) is based on numerous characteristics of the fish population. The percent relative abundance of darters is the percentage of darters to all the fish collected at a site (Angus, et al. 2002).

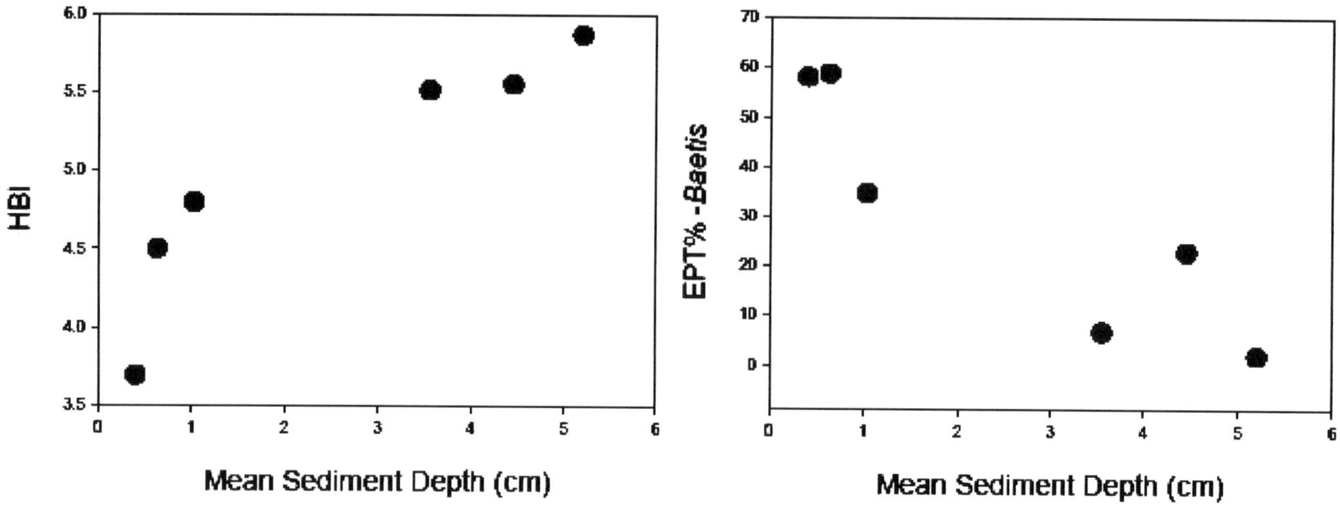

Figure 1.4. Associations between two macroinvertebrate metrics and the amount of stream sediment (Angus, et al. 2002).

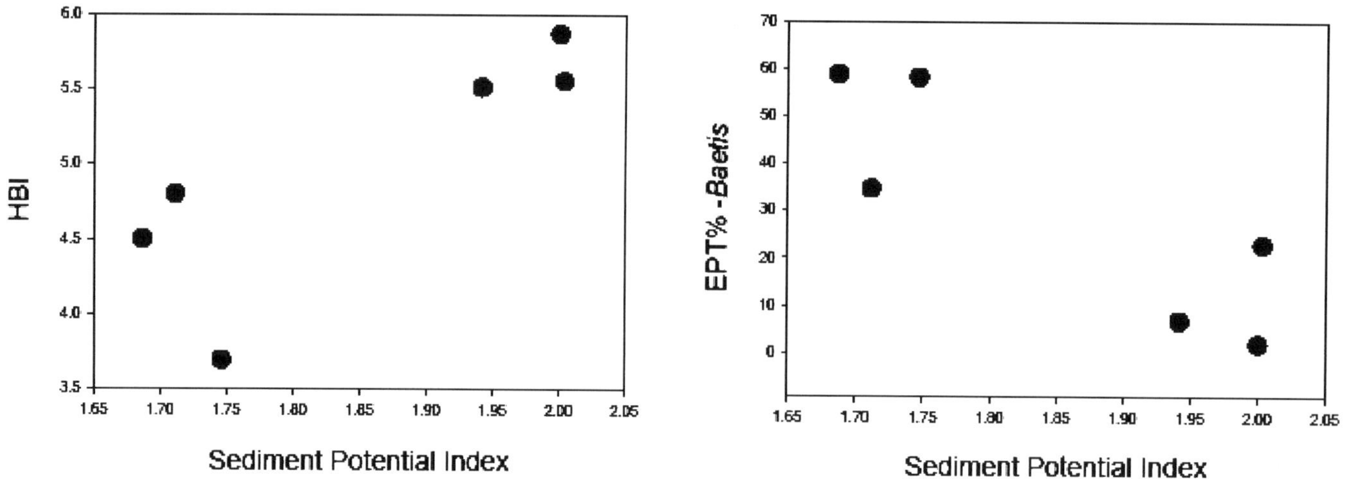

Figure 1.5. Associations between two macroinvertebrate metrics and the sedimentation potential in the same watershed (Angus, et al. 2002).

Receiving water problem investigations conducted at many locations throughout the country have led to increasing local and national regulations, and the development of new technologies and methods, for the reduction of construction site erosion. The rest of this introductory chapter outlines the Phase I and Phase II NPDES stormwater regulations affecting construction sites—the regulations that will affect almost all construction sites throughout the nation. Chapter 2 reviews erosion and sediment control tools that are described in some example state guidance handbooks for use in addressing specific elements of these regulations.

Trying to sell badly eroded land (difficult to sell lots and homes in these types of neglected areas).

Erosion Threatening Homes (This home is actually being constructed on 12 feet of fill soil. The foundation footers are 14 feet below the groundline. Note the rills draining down to the drainage swale) (Photographs by D. Lake).

Damaged from erosion requiring repairs (IECA photo).

Sediment Problems (WI DNR) (Natural streams alternate pools and riffles and have varying stream sediment textures. With erosion impacts, pools are filled and coarse material becomes covered with fines).

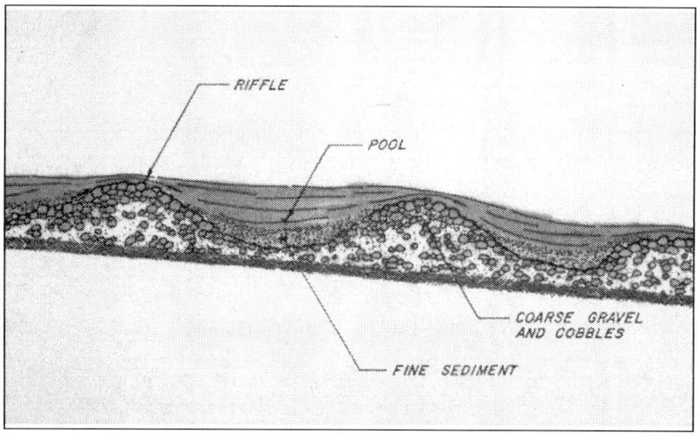

Typical stream showing riffle and pools (WI DNR).

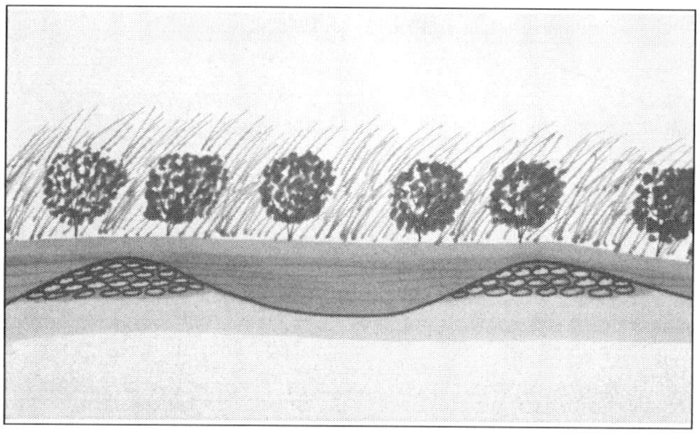

Sediment-laden channel bottom: unsupportive (WI DNR).

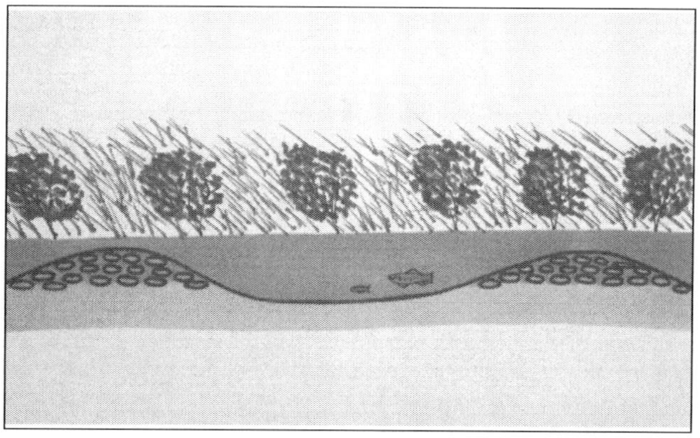

Clean gravel channel bottom: supportive of fish (WI DNR).

Natural stream showing coarse bottom material in riffle area, and impacted stream showing siltation in area of coarse bottom material.

Drainage Ditch Filled with Construction Sediment (J. Voorhees) (Decreased drainage capacity causes increased flooding).

Eroded Streams and Channels (WI DNR) (Eroding banks affect shoreline water quality).

Most stormwater has low turbidity unless affected by eroding soils.

Local erosion problem affecting turbidity of one drainage branch.

Eroding soils from bare ground can be responsible for much sediment loss.

A small utility trench can cause high turbidities in downstream runoff.

Buried debris and other material adversely affects soil structure and future drainage.

Vast amounts of bare ground exposed for extended periods at construction sites are responsible for most of the erosion problems, especially if on a slope.

Lack of Following a Good Plan

This site began, at least on paper, with all the promise of a well-planned, phased, and properly designed residential development. The consulting engineering firm had divided this 70 acre site into four construction phases, balancing the cuts and fills for each phase. They had incorporated appropriate erosion and sediment control practices in the stormwater pollution prevention plan that also contained an 8 acre pond for water quality and quantity control.

Unfortunately, once the project began construction in mid-August, the developer instructed the contractor to ignore the plan and build the entire site and infrastructure in one phase, with 65 aces disturbed at one time. In fact, the developer never signed nor submitted the Notice of Intent for the project as required under the state permit regulations. It began raining in mid-October and was still raining in late December. Unfortunately this site's outlet drains to a tributary only 3,800 feet from a high quality sport fishing recreational lake. It was later determined that the soils that washed off this site destroyed two acres of walleye spawning area. Soil analyses indicated that the site soil contained 75–98 percent material smaller than 0.074 mm, or 74 μm, and was therefore highly mobile when eroded.

The site was shut down by state authorities after the sediment plume into the lake was noticed in late December. Remediation included seeding and mulching the entire site for spring thaw conditions, placing stone check dams in all drainage conveyances, installing a rock dam to create a large sediment basin, construction of five rock chutes for gradient control, and six sediment traps at various locations on the site. The cost for this work was approximately $35,000. In addition, the developer paid a $10,000 fine and was placed on a graduated fine scale for any additional water quality violations.

Even though this site was relatively flat, the high content of fine particulates in the soil coupled with the total disregard for erosion control practices and lack of knowledge of the drainage area caused this disaster.

This exposed site is under a state shutdown order for destroying 2 acres of walleye spawning area in a nearby lake. 65 acres of the 70 acre site was stripped exposing soil containing 75–98 percent fines that could not settle out on site.

Sediment Sources

Continuous operations at a solid waste landfill require special precautions to prevent excessive erosion. This site has a large sediment pond, with pre-treatment forebays, plus a final sand filter, to meet their 50 NTU discharge permit requirement for turbidity.

End of season construction site shutdowns can also result in excessive erosion during late winter and early spring rains during periodic thaws unless the site is well-stabilized for the season.

Control of runoff is critical at the beginning of construction. Here the stormwater infrastructure is in place but the 24 inch storm sewer is 75% plugged with sediment. Note the large size of the material on the catch basin grate (Photograph by D. Lake).

Cleanup of excessive sediment on roads should not include rinsing the debris to the storm drainage inlet.

Stock Pile Problems and Working Close to Roads

Stock piles of material can be important sediment sources. (Especially when located on the road itself, directly connected to the drainage system and receiving water).

It is very difficult to work close to the road and prevent debris from entering the drainage system.

Improper Disposal of Construction Debris and Improper Equipment Maintenance

Engine repairs and other heavy equipment maintenance should not be allowed on construction sites, unless suitably protected from the elements.

Hazardous materials and other unsafe debris should never be left exposed at construction sites.

Improper waste concrete disposal.

Fuel spillage at re-fueling area is both hazardous and damaging.

Poor Drainage Construction

Poor grading directed runoff away from drain inlet and to the unprotected slope. Expensive repairs are now needed.

Another unfortunate example of poor grading allowing runoff to miss protected downslope channel.

Fugitive Dust Problems

In this commercial mall re-habilitation project, dust became a problem even though much of the site area was impervious. Complaints were received from homeowners beyond the work area in the direction of the prevailing winds.

Another example of fugitive dust causing potential traffic safety problems. Construction was halted this day due to high winds at this road-widening project, but unstabilized and exposed ground still allowed excessive dust losses.

Fugitive dust losses and traffic safety problem as heavy equipment was being driven on unprotected construction roads near existing roads during period of high winds.

CONSTRUCTION SITE EROSION AND SEDIMENT CONTROL REGULATIONS

The NPDES (National Pollutant Discharge Elimination System) was established as part of the Clean Water Act amendments of 1972. It was intended to control and regulate point sources of water pollution throughout the U.S., with the eventual objective of totally eliminating these discharges and ensuring all U.S. receiving waters were "fishable" and "swimmable." Over the years, these lofty objectives have been scaled back, but these regulations have done much to improve the quality of U.S. waters.

These initial regulations affected municipal sewage treatment plants (or "publicly owned treatment works," POTWs) and industrial discharges. Stormwater was initially considered an exempt point source and was not included in the initial regulations. After reviewing water quality data showing that stormwater caused problems, the EPA finally established separate regulations for stormwater in 1987. The original Phase I regulations for stormwater (implemented in 1990) applied to large municipalities (generally population >250,000) and certain industries. Medium-sized municipalities (100,000 to 250,000 in population, plus other industries) were regulated several years later. The recently implemented Phase II regulations are intended to be applied to all urban areas in the U.S. The Phase I regulations included construction activity as an industry and were applied to all construction sites greater than 5 acres. The Phase II regulations generally will apply to all construction sites larger than 1 acre.

Many municipalities and some states have had local regulations affecting construction sites for many years, independent of the federal regulations. Some features of these are included in Appendix 1A.

CWA 402(p)(6) Initial Phase II Rule (For Small Municipalities)

The purpose of the initial Phase II regulations was to designate additional sources of stormwater, beyond Phase I, that needed to be regulated to protect receiving water quality. These regulations required that all unregulated dischargers of stormwater apply for NPDES permits by March 10, 2003. According to the EPA, this regulation could apply to millions of industrial/commercial facilities and over 22,000 municipalities.

A Federal Advisory Committee (FACA) helped to develop the proposed Phase II rules. The membership in the FACA included a cross-section of interested stakeholders (private environmental groups, municipal representatives, trade associations, state regulators, and various other experts) from throughout the U.S. They held 14 meetings from 1995–1998 and prepared three preliminary drafts that were circulated for review and comment.

The proposed Phase II rule was published in the Jan. 9, 1998 edition of the *Federal Register* (40 CFR Parts 122 and 123, 63 FR 1563). There was a 90-day comment period and about 550 comments were received. The EPA held public hearings at six locations to explain the proposed Phase II rules and to obtain public comment. The final phase II rule was signed on December 8, 1999, after modifications based on these comments. Phase II NPDES permit applications were due starting March 10, 2003, but the specific compliance dates were set by each state regulatory agency.

Two new classes of facilities were established for automatic coverage on a nationwide basis:

1. Small municipal separate storm sewer systems located in urbanized areas (about 3,500 municipalities) [Phase I included medium and large municipalities]
2. Construction activities that disturb less than 5 acres of land (about 110,000 sites a year) [Phase I included construction sites larger than 5 acres]

A "no exposure" incentive for Phase I sites was also proposed for industrial activities (will exclude about 70,000 facilities).

Permit Requirements for Each Regulatory Agency

The following are the required elements for each plan to be prepared by the local regulatory agencies:

- Develop, implement, and enforce a program to reduce the discharge of pollutants and protect water quality to the "maximum extent practicable"
- Must include six minimum control measures:
  —Public education and outreach
  —Public involvement and participation
  —Illicit discharge detection and elimination
  —Construction site stormwater runoff control
  —Post-construction stormwater management in new development and redevelopment
  —Pollution prevention and good housekeeping for municipal operations
- Must submit a notice of intent (NOI), or permit application, and identify for each minimum control measure:
  —Best management practices to be used
  —Measurable goals
  —Timeframe for implementation
  —Responsible persons
- Must evaluate program and submit reports

The objective is to include greater flexibility in the Phase II rule by encouraging the use of general permits, encourage municipalities to determine appropriate stormwater controls, not require extensive monitoring by permittees, and recognize and contemplate the use of existing programs, including existing structures and mechanisms for public participation.

Construction Site Regulations

The Phase II regulations will extend existing Phase I regulations for construction coverage to:

- All sites that result in the disturbance of 1 acre or more, but less than 5 acres (designated nationwide)
- Sites that result in disturbance of less than 1 acre (potential designation by permitting authority).

The regulations will encourage the use of local regulations that control erosion and sediment to the "maximum extent practicable," control other waste at construction sites, allow the granting of waivers by the permitting authority, and to qualifying local and state programs.

The EPA allows the local agencies to waive coverage for construction sites that meet the following criteria:

- Rainfall erosivity factor (NRCS RUSLE rainfall factor "*R*") less than 2 (during the period of construction) ("low rainfall")
- Annual soil loss of less than 2 tons/acre/year ("low erosion potential")
- A watershed plan or TMDL assessment that addresses the pollutants of concern

The rule would require:

1. Control of other wastes at construction sites (discarded building materials, concrete truck washout, sanitary wastes, etc.)
2. Appropriate best management practices (such as silt fences, temporary detention ponds, etc.)
3. Pre-construction reviews of site management plans
4. Receipt and consideration of public information
5. Regular inspections during construction
6. Penalties to ensure compliance

If local regulations incorporate the following erosion-preventing principles and elements into its stormwater program, then it would be considered as a "qualifying" program that meets Federal requirements:

Five Principles:

1. Good site planning
2. Minimize soil movement
3. Capture sediment
4. Good housekeeping practices
5. Mitigation of post-construction stormwater discharges

Eight Elements:

1. Program description
2. Coordination mechanism
3. Requirements for nonstructural and structural BMPs
4. Priorities for site inspections
5. Education and training
6. Exemption of some activities due to limited impacts
7. Incentives, awards, and streamlining mechanisms
8. Description of staff and resources

Effluent Limit Guidelines Schedule

The following discussion is summarized from the EPA reports (USEPA 2002 and 2004) describing the proposed effluent guidelines, and the final ruling, for the construction and development NPDES categories.

Section 304(m) of the Clean Water Act (CWA) requires the U.S. Environmental Protection Agency (EPA) to publish a plan every two years that consists of three elements. First, under section 304(m)(1)(A), the EPA is required to establish a schedule for the annual review and revision of existing effluent guidelines in accordance with section 304(b). Section 304(b) applies to effluent limit guidelines (ELGs) for direct dischargers and requires the EPA to revise such regulations as appropriate. Second, under section 304(m)(1)(B), the EPA must identify categories of sources discharging toxic or nonconventional pollutants for which the EPA has not published Best Available Technology (BAT) ELGs under section 304(b)(2) or new source performance standards under section 306. Finally, under section 304(m)(1)(C), the EPA must establish a schedule for the promulgation of BAT and New Source Performance Standards (NSPS) for the categories identified under subparagraph (B) not later than three years after being identified in the 304(m) plan. Section 304(m) does not apply to pretreatment standards for indirect dischargers, which the EPA promulgates pursuant to sections 307(b) and 307(c) of the CWA.

On October 30, 1989, the Natural Resources Defense Council, Inc. (NRDC), and Public Citizen, Inc., filed an action against EPA in which they alleged, among other things, that EPA had failed to comply with section 304(m). Plaintiffs and the EPA agreed to a settlement of that action in a consent decree entered on January 31, 1992. (*Natural Resources Defense Council et al v. Whitman,* D.D.C. Civil Action No. 89-2980). The consent decree, which has been modified several times, established a schedule by which the EPA is to propose and take final action for eleven point source categories identified by name in the decree and for eight other point source categories identified only as new or revised rules, numbered 5 through 12. The EPA selected the Construction and Development (C&D) category as the subject for New or Revised Rule #10. The decree, as modified, called for the Administrator to sign a proposed ELG for the C&D category no later than May 15, 2002, and to take final action on that proposal no later than March 31, 2004. A settlement agreement between the parties, signed on

June 28, 2000, requires that the EPA develop regulatory options applicable to discharges from construction, development and redevelopment, covering site sizes included in the Phase I and Phase II NPDES stormwater rules (i.e., one acre, or greater). The EPA was required to develop options including numeric effluent limitations for sedimentation and turbidity; control of construction site pollutants other than sedimentation and turbidity (e.g. discarded building materials, concrete truck washout, trash, etc.); controls for reducing postconstruction runoff; controls for construction sites; and requirements to design stormwater controls to maintain pre-development runoff conditions, where practicable. The settlement also required the EPA to issue guidance to Municipal Separate Storm Sewer Systems (MS4s) and other permittees on maintenance of postconstruction controls identified in the proposed ELGs.

The EPA therefore proposed Effluent Limitation Guidelines for discharges associated with construction and development activities under the authority of Sections 301, 304, 306, 308, 402, and 501 of the Clean Water Act (CWA) (the Federal Water Pollution Control Act), 33 United States Code (U.S.C.) 1311, 1314, 1316, 1318, 1342, and 1361. The proposed rule contained three options for controlling stormwater discharges from construction sites (USEPA 2002):

> Option 1 would establish inspection and certification provisions to ensure proper implementation of controls. This option would apply to all construction sites disturbing one or more acres of land required to obtain a permit under the existing National Pollutant Discharge Elimination System (NPDES) stormwater regulations. This option would amend the NPDES regulations at 40 CFR Part 122, but would not create effluent limitation guidelines.
>
> Option 2 would add minimum requirements for preparation of a Stormwater Pollution Prevention Plan (SWPPP) as well as minimum requirements for sizing sediment basins, installing erosion and sediment controls, providing temporary stabilization to exposed soils, and conducting regular inspections. Option 2 would apply to all sites that disturb five or more acres of land, consistent with the permitting requirements of the Phase I NPDES stormwater regulations. This option would create a new effluent guidelines category at 40 CFR Part 450 and would also modify 40 CFR Part 122.
>
> Option 3 would not establish any new requirements.

The EPA estimated that Option 1 would cost approximately $130 million annually, while preventing the annual discharge of approximately 5.25 million tons of Total Suspended Solids (TSS) and associated turbidity to surface waters. The estimated annual monetized benefits of this option are $10.4 million. Option 2 was estimated to cost approximately $505 million annually, while preventing the discharge of approximately 11.1 million tons of TSS and associated turbidity to surface waters annually. The estimated annual monetized benefits of Option 2 are $22.0 million. Option 3 was not expected to have any costs or benefits.

Final Rule

On March 31, 2004, the EPA Administrator signed a Federal Register notice (published on April 4, 2004) opting for Option 3, basically to rely on the range of existing programs, regulations, and initiatives at the federal, state, and local levels for the control of runoff from construction sites rather than establish a new effluent guideline.

For additional information regarding the Construction & Development Effluent Guidelines project, the EPA listed the following main contacts:

Jesse W. Pritts
202-566-1038
pritts.jesse@epa.gov
Engineering and Analysis Division (4303T)
US EPA
1200 Pennsylvania Ave, NW
Washington, DC 20460

George Denning, Economist
202-566-1067
denning.george@epa.gov

Existing Regulations

The following is an excerpt from the *Construction and Development Fact Sheet: Final Action–Selection of Non-Regulatory Option,* EPA 821-F-04-001; March 2004.

> The EPA decided to rely on the range of existing programs, regulations, and initiatives at the federal, state, and local level for the control of stormwater runoff from construction sites rather than establish a national effluent guideline at this time. After careful study, they determined that almost every state has requirements in place that are equivalent to, or even more protective, than those contained in the proposed effluent guidelines (option 2). In addition, over 5,000 municipalities are currently developing, or upgrading, local programs and requirements for construction site runoff. The current system of federal requirements as outlined in the NPDES regulations allows states and local governments to develop programs that will both protect the environment and maintain flexibility to tailor requirements to meet local conditions.

The EPA wants local decision-makers to have maximum flexibility to develop control strategies that are tailored to the discharges of stormwater runoff from construction sites under their jurisdiction. They believed that the proposed regulatory options would have limited the flexibility permitting authorities currently have to use control strategies that reflect local conditions. Further, they felt that the costs of the proposed regulatory options would be very high, and these options would provide only marginal environmental improvements over regulations already in place. EPA's economic analysis indicated that the average incremental cost of construction and post-construction controls for a single family house would have ranged from about $1,000 to $2,200, depending on the degree of implementation of the

Phase II stormwater program. They concluded that the most stringent of the regulatory options would have reduced sediment loadings from construction sites by only about one percent more than the existing regulations (assuming, of course, adequate compliance and enforcement of these existing regulations).

Existing State Programs

In March 2003, Phase II of EPA's NPDES regulations for stormwater went into effect and required that permitting authorities establish programs to regulate runoff from construction sites of one to five acres in size. These new requirements are expected to affect approximately 200,000 construction sites annually. Larger construction sites have been regulated under the NPDES program since 1992. The authorized states and EPA are implementing these new requirements (Phase II) and they will result in significant reductions of pollutants from well-designed and maintained construction sites.

The EPA's analyses concluded that every state had regulations and programs in place that incorporate most of the provisions of the most stringent proposed option (#2). The following lists how states are already addressing these key requirements of the proposed effluent guideline:

Stormwater Pollution Prevention Plans—All 50 states require site managers to prepare a stormwater pollution prevention plan, erosion and sediment control plan, or an equivalent document.

Inspections by Construction Site Operator—All 50 states require construction site operators to inspect their sites on a regular basis.

Erosion and Sediment Control—All 50 states require site managers to implement a combination of erosion and sediment controls to prevent soil erosion and to manage construction site runoff. The EPA's proposed effluent guideline would have mandated sediment basins of a particular size across the country. Currently, states base their technical requirements for basins or other erosion control techniques on local rainfall patterns and other considerations.

Stabilization of Soils after Construction—All 50 states require stabilization of soils after construction activities have temporarily or permanently ceased. The EPA's proposed effluent guidelines would have mandated this step within 14 days. States currently set their own requirements based on local conditions. In dry areas, for instance, 14 days may not be necessary because of low rainfall. It may also be impractical due to slow growth of vegetation.

Existing Local Programs

Many local governments also have long-standing programs in place to control sediment and erosion from construction sites within their jurisdiction. EPA's stormwater regulations (Phase I and Phase II) set minimum requirements for these programs. Approximately, 6,000 municipalities are covered by these regulations. Many of the approximately 5,000 communities covered by Phase II are currently developing or upgrading their programs to meet these requirements. These are some of the minimum requirements for these programs:

- Ordinances or other regulatory mechanisms requiring the implementation of proper erosion and sediment controls
- Review of site plans to ensure proper design and installation of sediment and erosion controls
- Site inspections and enforcement of control measures
- Sanctions to ensure compliance
- Procedures for public review and comment
- Review of site plans

The NPDES regulations require that municipalities set up procedures for review of site plans to ensure proper implementation of sediment and erosion controls. The EPA's proposed effluent guideline would have required certification of the design and installation of sediment and erosion controls by a qualified professional (generally a third-party). The 5,000 communities covered under the Phase II requirements are starting to implement their programs for site plan review. States and communities are working together to define and develop effective programs. Communities have until 2008 to fully implement these requirements.

EPA Resources for Construction Site Stormwater Management

A range of regulatory programs and resources are currently in place and being implemented at the federal, state and local levels address construction site stormwater runoff.

Regulatory Programs

NPDES Regulations—The NPDES regulations for stormwater cover construction sites in two ways. First, authorized states and EPA (in non-authorized states) must develop programs and permits for sites disturbing one or more acres of land. Second, municipalities in urbanized areas must develop comprehensive programs to regulate stormwater from construction activities within their jurisdiction.

Construction—The NPDES Phase I and Phase II stormwater regulations require permits for construction sites that disturb one or more acres of land. Phase I became effective in 1992 and regulates construction sites five acres or larger in size. Authorized states and EPA developed detailed permit requirements for these sites and refined those requirements as permits are reissued (NPDES permits are reissued every 5 years). Effective in March 2003, Phase II extends these requirements to also cover sites of one to five acres.

Municipal—Approximately 6000 municipalities with separate storm sewer systems are covered by EPA's NPDES stormwater regulations (Phase I and II). They are required to develop programs to regulate stormwater from sites within their jurisdiction that are one acre or larger. Most municipalities have programs that cover construction sites.

The NPDES regulations outline a set of minimum controls and many cities are enhancing their current programs to meet these requirements. Municipal programs must include local enforceable ordinances, review of site plans, inspections, and enforcement procedures. Effective March 2003, the Phase II regulations cover municipalities in urban areas with populations up to 100,000 (the earlier Phase I regulations addressed larger municipalities). These communities have five years to develop and fully implement these programs.

EPA Resources for the Control of Construction Site Runoff

The following websites are listed by the EPA as main sources of information and technical assistance that they provide for state and local agencies, plus contractors and others involved in construction site erosion control:

State Water Pollution Control Program Grants Program
http://www.epa.gov/owm/cwfinance/pollutioncontrol.htm
(Section 106) provides funding to state programs to implement the programs under the Clean Water, including stormwater programs.

Stormwater Website
http://cfpub.epa.gov/npdes/home.cfm?program_id=6
Contains comprehensive reference and guidance materials for control of construction site runoff.

Construction Industry Compliance Assistance Center
http://cicacenter.org/
Contains information and links to a wide variety of information, including state regulatory programs and manuals for sediment and erosion controls.

Electronic Notice of Intent System
http://cfpub.epa.gov/npdes/stormwater/enoi.cfm
An online, electronic application system for obtaining coverage under EPA's Construction General Permit. This system also provides construction site operators with comprehensive information on controlling runoff and meeting permit requirements.

National Management Measures to Control Nonpoint Source Pollution from Urban Areas
http://www.epa.gov/owow/nps/urbanmm/
A technical guidance and reference document on best management practices to control urban runoff.

Smart Growth Program
http://www.epa.gov/livability/
Provides tools, technical and financial assistance, and training on complying with stormwater requirements while also encouraging innovation in land development.

Section 319 Nonpoint Source Management Program
http://www.epa.gov/owow/nps/cwact.html
Provides grants to states, territories and tribes to support a variety of nonpoint source implementation projects including those addressing stormwater runoff.

Copies of the final Federal Register notice and supporting materials are available at:
http://www.epa.gov/guide/ construction
Additionally, they can be requested by sending an email to center.water-resource@epa.gov. For further information pertaining to the final ruling, they list Ms. Pamela Barr at (202) 566-0430 or send her an email at barr.pamela@epa.gov, for further information.

Proposed EPA Effluent Guidelines for Construction and Development Category

The following discussion is summarized from the EPA's guidance document prepared for the proposed effluent guidelines (USEPA 2002) and from the fact sheet describing the final ruling (USEPA 2004). The proposed effluent guidelines contained three options. Option 2 would have required the permittee to prepare a stormwater pollution prevention plan (SWPPP) and implement the erosion and sediment controls contained in the EPA's Construction General Permit (CGP). In addition, the permittee would have been required to conduct periodic site inspections and provide certifications in a site log book. The final rule published in early April 2004 accepted the third option, which was to rely on the range of existing programs for the control of runoff from construction sites, rather than establish a new effluent guideline. Their rational was that provisions contained in the most demanding option (#2) were already included in almost all state and local regulations. Therefore, the originally proposed option 2 may possibly be considered a basic benchmark, and is summarized below.

General Erosion and Sediment Controls

Each SWPPP would have been required to include a description of appropriate controls designed to retain sediment on site to the extent practicable. These general erosion and sediment controls would be required to be included in the SWPPP described below. The SWPPP would be required to include a description of interim and permanent stabilization practices for the site, including a schedule of when the practices would be implemented. Stabilization practices could include the following:

1. Establishment of temporary or permanent vegetation;
2. Mulching, geotextiles, or sod stabilization;
3. Vegetative buffer strips;
4. Protection of trees and preservation of mature vegetation.

The EPA recommended that all controls be properly selected and installed in accordance with sound engineering practices and, manufacturer's specifications.

Sediment Controls

Operators would be required to design and install structural controls to divert flows from exposed soils, to store flows, or otherwise to limit runoff and the discharge of pollutants from exposed areas, and to describe controls in the SWPPP. The controls required are as follows:

1. For common drainage locations that serve an area with 10 or more acres disturbed at one time, the operator would be required to provide a temporary (or permanent) sediment basin that provides storage for a calculated volume of runoff from a 2 year, 24-hour storm from each disturbed acre drained, or equivalent control measures, where attainable, until final stabilization of the site. Where no such calculation has been performed, the operator would be required to provide a temporary (or permanent) sediment basin providing 3,600 cubic feet of storage per acre drained, or equivalent control measures, where attainable, until final stabilization of the site. When computing the number of acres draining into a common location, it would not be necessary to include flows from off-site areas and flows from on-site areas that are either undisturbed or have undergone final stabilization where such flows are diverted around both the disturbed area and the sediment basin.
2. In determining whether a sediment basin is attainable, the operator may consider factors such as site soils, slope, available area on site, etc. In any event, the operator would be required to consider public safety, especially as it relates to children, as a design factor for the sediment basin. Use of alternative sediment controls would be required where site limitations preclude a safe basin design.
3. For portions of the site that drain to a common location and have a total contributing drainage area of less than 10 acres, the operator would be required to consider installation of sediment traps or other sediment control devices.
4. Where neither a sediment basin nor equivalent controls are attainable due to site limitations, the operator would be required to install silt fences, vegetative buffer strips or equivalent sediment controls for all downslope boundaries of the construction area and for those side slope boundaries deemed appropriate for individual site conditions.

Pollution Prevention Measures

The operator would be required to implement the following pollution prevention measures:

1. The operator would be required to prevent litter, construction chemicals, and construction debris from becoming a pollutant source in stormwater discharges; and
2. The operator would be required to contain construction and building materials in appropriate storage areas and manage the materials to prevent contamination of stormwater runoff.

Stormwater Pollution Prevention Plan

Permittees would be required to develop and implement Stormwater Pollution Prevention Plans (SWPPPs) prior to groundbreaking at any construction site. In areas where EPA is not the permit authority, operators may be required to prepare documents that may serve as the functional equivalent of a SWPPP. Such alternate documents would satisfy the requirements for a SWPPP so long as they contain the necessary elements of a SWPPP. A SWPPP would be required to incorporate the following information:

1. A narrative description of the construction activity, including a description of the intended sequence of major activities that disturb soils on the site (Major activities include any clearing, grubbing, excavating, grading, soil stockpiling, and utilities and infrastructure installation, or any other activity that results in significant disturbance of soils.);
2. A general location map (e.g., portion of a city or county map) and a site map. The site map shall include descriptions of the following:
   a. Drainage patterns and approximate slopes anticipated after major grading activities;
   b. The total area of the site and the area of the site that is expected to be disturbed by excavation, clearing, grading and other construction activities during the life of the permit;
   c. Areas that will not be disturbed;
   d. Locations of erosion and sediment controls identified in the SWPPP;
   e. Locations where stabilization practices are expected to occur;
   f. Locations of off-site material, waste, borrow or equipment storage areas;
   g. Surface waters (including wetlands); and
   h. Locations where stormwater discharges to a surface water;
3. A description of available data on soils present at the site;
4. A description of the controls to be used to reduce pollutant discharges during construction
5. A description of the general timing (or sequence) in relation to the construction schedule when each erosion and sediment control is to be implemented;

6. An estimate of the pre-development and post-construction runoff coefficients of the site;
7. The name(s) of the receiving water(s);
8. Delineation of SWPPP implementation responsibilities for each site owner or operator;
9. Any existing data that describe the stormwater runoff characteristics at the site (such as data that may be collected during a site assessment).

Updating the SWPPP

The operator would be required to amend the SWPPP and corresponding erosion and sediment control practices whenever:

1. There is a change in design, construction, or maintenance that is expected to have a significant effect on the discharge of pollutants; or
2. Inspections or investigations by site operators, local, State, Tribal or Federal officials indicate that any erosion and sediment controls described in the SWPPP are ineffective in eliminating or significantly minimizing pollutant discharges.

Site Log Book/Certification

The operator would be required to maintain a record of site activities in a site log book, as part of the SWPPP. The site log book shall be maintained as follows:

1. A copy of the site log book would be required to be maintained on site and be made available to the permitting authority upon request. EPA recommends that the operator make a copy of the site log book available to the public upon request within a reasonable period;
2. In the site log book, the operator would be required to certify, prior to the commencement of construction activities, that the SWPPP meets all Federal, State and local erosion and sediment control requirements and is available to the permitting authority;
3. The operator would be required to have a qualified professional conduct an assessment of the site prior to groundbreaking and certify that the appropriate erosion and sediment controls described in the SWPPP have been adequately designed, sized and installed to ensure overall preparedness of the site for initiation of groundbreaking activities. The operator would be required to record the date of initial groundbreaking in the site log book. The operator would be required to certify that the site inspections, soil stabilization activities, and maintenance activities required by the proposed rule have been satisfied within 48 hours of actually meeting such requirements;
4. The operator would be required to post at the site, in a publicly-accessible location, a summary of the site inspection activities on a monthly basis. EPA recommends that the operator provide contact information for obtaining a copy of the SWPPP and a copy of the site inspection log book.

Site Inspections

The operator or designated agent of the operator (such as a consultant, subcontractor, or third party inspection firm) would be required to conduct regular inspections of the site and record the results of such inspection in the site log book. The specific activities that would require inspection and certification are:

1. After initial groundbreaking, operators would be required to conduct site inspections at least every 14 calendar days and within 24 hours of the end of a storm event of 0.5 inches or greater. These inspections would be required to be conducted by a qualified professional. During each inspection, the operator or designated agent would be required to record the following information:
   a. On a site map, indicate the extent of all disturbed site areas and drainage pathways. Indicate site areas that are expected to undergo initial disturbance or significant site work within the next 14-day period;
   b. Indicate on a site map all areas of the site that have undergone temporary or permanent stabilization;
   c. Indicate all disturbed site areas that have not undergone active site work during the previous 14-day period;
   d. Inspect all sediment control practices and note the approximate degree of sediment accumulation as a percentage of the sediment storage volume (for example 10 percent, 20 percent, 50 percent, etc.). Record all sediment control practices in the site log book that have sediment accumulation of 50 percent or more; and
   e. Inspect all erosion and sediment controls and record all maintenance requirements such as verifying the integrity of barrier or diversion systems (earthen berms or silt fencing) and containment systems (sediment basins and sediment traps). Identify any evidence of rill or gully erosion occurring on slopes and any loss of stabilizing vegetation or seeding/mulching. Document in the site log book any excessive deposition of sediment or ponding water along barrier or diversion systems. Record the depth of sediment within containment structures, any erosion near outlet and overflow structures, and verify the ability of rock filters around perforated riser pipes to pass water.
2. Prior to filing of the Notice of Termination, or the end of the permit term, a final site erosion and sediment control

inspection would be required to be conducted by the operator or designated agent. The inspector would be required to certify that the site has undergone final stabilization using either vegetative or structural stabilization methods and that all temporary erosion and sediment controls (such as silt fencing) not needed for long-term erosion control have been removed.

Stabilization

The operator would be required to initiate stabilization measures as soon as practicable in portions of the site where construction activities have temporarily or permanently ceased, but in no case more than 14 days after the construction activity in that portion of the site has temporarily or permanently ceased. This provision would not apply in the following instances:

1. Where the initiation of stabilization measures by the 14th day after construction activity temporarily or permanently ceased is precluded by snow cover or frozen ground conditions, the operator shall initiate stabilization measures as soon as practicable;
2. Where construction activity on a portion of the site is temporarily ceased, and earth disturbing activities will be resumed within 21 days, temporary stabilization measures need not be initiated on that portion of the site.
3. In arid areas (areas with an average annual rainfall of 0 to 10 inches), semi-arid areas (areas with an average annual rainfall of 10 to 20 inches), and areas experiencing droughts where the initiation of stabilization measures by the 14th day after construction activity has temporarily or permanently ceased is precluded by seasonably arid conditions, the operator shall initiate stabilization measures as soon as practicable.

Maintenance

The operator would be required to remove accumulated sediment from sediment traps and ponds identified as having sediment accumulations greater than 50 percent to restore the original design capacity,

State Regulations

States and municipalities have been regulating discharges of runoff from the construction and land development industry to varying degrees for some time. A compilation of state and selected municipal regulatory approaches was prepared by the EPA (USEPA 2002) to help establish the baseline for national and regional levels of control. They collect data by reviewing state and municipal web sites, summary references, state and municipal regulations, and stormwater guidance manuals. All states (and the selected municipalities) were contacted to confirm the data collected and to fill in data gaps. Eighty-seven percent of the state agencies, but a much smaller percentage of municipalities, responded. The state and municipal regulatory data are described below and the complete data summaries are included in Appendix 1A. Table 1A.1 lists example exemptions and waivers, Table 1A.2 shows some preferred practices, and Table 1A.3 lists allowed practices. These three tables include information for both local regulations and some state regulations. Table 1A.4 was prepared by the EPA (USEPA 2002) and lists some specific requirements (numeric standards, design storm frequency, soil stabilization requirements, and inspection frequencies). It is expected that all of the information on these tables may not be currently accurate, but they do show a good distribution of information. It is always necessary to contact the local planning departments and the regional NPDES authority to obtain the most recent compliance requirements.

Compilation of State and Municipal Existing Control Strategies, Criteria, and Standards

A summary of criteria and standards that are implemented by States and municipalities as of August 2000 are presented in Tables 1.3 and 1.4, respectively. The EPA (USEPA 2002) concluded that State requirements are generally equal to, or less stringent, than municipalities that are covered under the

TABLE 1.3. State or Regional Planning Authority Requirements for Water Quality Protection (USEPA 2002).

Standard	Number of States with Requirement*	Percent of National Developed Acreage with Requirement	Percent of National Developed Acreage without Requirement	Percent of National Developed Acreage without Information
Solids or sediment reduction	11	24	61	15
Numeric effluent limits for TSS, settleable solids, or turbidity	2	11	76	13
Numeric design depth or volume for water quality treatment	22	53	28	19
Habitat/biological measures	3	7	80	13
Physical in-stream condition controls	8	17	70	13
Water quality or effluent monitoring requirement	3	6	83	11

TABLE 1.4. Municipal Planning Authority Requirements (USEPA 2002).

Standard	Percent of Municipalities Reviewed with Requirement	Percent of Municipalities Reviewed without Requirement	Percent of Municipalities without Information
Design storm for peak discharge control	39	45	16
Solids or sediment percent reduction	7	77	16
Numeric design depth, storm, or volume for water quality treatment	NA	NA	NA
Design storm for flood control	39	16	23
Habitat/biological measures	3	65	32
Physical in-stream condition controls	10	58	32

NA = Not Available
Note: This table reflects data collected from 31 municipalities.

federal Clean Water Act NPDES Stormwater Program because State requirements apply to all developments within their boundaries including single site development and low-to-high density developments. NPDES Stormwater Program-designated municipalities generally have a population of 100,000 or more and can collect and fund the resources necessary to design, implement, and monitor separate and potentially more stringent stormwater management programs. Table 1.3 contains responses from 47 of the 54 State controlling agencies. The total is greater than 50 because Florida has 5 intrastate regional authorities. Some State data were uncertain and repeated contacts to the responsible State agencies to confirm the data were not returned. For the same reason, some of the data sought from municipal agencies also were not available for the summaries.

The data collected, as shown in Table 1A.4, reflect a cross section of the U.S. by location, but are representative mostly of municipalities that have a population of 100,000, or greater with relatively few municipalities of smaller populations represented. Thirty-one municipalities are included in the summary tables, which is a relatively small data set compared to the approximately 240 municipalities with NPDES programs and nearly 3,000 municipalities nationwide. The data presented for the States in Table 1.3 is fairly comprehensive, while data for the municipalities presented in Table 1.4 is not comprehensive, but does reflect the diversity of management techniques used at the municipal level.

Tables 1.3 and 1.4 indicate that the following key erosion and sediment control measures are being employed by States and municipal/regional authorities to implement the NPDES Stormwater Program:

- Stormwater controls designed for peak discharge control
- Stormwater controls designed for water quality control
- Stormwater controls designed for flood control
- Specified depths of runoff for water quality control
- Percent reduction of loadings for water quality control (primarily solids and sediments)
- Numeric effluent limits for water quality control (primarily total suspended solids, settleable solids, or turbidity)
- Control measures for biological or habitat protection
- Control measures for physical in-stream condition controls (primarily streambed and streambank erosion).

The water quantity control measures for peak discharge and runoff volume controls that apply to the post-development conditions typically are not applicable during the construction phase when the site is disturbed. Pollutant control measures are commonly required during the construction phase, though the requirements for post-development stormwater management are broader and potentially more stringent.

A variety of manuals and documents were used by the EPA (USEPA 2002) to obtain information on design and effectiveness of various erosion and sediment controls, including:

1. State design manuals such as the:
   - *Virginia Erosion and Sediment Control Handbook* http://www.dcr.state.va.us/sw/e&s-ftp.htm
   - *Maryland Stormwater Design Manual* http://www.mde.state.md.us/environment/wma/stormwatermanual
   - *Denver Urban Drainage Criteria Manual* http://www.udfcd.org
2. Guidance documents such as the
   - *Texas Nonpoint Source Book* http://www.txnpsbook.org
   - *EPA's National Menu of BMPs* http://www.epa.gov/npdes/menuofbmps/menu.htm
3. Consensus design manuals such as manuals of practice on stormwater design developed by ASCE and the Water Environment Federation (ASCE and WEF, 1992 and 1998) have been used to determine various management strategies.

Links to on-line manuals and guidance documents are provided on EPA's website at http://www.epa.gov/waterscience/guide/construction/.

State Erosion Control Handbooks Available on the Internet

Alabama
Alabama Handbook for Erosion Control
http://swcc.state.al.us/erosion_handbook.htm

California
California Storm Water BMP Construction Handbook
http://www.swrcb.ca.gov/stormwtr/index.html

Colorado
Denver Urban Drainage Criteria Manual
http://www.udfcd.org

Delaware
Delaware Erosion and Sediment Control Handbook
http://www.dnrec.state.de.us/dnrec2000/divisions/soil/stormwater/stormwater.htm

Florida
Florida Development Manual: A Guide to Sound Land and Water Management
http://www.dep.state.fl.us/water/nonpoint/urban2.htm

Georgia
Georgia Storm Water Management Manual
http://www.atlantaregional.com/water/waterquality/stormwatertaskforce.html

Idaho
Catalog of Storm Water BMPs for Idaho Cities & Counties
http://www2.state.id.us/deq/water/stormwater_catalog/index.asp

Louisiana
State of Louisiana Nonpoint Source Pollution Management Program—Construction
http://nonpoint.deq.state.la.us/manage10.html

Maryland
Maryland Stormwater Design Manual
http://www.mde.state.md.us/environment/wma/stormwatermanual

Maryland Storm Water Design Manual, Volumes I & II
http://www.mde.state.md.us/Programs/WaterPrograms/SedimentandStormwater/stormwater_design/index.asp

Massachusetts
Erosion and Sedimentation Control Guidelines: a guide for planners, designers, and municipal officials
http://www.mass.gov/dep/brp/stormwtr/stormpub.htm

Minnesota
Protecting Water Quality in Urban Areas: A Manual
http://www.pca.state.mn.us/water/pubs/sw-bmpmanual.html

Urban Small Sites Best Management Practice Manual
http://www.metrocouncil.org/environment/watershed/bmp/manual.htm

Missouri
Protecting Water Quality: A Construction Site Water Quality Field Guide
http://www.dnr.state.mo.us/wpscd/wpcp/wpcp-guide.htm

New Hampshire
Managing Storm Water as a Valuable Resource
http://www.des.state.nh.us/dwspp/stormwater.pdf

New Jersey
Revised Manual for New Jersey: BMPs for Control of Nonpoint Source Pollution from Storm Water
http://www.state.nj.us/dep/watershedmgt/bmpmanual.htm

New York
New York State Stormwater Management Design Manual
http://www.dec.state.ny.us/website/dow/toolbox/swmanual/

New York State Standards and Specifications for Erosion and Sediment Control
http://www.dec.state.ny.us/website/dow/toolbox/escstandards/index.html

Ohio
Storm Water Program—Factsheets, Forms, & Check Lists
http://www.epa.state.oh.us/dsw/storm/

Oregon
BMPs & Storm Water Pollution Control Plan
http://www.deq.state.or.us/wq/wqpermit/wqpermit.htm

Pennsylvania
Handbook of BMPs for Developing Areas
http://www.pacd.org/products/bmp/bmp_handbook.htm

South Carolina
Sediment, Erosion, & Storm Water Management
http://www.scdhec.net/water/html/erfmain.html

Tennessee
Tennessee Erosion and Sediment Control Handbook
http://www.state.tn.us/environment/wpc/

Knoxville BMP Manual
http://www.ci.knoxville.tn.us/engineering/bmp_manual/

Texas
Texas Nonpoint Sourcebook—Interactive BMP Selector
http://www.txnpsbook.org/SiteMap.htm

Utah
UPDES Storm Water Home Page
http://www.deq.state.ut.us/EQWQ/updes/stormwater.htm

Virginia
Virginia Erosion and Sediment Control Handbook
http://www.dcr.state.va.us/sw/e&s-ftp.htm

Northern Virginia BMP Handbook: A Guide to Planning and Designing BMPs in Northern Virginia
http://www.novaregion.org/pdf/NVBMP-Handbook.pdf

Washington
Storm Water Management Manual for Western Washington
http://www.ecy.wa.gov/programs/wq/stormwater/manual.html#copies

King County Storm Water Pollution Control Manual
http://dnr.metrokc.gov/wlr/Dss/Spcm.htm

Wisconsin
Wisconsin Construction Site Best Management Practice Handbook
http://dnr.wi.gov/org/water/wm/nps/stormwater/constrforms.htm#wicon

BASIC CONTROL OF CONSTRUCTION SITE RUNOFF

One of the main problems associated with the control of construction site runoff is that the actual monitored field performance of most construction site erosion controls has been disappointedly low. Excellent filter fence installations (well maintained and well constructed) provide only about 50% control, at a maximum. Typical monitored performance has shown negligible benefits due to installation and maintenance problems. The use of rock berms in channels are more robust, but still provide less than about 25% suspended solids control. Sediment ponds can be designed to provide good control (>50%) of suspended solids, but they would have to be very large (about 2% of the drainage area) to provide significant removal of fine sediment. The effluent turbidity from sediment control ponds at construction sites is still high, unless additional controls are used.

Prevention is therefore the best and typically least expensive control solution. Typical preventative measures include:

1. Divert flows around exposed soils
2. Schedule site activities to minimize amount of exposed soil
3. Use temporary mulch
4. Use erosion control blankets in sensitive areas (concentrated flow channels, steep slopes)

Basic Goals and Performance Standards for Erosion and Sediment Control

The most common goal of jurisdictions implementing an erosion and sediment control program is protection of public safety, water quality, or other aquatic related resources such as habitat or fisheries. A more realistic goal is minimization, "to the extent practical," of off-site impacts. That is because, even with the best designs, the process of site development with its associated earth disturbance can still create adverse downstream impacts because of the limited effectiveness of current erosion and sediment practices, especially when severe storm events exceed the design capacity for these practices. The intent of erosion and sediment control programs should be to minimize the potential for off-site impacts by reducing the aerial extent and time duration of impacts.

In defining how a program can minimize impacts, a dual strategy is recommended. The program should seek first to prevent erosion from occurring and also to seek to reduce the associated sedimentation. *Prevention* practices include sequencing construction to reduce areas of disturbance, conducting land disturbance during the dry season, establishing limits on areas of disturbance during the wet season, and timely stabilizing (temporary or permanent) disturbed areas. *Reduction* of impacts would follow using traditional erosion and sediment control practices such as stabilized construction entrances, silt fences, diversion dikes, sediment traps and basins. Reduction practices are most effective at removing coarser sediments, while preventive practices are more effective at controlling silt or clay particles by preventing their initial movement. In summary, a basic goal of erosion and sediment control programs should be to minimize off-site impacts by following a philosophy of first preventing erosion and then maximizing control of sedimentation on-site.

Once the program's goal is determined, it is necessary to establish an achievable performance standard which will form the basis for the development of design criteria for the various erosion and sediment control practices to be used. Performance standards can be either technology based or water-quality based. Technology-based standards are the most common. They typically are related to a reduction in the level of suspended solids (e.g. 80%) leaving a site, or may be expressed in terms of retaining sediment on-site. The former standard is appropriate because there is a good understanding of the processes involved in the reduction of suspended solids. The latter performance standard addresses potential adverse impacts beyond water quality such as public safety concerns associated with tracking sediments onto public streets or sediment clogging of runoff conveyances which can increase flooding. Water quality-based standards often are a "backstop" since most environmental laws prohibit violations of water quality standards. A common water-quality-based standard, for example, would be a requirement that discharges may not increase turbidity, measured in NTU, above background conditions by more than a specified amount (such as 50 NTU).

Design Criteria

Once a performance standard has been established, then design criteria need to be developed for the individual erosion and sediment controls. By providing both performance standards and design criteria, site planners and engineers can select those practices which will work best on a given site because of its specific soils, topography, slopes,

Diversion of Flows

A small berm and sodded swale divert flows from newly graded and mulched hillside.

A diversion downslope pipe at a highway construction site (during installation) to prevent erosive flows from damaging an unprotected slope.

Minimize Exposed Soil

Most construction sites are characterized with large expanses of unprotected soil, even after utilities are installed (WI DNR photo).

Large unprotected area at new commercial site.

Unprotected newly graded area at highway expansion project.

Temporary Mulch (Minimal tacking and no netting to retain material on site for short periods)

Spray mulch blown to protect exposed soil.

Complete ground cover after hydro-mulching.

Erosion Control Blankets

Vegetation starting to grow through erosion control netting at highway construction site.

Newly installed erosion control mats on steep highway embankment.

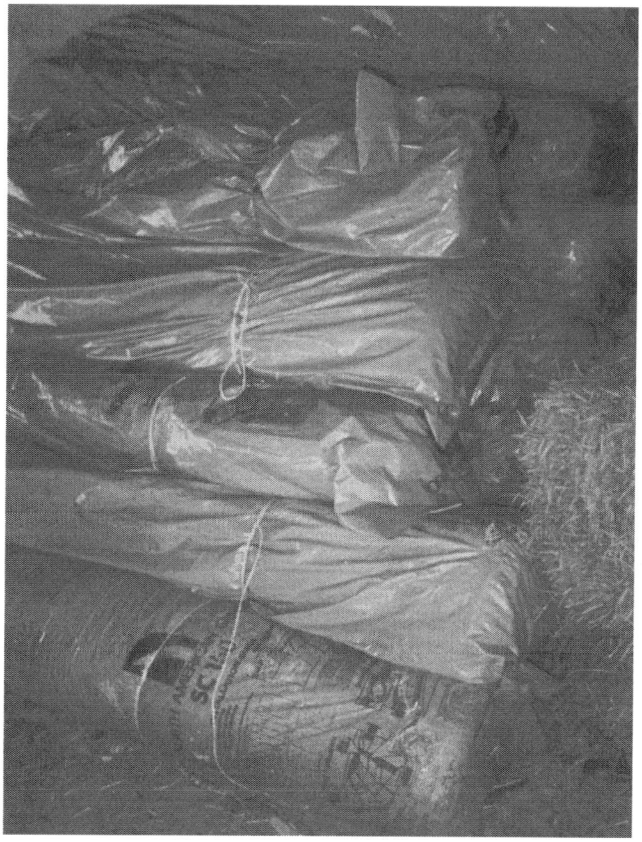

Stored erosion mats at construction site.

geology, and hydrology characteristics. Design criteria need to be specified for both prevention and reduction practices.

Specific design criteria should be included for at least two prevention practices. First, a maximum area of disturbance at any one time should be specified, with a variance provision for specific activities which cannot meet that limitation. Second, a maximum time frame for either temporary or permanent site stabilization upon cessation of grading needs to be set. As an example, Delaware's program limits site disturbance at any one time to a maximum of 20 acres and requires site stabilization within 14 days when an area is not being actively worked. The specific design criteria for prevention practices will depend largely on local rainfall patterns and associated runoff characteristics. If there is a defined seasonality to the rainfall, the criteria may be primarily directed towards activities conducted during the wetter seasons. This approach is used by the Puget Sound Water Quality Management Program which establishes seasonal limits for disturbed areas.

Design criteria for reduction practices often are based on sizing criteria, either in terms of contributing drainage area or storage volume, or both. Most programs establish a minimum size for sediment traps and basins, such as 1,800 cubic feet per acre of drainage area. This volume figure was developed years ago by the Natural Resources Conservation Service (then the Soil Conservation Service) to achieve a 70% reduction in suspended solids on a Piedmont hydrologic group C soil. This volume was then used as a design criteria as a minimum standard for site design.

Exemptions and Waivers

If the erosion and sediment control program is integrated with the stormwater management program, the exemptions and waivers should be consistent, but not necessarily identical. There are activities which, due to their limited size, should not be required to provide permanent stormwater management, but which should be required to implement erosion and sediment control. An example is single family home construction that is not part of a larger development.

The most common and simplest approach for establishing exemptions and waivers is based on the amount of disturbed area. This approach is easily implemented since determining the amount of disturbed area is simple. The size of the disturbed area for an exempt activity will depend to some extent on local conditions such as rainfall patterns, soil types, and topography. It is recommended that the threshold size of disturbance be relatively small, such as 5,000 square feet. This emphasizes that erosion and sediment control are integral components of site development. It also helps to minimize potential cumulative impacts if many construction projects are on-going within a watershed.

There also has to be some flexibility for unforeseen types of activities for which pre-construction review and approval would be an undue hardship and not be in the best public interest. These activities typically are of an emergency nature, such as those required after an extreme storm event which creates situations needing an immediate response. Such activities must still implement erosion and sediment controls, but implementation should be based on requirements defined on-site. Alternatively, a special process can be established which calls for submission and review of plans within an appropriate time frame.

Design Assistance and Guidance

To maximize program effectiveness and the proper use, design, construction, and maintenance of erosion and sediment controls, it is essential to have a design guidance document available for designers, developers, and contractors. Most areas of the country already have one available. To a large extent, the manuals are very similar to one another, either based on the early Virginia manual, or the manuals prepared by the SCS (NRCS) for various states. In some cases, special local practices have been developed and the manuals are more unique. For each practice, the design manual should specify the purpose, applicability in different site situations, sizing, materials, construction standards, maintenance needs, and operational information. The manual must include both structural and vegetative practices. Many of the structural practices, except for storage volumes of sediment traps or basins, tend to have universal design criteria. Vegetative practices must include local considerations such as the types of plant materials and how they are best established and maintained. It is critical in all locales that design manuals consider local conditions, especially rain characteristics, typical soils, and topography. This hinders the simple transfer of design manuals throughout the country.

Checklists to Ensure Plan Completeness and to Aid in Regulatory Review

Some regulatory agencies responsible for erosion and sediment control plan review have developed a series of checklists to aid in quickly determining whether the required plan components were included in the submitted package. Pennsylvania has developed two checklists to fulfill this purpose. Use of these checklists in plan development is useful to the designer to ensure that he/she has addressed the pertinent issues and demonstrated how the plan has met the regulations. According to the PA Department of Environmental Protection (DEP), "the Complete Plan Checklist is used to determine if an erosion and sediment control plan includes all required elements. This checklist is intended to serve as a tool to determine whether an erosion and sediment control plan addresses all eleven items required by Section 102.4(b)(5). It need not be included as part of the plan submittal."

The E&S Control Plan Technical Review Checklist is

COMPLETE PLAN CHECKLIST

Project: ______________

I. Existing topographic features of the project site.

A. The existing topographic features of the project site and the immediate surrounding area are shown on maps included in the drawings .... ☐

B. A location map has been provided (8½" x 11" copy of a USGS map with the outline of the project area) .... ☐

II. The Types, depth, slope, locations and limitations of the soils

A. A soils map with the project area outlined has been provided .... ☐

B. Physical characteristics of the soil types and their limitations are addressed in the narrative .... ☐

C. Construction techniques or special considerations to address the soil(s) limitations are noted on the drawings.. ☐

III. Characteristics of the earth disturbance activity

A. Limits of the project are shown on the plan map(s) .... ☐

B. Original and final contours are shown on the plan map(s).. ☐

C. Past, present and proposed land uses are addressed in the narrative .... ☐

IV. The amount of runoff from the project area and its upstream watershed area

A. Drainage areas to hydraulic BMPs are shown on plan map(s) .... ☐

B. Calculations are provided which show anticipated peak flows for the design storms ... ☐

V. The location of waters of the Commonwealth which may receive runoff within or from the project site.

A. The location(s) of streams or other waterbodies which may receive site runoff are shown on the plan map(s) .... ☐

B. The Chapter 93 classification of streams or other waterbodies which may receive site runoff is addressed in the narrative .... ☐

Figure 1.6. Complete Plan Checklist for Pennsylvania Erosion and Sediment Control Plans (PA DEP 2000).

VI. Locations and types of perimeter and on site BMPs

A. Plan map(s) show locations of proposed temporary BMPs to control runoff and provide sediment removal .... ☐

B. Plan map(s) show locations of proposed permanent BMPs to control erosion .... ☐

C. Construction details and specifications for all proposed BMPs are shown on the plan map(s) .... ☐

VII. Sequence of BMPs installation & removal

A. A construction sequence has been provided on the plan map(s) .... ☐

VIII. Supporting calculations

A. Supporting calculations for all proposed BMPs are included in the narrative .... ☐

IX. Plan drawings

A. Plan drawings are complete and legible .... ☐

X. Maintenance Program

A. A maintenance program has been provided .... ☐

XI. Measures for the recycling or disposal of materials from the project site.

A. A program for the recycling or disposal of materials associated with or from the project site has been provided .... ☐

Figure 1.6 (continued). Complete Plan Checklist for Pennsylvania Erosion and Sediment Control Plans (PA DEP 2000).

E & S CONTROL PLAN TECHNICAL REVIEW CHECKLIST

Project: ______________________ NPDES/Project No. ______________

Project Location: ______________________ Date: ______________________

Check-off: c = Complies, d = Deficient, na = Not applicable
Item Location: D = E&S Drawings, N = E&S Narrative, D&N = Drawings & Narrative

102.4(b)(3) **"The Erosion and Sediment Control Plan shall be prepared by a person trained and experienced in erosion and sediment control methods and techniques, and shall be designed to minimize the potential for accelerated erosion and sedimentation".**

Name ______________ Address ________________ Telephone No. __________ D&N
Qualifications __________ N

102.4(b)(5)(i) **"The existing topographic features of the project site and the immediate surrounding area."**

__________	Legible mapping	D
__________	Existing contours	D
__________	Existing improvements, i.e. roads, buildings, utilities , etc.	D
__________	Existing streams, wetlands, receiving watercourses, etc.	D
__________	Sufficient surrounding area	D
__________	Location map, i.e. USGS	D or N

102.4(b)(5)(ii) **"The types, depth, slope, locations and limitations of the soils"**

__________	Types, slopes, & locations of soil types	D
__________	Soil type use limitations and resolutions	N
__________	Hydric soils	N

102.4(b)(5)(iii) **"The characteristics of the earth disturbance activity, including the past, present, and proposed land uses and the proposed alteration to the project site."**

__________	Proposed NPDES boundary and limits of construction	D
__________	Proposed contours/grades	D
__________	Proposed waterways & stormwater management facilities	D
__________	Proposed improvements, i.e., roads, buildings, utilities, etc.	D
__________	Complete mapping symbols legend and north arrow	D
__________	Past, present and proposed land uses	N

102.4(b)(5)(iv) **"The amount of runoff from the project area and its upstream watershed area."**

__________	Maximum during construction drainage areas	D
__________	Offsite drainage area(s) on USGS quadrangle map	N
__________	Peak flow calculations for all channels	N

Figure 1.7. Pennsylvania E&S Control Plan Technical Review Checklist (PA DEP 2000).

102.4(b)(5)(v) "The location of waters of the Commonwealth which may receive runoff within or from the project site and their classification pursuant to Chapter 93 of this title."

________	Existing streams, wetlands, floodway, etc.	D
________	Receiving watercourses	D
________	Chapter 93 classification of streams or other waterbodies	N

102.4(b)(5)(vi) "A written depiction of the location and type of perimeter and on site BMPs used before, during, and after the earth disturbance activity."

102.4(b)(5)(vii) "A sequence of BMP installation and removal in relation to the scheduling of earth disturbance activities, prior to, during and after earth disturbance activities".

________	Complete and site specific sequence of BMPs installation	D
________	Activities planned to limit exposed areas	D&N
________	Removal of temporary BMPs	D&N

102.4(b)(5)(viii) "Supporting calculations"

102.4(b)(5)(ix) "Plan Drawings"

Channels

________	Locations ________ Contours and Grades ________ Complete details	D
________	Capacity & freeboard calculations	N
________	Protective lining calculations	N

Sediment Basins

________	Locations ________ Contours	D
________	Complete berm & outlet details ________ Cleanout information	D&N
________	Capacity info ________ Discharge calculations	N
________	Dewatering calculations	N
________	Discharge to waters of the Commonwealth or approved alternative	D
________	Structurally sound	D&N

Sediment Traps

________	Locations ________ Contours	D
________	Complete berm & outlet details ________ Cleanout information	D&N
________	Capacity information ________ Discharge calculations	N
________	Discharge to waters of the Commonwealth or approved alternative	D

Silt Fencing

________	Locations ________ Complete Details	D

Outlet Protection

________	Locations ________ Complete Details	D
________	Design Calculations	N

Figure 1.7 (continued). Pennsylvania E&S Control Plan Technical Review Checklist (PA DEP 2000).

Other BMPs (specify) ______________________

____________ Locations	____________ Complete Details		D
____________ Design Calculations			N

Temporary Stabilization

	Seed	Lime	Fertilizer	Mulch	Others	
Types	______	______	______	______	______	D
Rates	______	______	______	______	______	D

Permanent Stabilization

	Seed	Lime	Fertilizer	Mulch	Others	
Types	______	______	______	______	______	D
Rates	______	______	______	______	______	D

102.4(b)(5)(x) "A maintenance program which provides for inspection of BMPs on a weekly basis and after each measurable rainfall event, including the repair of the BMPs to ensure effective and efficient operation."

____________	Inspection schedule	D
____________	Maximum sediment storage elevation/level in BMPs	D
____________	Time frames for completing specific maintenance & repairs for each type of BMP proposed.	D
____________	Site stabilization repair parameters & directions	D
____________	Disposal directions for sediment removed from BMPs	D

102.4(b)(5)(xi) "Procedures which ensure that the proper measures for the recycling or disposal of materials associated with or from the project site will be undertaken in accordance with Department regulations."

____________	Project construction wastes are identified	N
____________	Directions for recycling/disposal of construction wastes	D
____________	Soil/rock disposal areas provided with BMPs	D&N

Figure 1.7 (continued). Pennsylvania E&S Control Plan Technical Review Checklist (PA DEP 2000).

used to determine the technical adequacy of an erosion and sediment control plan. "This checklist is to be used by the reviewing agency to ensure the erosion and sediment control plan meets the requirements of Chapter 102 and the standards of the Department's Erosion and Sediment Pollution Control Program Manual, No. 363-2134-008 (January 2000), as amended and updated. It should not be included as part of the plan submittal."

EXAMPLE CONSTRUCTION SITE EROSION CONTROL AND STORMWATER MANAGEMENT REQUIREMENTS[1]

Rationale and Purpose

The objective of an effective construction site erosion control and stormwater management ordinance is to protect the local water resources from water quality degradation from many potential sources and activities. Specific provisions of an ordinance may:

- Provide for treatment practices which promote the public health, safety, and general welfare, and
- Restrict or prohibit discharges which are dangerous to, or potentially may increase pollution of, the watershed and public water supply.

Standards and Specifications for Construction Site Erosion Control

Actual monitoring of construction sites (especially research on the yields and delivery of construction site erosion material) has shown that the type of development (i.e., final land use) has very little effect on erosion rates. Instead, construction site erosion losses vary with the amount of land disturbed, the duration of that disturbance, and the presence of effective erosion controls. A watershed protection ordinance, therefore, should require erosion control permits for all types of development and exclude only small construction projects (such as those disturbing less than 2,000 square feet, or involving excavation and/or filling of less than 500 cubic yards of material). Thus, projects such as home additions or household gardening activities will generally be too small to require permits, while construction of most individual homes and all larger types of development would require permits. Even small land disturbing activities should have erosion controls, even if formal permits are not required. In most cases, these small projects would only require simple good housekeeping provisions, good drainage, simple mulching, and a quick project period.

Construction site monitoring has also revealed that sediment delivery (the amount of sediment leaving its source compared to the amount entering the receiving water) is very close to 100 percent for typical urban construction sites in developing areas. Watershed monitoring has shown that almost all of the sediment from construction areas that disturb more than about ten percent of a watershed, and about one-half of that from construction areas that disturb less than ten percent, reach the receiving water. These very large delivery ratios probably result from the normal practice of installing the storm drainage system during the initial construction phase, because sediment travels much more efficiently in conventional storm drainage systems than in natural sheetflows or in small tributary streams. The early installation of storm drainage systems also apparently makes sediment yield and delivery insensitive to site slope. An erosion control ordinance, therefore, should not exempt construction projects on the basis of percentage disturbance of a watershed, or construction site slope.

Vague regulations and general criteria regarding erosion control sometimes found in many erosion control ordinances should be replaced by criteria that specify when and where specific control practices are to be used. Such guidance should help site engineers as well as site plan reviewers and inspectors. In addition, specific criteria should promote more uniform construction site erosion control throughout the watershed.

The main purpose of construction site erosion control requirements is to prevent sediment and other pollutants from leaving construction sites. The secondary purpose is to significantly reduce the quantity of any "escaped" material that reaches receiving waters. Past research projects that have characterized construction erosion discharges and transport processes have concluded that very large amounts of sediment, phosphorus, and other pollutants erode from most construction sites. Sediment yields from uncontrolled construction sites may, for example, be several hundred to several thousand times the annual sediment yields from most developed urban areas. Small areas of active construction may therefore contribute much more pollution to receiving waters than entire cities or surrounding agricultural lands. By requiring reasonable and effective construction site erosion controls for most developing areas, discharges of many pollutants to receiving waters can be greatly reduced.

Site Erosion Control Requirements

Site erosion control requires three main elements to protect downslope property, the storm drainage system, and receiving waters. The first main element involves diverting

[1]This discussion (and much of the preceding material) is based on material and experiences from a number of individuals and agencies besides the authors. Earl Shaver, currently of the Auckland Regional Council, New Zealand, was helpful in the preparation of some of the material reflecting his many years of experience in Maryland and Delaware. While working at the Wisconsin Department of Natural Resources, Bob Pitt was greatly influenced by his colleagues while preparing early versions of the WI model ordinance and later by environmental attorneys and other reviewers when he prepared an early version of the watershed protection ordinance for the Cahaba River watershed in Jefferson County, AL. These discussions therefore reflect a compilation of ideas that are presented to aid local agencies in meeting NPDES erosion control requirements.

The lack of an appropriate diversion structure to safely drain water down sensitive slopes can cause much damage and sediment loss.

water from upslope, undisturbed areas so that it does not flow across disturbed land. This preventive measure can reduce the volume of water and energy available to transport soil exposed by construction activity.

The second element requires mulching disturbed ground at time intervals that permit necessary grading but that also reduces erosion losses during intense rains. Of course, careful planning to decrease the amount of land disturbed at one time and to speed the entire construction process is assumed. Site erosion control, on-site mulch, or temporary vegetation is needed in order to control erosion from disturbed sites during periods of site inactivity or when the erosion potential is very high. In some areas of the country, storms having high erosion potential can occur at any time, so immediate on-site mulching is a very important aspect of effective construction site erosion control programs. A risk assessment of the erosion potential of Jefferson County, AL, rains showed that rains occur about every three days. Although about three rains could occur during any seven-day period, the probability of a rain with high erosion potential during any seven-day period is relatively low. The probability increases with longer periods of time, however. A time limit of 14 days of no activity before mulching is required on portions of the construction site is a compromise between potential erosion damage and construction scheduling problems. Unfortunately, many disturbed sites are commonly left inactive for periods much longer than 14 days, resulting in very high probabilities of severely erosive rains occurring when sites are left disturbed and inactive. Stabilization of these inactive but disturbed areas is needed, therefore, to prevent site erosion, to eliminate the cost of regrading severely eroded areas, and to protect off-site areas from erosion products. In many cases, better timing of grading operations could also reduce the time an area is left disturbed.

The third site erosion control element requires downslope controls to minimize the quantity of erosion products that leave the site. This element is necessary because significant exposed land will always occur at construction sites. Moreover, plantings can require several weeks to become established and capable of reducing erosion. For small sites (less than 10 acres) with no channelized flow, filter fences or other perimeter controls are probably adequate. These controls are fragile, however, and suitable only for sheetflows at low velocities. When larger flows can be expected, sedimentation basins are needed because high flow rates can quickly destroy filter fences.

Downslope controls alone cannot offer adequate protection from severely erosive rains that may occur at any time during the construction season. Because such rains could completely and quickly wash out a filter fence or silt-in a sedimentation basin if a site had no other protection, downslope controls should be installed in conjunction with above-site flow diversions and site mulching or plantings. Together, these three erosion control elements can significantly reduce potential erosion damage, which can be very expensive, if not impossible, to remedy once it has occurred. Nevertheless, occasional severe rains occurring at the "wrong time" in relation to site protection requirements may still cause downstream damage. The intent of an erosion control ordinance is to give site planners and engineers as much flexibility as possible in applying required specifications and standards to proposed projects. Although

Unattended severely eroded land causes great amounts of sediment loss and requires site regrading.

Filter Fencing for Small Drainage Areas

Sediment Ponds for Larger Construction Areas

construction site regulations may appear restrictive, they should allow many choices about matters such as location of storage piles, mulch types, timing of grading, etc.

Summary of Erosion Control Requirements

As included in many regulations, including the proposed EPA *Effluent Guidelines for the Construction and Development Industrial Category* (June 24, 2003 *Federal Register,* 40 CFR Parts 122 and 450), all erosion control efforts should consist of three basic elements:

1. Divert upslope water around the disturbed site, or pass it through the site along a protected channel,
2. Expose disturbed areas for the shortest possible time (allowing a maximum time limit of about 14 days for inactive disturbed land before required protection), either through improved construction phase scheduling, or through temporary or permanent mulching, and
3. Treat any runoff water before it leaves the site (by perimeter filter fencing, or if a "large" site, with a sediment pond).

This triple approach is needed because of the potential failure of any one system due to random rains that may cause severe site and erosion damage. As an example, if a temporary seeding is not fully established, a moderate rain of greater than 0.5 inch (which may occur about every 10 days in the Birmingham, AL, area) can easily wash it away. In addition, special considerations are also necessary, such as the following examples:

- Construction wastes (do not allow their burial on the site),
- Tracking restrictions (all main site roads, which have greater than about 25 vehicles per day traffic, and all site entranceways have to be graveled, and travel is restricted off these graveled areas),
- Treat dewatering wastes before discharge,
- Protect storm drain inlets (such as with straw bale or filter fence barriers),
- Locate material storage piles away from storm drain inlets (by at least 50 feet), and if left for a long time (greater than 14 days), then they must be covered, mulched, or surrounded with a perimeter filter fence or straw bale barrier,
- Direct all on-site concentrated runoff (especially down steep slopes) along protected channels, or in flexible down drains,
- Require contractor to inspect all erosion controls on the site and make necessary repairs at least weekly and after large rains (greater than about 0.5 inch),
- Perform construction vehicle maintenance in special protected areas.

Barrier fencing setting outer limits of disturbance at construction site.

Preventative measures and "good housekeeping" controls should also be used at construction sites.

Truck being cleaned as it leaves construction site for public right-of-way.

Vehicle being cleaned as it leaves a construction site to prevent debris from affecting pubic roads.

NEED FOR ADEQUATE DESIGN AND INSPECTION

Adequate design specifications, especially those based on local experience, can minimize potential construction site erosion problems. Construction site erosion controls may fail for several reasons. Unusual rains that exceed the design capacities of even correctly constructed and maintained control facilities may cause their failure. Most construction erosion controls are relatively fragile and cannot survive large rains. However, a wet detention basin installed early during the construction period will act as a good sediment trap during a wide range of rains. In-stream detention facilities that receive large amounts of runoff from above a construction project can be easily damaged during large rains. The basin must be cleaned (dredged) often during construction and after final landscaping, for the construction period can produce as much sediment as many years of "normal" urban runoff. Large rains can also damage filter fences and other barriers and can severely erode culverts and waterway diversions. Failed controls are not only unable to reduce expected large amounts of erosion materials during severe rains but also may discharge previously retained sediment.

Obviously, downslope controls (filter fences and sediment ponds) must be installed first, followed by upslope diversions and then any on-site channel protection measures. Construction limit barriers may also need to be installed. Only when these controls are suitably installed should actual construction begin.

Improperly located, designed, constructed, or maintained control devices produce little benefit. A common example of a poor location for a control device is the placement of filter fences in established waterways that drain large areas. Filter fences slow down water passing through them and create small detention areas. Particles then settle from the ponded water. They can be designed as small wet detention basins, based on their allowable water seepage rates (outfall velocities), and not as filtration devices. They are supposed to be used to control shallow sheetflows. When placed in channels draining areas that are too large, backed up water may topple the filter fence, or the stream may increase in elevation and collapse the fencing, or the water may flow around the filter fence edges. Similar problems exist when straw bales are placed in large waterways. These devices are best used to control sheetflows before they enter the drainage channels. If large drainage channels cannot be diverted and must pass through a project, filter fencing must be placed appropriately to control sheetflows entering the channel. Well designed wet detention (sediment) basins may also be needed below the site.

Probably the most common reason for failure of construction site erosion control devices is inadequate maintenance. These devices are often reluctantly installed and then ignored. If control devices are properly constructed, but not properly or frequently maintained, very little benefit may be expected. Newly installed devices will perform as initially expected until their "capacity" is exceeded. Filter fences, for example, should be maintained before the material that accumulates behind them becomes excessive. More importantly, the integrity of the fence also needs to be checked frequently. Many filter fences at construction sites are undermined or bypassed because of large flows or large sediment accumulations. Sedimentation basins, silt traps, catchbasins, etc., also need to be cleaned frequently. The cleaning frequency of these devices located in areas undergoing construction can be quite high because of the very large discharges of sediment from construction sites. Rill or gully erosion must be corrected immediately when first observed. Similarly, mulched or planted areas need frequent inspections and repairs before large amounts of material are lost. Proper plan reviews and adequate inspections by administrative officials can prevent many of the problems caused by improper location, construction, and maintenance of construction erosion and stormwater control devices.

Inspection During Construction[2]

During construction, inspections need to be made of both erosion and sediment controls and stormwater management facilities. Erosion and sediment controls must be inspected periodically throughout the construction process, especially after storms. Stormwater management systems need to be inspected at critical times during construction of the individual practices.

Inspection frequency needs to be flexible, corresponding to shifts in the intensity of activity occurring at the site. When active construction is occurring, erosion and sediment control inspections should be conducted on a specified, appropriate frequency. When work on the site stops temporarily, inspections should be done periodically to assure that erosion and sediment controls are being maintained and still working, and to ensure that work has not resumed. Ideally, inspections should be done at a specified regular time interval and after significant storm events. This allows any changes in site conditions to be observed, and ensures that erosion and sediment controls are still functioning as designed and approved. It is recommended that inspections be conducted by a public agency representative at least once every two weeks.

Inspection staff resources typically are insufficient to visit all active construction sites as frequently as needed. An implementation strategy decision must be made whether to visit fewer sites and completely follow the inspection

[2]Earl Shaver, Auckland Regional Council, New Zealand, prepared the following comments on construction site inspections based on his many years of developing and managing erosion and sediment control programs in Maryland and Delaware.

Improper use of erosion controls must be corrected before excessive damage occurs (J. Voorhees photo).

Inspections should require replacement of damaged mulch, or preferably the use of appropriate materials that are suitable for the site conditions.

Inspections must require replacement of damaged erosion controls (J. Voorhees photo).

Inspections must enforce needed maintenance before failure. This silt fence is retaining massive amounts of sediment and is near its limit and needs to be maintained soon.

Inspections need to monitor sediment accumulations. This dry sediment pond is almost full of accumulated material.

Lack of supplemental irrigation jeopardizes sodded areas.

Poorly covered mulched areas need to be remulched.

Excessive tracking due to insufficient or non-maintained graveled access needs to be corrected.

Damaged erosion controls need to be repaired or replaced as soon as possible.

Necessary Enforcement and Education

Needed enforcement actions need to be obvious (WI DNR photo).

Education of erosion control contractors is mandatory (Maryland DNR, Earl Shaver, photo). These straw bales are located on the dashes on the site erosion control plan map designating the site boundary controls.

procedures, or to conduct less comprehensive inspections at more sites. It is recommended that the inspection procedures be followed completely at sites which are inspected. Inspections need to be prioritized based on potential impacts, helping to assure compliance on tougher sites. Following the prescribed procedures also is important should legal enforcement action become necessary.

Inspectors should always attempt to contact an on-site individual who is responsible for the site grading activities. The contractor should be aware that the inspector is visiting the site even if the contractor does not accompany the inspector. This improves the dialogue that is important between the inspector and contractor. Highly visible inspections reinforce the commitment and importance a jurisdiction places on effective implementation of site controls. By knowing that the site will be inspected periodically, contractors are more likely to be aware of, and meet, site control responsibilities.

After completing the inspection, the inspector should leave an inspection report with the contractor, and should send a copy to the developer and possibly the property owner. The report should serve as a site report card, clearly documenting proper installation and maintenance of site controls as well as any deficiencies in site control implementation. If there is a violation, the inspection report initiates a "paper trail" which is integral to successful enforcement actions.

It is unlikely that public agencies will ever have enough inspectors, simply due to the large number of active construction projects at any time and to the resource limitations of stormwater management programs. A creative innovation to solve this problem is a partnership between the stormwater management program agency and the development community. This concept is being used in Delaware where the contractor or developer supplies their own inspectors. This person must attend and pass a State sponsored training course for inspectors. They are then responsible for inspecting the site at least once a week, completing an inspection form, and providing a copy of the form to the contractor, developer, and appropriate inspection agency. Having a "certified" private inspector on the site weekly can reduce the inspection frequency by the appropriate agency.

To improve the effectiveness of inspections, it is important to establish standard, well-documented inspection procedures. These procedures should specify in detail the actions an inspector conducts at a site, set out options and list steps to be taken when site compliance is inadequate, and establish an appeals process, should the inspector and with suitable developer disagree on matters. The procedures need to be developed in conjunction with available legal authorities and with suitable penalty provisions. Inspection of the stormwater management system during construction typically is not done on a regular schedule, but at certain stages of construction. For each type of construction site control practice, there are certain stages of construction where inspection is essential to assure proper construction and performance.

IMPORTANT INTERNET LINKS

The following are the main Internet links referenced in this chapter and provide much additional information, especially concerning the federal programs and resources. These are likely to change with time, but current linkage addresses can usually be found by using an Internet search tool.

EPA Office of Wastewater Management (OWM) information:
http://www.epa.gov/owm/

EPA Stormwater Program information, Final Phase II NPDES rule:
http://cfpub.epa.gov/npdes/home.cfm?program_id=6

Final Federal Register notice and supporting materials for Effluent Limits Guidelines for Erosion and Sediment control:
http://www.epa.gov/guide/construction

EPA Fact Sheet Series:
http://cfpub.epa.gov/npdes/stormwater/swfinal.cfm

EPA stormwater regulations:
http://cfpub.epa.gov/npdes/regs.cfm?program_id=6

EPA information on discharges from construction activities:
http://cfpub.epa.gov/npdes/stormwater/const.cfm

EPA National Menu of stormwater, and erosion and sediment control practices:
http://www.epa.gov/npdes/menuofbmps/menu.htm

EPA links to on-line manuals and guidance documents:
http://www.epa.gov/waterscience/guide/construction/.

State Water Pollution Control Program Grants Program
http://www.epa.gov/owm/cwfinance/pollutioncontrol.htm

Stormwater website
http://cfpub.epa.gov/npdes/home.cfm?program_id=6

Electronic Notice of Intent System
http://cfpub.epa.gov/npdes/stormwater/enoi.cfm

National Management Measures to Control Nonpoint Source Pollution from Urban Areas
http://www.epa.gov/owow/nps/urbanmm/

Smart Growth Program
http://www.epa.gov/livability/

Section 319 Nonpoint Source Management Program
http://www.epa.gov/owow/nps/cwact.html

The Construction Industry Compliance Assistance Center (http://cicacenter.org/) contains information and links to a wide variety of information, including state regulatory programs and manuals for sediment and erosion controls.

REFERENCES

ASCE and WEF. *Design and Construction of Urban Stormwater Management Systems.* 1992.

ASCE and WEF. *Urban Runoff Quality Management.* ASCE Manual and Report on Engineering Practice No. 87; WEF Manual of Practice No. 23. American Society of Civil Engineers, Reston, VA. Water Environment Federation, Alexandria, VA. http://www.asce.org and http://www.wef.org, 1988.

Chesters, G., J. Konrad and G. Simsiman. "Menomonee River Pilot Watershed Study- Summary and Recommendations", EPA-905/4-79-029. U.S. Environmental Protection Agency, Chicago, Ill., 1979.

Madison, F., J. Arts, S. Berkowitz, E. Salmon, and B. Hagman. "Washington County Project". EPA 905/9-80-003, U.S. Environmental Protection Agency, Chicago, Ill., 1979.

Nelson, J. *Characterizing Erosion Processes and Sediment Yields on Construction Sites.* MSCE thesis. Dept. of Civil and Environmental Engineering, University of Alabama at Birmingham. 94 pgs. 1996.

Novotny, V. and G. Chesters. *Handbook of Nonpoint Pollution Sources and Management.* Van Norstrand Reinhold Company, New York, 1981.

Southeastern Wisconsin Planning Commission (SEWRPC). "Sources of Water Pollution in Southeastern Wisconsin: 1975". Technical Report No. 21. Waukesha, Wisconsin, 1978.

USEPA. *Development Document for Proposed Effluent Guidelines and Standards for the Construction and Development Category.* United States Environmental Protection Agency, Office of Water (4303T). EPA-821-R-02-007. Washington, DC 20460 (www.epa.gov/waterscience/guide/). June 2002.

USEPA. "Effluent guidelines for the construction and development industrial category." *Federal Register,* 40 CFR Parts 122 and 450. June 24, 2003.

USEPA. *Construction and Development Fact Sheet: Final Action—Selection of Non-Regulatory Option,* EPA 821-F-04-001; March 2004.

Virginia. Erosion and Sediment Control Handbook. Second Edition. Division of Soil and Water Conservation. Virginia Dept. of Conservation and Historic Resources. Richmond, Virginia. 1980.

Willett, G. Urban Erosion, in National Conference on Urban Erosion and Sediment Control; Institutions and Technology. EPA 905/9-80-002. U.S. Environmental Protection Agency, 1980.

PROBLEMS

1. Conduct a search in a newspaper database of articles related to sediment and erosion problems in your area (limit this search to the last three years). If no local articles are found, broaden your search to the state or national region. What percentage of these articles focused on agricultural erosion and what percentage were focused on erosion during land development for urban uses?
2. Determine which state agency is responsible for sediment and erosion control. Review that agency's website relating to sediment and erosion control. Is the authority for plan review and enforcement retained at the state level? County level? Local level?
3. Determine which public agency is responsible for plan review. Find out how to obtain a copy of an approved erosion and sediment control plan (do not ask for one unless requested by the instructor). Find out if the plans are available for review in the office of the review agency. Who is responsible for writing project-specific erosion and sediment control plans?
4. Determine if the state and local authority's sediment erosion control regulations are available on the Internet. If not, find out where you can locate them.
5. Find three construction sites near your home, school or office. Answer the following questions regarding each site:
   a. Are there noticeable erosion problem on the site?
   b. Are these resulting in off-site problems?
   c. Do the perimeter erosion-control measures look well maintained? (*Note:* Do not enter any part of an active job site without the owner's permission, preferably in writing).
   d. If they are in the early stages of construction, were the minimum controls followed in setting up the work area for construction?

Appendix 1A State Regulations on the Control of Construction Phase Stormwater

TABLE 1A.4. State Regulations on the Control of Construction Phase Stormwater (USEPA, 2002).

Geographic Area Name	Disturbed Area Limit for Permit Coverage (square feet)[1]	Numeric standard or Pollutant Reduction Requirement	Minimum Depth or Runoff or Storm Return Frequency to Treat for Water Quality Management (per acre)	Maximum Allowed Denuded Acreage or Soil Stabilization Requirement	Visual Inspection Frequency	Notes
Clean Water Act NPDES Storm Water program for Phase I and Phase II MS4s	43,560				After 0.5 Inch rainfall and every 14 days	Phase II compliance date is March 10, 2003.
CZARA	5,000					Must prepare and implement an dapproved erosion and sediment control plan or similar document that contains erosion and sediment control provisions.
Alabama	217,800	Turbidity <50 NTU				
Alaska	217,800	TSS > 20 microns	2 year/6 hour		After 0.5 inch rainfall and every 7 days	Inspector mus be qualified personnel provided by the discharger.
Arizona	217,800					
Arkansas	217,800		10 year/24 hour		Every 7 days	Developers must submit erosion and sediment control plan and storm water pollution prevention plan before filing a notice of intent. Sites 10 acres or more need temporary or permanent sediment basin. Sites less than 10 acres need sediment traps and silt fences
California	217,800		2 year/24 hour		After 0.5 inch rainfall	Inspections will be performed before anticipated storm events, during extended storm events, and after storm events, and once each 24-hour period during extended storm events to identify BMP effectiveness and implement repairs or design changes as soon as feasible depending on field conditions. Discharger is also responsible for inspecting and cleaning all public and private roads for sediment. Construction activities that fall under the jurisdiction of the California Department of Transportation (CALTRANS) have separate permit and regulations.
Colorado	217,800				Any percipitation or snowmelt event that causes erosion and every 14 days	Storm water management plan must be submitted to state for a 10 day review, as well as be retained on site.
Connecticut	217,800	80% TSS reduction				
Delaware	5,000	80% TSS reduction	0.5 inch			
Florida, DEP, Northern Discrict (only applies in NW Florida)	217,800	80% TSS reduction	0.5 inch*			*>100 acres, 1 inch rainfall, <100 acres, 0.5 inch rainfall.

(continued)

TABLE 1A.4 (continued). State Regulations on the Control of Construction Phase Stormwater (USEPA, 2002).

Geographic Area Name	Disturbed Area Limit for Permit Coverage (square feet)[1]	Numeric standard or Pollutant Reduction Requirement	Minimum Depth or Runoff or Storm Return Frequency to Treat for Water Quality Management (per acre)	Maximum Allowed Denuded Acreage or Soil Stabilization Requirement	Visual Inspection Frequency	Notes
Florida, South Florida Water Management District (General, Standard General, Noticed General and Individual Permits)	435,600		1 inch			
Florida, Southwest Florida Water Management District	217,800		0.5 inch			
Florida, St. Johns River Water Management Disctict	217,800	Turbidity <29 NTU				
Florida, Suwannee River Water Management Disctrict	43,560	80% TSS reduction	1 inch			
Georgia	47,916	Turbidity = 10–25 NTU*	25 year/24 hour			*<25 nephelometric turbidity units for waters supporting warm water fisheries, or <10 nephelometric turbidity units for waters classified as trout waters.
Hawaii	217,800				After 0.5 inch rainfall and every 7 days during dry season, every day during rainy season	
Idaho	217,800				After 0.5 inch rainfall and every 14 days	
Illinois	217,800		3,600 cubic feet per acre		Every 7 days	
Indiana	217,800				Every 7 calendar days and within 24 hours of 0.5 inch of precipitation	
Iowa	217,800	80% TSS reduction			Every 7 days	
Kansas	217,800				At least once per week	
Kentucky	217,800	Goal of 80% TSS reduction (compared to pre-construction conditions)				
Louisiana	217,800					
Maine	217,800	40–80% TSS reduction	2 year			
Maryland	5,000	80% TSS reduction*	2 year/24 hour			*Based on the average annual TSS loading from all storms less than or equal to the 2 year/24 hour storm.

(continued)

TABLE 1A.4 (continued). State Regulations on the Control of Construction Phase Stormwater (USEPA, 2002).

Geographic Area Name	Disturbed Area Limit for Permit Coverage (square feet)[1]	Numeric standard or Pollutant Reduction Requirement	Minimum Depth or Runoff or Storm Return Frequency to Treat for Water Quality Management (per acre)	Maximum Allowed Denuded Acreage or Soil Stabilization Requirement	Visual Inspection Frequency	Notes
Montana	217,800		2 year/24 hour		After 0.5 inch rainfall and every 7 days	Dischargers must submit with the state application form a stormwater erosion control plan (SWECP) that resembles EPA's construction site SWPP. Permit coverage begins only when Montana DEQ reviews and approves SWECP. Must also inspect everyday during prolonged precipitation or snowmelt periods. A registered PE must prepare the ESC plan if site is greater than 20 acres. Also regulate down to 1 acre if construction site within 100 feet of a surface water body. Montana has a sediment and erosion control guidance manual that lists standard use BMPs. If other BMPs are used, they need to be submitted with ESC plan to the state for approval. For slopes steeper than 3:1 and greater than 5 vertical feet, surface roughening is required. Filter fences should be used on drainage areas >1 acre; sediment traps should only be used on drainage areas >3 acres; and temporary sediment ponds should only be used on drainage areas >10 acres.
North Carolina	43,560	Y		20 acres total disturbance at any given time for areas discharging to high quality waters	Every 7 days	
Nebraska	217,800				Once a month	
Nevada	217,800					
New Hampshire	100,000					
New Jersey	5,000					
New Mexico	217,800				Y	
New York	217,800	0.5 inch				
North Dakota	217,800				Y	
Ohio	217,800					
Oklahoma	217,800		3,600 cubic feet per acre		Y	A vegetated buffer zone of at least 100 ft must be retained or successfully established between the area dusturbed during construction and all perennial or intermittent streams on or adjacent to the construction site. A vegetated buffer zone at least 50 ft wide must be retained or established between the area disturbed during construction and all ephemeral streams or drainages. Treatment volume is the lesser of 3,600 ft^3 or the runoff volume of a 2 year/24 hour storm.

(continued)

TABLE 1A.4 (continued). State Regulations on the Control of Construction Phase Stormwater (USEPA, 2002).

Geographic Area Name	Disturbed Area Limit for Permit Coverage (square feet)[1]	Numeric standard or Pollutant Reduction Requirement	Minimum Depth or Runoff or Storm Return Frequency to Treat for Water Quality Management (per acre)	Maximum Allowed Denuded Acreage or Soil Stabilization Requirement	Visual Inspection Frequency	Notes
Oregon	217,800				Every 7 days, and daily during periods of stormwater runoff and snowmelt runoff, only every 14 days during periods of 7 days or more of non-construction acitivity	If site is >20 acres, erosion and sediment control plan must be prepared by a Professional Engineer, or Registered Landscape Architect, or Certified Professional in Erosion and Sediment Control, and plan must be submitted 90 days before construction begins. All permitees must submit an Oregon Land Use Compatibility Statement if they do not already have one on file with Oregon DEQ.
Pennsylvania	217,800		5 year		Y	Basins should drain no quicker than 4 days and no longer than 7 days.
Rhode Island	217,800	80–90% TSS reduction	10 year			
South Carolina	217,800		5 year			
Tennessee	217,800				Y	The permitee shall maintain records of checks and repairs.
Texas	217,800		3,600 cubic feet per acre			
Utah	217,800		24 hour or 1 inch storm event		Once every 14 days, before anticipated storm events expected to cause significant runoff, and within 24 hours of the end of a storm that is 0.5 inch or greater.	Where sites have been finally or temporarily stabilized, or when runoff is unlikely due to winter conditions, or during seasonal arid periods in arid areas and semi-arid areas inspections shall be conducted at least once every 30 days, 10 yr, 24 hr strom event for water quality is for 10 acres or greater. For areas less than 10 acres, or where calculations for volume of runoff for disturbed acres is not performed, a sediment basin providing 3600 cubic feet of storage per acre drained or equivalent control measures shall be provided. 1) Where the initiation of stabilization measures by the 14th day after construction activity temporary or permanently cease is preclude by snow cover or frozen ground conditions, stabilization measures shall be initiated as soon as possible. 2) In arid areas, semi-arid areas, and areas experiencing droughts where the initiation stabilization measures by the 14th day after construction activity has temporariliy or permanently ceased is precluded by seasonal arid conditions, stabilization measures shall be initiated as soon as possible.
Vermont	217,800					

(continued)

TABLE 1A.4 (continued). State Regulations on the Control of Construction Phase Stormwater (USEPA, 2002).

Geographic Area Name	Disturbed Area Limit for Permit Coverage (square feet)[1]	Numeric standard or Pollutant Reduction Requirement	Minimum Depth or Runoff or Storm Return Frequency to Treat for Water Quality Management (per acre)	Maximum Allowed Denuded Acreage or Soil Stabilization Requirement	Visual Inspection Frequency	Notes
Virginia	217,800		3,600 cubic feet per acre		Y	Sediment basins required for sites of 10 acres or more (except those with final stabilization); for sites <10 acres, filter fences required but only for sideslope and downslope boundaries of construction sites.
Washington, Large Parcel	>43,560		24 hour/6month	2 days between October 1 and April 30 (i.e., the wet season); 7 days between May 1 to September 30 (dry season)		
Washington, Small Parcel	<43,560		24 hour/6month	2 days between October 1 and April 30 (i.e., the wet season); 7 days between May 1 to September 30 (dry season)		
West Virginia	130,680		2 year		Y	
Wisconsin	*217,800				Y	
Wyoming	217,800	Turbidity <10-15 NTU			Inspect every 7 days, except during seasonal shutdowns and during the period following completion of constructin but prior to return of the site to "finally stabilized" conditions and termination of coverage, then the site must be inspected every quarter.	

Selection of Controls and Site Planning

INTRODUCTION

THIS chapter outlines some of the available guidance for selecting erosion controls for construction sites. There are many manuals available for throughout the U.S., some have been in use for more than 25 years. One example is the *Alabama Handbook for Erosion Control, Sediment Control, and Stormwater Management on Construction Sites and Urban Areas,* that was originally produced for the Alabama Soil and Water Conservation Committee in 1993 by the SCS (now the NRCS). This local manual was revised in 2003 as guidance for the EPA Phase II stormwater regulations (http://swcc. state.al.us/erosion_handbook. htm). An earlier Alabama manual was prepared by the Birmingham Regional Planning Commission as part of their "208" project in 1980: *Best Management Practices for Controlling Sediment and Erosion from Construction Activities.* Chapter 1 lists many other handbooks that are available for other areas of the country.

This chapter organizes some of the major control categories according to site erosion control issues that are listed in the Phase II stormwater NPDES regulations that will affect construction sites. In addition, steps are provided to guide a user in preparing an erosion control plan for local construction sites. Later chapters discuss how local rains, soils, and objectives need to be considered when designing the selected controls for site specific conditions.

Also included in this chapter is Appendix 2A which lists some costs for on-site erosion and sediment controls, summarized from *Costs of Urban Nonpoint Source Water Pollution Control Measures* prepared by the Southeast Wisconsin Regional Planning Commission (June 1991). These costs can be multiplied by 1.6 to estimate 2005 cost values, according to the Engineering News Record's *Construction Site Index.* Obviously, different regions of the country have different labor rates and material costs, so these should only be used as initial estimates.

EXAMPLE CONSTRUCTION SITE CONTROL REQUIREMENTS

Construction site control requirements can be divided into two major categories, primary controls and supporting controls, as described in Chapter 1. It is also possible to categorize the controls into preventative measures (much preferred), usually termed erosion control practices, and treatment measures (typically not as effective), usually termed sediment controls.

Over the years, two general family "trees" of construction site erosion manuals have evolved. The State of Virginia produced one of the earliest manuals in 1980, and is widely copied by many states and local governments throughout the country. Another type of manual has been produced by the SCS, (now NRCS), and has been modified by them for a number of states. In recent years, there also have been a number of independently-produced local manuals that reflect local conditions and include some emerging procedures and techniques. These design standards for the needed practices obviously can be supplemented and many need to be modified to reflect local conditions, based on site-specific hydrology and erosion conditions, as described in the later chapters of this book.

The requirement categories are summarized in the following sections, along with a list of example controls that can be used to help meet each requirement, as referenced to the SCS (NRCS) standards (as modified for New Jersey or Alabama 1993) and selected Virginia standards (as modified by the Wisconsin Department of Natural Resources, 1994). This list is not comprehensive, but does indicate the range of available tools to address these issues. Many of the Phase II NDPES requirements are similar to these categories.

Primary Construction Site Control Requirements

The following discussion lists available construction site controls that can be applied to different categories of site issues. The listed names are followed by the section number in the manual referenced. Obviously, these lists are not comprehensive, but do illustrate the diversity and number of practices that can be used. These practices are organized by the different site issues that should be addressed for all construction sites. Typical applications would require that each category be addressed for all construction sites, but the specific controls need to be selected based on site-specific conditions.

Minimize Upslope Water Contributions

Upslope water must be diverted around disturbed areas, and existing large channels passing through the site must be protected from erosion runoff. These controls must be installed before any other site disturbance in order to minimize the amount of water flowing across disturbed areas, contributing to site erosion and placing a greater burden on sediment control practices. These controls are all preventative erosion control practices.

General Diversion Structures

- Diversions (Virginia standards III-51)
- Diversions (SCS/NJ standards 4.2.1)
- Level spreaders (Virginia standards III-161)
- Diversions (SCS/AL standards III-DV-1)
- Diversion design (SCS/AL standards III-DN-6)

Temporary Diversion Structures

- Temporary diversion (WDNR standards 4-7)
- Temporary diversion dike (Virginia standards III-139)
- Temporary fill diversion (Virginia standards III-43)
- Temporary right-of-way diversion (Virginia standards III-47)

Permanent Diversion Structures

- Permanent diversion (WDNR standards 4-4)

General Channel Stabilization

- Permanent channel stabilization (WDNR standards 4-48)
- Structural streambank stabilization (Virginia standards III-175)
- Rock and concrete lined waterways (WDNR standards 4-59)
- Channel stabilization (SCS/NJ standards 4.6.1)
- Lined waterway (SCS/NJ standards 4.11.1)
- Channel stabilization (SCS/AL standards III-CS-1, III DV-6)
- Gabion (SCS/AL standards III-GB-1)

Check Dams

- Check dams (Virginia standards III-151)
- Temporary sediment trap (Virginia standards III-55)
- Sediment traps (WDNR standards 4-35)
- Check dams (SCS/AL standards III-CD-1)

Riprap

- Riprap (SCS/NJ standards 4.12.1)
- Riprap (Virginia standards III-137)
- Riprap (SCS/AL standards III-RR-1)

Waterway Drops

- Grade stabilization structure (SCS/NJ standards 4.17.1)
- Waterway drop structure (Virginia standards III-155)
- Drop structure (SCS/AL standards III-DS-1)
- Gabion (SCS/AL standards III-GB-1)

Stream Crossing

- Temporary stream crossing (Virginia standards III-183)
- Stream crossing (SCS/AL standards III-SX-1)

Grassed Waterways

- Vegetative streambank stabilization (Virginia standards III-165)
- Grassed waterways (SCS/NJ standards 4.3.1)
- Sodding (WDNR standards 4-52)
- Grassed waterway (WDNR standards 4-55)
- Geotextile reinforced grassed waterway (WDNR standards 4-57)
- Waterway or stormwater channels (SCS/AL standards III-WW-1)

Slope Protection

- Slope protection structures (SCS/NJ standards 4.5.1)
- Temporary slope drain Virginia standards III-89)
- Paved flume (Virginia standards III-95)
- Paved flume (SCS/AL standards III-PF-1)
- Retaining wall (SCS/AL standards III-RW-1)
- Down drain structure (SCS/AL standards III-DN-1)
- Gabion (SCS/AL standards III-GB-1)

Provide Downslope Controls

In general, wet detention (sediment) ponds are required to treat all runoff leaving construction sites for drainage areas greater than about 10 acres. If the drainage area is less than 10 acres, then filter fences, or equivalent perimeter sediment controls, may be used at all side slope and downslope edges of the construction site, depending on the site hydraulics. These controls must also be installed before any other site disturbance. These controls are all treatment, or sediment control, practices, as they are intended to remove sediment from the flowing water before it leaves the construction site. Erosion control (prevention) practices must always be emphasized, but sediment controls will always be needed as

Slope Diversions

Large diversion berm and swale to divert water from downslope area at an abandoned mine site (SCS photo).

Temporary slope diversion at highway construction site.

Highway slope diversion during initial construction.

This photo of U.S. Route 1 being relocated around Dover, Delaware shows pipe slope drains that carry sediment laden runoff downslope to a collector swale. Note that the side slopes are graded and seeded as the work progresses to keep soil from washing down (Photograph by D. Lake).

Large slope diversions carrying water from upslope benches (IECA photo).

Downslope side of perimeter filter fence intercepting sheetflows.

Filter fabric fence on mulched slope (SCS photo).

it will not be possible to prevent all erosion from occurring in the first place.

General Sediment Fence

- Sediment barrier (SCS/NJ standards 4.13.1)
- Sediment barrier/fence (SCS/AL standards III-SF-1)
- Retrofitting (SCS/AL standards III-RT-1)

Filter Fabric Fences

- Silt fence (Virginia standards III-17)
- Filter fabric fences (WDNR standards 4-11)
- Filter fabric barriers (WDNR standards 4-25)
- Temporary right-of-way diversion (Virginia standards III-47)
- Sediment barrier/fence (SCS/AL standards III-SF-1)

Straw Bale Fences

- Straw bale fences (WDNR standards 4-15)
- Straw bale barriers (WDNR standards 4-30)
- Straw bale barriers (Virginia standards III-9)
- Brush barrier (Virginia standards III-25)
- Sediment barrier/fence (SCS/AL standards III-SF-1)

Sediment Basins

- Temporary sediment basin (Virginia standards III-59 and III-87)
- Sediment basins (SCS/NJ standards 4.4.1)
- Sediment basins (WDNR standards 4-39)
- Minimum area for sedimentation basins (SCS undated)
- Sediment basin (SCS/AL standards III-SB-1)
- Storm water retention structure (SCS/AL standards III-RS-1)

Outlet Protection

- Outlet protection (Virginia standards III-127)
- Conduct outlet protection (SCS/NJ standards 4.14.1)
- Outlet protection (SCS/AL standards III-OP-1, III-DN-6)

Protect Disturbed Areas

Disturbed areas exposed for extended periods (14 days is a typical limit) without any activity must be stabilized with mulches, temporary vegetation, permanent vegetation, or by other equivalent control measures. These controls would all be considered preventative, or erosion control, practices, and

Large expanses of unprotected soils left exposed for long periods cause most of the sediment losses from construction sites.

Various slope protection treatments and tree conservation (Photograph by D. Lake).

are usually considered the most effective, especially when used in conjunction with a good phasing plan to minimize the amount of land being disturbed at any one time.

Mulching

- Mulching (Virginia standards M-247)
- Mulching (WDNR standards 4-19)
- Stabilization with mulch only (SCS/NJ standards 3.3.1)
- Guide to mulching materials (King Co. Wash. 1989)
- Mulching (SCS/AL standards IV-MU-1)

Local Vegetation Information

- Vegetative BMPs to protect exposed surfaces (Birmingham Regional Planning Commission, BRPC, temporary and permanent covers)
- Lime and fertilizer requirements for plant growth (BRPC Appendix 1)
- Planting guide (SCS, Jefferson County, AL-6, 1975)
- Seed, fertilizer, and lime requirements for cost-share rates (SCS, Jeff. Co., Exhibit 1)
- Selection of vegetation (SCS/AL standards IV-7)
- Information on installing vegetative measures (SCS/AL standards Appendix A4)

General Seeding

- Surface roughening (Virginia standards III-201)
- Topsoiling (Virginia standards III-207)
- Topsoiling (SCS/NJ standards 3.5.1)
- Seeding (WDNR standard 4-22)
- Topsoil (SCS/AL standards III-TS-1)
- Surface roughening (SCS/AL standards III-SR-1)

Temporary Seeding

- Temporary seeding (Virginia standards M-211)
- Temporary vegetative cover for soil stabilization (SCS/NJ standards 3.1.1)
- Temporary vegetation-seeding (SCS/AL standards IV-TV-1)

Permanent Seeding

- Permanent seeding (Virginia standards III-215)
- Permanent vegetative cover for soil stabilization (SCS/NJ standards 3.2.1)
- Permanent seeding (SCS/AL standards IV-PS-1)

Sodding

- Sodding (Virginia standards M-231)
- Permanent stabilization with sod (SCS/NJ standards 3.4.1)
- Bermudagrass establishment Virginia standards III-241)
- Sodding (SCS/AL standards IV-SD-1)

Trees and Shrubs

- Trees, shrubs, vines, and ground covers (Virginia standards III-257)
- Shrub, vine, and ground cover planting (SCS/AL standards IV-SVG-1)

Maintenance of Vegetation

- Maintaining vegetation (SCS/NJ standards 3-6.1)
- Tree preservation and protection (Virginia standards III-279)
- Tree protection during construction (SCS/NJ standards 3.9.1)
- Tree preservation and protection (SCS/AL standards IV-TPP-1)
- Irrigation (SCS/AL standards IV-IR-1)

Supporting Construction Site Controls

A number of construction site controls are also typically specified in local ordinances. The following are examples of some of these controls, some of which are preventative (represented by the "good-housekeeping" controls) while others are treatment practices (such as inlet filters). The *Alabama Handbook*, along with other erosion control manuals, contains descriptions of many "structural" practices that can be used on construction sites to prevent erosion, or to capture sediment that has already eroded. The following excerpts from the *Alabama Handbook* are only a few that are included in this chapter, but are the most basic controls that should be considered: construction site exits, stormdrain inlet protection, use of riprap, check dams in channels, and protection of outlets from ponds. These sections contain recommendations for the use of these controls for Alabama conditions. Other jurisdictions have developed their own list of mandatory and recommended controls. These handbooks are periodically revised, and local regulatory agencies and/or the local USDA extension offices should be consulted for updated recommendations. The erosion and sediment control benefits of most of these controls have not been measured in the field, but these controls are generally acknowledged as essential elements of construction site erosion control programs.

Unsafe storage disposal of empty oil containers at construction site.

Control Wastewater from Dewatering Operations

Wastewater from site dewatering operations should be controlled to limit the discharge of sediment. Typical criterion restricts particles greater than 50 μm from being discharged during dewatering operations. This level of control can be obtained by using simple sedimentation devices sized according to the maximum dewatering pumping rates.

- Dewatering settling basin (WDNR standards 4-72)
- Dewatering sediment basin (SCS/AL standards III-RS-5)

Properly Dispose of Construction Debris

All building material and other wastes need to be removed from the site and disposed of in licensed disposal facilities. No wastes or unused building materials may be buried, dumped, or discharged at construction sites.

Control Tracking of Sediment Off-Site

Each site needs to have graveled access drives and parking areas to reduce the tracking of sediment onto public or private roads. An example regulation would require that all unpaved roads on the site carrying more than 25 vehicles per day also be graveled. Any sediment or debris tracked onto public or private roads needs to be removed daily by street cleaners (and not by washing it down the storm drain system).

Entrance Controls

- Temporary gravel construction entrance (Virginia standards III-1)
- Stabilized construction entrance (SCS/NJ standards 4.15.1)
- Construction exit (SCS/AL standards III-CE-1)

Site Road Controls

- Construction road stabilization (Virginia standards III-5)
- Temporary graveled access roads and parking areas (WDNR standards 4-74)
- Traffic control (SCS/NJ standards 4.9.1)
- Construction exit (SCS/AL standards III-CE-1)

Dust Control

- Dust control Virginia standards III-299)
- Dust control (SCS/NJ standards 4.10.1)
- Dust control (SCS/AL standards IV-DU-1)

Proper construction site entrance or graveled driveway can eliminate much tracking of sediment onto public roads.

WI DNR photo.

Protect Construction Site Entrances and Exits

The following discussion is from the *Alabama Handbook* (USDA, 2003) and is an example of the guidance provided by different state agencies for the control of tracking from construction exits.

Construction Site Exit-CE

Definition

A stone or rock stabilized pad located at points of vehicular ingress or egress to a construction site.

Purpose

To reduce or eliminate the transport of mud from the construction area onto public right-of-ways by motor vehicles or by runoff.

Conditions Where Practice Applies

This practice is applied where vehicular traffic will be leaving a construction site and move directly onto a public road or street.

Planning Considerations

Roads and streets adjacent to construction sites should be kept clean for the general safety and welfare of the public. A construction exit (Figure 2.1) should be provided where mud can be removed from construction vehicle tires before they enter a public road. If traveling over the rock stabilized pad does not remove the mud from construction vehicles, a wash area should be provided for that purpose. Whenever washing is used, the wash water needs to be collected in a sediment basin before leaving the site.

Construction of stabilized roads throughout the development site should be considered to lessen the amount of mud transported by vehicular traffic. The rock pad should be located to provide for maximum use by all construction vehicles. Consideration should be given to limiting construction vehicles to only one ingress and egress point. Measures may be necessary to make existing traffic use the construction exit.

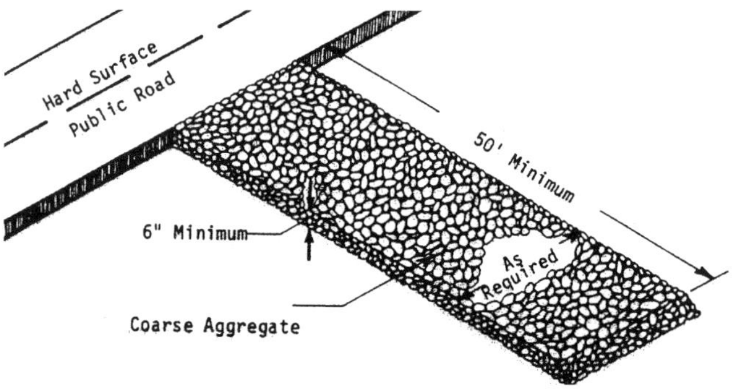

Figure 2.1. Gravel construction exit (USDA, 2003).

Design Criteria

Aggregate Size—Aggregate should be Alabama Highway Department coarse aggregate gradation No. 1, or equivalent. AL DOT coarse aggregate No. 1 has the following size specifications:

	Percent Passing
100 mm (4 in)	100%
90 mm (3-1/2 in)	90–100%
63 mm (2-1/2 in)	25–60%
37.5 mm (1-1/2 in)	0–15%
19 mm (3/4 in)	0–5%

Entrance Dimensions—The rock pad shall be a minimum of six inches thick. It shall be at least 50 feet long or the length required to enter and park the longest anticipated construction vehicles. The width shall be at least 20 feet.

Geotextiles—A non-woven geotextile meeting the requirements of Soil Conservation Service Material Specification 592, Class IV should be used under the rock when the subgrade is soft or the blow count is less than 10.

Washing—A wash rack shall be provided as necessary to prevent mud from being transported to public streets and highways. It shall be constructed of concrete and/or other durable materials. Provisions shall be provided for the mud and other material to be carried away from the wash rack to a sediment basin to remove the mud from the water before release from the site.

Maintenance

The construction exit shall be maintained in such a way to prevent the movement of mud into public travel ways. Aggregate should be added to the pad whenever it will not serve as an all weather travel way for the construction vehicles. Sediment basins shall be cleaned out whenever one-half of the design storage volume is depleted.

Construction Specifications for Construction Exit

1. Remove all vegetation, roots and other objectionable material from the stone pad area.
2. Smooth the area to an even grade and fill in and recompact material in holes low places or over excavated areas. Recompacted material shall be as dense as the surrounding material.
3. Place any required geotextile over the area to be protected. Take care not to pull the geotextile tight, but leave sufficient slack for the fabric to conform to the ground after rock is placed and loaded with vehicles. The fabric shall be unrolled parallel to the roadway centerline. The recommended geotextile overlap is 24 inches when the blow count is 10, 36 inches when the blow count is four to nine and 48 inches when the blow count is three or less. Geotextiles that are the full width of the roadway are needed.

4. The stone pad should be dumped and spread in a full uniform thickness before vehicular traffic is permitted to travel on it.
5. Wash racks shall be installed in accordance with manufacturers recommendations.
6. Sediment basins or other related facilities constructed in conjunction with the wash rack shall be constructed in accordance with the plans and specifications. Sediment basins for wash racks shall be constructed before the wash rack is put into service.

Protect Storm Drain Inlets

All storm drain inlets need to be protected from erosion materials.

- Storm drain inlet protection (Virginia standards III-29)
- Inlet protection barriers (WDNR standards 4-64)
- Storm sewer inlet protection (SCS/NJ standards 4.16.1)
- Inlet insert baskets (WDNR standards 4-66)
- Inlet protection (SCS/AL standards III-NP-1)

The following discussion is from the *Alabama Handbook* (USDA, 2003) and is an example of the guidance provided by different state agencies for the protection of stormdrain inlets.

Stormdrain Inlet Protection (NP)

Definition

A sediment filter installed around a storm drain drop inlet or curb inlet to reduce sediment discharge.

Purpose

To prevent sediment from entering storm drainage systems during construction and prior to permanent stabilization of the disturbed area.

Conditions Where Practice Applies

Where storm drain inlets are to be made operational before permanent stabilization of the disturbed drainage area. Different types of structures are applicable to different situations.

Planning Considerations

Storm sewers which are made operational before their drainage area is stabilized can convey large amounts of sediment to natural drainageways. In cases of extreme sediment loadings, the storm sewer itself may clog and lose a major portion of its capacity. To avoid these problems, it is necessary to minimize that amount of sediment that enters the system at the inlets.

This practice contains several types of inlet filters and traps which have different applications dependent upon site conditions and type of inlet. These inlet protection devices are for drainage areas of less than one acre. Runoff from large disturbed areas should be routed through a sediment basin.

The best way to prevent sediment from entering the storm sewer system is to stabilize the site as quickly as possible, preventing erosion and stopping sediment at its source. Inlet protection devices likely have limited benefits for most of the eroding sediment, although they are more effective for the larger materials that may clog inlets and drainage systems. Sediment is best treated by preventing erosion. Leave as much of the site undisturbed as possible in the total site plan. Clear and disturb the site in small increments, if possible.

Design Criteria

1. The drainage area shall be no greater than 1 acre.
2. The inlet protection device shall be constructed in a manner that will facilitate cleanout and disposal of trapped sediment and minimize interference with construction activities.
3. The inlet protection devices shall be constructed in such

Cinder block and gravel barrier to protect inlet (SCS photo).

Proprietary filter fabric storm drain inlet covers.

There have been many inlet barriers used over the years, with poor to moderate success. Most have suffered from lack of proper maintenance or poor construction.

Typical filter fabric enclosure surrounding inlet (J. Voorhees photo).

Inlet protected by filter fabric, thick matting to protect new grass, and chemically stabilized soil (Illinois).

Large accumulation of debris surrounding filter fabric inlet barrier, requiring maintenance.

Typical reinforced filter fabric barrier surrounding elevated inlet.

(continued)

There have been many inlet barriers used over the years, with poor to moderate success. Most have suffered from lack of proper maintenance or poor construction *(continued)*.

Older concrete block, lumber, and stone inlet barrier (historical SCS photo).

Older concrete block, lumber, and stone inlet barrier (historical SCS photo).

Older concrete block and stone inlet barrier (historical SCS photo).

Netted stone barrier to attempt to divert bypassing gutter flows into inlet.

Typical straw bale barrier surrounding inlet (notice tight bales and large amount of sediment collected around outside of bales, needing removal).

Straw bale barrier showing large gaps between bales and decomposing bales, needing replacement (SCS photo).

(continued)

There have been many inlet barriers used over the years, with poor to moderate success. Most have suffered from lack of proper maintenance or poor construction *(continued)*.

Temporary inlet filter fabric bag placed under inlet at redevelopment construction site.

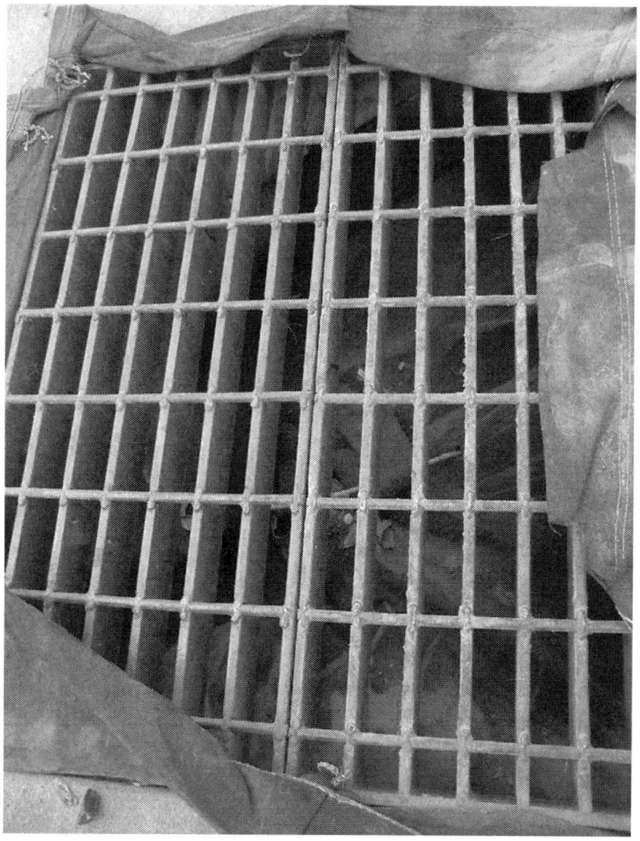

Temporary inlet filter fabric bag placed under inlet at redevelopment construction site.

a manner that any resultant ponding of stormwater will not cause excessive inconvenience or damage to adjacent areas or structures.

4. Design criteria more specific to each particular inlet protection device will be found with that construction specification.
5. Ponding of water or deposition of sediment on roadways that will create traffic hazards will be prevented.

Maintenance

1. The structure shall be inspected after each rain and repairs made as needed.
2. Sediment shall be removed and the trap restored to its original dimensions when the sediment has accumulated to 1/2 the design depth of the trap. Removed sediment shall be deposited in a suitable area and in such a manner that it will not erode. Stabilize all sediment disposal areas with appropriate vegetation.
3. Structures shall be removed and the area stabilized when the contributing drainage area has been properly stabilized.

Construction Specifications for Inlet Protection

1. Straw bale drop inlet structure. (Figure 2.2). This method of inlet protection is applicable where the inlet drains a relatively flat area (slopes no greater than 5 percent) where sheet or overland flows (not exceeding 0.5 cfs) are typical. The method shall not apply to inlets receiving concentrated flows, such as in street or highway medians.

   a. Bales shall be either wire-bound or string-tied with the bale oriented so that the bindings are around the sides rather than over and under the bales. Bales will be laid on edge.
   b. Bales shall be placed lengthwise in a single row surrounding the inlet, with the ends of adjacent bales pressed together.
   c. The filter barrier shall be entrenched and backfilled. A trench shall be excavated around the inlet the width of a bale to a minimum depth of 4 inches. After the bales are staked, the excavated soil shall be backfilled and compacted against the filter barrier.
   d. Each bale shall be securely anchored and held in place

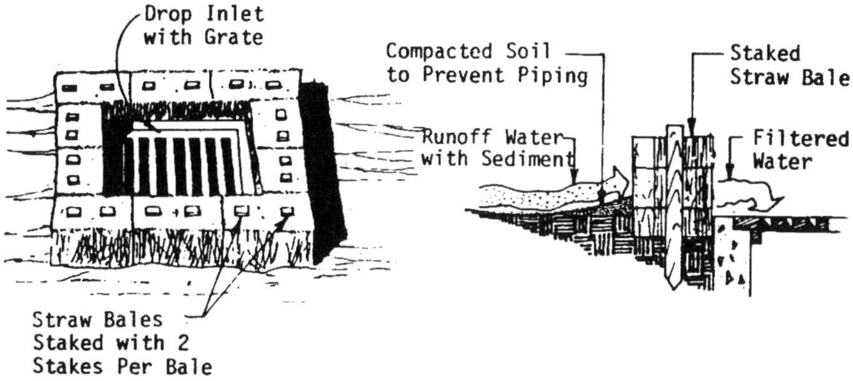

Figure 2.2. Straw bale drop inlet sediment trap (USDA, 2003).

by at least two stakes or rebars driven through the bale.

e. Loose straw shall be wedged between bales to prevent water from entering between bales.

f. Stakes for anchorage shall be nominal 1″ × 2″ durable wood or equivalent. The wood shall be sound with a minimum actual dimension of 1/2″. The minimum embedment into the ground shall be 12 inches.

2. Gravel and wire mesh drop inlet sediment filter. (Figure 2.3). This method of inlet protection is applicable where heavy concentrated flows are expected, but not where ponding around the structure might cause excessive inconvenience or damage to adjacent structures and unprotected areas.

   a. Wire mesh shall be laid over the drop inlet so that the wire extends a minimum of 1 foot beyond each side of the inlet structure. Hardware cloth or comparable wire mesh with 1/2-inch openings shall be used. If more than one strip of mesh is necessary, the strips shall be overlapped and securely tied or wired together.

   b. Alabama Highway Department No. 1 Coarse Aggregate, or equivalent, shall be placed over the wire mesh as indicated on Figure 2.3. The depth of stone shall be at least 12 inches over the entire inlet opening. The stone shall extend beyond the inlet opening at least 18 inches in all directions.

   c. If the stone filter becomes clogged with sediment so that it no longer adequately performs its function, the stones must be pulled away from the inlet, cleaned and replaced.

Warning: This filtering device has no overflow mechanism, therefore, ponding is likely, especially if sediment is not removed regularly. This type of device must never be used where overflow may endanger an exposed embankment slope. Consideration should also be given to the possible effects of ponding on traffic routes, nearby structures, working areas, adjacent property, etc.

3. Gravel curb inlet sediment filter. (Figure 2.4). This method of inlet protection is applicable at curb inlets where ponding in front of the structure is not likely to cause inconvenience or damage to adjacent structures and unprotected areas.

   a. Hardware cloth or comparable wire mesh with 1/2 inch openings shall be placed over the curb inlet opening so that at least 12 inches of wire extends across the inlet cover and at least 12 inches of wire extends across the concrete gutter from the inlet opening.

   b. Stone shall be piled against the wire so as to anchor it against the gutter and inlet cover and to cover the inlet opening completely. Alabama Highway Department No. 1 Coarse Aggregate, or equivalent, shall be used.

   c. If the stone filter becomes clogged with sediment so

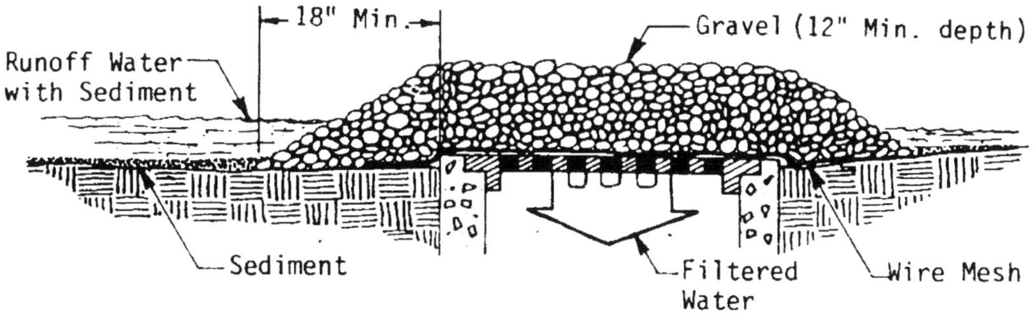

Figure 2.3. Gravel and wire mesh drop inlet filter (USDA, 2003).

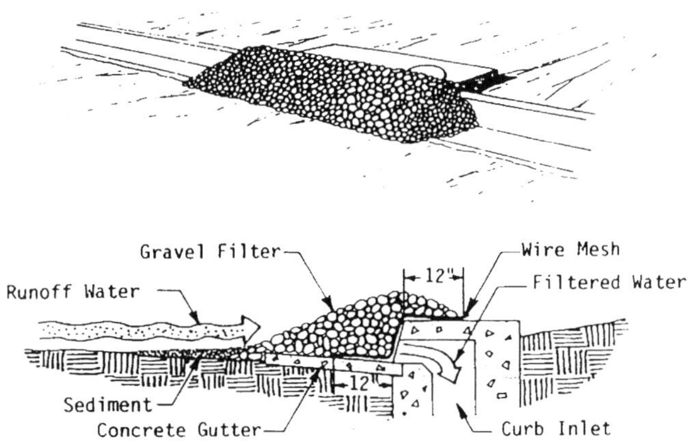

Figure 2.4. Gravel curb inlet sediment filter (USDA, 2003).

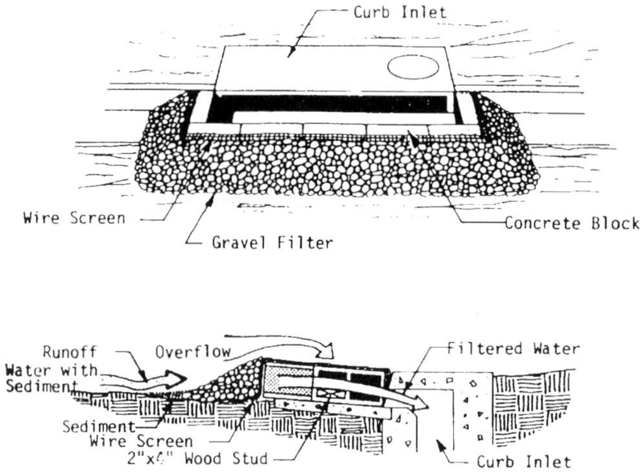

Figure 2.5. Block and gravel inlet filter (USDA, 2003).

that it no longer adequately performs its function, the stone must be pulled away from the block, cleaned and replaced. Do not hose the debris into the curb inlet.

4. Block and gravel curb inlet sediment filter. (Figure 2.5). This method of inlet protection is applicable at curb inlets where an overflow capability is necessary to prevent excessive ponding in front of the structure.
   a. Two concrete blocks shall be placed on their sides abutting the curb at either side of the inlet opening.
   b. A 2-inch by 4-inch stud shall be cut and placed through the outer holes of each spacer block to help keep the front blocks in place.
   c. Concrete blocks shall be placed on their sides across the front of the inlet and abutting the spacer blocks as illustrated in Figure 2.5.
   d. Wire mesh shall be placed over the outside vertical face (webbing) of the concrete blocks to prevent stone from being washed through the holes in the blocks. Chicken wire or hardware cloth with 1/2-inch openings shall be used.
   e. Alabama Highway Department No. 1 Coarse Aggregate, or equivalent, shall be piled against the wire to the top of the barrier as shown in Figure 2.5.
   f. If the stone filter becomes clogged with sediment so that it no longer adequately performs its function, the stone must be pulled away from the blocks, cleaned and replaced.
5. Block and gravel drop inlet sediment filter. (Figure 2.6). This method of inlet protection is applicable where heavy flows are expected and where an overflow capacity is necessary to prevent excessive ponding around the structure.
   a. Place concrete blocks lengthwise on their sides in a single row around the perimeter of the inlet, with the ends of adjacent blocks abutting. The height of the barrier can be varied, depending on design needs, by stacking combinations of 4-inch, 8-inch and 12-inch wide blocks. The barrier of blocks shall be at least 12 inches high and no greater than 24 inches high.
   b. Wire mesh shall be placed over the outside vertical face of the concrete blocks to prevent stone from being washed through the holes in the blocks. Hardware cloth or comparable wire mesh with 1/2-inch openings shall be used.
   c. Stone shall be piled against the wire to the top of the block barrier, as shown in Figure 2.6. Alabama Highway Department No. 1 Coarse Aggregate, or equivalent, shall be used.
   d. If the stone filter becomes clogged with sediment so that it no longer adequately performs its function, the stone must be pulled away from the blocks, cleaned and replaced.

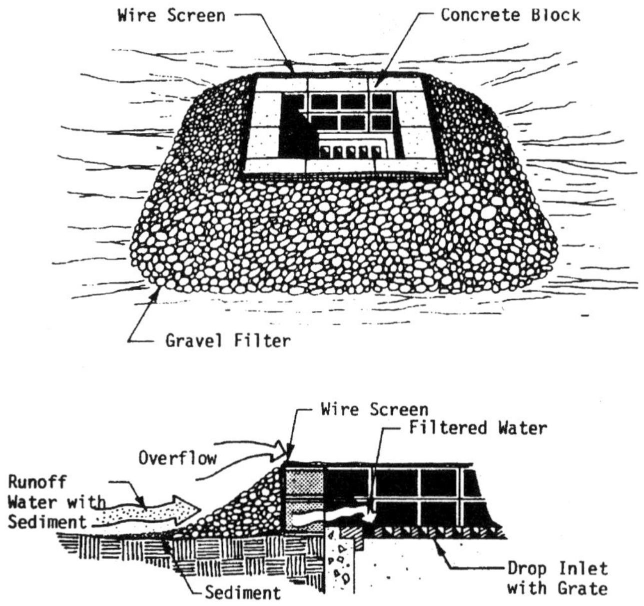

Figure 2.6. Block and gravel drop inlet filter (USDA, 2003).

Small-scale slope diversion to safely carry roof runoff away from building and down sensitive adjacent slope.

Example Proprietary Inlet Protection Devices
Nutec Supply:
http://www.nutec-supply.com/erosion/inlet/#post

Crow Company:
http://www.geosyntheticproducts.com/Erosion_Control/erosion_control.html

EarthSaver Company:
http://www.earth-savers.com/index.html?Main%20Window=applictn.html

EPA discussion on inlet protection:
http://www.epa.gov/npdes/menuofbmps/site_17.htm

Minimize Area Disturbed

One of the most effective erosion controls would require that all construction activities be conducted in a logical sequence to minimize the area of bare soil disturbed at any one time.

- Land grading (SCS/NJ standards 4.1.1)

Control Erosion Scour from Roof Runoff

Roof runoff must be directed to stabilized surfaces.

- Down drain structure (SCS/AL standards III-DN-1)

Control Erosion from Storage Piles

All uncovered soil or dirt storage piles also need to be controlled to prevent erosion. An example regulation may contain the following restrictions.

> An uncovered storage pile, containing more than 10 cubic yards of material, should be located more than 25 feet from a roadway or drainage channel. If these piles remain for 14 or more days, then their surfaces must be stabilized. If the piles will be in place for less than 14 days, then their perimeters must be surrounded by filter fencing or straw bales. Dirt or soil storage piles located less than 25 feet from the road, containing more than 10 cubic yards of material, and in place for 14 or more days must be covered with tarps or other control. If the piles will be in place for less than 14 days, then their perimeters must be surrounded by filter fencing or straw bales. Storm drain inlets must be protected from potential erosion from near-street storage piles by filter fencing or other appropriate barriers.

Soil stockpile next to road needing protection.

Many of the above practices may be applicable for erosion control of storage piles, such as filter fabric and straw bale fences for perimeter protection, plus temporary mulching and seeding practices to reduce direct erosion of material from the storage piles.

PLANNING STEPS AND COMPONENTS FOR CONSTRUCTION SITE CONTROL

Most state guidance provides for the incorporation of newly-developed control practices for erosion and sediment control, provided that the new control's performance is known. For example, Pennsylvania provides detailed guidance for the specification and use of controls that are not contained in the Erosion & Sediment Pollution Control Manual. This section of the state manual is quoted below. The interesting point of interest for persons planning to use a novel control practice is the requirement that a conventional control practice of known performance must be specified to be installed if the novel control practice fails (PA DEP 2000).

> **NEW PRODUCTS AND PROCEDURES**
>
> The BMPs set forth in this manual shall be appropriately incorporated into all erosion and sedimentation control plans unless the designer shows that alteration of these BMPs or inclusion of other BMPs shall effectively minimize accelerated erosion and sedimentation. Since the burden of proof for whether a proposed new product or procedure will be effective lies with the designer, all necessary information required to approve the use of the new product or procedure must be submitted as part of the application. At a minimum, this should include:
>
> 1. The name of the product (and type of control if a brand name is used).

2. Proposed use (e.g. storm sewer inlet protection). If this product or procedure has the potential to minimize accelerated erosion and sedimentation more effectively or efficiently than current methods, this should be stated and the reason given (e.g. same protection for less cost, less maintenance required, etc.). It should be demonstrated that the proposed use meets with any manufacturer's recommendations (e.g. manufacturer's recommendations showing such use, test data, limitations, etc.).
3. Where the proposed use is in a protected watershed (HQ* or EV*) or a critical area (e.g. adjacent to a stream channel or wetland), an alternative conventional BMP should be specified for installation should the innovative product or procedure fail. The definition of a product failure must be clearly stated.
4. Sufficient installation information must be provided to ensure its proper use. This should include a clear, concise sequence as well as a typical detail showing all critical dimensions and/or elevations.
5. The plan maps must show all locations where the proposed new product or procedure will be used. All receiving waters must be identified. Any downstream public water supplies, fish hatcheries, or other environmentally sensitive facilities must be noted.
6. A suitable maintenance program must be provided. Specific instructions, which identify potential problems and recommended remedies must be included.

New products and procedures which meet the above criteria will be reviewed on a case-by-case basis until their effectiveness has been sufficiently demonstrated by successful use in the field.

**Note:* HQ: high quality and EV: exceptional value.

Most construction site control handbooks and design manuals include some information pertaining to the selection of controls needed for construction sites, and guidance on submitting acceptable control plans. As an example, the following discussion lists the minimum standards applicable for all construction sites in Virginia. Also included is planning guidance from the 2003 Alabama Handbook for erosion control.

Virginia Erosion and Sediment Control Regulations, Minimum Standards

The following is the list of the 19 "minimum standards" for erosion and sediment control as required in Section 4VAC50-30-40 of the Virginia Erosion and Sediment Control Regulations. This is a typical listing representative of most erosion and sediment control regulations and indicates which controls need to be considered for construction-site activities.

(1) Soil Stabilization

Permanent or temporary soil stabilization shall be applied to denuded areas within seven days after final grade is reached on any portion of the site. Temporary soil stabilization shall be applied within seven days to denuded areas that may not be at final grade but will remain dormant for longer than 30 days, but less than one year. Permanent stabilization shall be applied to areas that are to be left dormant for more than one year

(2) Soil Stockpile Stabilization

During construction, soil stockpiles and borrow areas shall be stabilized or protected with sediment trapping measures. Temporary protection and permanent stabilization shall be applied to all soil stockpiles on site and borrow areas or soil intentionally transferred off site.

(3) Permanent Stabilization

Permanent vegetative cover shall be established on denuded areas not otherwise permanently stabilized. Permanent vegetation shall not be considered established until a ground cover is achieved that is: uniform, mature enough to survive, and will inhibit erosion.

(4) Sediment Basins & Traps

Sediment basins, sediment traps, perimeter dikes, sediment barriers, and other measures intended to trap sediment shall be constructed as a first step in any land-disturbing activity and shall be made functional before upslope land disturbance takes place.

(5) Stabilization of Earthen Structures

Stabilization measures shall be applied to earthen structures such as dams, dikes, and diversions immediately after installation.

(6) Sediment Traps and Sediment Basins

Sediment traps and basins shall be designed and constructed based upon the total drainage area to be served by the trap or basin as follows:

Sediment Traps—Only control drainage areas less than three acres. Minimum storage capacity of 134 cubic yards per acre of drainage area.

Sediment Basins—Control drainage areas greater than or equal to three acres. Minimum storage capacity of 134 cubic yards per acre of drainage area. The outfall system shall, at a minimum, maintain the structural integrity of the basin during a 25 year storm of 24-hour duration.

(7) Cut and Fill Slopes Design and Construction

Cut and fill slopes shall be designed and constructed in a manner that will minimize erosion. Slopes found to be eroding excessively within one year of permanent

Permanent Stabilization Solutions (Illinois roadside).

Thick netting and fiber mulch protection for new grass.

Netting and mulch along with cemented soil for roadside stabilization.

stabilization shall be provided with additional slope stabilizing measures until the problem is corrected.

(8) Concentrated Runoff Down Slopes

Concentrated runoff shall not flow down cut or fill slopes unless contained within an adequate temporary or permanent channel, flume, or slope drain structure.

(9) Slope Maintenance

Whenever water seeps from a slope face, adequate drainage or other protection shall be provided.

(10) Storm Sewer Inlet Protection

All storm sewer inlets made operable during construction shall be protected so that sediment-laden water cannot enter the stormwater conveyance system without first being filtered/treated to remove sediment.

(11) Stormwater Conveyance Protection

Before newly constructed stormwater conveyance channels or pipes are made operational, adequate outlet protection and any required temporary or permanent channel lining shall be installed in both the conveyance channel and the receiving channel.

(12) Work in Live Watercourse

When work in a live watercourse is performed, precautions shall be taken to minimize encroachment, control sediment transport, and stabilize the work area to the greatest extent possible during construction; nonerodible material shall be used for the construction of causeways and cofferdams; and earthen fill may be used for these structures if armored by nonerodible cover materials.

(13) Crossing Live Watercourse

When a live watercourse must be crossed by construction vehicles more than twice in any six-month period, a temporary vehicular stream crossing constructed of nonerodible material shall be provided.

(14) Regulation of Watercourse Crossing

All applicable federal, state and local regulations pertaining to working in or crossing live watercourses shall be met.

(15) Stabilization of Watercourse

The bed and banks of a watercourse shall be stabilized immediately after work in the watercourse is completed.

(16) Underground Utility Line Installation

Underground utility lines shall be installed in accordance with the following standards in addition to other applicable criteria: no more than 500 linear feet of trench may be opened at one time; excavated material shall be placed on the uphill side of trenches; effluent from dewatering operations shall be filtered or passed through an approved sediment trapping device, or both, and discharged in a manner that does not adversely affect flowing streams or off-site property; material used for backfilling trenches shall be properly compacted in order to minimize erosion and promote stabilization; restabilization shall be accomplished in accordance with these regulations; and all work shall comply with applicable safety regulations.

Working in rivers, streams, or lake shorelines requires special consideration (none of these examples have any erosion controls).

Use of Cofferdams to Protect Waterbodies to Support Near-Shore Construction

The state power authority built a hydraulic turbine generating plant on the eastern shore of a pristine lake in 1918. After almost 80 years of operation, a crack was noticed in the supporting structural carriage for one of the turbines at the plant's outlet on the lake. In order to gain access to the turbine area, the authority needed to construct a coffer dam out into the lake, and then dewater the area, construct access and complete repairs. The neighborhood along the shore and inhabitants of the area had changed dramatically in the decades since the plant was first built. There were many concerns about construction impacts such as noise, access and disruption of traffic as well as environmental impacts to the lake, even with the construction of a coffer dam to isolate the work area. The project engineer for the authority investigated constructing an earthen coffer dam with a rock riprapped face for wave protection. The length would be only 150 feet and the height needed about 4 feet. He found that this would take a week to construct at a cost of about $27,000 and then another week to remove after the turbine repairs were made. However, this activity would also cause disturbance to the lake bed.

After consulting with an erosion control expert, he decided to use a coffer dam of two polyethylene tubes filled with water and wrapped with a durable geotextile. The system cost $2,100 and was installed in just four hours, including dewatering. At the completion of the work, the system was drained and removed in an hour and a half. This system was floated into position then filled with water and had essentially no disturbance to the lake bed. Its height extended well above lake level to allow protection from wave action by wind or watercraft. Although this structure is not bullet proof and can freeze solid if used in cold climate applications, it is an excellent system for isolating work areas that require small depth control with low environmental impact.

Water-Filled Coffer Dams Allowing Near-Shore Work

These two photos at the New York State Electric & Gas Corporation hydroelectric plant on Keuka Lake demonstrate the use of water structures as cofferdams. Two tubes of polyethylene wrapped with a geotextile are filled with water to act as a low ground pressure, environmentally friendly cofferdam. They can be installed and removed quickly (Photograph by D. Lake).

This is another example of how utility line crossings can be accomplished without routing construction equipment through the stream. This Aqua-Barrier allows one side to be completed then the set up can be moved to the opposite bank for completion of the crossing.

Use of Cofferdams to Support In-Lake Dam Rehabilitation

Not 60 miles from a major U.S. city, a nuclear fuel processing facility operated on this 54-acre lake from 1946–1972. Dam safety inspections statewide found a number of dams that did not meet safety standards. This particular project required the installation of a reservoir-drain system in the high-hazard dam in order to assist in meeting current safety standards. Since the existing 100 foot long dam had a competent concrete core in the center of the dam beginning about four feet below the top of dam and extending down to the rock foundation, breaching the dam to install a conventional pipe/gate system was not feasible.

It was decided to install a siphon system for reservoir drawdown. To do this and maintain the integrity of the ecosystem of the lake, a coffer dam system was constructed of structural steel A-frames with a geo-membrane that was placed on the frame and extended out into the pool area. The entire system was put in place by divers. This coffer dam was about 125 feet long and about 8 feet high. Once the dam was in place, the work site was dewatered by pumps whose intakes were located well away from the base of the structural frame.

Once the construction was complete, the area was cleaned up of excess materials and some fish habitat structures placed in the area. The water was then pumped back into the work area and the divers removed the coffer dam with minimal disturbance to the lake bottom.

Near-Shore Barrier Dams

These photos show the use of a Port-A-Dam system in use at Nuclear Lake in Dutchess County, New York. This 54 acre lake is about 12 feet deep and would have had to be pumped dry to fix the dam. This system was installed by divers, then the interior pumped dry for working, protecting lake ecosystem (Photograph by D. Lake).

(17) Vehicular Sediment Tracking

Where construction vehicle access routes intersect paved or public roads: provisions shall be made to minimize the transport of sediment by vehicular tracking onto the paved surface; where sediment is transported onto a paved or public road surface, the road surface shall be cleaned thoroughly at the end of each day; and sediment shall be removed from the roads by shoveling or sweeping and transported to a sediment control disposal area. Street washing shall be allowed only after sediment is removed in this manner.

(18) Removal of Temporary Measures

All temporary erosion and sediment control measures shall be removed within 30 days after final site stabilization, or after the temporary measures are no longer needed, unless otherwise authorized by the program authority. Trapped sediment and the disturbed soil areas resulting from the disposition of temporary measures shall be permanently stabilized to prevent further erosion and sedimentation.

(19) Stormwater Management

Properties and waterways downstream from development sites shall be protected from sediment deposition, erosion, and damage due to increases in volume, velocity, and peak flow rate of stormwater runoff for the stated frequency storm of 24-hour duration in accordance with the following standards and criteria:

- Concentrated stormwater runoff leaving a development site shall be discharged directly into an adequate natural or man-made receiving channel, pipe, or storm sewer system. For those sites where runoff is discharged into a pipe or pipe system, downstream stability analyses at the outfall of the pipe or pipe system shall be performed.
- Adequacy of all channels and pipes shall be verified:
  —*Natural Channels*—use 2-year storm event
  —*Manmade Channels*—use 2- and 10-year storm events
  —*Pipe and Pipe Systems*—use 10-year storm event
- If existing natural receiving channels or previously constructed man-made channels or pipes are not adequate, the applicant shall provide channel, pipe, or pipe system improvement or provide a combination of channel improvement, site design, stormwater detention, or other measures that is satisfactory to the program authority to prevent downstream erosion.
- Provide evidence of permission to make the improvements.
- If the applicant chooses an option that includes stormwater detention, he shall obtain approval from the locality of a plan for maintenance of the detention facilities. The plan shall set forth the maintenance requirements of the facility and the person responsible for performing the maintenance.
- Outfall from a detention facility shall be discharged to a receiving channel, and energy dissipators shall be placed at the outfall of all detention facilities as necessary to provide a stabilized transition from the facility to the receiving channel.
- Increased volumes of sheetflows that may cause erosion or sedimentation on adjacent property shall be diverted to a stable outlet, adequate channel, pipe or pipe system, or to a detention facility.
- In applying these stormwater runoff criteria, individual lots or parcels in a residential, commercial or industrial development shall not be considered to be separate development projects. Instead, the development as a whole shall be considered to be a single development project.
- All measures used to protect properties and waterways shall be employed in a manner that minimizes impacts on the physical, chemical and biological integrity of rivers, streams and other waters of the state.

The complete, unedited version of the Virginia Erosion and Sediment Control Regulations (4VAC50-30) as codified in the Virginia Administrative Code is available through the Commonwealth of Virginia website at www.vipnet.org/vipnet/portal/government.

Alabama Procedures for Developing Plans for Erosion and Sediment Control

The following discussion is excerpted from the *Alabama Handbook for Erosion Control, Sediment Control, and Stormwater Management on Construction Sites and Urban Areas,* produced by the Alabama Soil and Water Conservation Committee Montgomery, AL, July 2003 (http://swcc.state.al.us/erosion_handbook.htm).

An erosion and sediment control plan is a working document which explains and stipulates the measures and actions which are to be taken to control potential erosion and sedimentation problems. The plan has a written narrative and a graphic portion known as a treatment map or site map. It contains specifications that describe how the measures are to be installed to meet the appropriate criteria. Also, it contains enough information to ensure that the party responsible for development of a site can install the measures in the correct sequence at the appropriate season of the year. The plan may contain a description of the potential erosion and sedimentation problems.

The purpose of an erosion and sediment control plan is to establish clearly which control measures are intended to prevent erosion and off-site sedimentation. The plan should serve as a blueprint for the location, installation, and maintenance of practices to control all anticipated erosion, and prevent sediment from leaving the site.

Developers and others can minimize erosion, sedimentation, and other construction problems by selecting areas appropriate for the intended use. Tracts of land vary in suitability for development. Knowing the soil type, topography, natural landscape values, drainage patterns, flooding potential, and other pertinent data helps identify both beneficial features and potential problems of a site.

The planner should have a sound understanding of the requirement of any state and/or local erosion and sediment control laws, erosion and sedimentation control principles, and vegetative, structural, and management type measures and their role in erosion and sediment control before preparing an erosion and sedimentation control plan on a selected site.

The erosion and sediment control plan should be a separate document. The plan should include, at a minimum, the erosion and sediment control layout, measure details and specifications. The approved plan should be included in the general construction contract.

An erosion and sediment control plan must contain sufficient information to describe the site development and the system intended to control erosion and off-site sedimentation. If regulations exist, the plan must satisfy the approving authority that the potential problems of erosion and sedimentation will be adequately addressed. The length and complexity of the plan should be commensurate with the size and importance of the project, severity of site conditions, and the potential for off-site damage.

Obviously, a plan for constructing a house on a single subdivision lot may not need to be as complex as a plan for a shopping center development. Plans for projects undertaken on flat terrain will generally be less complicated than plans for projects constructed on steep slopes with higher erosion potential. The greatest level of planning and detail should be evident on plans for projects which are adjacent to flowing streams, dense population centers, high value properties, etc., where damage may be particularly costly or detrimental to the environment.

The owner or lessee of the land being developed has the responsibility for plan preparation and submission. The owner or lessee may designate someone (i.e, an engineer, architect, contractor, etc.) to prepare and implement the plan, but the owner or lessee retains the ultimate responsibility. The following outline of the procedures can be used by planners and reviewers as a checklist for plan content and format.

Components of the Plan

As a minimum, include the following components in the plan:

- *A location or vicinity map, a narrative written in a clear, concise manner that describes the type of proposed development, existing conditions at the site and adjacent areas, proposed erosion and sediment control measures, and rationale or justification for those decisions.* Adequate information provided by the narrative is important for the plan reviewer who may not be familiar with the site and to the construction superintendent and inspector who are responsible for plan installation. Details of the narrative can save time and insure that erosion and sediment control measures are properly installed.
- *Specifications for planned erosion and sediment control measures.* These should include standards and specifications for both vegetative and structural measures. The specific name and number of planned measures should be identified in the narrative and marked on the site plan or treatment map. By properly referencing the state *Handbook,* the planner could reduce the need for detailed drawings and lengthy conservation practice descriptions. New innovative conservation measures or modifications to the State standard measures may be used, but only after being thoroughly described, detailed designs developed, and concurred by the approving authority.
- *Site plan or treatment map.* This map may include a site development drawing and a site erosion and sediment control drawing depicting type and, to the extent possible, locations of planned conservation practices. Map scales and drawings should be appropriate for clear interpretation.

Site planners are urged to use the standard coding system for conservation practices contained in the *Alabama Handbook.* Use of the coding system will result in increased uniformity of plans and better readability for plan reviewers, job superintendents, and inspectors statewide.

The following components should be separate or included as applicable in the written narrative or site plan:

- Supporting material such as sketches and calculations for design of conservation practices, construction schedule, other maps (e.g., soils maps, charts, or other materials) as applicable.

Step-By-Step Procedures for Plan Development

Step 1—Data Collection

Inventory the existing site conditions to gather information which will help the planner develop the most effective erosion and sediment control plan. The information obtained should be shown on a map and verbally explained in the narrative portion of the plan.

A. *Topography*—A small-scale topographic map of the site

should be prepared to show the existing contour elevations. The suggested interval is usually 1 to 5 feet, depending upon the slope of the terrain. However, the contour interval may be increased on steep slopes.

B. *Drainage Patterns*—All existing drainage swales and patterns on the site should be located and clearly marked on the topographic map.

C. *Soils*—Major soil type(s) on the site should be determined and shown on the topographic map. Soils information can be obtained from the County Soil Survey, available from the local Soil Conservation District Office. Commercial soils evaluations are also available from consultants. For ease of interpretation, soils information should be plotted directly onto the map, or an overlay of the same scale. Chapter 4 describes soil characteristics as contained in the County Soil Surveys that are relevant to erosion control plans.

D. *Ground Cover*—The existing vegetation on the site should be shown. Such features as trees and other woody vegetation, grassy areas, and unique vegetation should be shown on the map. In addition, existing bare or exposed soil areas should be indicated.

E. *Adjacent Areas*—Areas adjacent to the site should be delineated on the topographic map. Applicable features such as streams, roads, houses, and utilities or other buildings and wooded areas should be shown.

Step 2—Data Analysis

When all of the data in Step 1 are considered together, a picture of the site potentials and limitations should begin to emerge. The site planner should be able to determine those areas which have potentially critical erosion hazards. The following are some important points to consider in site analysis:

A. *Topography*—The primary topographic considerations are slope steepness and slope length. The longer and steeper the slope, the greater the erosion potential from surface runoff. When the percent of slope has been determined, areas of similar steepness should be outlined. Slope gradients can be grouped into three general ranges of soil erodibility:

0–2%—Low erosion hazard potential
2–5%—Moderate erosion hazard potential
over 5%—High erosion hazard potential

Within these slope gradient ranges, longer slope lengths further increase the erosion hazard. Therefore, in determining potential critical areas, the site planner should be aware of excessively long slopes. As a general rule, the erosion hazard will become critical if slope lengths exceed these combined values:

0–2%—300 feet
2–5%—150 feet
over 5%—75 feet

Figures 2.7 and 2.8 are examples of pre-development and final grading site topography evaluations for these slope erosion hazards. The pre-development topography shows much of the site having steep slopes and critical erosion hazards because many of the steep slopes were greater than 75 feet long. The site was originally heavily wooded, with little observed erosion problems. However, the site clearing operations left these soils exposed at these slopes while the site was slowly graded to the final site contours, as shown in Figure 2.8. Because the site was located at the top of the local drainage area and was surrounded by major roads on the upslope sides, little off-site drainage flowed across the site as it was being developed. Diversion structures were therefore not needed, but downslope controls were critical during the grading operation to minimize sediment transport off the site. Because the site was relatively small (between 5 and 10 acres), with concurrently small subdrainage areas, only filter fabric fences were used, and not a sediment pond. However, a pond would have been more suitable due to most of the site draining towards one area. The final grading contours shown on Figure 2.8 show that most of the site was graded flat for building pads and therefore had low erosion hazards. The slopes on the bottom edges of the terraces, however, are quite steep and have high erosion hazard potentials. The final slope lengths are all relatively short, so the only critical erosion hazard is near the bottom outlet area. These steep slopes require protection, as described in Chapter 5.

B. *Drainage Patterns*—Natural drainage patterns exist on the land. These patterns, known as swales, depressions, and natural watercourses, should be identified in order to plan around critical areas where water will concentrate. Where it is possible, natural drainage ways should be used to convey runoff over and off the site to avoid the expense and problems of constructing an artificial drainage system. Man-made ditches and waterways will become part of the erosion problem if they are not properly stabilized. Care should also be taken to be sure that increased runoff from the site will not erode or flood the existing natural drainage system; this includes locating possible sites for stormwater detention. Chapter 3 presents examples for determining site drainage evaluations.

C. *Soils*—Such soils' properties as natural drainage, depth to bedrock, depth to seasonal water table, permeability, shrink-swell potential, texture, and erodibility should exert a strong influence on land development decisions. Also, the flood hazard related to the soils can be determined based on the relationship between soils and

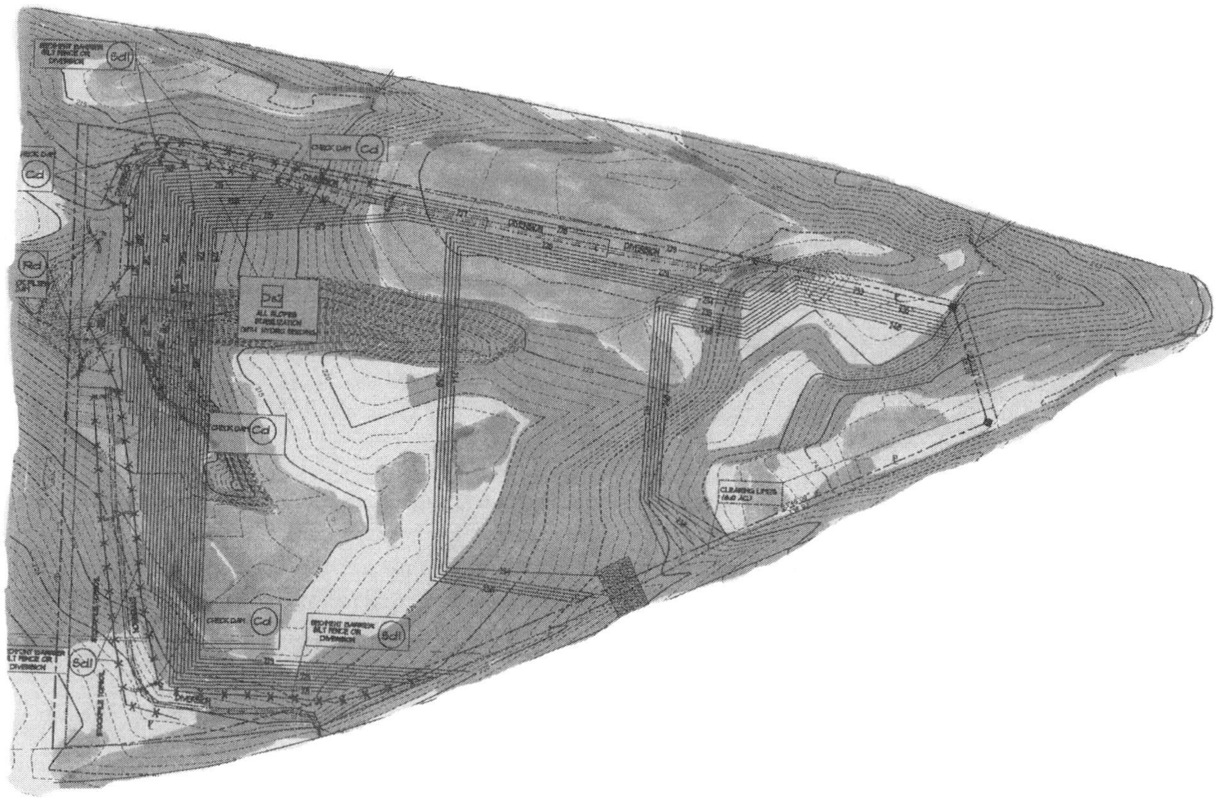

Figure 2.7. Evaluation of pre-development topography (dashed contour lines) for erosion hazards (orange: >10% slopes and high hazard; yellow: 5 to 10% slopes and high hazard; blue: 2 to 5%slopes and moderate hazard; pink: <2% slopes and low hazard). Publishers note: to see the figure in color please contact the authors.

Figure 2.8. Evaluation of final grading plan topography (solid contour lines) for erosion hazards (orange: >10% slopes and high hazard; yellow: 5 to 10% slopes and high hazard; blue: 2 to 5% slopes and moderate hazard; pink: <2% slopes and low hazard). Publishers note: to see the figure in color please contact the authors.

flooding. A discussion of soils, along with interpretations for developmental uses, is included in the County Soil Maps, from the local NRCS (SCS) offices. Chapters 3 and 4 both discuss important soil considerations.

D. *Ground Cover*—Ground cover is the most important factor in terms of preventing erosion. Any existing vegetation which can be saved will help prevent erosion. Trees and other vegetation protect the soil as well as beautify the site after construction. If the existing vegetation cannot be saved, the planner should consider staging construction, temporary mulching and/or vegetation. Staging of construction involves stabilizing one part of the site before disturbing another. In this way, the entire site is not disturbed at once, minimizing the time ground is left bare. Temporary mulching and/or vegetation involves seeding or mulching areas which would otherwise be left bare for long periods of time; therefore, time of exposure is limited and the erosion hazard is reduced.

E. *Adjacent Areas*—Generally, the analysis of adjacent properties should focus on areas downslope or downstream from the construction project. Watercourses which will receive direct runoff from the site should be of major concern; these streams should be analyzed to determine their carrying capacity. The potential for sediment pollution of these watercourses should be considered as well as the potential for downstream channel erosion due to increased velocity and peak flow rate of storm water runoff from the site. The potential for sediment deposition on adjacent properties due to sheet and rill erosion should also be analyzed so that appropriate sediment retention measures can be planned.

Step 3—Facility Plan Development

This step does not apply to established developments. On the other hand, this step is relevant to those situations where facilities are being planned and there is flexibility in their extent and location. After analyzing the data about the site and determining any site limitations, the planner can then develop a site plan that is in harmony with the landscape. An attempt should be made to locate the buildings, roads, and parking lots and develop landscaping plans to exploit the strengths and overcome the limitations of the site. The following are some points to consider in making these decisions:

A. *Fit development to terrain*—The development of an area should be tailored, as much as possible, to existing site conditions. This will avoid unnecessary land disturbance, while minimizing the erosion hazards and development costs.

B. *Confine construction activities to the least critical areas*—Any land disturbance in the critically erodible areas will necessitate the installation of more costly erosion and sediment control measures.

C. *Cluster buildings together*—This minimizes the amount of disturbed area, concentrates utility lines and connections in one area while leaving more open natural space. The cluster concept not only lessens the erodible area, but it generally reduces runoff and development costs.

D. *Minimize impervious areas*—Keep paved areas, such as parking lots and roads, to a minimum. This goes hand in hand with cluster developments in eliminating the need for duplicating parking areas, access roads, etc. The more land that is kept in vegetative cover, the more water will infiltrate, thus minimizing runoff and erosion. Consider the use of special pavements which will allow water to infiltrate, or cellular blocks which have soil and vegetation components.

E. *Utilize the natural drainage system*—If the natural drainage system of a site can be preserved instead of being replaced with storm sewers or concrete channels, the potential for downstream damages due to increased runoff can be minimized, making compliance with stormwater management criteria much easier.

Step 4—Planning for Erosion and Sediment Control

When the site facility plan layout has been developed, a plan to control erosion and sedimentation from the disturbed areas then is formulated. The following general procedure is recommended for erosion and sediment control planning:

A. *Divide the site into drainage areas*—Determine how runoff will travel over the site. Consider how erosion and sedimentation can be controlled in each small drainage area before looking at the entire site. Remember, it is easier to control erosion than to contend with sediment after it has been carried downstream.

B. *Determine the limits of clearing and grading*—Decide exactly which areas must be disturbed in order to accommodate the proposed construction. Pay special attention to critical areas which must be disturbed. The important point in this activity is to minimize the areas to be disturbed.

C. *Select erosion and sediment control measures*—Erosion and sediment control practices can be divided into 3 broad categories: vegetative measures, structural measures, and management measures. The *Alabama Handbook* should be used for the selection and design of vegetative and structural measures. Management measures include items such as construction management techniques which, if properly utilized, can minimize the need for more costly vegetative/structural erosion and sediment control measures.

1. *Vegetative Controls*—Vegetative controls should generally be considered first, because of economics. Usually, vegetation should be established on a temporary basis to minimize offsite impacts at the beginning of land disturbances. Vegetation protects the soil surface from raindrop impact and overland flow of runoff water. Vegetative measures should be maximized to provide as much erosion and sediment control as possible, with a minimum of structural measures. One of the simplest ways to protect the soil surface is by preserving existing ground cover where protective cover already exists. Where existing ground cover must be removed and land disturbance is necessary, temporary seeding or mulching can be used on areas that are to be exposed for long periods. Erosion and sediment control plans must contain provisions for permanent stabilization of disturbed areas. Selection of permanent vegetation should include the following considerations:
   a. adaptability to site conditions
   b. establishment requirements
   c. aesthetics
   d. maintenance requirements
2. *Structural Controls*—Structural measures are generally more costly than vegetative controls. However, they are necessary on areas where vegetation alone will not control erosion. In addition, structural measures are often needed in combination with vegetative measures as a second or third line of defense to capture sediment before it leaves the site. It is very important that structural measures be selected, designed, constructed, and maintained according to the standards and specifications in the *Alabama Handbook.* Poorly planned or constructed structural measures can increase development costs and create maintenance problems. Structural measures that fail may increase erosion and sedimentation. Therefore, it is very important that structural measures be designed and installed properly.
3. *Management Measures*—Good construction management is as important as physical measures for erosion and sediment control and there is generally little or no cost involved. Following are some management considerations which should be included in the erosion and sediment control plan:
   a. Sequence construction so that no area remains exposed for unnecessarily long periods of time.
   b. When possible, avoid grading activities during months such as July and November through February, because these months are unsuitable for seeding and the potential for erosion and sedimentation is high.
   c. Temporary seedings should be done immediately after grading.
   d. On large projects, stage the construction if possible, so that one area can be stabilized before another is disturbed.
   e. Develop and carry out a regular maintenance schedule for erosion and sediment control measures.
   f. Physically mark off limits of land disturbance on the site with tape, signs or other methods, so the workers can see areas to be protected.
   g. Make sure that all workers understand the major provisions of the erosion and sediment control plan.
   h. Responsibility for implementing the erosion and sediment control plan should be designated to one individual (preferably the job superintendent or foreman).
4. Plan for stormwater management. Where increased runoff will cause the carrying capacity of a receiving channel to be exceeded (for a 2-year storm), the site planner must select appropriate stormwater management measures.

Step 5—Plan Assembly

The necessary planning work was done in steps 1 through 4; therefore, this final step consists of consolidating the pertinent information and developing it into a specific erosion and sediment control plan for the project. The two major plan components are a narrative and a site plan. The narrative verbally explains the problems and their solutions with all necessary documentation. The site plan is one, or a series, of maps or drawings pictorially explaining information contained in the narrative. The following checklists may be used in completing the narrative and site plan. These checklists can be used as a guide by the site planner as a ready reference to be sure all major items are included in the erosion and sediment control plan.

Checklist for Erosion and Sediment Control Plans

Narrative

- *Project description*—Briefly describe the nature and purpose of the land disturbing activity and the amount of grading involved.
- *Existing site conditions*—A description of the existing topography, vegetation, and drainage.

- *Adjacent areas*—A description of neighboring areas such as streams, lakes, residential areas, roads, etc., which might be affected by the land disturbance.
- *Soils*—A brief description of the soils on the site giving such information as soil names, mapping unit, erodibility, permeability, depth, texture, soil structure, and any other limitations.
- *Critical areas*—A description of areas on the site which have potentially serious erosion problems.
- *Erosion and sediment control measures*—A description of the methods which will be used to control erosion and sedimentation on the site.
- *Permanent stabilization*—A brief description, including specifications, of how the site will be stabilized after construction is completed.
- *Stormwater management considerations*—Will the development of the site result in increased peak rates of runoff? Will this result in flooding or channel degradation downstream? If so, considerations should be given to stormwater control structures on the site. Local ordinances must be considered and met.
- *Maintenance*—A schedule of regular inspections and repair of erosion and sediment control measures should be set forth.

Authors' Note: A review of other state regulations indicates that this list is similar to the elements required of the narrative section of the erosion control plans in other states. Items addressed by other states include the additional protections required if the construction site is adjacent to or draining to a previously-defined sensitive stream. For example, Pennsylvania requires additional protective measures for its trout-fishing streams and other "outstanding" waterways.

TABLE 2.1. Legend of Measures for Erosion and Sediment Control Plans (Alabama 1993).

Vegetative Measures	
BZ	Buffer Zone
DU	Dust Control
IR	Irrigation
MU	Mulching
PS	Permanent Seeding
TP	Tree Planting on Disturbed Sites
SD	Sodding
SVG	Shrub, Vine and Groundcover Planting
TV	Temporary Vegetation—Seeding
TPP	Tree Preservation and Protection
Coastal Dune Measures	
DC	Dune Crosswalk
DSF	Dune Sand Fence
DV	Dune Vegetation
Structural Measures	
CD	Check Dam
CS	Channel Stabilization
CE	Construction Exit
DV	Diversion
DN	Downdrain Structure
DS	Drop Structure
GB	Gabion
NP	Inlet Protection
OP	Outlet Protection
PF	Paved Flume
RW	Retaining Wall
RT	Retrofitting
RR	Riprap
SF	Sediment Barrier/Fence
SB	Sediment Basin
RS	Storm Water Retention Structure
SX	Stream Crossing
SR	Surface Roughening
TS	Topsoil
WW	Waterway or Storm Water Conveyance Channel

Site Plan

- *Vicinity map*—A small map locating the site in relation to the surrounding area.
- *Existing contours*—The existing contours of the site should be shown on a map.
- *Existing vegetation*—The existing tree lines, grassy areas, or unique vegetation should be shown on a map.
- *Soils*—The boundaries of the different soil types should be shown on a map.
- *Indicate north*—The direction of north in relation to the site should be shown. The top of all maps should be north, if possible and practical.
- *Critical erosion areas*—Areas with potentially serious erosion problems should be shown on a map.
- *Existing drainage patterns*—The dividing lines and the direction of flow for the different drainage areas should be shown on a map.
- *Final contours*—Changes to the existing contour should be shown on a map.
- *Development features*—The outline of buildings, roads, drainage appurtenances, utilities, landscaping features, parking areas, improvements, impervious areas, topographic features, and similar man-made installations should be shown to scale and relative location.
- *Limits of clearing and grading*—Areas which are to be cleared and graded should be outlined on a map.
- *Location of measures*—The locations of the erosion and sediment control and stormwater management practices used on the site should be shown on a map, using the notation shown on Table 2.1.
- *Detail drawings*—Any structural measures used that are not referenced to the manual or other local manuals should be explained and illustrated with detailed drawings.

AMOUNTS OF CONSTRUCTION ACTIVITY SUBJECT TO EROSION AND SEDIMENT CONTROL AND THEIR COSTS[1]

Historic Trends

The Department of Commerce (DOC) began collecting detailed information on housing starts in 1963. Data on housing permits and starts are published monthly by the DOC and are viewed by economists as leading indicators of economic activity. More detailed industry information is collected through the Census of Construction Industries (CCI), which is conducted every 5 years (in years ending in a 2 or a 7) as part of the Census Bureau's Economic Census program. These data provide the most detailed snapshot of the status of the construction industry. The CCI covers all employer establishments primarily engaged in construction. Table 2.2 summarizes housing starts for the period from 1979 to 1999. In this table, the number of construction starts is shown by regional location and type of structure. The table also provides national totals for both single- and multifamily housing starts (BOC, 2001). As shown in the table, single-family housing accounts for the majority of housing construction starts. The number of construction starts annually for privately-owned housing units has decreased from approximately 1.7 million starts in 1979 to roughly 1.6 million starts in 1999 (BOC, 2001). At the regional level, growth rates have varied to a large degree. Construction of housing has increased by nearly 40 percent in the South, while construction starts in 1999 in the Northeast actually decreased by almost 13 percent from 1989 levels. Housing starts in the Midwest also increased significantly over 1989 levels, while housing starts in the West remained at about the same level as a decade earlier. Tables 2.3 and 2.4 list the markets having the most housing starts in 1999.

Construction Site Size Categories and Estimates of Amount of Disturbed Land

The Phase I and Phase II NPDES stormwater permit requirements apply to construction sites of all types (i.e., residential, commercial, and industrial) of more than one acre. Because the costs of erosion and sediment control are largely driven by site size, the EPA estimated the distribution of construction sites by size category, land use type, and geographic region in order to estimate the cost of erosion controls (USEPA 2002).

National Estimates of Disturbed Acreage

The EPA used the U.S. Department of Agriculture's (USDA's) 1997 National Resources Inventory (NRI) (USDA, 2000) to estimate the level of new U.S. development each year. The NRI is designed to track changes in land cover and land use over time. The inventory, conducted every five years, covers all non-federal lands in the U.S., which are 75

TABLE 2.2. Annual Housing Construction Starts by Type and Region (Starts are in thousands of units) (EPA, 2002).

		Northeast		Midwest		South		West	
Year	United States	Single-family	Multi-family	Single-family	Multi-family	Single-family	Multi-family	Single-family	Multi-family
1979	1,745	123	55	243	106	522	225	306	165
1980	1,292	87	38	142	76	428	215	196	110
1981	1,084	84	33	110	55	363	198	148	92
1982	1,062	79	37	99	50	357	234	127	78
1983	1,703	123	45	153	65	557	378	234	148
1984	1,750	158	46	167	76	528	338	230	206
1985	1,742	182	70	148	92	504	278	239	230
1986	1,805	228	66	188	108	504	229	261	222
1987	1,621	204	65	203	95	485	149	255	165
1988	1,488	181	54	194	80	443	132	264	140
1989	1,376	132	47	190	76	409	127	272	124
1990	1,193	104	27	193	60	371	108	226	103
1991	1,014	99	14	191	42	353	62	197	57
1992	1,200	112	15	236	52	439	58	244	45
1993	1,288	116	11	251	47	498	63	261	41
1994	1,457	123	16	268	61	522	117	286	65
1995	1,354	102	16	233	57	485	130	256	76
1996	1,447	112	20	254	68	524	138	271	90
1997	1,474	111	26	238	66	507	164	278	86
1998	1,617	122	26	223	58	573	169	303	92
1999	1,641	126	29	289	59	580	167	308	84

[1]This section is summarized from the USEPA's *Development Document for Proposed Effluent Guidelines and Standards for the Construction and Development Category*, United States Environmental Protection Agency, Office of Water (4303T), EPA-821-R-02-007, Washington, DC 20460 (www.epa.gov/waterscience/guide/), June 2002.

TABLE 2.3. Busiest Markets for Single-Family Housing Permits for 1999.

Market Area	Single-family Housing Permits (1999)	Percent Change from 1998
Atlanta	25,066	+11%
Phoenix	21,290	+13%
Dallas-Ft. Worth	17,434	+6%
Chicago	14,954	+7%
Washington, D.C.	14,703	+0.07%

Source: U.S. Housing Markets, 1999a.

percent of the U.S. total land area, using land use information from about 800,000 statistically-selected locations. From 1992 to 1997, about 2.2 million acres per year were converted from non-developed to developed status. Table 2.5 shows the allocation of this converted land area by type of land or land cover, while Table 2.6 shows the amount of new land areas developed per year. Table 2.7 also lists the distribution of the construction activities by site area and land use.

Costs of Erosion and Sediment Controls for Construction Sites

The following discussion summarizes some of the expected costs associated with erosion and sediment control practices at construction sites, as prepared by the EPA (USEPA 2002). Appendix 2A also includes selected data from the comprehensive report prepared by the Southeast Wisconsin Regional Planning Commission (SEWRPC), *Costs of Urban Nonpoint Source Water Pollution Control Measures* (SEWRPC 1991). In the data sets given in this section, the SEWRPC costs are for 1988 and 1989, while the R.S. Means costs are for 2000. In order to convert these costs to estimated 2005 U.S. dollars, the SEWRPC costs should be multiplied by about 1.6, while the R.S. Means costs should be multiplied by about 1.1 (based on historical and projected Consumer Price Index values).

Vegetative Stabilization

Grass-Lined Channels—Grassed channel construction costs can be estimated using unit cost values. Shallow trenching (1 to 4 feet deep) with a backhoe in areas not requiring dewatering can be performed for $4 to $5 per cubic yard of removed material (R. S. Means 2000). Assuming no disposal costs (i.e., excavated material is placed on either side of the trench), only the cost of fine grading, soil treatment, and grassing (approximately $2 per square yard of earth surface area) should be added to the trenching cost to approximate the total construction cost. Site-specific hydrologic analysis of the construction site is necessary to estimate the channel conveyance requirement; however, it is not unusual to have flows on the order of 2 to 4 cfs per acre served. For channel velocities between 1 and 3 feet per second, the resulting range in the channel cross-section area can be as low as 0.67 square foot per acre drained to as high as 4 square feet per acre. If the average channel flow depth is 1 foot, then the low estimate for grassed channel installation is $0.27 per square foot of channel bottom per acre served per foot of channel length. The high estimate is $1.63 per square foot of channel bottom per acre served per foot of channel length.

Seeding—Seeding costs range from $200 to $1,000 per acre and average $400 per acre. Maintenance costs range from 15 to 25 percent of initial costs and average 20 percent (USEPA 1993). R. S. Means (2000) indicates the cost of mechanical seeding to be approximately $900 per acre, and

TABLE 2.5. Acres Converted from Undeveloped to Developed State, 1992–1997 (EPA, 2002).

Type of Land	Acres Converted to Development 1992–1997 (thousands), Annual Average	Percent Contribution by Type of Land
Cropland	574.8	26.6%
Conservation Reserve Program land	1.5	0.1%
Pastureland	391.2	17.4%
Rangeland	245.9	11.0%
Forest land	939.0	41.9%
Other rural areas	89.1	4.0%
Water areas and federal land	1.8	0.1%
Total	2,243.4	100%

TABLE 2.4. Busiest Markets for Multifamily Housing Permits for 1999.

Market Area	Multifamily Housing Permits (1999)	Percent Change from 1998
Dallas-Ft. Worth	8,488	–15%
Orlando	7,303	+46%
New York-Long Island	6,255	+55%
Puget Sound	6,122	+19%
Houston	5,900	–50%

Source: U.S. Housing Markets, 1999b.

TABLE 2.6. National Estimates of Land Area Developed Per Year (EPA, 2002).

Type of Construction		Total NRI Acreage[a]	Acres Waived or not Covered	Adjusted NRI Acreage[b]
Residential	Single-family	546,783	12,905	533,878
	Multifamily	258,616	6,434	252,182
Nonresidential	Commercial	1,377,070	44,594	1,332,476
	Industrial	60,932	3,412	57,523
Total		2,243,400	67,345	2,176,058

TABLE 2.7. Distribution of National Construction by Site Size and Development Type.

Site Size (Acres)	No. of Permits	Acres by Size	Pct. Acres by Size
Single-Family Residential			
1	12,392	12,392	2.3%
3	10,622	31,865	6.0%
7.5	6,429	48,217	9.0%
25	8,153	203,815	38.2%
70	1,398	97,831	18.3%
200	699	139,759	26.2%
Total	39,691	533,878	100.0%
Multifamily Residential			
1	3,120	3,120	1.2%
3	7,256	21,768	8.6%
7.5	4,426	33,196	13.2%
25	5,152	128,794	51.1%
70	726	50,792	20.1%
200	73	14,512	5.8%
Total	20,752	252,182	100.0%
Commercial			
1	64,280	64,280	4.8%
3	86,029	258,086	19.4%
7.5	20,782	155,866	11.7%
25	21,990	549,761	41.3%
70	4,350	304,483	22.9%
200	0	0	0.0%
Total	197,431	1,332,476	100.0%
Industrial			
1	3,256	3,256	5.7%
3	4,592	13,775	3.9%
7.5	835	6,262	10.9%
25	668	16,698	29.0%
70	250	17,532	30.5%
200	0	0	0.0%
Total	9,601	57,523	100.0%
Totals			
1	83,048	83,048	3.8%
3	108,498	325,494	15.0%
7.5	32,472	243,541	11.2%
25	35,963	899,067	41.3%
70	6,723	470,638	21.6%
200	771	154,271	7.1%
Grand Total	267,475	2,176,059	100.0%

Based on permitting data from the following municipalities or counties: Austin, TX; Baltimore County, MD; Cary, NC; Ft. Collins, CO; Lacey, WA; Loudoun County, VA; New Britain, CT; Olympia, WA; Prince George's County, MD; Raleigh, NC; South Bend, IN; Tallahassee, FL; Tuscon, AZ; and Waukesha, TI. Source: USEPA, 1999.

demonstrates that the coverage cost varies with the seed type, seeding approach and scale (total acreage to be seeded). For example, hydro or water-based seeding for grass is estimated to be $700 per acre but seeding of "field" grass species is only $540 per acre (Costs include materials, labor, and equipment, with profit and overhead). If surface preparation is required, then the installation costs increase. R. S. Means suggests the cost of fine grading, soil treatment, and grassing is approximately $2 per square yard of earth surface area.

Sodding—Average construction costs of sod average $0.20 per square foot and range from $0.10 to $1.10 per square foot; maintenance costs are approximately 5 percent of installation costs (USEPA 1993). R. S. Means (2000) indicates the sodding ranges between $250 and $750 per 1000 square feet for 1″ deep bluegrass sod on level ground, depending on the size of the area treated (unit costs value are for orders over 8,000 square feet and less than 1,000 square feet, respectively). Bent grass sod values range between $350 and $500 per 1000 square feet, again the lower value is more likely for most construction sites because it is for large area applications. (Costs include materials, labor, and equipment, with profit and overhead).

Mulching—The costs of seed and mulch average $1,500 per acre and range from $800 to $3,500 per acre (USEPA 1993). R. S. Means (2000) estimates the cost of power mulching to be $22.50 per 1000 square feet, for large volume applications. In addition, hydro- and mechanical seeding are approximately $700 to $900 per acre. Coverage cost varies with the seed type, seeding approach, and scale (total acreage to be seeded). For example, hydro or water-based seeding for grass is estimated to be $700 per acre, but seeding of "field" grass species is only $540 per acre. Costs include materials, labor, and equipment, with profit and overhead. If surface preparation is required, then the installation costs increase. R. S. Means (2000) suggests the cost of fine grading, soil treatment, and grassing is approximately $2 per square yard of earth surface area.

Geotextiles (Netting Covering Planted Area)—Costs for geotextiles range from $0.50 to $10.00 per square yard depending on the type chosen (SWRCP 1991). Geosynthetic turf reinforcement mats (TRMs) are widely used for immediate erosion protection and long-term vegetative reinforcement, usually for steeply sloped areas or areas exposed to runoff flows. The Erosion Control Technology Council (a geotextile industry support association) estimates TRMs cost approximately $7.00 per square yard (installed) for channel protection. Channel protection is one of the most demanding of installations (much more demanding than general coverage of denuded area). The Erosion Control Technology Council estimates the cost to install a simple soil blanket (or rolled erosion control product), seed, and fertilizer to be $1.00 per square yard.

Vegetated Buffer Strips—Cost estimates for grassed buffer strips can be made based on square footage using unit cost values. R. S. Means (2000) estimates the cost of fine grading, soil treatment, and grassing to be $2 per square yard of earth surface area. This cost estimate is based on application of traditional lawn seed. The cost for field seed is lower than lawn seed, reducing the coverage price. Where gently sloping areas simply need to be grassed with acceptable species, the cost can be as low as $0.38 per square yard.

Topsoiling—Topsoiling costs are a function of the price of topsoil, the hauling distance, and the method of application. R. S. Means (2000) report unit cost values of $3 and $4 per square yard for 4 and 6 inches of top soil cover, respectively. This price is for furnishing and placing of top soil, and includes materials, labor, and equipment, with profit and overhead.

Water Handling Practices

Earth Dike—The cost of an earth dike depends on the design and materials used. Small dikes can cost approximately $2.00 per linear foot, while larger dikes can cost approximately $2.00 per cubic yard. The EPA states that an earth dike can cost approximately $4.50 per linear foot (NAHB undated). An alternative means to estimate conceptual costs for earthen dikes is to use unit cost values and a rough estimate of the quantities needed. Shallow trenching (1 to 4 feet deep) with a backhoe in areas not requiring dewatering can be performed for $4 to $5 per cubic yard of removed material (R. S. Means 2000). Based on this value, $2 per linear foot provides for 11 square feet of flow area and $4.50 per linear foot provides for 24 square feet of flow area. Based on standards for Virginia (VDCR 1995), most small drainage areas (made up of 5 acre or less), diversion dikes are approximately 18-in tall, with a 4.5-ft base. Assuming the excavation volume equals the volume of the dike, the resulting excavation volume is approximately 7 cubic feet per linear foot, which (conservatively) equates to $1.03 to $1.30 per linear foot for construction costs.

If the earthen dikes are to be permanent, then additional costs are incurred to vegetate the dike. R. S. Means (2000) estimates the cost of fine grading, soil treatment, and grassing is approximately $2 per square yard of earth surface area. This adds approximately $6 per linear foot of dike. Where gently sloping areas only need to be grassed with acceptable species, the cost can be as low as $0.38 per square yard.

Temporary Swale—Grassed-channel construction costs can be estimated using unit cost values. Shallow trenching (1 to 4 feet deep) with a backhoe in areas not requiring dewatering can be performed for $4 to $5 per cubic yard of removed material (R. S. Means 2000). Assuming no disposal costs (i.e., excavated material is placed on either side of the trench), only the cost of fine grading, soil treatment, and grassing (approximately $2 per square yard of earth surface area) should be added to the trenching cost to approximate the total construction cost. It is not unusual to have flows on the order of 2 to 4 cfs per acre served. For a design channel velocity of 1 foot per second, the resulting range in the channel cross-sectional area can be as low as 2, but as high as 4 square feet per acre drained. If the average channel flow depth is 1 foot, then the low estimate for grassed channel installation is $0.74 per square foot of channel bottom per acre served per foot of channel length. The high estimate is $1.48 per square foot of channel bottom per acre served per foot of channel length. Table 2.8 summarizes additional costs of grass swales.

Temporary Storm Drain Diversions—Depending on the size of the construction site, a temporary storm drain diversion's costs can include those associated with materials needed to construct the diversion and sediment trap or basin (mainly piping, concrete, and gravel), and also labor costs for installation and removal of the system, all of which may involve excavation, regrading, and inspections. Cost estimates can be based on unit cost values along with site-specific quantity estimates. R. S. Means (2000) indicates a range of pipe costs for surface placement, between $5.00 per linear foot for 4″ diameter PVC piping, and $9.20 per linear foot for 10″ diameter PVC piping. On construction

TABLE 2.8. Average Annual Operation and Maintenance Costs for a Grass Swale.

Component	Estimated Unit Cost ($)	$ for Swale Size: 0.5 m Deep 0.3 m Bottom Width 3 m Top Width	$ for Swale Size: 1 m Deep 1 m Bottom Width 7 m Top Width	Comments
Mowing	0.89/100 m^2	145.0	241.0	Mow 2–3 times per year
General grass care	8.8/100 m^2	162.98	274.0	Grass maintenance is (top width + 3 m) x length
Debris/litter removal	0.51/m^2	93.0	93.0	
Reseeding/ fertilization	0.35/m^2	5.9	10.37	Area revegetated is 1% of maintenance area per year
Inspection and general administration	0.74/m^2	231.0	231.0	Inspection once per year
Total		638.0	850.0	

Source: Ellis, 1998

sites, temporary inlets and outlets are usually formed by small rock-lined depressions. Assuming 4 cubic yards of crushed rock (1.5″ mean diameter) per opening, an inlet and outlet combine to add approximately $200 per pipe installation, based on $25 per cubic yard of stone (R. S. Means 2000).

Stone Check Dam—The cost of check dams varies based on the material used for construction and the width of the channel to be dammed. In general, it is estimated that check dams constructed of rock cost about $100 per dam (USEPA, 1992). Brown and Schueler (1997) estimated rock check dams would cost approximately $62 per installation, including the cost for filter fabric bedding. Other materials, such as logs and sandbags, may be a less expensive alternative, but they might require higher maintenance costs.

Sediment Trapping Devices

Silt Fence—There is a wide range of data on installation costs for silt fences. The EPA estimates these costs at approximately $6.00 per linear foot (USEPA, 1992) while SWRPC estimates unit costs between $2.30 and $4.50 per linear foot (SWRPC, 1991). Silt fences have an annual maintenance cost that is 100 percent of the installation cost (Brown and Schueler, 1997). These values are significantly greater than that reported by R. S. Means (2000), which indicates a 3 foot tall silt fence installation cost between $0.68 and $0.92 per linear foot (for favorable and challenging installations). It should be noted that the R. S. Means value covers just a single installation, without the expected costs of maintenance (e.g., removal of collected sediment). In addition, the type of silt fence fabric employed will also affect the total installation costs.

Sediment Trap—The cost of installing temporary sediment traps ranges from $0.20 to $2.00 per cubic foot of storage (about $1,100 per acre of drainage). For a recent national assessment, USEPA (1999) estimated the following costs for sediment traps, which vary as a function of the volume of storage: $513 for 1,800 cubic yards, $1,670 for 3,600 cubic yards, and $2,660 for 5,400 cubic yards. In addition, it has been reported that a sediment trap has an annual maintenance cost of 20 percent of the installation cost (Brown and Schueler 1997).

Sediment Basins—Sediment basins have an estimated 25 percent annual maintenance cost as a percentage of installation (Brown and Schueler, 1997). If constructing a sediment basin with less than 50,000 cubic feet of storage space, the cost of installing the basin ranges from $0.20 to $1.30 per cubic foot of storage (about $1,100 per acre of drainage). The average cost for basins with less than 50,000 cubic feet of storage is approximately $0.60 per cubic foot of storage (USEPA, 1993). If constructing a sediment basin with more than 50,000 cubic feet of storage space, the cost of installing the basin ranges from $0.10 to $0.40 per cubic foot of storage (about $550 per acre of drainage). The average cost for basins with greater than 50,000 cubic feet of storage is approximately $0.30 per cubic foot of storage (USEPA, 1993).

As an alternative costing method, designers can use cost curves developed for permanent basins used to manage stormwater from urban areas. However, since permanent stormwater basins typically include design features that would not be included in temporary sediment basins, this approach is expected to greatly overestimate the actual costs to construct sediment basins. For many sites, sedimentation basins installed for erosion and sediment control during the construction phase are retained/modified to meet other runoff management requirements. As a result, the sedimentation basins' installation costs are partially offset by a later cost reduction or savings. Work by the Center for Watershed Protection (CWP, 1996) provides capital cost equations for different types of sediment basins for permanent installations. For example, dry extended detention ponds:

$$CC = 8.16\,(V_s)^{0.78}$$

and for all ponds regardless of type (including wet ponds):

$$CC = 20.18\,(V_s)^{0.70}$$

where,

CC = base construction cost, not including design, engineering, and contingencies
V_s = Storage volume below the crest of the emergency spillway, in cubic feet

Design, engineering, and contingency costs are given as approximately 32 percent of the base construction costs. Base construction costs for permanent ponds are composed of approximately 48 percent excavation/grading cost, 36 percent control structure cost, and 16 percent appurtenances cost. R. S. Means (2000) suggests the cost to remove the eroded sediment collected in a small basin during construction is approximately $4 per cubic yard (value includes a 100 percent surcharge for wet excavation). The cheapest management of dredge material is application to land areas adjacent to the basin, followed with application of a vegetative cover.

Other Control Practices

Rock Outlet Protection—R. S. Means indicates machine-placed riprap costs of approximately $40 per cubic yard. For a riprap maximum size between 15 and 24 inches, a cubic yard of riprap will cover between 13.5 and 17 square feet of a channel bed. This suggests that riprap lining will be between $21 and $27 per square foot of outlet (includes materials, labor, and equipment, with overhead and profit). R. S. Means (2000) provides a cost range for gabions ($2.80 to $9 per square foot of coverage) for stone fill depths of 6 inches to 36 inches, respectively. These costs include all costs of materials, labor, and installation.

Sump Pit—R. S. Means (2000) provides information appropriate for assessment of a wide range of dewatering scenarios (i.e., different sump sizes, dewatering durations, and discharge conditions). In general, installation of earthen sump pits is estimated as costing approximately $1.50 per cubic foot of sump volume. Costs for piping to and from the sump ranges from $30 to $60 per linear foot. Pump rentals and operation range between $150 and $500 per day of pumping, depending on the rate of dewatering. All costs include material, labor, and equipment, with overhead and profit.

Stabilized Construction Entrance—Without a wash rack, construction site entrance stabilization costs range from $1,000 to $4,000. On average, the initial construction cost is around $2,000 per entrance. When maintenance costs are included, the average total annual cost for a 2-year period, is approximately $1,500. If a wash rack is included in the construction site entrance stabilization, the initial construction costs range from $1,000 to $5,000, with an average initial cost of $3,000 per entrance. Total annual cost, including maintenance for an estimated 2-year life span, is approximately $2,200 per year (USEPA, 1993).

Temporary Access Waterway Crossing—In general, temporary bridges are more expensive to design and construct than culverts. Bridges are also associated with higher maintenance and repair costs should they fail. Temporary bridging costs range as a function of the width of the bridge span and the duration of application. If the bridging is permanent, a mean cost of $50 per square foot for an 8-foot wide steel arch bridge (no foundation costs included) can be used for conceptual cost estimation (R. S. Means, 2000). If rental bridging is employed, then rates are probably on the order of 20 to 50 percent of the bridge (permanent) cost, but will range based on the rental duration and mobilization distance.

Storm Drain Inlet Protection—The cost of implementing storm drain drop inlet protection measures will vary depending on the control measure chosen. Generally, initial installation costs range from $50 to $150 per inlet, with an average cost of $100 (USEPA, 1993). Maintenance costs can be high (annually, up to 100 percent of the initial construction cost) because of frequent inspection and repair needs. The Southeastern Wisconsin Regional Planning Commission has estimated that the cost of installation of inlet protection devices ranges from $106 to $154 per inlet (SEWRPC, 1991).

Polyacrylamide (PAM)—The cost of PAM ranges from $1.25 per pound to $5.00 per pound (Entry, *et al.* 1999). The cost of PAM application depends on the system employed. PAM can be used in a centralized treatment system (e.g., at a sedimentation basin) to treat larger areas, or dispersed in granular or liquid form. In Tobiason, et al. (2000), the startup costs for the batch treatment system amounted to $90,000. Monthly expenses averaged $18,000 for operations and maintenance and $13,000 for materials and equipment. The total costs for this phase totaled about $245,000, less than 1 percent of total construction costs. If dispersed through irrigation systems, the seasonal cost of PAM treatment is $9 to $15 per acre (Kay-Shoemake, *et. al.* 2000), where a season probably requires between 5 and 10 applications.

For construction sites, it is more likely that PAM would be applied as an additive to the hydroseed mix and applied when final grade is established and cover vegetation is installed. There are numerous suppliers who provide PAM as a low cost additive for hydroseeding, suggesting PAM application costs can be incorporated into that of hydroseeding ($540 to $700 per acre depending on which seed is applied). An additional cost would be incurred to sample site soils to customize the dosage and delivery mechanisms for individual sites. In addition, re-application of PAM in granular or liquid form to areas with rill development (poor vegetation cover) may be necessary. Where re-application of granular PAM is used, R. S. Means (2000) suggests a cost of approximately $5 per 1,000 square feet for spreading soil admixtures by hand.

Extent of Erosion Control Effort for Different Development Types

The EPA estimated a reference or standard application effort for each erosion and sediment control that could be applied for different types of land development (e.g., 621 feet of silt fence for a 3-acre single-family residential construction site) to meet their proposed option 2 discharge limits. Reference quantities of various erosion and construction controls are listed in Tables 2.9 through 2.17, along with unit costs and the assumptions used in EPA's compliance cost assessment. Note that for some controls, reference quantities are given in terms of the number of units that will be constructed (i.e., the number of construction entrances anticipated for a certain size site). In addition, where unit costs are nonlinear (i.e., the unit cost varies with the size of the unit), both a design quantity and a number of units per site size class are required to estimate costs. An example of this is for sediment basins, where the total volume (the site size in acres times 3,600 cubic feet per acre) is apportioned into a number of installations (i.e., a 70-acre site is estimated to have 2 installations). This process helps ensure that any economies of scale in the calculation of compliance costs are reasonable.

TABLE 2.9. Quantities of Sit Fencing and Diversion Dikes for Different Land Development Scenarios (USEPA, 2002).

Site Size (acres)	Feet of Silt Fence: Single-family	Multi-family	Commercial	Industrial	Feet of Diversion Dike: Single-family	Multi-family	Commercial	Industrial
1	–	–	–	–	–	–	–	–
3	621	722	361	361	621	722	361	361
7.5	1,553	1,143	600	600	1,553	1,143	600	600
25	5,175	3,129	2,087	2,087	5,175	3,129	2,087	2,087
50	14,490	5,238	3,492	3,492	14,490	5,238	3,492	3,492
200	41,400	8,853	5,902	5,902	41,400	8,853	5,902	5,902

Both silt fencing and diversion dike lengths were based on 207 feet per acre on the site.
Costs for new installation of silt fence are based on \$0.92/ft length, excluding profit and overhead (R.S. Means, 2000).
Costs for new installation of diversion ditch are based on \$0.55/ft length installation, excluding profit and overhead (R.S. Means, 2000).

TABLE 2.10. Quantities of Mulched Area for Different Land Development Scenarios (USEPA, 2002).

Site Size (acres)	Mulched Acreage to Control: Single-family	Multi-family	Commercial	Industrial
1	0.0	0.0	0.0	0.0
3	0.8	0.8	0.8	0.8
7.5	1.9	1.9	1.9	1.9
25	6.3	6.3	6.3	6.3
50	17.5	17.5	17.5	17.5
200	50.0	50.0	50.0	50.0

For sites larger than 1 acre, mulching is limited to the site acreage times half the percentage of ultimate impervious area as a temporary means to stabilize denuded surfaces. The maximum coverage is set to 25% of the total site acreage. Cost to mulch is set to \$0.20 per square yard for materials/installation without overhead and profit (R.S. Means 2000).

TABLE 2.11. Amount of Land Treated with PAM for Different Land Development Scenarios (USEPA, 2002).

Site Size (acres)	Acres Treated with PAM: Single-family	Multi-family	Commercial	Industrial
1	0.00	0.00	0.00	0.00
3	0.84	1.32	1.50	1.50
7.5	2.10	3.29	3.75	3.75
25	7.00	10.96	12.50	12.50
50	19.60	30.70	35.00	35.00
200	56.00	87.72	100.0	100.0

PAM is costed at \$200 per acre per treatment based on a survey of commercial vendors and assuming costs are similar to herbicide for soil treatment (\$0.04 per square yard without profit and overhead based on spraying from truck). The acreage treated is equal to the site size times the ultimate impervious percentage, to a maximum of 50% of the site size.

TABLE 2.12. Numbers of Stone Check Dams and Sediment Traps for Different Land Development Scenarios (USEPA, 2002).

Site Size (acres)	The Number of Equal Size Units Installed to Provide Required Protection — Number of Stone Check Dams: Single-family	Multi-family	Commercial	Industrial	Number of Sediment Trap: Single-family	Multi-family	Commercial	Industrial
1	0	0	0	0	0	0	0	0
3	0	0	0	0	1	1	1	1
7.5	10	10	10	10	1	1	1	1
25	35	35	35	35	0	0	0	0
50	50	50	50	50	0	0	0	0
200	100	100	100	100	0	0	0	0

TABLE 2.13. Numbers of Sediment Basins for Different Land Development Scenarios (USEPA, 2002).

	Number of Sediment Basins			
Site Size (acres)	Single-family	Multi-family	Commercial	Industrial
1	0	0	0	0
3	0	0	0	0
7.5	1	1	1	1
25	2	2	2	2
50	2	2	2	2
200	4	4	4	4

Sediment pond of 3,600 cubic feet per acre served. Cost in dollars is computed from the equation: [0.76 × 7.47 × (volume required, cubic feet/number of ponds per site size)0.78]. The value of 0.76 removes overhead and profit from cost estimate.

TABLE 2.14. Numbers of Construction Entrances for Different Land Development Scenarios (USEPA, 2002).

	Number of Construction Entrances			
Site Size (acres)	Single-family	Multi-family	Commercial	Industrial
1	0	0	0	0
3	1	1	1	1
7.5	1	1	1	1
25	1	1	1	1
50	2	2	2	2
200	4	4	4	4

Costs for construction entrance based on $6.92 per square yard (gravel installed) for a footprint covering 100 square yards, excluding profit and overhead (R.S. Means, 2000).

TABLE 2.15. Numbers of Site Inspections for Different Land Development Scenarios (USEPA, 2002).

Administrative BMPs for Erosion and Sediment Control Management				
	E&S Site Inspection			
Site Size (acres)	Single-family	Multi-family	Commercial	Industrial
1	0	0	0	0
3	1	1	1	1
7.5	1	1	1	1
25	2	2	2	2
50	7	7	7	7
200	20	20	20	20

E&S Inspection includes multiple site visits by a certified inspector to verify the proper installation and operation of ESC BMPs. Values above are the number of half-day site inspections. Costs are based on 16 hours of inspection/documentation time per 10-acre-unit of a site, at a rate of $28.44 per hour.

TABLE 2.16. Numbers of Site Certifications for Sediment Basins for Different Land Development Scenarios (USEPA, 2002).

Administrative BMPs for Erosion and Sediment Control Management				
	E&S Site Certification of Sedimentation Basins			
Site Size (acres)	Single-family	Multi-family	Commercial	Industrial
1	0	0	0	
3	1	1	1	1
7.5	1	1	1	1
25	1	1	1	1
50	2	2	2	2
200	4	4	4	4

E&S Site Certification includes multiple site visits by a certified inspector to verify the proper installation of sedimentation basins. Costs based on 2 hours of inspection/documentation by a licensed engineer per 10-acre-unit of a site, at a rate of $56.74 per hour.

TABLE 2.17. Phasing Activities for Different Land Development Scenarios (USEPA, 2002).

	Phasing of Construction			
Site Size (acres)	Single-family	Multi-family	Commercial	Industrial
1	0	0	0	0
3	0	0	0	0
7.5	0	0	0	0
25	2	2	2	2
50	6	6	6	6
200	19	19	19	19

For sites larger than 10 acres, the number of remobilizations required is based on a maximum of 10 acres denuded at any single time to prevent large unstabilized construction sites. Costs are based on $1,000 per remobilization.

REFERENCES

Alabama. *Alabama Handbook For Erosion Control, Sediment Control, And Stormwater Management On Construction Sites And Urban Areas*, produced by the Alabama Soil and Water Conservation Committee Montgomery, AL, July 1993.

BOC. *New Privately Owned Housing Units Started: Annual Data, 2001.* U.S. Bureau of the Census, Washington, DC. http://www.census.gov/const/startsan.pdf. Accessed May 23, 2002. 2001.

Brown, W. and T. Schueler. *The Economics of Stormwater BMPs in the Mid-Atlantic Region.* Prepared for: Chesapeake Research Consortium. Edgewater, MD. Center for Watershed Protection, Ellicott City, MD. 1997.

BRPC (Birmingham Regional Planning Commission). *Best Management Practices for Controlling Sediment and Erosion from Construction Activities.* Birmingham, Alabama. 1980.

CWP (Center for Watershed Protection). *Design of Stormwater Filtering Systems.* Prepared for Chesapeake Research Consortium, Solomons, MD, and U.S. EPA Region 5, Chicago, IL, by Center for Watershed Protection, Ellicott City, MD. 1996.

Entry, J.A., and R.E. Sojka. Polyacrylamide Application to Soil Reduces the Movement of Microorganisms in Water. In *1999 Proceedings of the International Irrigation Show,* Irrigation Association, Orlando, FL, November 9, 1999, pp. 93–99. 1999.

Kay-Shoemake, J, M. Watwood, R. Sojka, and R. Lentz. "Soil Amidase Activity in Polyacrylamide (PAM) Treated Soils and Potential Activity toward Common Amide Containing Agrochemicals". *Biology and Fertility of Soils* 31(2): 183–186. 2000.

R.S. Means. *Site Work & Landscape Cost Data, 19th Edition.* R.S. Means Co., Kingston, MA, 2000.

SEWRPC (Southeast Wisconsin Regional Planning Commission). *Costs of Urban Nonpoint Source Water Pollution Control Measures,* Technical Report #31. Waukesha, WI. June 1991.

Tobiason, S., D. Jenkins, E. Molash, and S. Rush. "Polymer Use and Testing for Erosion and Sediment Control on Construction Sites: Recent experience in the Pacific Northwest". In: *Proceedings of Conference 31.* International Erosion Control Association. Palm Spring, CA, February 21–25, 2000, pp. 41–52. 2000.

U.S. Housing Markets. *Busiest Markets, Single-family Building.* Meyers Real Estate Information, Inc. (http://www.housingusa.com/ushm/pmts/mulbuzz.html). Accessed October 25, 2000. 1999a.

U.S. Housing Markets. *Busiest Markets, Multifamily Construction.* Meyers Real Estate Information, Inc. (http://www.housingusa.com/ushm/pmts/mulbuzz.html). Accessed October 25, 2000. 1999b.

USDA. *1997 National Resources Inventory.* U.S. Department of Agriculture, National Resources Conservation Service, Washington, DC. (www.nrcs.usda.gov/technical/NRI/). 2000.

USEPA. *Stormwater Management for Construction Activities: Developing Pollution Prevention Plans and Best Management Practices.* EPA 832-R-92-005. U.S. Environmental Protection Agency, Office of Water, Washington, DC. 1992.

USEPA. *Stormwater Management for Industrial Activities: Developing Pollution Prevention Plans and Best Management Practices.* EPA 832-R-92-006. U.S. Environmental Protection Agency, Office of Water, Washington, DC. 1992.

USEPA. *Guidance Specifying Management Measures for Sources of Nonpoint Pollution in Coastal Waters.* EPA 840-B-92-002. U.S. Environmental Protection Agency, Office of Water, Washington, DC. 1993.

USEPA. *Economic Analysis of the Final Phase II Stormwater Rule.* U.S. Environmental Protection Agency, Office of Wastewater Management. Washington, DC. 1999.

USEPA. *Development Document for Proposed Effluent Guidelines and Standards for the Construction and Development Category.* United States Environmental Protection Agency, Office of Water (4303T). EPA-821-R-02-007. Washington, DC 20460 (www.epa.gov/waterscience/guide/). June 2002.

VDCHR (Virginia Dept. of Conservation and Historic Resources). *Erosion and Sediment Control Handbook. Third Edition. Division of Soil and Water Conservation.* Richmond, Virginia. 1992.

VDCR (Virginia Department of Conservation and Recreation). *Virginia Erosion & Sediment Control Field Manual.* Second Edition, Division of Soil and Water Conservation, Richmond, VA. 1995.

VDCR. *Erosion and Sediment Control Handbook.* Second Edition. Division of Soil and Water Conservation. Richmond, Virginia. 1980.

WI DNR (Wisconsin Department of Natural Resources). *Wisconsin Construction Site Best Management Practice Handbook, revised.* Madison, WI. 1994.

PROBLEMS

1. After locating a site for project study, obtain a copy of the erosion and sediment control regulations that govern the on-site practices. Who has the authority to approve the plans? Who has the authority to enforce the plans? Are they the same entity? What is the scheduled inspection frequency for the site?

2. Find out from the regulatory agency that approved the plan who is the responsible party. Contact the responsible party and obtain agreement to use the site for a class project. If agreement cannot be obtained, repeat problems 1 and 2, or obtain permission from instructor to still use the site, but without the specific information. The site will need to be highly visible from public access areas. It may be possible to obtain copies of the erosion control plan and site maps from the regulatory agency.

3. On your site, perform a preliminary inventory of the erosion-control measures that have been installed. Are the erosion and sediment control categories discussed in this chapter considered?

4. Compare the approved erosion and sediment control plan with actual site conditions (try to find a site that will release a copy of the erosion-control measures map/plan, or where a plan is available from the review agency). Are the measures located where the plan writer described for each measure? If not, speculate why not. For example, did site conditions require revision of the plan and the "final" location of these structures/measures?

5. Given your site plan, estimate the cost of the erosion control measures listed/described on the plan. If the data is available, compare the cost of erosion and sediment control to the overall cost of the project. What percentage of the project is represented by the erosion control costs?

IMPORTANT INTERNET LINKS

The following Internet Links are referenced in Chapter 2. These sites should be visited to obtain additional information. Some of the locator addresses will likely change, but the material can still likely be located using a search tool.

Alabama Department of Environmental Management (ADEM), Nonpoint Program
http://www.adem.state.al.us/Education%20Div/Nonpoint%20Program/WSNPSProgram.htm

Jefferson County Stormwater Management, Inc.
http://www.swma.com/

Alabama Soil and Water Conservation Committee
http://www.swcc.state.al.us/

Alabama Soil and Water Conservation Committee (Sediment and Erosion Control Handbook):
http://swcc.state.al.us/erosion_handbook.htm

Geological Survey of Alabama
http://www.gsa.state.al.us/

Natural Resources Conservation Service, USDA
http://www.nrcs.usda.gov/

Alabama on-line soil surveys available to download (only a few counties):
http://soils.usda.gov/survey/online_surveys/alabama/

EPA Region 4 Nonpoint Source Information
http://www.epa.gov/region4/water/nps/

EPA "Surf you Watershed" (compiled water and watershed information for your watershed)
http://www.epa.gov/surf/

USGS "Science in your Watershed" (additional water and watershed information)
http://water.usgs.gov/wsc/index.html

Microsoft TerraServer maps (maps and aerial photographs for most of US)
http://terraserver.microsoft.com/

NOAA Data and Information Server (many linked environmental databases)
http://www.esdim.noaa.gov/NOAAServer/

Code of Federal Regulations, Title 40 (environmental regulations)
http://www.access.gpo.gov/cgi-bin/cfrassemble.cgi?title=199840

Example Site Description for Final Plan

The following is excerpted from a homework assignment prepared by Heather Hill, a student at the University of Alabama at Birmingham, as part of the Construction Site Erosion and Sediment Control Class taken during the summer of 2005. This assignment was to prepare a site description for a construction site erosion control plan for a construction site that has been studied during the class term.

Site Description

The Cahaba Village development is located on the north side of Highway 280 in the city of Mountain Brook in Jefferson County, Alabama. It is bordered to the south by Highway 280, the west and north by Green Valley Road and to the east by woods and housing. The site has been a dumping ground for numerous soil and rubble materials. There is a mound of dirt and debris on the east side of the site. A dirt/gravel road runs through the middle of the site. Two-thirds of the way into the site going north is a tributary to the Little Shades Creek, which flows west to east.

The Cahaba Village is a 16-acre development and will consist of a large grocery store with parking on the east side and a strip mall and high-rise condominiums on the west side. The site will be leveled to approximately Elevation 740′. The creek will be rerouted through a 13 foot culvert and covered with approximately 20 feet of fill in order to move Green Valley Road. Green Valley Road will be moved to the center of the site basically where the existing dirt/gravel road is and extend north to join the existing Green Valley Road. The west end of the site where Green Valley Road is going to be removed will be part of the final development.

Topography and Soils

The current topography is not the original topography of the site. For years, material ranging from soil to trees to concrete rubble has been disposed of on this site. The Jefferson County Soil Survey describes the soils in this area as the Nauvoo-Townley-Montevallo Association. The Leesburg Series is cobbly fine sandy loam on the surface, and the subsoil is clay loam. Rock outcrops consist of Sandstone. Nauvoo-Montevallo Series consists of numerous linear, roughly parallel, low mountains and ridges that extend from the southwest to the northeast across the county. The Nauvoo-Montevallo is underlain by sandstone and shale tilted and dipping to the southeast with 10% to 40% slopes. The north side of the site is made of a rock ridge of shale and sandstone. The natural soils at the site are silt and clay and residuum from the shale and sandstone. Water drains from the rock ridge down to the creek and off of the mound towards the creek. The creek enters the site on the west end and flows to the east and exits the site north of the pile.

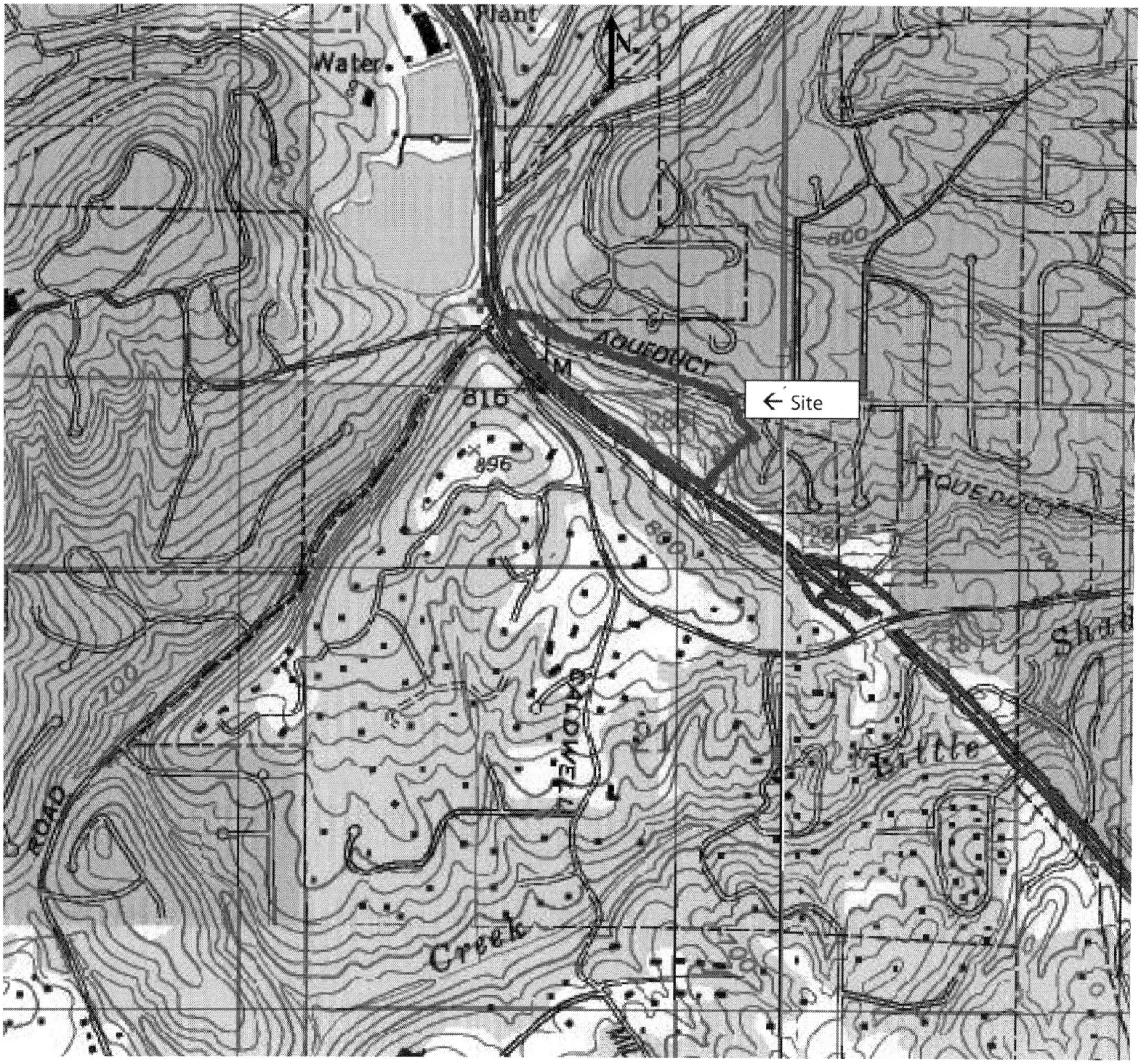

USGS topographic map.

Drainage Patterns

The drainage basin for the site is approximately 675 acres. The site is at the bottom of the drainage basin. Water flows off of the large ridge to the north about a half mile away and into the Little River Creek. Drainage on the site flows from the north into the creek and from the south into the creek. None of the runoff flows onto Highway 280.

Ground Cover

The ground cover over the site ranges from bare soil and rock to densely vegetated slopes. The mound of dirt is sparsely vegetated, as well as the entrance road. The creek area has dense vegetation growing on the slopes. The western side has been undisturbed and has mature trees and shrubs over it. The rock ridge on the north side is sporadically vegetated with kudzu and small trees able to grow on the steep slope.

Adjacent Property

To the north of the site, on the north side of Green Valley Road is a residential area, and also on the east side of the site. The south side of the site is adjacent to Highway 280, and the west side is adjacent to Green Valley Road. An old (100 years +) water supply line runs east-west across the site and almost parallel to the creek.

Construction Phases

Phase I

The first phase of construction consists of constructing a 13′ diameter corrugated metal pipe to channel the flow of the tributary of Little Shades Creek that runs through the site. In order to do this, a culvert pipe will be placed in the stream to allow access to the other side of the creek, and a series of pump systems have to be installed for water quality. A road will then be cut on the north side to access the creek. Another road will be cut from Green Valley Road to access the stream and install the pump.

The appropriate erosion control measures will be installed at each of the areas to maintain sediment runoff. Silt fences will be installed on each side of the new cut roads. A holding pond will receive all of the runoff from the active construction areas. Disturbed areas not under active construction will be seeded and temporarily mulched.

A culvert will be placed in the creek to allow traffic to reach the north side of the creek. A road will be cut parallel to the creek to access the creek and install a pump and a 12 inch pipe to divert the water from where it enters the site back to where it exits. A road will be cut from Green Valley Road to the south along the ridge to install the pump to divert the water. An impermeable diversion dam was installed where the creek enters the site in order to prevent the flow of water onto the site. A holding pond will be installed to the north of the mound and to the south of the creek to collect site runoff water. The water in the holding pond will then be pumped to the detention pond to settle the sediments out and then released into the natural drainage area.

A turning lane and curbs will be installed on Highway 280. The areas that are disrupted will be seeded and mulched temporarily. The area will later be landscaped with sod, shrubs, and trees.

Silt fences will be installed around the outside of the mound to prevent sediment runoff onto Highway 280. Silt Fences will also be installed on the north side where the road will be cut to prevent runoff into the creek bed.

Phase II

Phase II will be site grading. While the 13′ diameter culvert is being installed, the mound of dirt is being excavated and sieved to acquire suitable backfill for the site. The final site grading will consist of covering the 13′ diameter culvert with approximately 25′ of backfill and taking the mound down to original or near original grade at approximately Elevation 740′. The entire site will be fairly flat with mainly parking lots, roads and buildings. Green Valley Road will be rerouted to come down the center of the site and the current Green Valley Road will become part of the commercial development.

The site will be graded to drain to the southeast to the detention pond. The site will be almost level, with an approximately 1% grade. The material in the mound will be sieved to obtain the appropriate structural fill. An area of approximately one acre will be undisturbed on the west side

near the existing Green Valley Road-Highway 280 intersection. Also, the majority of the area to the east of the detention pond and continuing to the north will be left undisturbed and the site contours will be graded to them.

Phase III

Phase III of construction will be fine-grading the site to prepare for the foundations of the building, road and parking lot base, and areas to be landscaped.

The final stage of development will leave very little area to be vegetated. The majority of the site will be buildings and asphalt roads and parking lots. The front entrance will be landscaped and the northern extents will be graded and sodded.

Data Analysis

The site is sparsely vegetated near the road and mound of dirt. Outside of that area, the site is moderately to densely vegetated. The road and mound are actually fairly stable with a small amount of erosion features. However, the steepness of the mound and the northern ridge line, as well as the channel walls of the creek makes these areas a high erosion hazard potential. The rest of the area is fairly flat, thus having a low erosion hazard potential.

However, once the site construction for Phase I starts, the area beneath the ridge will be cleared to create a road and a road was cut to access the top of the mound. Silt fence will be installed along the south edge of the road on the north side of the creek to divert flow toward the collection pond. Silt fence will also be installed around the mound to divert flow from going to Highway 280. All runoff will be diverted to the detention pond on the east side of the site.

The final grading contours are relatively flat for the buildings, parking lots, and roads and therefore have low erosion hazard potential. The slopes on the north side of the site will be steep at a 2:1 slope and therefore have high erosion hazard potential.

The majority of the soils at the site are not the native soils. The soil in the mound will be sieved and used as backfill for the site. The backfill soil is mainly silty clay. All debris will be removed from the soil and disposed of off site.

Approximately 10% of the site will be undisturbed. Approximately 70% of the site will be graded and paved or have a building on it. The remaining 20% will consist of landscaping areas, vegetated/protected slopes, and the detention pond. Temporary mulching and seeding will be used during the interim to reduce erosion potential.

Adjacent areas that may be affected are Highway 280 and the wooded area to the east. Highway 280 should not be affected due to the grading of the area adjacent to 280, but during construction, heavy rain may cause runoff onto Highway 280. A drainage system will be installed along Highway 280 to collect runoff from the road and thus any potential runoff from the site onto the road. The wooded area to the east should not be impacted. The water from the detention basin will be released toward this area, but will travel over a jute mesh and then into the creek bed. The water will be released slowly into the creek.

Facility Plan Development

The creek on the site will flow through a 13-foot culvert pipe generally along the original creek flow path. This area will then be covered to allow for the site construction grading. The site will be flat with approximately a 1% slope to allow drainage to the detention pond. The creek will release at the same elevation it previously did and the water from the detention pond will drain down the original slope back into the creek. The water from the detention pond will have reduced sediment.

Erosion and Sediment Control Plan

Phase I—Preparation for Culvert Installation

A small culvert will be installed in the creek to access the north side of the property and silt fences will be installed on each side to prevent sediment runoff into the creek. Once the road is cut on the north side of the stream, silt fences will be installed along the south side of the road/the north side of the

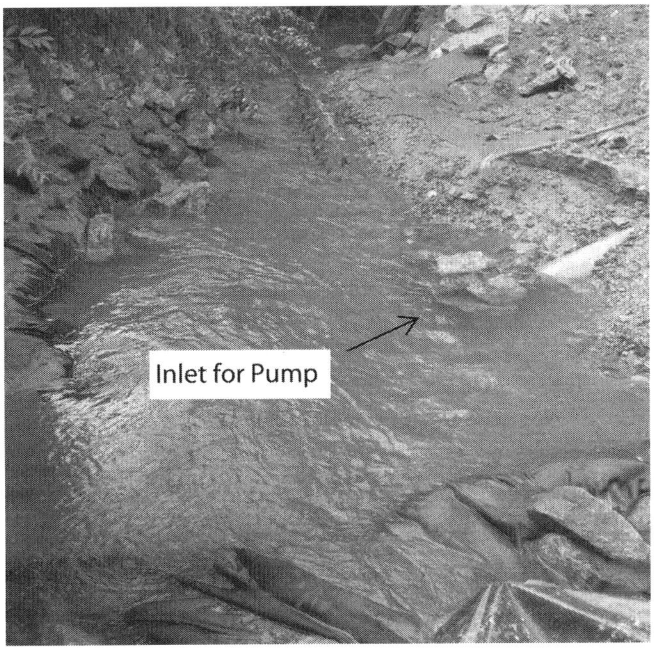

creek. An impermeable diversion dam, 15′ wide, 29′ long and 5′ high, will be installed in the creek to back up flow for the pump to divert water to the other end. The diverted water will then be released into a riprap area, flow into a small settling basin, and then released to the existing creek bed.

Once the turn lanes are installed on Highway 280, the disrupted area will be graded, seeded, and mulched temporarily. The grassed area in the median of Highway 280 will have an excelsior blanket placed on it to assist in permanent vegetation. This area on the site will be regraded again at the end of the project.

The detention pond will be constructed on the east side of the site and seeded to reduce erosion. A split pipe flocculation system will be installed at the head of the pond to reduce the sediment flow into the pond and protect the slope from scour.

Phase II—Site Grading

Phase II will coincide with Phase I, and the same erosion control measures will be in place. The material in the mound will be sieved and the sieved material will be used for backfill. The sieved material will be placed on site and silt fences will be placed around the piles to control runoff. The site will be graded to almost level and runoff will be directed to the detention pond.

Phase III—Final Grading

The site will be graded to accommodate buildings, parking lots, and roads mainly. These areas will be graded to drain to sewer systems that dump into the detention pond. The landscaped area in the front, adjacent to Highway 280 will be landscaped with sod, shrubs, and trees. Silt fences will be installed at the eastern end where the slope splits and the water leaves the site instead of flowing into the detention pond. At the north side of the site, the slopes will be graded to 2:1 and armored with North American Green SC150B. The toe of the slope will be lined with silt fences. The surfaces will be graded to drain to a sewer inlet that dumps into the detention pond.

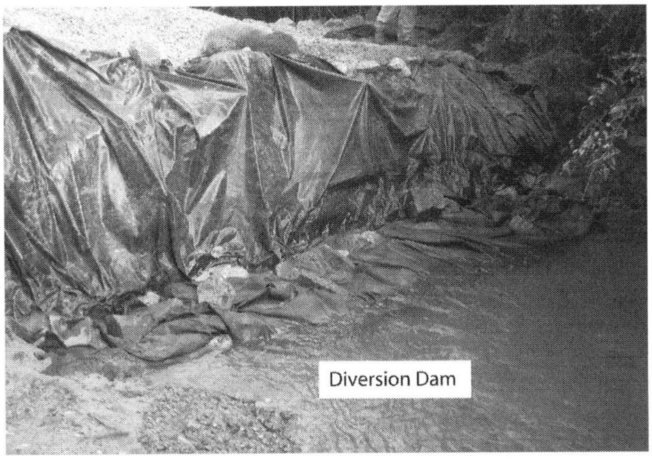

APPENDIX 2A. COSTS OF CONTROL OPTIONS AT CONSTRUCTION SITES

The following tables are from Technical Report #31, *Costs of Urban Nonpoint Source Water Pollution Control Measures.* Southeast Wisconsin Regional Planning Commission, Waukesha, WI. June 1991. The costs reported in the tables in this appendix are mostly from the 1988 and 1989 construction period. To adjust for 2005 costs, multiply by approximately 1.6 (based on long-term inflation factors, but not different locations).

TABLE 2A.1. Reported Costs of Selected Construction Erosion Control Measures (SEWRPC, 1991; multiply by 1.6 for 2005 costs).

Measure	Unit	Number of Reported Costs	Minimum	Maximum	Mean
Temporary Seeding	Square yard	8	$0.08	$0.12	$0.10
Mulching	Square yard	91	0.10	1.00	0.30
Sodding	Square yard	117	1.40	10.10	2.40
Filter Fabric Fence	Lineal foot	290	0.60	8.00	3.40
Straw Bale Barrier	Bale	136	5.00	12.00	9.20

Source: City of Madison, Wisconsin; City of West Bend, Wisconsin, Crispell-Snyder Consulting Engineers; GeoSynthetics, Inc.; Hornburg Contractors; Ruekert & Mielke, Inc; Terra Engineering; Wisconsin Department of Transportation; and SEWRPC.

TABLE 2A.2. Unit Capital Costs for Selected Construction Erosion Control Measures (SEWRPC, 1991; multiply by 1.6 for 2005 costs).

		Unit Cost		
Component	Unit	Low	Moderate	High
Temporary Seeding	Square yard	$0.05	$0.10	$0.20
Temporary Seeding	Pound	1.80	4.60	7.40
Mulching	Square yard	0.10	0.30	0.50
Sodding	Square yard	1.20	2.40	3.60
Filter Fabric Fence	Lineal foot	2.30	3.40	4.50
Straw Bale Barrier	Bale	7.80	9.20	10.60
Inlet Protection Device	Inlet	106.00	130.00	154.00

TABLE 2A.3. Estimated Capital Cost of a 1.5-Foot-Deep Diversion Swale (SEWRPC, 1991; multiply by 1.6 for 2005 costs).

			Unit Cost			Total Cost		
Component	Unit	Extent	Low	Moderate	High	Low	Moderate	High
Site Preparation								
Excavation	Cubic yard	11.81	$2.10	$3.70	$5.30	$24.80	$43.70	$62.59
Place and Compact Fill	Cubic yard	11.81	0.60	1.10	1.60	709	12.99	18.90
Grading	Square yard	144.4	0.10	0.20	0.30	14.44	28.88	43.32
Site Development								
Salvaged Topsoil Seed, and Mulch	Square yard	72.2	$0.40	$1.00	$1.60	$28.88	$72.20	$115.52
Sod	Square yard	72.2	1.20	2.40	3.60	86.64	173.28	259.92
Subtotal	—	—	—	—	—	$162.00	$331.00	$500.00
Contingencies	Swale	1	25 percent	25 percent	25 percent	$41.00	$83.00	$125.00
Total	—	—	—	—	—	$202.00	$414.00	$625.00

Note: Swale height is from top of dike to bottom of channel. Dike top width equals channel bottom width of two feet. Swale has an assumed length of 100 feet with 3:1 side slopes.

TABLE 2A.4. Estimated Capital Cost of a 3.0-Foot-Deep Diversion Swale (SEWRPC, 1991; multiply by 1.6 for 2005 costs).

			Unit Cost			Total Cost		
Component	Unit	Extent	Low	Moderate	High	Low	Moderate	High
Site Preparation								
Excavation	Cubic yard	36.11	$2.10	$3.70	$5.30	$75.83	$133.61	$191.38
Place and Compact Fill	Cubic yard	36.11	0.60	1.10	1.60	21.67	39.72	57.78
Grading	Square yard	244.4	0.10	0.20	0.30	24.44	48.88	73.32
Site Development								
Salvaged Topsoil								
Seed, and Mulch	Square yard	122.2	$0.40	$1.00	$1.60	$48.88	$122.20	$195.52
Sod	Square yard	122.2	1.20	2.40	3.60	146.64	293.28	439.92
Subtotal	—	—	—	—	—	**$162.00**	**$331.00**	**$500.00**
Contingencies	Swale	1	25 percent	25 percent	25 percent	**$79.00**	**$159.00**	**$240.00**
Total	—	—	—	—	—	**$397.00**	**$79700**	**$1,198.00**

Note: Swale height is from top of dike to bottom of channel. Dike top width equals channel bottom width of two feet. Swale has an assumed length of 100 feet with 3:1 side slopes.

TABLE 2A.5. Estimated Capital Cost of a 3.0-Foot-Deep Sediment Trap (SEWRPC, 1991; multiply by 1.6 for 2005 costs).

			Unit Cost			Total Cost		
Component	Unit	Extent	Low	Moderate	High	Low	Moderate	High
Site Preparation								
Excavation	Cubic yard	117	$2.10	$3.70	$5.30	$246.00	$433.00	$620.00
Site Development								
Outlet								
Crushed Stone Fill	Cubic yard	1.8	$14.80	$19.40	$24.00	$26.60	$34.90	$43.20
Filter Fabric	Square yard	6.7	1.00	2.00	3.00	6.70	13.40	20.10
Subtotal	—	—	—	—	—	**$279.00**	**$481.00**	**$683.00**
Contingencies	Sediment trap	1	25 percent	25 percent	25 percent	**$70.00**	**$121.00**	**$171.00**
Total	—	—	—	—	—	**$349.00**	**$602.00**	**$854.00**

Note: Trap has an assumed surface area of 1,000 square feet.

TABLE 2A.6. Estimated Capital Cost of a 5.0-Foot-Deep Sediment Trap (SEWRPC, 1991; multiply by 1.6 for 2005 costs).

			Unit Cost			Total Cost		
Component	Unit	Extent	Low	Moderate	High	Low	Moderate	High
Site Preparation								
Excavation	Cubic yard	926	$2.10	$3.70	$5.30	$1,954.00	$3,426.00	$4,908.00
Site Development								
Outlet								
Crushed Stone Fill	Cubic yard	3	$14.80	$19.40	$24.00	$44.40	$58.20	$7200
Filter Fabric	Square yard	11	1.00	2.00	3.00	11.00	22.00	33.00
Subtotal	—	—	—	—	—	**$2,400.00**	**$4,383.00**	**$5,013.00**
Contingencies	Sediment trap	1	25 percent	25 percent	25 percent	**$600.00**	**$877.00**	**$1,253.00**
Total	—	—	—	—	—	**$3,000.00**	**$4,383.00**	**$6,266.00**

Note: Trap has an assumed surface area of 5,000 square feet.

TABLE 2A.7. Estimated Capital Cost of a 0.1-Acre Sedimentation Basin (SEWRPC, 1991; multiply by 1.6 for 2005 costs).

			Unit Cost			Total Cost		
Component	Unit	Extent	Low	Moderate	High	Low	Moderate	High
Site Preparation								
Excavation	Cubic yard	462	$2.10	$3.70	$5.30	$970.00	$1,709.00	$2,449.00
Place and Compact Fill	Cubic yard	310	0.60	1.10	1.60	186.00	341.00	496.00
Site Development								
Basin Inlet	Basin	1	$1,310.00	$2,870.00	$4,430.00	$1,310.00	$2,870.00	$4,430.00
Basin Outlet	Basin	1	1,320.00	3,380.00	5,440.00	1,320.00	3,380.00	5,440.00
Riprap	Cubic yard	2.42	16.40	29.60	42.80	39.70	71.60	104.00
Subtotal	—	—	—	—	—	**$3,826.00**	**$8,372.00**	**$12,919.00**
Contingencies	Basin	1	25 percent	25 percent	25 percent	**$956.00**	**$2,093.00**	**$3,230.00**
Total	—	—	—	—	—	**$4,782.00**	**$10,465.00**	**$16,149.00**

Note: Basin has side slopes of 3:1 and a depth of five feet.

TABLE 2A.8. Estimated Capital Cost of a 0.25-Acre Sedimentation Basin (SEWRPC, 1991; Multiply By 1.6 For 2005 Costs).

			Unit Cost			Total Cost		
Component	Unit	Extent	Low	Moderate	High	Low	Moderate	High
Site Preparation								
Excavation	Cubic yard	1,509	$2.10	$3.70	$5.30	$3,169	$5,583	$7,998
Place and Compact Fill	Cubic yard	1,011	0.60	1.10	1.60	607	1,112	1,618
Site Development								
Basin Inlet	Basin	1	$1,310.00	$2,870.00	$4,430.00	$1,310	$2,870	$4,430
Basin Outlet	Basin	1	1,320.00	3,380.00	5,440.00	1,320	3,380	5,440
Riprap	Cubic yard	6.1	16.40	29.60	42.80	100	181	261
Subtotal	—	—	—	—	—	**$6,506**	**$13,126**	**$19,747**
Contingencies	**Basin**	**1**	**25 percent**	**25 percent**	**25 percent**	**$1,627**	**$3,282**	**$4,937**
Total	—	—	—	—	—	**$8,133**	**$16,408**	**$24,684**

Note: Basin has side slopes of 3:1 and a depth of five feet.

TABLE 2A.9. Estimated Capital Cost of a 1.0-Acre Sedimentation Basin (SEWRPC, 1991; multiply by 1.6 for 2005 costs).

			Unit Cost			Total Cost		
Component	Unit	Extent	Low	Moderate	High	Low	Moderate	High
Site Preparation								
Excavation	Cubic yard	7,252	$2.10	$3.70	$5.30	$15,229	$26,832	$38,436
Place and Compact Fill	Cubic yard	4,859	0.60	1.10	1.60	2,915	5,345	7,774
Site Development								
Basin Inlet	Basin	1	$1,310.00	$2,870.00	$4,430.00	$1,310	$2,870	$4,430
Basin Outlet	Basin	1	1,320.00	3,380.00	5,440.00	1,320	3,380	5,440
Riprap	Cubic yard	24.2	16.40	29.60	42.80	397	716	1,036
Subtotal	—	—	—	—	—	**$26,464**	**$48,929**	**$71,395**
Contingencies	**Basin**	**1**	**25 percent**	**25 percent**	**25 percent**	**$5,293**	**$9,786**	**$14,279**
Total	—	—	—	—	—	**$26,464**	**$48,929**	**$71,395**

Note: Basin has side slopes of 3:1 and a depth of five feet.

TABLE 2A.10. Annual Maintenance Unit Costs for Construction Erosion Control Measures (SEWRPC, 1991; multiply by 1.6 for 2005 costs).

Measure	Annual Maintenance Cost Expressed As Percentage of Capital Cost	Unit Annual Maintenance Cost
Temporary Seeding and Mulching	25	$0.12/square yard
Sodding	5	$0.12/square yard
Filter Fabric Fence	100	$3.40/lineal foot
Straw Bale Barrier	100	$9.20/bale
Inlet Protection Device	100	$123/inlet
Diversion Swale	20	$1.50–5.20/lineal foot
Sediment Trap	20	$1.00–1.80/lineal foot
Sedimentation Basin	25	$1.50–3.25/cubic yard

TABLE 2A.11. Construction Component Unit Costs for Urban Nonpoint Pollution Control Measures (SEWRPC, 1991; multiply by 1.6 for 2005 costs).

Description	Unit	Installation Costs: Material	Labor	Equipment	Indirect Cost	Total Cost	Year of Cost	Comments
Site Clearing								
Clear and Grub								
Light	Acre	—	$ 865.00	$ 850.00	$ 510.00	$2,225.00	January 1989	All trees are cut and chipped
Medium	Acre	—	1,225.00	1,225.00	750.00	3,200.00		
Heavy	Acre	—	2,875.00	2,850.00	1,725.00	7,450.00		
Clear and Grub								
Light	Acre	—	$ 283.12	$ 365.00	$ 256.01	$904.13	Mid-1988	—
Medium	Acre	—	943.73	1,216.67	853.36	3,013.76		
Heavy	Acre	—	3,147.20	4,144.00	2,863.99	10,155.19		
Clear Brush								
By Hand	Acre	—	$ 1,125.00	$ 430.00	$ 620.00	$2,175.00	January 1989	—
With Brush Saw	Acre	—	540.00	205.00	305.00	1,050.00		
Clear Trees								
<24 Inches	Each	—	$ 118.87	$52.17	$86.25	$257.29	Mid-1988	—
>24 Inches	Each	—	178.30	78.25	129.39	385.94		
Earthwork								
Grading								
By Hand	Cubic yard	—	$ 40.64	—	$30.66	$71.30	Mid-1988	—
Dozer	Cubic yard	—	0.46	$0.91	0.53	1.90		
Grading								
≤1,000 Foot Haul	Cubic yard	—	0.30	0.82	0.41	1.53		
>1,000 Foot Haul	Cubic yard	—	0.30	0.88	0.41	1.59		
Excavating								
To Five-Foot Depth	Cubic yard	$0.50	$ 5.33	$ 2.77	$4.70	$13.30	Mid-1988	—
To 10-Foot Depth	Cubic yard	0.50	3.04	1.58	2.74	7.86		
By Hand	Cubic yard	—	—	40.64	30.66	71.30	Mid-1988	—
Common Excavation	Cubic yard	—	—	—	—	$2.82 2.00–5.00	1983	Average Typical range
Excavation								
Loam, Sand, and Gravel	Cubic yard	—	$ 0.24	$0.30	$0.24	$0.78	Mid-1988	Two-and-one-half cubic yard power shovel
Compacted Gravel and Till	Cubic yard	—	0.26	0.33	0.27	0.86		
Hard Clay and Shale	Cubic yard	—	0.32	0.40	0.33	1.05		
Excavation								
Loam, Sand, and Gravel	Cubic yard	—	$ 0.17	$0.07	$0.07	$0.38	Mid-1988	Two-cubic-yard front end loader
Compacted Gravel and Till	Cubic yard	—	0.18	0.07	0.16	0.41		
Hard Clay and Shale	Cubic yard	—	0.23	0.09	0.19	0.51		
Excavation: Structure								
Backhoe								
Common Earth	Cubic yard	—	$ 3.63	$4.78	$2.24	$10.65	January 1989	3/4 Cubic Yard Bucket
Common Earth	Cubic yard	—	3.03	4.97	1.95	9.95		One-Cubic-Yard Bucket
Common Earth	Cubic yard	—	2.27	4.44	1.54	8.25		1-1/2-Cubic-Yard Bucket
Common Earth	Cubic yard	—	1.63	4.46	1.26	7.35		Two-Cubic-Yard Bucket

(continued)

TABLE 2A.11 (continued). **Construction Component Unit Costs for Urban Nonpoint Pollution Control Measures (SEWRPC, 1991; multiply by 1.6 for 2005 costs).**

Description	Unit	Installation Costs			Indirect Cost	Total Cost	Year of Cost	Comments
		Material	Labor	Equipment				
Backfill By Hand								
Light Soil	Cubic yard	—	$ 9.65	—	$ 4.85	$14.50	January 1989	No compaction
Heavy	Cubic yard	—	12.25	—	6.20	18.45		
Backfill: Compaction								
Light Soil Six Inches Deep								
Hand Tamp	Cubic yard	—	$16.20	—	$8.15	$24.35	January 1989	—
Roller Compaction	Cubic yard	—	12.06	$0.67	6.10	18.83		
Light Soil 12 Inches Deep								
Hand Tamp	Cubic yard	—	13.61	—	6.84	20.45		
Roller Compaction	Cubic yard	—	11.25	0.45	5.68	17.38		
Heavy Soil Six Inches Deep								
Hand Tamp	Cubic yard	—	18.80	—	9.50	28.45		
Roller Compaction	Cubic yard	—	14.66	0.67	7.45	22.78		
Heavy Soil 12 Inches Deep								
Hand Tamp	Cubic yard	—	16.21	—	8.19	24.40		
Roller Compaction	Cubic yard	—	13.85	0.45	7.03	21.33		
Earth Fill								
Borrow Fill								
One Mile Haul	Cubic yard	$3.67	$0.44	$1.17	$0.72	$6.00	January 1989	Compact and Shape
Select Fill								
One Mile Haul	Cubic yard	5.75	0.44	1.17	$0.89	8.25		
>One Mile Haul	Cubic yard Mile	—	—	—	—	0.60		
Compacted Gravel Fill								
Four Inches Deep	Square feet	$0.11	$ 0.09	$0.01	$0.05	$0.26	January 1989	—
Six Inches Deep	Square feet	0.17	0.10	0.01	0.0.7	0.35		
Nine Inches Deep	Square feet	0.25	0.12	0.02	0.09	0.48		
12 Inches Deep	Square feet	0.33	0.14	0.02	0.10	0.59		
Crushed Stone Fill								
1-1/2 Inches	Cubic yard	$14.00	$ 0.87	$ 2.34	$2.04	$19.25	January 1989	—
3/4 Inches to 1-1/2 Inches	ton	9.10	—	—	0.90	10.00		
Stone Fill								
One to Two Inches Deep	Cubic yard	—	—	—	—	$22.50	1983	Average
						15.00–25.00		Typical range
Stone Tamping	Cubic yard	—	—	—	—	$2.00	1983	Average
Pea Gravel Fill	Cubic yard	$14.40	$16.00	—	$8.60	$38.00	January 1989	—
	Cubic yard	—	—	—	—	7.50	1983	Average
Clean Washed Sand Fill	Square feet	$12.95	$0.87	$2.34	$1.94	$18.10	January 1989	—
	Square feet	—	—	—	—	14.00	1983	Average
Hauling								
Off-Road								
1,000 Feet One Way	Cubic yard	—	$0.23	$0.52	$0.29	$1.04	Mid-1988	—
2,000 Feet One Way	Cubic yard	—	0.27	0.67	0.54	1.28		
Over-Road								
1,000 Feet One Way	Cubic yard	—	0.44	0.74	0.49	1.67		
2,000 Feet One Way	Cubic yard	—	0.47	0.82	0.52	1.81		
Six Cubic Yard Dump Truck								
1/4 Mile Round Trip	Cubic yard	—	0.59	1.12	0.04	2.11	January 1989	—
1/2 Mile Round Trip	Cubic yard	—	0.71	1.37	0.49	2.57		

(continued)

TABLE 2A.11 (continued). Construction Component Unit Costs for Urban Nonpoint Pollution Control Measures (SEWRPC, 1991; multiply by 1.6 for 2005 costs).

Description	Unit	Installation Costs: Material	Labor	Equipment	Indirect Cost	Total Cost	Year of Cost	Comments
Mobilization/Demobilization								
Shovel. Backhoe. or Dragline								
3/4 Cubic Yard	Each	—	$39.00	$175.00	$41.00	$255.00	January 1989	—
1-1/2 Cubic Yard	Each	—	47.00	210.00	48.00	305.00		
Construction								
Pond Linings								
Plain PVC Sheets								
10 mils Thick	Square feet	$0.10	$0.54	—	$0.35	$0.99	January 1989	—
20 mils Thick	Square feet	0.21	0.55	—	0.37	1.13		
30 mils Thick	Square feet	0.32	0.56	—	0.39	1.27		
PVC Mineral Fiberback								
45 mils Thick	Square feet	$0.70	$0.57	—	$0.43	$1.70	January 1989	—
Waterproof Membrane								
Two-Ply	Square yard	$5.41	$10.16	—	$8.80	24.37	Mid-1988	—
Three-Ply	Square yard	5.82	12.70	—	10.80	29.32		
Filter Fabric								
Minimum	Square feet	$0.26	—	—	$0.03	$0.29	January 1989	—
Maximum	Square feet	0.30	—	—	0.03	0.33		
Filter Cloth	Square feet	—	—	—	—	$2.71	1983	Average
						2.00–5.00		Typical range
PVC Pipe								
10-Foot Length								
Six-Inch Diameter	Lineal foot	$1.22	$1.32	—	$0.79	$0.33	January 1989	
Eight-inch Diameter	Lineal foot	1.75	1.38	—	0.87	4.00		
10-Inch Diameter	Lineal foot	2.80	1.67	$0.26	1.12	5.85		
Six-Inch Diameter	Lineal foot	—	—	—	—	$10.00	1983	Average
						8.00–12.00		Typical range
Eight-inch Diameter	Lineal foot	—	—	—	—	10.50		Average
10-Inch Diameter	Lineal foot	—	—	—	—	15.00		Average
Six-Inch Diameter	Lineal foot	$2.65	$0.79	—	$1.15	$4.59	Mid-1988	
Eight-inch Diameter	Lineal foot	4.48	0.83	—	1.57	6.88		
10-Inch Diameter	Lineal foot	7.19	0.89	—	2.17	10.25		
Perforated PVC Pipe								
10 Foot Length								
Four-Inch Diameter	Lineal foot	$0.57	$1.23	—	$0.68	$2.48	January 1989	—
Six-Inch Diameter	Lineal foot	1.22	1.32	—	0.79	3.33		
Eight-inch Diameter	Lineal foot	1.75	1.38	—	0.87	4.00		
10-Inch Diameter	Lineal foot	2.80	1.67	$0.26	1.12	5.85		
Six-Inch Diameter	Lineal foot	$2.65	$0.79	—	$1.15	$4.59	Mid-1988	
Eight-inch Diameter	Lineal foot	4.48	0.83	—	1.57	6.88		
10-Inch Diameter	Lineal foot	7.19	0.89	—	2.17	10.25		—
Six-Inch Diameter	Lineal foot	—	—	—	—	$10.00	1983	Average
						8.00–12.00		Typical range
Eight-inch Diameter	Lineal foot	—	—	—	—	10.50		Average
10-Inch Diameter	Lineal foot	—	—	—	—	15.00		Average

TABLE 2A.11 (continued). Construction Component Unit Costs for Urban Nonpoint Pollution Control Measures (SEWRPC, 1991; multiply by 1.6 for 2005 costs).

Description	Unit	Installation Costs			Indirect Cost	Total Cost	Year of Cost	Comments
		Material	Labor	Equipment				
Reinforced ConcretePipe (Class 111)								
15-Inch Diameter	Lineal foot	$7.31	$2.06	$1.17	$3.32	$13.86	Mid-1988	Gasket joints eight-foot lengths
18-Inch Diameter	Lineal foot	9.08	3.31	1.88	4.81	19.08		
21-Inch Diameter	Lineal foot	10.73	3.42	1.95	5.24	21.34		
24-Inch Diameter	Lineal foot	15.20	4.14	2.36	6.81	28.51		
Reinforced ConcretePipe (Class 111)								
15-Inch Diameter	Lineal foot	$7.50	$3.15	$0.48	$2.37	$13.50	January 1989	Gasket joints
18-Inch Diameter	Lineal foot	9.45	3.67	0.56	2.87	16.55		
24-Inch Diameter	Lineal foot	14.80	5.50	0.85	3.85	25.00		
Riprap Broken Stone								
Random Placement	Cubic yard	$9.20	$5.25	$6.35	$4.20	$25.00	January 1989	Machine placed for protection
3/8-1/4 Cubic Yard Pieces	Square yard	16.10	12.70	5.75	8.45	43.00		Grouted
18-Inch MinimumThickness	Square yard	11.50	19.20	8.70	11.60	51.00		Not grouted
Porous Pavement								
Two-Inch-Thick Surface	Square yard	—	—	—	—	$6.60	1976	12-inch sub-base
Two to Four Inches Thick	Square yard	$1.58	—	—	—	—	1983	
Grassed Driveways								
(porous surfaces)	Cubic yard	—	—	—	—	$70.00	1976	Brick lattices, gravel filled, covered with top soil
Landscaping								
Sodding								
Level								
>400 Square Yards	Square yard	$0.98	$0.85	$0.17	$0.56	$2.56	January 1989	—
100 Square Yards	Square yard	1.36	1.07	0.22	0.70	3.35		
50 Square Yards	Square yard	1.95	1.14	0.23	0.80	4.12		
Slopes								
400 Square Yards	Square yard	1.03	1.19	0.24	0.72	3.18		
Seeding								
Mechanical Seeding	Acre	$410.00	$435.00	$165.00	$290.00	$1,300.00	January 1989	—
	Square yard	0.08	0.09	0.03	0.06	0.26		Includes fertilizer and lime
Fine Grade/Seed	Square yard	0.15	0.85	0.17	0.48	1.65		
Push Spreader								
Grass Seed	1,000 square feet	$8.60	$0.67	$0.26	$1.22	$10.75	January 1989	—
Limestone	1,000 square feet	2.05	0.67	0.26	0.58	3.56		
Fertilizer	1,000 square feet	5.40	0.67	0.26	0.92	7.25		
Level Areas	Acre	578.21	149.30	80.63	251.00	1,059.14	Mid-1988	—
Sloped Areas	Acre	578.21	238.88	129.00	328.75	1,274.84		
Mulching								
Hay	Acre	$255.76	$74.65	$40.31	$118.50	$489.22	Mid-1988	—
	Square yard	—	—	—	—	0.58	1983	Average
						0.25–1.00		Typical range

Note: Total cost includes operation and maintenance, taxes, insurance, and other contingencies.

Regional Rainfall Conditions and Site Hydrology for Construction Site Erosion Evaluations

INTRODUCTION: HYDROLOGY FOR THE DESIGN OF CONSTRUCTION EROSION CONTROLS

THIS chapter provides an overview of hydrology analysis techniques appropriate for the design of construction site erosion controls. The NRCS's TR-55 procedure will be used in this chapter, as it provides most of the needed information and is generally applicable to conditions found on most construction sites.

The reference list contains the URL for an on-line copy of TR-55, *Urban Hydrology for Small Watersheds* by the U.S. Dept. of Agriculture/Soil Conservation Service (now NRCS) (1986). Recently, a Windows version of TR-55 (WinTR55) has become available (beta version) that can be used to greatly simplify these calculations, and that appropriate URL is also given. TR-55 provides a good set of tools to determine a number of hydrology parameters needed for effective design of construction site erosion controls. The following list shows typical controls and the types of hydrology information needed for complete evaluations and design (later chapters will review and present examples of how this information is used in these designs):

- Mulches—water velocities and water depth
- Ditch liners—water velocities and water depth
- Slope down shoots—peak flow rates
- Diversion dikes and swales—peak flow rates
- Filter fabric fences—water velocities and hydrographs
- Sediment ponds—water volume and hydrographs

Factors Affecting Runoff

Rainfall

The temporal extent of the storm and the distribution of rainfall during the storm are two major factors which affect the peak rate of runoff. The storm distribution can be thought of as a measure of how the rate of rainfall (intensity) varies within a given time interval. If a certain amount of precipitation was measured in a given 24-hour period, this precipitation may have occurred over the entire 24-hour period or in just one hour. The duration of the rain (and the peak intensity) directly affect the runoff rates.

The size of the storm is often described by the length of time over which precipitation occurs, the total amount of precipitation occurring and how often this same storm might be expected to occur or be exceeded (frequency). Thus, a 10-year, 24-hour storm can be thought of as a storm producing the amount of rain in 24 hours with a 10% chance of occurrence in any given year.

Antecedent Moisture Content

The runoff from a given storm is affected by the existing soil moisture content resulting from the precipitation preceding the event of interest (defined as a five day period by the NRCS). This has a much smaller effect in areas having mostly paved surfaces. On construction sites, this factor can be important, at least in areas where substantial soil compaction has not occurred.

Surface Cover

The type of cover and its condition affects the runoff volume through its influence on the infiltration rate of soil. Bare soil at a construction site generates more runoff than forested or grass land for a given soil type. As a site develops, paving areas reduce the surface storage and infiltration capacity of the area and thus increases the amount of runoff.

The foliage and its litter maintain the soils infiltration potential by preventing the sealing of the soil surface from the impact of the raindrops. Some of the raindrops are retained on the surface of the foliage, increasing their chance of being evaporated back to the atmosphere. Some of the intercepted moisture is so long draining from the plant down to the soil that it is withheld from the initial period of runoff. Foliage also transpires moisture into the atmosphere, thereby creating a moisture deficiency in the soil which must be replaced by rainfall before runoff occurs. Vegetation, including its ground litter, forms numerous barriers along the path of the water flowing over the land surface, which slows the water down and reduces its peak rate of runoff.

Soils

In general, the higher the rate of infiltration, the lower the quantity of stormwater runoff. Fine textured soils, such as clay, produce a higher rate of runoff than do coarse textured soils, such as sand. In addition, compacted soils also produce much more runoff than natural soils (Pitt, *et al.* 1999). Sites having clay soils are much more susceptible to compaction problems than most other soils.

Time of Concentration (T_c or t_c)

The time of concentration (T_c) is the minimum time needed for runoff originating from the complete project site to arrive at the outlet. By definition, T_c is the time required for water to flow from the hydraulically most-distant point in the watershed to the outlet. When rain events last at least as long as the T_c, the outlet is receiving runoff from the entire watershed. The time of concentration affects the peak and shape of the hydrograph. With land clearing and subsequent development, the drainage efficiency usually dramatically increases, resulting in much greater peak runoff values that occur earlier in the storm. In addition, land development (and soil compaction) decease the infiltration capacity of the site, further increasing the runoff volume and the peak runoff rate.

Important aspects of T_c to remember include the following:

- The design storm duration must be equal to the time of concentration for the drainage area.
- The time of concentration (T_c) is equal to the longest flow path (by time).
- If the T_c is 5 min for a storm having a return period of 25 years, the associated peak intensity (which has a duration of 5 min) would be about 8.6 in/hr for Birmingham, AL.
- If the T_c for this same return period was 40 min, the peak rain intensity would be "only" 3.8 in/hr.

Figure 3.1 illustrates the relationships between watershed topography, slopes, and drainage times (McCuen 1989). The "iso-time" plot indicates the times for water to travel to the watershed outlet from all locations in the watershed. This is a complete, but tedious, method to determine T_c. The T_c for this watershed is seen to be 13 minutes.

An area-time plot for this watershed example is shown in Figure 3.2 (McCuen 1989). In this example, 13 minutes is the watershed time of concentration, but almost all of the watershed area is contributing runoff at 9 or 10 minutes. The very small additional area contributed by the increased travel time would normally not compensate for the increased T_c used in calculating the peak flow rate for this watershed.

Generally, only a rain duration equal to the T_c produces the maximum peak runoff rate at the critical rain intensity. Shorter duration rains do not produce runoff from the complete area, while longer duration rains do not have any additional contributing areas, as shown on Figure 3.3.

Rains having durations equal to the T_c must be used in drainage designs as they produce the critical intensity for the area and the level of service (likelihood of failure in any one year), as indicated on Figure 3.4. Longer duration rains have lower intensities for the same level of service, while shorter duration rains do not have the complete drainage area contributing flows during that time period. It is important that the same rain frequency (level of service associated with the acceptable failure rate) be used when examining alternative durations and rain intensities.

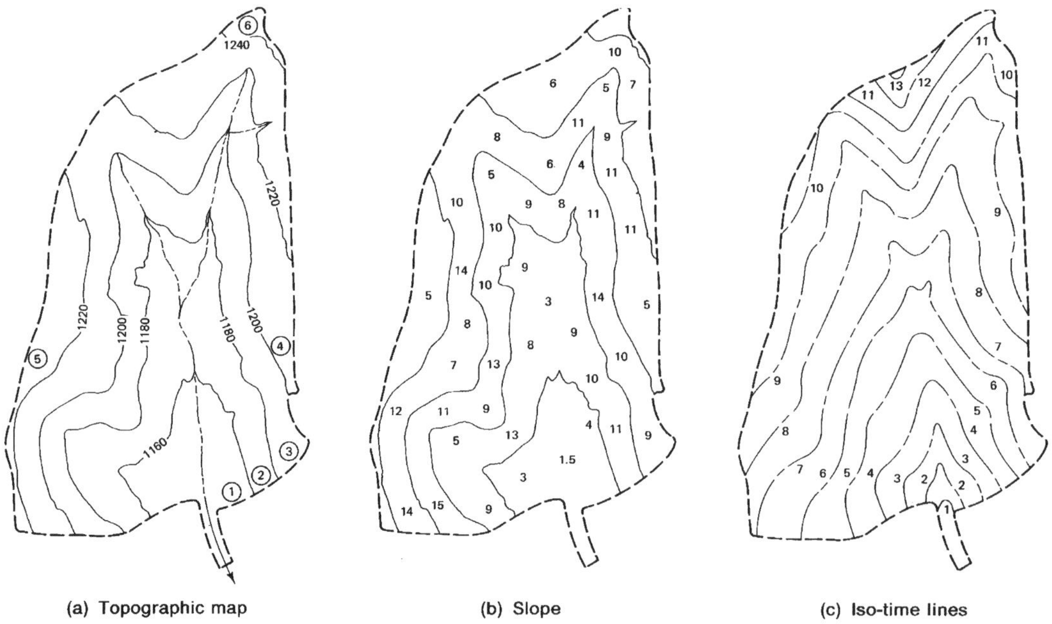

(a) Topographic map (b) Slope (c) Iso-time lines

Figure 3.1. Relationships between water topography, slope, and drainage times (McCuen 1989, with permission).

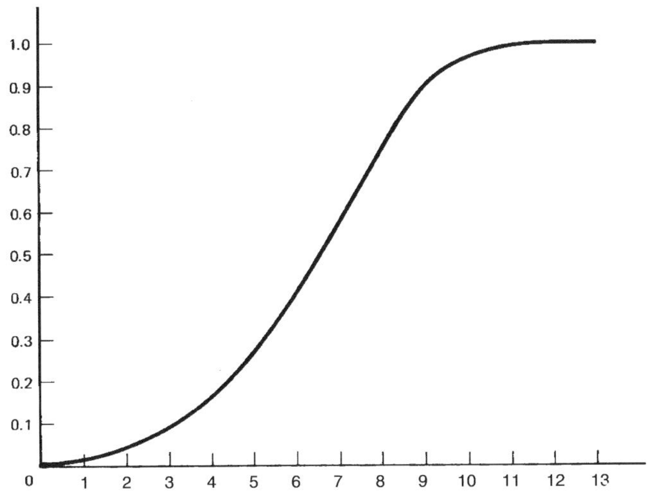

Figure 3.2. Area of watershed contributing runoff as a function of flow travel time (T_t) (McCuen 1989, with permission).

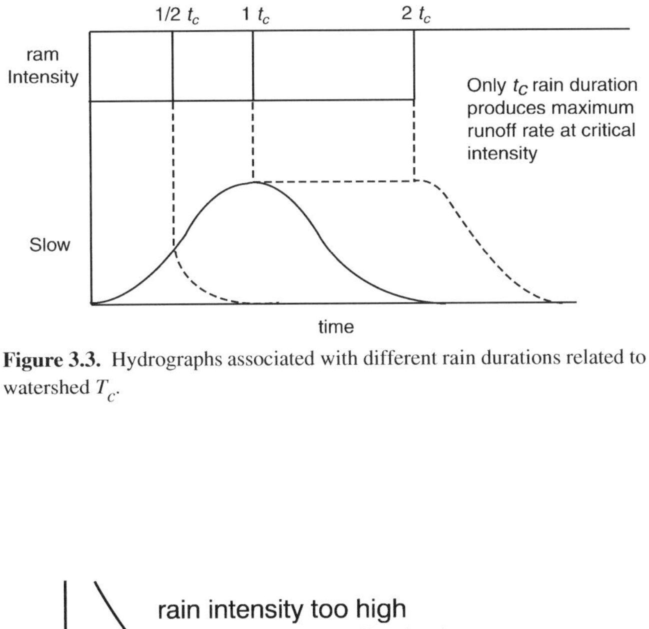

Figure 3.3. Hydrographs associated with different rain durations related to watershed T_c.

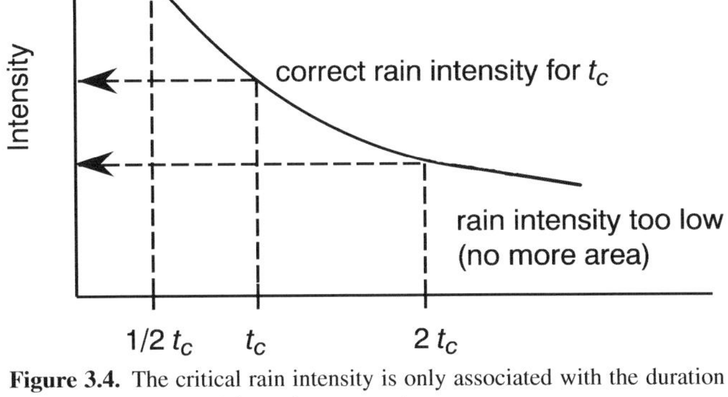

Figure 3.4. The critical rain intensity is only associated with the duration equal to the watershed time of concentration.

LOCAL RAINFALL CONDITIONS RELEVANT TO CONSTRUCTION SITE EROSION AND SEDIMENT CONTROL DESIGN

The following discussion is an example assessment of typical Alabama rain conditions to determine the frequency of highly erosive rains and the relative importance of various rains in generating construction site erosion yields. Figures 3.5, 3.6 and 3.7 show the general variations of rain conditions over Alabama. These figures were prepared by Pitt and Durrans (1995) as part of a research project for the Alabama Dept. of Transportation. These analyses used data from the 1976 and 1977 rain period. These two years were determined to be representative of the average conditions from 1948 through 1994 based on total rain depth and the monthly distribution of rains. These data were obtained from EarthInfo (Golden, CO) CD-ROMs, which archive the official NOAA data. Figure 3.5 is a contour map of the total annual rain depth throughout Alabama, based on analyses at more than 120 rain gage stations located in Alabama and in surrounding states (rain gauges represented on Figure 3.5 by dots). There is little variability in rain conditions over most of the state (50 to 56 inches per year). The northwest corner has less rain (down to about 46 inches), while the rain depth increases substantially moving towards the Gulf Coast (as high as 66 inches per year). There are usually slightly more than 100 separate rain events per year in Alabama (defined

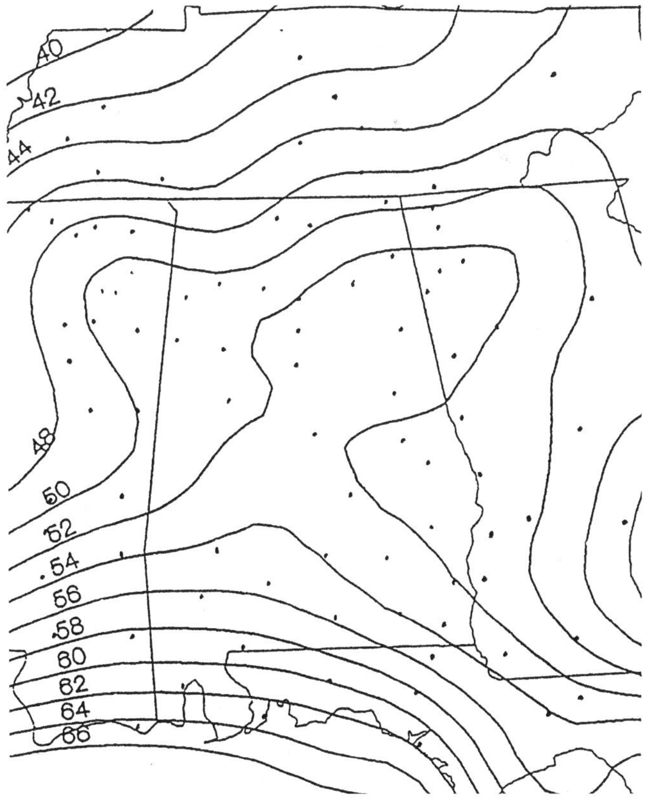

Figure 3.5. Annual rainfall depths throughout Alabama in inches (Pitt and Durrans, 1995).

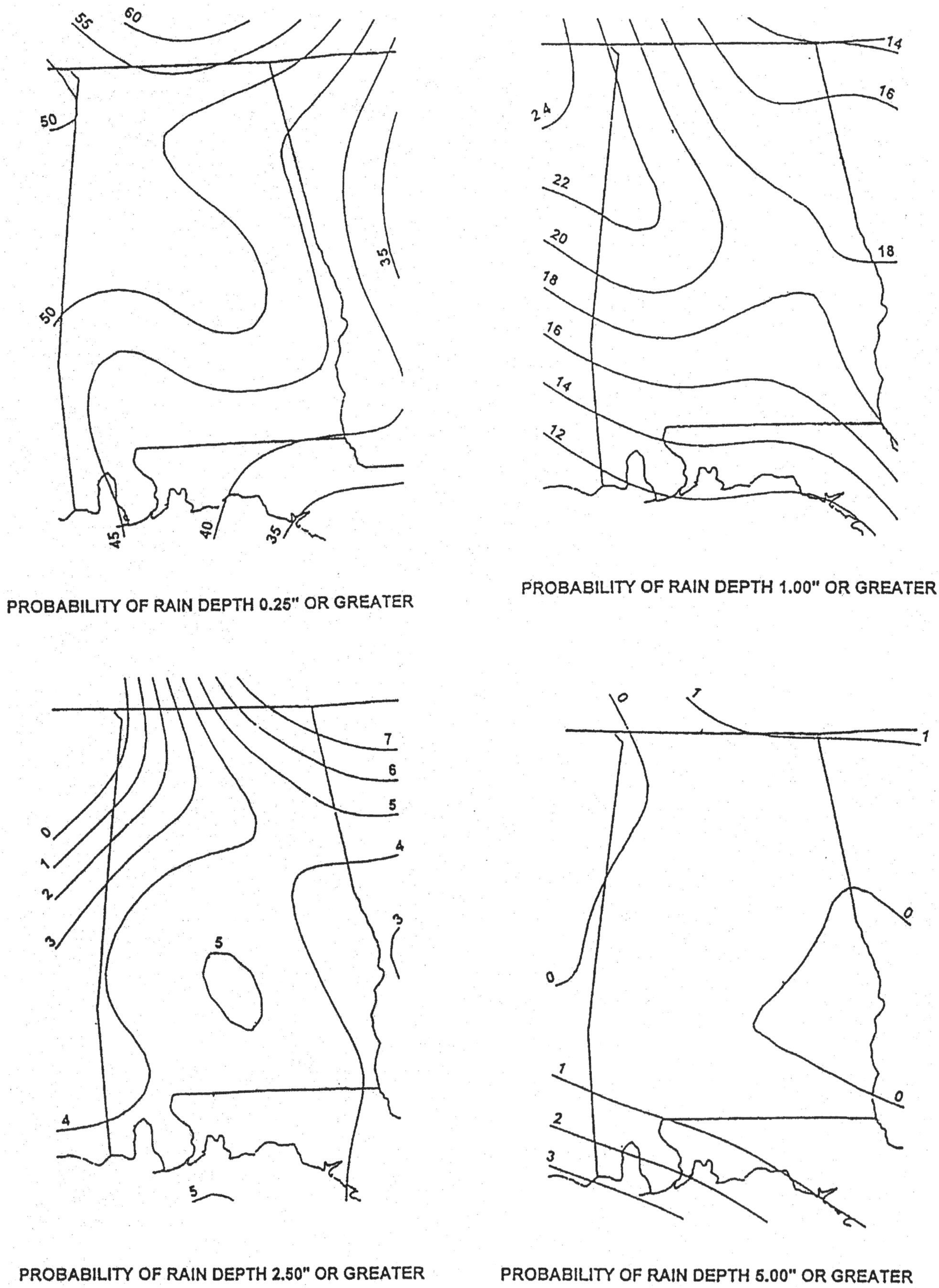

Figure 3.6. Probabilities (expressed as percents) of individual rain storms having various rain depths in Alabama (Pitt and Durrans, 1995).

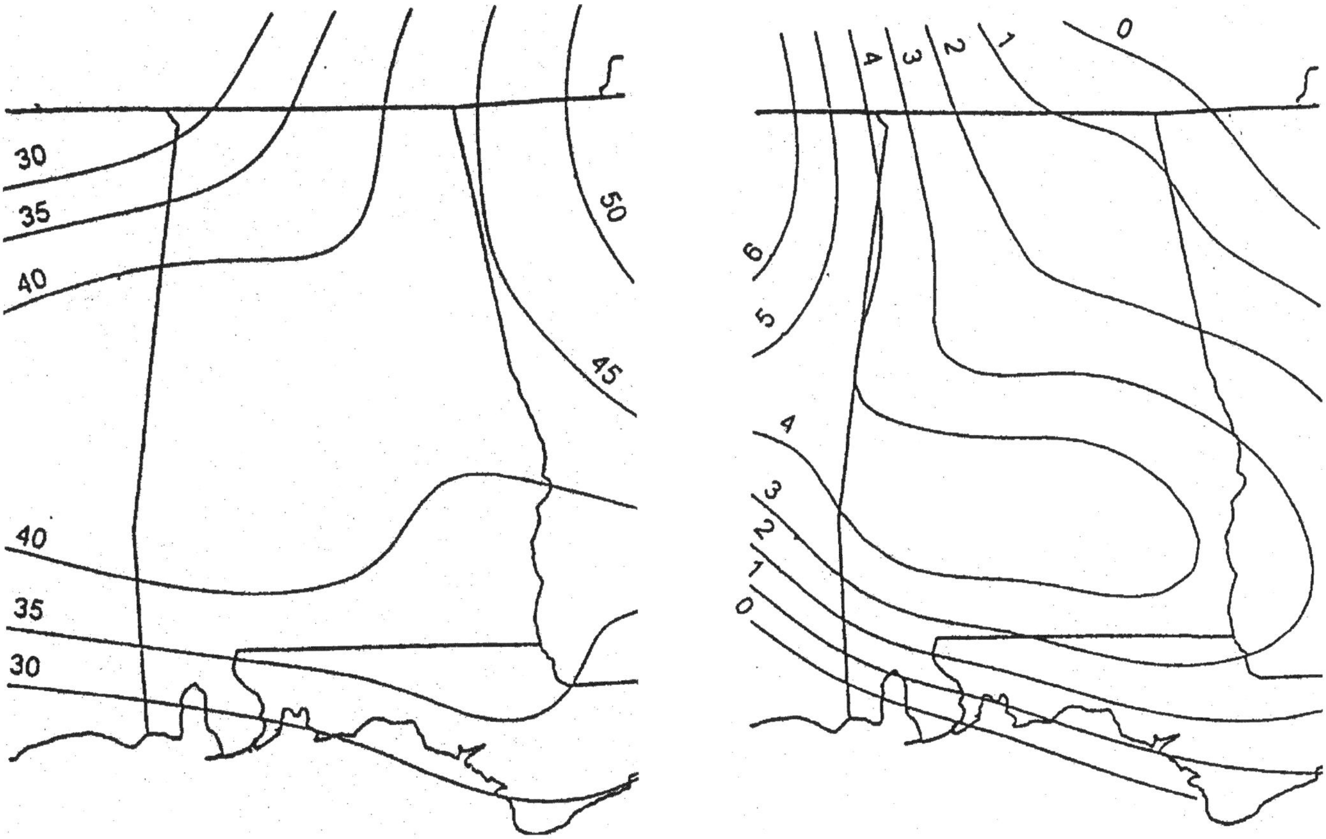

(a) **Probabilities of rains having at least 3-day antecedent dry periods.**

(b) **Probabilities of rains having at least 15-day antecedent dry**

Figure 3.7. Probabilities (expressed as percents) for average rain storm interevent periods for Alabama (Pitt and Durrans, 1995).

using a minimum of 6 hours for the interevent period), with the smallest rains being 0.01 inches and the largest approaching 10 inches. Figure 3.6 presents the percentages of these annual rains having at least 0.25, 1.00, 2.5, and 5.00 inches. Few, if any, of the rains are likely greater than 5 inches in the central and northern portions of the state, but several rains greater than this amount likely occur each year near the coast. At least 40 to 50% of all rains are at least 0.25 inches in depth throughout the state. Figure 3.7 shows the percentages of all storm interevent periods that are at least 3 and 15 days. Most interevent periods are about 3 days throughout the state, but few last as long as 2 weeks, especially near the gulf coast.

Typical Birmingham Rain Conditions

Monthly rain depths from 1955 to 1986 were examined to identify a single rain year that had total depths and rain distributions similar to the longterm average conditions. The years 1975 and 1976 both were found to have similar rain conditions that were close to these average conditions. Individual events in these years were identified using hourly rain records. A rain event was defined as a series of hourly observations containing no more than six adjacent hours having no rain. This definition has been commonly used in many urban runoff studies as it produces discrete runoff hydrographs. The six-hour period of no rain also almost always allows urban streams to return to near baseflow conditions. Tables 3.1 and 3.2 summarize these rains.

Table 3.1 lists the expected rainfall distribution for typical Birmingham conditions. There are about 100 individual rains per year in Birmingham, ranging from 0.01 to about 4 inches in depth. Most of the rains are less than 0.5 inches in depth, but more than onehalf of the total annual rain depth is associated with rains greater than one inch. Rain interevent periods are important when determining the periods of time that bare ground may remain unprotected at construction sites. The interevent periods shown on this table are for all rains greater than the minimum rain in the range. As an example, rains greater than 2 inches occur about every 56 days, while rains greater than 0.5 inch occur about every 10 days.

Table 3.2 summarizes the runoff quantities that may be expected for each rain depth class, for a typical construction site area, without significant compaction. More than half of the runoff from this area is associated with rains less than 1.5 inches in depth. Less than 20 percent of the runoff is associated with rains greater 2.5 inches in depth. Only rains

TABLE 3.1. Birmingham Rain Depth Distributions (average for 1975 and 1976).

Rain Depth Range (inches)	Interevent Period (days)	Annual Number of Rains in Range (out of 100 rains per year)	Total Rain in Range (inches)	% of Annual Rain in Range	Accumulative % of Rain in Range
0.01 to 0.5	4	62	15.5	25	25
0.5 to 1.0	10	19	14.3	23	48
1.0 to 1.5	21	9	11.3	17	65
1.5 to 2.0	41	3	5.3	8	73
2.0 to 2.5	56	3	6.8	10	83
2.5 to 3.0	122	2	5.5	8	91
3.0 to 3.5	183	1	3.5	3	94
3.5 to 4.0	365	1	3.8	6	100

greater than about 1.25 inches will contribute runoff quantities greater than 0.5 inches, a commonly used detention criterion contained in runoff control ordinances. The first 0.5 inch of runoff from all rains therefore includes all rains smaller than about 1.25 inches, plus portions of larger rains. The remaining runoff, after the first 0.5 inch, totals about 5.5 inches for typical construction areas using the 1975 and 1976 Birmingham rains.

Erosion Yields for Different Alabama Rain Categories

It is possible to estimate the relative erosion contributions of different rains, as shown in Tables 3.3 through 3.5. Thronson (1973) presented the following equation to estimate the erosion potential for individual rains, when complete intensity information is not available:

$$R = \frac{19.25(P)^{2.2}}{(dur)^{0.4672}}$$

where,

P = rain depth (inches)
dur = rain duration (hours)

This equation was proposed for the original SCS type II rain category which was applicable for the complete U.S., except for the extreme west coast. Long-term rain series data for Huntsville, Birmingham, Tuscaloosa, Montgomery, and Mobile were extracted from EarthInfo CD-ROMS (Golden, CO) and processed in WinSLAMM (www.winslamm.com) to combine the hourly data into individual rain records. Each rain was defined as having at least a 6-hour-dry interevent period. About 50 years of data were available for each city, although some of the records were incomplete. The number of events evaluated for each city ranged from about 2500 to 5200 separate rains. The calculations were made for each of 12 rain categories and the total annual R was estimated by multiplying the partial R for each category by the number of events in each category. The calculated annual R values for these 5 cities were slightly larger (differences of 6 to 34%) than the published annual R values. The main reason for these differences is that the published annual R values are median values based on many years of record where R values were calculated for individual years, while the R values used here were averaged values, which would be larger. The calculated R values for each category were therefore adjusted to indicate the approximate portion of the total annual R associated with the different rain categories.

Figure 3.8 is a plot of the accumulative total R associated with the rains. The larger rains contribute most of the erosion potential for Alabama conditions. For all of these cities, except Mobile, the rain depth associated with the median of the annual R is about 2 inches, while it is about 2.5 inches for Mobile. About 5% of the annual rains are therefore responsible for about half of the annual erosion potential. Rains less than about 0.75 to 1 inches in depth are

TABLE 3.2. Birmingham Runoff Volume Distributions for Typical Construction Site.

Rain Depth Range (inches)	Volumetric Runoff Coefficient (R_v)	Annual Runoff in Range (inches)	% of Runoff in Range	Accumulative % of Runoff in Range
0.01 to 0.5	0.27	4.2	19	19
0.5 to 1.0	0.34	4.9	22	41
1.0 to 1.5	0.36	4.1	17	58
1.5 to 2.0	0.39	2.0	9	67
2.0 to 2.5	0.41	2.8	11	78
2.5 to 3.0	0.44	2.4	10	88
3.0 to 3.5	0.45	1.5	4	92
3.5 to 4.0	0.48	1.8	8	100
Total, or weighted average	0.36	23.7	100	

TABLE 3.3. Erosion Potential Analysis for Birmingham Rains Occurring from 1948 through 1999.

Rain range (inches)	Mid Point Rain (inches)	Average Duration (hours)	Average Intensity (in/hr)	#/year in Range Category	% of Rains in Category	Thronson *R*	% of Annual *R* in Category	Accumulative % of Total *R*
0.01 to 0.05	0.03	3	0.01	22.9	20.7	0.1	0.0	0.0
0.06 to 0.10	0.08	7	0.01	17.4	15.8	0.4	0.1	0.1
0.11 to 0.25	0.18	8	0.02	17.3	15.6	2.4	0.7	0.8
0.26 to 0.50	0.38	10	0.04	19.5	17.6	12.4	3.5	4.4
0.51 to 0.75	0.63	12	0.05	9.4	8.5	16.6	4.8	9.1
0.76 to 1.00	0.88	14	0.06	8.3	7.5	28.6	8.2	17.3
1.01 to 1.50	1.26	16	0.08	7.9	7.2	56.4	16.1	33.4
1.51 to 2.00	1.76	18	0.10	3.8	3.5	53.9	15.4	48.8
2.01 to 2.50	2.26	20	0.11	1.6	1.5	38.0	10.9	59.7
2.51 to 3.00	2.76	24	0.12	0.8	0.7	26.3	7.5	67.2
3.01 to 4.00	3.5	30	0.12	1.1	1.0	57.0	16.3	83.5
over 4.01	5.67	36	0.16	0.4	0.4	57.9	16.5	100.0
4583 events	41.5 years	13.58 in. max rain	Totals	110.5	100.0	350.0	100.0	

responsible for only about 10% of the total erosion potential. About 20 to 30% of the rains (generally between 0.75 and 4 inches) are associated with about 80% of the erosion potential. Because of the long rain record used here, these rain series include several rare events, including the "50-year" event. It may be impractical to design erosion controls that can effectively withstand the very large events. Except for Mobile, rains greater than 4 inches occur less than once a year in most parts of the state. If a "typical" rain year was examined, the effects of these very large rains would be somewhat diminished. When only the 1976 rain year for Birmingham was examined (a typical year for local rains), for example, the rain depth associated with the median erosion potential was reduced to about 1.75 inches. These calculations are repeatable for any location, provided that sufficient rainfall records are available. The longer rain records typically contain "rare" events that, while uncommon and difficult to plan for, may affect the erosion yield and cause damage to the site that would require substantial regrading.

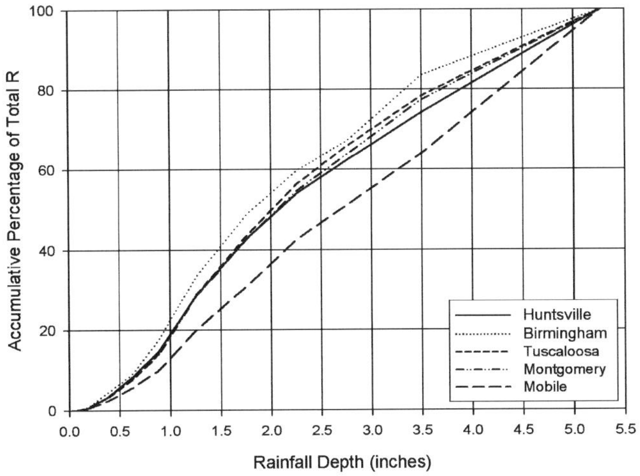

Figure 3.9. Distribution of erosion potential associated with different rains for major Alabama cities.

Table 3.4 shows the variation in frequency of these large rains for the 1948 through 1999 rain period for Birmingham (41.5 years of data due to some missing data periods). Between 1 and 8 (an average of 4.1) of these large rains occur each year, but no obvious pattern is indicated in the group in terms of predicting the number of large rains in any given year. Table 3.5 examines these highly erosive rains for each month of the year for this same Birmingham rain period. May through November appears to have fewer of these rains. However, September had the largest number of any month, which is not unexpected for any area whose rainfall distribution is influenced by tropical storms. August and September are considered the most-active months for the development and sustenance of tropical weather (the Atlantic hurricane season is considered to peak in September).

TABLE 3.4. Number of Large Rains (>2 inches) per Year for Birmingham.

Year	#/Year	Year	#/Year	Year	#/Year
1948	4	1962	4	1976	7
1949	2	1963	6	1977	8
1950	7	1964	8	1978	3
1951	6	1965	2	1979	2
1952	2	1966	5	1980	3
1953	4	1967	6	1981	3
1954	3	1968	5	1982	5
1955	1	1969	6	1983	1
1956	3	1970	5	1984	4
1957	8	1971	4	1985	4
1958	2	1972	3	1986	5
1959	2	1973	5	1987	1
1960	1	1974	3	1988	6
1961	6	1975	5	1999	2

total = 172 large storms from 1948 through 1999
average = 4.1 large storms/year
minimum = 1 large storms/year
maximum = 8 large storms/year
standard deviation = 2.0
COV = 0.49

TABLE 3.5. Birmingham Rains by Months.

	2.00 to 2.50	2.51 to 3.00	3.01 to 4.00	over 4.01	Total
January	7	2	4	4	17
February	7	2	4	1	14
March	9	5	5	2	21
April	5	1	5	1	12
May	7	4	4	1	16
June	6	0	5	0	11
July	5	2	2	2	11
August	4	5	1	1	11
September	9	7	5	1	22
October	0	3	5	1	9
November	8	1	1	1	11
December	6	2	6	3	17
Total for 41.5 years of record	73	34	47	18	172
Average (#/year)	1.8	0.8	1.1	0.4	4.1

Intensity, Duration and Frequency (IDF) Information for Rains Used to Design Erosion Controls

As noted above, rains having high intensities typically contribute the highest erosion yields. Individual rains that may occur at any time of the year can contribute excessive erosion losses. Very rare rains, occurring at most only once every year and usually much less frequently, typically receive the most attention for flooding and drainage studies. When these rare rains do occur, great erosion yields will occur and most erosion and sediment control devices will fail. As an example, Figure 3.9 (the IDF curve for Birmingham, AL) shows the relationship between rainfall duration, peak intensity, and return period. (*Note:* The return period of a storm is defined as the inverse of the probability [expressed as a decimal fraction] of a storm of a specific depth and duration being equaled or exceeded within a pre-specified time frame, typically one year. The IDF curve for Birmingham, as displayed in Figure 3.9, shows this relationship for durations up to 60 min. As seen in this figure, rains having average intensities of almost 3 inches per hour lasting for 30 minutes are expected to occur with a 50 percent probability every year. Five minute peak rain intensities of more than 6 inches per hour also occur with a probability of at least 50 percent every year. Table 3.6 lists the approximate rain depths (inches) and average rain intensities (inches per hour) associated with rain, durations from 1 to 24 hours and return frequencies of 1 to 100 years for Birmingham. Also shown on this table are three maximum probable events, associated with 6, 12, and 24 hour rain durations. A review of the extreme-event data demonstrates that it would be very

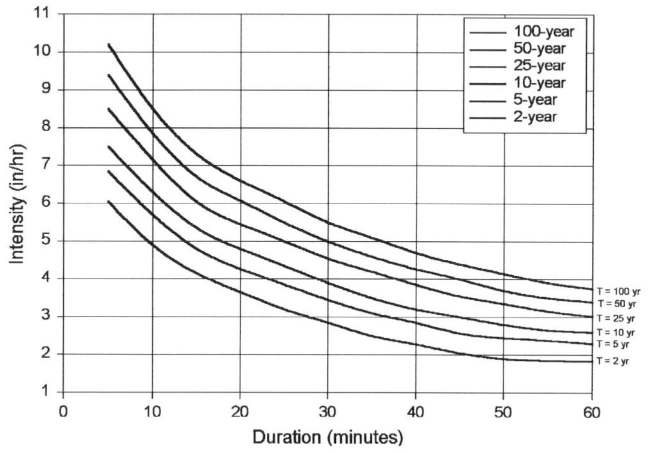

Figure 3.9. Intensity, duration, and frequency (IDF) curve for Birmingham, AL (from National Weather Service, Hydro-35).

TABLE 3.6. Rare Birmingham Rain Conditions.

Duration (hours)	Probability (*P*, % occurrence per year)	Frequency (1/*P*, years)	Rain Depth (inches)	Rain Intensity (inches per hour)
1	100	1	1.5	1.5
2	100	1	1.9	1.0
3	100	1	2.1	0.7
6	100	1	2.5	0.4
12	100	1	3.0	0.3
24	100	1	3.5	0.1
1	20	5	2.3	2.3
2	20	5	2.8	1.4
3	20	5	3.1	1.0
6	20	5	3.8	0.6
12	20	5	4.5	0.4
24	20	5	5.3	0.2
1	10	10	2.6	2.6
2	10	10	3.3	1.7
3	10	10	3.5	1.2
6	10	10	4.3	0.7
12	10	10	5.1	0.4
24	10	10	6.0	0.3
1	4	25	3.1	3.1
2	4	25	3.6	1.8
3	4	25	4.0	1.3
6	4	25	5.0	0.8
12	4	25	6.0	0.5
24	4	25	6.9	0.3
1	2	50	3.4	3.4
2	2	50	4.0	2.0
3	2	50	4.4	1.5
6	2	50	5.5	0.9
12	2	50	6.6	0.6
24	2	50	7.6	0.3
1	1	100	3.8	3.8
2	1	100	4.4	2.2
3	1	100	4.9	1.6
6	1	100	6.0	1.0
12	1	100	7.2	0.6
24	1	100	8.4	0.4
6	Maximum probable event		31	5.2
12	Maximum probable event		37	3.1
24	Maximum probable event		42	1.8

difficult to design effective erosion and sediment control practices that can withstand the high runoff rates than may occur during many of these "design storm" events.

In some states, IDF information is compiled by region and is available through one of the state agencies, typically the state Department of Transportation. Pennsylvania is an example of one such state. Pennsylvania is divided into five regions based on the region's characteristic patterns. Pennsylvania design rain information also includes a figure showing rainfall depth based on storm duration and return period (Figure 3.10). The current Pennsylvania manual (Field Manual of the Pennsylvania Department of Transportation 1986) has been updated and has replaced the older information originally obtained from TP-40 with newer calculated design curves. These regional rainfall design curves in this Pennsylvania field manual were developed from frequency analyses based on hourly records from 153 climatological stations and 15-minute records from 45 stations in Pennsylvania. The analysis leading to the design curves is fully described in the project report "Pennsylvania Department of Transportation Rainfall Intensity-Duration-Charts," submitted to the Pennsylvania Department of Transportation.

Appendix 3B contains rainfall distribution maps for the whole country from the NRCS TR-55 manual (SCS 1986). In these maps, the return period of the storm is given and the rainfall duration is set at 24-hours. The approximate rainfall depth is read from the map based on the site location. These maps were prepared by the NRCS (SCS) as part of the runoff peak flow rate calculation procedure contained in Technical Release 55 (TR-55).

Similar to Pennsylvania's update, rainfall data collected since the publication of NWS' Technical Paper 40 and HYDRO-35, and NOAA's Atlas 2 and Atlas 14, and the development of improved statistical methods, motivated several states to initiate update studies of precipitation distributions (Durrans and Brown 2001).

The study by Durrans and Brown (2001) is an interesting one to highlight for three reasons. One, it uses a substantially longer period of record to perform the statistical calculations. Second, it was based on extreme-event probability calculations. Third, the results are widely disseminated on the Internet and information for several smaller cities is available, based on their historical rainfall record. Since much of the interest in precipitation records has come from state Departments of Transportation as part of their need to calculate runoff peak flow rates for design purposes, it is logical that this study was funded by the Alabama Department of Transportation.

The Alabama Rainfall Atlas is available at: http://www.bama.ua.edu/~rain/. This web site, prepared by Dr. Rocky Durrans of the University of Alabama for the Alabama Dept. of Transportation, calculates and presents IDF curves for any location in the state of Alabama. IDF equation coefficients were calculated based on long term rain records for many state locations. This web site then

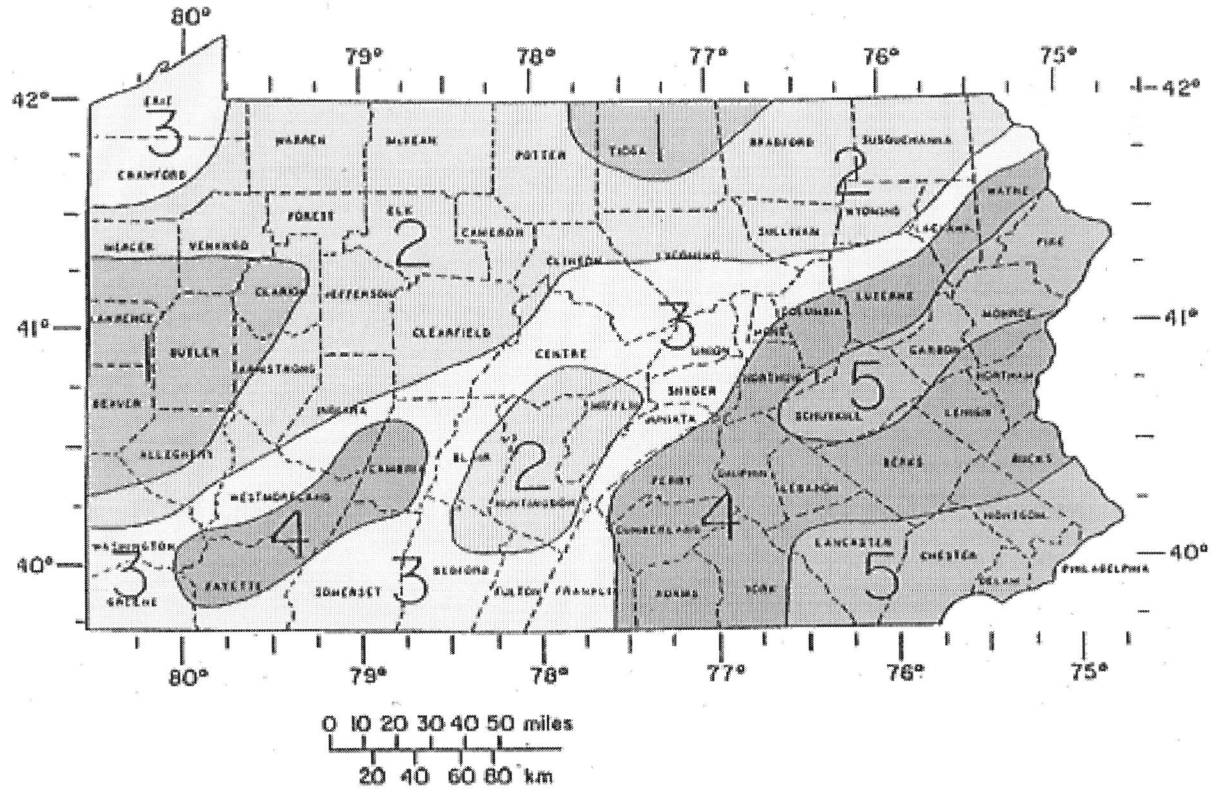

Figure 3.10. Rainfall zones for the State of Pennsylvania (Field Manual of the Pennsylvania Department of Transportation, 1986).

interpolates the coefficients for any location on the state map and presents graphical and tabular IDF information. The IDF information is presented for 2 to 500 year rains and for 5 minutes to 48 hours durations. The web site will also produce SCS design hyetographs. Figure 3.12 is the main map that is displayed for the Atlas. The user simply clicks the mouse anywhere an IDF calculation is desired, and selects if a map or table (or both) is desired. In most cases, the "partial duration" option is probably desired in order to be more consistent with historical NOAA IDF curves (not a significant difference for the large, rare, rains, but more of an effect on the smaller events). These IDF curves are likely to vary from the "official" older NOAA IDF curves as they are obtained from more recent data (the Alabama Rainfall Atlas values seem to be slightly smaller than the NOAA values). The bottom button is then clicked to accept the choices and the desired outputs are produced. Figure 3.13 is an example for Mobile, AL, showing both an IDF graph and a table. This is a preliminary product and the "print" options indicated are not yet functioning. However, it is possible to use a simple print screen utility to capture the calculated IDF information.

Figures 3.14 and 3.15 refer to the SCS rain distribution types that are commonly used in urban drainage design. The cumulative rain distributions in Figure 3.14 shows how the rain intensities vary throughout these hypothetical events. The slope of this curve, averaged over the time of concentration (described later) and multiplied by the rainfall depth, equals the rain intensity that would be plotted on an IDF curve for each hypothetical distribution. Figure 3.15 shows which of these rain types are applicable for different southeastern U.S. areas. Most of the U.S. uses Type II rains, but the gulf coast and eastern seaboard use Type III rains. Types I and IA are used in some parts of the western states. Appendix 3B includes a map showing the rainfall distribution types for the entire U.S.

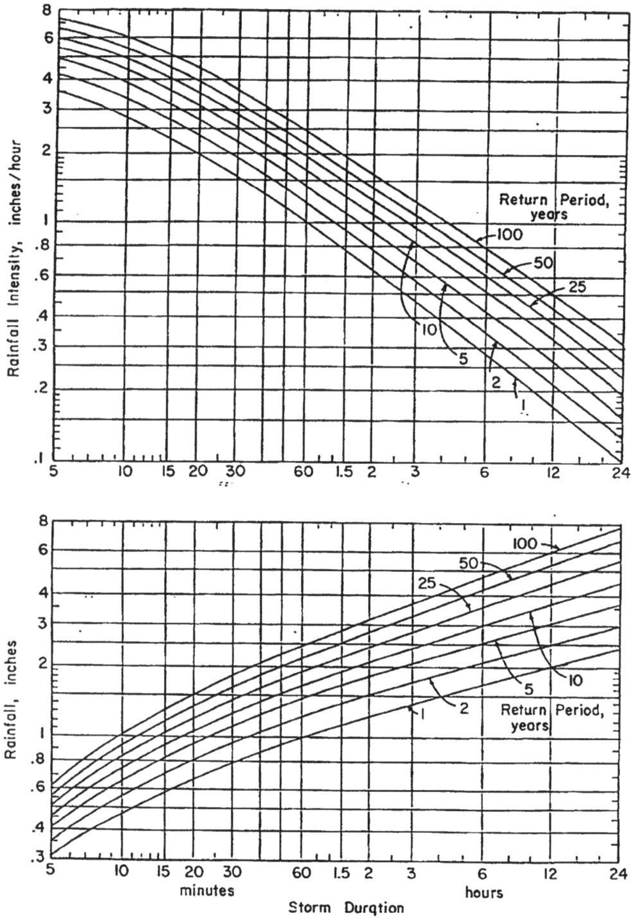

Figure 3.11. IDF Curves for Pennsylvania Region 4 (Field Manual of the Pennsylvania Department of Transportation 1986).

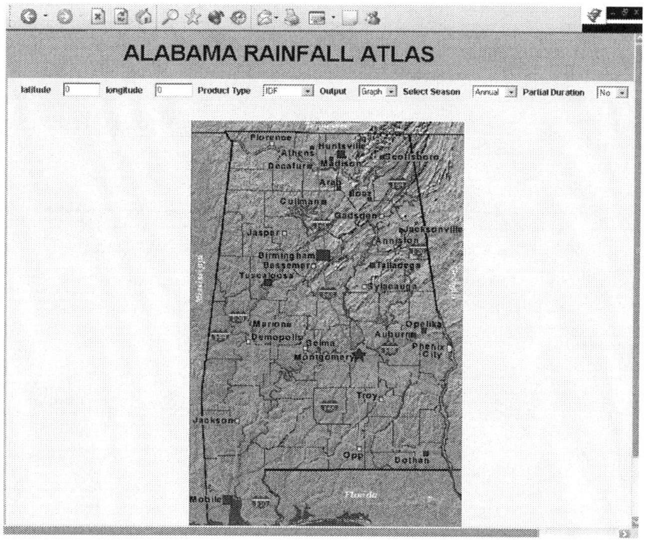

Figure 3.12. Opening map for the Alabama Rainfall Atlas.

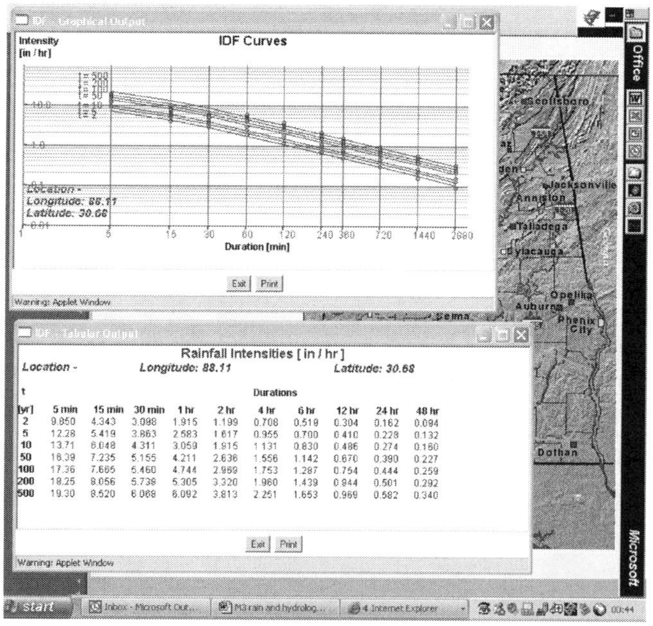

t [yr]	5 min	15 min	30 min	1 hr	2 hr	4 hr	6 hr	12 hr	24 hr	48 hr
2	9.850	4.343	3.098	1.915	1.199	0.708	0.519	0.304	0.162	0.094
5	12.28	5.419	3.863	2.583	1.617	0.955	0.700	0.410	0.228	0.132
10	13.71	6.048	4.311	3.059	1.915	1.131	0.830	0.486	0.274	0.160
50	16.39	7.235	5.155	4.211	2.636	1.556	1.142	0.670	0.390	0.227
100	17.36	7.665	5.460	4.744	2.969	1.753	1.287	0.754	0.444	0.259
200	18.25	8.056	5.738	5.305	3.320	1.960	1.439	0.844	0.501	0.292
500	19.30	8.520	6.068	6.092	3.813	2.251	1.653	0.969	0.582	0.340

Figure 3.13. IDF information produced by the Alabama Rainfall Atlas for Mobile, AL.

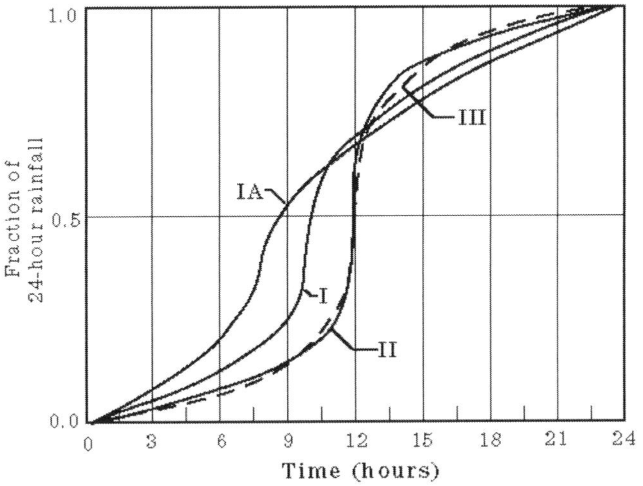

Figure 3.14. Cumulative distribution curves for different SCS rain types (SCS 1986).

Selection of Design Storms for Varying Risks and Project Durations

The selection of appropriate control practices must consider potentially high runoff flow rates corresponding to relatively large rains. As an example, the use of filter fences is not recommended in channels that drain large areas. Filter fences are most suitable for controlling sheet flows originating from relatively small areas. More robust sediment control practices, such as wet detention ponds, are needed for treating runoff from large areas. Similarly, the use of unreinforced mulches can only be used on flat slopes with small contributing areas. The following paragraphs describe how to select an appropriate "design storm" based on acceptable failure rates and exposure periods.

The following equation (from McGhee, 1991) can be used to calculate the probability that a rain having a return period of "n" years, will occur at least once in the next "y" years:

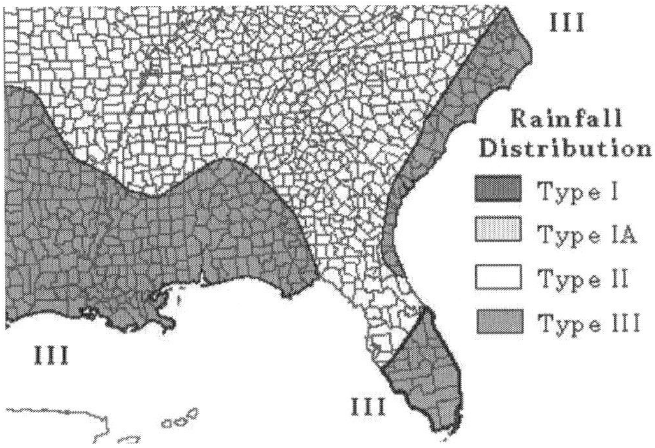

Figure 3.15. SCS rain distribution types for southeastern U.S. (NRCS, 2002b).

$$P = 1 - \left(1 - \frac{1}{n}\right)^{y}$$

This equation can be reworked to relate the service life to the needed design return period and probability of exceedence (or failure).

$$P = 1 - \left(1 - \frac{1}{T_{needed}}\right)^{\text{design life}}$$

$$1 = P - \left(1 - \frac{1}{T_{needed}}\right)^{\text{design life}}$$

$$(1 - P)^{(1/\text{design life})} = 1 - \frac{1}{T_{needed}}$$

$$(1 - P)^{(1/\text{design life})} - 1 = -\frac{1}{T_{needed}}$$

$$1 - (1 - P)^{(1/\text{design life})} = \frac{1}{T_{needed}}$$

$$T_{needed} = \frac{1}{1 - [(1 - P)^{(1/\text{design life})}]}$$

Figure 3.16 is a plot illustrating this relationship, but modified to show the probability of an event not being exceeded during the design period.

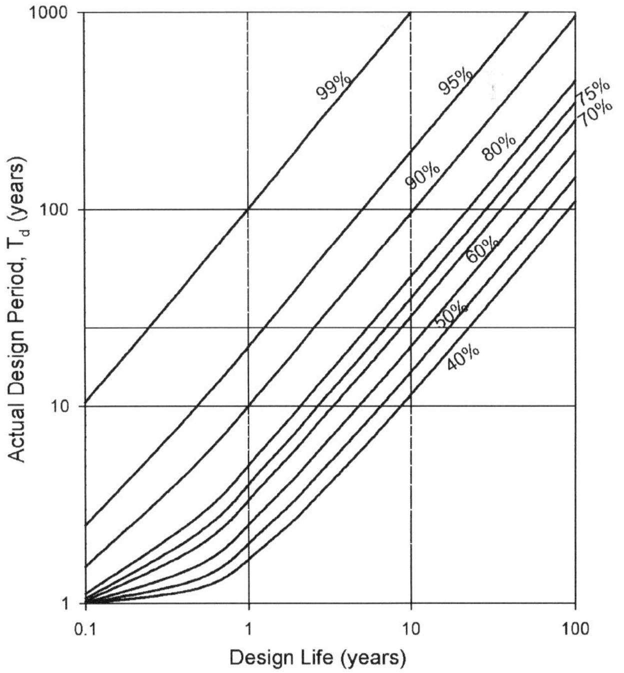

Figure 3.16. Probability, expressed as a percent, of a design storm (design return period) not being exceeded during the project life (design period) (from McGhee, 1991, with permission).

TABLE 3.7. Design Storm Return Periods Associated with Different Probability Levels for a 1-year Construction Period.

Probability of Storm Not Being Exceeded in a One Year (T_d on Figure 3.16) Construction Period	Design Storm Return Period (T on Figure 3.16) (yr)
50%	2
75%	6.5
90%	10
95%	20

As an example, one needs to be certain, with a 90% probability that a failure would not occur during a 5-year project period (the exposure period, or T_d). A storm having a 50 year return period (T) would be the appropriate design storm frequency for this condition.

Obviously, if failure could possibly lead to serious property damage or loss of life, then the probability of an event that may cause such failure not occurring during the project design life will need to be very large. Similarly, if only minor inconvenience will be associated with a failure, then the probability of that event not occurring during the design period can be much less. Table 3.7 illustrates several examples for a typical construction period of one year. The design storms could therefore vary greatly for different elements on the same project site. A filter fence failure may not be very serious if the site runoff is also being captured by a downstream sediment pond. However, the failure of the pond could cause much greater problems. Similarly, the slope along a filled embankment near a building foundation could cause structural failure if massive erosion occurred on the slope. In these cases and for a one year construction period, the filter fence may be designed using a 2-year design storm (acceptable failure probability of 50% in the one year period), the pond may require a 10-year design storm (acceptable failure probability of 10% in the one year period), while the slope near the building may need a 20+-year design storm (acceptable failure probability of <5% in the one year period).

METHODS OF DETERMINING RUNOFF

Many different methods of computing runoff have been developed. Some of the methods and limitations of each are summarized on Table 3.8 and in the paragraphs below (from Illinois 1989).

(1) The Rational Method

The Rational Method is an empirical formula used for computing peak rates of runoff that has been used in urban areas for over 100 years ($Q = CiA$). It is useful for estimating runoff on relatively small areas such as roof tops, parking lots, or other homogeneous areas. Use of the Rational equation should be limited to drainage areas less than 20 acres that do not vary in surface character and do not have branched drainage systems. The most serious drawback of the Rational Method is that it gives only the peak discharge and provides no information on the time distribution of the storm runoff, disallowing routing of hydrographs through the drainage system or storage structures. Newer methods that would allow runoff hydrographs to be developed based on a modified Rational Method have been proposed, but are not in wide public use. Furthermore, the choice of "C" and "T_c" when choosing "i" in the rational method is more an art of judgment than a precise account of the antecedent moisture condition. It also is not an aerial distribution of rainfall intensity. Many errors have been reported in the use of the Rational Method, and it cannot be easily verified. Modifications of the rational method have similar limitations. The rational method may be applicable in small, isolated sections of construction sites. The rational method will be used later in this chapter, and in the next chapter, for predicting sheetflow runoff depth needed for shear stress calculations for isolated slopes.

(2) SCS TR-20 Method

The SCS-TR-20 computer program uses hydrologic soil and cover runoff curve numbers to determine runoff

TABLE 3.8. Selection Criteria for Runoff Calculation Methods (Illinois 1988).

Output Requirements	Drainage Area	Rational Method	SCS TR-20 Method	SCS TR-55 Tabular Method	SCS TR-55 Graphical Peak Discharge Method	COE HEC-1 Method (now replaced with the HEC-HMS method)
Peak Discharge Only	Up to 20 acres	X		X	X	X
	Up to 2,000 acres		X	X	X	X
	Up to 5 square miles		X	X		X
	Up to 20 square miles		X	X		X
Peak Discharge and Total Runoff Volume	Up to 2,000 acres		X	X	X	X
	Up to 5 square miles		X	X		X
	Up to 20 square miles		X	X		X
Runoff Hydrograph	Up to 5 square miles		X	X		X
	Up to 20 square miles		X	X		X

volumes, and it uses synthetic unit hydrographs to determine peak rates of discharge and combined hydrographs. Factors needed to use the method are the 24-hour rainfall amount, a given rainfall distribution, runoff curve numbers, time of concentration, travel time, and drainage area. This procedure probably should not be used for drainage areas less than 50 acres or more than 20 square miles. It is very useful for larger drainage basins, especially when there are a series of structures or several tributaries to be studied. Recently, a preliminary Windows version of TR-20 has become available, making the method easier to use.

(3) SCS TR-55 Tabular Hydrograph Method

The SCS TR-55 Tabular hydrograph is an approximation of the more detailed SCS TR-20 method. The Tabular Method divides the watershed into subareas, computes an outflow hydrograph for each, and then combines and routes each subarea hydrograph to the outlet. It is especially useful for measuring the effects of changing land use in a part of a watershed. It can also be used to determine the effects of hydraulic structures and combinations of structures, including channel modifications, at different locations in a watershed. The Tabular Method should not be used when large changes in the curve number occur among subareas within a watershed and when runoff volumes are less than about 1.5 inches for curve numbers less than 60. For most watershed conditions, however, this procedure is adequate to determine the effects of urbanization on peak rates of discharge for subareas up to approximately 20 square miles in size. The recent preliminary Windows version of TR-55 has many improvements and is much easier to use than the older manual method or the original computer version. It is applicable for many conditions at construction sites and will be described later in this chapter.

(4) SCS TR-55 Graphical Method

The SCS TR-55 Graphical Method calculates peak discharge using an assumed unit hydrograph and an evaluation of the soils, slope, and surface cover characteristics of the watershed. The assumed unit hydrograph is based on design considerations rather than meteorological factors. Correction factors for swampy or ponding conditions can be used. This method is a component of the older TR-55 procedures and is not included in the new Windows version of TR-55. It is not a very suitable tool, as it has most of the same limitations as the rational method (specifically no hydrograph routing capabilities).

(5) U.S. Army Corps of Engineers HEC-1/HEC-HMS

The COE-HEC 1 provides similar site evaluations as the SCS TR-20. It is a rainfall-runoff model that can be calibrated to gauge records. Like TR-20, it can be used on both simple and complex watersheds. Several years ago, the older HEC-1 was superseded by the HEC-HMS (Hydrologic Modeling System) that is a Windows-based program and much easier to use. Because of its complexity, it is not a very suitable tool for use at most construction sites. However, if complex conditions exist, like at some highway sites where relatively large streams are crossed by the construction activities, its use may be warranted.

WATERSHED DELINEATION

One of the first steps in conducting a hydrologic evaluation of an area is to delineate the watershed draining to the location of concern. For construction sites, this may include determining the area draining to a sediment pond, the area draining to a filter fence, the area draining to a diversion channel, etc. The following discussion outlines a general approach in determining the watershed boundaries.

Topographic Map Data Sources

The fundamental source of data for delineating and studying watersheds is the U.S. Geological Survey Quadrangle map. Each "Quad Sheet" map covers 7.5 minutes of longitude and latitude. These maps give a wealth of information including topographic contour lines, locations of cities, buildings, roads, road types, railroads, pipelines, water bodies, forested land, stream networks, and USGS stream gauging stations and benchmarks. The quad sheets typically have a scale of 1:24,000 (i.e., 1 inch on the map = 24,000 inches on the land). Depending on the age of the map, elevation data may be in U.S. Customary or Metric units. Typically, in the Midwest, the contour intervals of the elevation data are 5 feet or 1.5 meter. In the south, the contour intervals may be 20 ft. For watershed delineation, quad sheets offer an important starting point. However, for detailed investigations, especially for small areas, more detailed site maps having 1 to 5 ft contour intervals are usually required for final analyses. Many of the quad sheets are available on the Internet, although at relatively low resolution and for small areas at a time. Internet aerial photographic sources are also valuable to understand cover and development conditions. Some of these available aerial photographic sources are quite dramatic, with increasing resolution and coverage being constantly added. Detailed site maps are usually produced by the site developer. These may be available to others from the regulatory reviewing agency.

Steps in Determining the Watershed Boundaries

The following is a brief outline of the steps that can be followed to determine the watershed boundaries of a drainage area affecting a specific location.

1. Trace out the main drainage pathways upstream from the point of concern. It is suggested that a medium point marker trace the blue line on the quad sheet upstream from the point of concern, as in the following map:

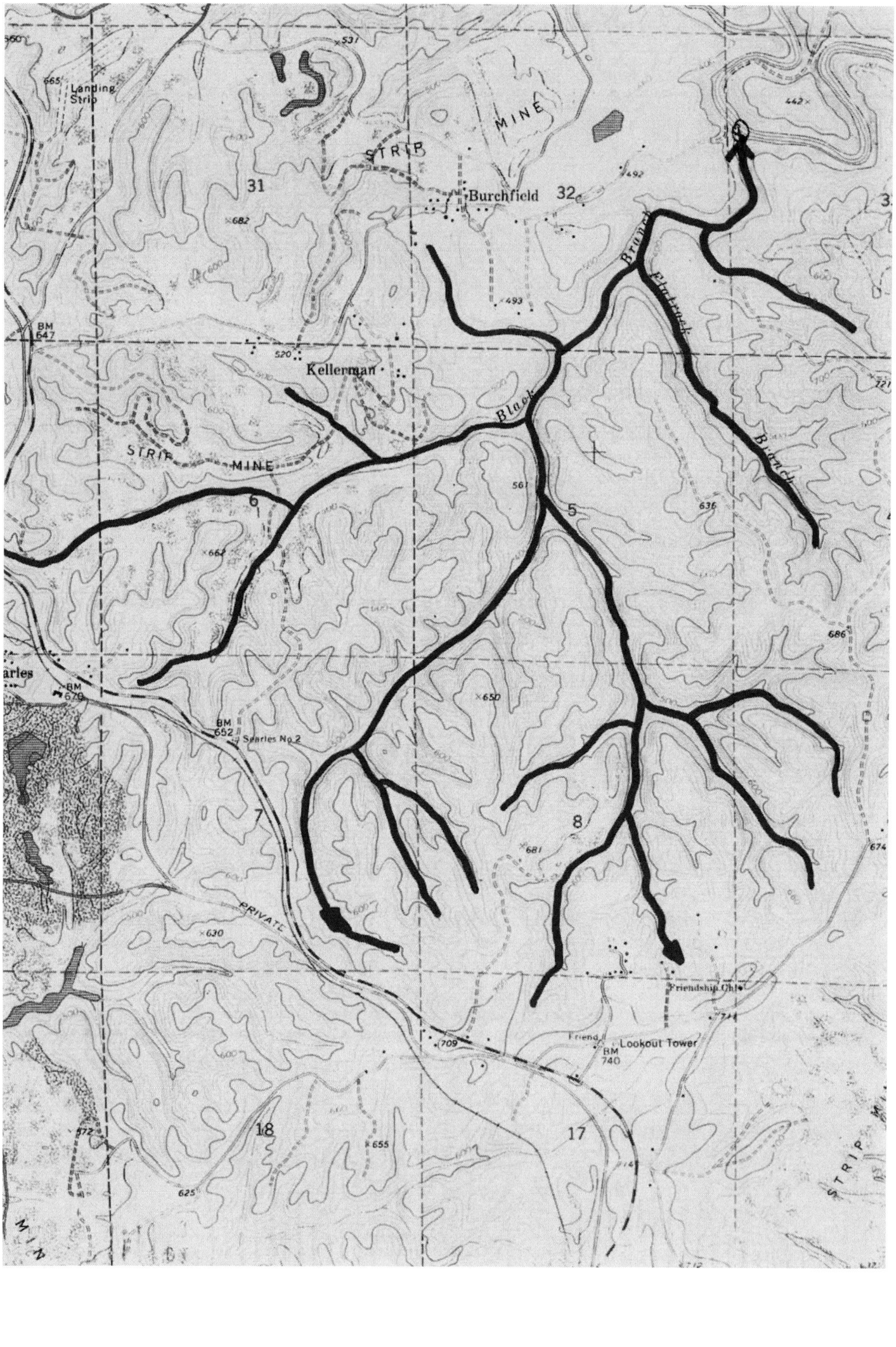

2. Using a different color, trace the drainage pathways marked on the quad sheet draining away from the area, as shown on the following map:

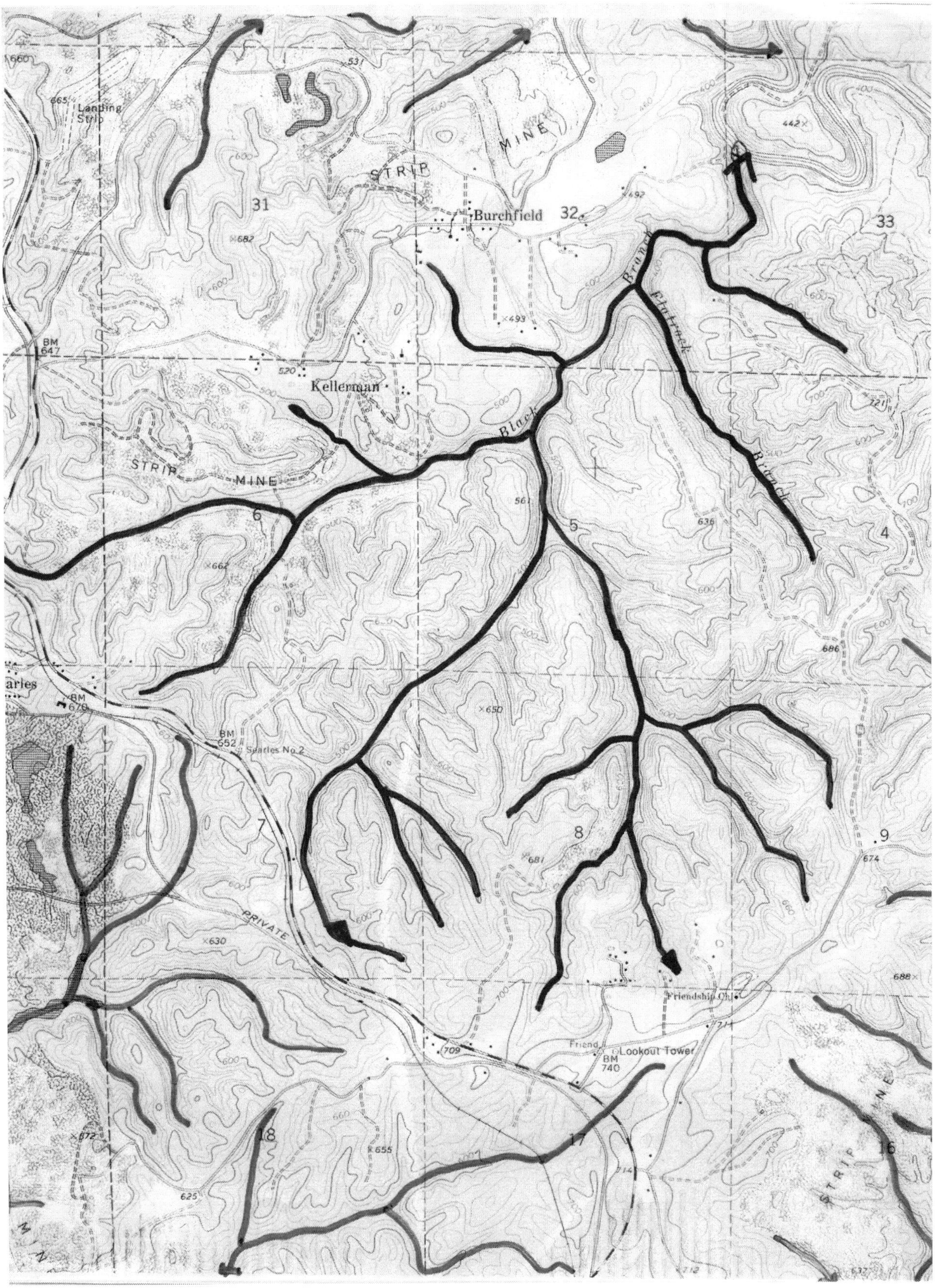

3. Extend the drainage way highlights along obvious drainage pathways, such as gullies/ravines. Also, locate the peaks along the ridges between these drainage systems with a large dot in the center of enclosed contours:

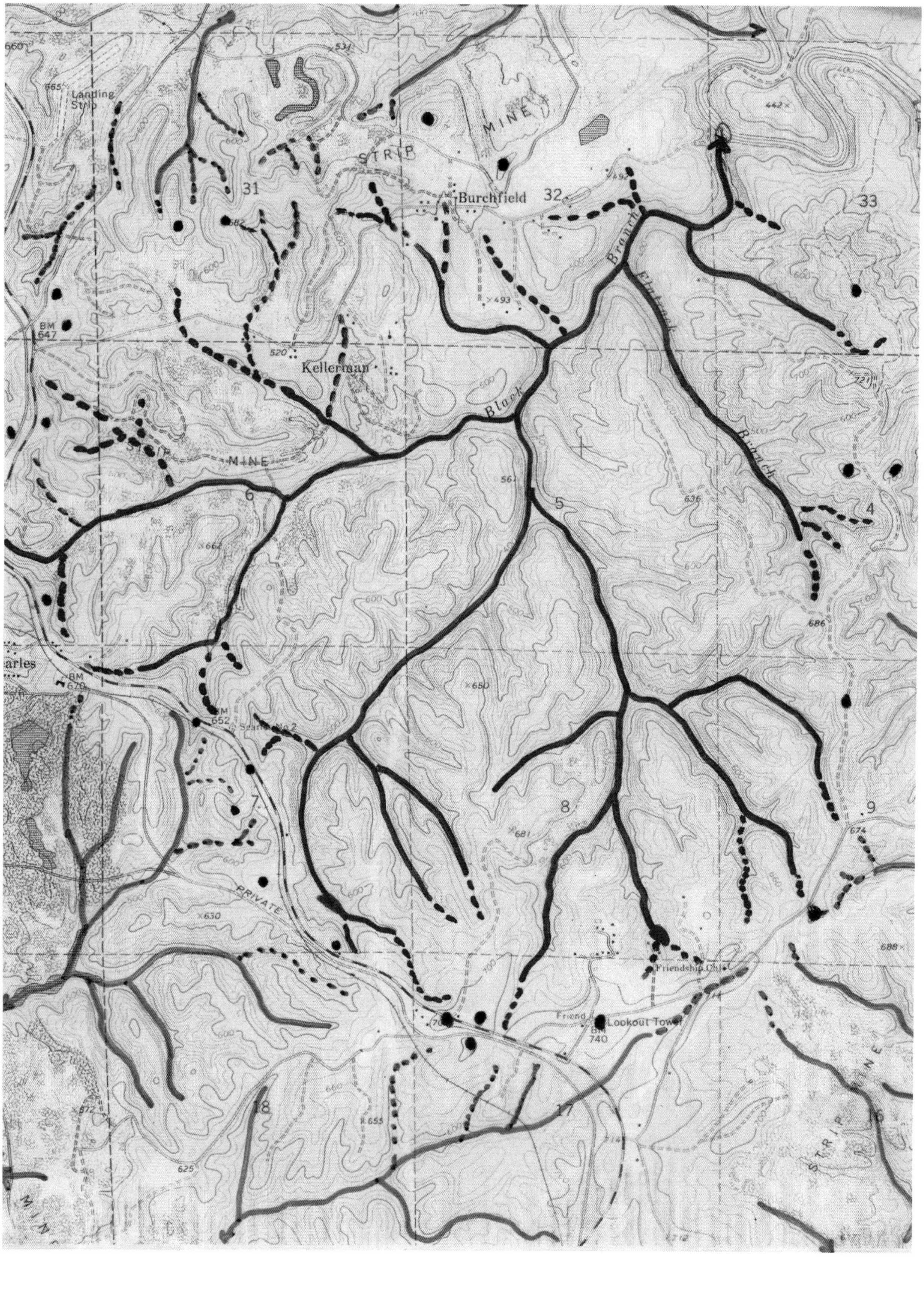

4. Starting at the bottom of the area at the location of interest, connect the peaks between the drainage systems along the ridges to delineate the watershed boundary. Make sure the watershed boundary line only crosses the topographic lines at 90 degree angles.

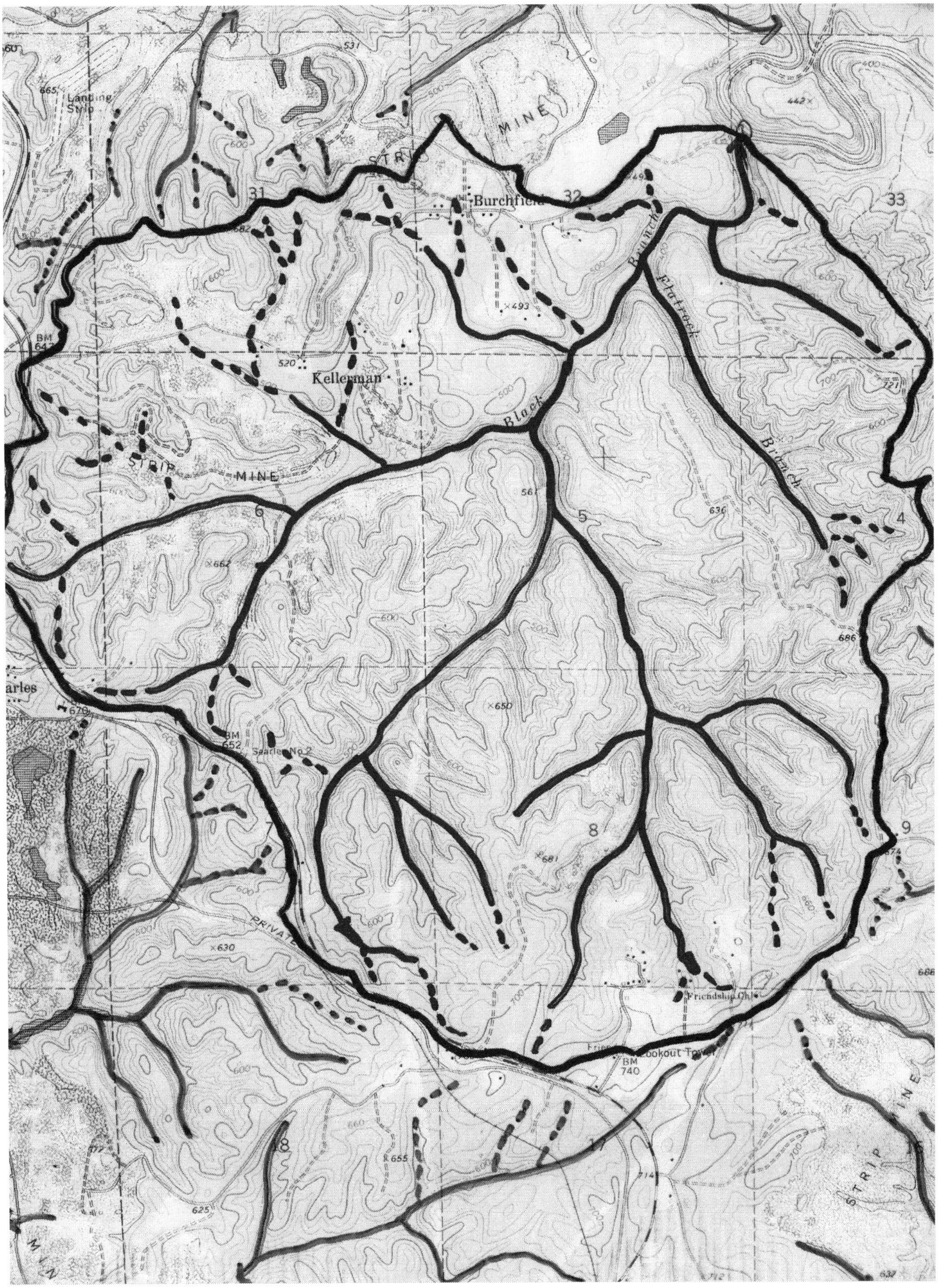

5. Make modifications to the watershed boundary to consider anthropogenic modifications to the landscape. A site survey should identify locations that are different than described on the (usually outdated) quad sheet. In the above example, the site has been extensively strip mined. This example also has roads that are near the ridges that serve as watershed boundaries. Roads are notorious in affecting the local drainage patterns. Roadside ditches commonly collect water from the watershed of interest, but divert it alongside the road and then let it drain into an adjacent watershed. Also, culverts may collect water from parts of an adjacent drainage area and discharge the water into the watershed of interest. Finally, buildings may be constructed on the watershed divide itself (fairly common in small urban drainage areas). Roof drains, graded paved parking lots, and other disturbances can frequently divert small fractions of adjacent watersheds back and forth. In these areas, it is best to carefully examine the expected watershed boundary and account for these modifications, depending on the needed accuracy of the area calculations.

USE OF THE SCS (NRCS) TR-55 METHOD FOR CONSTRUCTION SITE HYDROLOGY EVALUATIONS

General Description of TR-55 for Small Watersheds

The complete User Guide for TR-55 (1986 version) can be downloaded from:

http://www.wcc.nrcs.usda.gov/hydro/hydro-tools-models-tr55.html

According to the NRCS (2002), Technical Release 55 (TR-55) Urban Hydrology for Small Watersheds was first issued in January 1975 as a simplified procedure to calculate the storm runoff volume, peak rate of discharge, hydrographs and storage volumes required for storm water management structures (SCS, 1975). This initial version involved manual methods and assumed the Type II rainfall distribution for all calculations. In June 1986, major revisions were made in TR-55 by adding three additional rainfall distributions (Type I, IA and III) and developing a DOS-based computer program. Time of concentration was estimated by splitting the hydraulic flow path into separate flow phases (SCS, 1986). This 1986 version is the last non-computerized version and has been widely used for drainage design in urban areas.

Even though the manual version of TR-55 is currently being phased out, its use may still be of interest when examining construction sites. In addition, the User Guide for TR-55 (SCS 1986) contains a more through description of the basic processes included in the model. A later discussion presents a description and example of the Windows version of the program.

Only the following site characteristics are needed to use TR-55: drainage area, curve number (CN), and time of concentration (T_c). With this information, it is possible to develop a hydrograph for a specific design storm. In a complex drainage area, the watershed should be subdivided into relatively-homogeneous subwatersheds for routing the flows through the system. The following paragraphs describe the elements of TR-55 that are of most interest for use on construction sites, and present examples for its use.

Selection of the Curve Number

The first part of using TR-55 is to select the curve number. The curve number is simply the single parameter that relates runoff to rainfall. This is illustrated in Figure 3.17. The following equation shows how the CN is used to calculate the runoff depth, Q (in inches), from the precipitation depth, P (in inches), and the curve number, CN (dimensionless):

$$Q = \frac{\left[P - 0.2\left(\frac{1000}{CN} - 10\right)\right]^2}{P + 0.8\left(\frac{1000}{CN} - 10\right)}$$

Tables 3.9 and 3.10 are used to select the most appropriate curve numbers for an area. For construction sites, Table 3.6 shows that newly graded areas have curve numbers ranging from 77 for A type soils to 94 for D type soils. These are relatively high compared to typical pre-development conditions (woods ranging from 30 to 77), reflecting the increase in runoff volume during the period of construction and the associated increased runoff rate.

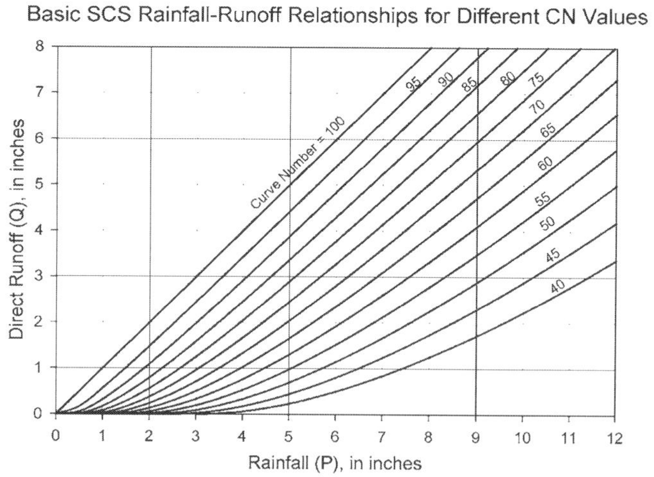

Figure 3.17. Basic SCS rainfall-runoff relationship for different CN values (SCS 1986).

TABLE 3.9. Typical Curve Number Values for Urban Areas (SCS 1986)[1].

Land Use Description/Treatment	Hydrologic Condition	Hydrologic Soil Group A	B	C	D
Residential[2]					
Average lot size:	Average Percent Imperviousness[3]				
1/8 acre or less	65	77	85	90	92
1/4 acre	38	61	75	83	87
1/3 acre	30	57	72	81	86
1/2 acre	25	54	70	80	85
1 acre	20	51	68	79	84
Paved parking lots, roofs, driveways, etc.[4]		98	98	98	98
Streets and roads					
Paved with curbs and storm sewers[3]		98	98	98	98
Gravel		76	85	89	91
Dirt		72	82	87	89
Commercial and business areas (85 percent imperviousness)		89	92	94	95
Industrial districts (72 percent imperviousness)		81	88	91	93
Open spaces, lawns, parks, golf courses, cemeteries, etc.					
Good condition: grass cover on 75 percent or more of the area		39	61	74	80
Fair condition: grass cover on 50 to 75 percent of the area		49	69	79	84
Poor condition: grass cover on less than 50 percent		68	79	86	89
Western Desert Urban Areas					
Natural desert landscaping (pervious areas only)[5]		63	77	85	88
Artificial desert landscaping (impervious weed barrier, desert shrub with 1- to 2-inch sand or gravel mulch and basin borders)		96	96	96	96
Developing Urban Areas					
Newly developing areas (pervious areas only, no vegetation)		77	86	91	94

[1]Average runoff condition, and $Ia = 0.2S$.
[2]Curve numbers are computed assuming the runoff from the house and driveway is directed toward the street with a minimum of roof water directed to lawns where additional infiltration could occur. Impervious areas have a *CN* of 98 and pervious space considered equivalent to open space in good hydrologic condition.
[3]The remaining pervious areas (lawn) are considered to be in good pasture condition for these curve numbers.
[4]In some warmer climates of the country, a curve number of 95 may be used.
[5]Composite curve numbers for natural desert landscaping should be computed using the following figures based on the impervious area percentage and the pervious area *CN*. The pervious area *CN*s are assumed equivalent to desert shrub in poor hydrologic condition.
[6]Composite *CN*s to use for the design of temporary measures during grading and construction should be computed using the following figures based on the degree of development (impervious area percentage) and the *CN*s for the newly graded pervious areas.

TABLE 3.10. Typical Curve Number Values for Non-Urban Areas (SCS 1986)[1].

Cover Description	Hydrologic Condition	Curve Numbers for Hydrologic Soil Group A	B	C	D
Pasture, grassland, or range—continuous forage for grazing[2]	Poor	68	79	86	89
	Fair	49	69	79	84
	Good	39	61	74	80
Meadow—continuous grass, protected from grazing and generally mowed for hay	–	30	58	71	78
Brush—brush-weed-grass mixture with brush the major element[3]	Poor	48	67	77	83
	Fair	35	56	70	77
	Good	30	48	65	73
Woods-grass combination (orchard or tree farm)[5]	Poor	57	73	83	86
	Fair	43	65	76	82
	Good	32	58	72	79
Woods[6]	Poor	45	66	77	83
	Fair	36	60	73	79
	Good	34	55	70	77
Farsteads—buildings, lanes, driveways, and surrounding lots	–	59	74	82	86

[1]Average runoff condition, and $Ia = 0.2S$.
[2]Poor: < 50% ground cover or heavily grazed with no mulch.
Fair: 50 to 75% ground cover and not heavily grazed.
Good: > 75% ground cover and lightly or only occasionally grazed.
[3]Poor: <50% ground cover.
Fair: 50 to 75% ground cover.
Good: >75% ground cover.
[4]Actual curve number is less than 30; use $CN = 30$ for runoff computations.
[5]CNs shown were computed for areas with 50% woods and 50% grass (pasture) cover. Other combinations of conditions may be computed from the *CN*s for woods and pasture.
[6]Poor: Forest litter, small trees, and brush are destroyed by heavy grazing or regular burning.
Fair: Woods are grazed, but not burned, and some forest litter covers the soil.
Good: Woods are protected from grazing, and litter and brush adequately cover the soil.

Soil Characteristics

The hydrologic soil groups (HSG) shown on the curve number tables greatly affect the selected curve number for a specific cover type or landuse type. The following are the descriptions for the four soil categories, as given by the SCS (1986):

> Group A soils have low runoff potential and high infiltration rates, even when thoroughly wetted. They consist chiefly of deep, well to excessively drained sands or gravels and have a high rate of water transmission (greater than 0.30 in/hr).
>
> Group B soils have moderate infiltration rates when thoroughly wetted and consist chiefly of moderately deep to deep, moderately well to well drained soils, with moderately fine to moderately coarser textures. These soils have a moderate rate of water transmission (0.15 to 0.30 in/hr).
>
> Group C soils have low infiltration rates when thoroughly wetted and consist chiefly of soils with a layer that impedes downward movement of water and soils with moderately fine to fine textures. These soils have a low rate of water transmission (0.05 to 0.15 in/hr).
>
> Group D soils have high runoff potential. They have very low infiltration rates when thoroughly wetted and consist chiefly of clay soils with a high swelling potential, soils with a permanent high water table, soils with a claypan or clay layer at or near the surface, and shallow soils over nearly imperious material. These soils have a very low rate of water transmission (0 to 0.05 in/hr).

The transmission/percolation rates noted above are the rates that water moves within the soil and are controlled by the soil profile. These are not the same as the water infiltration rates which are the rates that water enters the soil at the soil surface and are therefore controlled by surface conditions. For undisturbed natural conditions, the soil characteristics are usually obtained from local county soil maps that are available from the county USDA offices for all areas of the U.S. Consider the following example from a local county soil survey. Figure 3.18 is a small section of the soil survey map for the Cripple Creek Church area, adjacent to Cripple Creek and North River, in Tuscaloosa County, AL. The maps are also aerial photographs (usually several decades old) that show the presence of woods, agricultural operations, and land development features, along with waterways. The large numbers (15 and 22) are the county survey/deed record section numbers. For example, these sections are located in R. 10 W. and T. 18 S. The small numbers (21, 23, and 33) refer to the soil types within the

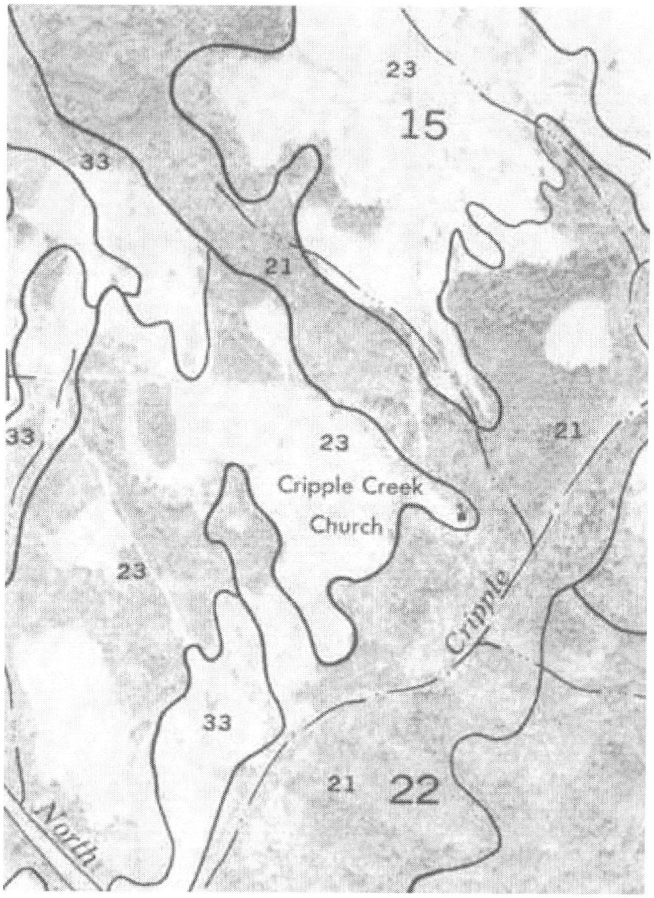

Figure 3.18. Cripple Creek Church, Tuscaloosa County, AL, soil survey.

TABLE 3.11. Soil Survey Characteristics for Area near Cripple Creek Church, Tuscaloosa County, AL.

Soil Number (name) and Depth	Hydrologic Soil Group	Depth to Bedrock (inches)	Permeability (in/hr)	Erosion Factor, *k*	Tolerable Soil Loss, *T* (tons/ac/yr)	Organic Matter (%)
21 (Montevallo)	D	10–20			2	0.5–2
0–7			0.6–2.0	0.37		
7–12			0.6–2.0	0.32		
12–20			–	–		
23 (Nauvoo)	B	40–60			3	0.5–2
0–17			2.0–6.0	0.28		
17–35			0.6–2.0	0.32		
35–41			0.6–2.0	0.32		
41–60			–	–		
33 (Smithdale)	B	>60			5	0.5–2
0–5			2.0–6.0	0.28		
5–42			0.6–2.0	0.24		
42–72			2.0–6.0	0.28		

TABLE 3.12. Particle-Size Distribution for Smithdale Soil (percent in size category, less than 2 mm).

Sample Number	Depth (inches)	Horizon	Clay (<0.002 mm)	Silt (0.002–0.05 mm)	Sand (0.05–2.0 mm)	Cation Exchange Capacity (meq/100 mL)
S77AL-125-11-1	0–5	Ap	2.8	29.2	68.0	3.65
S77AL-125-11-2	5–20	B21t	22.2	34.9	42.9	9.02
S77AL-125-11-3	20–42	B22t	20.2	29.1	50.7	5.36
S77AL-125-11-4	42–52	B23t	12.3	26.5	61.2	4.06
S77AL-125-11-5	52–72	B2t	21.2	12.8	66.0	3.52

dark outlines. These are the soils of interest for this area. About two soil samples per square mile were obtained and analyzed by USDA soil scientists in the preparation of these maps, so they are not absolutely accurate for small areas. They were able to extend the likely areas associated with each soil type based on surface features and aerial photographs. As an example, soil 21 (Montevallo) is generally in the bottom lands along the creeks. Table 3.11 lists some of the characteristics of these soils pertaining to erosion and runoff considerations, while Table 3.12 shows detailed particle-size information for samples obtained at different depths for Smithdale soil (the only one of these 3 with this information complete in the soil survey) and Table 3.13 lists some potential problems that may be encountered if the site is to be used for building development.

The information summarized on these tables is only a small fraction of the tremendous amount of information in the soil surveys. Unfortunately, not all of this information can be used for developed areas, or for areas undergoing development. Soils are dramatically altered during construction projects. These changes range from stripping off the topsoil and compacting the remaining soil, to removing large amounts of native soils in cut operations, to bringing in large amounts of new material if fill is needed. The surface soils exposed to potential erosion and which affects the amount of runoff at the site can therefore vary for different construction phases.

Therefore, it is important to determine the native soils on the proposed construction site (an overlay of soil types is usually required for most erosion control plans). Widely varying soil characteristics on the site should be especially noted. Descriptions of how the soils (and topography) will be affected and changed are also needed, as is the description of the fill soil, if a fill soil will be used and if the description is known. The excavations and fills during different construction phases should be described by the depth of material to be removed, or brought in, and the resulting surface soils. The SCS (1986) notes that due to urbanization, the soil profile may be considerably altered and the soil survey data may not be applicable for final surface soil conditions. They recommend that the hydrologic soil group be estimated based on the soil texture. They provide the following list to estimate the soil groups, based on texture, provided that significant compaction has not occurred:

HSG	Soil Textures
A	Sand, loamy sand, or sandy loam
B	Silt, silt loam or loam
C	Sandy clay loam
D	Clay loam, silty clay loam, sandy clay, silty clay, or clay

Figure 3.19 shows the standard USDA soil triangle with the hydrologic soil groups marked, based on the above categories. Soil compaction can have severe effects on the runoff potential of soils and needs to be considered. As reported by Pitt, *et al.* (1999), unpublished double-ring infiltration tests conducted by the Wisconsin Department of Natural Resources (DNR) in Oconomowoc, Wisconsin, indicated highly variable infiltration rates for soils that were generally sandy (Natural Resources Conservation Service (NRCS) A/B hydrologic group soils) and dry. The median initial rate was about 75 mm/hr (3 in/hr), but ranged from 0 to 640 mm/hr (0 to 25 in/hr). The final rates also had a median value of about 75 mm/hr (3 in/hr) after at least 2 hr of testing, but ranged from 0 to 380 mm/hr (0 to 15 in/hr). Many infiltration rates actually increased with time during these tests. In about 1/3 of the cases, the infiltration rates remained very close to zero, even for these sandy soils. Areas that experienced substantial disturbances or traffic (such as school playing fields), and siltation (such as in some grass swales) had the lowest infiltration rates.

This data indicated that a potential problem existed in terms of estimating the infiltration rate for typical urban soils. Therefore, the research team performed more than 150 infiltration tests (as a full factorial experimental design that

TABLE 3.13. Building Site Development Limitations.

Soil	Shallow Excavations	Local Streets and Roads	Dwellings with Basements	Lawns and Landscaping
21 (Montevallo)	Severe (depth to rock, slope)	Severe (slope)	Severe (depth to rock, slope)	Severe (droughty, slope, thin soil layer)
23 (Nauvoo)	Slight	Moderate (low strength)	Slight	Slight
33 (Smithdale)	Moderate (slope)	Moderate (slope)	Moderate (slope)	Moderate (slope)

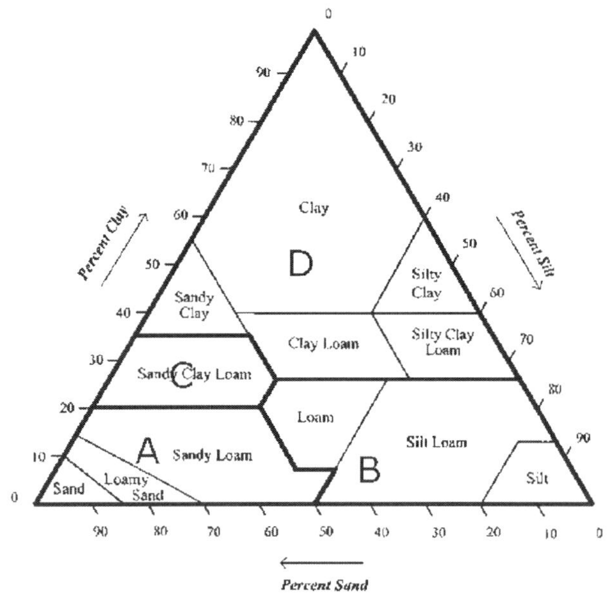

Figure 3.19. USDA standard soil triangle, with hydrologic soil groups for disturbed soils.

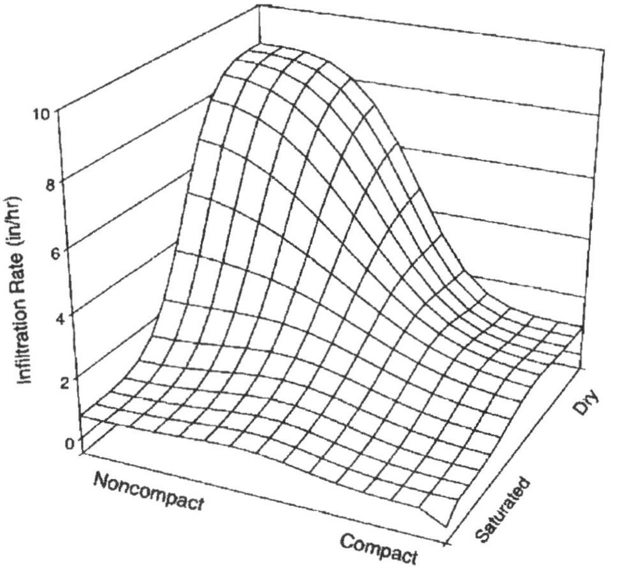

Figure 3.21. Three-dimension plots of infiltration rates for clayey soil (Pitt, *et al.* 1999).

allowed the researchers to investigate the effects of soil type, compaction, moisture content and age since development) on disturbed urban soils. Compaction had dramatic effects on infiltration rates through sandy soils, while compaction and moisture affected the infiltration rates in clayey soils. Moisture was not a factor controlling infiltration rates in the sandy soils. Figures 3.20 and 3.21 show the impacts of both compaction and moisture on the infiltration rates of sandy and clayey soils, respectively.

Table 3.14 shows the results of controlled laboratory tests measuring the water transmission rates for different soil mixtures with varying levels of compaction. Also shown are

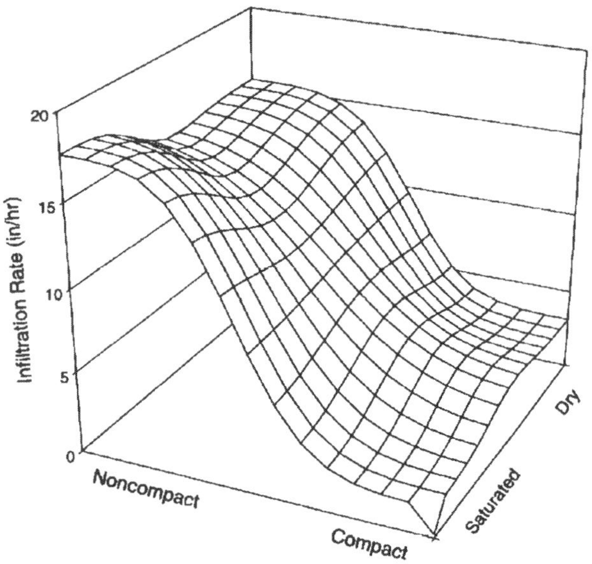

Figure 3.20. Three-dimension plots of infiltration rates for sandy soil (Pitt, *et al.* 1999).

the effects of duration for some of the test conditions. In all cases, except for the clay loam, the uncompacted soils behaved as predicted and as shown on the USDA soil triangle, Figure 3.19. Clay loam had a unexpectedly high water transmission rate for the uncompacted soil. In all cases, except for 100% sand, compaction resulted in significantly reduced water transmission rates, resulting in a different HSG than if uncompacted. All severely compacted soils, except for 100% sands, are in the D category. Sands remain in the A category for all compaction conditions. During the tests, the transmission rates for sands dropped significantly, but still remained in the HSG A category.

Time of Concentration (T_c or t_c) Calculations

The time of concentration needs to be determined for each subwatershed in the study area. It is usually necessary to investigate several candidate flow paths in order to be relatively certain of the one that takes the longest time to reach the end of the subwatershed area. There are many different time-of-concentration formulas typically presented in hydrology textbooks, usually for different conditions and locations. The SCS/NRCS method has become relatively common recently. It is necessary to use this method when using TR-55 (and TR-20). This method separates the flow path into three segments: sheetflow, shallow concentrated flow, and channel flow. The time of concentration is equal to the sum of travel times in each of these flow segments for the critical flow path. In some cases, especially for small sites, only sheetflow and possibly shallow concentrated flow may be evident. Sheetflow is usually limited to less than 150 ft.

The candidate flow paths are drawn on a site topographic map, originate on the subwatershed boundary, and proceed all the way to the bottom of the subwatershed. (*Note:* In rare

TABLE 3.14. Laboratory Water Transmission Tests for Various Soil Textures and Densities (densities and observed infiltration rates for different durations) (Pitt, *et al.* 2002).

	Hand Compaction	Standard Compaction	Modified Compaction
Sand (100% sand)	Density: 1.36 g/cc (ideal for roots) 0 to 1.6 hrs: A	Density: 1.71 g/cc (may affect roots) 0 to 2.7 hrs: A	Density: 1.70 g/cc (may affect roots) 0 to 2.7 hrs: A
Silt (100% silt)	Density: 1.36 g/cc (close to ideal for roots) 0 to 35 hrs: B	Density: 1.52 g/cc (may affect roots) 0 to 48 hrs: D	Density: 1.75 g/cc (will likely restrict roots) 0 to 48 hrs: D
Clay (100% clay)	Density: 1.45 g/cc (may affect roots) 0 to 48 hrs: D	Density: 1.62 g/cc (will likely restrict roots) 0 to 100 hrs: D	Density: 1.88 g/cc (will likely restrict roots) 0 to 100 hrs: D
Sandy Loam (70% sand, 20% silt, 10% clay)	Density: 1.44 g/cc (close to ideal for roots) 0 to 7.5 hrs: A	Density: 1.88 g/cc (will likely restrict roots) 0 to 3.82 hrs: A 3.82 to 24.32 hrs: B	Density: 2.04 g/cc (will likely restrict roots) 0 to 175 hrs: D
Silty Loam (70% silt, 20% sand, 10% clay)	Density: 1.40 g/cc (may affect roots) 0 to 7.22 hrs: B 7.22 to 47 hrs: C	Density: 1.64 g/cc (will likely restrict roots) 0 to 144 hrs: D	Density: 1.98 g/cc (will likely restrict roots) 0 to 144 hrs: D
Clay Loam (40% silt, 30% sand, 30% clay)	Density: 1.48 g/cc (may affect roots) 0 to 6.1 hrs: A	Density: 1.66 g/cc (will likely restrict roots) 0 to 93 hrs: D	Density: 1.95 g/cc (will likely restrict roots) 0 to 93 hrs: D

circumstances, it is possible for the T_c flow path to originate at an internal elevated location and not along the subwatershed boundary. This should be investigated for all sites to confirm that the T_c pathway does not have an internal "starting point"). Sheetflow is usually the first element considered and normally is assumed to last for a maximum of 300 ft (300 ft for very smooth surfaces; usually 50–150 ft for surfaces with natural ground cover; WinTR-55 currently limits the sheetflow length to a maximum of 100 ft). The travel time for sheetflow is calculated using a kinematic solution to Manning's equation. Sheetflow ends when it is assumed that the depth of flow exceeds 0.1 ft (SCS 1986). The flow path is then assumed to occur as shallow concentrated flow, until a designated channel on the topographic map is reached (usually taken as a designated creek or stream on a USGS quadrangle map). When several candidate flow paths are evaluated, the one with the longest travel time is assumed to represent the time of concentration for the subwatershed. If a rain lasts for at least that time period, the runoff at the outlet will contain water from the complete area, resulting in maximum runoff rates.

The following discussions show how the travel times are calcualted for each flow path element.

Sheetflow

The following equation (a kinematic solution to the Manning's equation) is used in the SCS procedures to calculate the travel time along the sheetflow path segment:

$$T_t = \frac{0.007(nL)^{0.8}}{(P_2)^{0.5}\, S^{0.4}}$$

where,

T_t = travel time (hr)
n = Manning roughness coefficient (for sheet flow)
L = flow length (ft) (maximum of 300 ft; WinTR55 only allows a maximum length of 150 ft)
P_2 = 2-year, 24-hour rainfall depth (in), and
s = slope of hydraulic grade line (land slope, ft/ft)

The sheetflow Manning's n roughness coefficient values are different from the channel lining roughness coefficients. Table 3.15 lists these sheetflow values. These are all greater than the channel lining n values for the rougher surfaces, due to the shallow nature of the flows, which results in friction affecting more of the flow depth. As an example, a common channel-lining n value for grass is 0.024, while the sheetflow n value for grass is 0.24, or 10 times higher. The grass has a much greater effect on flow when the flow is shallow than when the flow is deep. However, the smooth surface sheetflow n values (0.011) are very similar to the values that would be used for these surfaces in channels. This is because these smooth surfaces have a minimal effect on both shallow and deeper flows due to their relatively low effective roughness heights. An important factor for construction sites is the roughness coefficient of 0.011 for bare soils, compared to cultivated soils (with mulch covers of >20%) of 0.17, and dense grasses of 0.24. Natural woods can have n coefficients of 0.4 to 0.8, depending on the height of the underbrush. Figure 3.22 includes graphs that can be used to estimate the travel time for different sheetflow conditions, calculated using the above SCS sheetflow formula, using a P_2 value of 4.2 inches (appropriate for Birmingham, AL). If the P_2 ratio is not 4.2 inches, the Figure 3.22 values can be adjusted using the above sheetflow equation and the following factors:

Actual P_2 Value (inches)	Multiplier for Sheetflow Travel Times (if P_2 is not 4.2 inches)
1.0	2.0
1.5	1.7
2.0	1.4
2.5	1.3
3.0	1.2
3.5	1.1
4.0	1.0
4.5	1.0
5.0	0.9
5.5	0.9
6.0	0.8

Shallow Concentrated Flow

After a maximum of 300 ft, sheetflow usually becomes shallow concentrated flow which is characterized by much narrower flow paths and faster flows. The flow depth also is greater than 0.1 ft, and therefore friction effects of the surface cover are not as dramatic. The following equations are used to calculate the velocities of this flow segment, based on the nature of the surface (paved or unpaved). Figure 3.13 contains graphical solutions for these equations.

$$V = 16.1\sqrt{s} \quad \text{(Unpaved)}$$

$$V = 20.3\sqrt{s} \quad \text{(Paved)}$$

where,

V = average velocity (ft/s), and
s = slope of hydraulic grade line (watercourse slope, ft/ft)

TABLE 3.15. Sheetflow Manning's Equation Roughness Coefficients (SCS, 1986).

Surface Description	Sheetflow Roughness Factor, n
Smooth surfaces (concrete, asphalt, gravel, or bare soil)	0.011
Fallow (no residue)	0.05
Cultivated soils:	
Residue cover ≤20%	0.06
Residue cover >20%	0.17
Grass:	
Short grass prairie	0.15
Dense grass	0.24
Bermudagrass	0.41
Range (natural)	0.13
Woods[2]	
Light underbrush	0.40
Dense underbrush	0.80

[1]Includes species such as weeping lovegrass, bluegrass, buffalo grass, blue gama grass, and native grass mixtures

[2]When selecting n for woods, consider cover to a height of about 0.1 ft. This is the only part of the plant cover that will obstruct sheet flow.

These two equations are based on a solution of the Manning equation with different assumptions for n (Manning roughness coefficient) and R (hydraulic radius, ft). For unpaved areas, n is 0.05 and R is 0.4 ft; for paved areas, n is 0.025 and R is 0.2 ft. The travel time associated with the shallow-concentrated flow segment is calculated using this velocity and the flow-path length.

The following empirical formula is given by CA DOT (http://www.dot.ca.gov/hq/oppd/hdm/pdf/chp0810.pdf) in

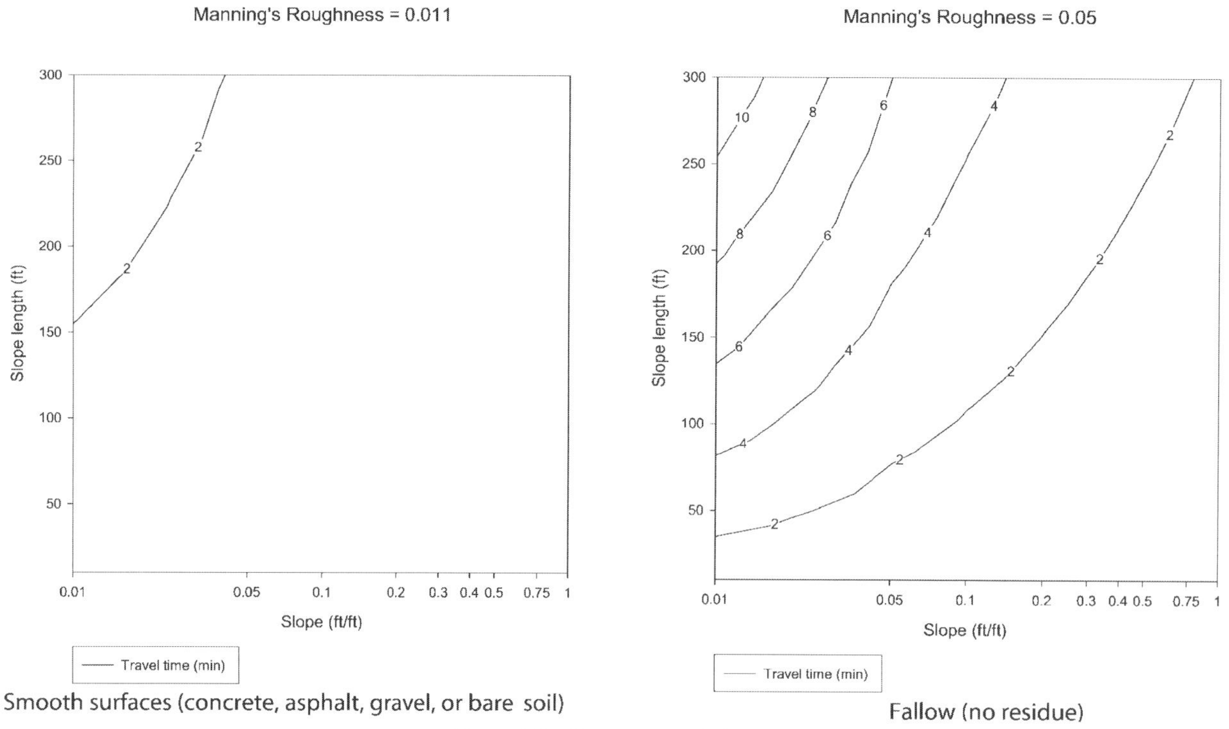

Figure 3.22. Sheetflow travel times.

Manning's Roughness = 0.06

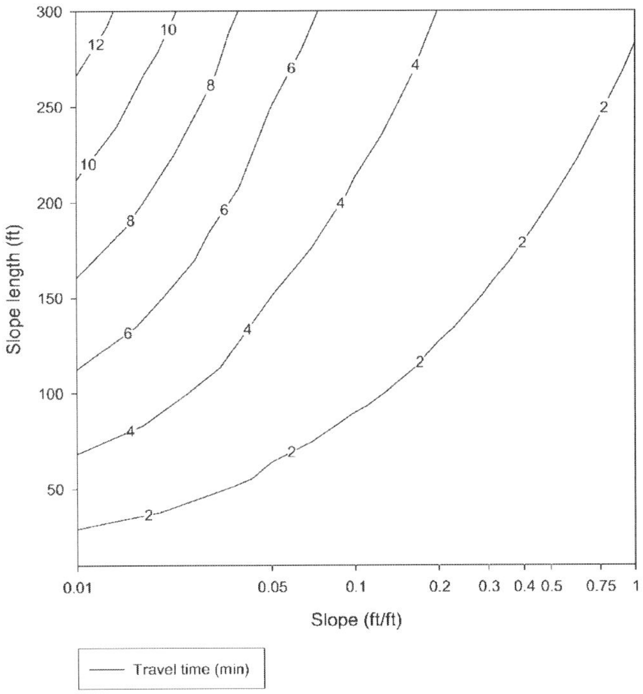

Cultivated soils: residue cover ≤ 20%

Manning's Roughness = 0.17

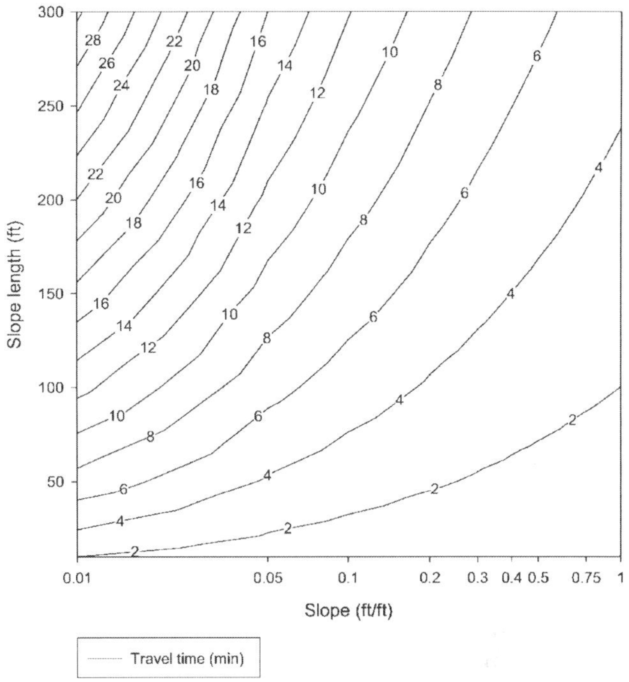

Cultivated soils: residue cover > 20%

Manning's Roughness = 0.15

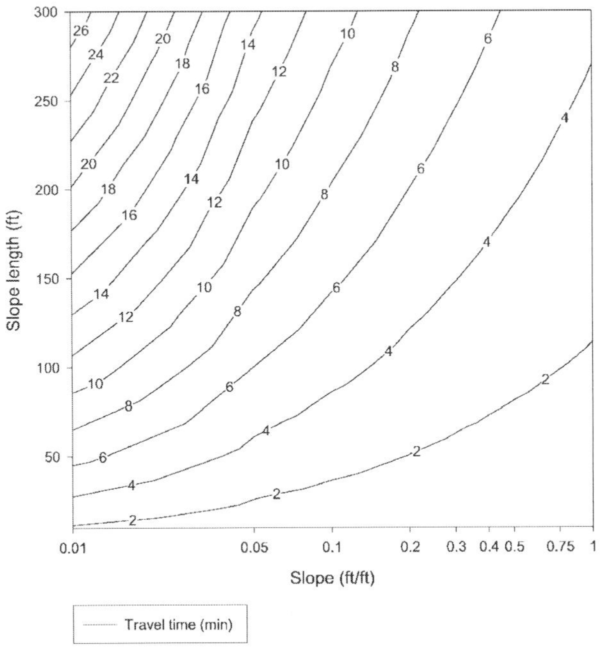

Grass: short grass prairie

Manning's Roughness = 0.24

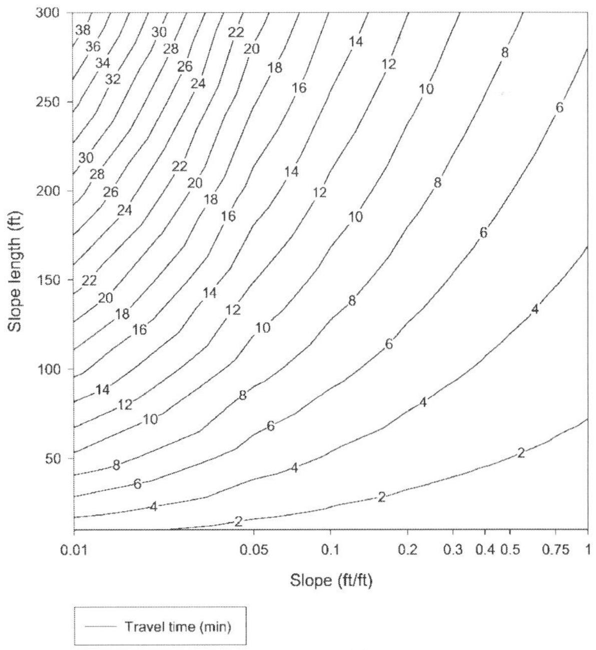

Grass, dense (weeping lovegrass, bluegrass, buffalo grass, blue gama grass, and native grass mixtures

Figure 3.22 (continued). Sheetflow travel times.

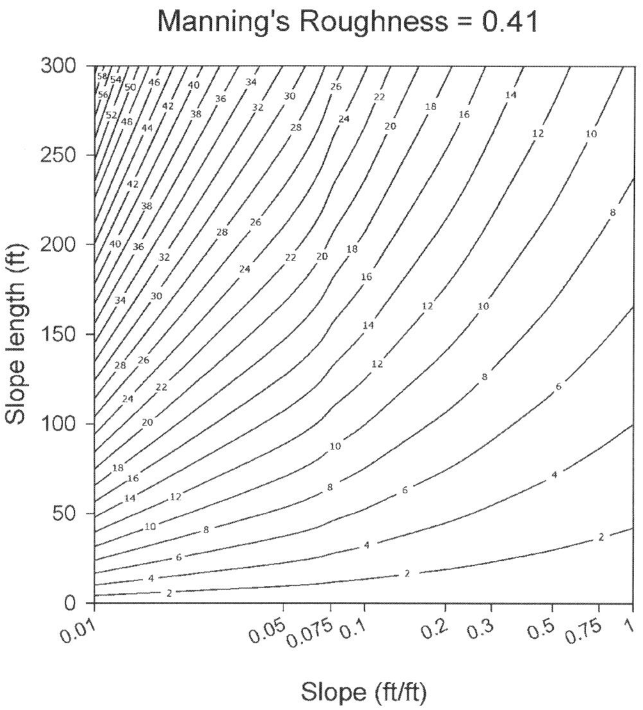

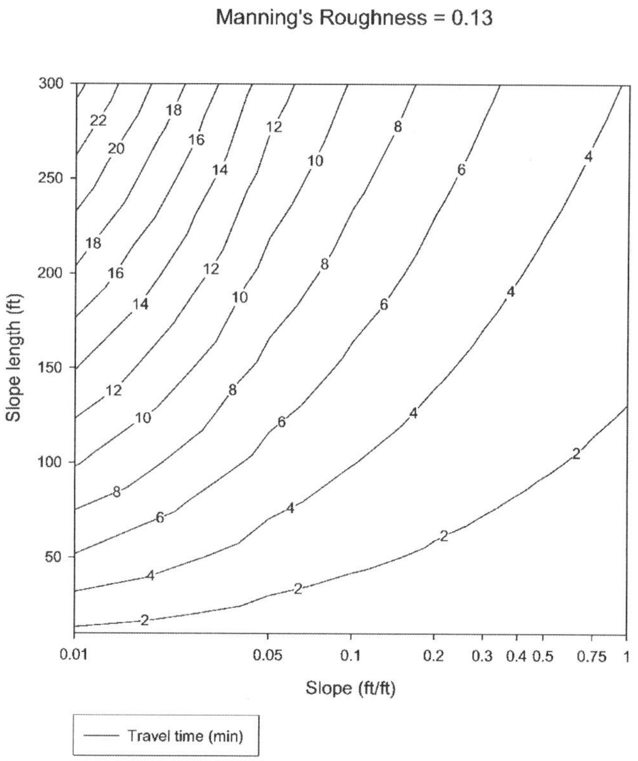

Range (natural)

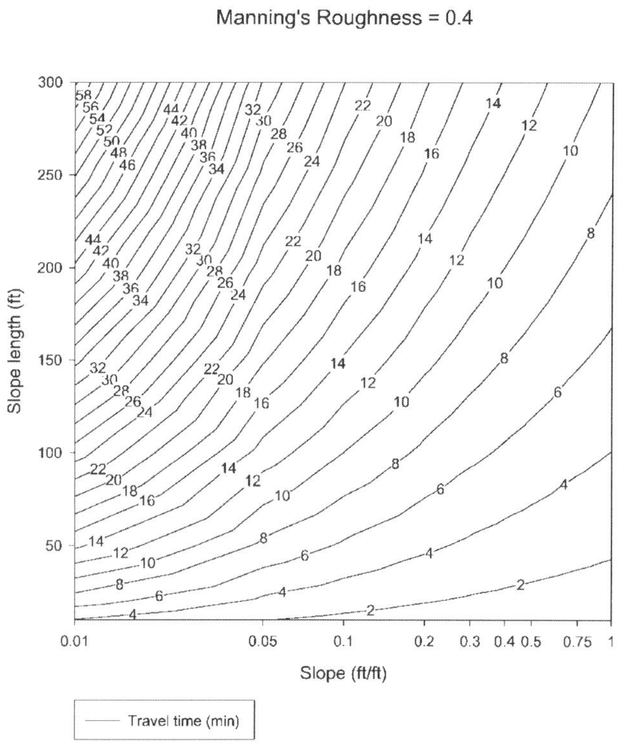

Woods, light underbrush (considering cover to height of about 0.1 ft)

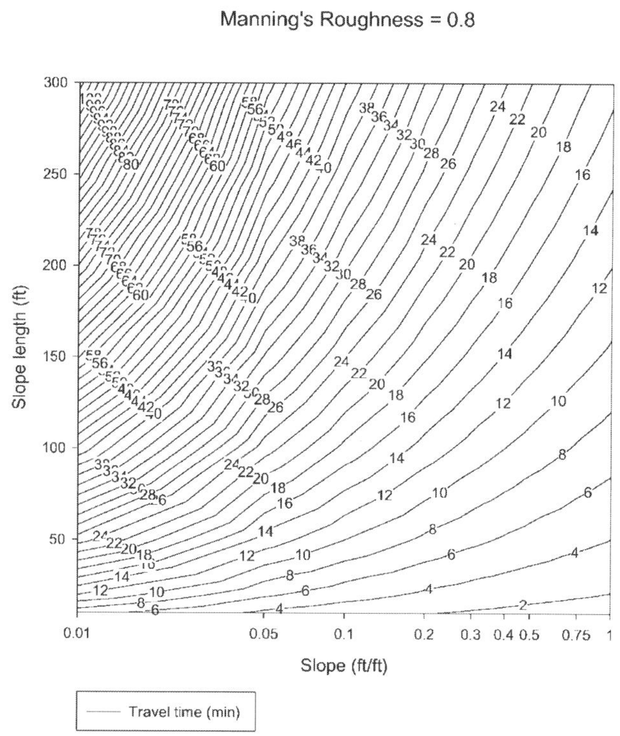

Woods: dense underbrush (considering cover to height of about 0.1 ft)

Figure 3.22 (cotinued). Sheetflow travel times.

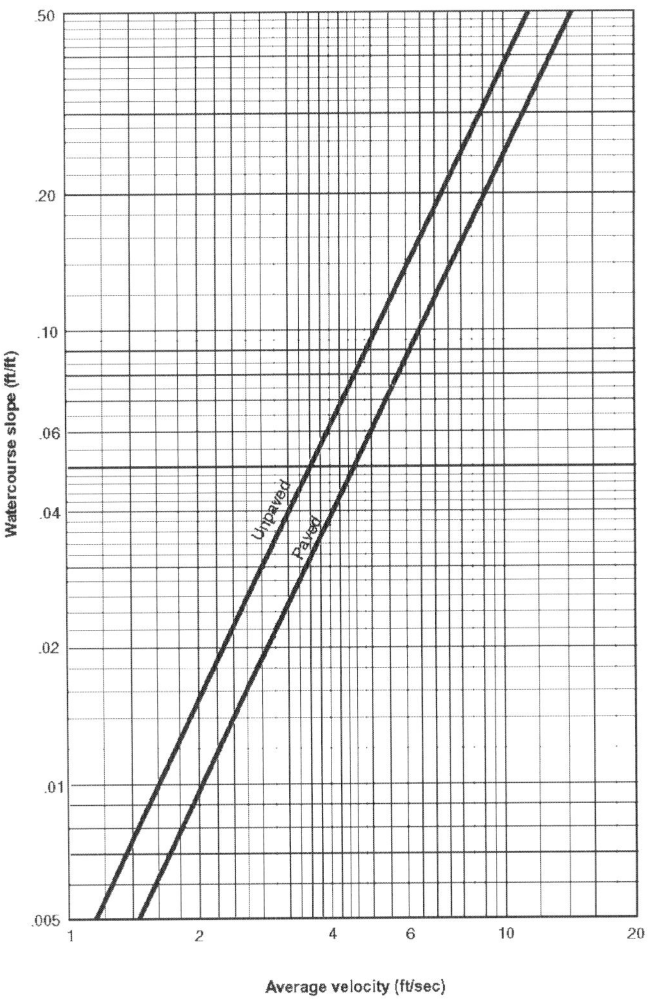

Figure 3.23. Shallow concentrated flow velocities (SCS 1986).

their Hydrology Design Manual, Chapter 810, as an alternative to estimate the flow velocity (in m/sec):

$$V = kS^{1/2}$$

Where S is the slope in percent and k (m/s) is an intercept coefficient depending on land surface cover as shown below:

Forest with heavy ground litter; hay meadow (overland flow): 0.076

Trash fallow or minimum tillage cultivation; contour or strip cropped; woodland (overland flow): 0.152

Short grass pasture (overland flow): 0.213

Cultivated straight row (overland flow): 0.274

Nearly bare and untilled (overland flow); alluvial fans: 0.305

Grassed waterway (shallow concentrated flow): 0.457

Unpaved (shallow concentrated flow): 0.491

Paved area (shallow concentrated flow); small upland gullies: 0.619

Channel Flow

If the flow path includes a designated channel shown on a USGS quadrangle map, the Manning's equation is used to calculate the velocity in the channel reach. The travel time in the reach is then calculated using this channel-full velocity and the length of the channel.

$$V = \frac{1.49R^{2/3}\sqrt{s}}{n}$$

where,

V = average velocity (ft/s), and
R = hydraulic radius (ft) and is equal to a/p_w
a = cross sectional flow area (ft^2)
p_w = wetted perimeter (ft)
s = slope of hydraulic grade line (channel slope, ft/ft)
n = Manning roughness coefficient (for open channel flow)

This is the conventional Manning's equation, and appropriate channel lining n coefficients are used. The depth of water in the channel is assumed to be equal to the depth at bankfull conditions, assumed by TR-55 to be the 2-year storm (to be consistent with the sheetflow calculations) (SCS, 1986).

The hydraulic radius (R) in the equation is the ratio of the cross-sectional flow area to the wetted perimeter length (the wet edge of the channel). For a fully-flowing circular pipe, this is equal to the diameter divided by 4, while for sheetflows (where the depth is less than about 10 times the flowwidth) the hydraulic radius is close to the depth of flow.

The Manning's roughness coefficients, n, for channel conditions where deep flow is typical, are substantially different than for the previously presented values for sheetflow. Table 3.16 is a set of typical Manning's n values for different channel (and conduit) conditions (Chow, 1959):

Table 3.16 presents reasonable values for simple channels that are likely to be constructed at construction sites, including downslope pipe diversions. The USGS (Arcement and Schneider, 1984) presents the following summary for determining Manning n values for the natural channels that may also be present on construction sites:

The most important factors that affect the selection of channel n values are the type and size of the materials that compose the bed and banks of the channel, and shape of the channel. Cowan (1956) developed a procedure for estimating the effects of these factors to determine the value of n for a channel. The value of n may be computed by

$$n = (n_b + n_1 + n_2 + n_3 + n_4)m$$

where,

n_b = a base value of n for a straight, uniform, smooth channel in natural materials
n_1 = a correction factor for the effect of surface irregularities

n_2 = a value for variations in shape and size of the channel cross section
n_3 = a value for obstructions
n_4 = a value for vegetation and flow conditions
m = a correction factor for meandering of the channel

The following discussion on the basic n values and modifications for channels is summarized from Arcement and Schneider (1984).

Base n *Values* (n_b) *for Channels*

In the selection of a base n value for channel subsections, the channel must be classified as either a stable channel or as a sand channel. A stable channel is defined as a channel in which the bed is composed of firm soil, gravel, cobbles, boulders, or bedrock and the channel remains relatively unchanged throughout most of the range in flow. The following table is modified from Aldridge and Garrett, 1973) and lists base n_b values for stable channels and sand channels. The base values of Benson and Dalrymple (1967) in Table 3.17 apply to conditions that are close to average, while Chow's (1959) base values are for the smoothest reach attainable for a given bed material.

TABLE 3.16. Manning's *n* Values for Different Channel Conditions (Chow, 1959).

Type of Channel and Description of Closed Conduits	Minimum
Concrete Pipe:	
Culverts with bends, connections & debris	0.013
Storm sewer	0.013
Subdrain with open joints	0.016
PVC Pipe	0.011
Concrete Surfaces (bottom & sides):	
Smooth finish	0.015
Unfinished	0.017
Concrete Bottom (with sides made of):	
Mortared stone	0.020
Dry rubble or riprap	0.030
Gravel Bottom (with sides made of):	
Formed concrete	0.020
Dry rubble or riprap	0.040
Excavated or Dredged Channels and Ditches:	
Earthen, straight & uniform, no brush or debris:	
Grassed, less than 6″ high with:	
Depth of flow 2.0 ft	0.035
Depth of flow 2.0 ft	0.030
Grassed, approximately 12″ high with:	
Depth of flow 2.0 ft	0.060
Depth of flow 2.0 ft:	0.035
Grassed, approximately 24ö high with:	
Depth of flow 2.0 ft	0.070
Depth of flow 2.0 ft	0.035
Earth bottom with riprap on sides	0.040
Rock or shale cuts:	
Smooth and uniform	0.035
Jagged and irregular	0.040
Curb and Gutter (Concrete)	0.016

TABLE 3.17. Base *n* values for channels.

Bed Material	Median Size of Bed Material (mm)	Base *n* value: Straight Uniform Channel[1]	Base *n* value: Smooth Channel[2]
Sand Channels	2.00 to 2.50	2.51 to 3.00	Total
Sand[3]	0.2	0.012	–
	0.3	0.017	–
	0.4	0.020	–
	0.5	0.022	–
	0.6	0.023	–
	0.8	0.025	–
	1.0	0.026	–
Stable Channels and Flood Plains			
Concrete	–	0.012.018	0.011
Rock Cut	–	–	0.025
Firm Soil	–	0.025–0.032	0.020
Coarse Sand	1–2	0.026–0.035	–
Fine Gravel	–	–	0.024
Gravel	2–64	0.028–0.035	–
Coarse Gravel	–	–	0.026
Cobble	64–256	0.030–0.050	–
Boulder	>256	0.040–0.070	–

Modified from Aldridge and Garret, 1973.
–No data
[1]Benson and Dalrymple .
[2]For indicated material (Chow 1959).
[3]Only for upper regime flow where grain roughness is predominant.

Barnes (1967) cataloged verified n values for stable channels having roughness coefficients ranging from 0.024 to 0.075. In addition to a description of the cross section, bed material, and flow conditions during the measurement, color photographs of the channels were provided.

A sand channel is defined as a channel in which the bed has an unlimited supply of sand. By definition, sand ranges in grain size from 0.062 mm (62 μm) to 2 mm. Resistance to flow varies greatly in sand channels because the bed material moves easily and takes on different configurations or bed forms. Bed form is a function of velocity of flow, grain size, bed shear, and temperature.

The flows that produce the bed forms are classified as lower regime flow and upper regime flow, according to the relation between depth and discharge. The lower regime flow occurs during low discharges, and the upper regime flow occurs during high discharges. An unstable discontinuity, called a transitional zone, appears between the two regimes in the depth to discharge relationship. In lower regime flow, the bed may have a plane surface and no movement of sediment, or the bed may be deformed and have small uniform waves or large irregular saw-toothed waves formed by sediment moving downstream. The smaller waves are known as ripples, and the larger waves are known as dunes. In upper regime flow, the bed may have a plane surface and sediment movement or long, smooth sand waves that are in phase with the surface waves.

TABLE 3.18. Adjustment Values for Factors that Affect the Roughness of a Channel [modified from Aldridge and Garrett, 1973].

Channel Conditions	*n* Value Adjustment	Example
Smooth	0.000	Compares to the smoothest channel attainable in a given bed material.
Minor	0.001–0.005	Compares to slightly degraded channels in good condition but having slightly eroded or scoured side slopes.
Moderate	0.006–0.010	Compares to dredged channels having moderate to considerable bed roughness and moderately sloughed or eroded side slopes.
Severe	0.011–0.020	Badly sloughed or scalloped banks of natural streams; badly eroded or sloughed sides of canals or drainage channels; unshaped, jagged, and irregular surfaces of channels in rocks.

Irregularity (n_1)

Where the ratio of width to depth is small, roughness caused by eroded and scalloped banks, projecting points, and exposed tree roots along the banks must be accounted for by fairly large adjustments, Table 3.18, Chow (1959) and Benson and Dalrymple (1967) showed that severely eroded and scalloped banks can increase *n* values by as much as 0.02. Larger adjustments may be required for very large, irregular banks that have projecting points.

Variation in Channel Cross Section (n_2)

The value of n is not affected significantly by relatively large changes in the shape and size of cross sections if the changes are gradual and uniform. Greater roughness is associated with alternating large and small cross sections and sharp bends, constrictions, and side-to-side shifting of the low-water channel. The degree of the effect of changes in the size of the channel depends primarily on the number of alternations of large and small sections and secondarily on the magnitude of the changes. The effects of abrupt changes may extend downstream for several hundred meters. The n value for a reach below a disturbance may require adjustment, even though none of the roughness-producing factors are apparent in the study reach, Table 3.19. A maximum increase in n of 0.003 will result from the usual amount of channel curvature found in designed channels and in the reaches of natural channels used to compute discharge (Benson and Dalrymple 1967).

Obstructions (n_3)

Obstructions, such as logs, stumps, boulders, debris, pilings, and bridge piers, disturb the flow pattern in the channel and increase roughness. The amount of increase depends on the following: the shape of the obstruction; the size of the obstruction in relation to that of the cross section; and the number, arrangement, and spacing of obstructions. The effect of obstructions on the roughness coefficient is a function of the flow velocity. When the flow velocity is high, an obstruction exerts a sphere of influence that is much larger than the obstruction because the obstruction affects the flow pattern for considerable distances on each side. The sphere of influence for velocities that generally occur in channels that have gentle to moderately steep slopes is about three to five times the width of the obstruction. Several obstructions can create overlapping spheres of influence and may cause considerable disturbance, even though the obstructions may occupy only a small part of a channel cross section. Chow (1959) assigned adjustment values to four levels of obstruction: negligible, minor, appreciable, and severe (Table 3.20).

Vegetation (n_4)

The extent to which vegetation affects n depends on the following: the depth of flow; the percentage of the wetted perimeter covered by the vegetation; the density of vegetation below the high-water line; the degree to which the vegetation is flattened by high water; and the alignment of vegetation relative to the flow. The adjustment values given in the following table apply to constricted channels that are narrow in width. In wide channels having small depth-to-width ratios and no vegetation on the bed, the effect of bank vegetation is small, and the maximum adjustment is about 0.005. If the channel is relatively narrow and has steep banks covered by dense vegetation

TABLE 3.19. n_2 Adjustment Factor.

Channel Conditions	*n* Value Adjustment	Example
Gradual	0.000	Size and shape of channel cross sections change gradually.
Alternating occasionally	0.001–0.005	Large and small cross sections alternate occasionally, or the main flow occasionally shifts from side to side owing to changes in cross-sectional shape.
Alternating frequently	0.010–0.015	Large and small cross sections alternate frequently, or the main flow frequently shifts from side to side owing to changes in cross-sectional shape.

TABLE 3.20. n_3 Adjustment Factors.

Channel Conditions	*n* Value Adjustment	Example
Negligible	0.000–0.004	A few scattered obstructions, which include debris deposits, stumps, exposed roots, logs, piers, or isolated boulders, that occupy less than 5 percent of the cross-sectional area.
Minor	0.005–0.015	Obstructions occupy less than 15 percent of the cross-sectional area, and the spacing between obstructions is such that the sphere of influence around one obstruction does not extend to the sphere of influence around another obstruction. Smaller adjustments are used for curved smooth-surfaced objects than are used for sharp-edged angular objects.
Appreciable	0.020–0.030	Obstructions occupy from 15 percent to 50 percent of the cross-sectional area, or the space between obstructions is small enough to cause the effects of several obstructions to be additive, thereby blocking an equivalent part of a cross-section.
Severe	0.040–0.050	Obstructions occupy more than 50 percent of the cross-sectional area, or the space between obstructions is small enough to cause turbulence across most of the cross section.

that hangs over the channel, the maximum adjustment is about 0.03. The larger adjustment values given in Table 3.21 apply only in places where vegetation covers most of the channel.

Meandering (m)

The degree of meandering, *m*, depends on the ratio of the total length of the meandering channel in the reach being considered to the straight length of the channel reach, Table 3.22. The meandering is considered minor for ratios of 1.0 to 1.2, appreciable for ratios of 1.2 to 1.5, and severe for ratios of 1.5 and greater. According to Chow (1959), meanders can increase the *n* values by as much as 30 percent where flow is confined within a stream channel. The meander adjustment should be considered only when the flow is confined to the channel. There may be very little flow in a meandering channel when there is flood-plain flow.

Example (Manning's n *Adjustment):*

Consider the following:

Basic *n* value for channel in earth (straight uniform channel in firm soil), $n_b = 0.030$; Modification for channel irregularity (minor), $n_1 = 0.002$; Modification for channel cross section (alternating occasionally), $n_2 = 0.003$; Modification for obstructions (negligible), $n_3 = 0.002$; Modification for vegetation (small, grass), $n_4 = 0.005$.

No meander correction

$$n = n_b + n_1 + n_2 + n_3 + n_4, \quad n = 0.042$$

Chow (1959) would indicate a value between 0.030 and 0.050 for this channel.

For most streams, a field survey is needed to determine the appropriate Manning's roughness and hydraulic radius values for a site, as it is not possible to estimate these from a map.

TABLE 3.21. n_4 Vegetation Adjustment Factors.

Channel Conditions	n Value Adjustment	Example
Small	0.002–0.010	Dense growths of flexible turf grass, such as Bermuda, or weeds growing where the average depth of flow is at least two times the height of the vegetation; supple tree seedlings such as willow, cottonwood, arrowhead, or saltcedar growing where the average depth of flow is at least three times the height of the vegetation.
Medium	0.010–0.025	Turf grass growing where the average depth of flow is from one to two times the height of the vegetation; moderately dense stemy grass, weeds, or tree seedlings where the average depth of flow is from two to three times the height of the vegetation; brushy, moderately dense vegetation, similar to 1-to-2-year-old willow trees in the dormant season, growing along the banks, and no significant vegetation is evident along the channel bottoms where the hydraulic radius exceeds 0.61 meters.
Large	0.025–0.050	Turf grass growing where the average depth of flow is about equal to the height of the vegetation; 8-to-10-year-old willow or cottonwood trees intergrown with some weeds and brush (none of the vegetation in foliage) where the hydraulic radius exceeds 0.60 m; bushy willows about 1 year old intergrown with some weeds along side slopes (all vegetation in full foliage), and no significant vegetation exists along channel bottoms where the hydraulic radius is greater than 0.61 m.
Very Large	0.050–0.100	Turf grass growing where the average depth of flow is less than half the height of the vegetation; bushy willow trees about 1 year old intergrown with weeds along side slopes (all vegetation in full foliage), or dense cattails growing along channel bottom; trees intergrow with weeds and brush (all vegetation in full foliage).

TABLE 3.22. Meander Adjustment Multiplier.

Channel Conditions	*n* Value Adjustment	Example
Minor	1.00	Ratio of the channel length to valley length is 1.0 to 1.2
Appreciable	1.15	Ratio of the channel length to valley length is 1.2 to 1.5
Severe	1.30	Ratio of the channel length to valley length is greater than 1.5

[1]Adjustments for degree of irregularity, variation in cross section, effect of obstructions, and vegetation are added to the base n value before multiplying by the adjustment for meander.
[2]Adjustment values apply to flow confined in channel and do not apply where downvalley flow crosses meanders.

Example (Travel Time Calculation):

The TR-55 User Guide (SCS 1986) includes the following example. Figure 3.24 shows a watershed in Dyer County, which is located in northwestern Tennessee. The problem is to compute T_c at the outlet of the watershed (point D). The 2-year 24-hour rainfall depth is 3.6 inches. All three types of flow occur from the hydraulically most distant point (A) to the point of interest (D). To compute T_c, first determine T_t for each segment from the following information:

Segment AB: Sheetflow; dense grass; slope (s) = 0.01 ft/ft; and length (L) = 100 ft.

Segment BC: Shallow concentrated flow; unpaved; s = 0.01 ft/ft; and L = 1400 ft.

Segment CD: Channel flow; Manning's n = 0.05; flow cross-sectional area (a) = 27 ft^2; wetted perimeter (p_w) = 28.2 ft; s = 0.005 ft/ft; and L = 7300ft.

Figure 3.25 is the SCS worksheet showing the calculations for the above problem. In this case, each flow segment is comprised of a single condition of slope and cover. In many cases, the individual flow segments may need to be broken up into subunits to represent different slopes or roughness coefficients. The travel times for each of the segments are added. For the sheetflow segment, however, the total travel length must still be less than 300 ft, not 300 ft for each calculation interval. Worksheet 3 has two columns to facilitate two segments for each portion. Additional segments may be needed. In this example, the total travel time for this flow path from A to D is 1.53 hours, with almost 1 hour associated with the channel flow time. For small sites, including most construction sites, the sheetflow segment will likely comprise the largest portion of the total flow time.

Again, in order to determine the time of concentration for the watershed, several different candidate flow paths are usually needed to be evaluated and the one with the longest travel time is used as the time of concentration. This may not be the path with the longest travel distance, but may be a shorter path affected by shallower slopes and rougher covers.

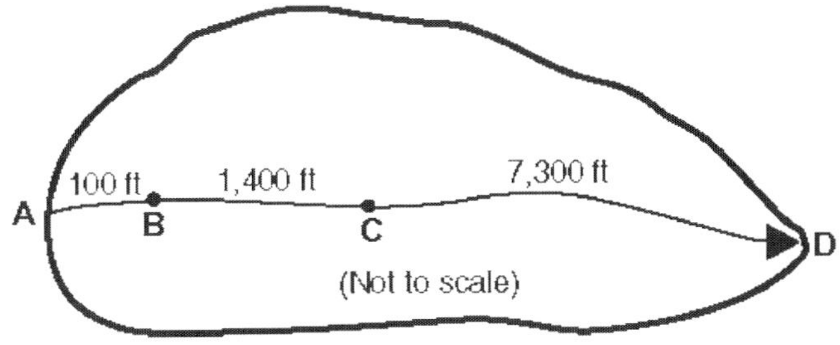

Figure 3.24. Watershed for TR-55 Tt calculation example (SCS, 1986).

Tabular Hydrograph Method

The SCS TR-55 tabular hydrograph method (SCS, 1986) can be used to develop a hydrograph for each subwatershed area that can then be routed through the downstream project segments. This method will also produce the total runoff volume and the peak flow rate. This method is not used in the new WinTR-55; this computerized version uses the more complete routing procedures from TR-20. However, the following is still presented as an optional method and to illustrate the sensitivity of T_c and CN selections. Appendix 3A includes all of the tabular hydrograph tables that can be used to calculate hydrographs for all locations in the U.S.

Example (Tabular Hydrograph Calculation)

The following example is from the TR-55 manual (SCS 1986) and illustrates how the T_c, CN, and other site characteristics are used to develop and route hydrographs for a complex watershed.

This example computes the 25-year frequency peak discharge at the downstream end of subarea 7 shown in Figure 3.26. This example is for present conditions and uses the worksheets presented in SCS (1986). The CN, T_c, and T_t for each subarea must be determined or calculated using the procedures in TR-55 Chapters 2 and 3. These values are entered on worksheet 5a (Figure 3.27). Then, the tabular hydrograph tables are used to determine the normalized hydrograph for downstream locations.

The hydrograph tables are presented in SCS (1986) according to rain type (there are sections of tables for types I, Ia, II, and III rain distributions). The first step is to find the table section pertaining to the rain distribution for the study area. In this case, the area has type II rains. The type II rain hydrograph tables are further grouped according to the T_c for the subarea, ranging from 0.1 to 2 hours. In the case for subarea #1, the T_c is 1.5 hours, so pg 5-37 from SCS (1986) is used (Table 3.24). Each page is further divided into three segments, corresponding to Ia/P ratios of 0.10, 0.30, and 0.50. The Ia is the initial abstractions for the area (not to be

Worksheet 3: Time of Concentration (T_c) or travel time (T_t)

Project: Heavenly Acres	By: DW	Date: 10/6/85
Location: Dyer County, Tennessee	Checked: NM	Date: 10/8/85

Check one: ☐ Present ☒ Developed

Check one: ☒ T_c ☐ T_t through subarea

Notes: Space for as many as two segments per flow type can be used for each worksheet. Include a map, schematic, or description of flow segments.

Sheet flow (Applicable to T_c only)

Segment ID		AB	
1. Surface description (table 3-1)		Dense Grass	
2. Manning's roughness coefficient, n (table 3-1)		0.24	
3. Flow length, L (total L ≤ 300 ft)	ft	100	
4. Two-year 24-hour rainfall, P_2	in	3.6	
5. Land slope, s	ft/ft	0.01	
6. $T_t = \frac{0.007\,(nL)^{0.8}}{P_2^{0.5}\,s^{0.4}}$ Compute T_t	hr	0.30	+ = 0.30

Shallow concentrated flow

Segment ID		BC	
7. Surface description (paved or unpaved)		Unpaved	
8. Flow length, L	ft	1400	
9. Watercourse slope, s	ft/ft	0.01	
10. Average velocity, V (figure 3-1)	ft/s	1.6	
11. $T_t = \frac{L}{3600\,V}$ Compute T_t	hr	0.24	+ = 0.24

Channel flow

Segement ID		CD	
12. Cross sectional flow area, a	ft^2	27	
13. Wetted perimeter, p_w	ft	28.2	
14. Hydraulic radius, $r = \frac{a}{p_w}$ Compute r	ft	0.957	
15. Channel slope, s	ft/ft	0.005	
16. Manning's roughness coefficient, n		0.05	
17. $V = \frac{1.49\,r^{2/3}\,s^{1/2}}{n}$ Compute V	ft/s	2.05	
18. Flow length, L	ft	7300	
19. $T_t = \frac{L}{3600\,V}$ Compute T_t	hr	0.99	+ = 0.99
20. Watershed or subarea T_c or T_t (add T_t in steps 6, 11, and 19)	Hr		1.53

Figure 3.25. Calculation example for travel time problem (SCS, 1986).

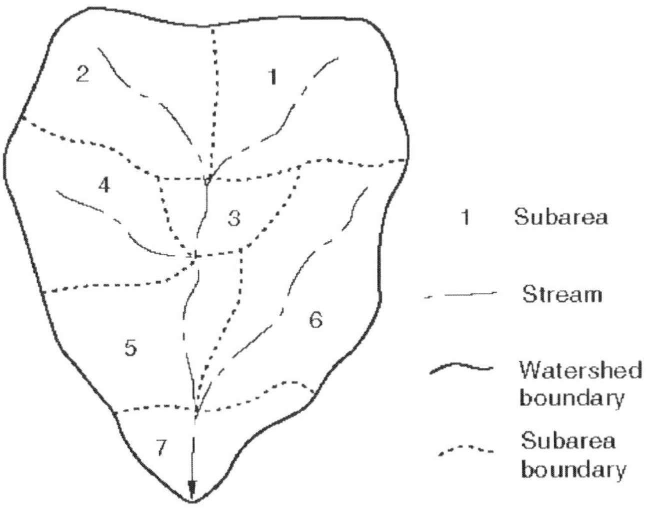

Figure 3.26. Example watershed for tabular hydrograph calculations (SCS 1986).

confused with rain distribution type *Ia*) and are a direct function of the *CN* value. These are given in the User Guide (SCS Table 5-1), and on Table 3.23. The *P* is the total rain depth being evaluated. The top set of values are used for *Ia*/*P* ratios of ≤0.2, the middle set for ratios from 0.2 to 0.4, while the bottom set is used for ratios of >0.4 (interpolation is not used; WinTR-55 and TR-20 calculate more precise values based on actual site conditions). In this case, the #1 subarea *Ia*/*P* is 0.18, so the top set of values are used. Finally, each segment has 12 lines representing different travel times from the bottom of the subwatershed area to the location of interest (typically the outlet). The largest unit peak runoff rate values (csm/in, or cubic feet per second of runoff per square mile of drainage area, per inch of direct runoff) on

TABLE 3.23. *Ia* Values for Runoff Curve Numbers (SCS, 1986).

Curve Number	*Ia* (inch)	Curve Number	*Ia* (inch)	Curve Number	*Ia* (inch)
40	3.000	60	1.333	80	0.500
41	2.878	61	1.279	81	0.469
42	2.762	62	1.226	82	0.439
43	2.651	63	1.175	83	0.410
44	2.545	64	1.125	84	0.381
45	2.444	65	1.077	85	0.353
46	2.348	66	1.030	86	0.326
47	2.255	67	0.985	87	0.299
48	2.167	68	0.941	88	0.273
49	2.082	69	0.899	89	0.247
50	2.000	70	0.857	90	0.222
51	1.922	71	0.817	91	0.198
52	1.846	72	0.778	92	0.174
53	1.774	73	0.740	93	0.151
54	1.704	74	0.703	94	0.128
55	1.636	75	0.667	95	0.105
56	1.571	76	0.632	96	0083
57	1.509	77	0.597	97	0.062
58	1.448	78	0.564	98	0.041
59	1.390	79	0.532		

Worksheet 5a: Basic watershed data

Project: Fallswood | Location: Dyer County, Tennessee | By: DW | Date: 10/1/85

Check one: ☒ Present ☐ Developed | Frequency (yr): 25 | Checked: NM | Date: 10/3/85

Subarea name	Drainage area A_m (mi^2)	Time of concentration T_c (hr)	Travel time through subarea T_t (hr)	Downstream subarea names	Travel time summation to outlet ΣT_t (hr)	24-hr rainfall P (in)	Runoff curve number CN	Runoff Q (in)	A_mQ (mi^2–in)	Initial abstraction I_a (in)	I_a/P
1	0.30	1.50	--	3, 5, 7	2.50	6.0	65	2.35	0.71	1.077	0.18
2	0.20	1.25	--	3, 5, 7	2.50	6.0	70	2.80	0.56	0.857	0.14
3	0.10	0.50	0.50	5, 7	2.00	6.0	75	3.28	0.33	0.667	0.11
4	0.25	0.75	--	5, 7	2.00	6.0	70	2.80	0.70	0.857	0.14
5	0.20	1.50	1.25	7	0.75	6.0	75	3.28	0.66	0.667	0.11
6	0.40	1.50	--	7	0.75	6.0	70	2.80	1.12	0.857	0.14
7	0.20	1.25	0.75	--	0	6.0	75	3.28	0.66	0.667	0.11

From worksheet 3 | From worksheet 2 | From table 5-1

Figure 3.27. Worksheet 5a for showing basic watershed data (SCS, 1986).

TABLE 3.24. Tabular Hydrograph Table for Example Problem (SCS, 1986, pg 5-37).

Exhibit 5-II: Tabular hydrograph unit discharges (csm/in) for type II rainfall distribution—continued

TRVL TIME (hr)	11.0	11.3	11.6	11.9	12.0	12.1	12.2	12.3	12.4	12.5	12.6	12.7	12.8	13.0	13.2	13.4	13.6	13.8	14.0	14.3	14.6	15.0	15.5	16.0	16.5	17.0	17.5	18.0	19.0	20.0	22.0	26.0
IA/P = 0.10 — * * * TC = 1.5 HR * * * — IA/P = 0.10																																
0.0	9	11	15	21	25	31	41	58	82	112	147	184	216	255	275	236	198	159	129	98	76	57	43	35	30	25	23	21	18	16	12	1
.10	8	10	13	18	20	23	28	37	51	72	98	131	166	226	265	254	226	187	151	113	86	63	46	37	31	26	23	21	19	16	13	2
.20	8	10	13	17	19	22	26	33	45	63	87	116	149	212	259	259	233	197	160	119	90	66	48	38	32	27	24	22	19	16	13	2
.30	7	9	12	16	18	21	24	30	40	55	76	103	134	197	244	255	238	206	169	125	95	68	49	38	32	27	24	22	19	17	13	2
.40	7	8	11	14	15	17	19	23	28	36	49	67	91	151	208	247	252	230	196	146	109	77	54	41	34	29	25	22	19	17	13	3
.50	6	8	10	13	15	16	18	21	26	33	43	59	80	136	194	238	249	235	204	154	115	81	56	42	34	29	25	23	20	17	13	3
.75	5	7	8	11	12	13	14	16	18	21	25	32	42	76	125	179	222	240	233	193	148	102	67	48	38	32	27	24	20	18	13	5
1.0	4	5	7	8	9	10	11	12	13	14	16	18	22	34	59	101	152	201	236	230	193	135	86	59	44	35	30	26	21	18	14	7
1.5	3	4	5	6	6	7	8	8	9	10	11	12	13	16	22	34	58	95	141	203	226	197	131	84	58	43	35	29	23	20	15	10
2.0	1	2	3	4	4	5	5	6	6	7	7	8	9	10	12	16	22	34	56	110	172	218	187	126	82	57	43	34	25	21	16	11
2.5	1	1	2	2	3	3	3	4	4	4	5	5	6	7	8	9	11	14	18	34	69	141	210	190	133	87	60	44	30	23	17	12
3.0	0	0	1	1	2	2	2	2	3	3	3	3	4	5	5	6	8	9	11	16	27	66	149	204	181	128	85	58	35	25	18	12
IA/P = 0.30 — * * * TC = 1.5 HR * * * — IA/P = 0.30																																
0.0	0	0	0	0	0	1	6	15	31	53	80	112	144	193	225	208	186	157	134	108	89	70	56	48	42	37	34	31	28	25	20	2
.10	0	0	0	0	0	0	1	4	12	25	43	68	97	157	198	219	203	178	151	120	98	77	60	50	44	38	35	32	28	25	20	3
.20	0	0	0	0	0	0	0	1	3	9	19	35	57	114	168	201	213	196	171	135	108	84	64	53	46	40	36	33	29	26	20	4
.30	0	0	0	0	0	0	0	1	2	7	15	29	48	100	155	193	210	200	177	140	113	87	66	54	46	41	36	33	29	26	20	5
.40	0	0	0	0	0	0	0	0	2	5	12	23	39	87	141	184	207	202	182	146	117	89	68	55	47	41	36	33	29	26	20	5
,50	0	0	0	0	0	0	0	0	0	1	4	9	18	51	101	153	190	205	197	164	131	99	73	58	49	43	38	34	30	26	20	7
.75	0	0	0	0	0	0	0	0	0	0	2	4	9	30	68	116	160	189	197	179	147	110	80	62	52	45	39	35	30	27	21	8
1.0	0	0	0	0	0	0	0	0	0	0	0	0	1	5	20	49	92	138	175	195	178	137	97	72	57	48	42	37	31	28	21	12
1.5	0	0	0	0	0	0	0	0	0	0	0	0	0	0	1	7	21	47	85	145	187	178	133	95	71	57	48	42	34	29	23	16
2.0	0	0	0	0	0	0	0	0	0	0	0	0	0	0	0	0	1	4	13	45	97	162	180	138	99	74	58	49	38	32	25	18
2.5	0	0	0	0	0	0	0	0	0	0	0	0	0	0	0	0	0	0	1	8	31	89	161	174	133	97	72	58	42	34	26	18
3.0	0	0	0	0	0	0	0	0	0	0	0	0	0	0	0	0	0	0	0	1	5	29	98	160	169	129	95	71	48	37	28	19
IA/P = 0.50 — * * * TC = 1.5 HR * * * — IA/P = 0.50																																
.0	0	0	0	0	0	0	0	3	8	16	27	42	59	92	116	128	130	121	112	100	90	78	67	60	55	50	46	43	39	35	29	4
.10	0	0	0	0	0	0	0	2	6	12	22	35	51	84	110	125	128	123	114	102	91	79	68	61	55	50	46	43	39	35	29	4
.20	0	0	0	0	0	0	0	0	1	4	10	18	29	60	91	114	126	128	120	108	97	83	71	63	57	52	47	44	40	36	29	5
.30	0	0	0	0	0	0	0	0	1	3	8	14	24	52	83	108	123	126	122	110	98	85	72	63	57	52	48	44	40	36	29	6
.40	0	0	0	0	0	0	0	0	0	1	2	6	12	31	60	90	112	124	126	116	104	90	75	66	59	54	49	45	41	37	29	8
.50	0	0	0	0	0	0	0	0	0	0	2	4	9	26	53	83	106	121	125	118	106	91	77	67	60	54	49	46	41	37	29	8
.75	0	0	0	0	0	0	0	0	0	0	1	2	5	16	36	62	88	108	119	122	112	97	81	69	62	56	51	47	42	38	30	11
1.0	0	0	0	0	0	0	0	0	0	0	0	0	0	3	10	26	49	75	98	118	121	108	90	76	66	59	54	49	43	39	31	16
1.5	0	0	0	0	0	0	0	0	0	0	0	0	0	0	1	3	11	25	45	80	107	118	106	89	75	65	59	53	45	41	32	23
2.0	0	0	0	0	0	0	0	0	0	0	0	0	0	0	0	0	1	4	11	32	63	100	115	104	87	74	65	58	48	42	34	26
2.5	0	0	0	0	0	0	0	0	0	0	0	0	0	0	0	0	0	0	1	4	16	48	94	113	105	89	76	66	53	45	36	27
3.0	0	0	0	0	0	0	0	0	0	0	0	0	0	0	0	0	0	0	0	0	3	15	54	96	111	103	88	75	58	48	38	28

RAINFALL TYPE = II * * * TC = 1.5 HR * * * SHEET 9 OF 10

(210-VI-TR-55, Second Ed., June 1986) 5-37

TABLE 3.25. Worksheet 5b for Example Hydrograph Calculation (SCS, 1986).

Worksheet 5b: Basic watershed data

Project Fallswood	Location Dyer County, Tennessee	By DW	Date 10/1/85
Check one: ☒ Present ☐ Developed	Frequency (yr) 25	Checked NM	Date 10/3/85

Subarea name	Subarea T_c (hr)	ΣT_t to outlet (hr)	I_a/P	A_mQ (mi²–in)	12.7	12.8	13.0	13.2	13.4	13.6	13.8	14.0	14.3	14.6	15.0	15.5
	Basic watershed data used 1/				Select and enter hydrograph times in hours from exhibit 5-II 2/ — Discharges at selected hydrograph times 3/ (cfs)											
1	1.50	2.50	0.10	0.71	4	4	5	6	6	8	10	113	24	49	100	149
2	1.25	2.50	0.10	0.56	3	4	4	6	7	8	11	16	32	64	110	127
3	0.50	2.00	0.10	0.33	5	5	6	8	12	21	41	67	98	92	60	29
4	0.75	2.00	0.10	0.70	8	9	11	14	20	34	62	106	172	192	149	81
5	1.50	0.75	0.10	0.66	21	28	50	83	118	147	158	154	127	98	67	44
6	1.50	0.75	0.10	1.12	36	47	85	140	200	249	269	261	216	166	114	75
7	1.25	0	0.10	0.66	169	187	205	176	140	108	85	69	51	40	31	24
Composite hydrograph at outlet					246	284	366	433	503	575	636	686	720	701	631	529

1/ Worksheet 5a. Rounded as needed for use with exhibit 5.
2/ Enter rainfall distribution type used.
3/ Hydrograph discharge for selected times is A_mQ multiplied by tabular discharge from appropriate exhibit 5.

each line are close to 12 hours for the top time, and shift to the right as the travel time increases. The shift between the largest values for each row is equal to the differences in the travel times between each line, representing routing of the hydrographs as they travel downstream. For the #1 subarea, the T_t is 2.5 hours. Therefore, the line near the bottom of the top segment, representing 2.5 hours, is used. The values in the table represent normalized hydrographs and are multiplied by A_mQ (the factor of the watershed area, in mi^2, and the direct runoff in inches) to obtain the flow values in traditional units of ft^3/sec, or cfs. These final cfs values are written on worksheet 5b (Table 3.25). As an example, the appropriate values for the peak discharge (q) for subarea 4 at 14.6 hr is:

$$q = q_t(A_mQ) = (274)(0.70) = 192 \text{ cfs}$$

Once all the prerouted subarea hydrographs have been tabulated on worksheet 5b, they are summed to obtain the composite hydrograph. The resulting 25-year frequency peak discharge is 720 cfs at 14.3 hr, as shown on Table 3.25.

Example (Tabular Hydrograph for Urban Watershed)

The following example is for a typical urban watershed, having four subareas that are quite different in their development characteristics. The following lists the procedure for evaluating this area:

1. Subdivide the watershed into relatively homogeneous subareas (as shown in Figure 3.28).
2. Calculate the drainage for each subarea.

I	0.10 mi^2
II	0.08
III	0.6
IV	0.32
Total	1.12 mi^2

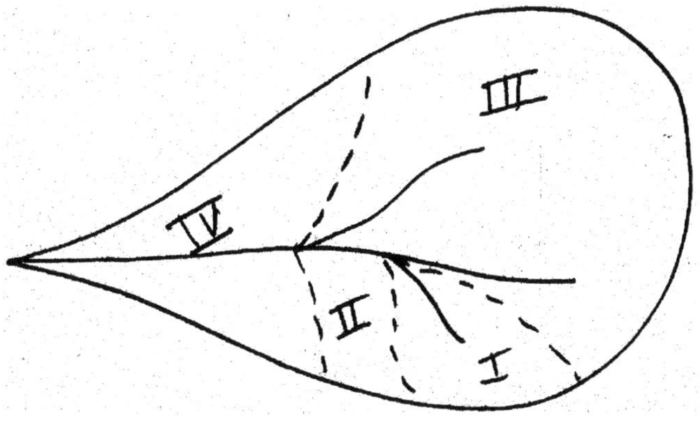

Figure 3.28. Relatively homogeneous subareas in example urban

3. Calculate the time of concentration (T_c) for each subarea (TR-55 chapter 3).

I	0.2 hrs
II	0.1
III	0.3
IV	0.1

4. Calculate the travel time (T_t) from each subarea discharge location to the location of interest (outlet of total watershed in this example) (TR-55 chapter 3).

I	0.1 hrs
II	0.05
III	0.06
IV	0.0

5. Select the curve number (CN) for each subarea, Tables 3.9 and 3.10, or local data as in this example.

I	Strip commercial, all directly connected	$CN = 97$
II	Medium density residential area, grass swales	$CN = 46$
III	Medium density residential area, curbs and gutters	$CN = 72$
IV	Low density residential area, grass swales	$CN = 40$

6. Determine the appropriate rainfall distribution (Type II for all areas in this example).
7. Find the 24-hour rainfall depth for storm, equal to 4.1 inches for this example.
8. Calculate total runoff (inches) from CN and rain depth (from SCS Figure 2-1), Figure 3.17.

I	$CN = 97$	$P = 4.1$ in.	$Q = 3.8$ in.
II	$CN = 46$	$P = 4.1$ in.	$Q = 0.25$
III	$CN = 72$	$P = 4.1$ in.	$Q = 1.5$
IV	$CN = 40$	$P = 4.1$ in.	$Q = 0.06$

9. Determine Ia for each subarea (SCS assumes $Ia = 0.2\,S$, where S is the total rainfall abstractions) (SCS table 5-1), Table 3.23.

I	$CN = 97$	$Ia = 0.062$ in.
II	$CN = 46$	$Ia = 2.348$ in.
III	$CN = 72$	$Ia = 0.778$ in.
IV	$CN = 40$	$Ia = 3.000$ in.

10. Calculate the ratio of Ia to P.

I	$Ia/P = 0.062/4.1 = 0.015$
II	$Ia/P = 2.348/4.1 = 0.57$
III	$Ia/P = 0.778/4.1 = 0.19$
IV	$Ia/P = 3.000/4.1 = 0.73$

11. Use worksheets SCS 5a and 5b to summarize above data and to calculate the composite hydrograph. These are shown in Tables 3.26 and 3.27.

TABLE 3.26. SCS Worksheet 5a for Urban Example.

Worksheet 5a: Basic watershed data

Project	Location	By	Date
EXAMPLE - URBAN	JEFFERSON COUNTY, AL	XYZ	
Check one: ☐ Present ☒ Developed	Frequency (yr) 2 YR	Checked	Date

Subarea name	Drainage area A_m (mi^2)	Time of concentration T_c (hr)	Travel time through subarea T_t (hr)	Downstream subarea names	Travel time summation to outlet ΣT_t (hr)	24-hr rainfall P (in)	Runoff curve number CN	Runoff Q (in)	A_mQ (mi^2—in)	Initial abstraction I_a (in)	I_a/P
I	0.10	0.2	–	–	0.1	4.1	97	3.8	0.38	0.062	0.015
II	0.08	0.1	–	–	0.05	4.1	46	0.25	0.02	2.348	0.57
III	0.62	0.3	–	–	0.05	4.1	72	1.5	0.93	0.778	0.19
IV	0.32	0.1	–	–	Ø	4.1	40	0.06	0.019	3.000	0.73
Σ =	1.12										
		From worksheet 3	From worksheet 3				From worksheet 2			From table 5-1	

TABLE 3.27. SCS Worksheet 5b for Urban Example.

Worksheet 5b: Basic watershed data

Project	Location	By	Date
EXAMPLE - URBAN	JEFFERSON CTY., AL	XYZ	
Check one: ☐ Present ☒ Developed	Frequency (yr) 2 YR.	Checked	Date

Basic watershed data used 1/ — Select and enter hydrograph times in hours from exhibit 5-II 2/ — Discharges at selected hydrograph times 3/ (cfs)

Subarea name	Subarea T_c (hr)	ΣT_t to outlet (hr)	I_a/P	A_mQ (mi^2—in)		11.0	11.6	12	12.2	12.3	13	14	16	18	26	
I	0.2	0.1	~~0.015~~ 0.10	0.38	csm/in:	19	39	168	601	733	83	43	25	18	Ø	p.5-30
					cfs:	7.2	14.2	63.8	228	278	31.5	16.3	9.5	6.8	Ø	
II	0.1	~~0.05~~ Ø	~~0.57~~ 0.50	0.02	csm/in:	Ø	Ø	70	377	196	99	67	46	38	Ø	p.5-29
					cfs:	Ø	Ø	1.4	7.5	3.9	2.0	1.3	0.9	0.8	Ø	
III	0.3	~~0.05~~ Ø	~~0.19~~ 0.10	0.93	csm/in:	20	41	235	676	676	80	42	24	18	Ø	p.5-31
					cfs:	18.6	38.1	218	628	628	74.4	39.1	22.3	16.7	Ø	
IV	0.1	Ø	~~0.73~~ 0.50	0.019	csm/in:	too small for calculation/contribution										
Composite hydrograph at outlet						25.8	52.3	283	863	910	108	56.7	32.7	24.3	Ø	

1/ Worksheet 5a. Rounded as needed for use with exhibit 5.
2/ Enter rainfall distribution type used.
3/ Hydrograph discharge for selected times is A_mQ multiplied by tabular discharge from appropriate exhibit 5.

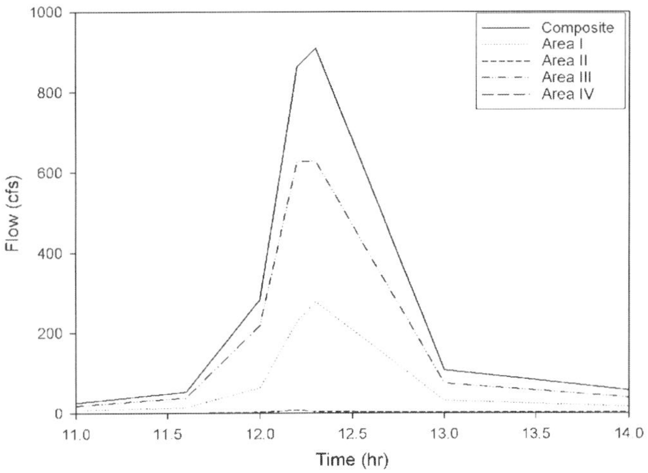

Figure 3.29. Plot of individual and composite hydrograph for urban example.

The peak flow is seen to be 910 cfs, occurring at 12.3 hours. Figure 3.29 is a plot of the 3 main components, plus the total hydrograph. Subarea III contributed most of the peak flow to the total hydrograph, while subareas II and IV contributed insignificant flows. The following chapter section introduces WinTR-55 and presents this same example. The main differences is that WinTR-55 requires a description of the channel as it calculates the travel times and conducts the channel routing using a more precise procedure. In addition, the hydrograph development uses TR-20, instead of the tabular hydrograph method.

WinTR-55

The following discussion is summarized from the WinTR-55 user guide information, while the example uses the previously described information.

A WinTR-55 work group was formed in the spring of 1998 to modernize and revise TR-55 and the computer software. The current changes included the following: upgrading the source code to Visual Basic, changing the philosophy of data input, developing a Windows interface and output post-processor, enhancing the hydrograph-generation capability of the software, and improving the generated flood-route hydrographs through stream reaches and reservoirs.

The availability and technical capabilities of the personal computer have significantly changed the philosophy of problem-solving for the engineer. Computer availability eliminated the need for TR-55 manual methods, thus the manual portions (graphs and tables) of the user document have been eliminated. The WinTR-55 user manual (NRCS 2002a) covers the procedures used in and the operation of the WinTR-55 computer program. Part 630 of the Natural Resources Conservation Service (NRCS) National Engineering Handbook provides detailed information on NRCS hydrology and is the technical reference for WinTR-55.

Program Description

WinTR-55 is a single-event rainfall-runoff small watershed hydrologic model. The model generates hydrographs from both urban and agricultural areas and at selected points along the stream system. Hydrographs are routed downstream through channels and/or reservoirs. Multiple sub-areas can be modeled within the watershed.

Model Overview

A watershed is composed of subareas (land areas) and reaches (major flow paths in the watershed). Each subarea has a hydrograph generated from the land area based on the land and climate characteristics provided. Reaches can be designated as either channel reaches where hydrographs are routed based on physical reach characteristics or as storage reaches where hydrographs are routed through a reservoir based on temporary storage and outlet characteristics. Hydrographs from sub-areas and reaches are combined as needed to accumulate flow as water moves from the upland areas down through the watershed reach network. The accumulation of all runoff from the watershed is represented at the watershed outlet. Up to ten sub-areas and ten reaches may be included in the watershed.

WinTR-55 uses the TR-20 (NRCS 2002b) model for all of the hydrograph procedures: generation, channel routing, storage routing, and hydrograph summation. Figure 3.30 is a diagram showing the WinTR-55 model, its relationship to TR-20, and the files associated with the model.

Capabilities and Limitations

WinTR-55 hydrology has the capability to analyze watersheds that meet the criteria listed in Table 3.28.

Model Input

The various data used in the WinTR-55 procedures are user entered via a series of input windows in the model. A description of each of the input windows follows the figure. Data entry is needed only on the windows that are applicable to the watershed being evaluated.

Minimum Data Requirements

While WinTR-55 can be used for watersheds with up to

TABLE 3.28. WinTR-55 Capabilities & Limitations (NRCS 2002a).

Variable	Limits
Minimum area	No absolute minimum is included in the software. However, carefully examine results from sub-areas less than 1 acre.
Maximum area	25 square miles (6,500 hectares)
Number of Subwatersheds	1–10
Time of concentration for any sub-area	0.1 hour ≤ T_c ≤ 10 hour
Number of reaches	0–10
Types of reaches	Channel or Structure
Reach Routing	Muskingum-Cunge
Structure Routing	Storage-Indication
Structure Types	Pipe or Weir
Structure Trial Sizes	1–3
Rainfall Depth[1]	Default or user-defined 0–50 inches (0–1,270 mm)
Rainfall Distributions	NRCS Type I, IA, II, III, NM60, NM65, NM70, NM75, or user-defined
Rainfall Duration	24-hour
Dimensionless Unit Hydrograph	Standard peak rate factor 484, or user-defined (e.g. Delmarva–see Example 3)
Antecedent Moisture Condition	2 (average)

[1]Although no minimum rain depth is listed by the NRCS in the above table, it must be recognized that the original SCS curve number methods, incorporated in this newer version, are not accurate for small storms. In most cases, larger storms used for drainage design are reasonably well suited to this method. Pitt (1987) and Pitt, *et al.* (2002) showed that rain depths less than 2 or 3 inches can have significant errors when using the *CN* approach.

ten sub-areas and up to ten reaches, the simplest run involves only a single sub-area. Data required for a single sub-area run can be entered on the TR-55 Main Window. These data include: identification data-user, state, county, project, and subtitle; dimensionless unit hydrograph; storm data; rainfall distribution; and subarea data. The subarea data can be entered directly into the subarea entry and summary table: subarea name, subarea description, subarea flows to reach/outlet, area, runoff curve number (CN), and time of concentration (T_c). Detailed information for the subarea CN and T_c can be entered here or on other windows; if detailed information is entered elsewhere the computational results are displayed in this window.

Watershed Subareas and Reaches

To properly route stream flow to the watershed outlet, the user must understand how WinTR-55 relates watershed subareas and stream reaches. Figure 3.31 and Table 3.29 show a typical watershed with multiple sub-areas and reaches.

Reaches define flow paths through the watershed to its outlet. Each subarea and reach contribute flow to the upstream end of a receiving reach or to the outlet. Accumulated runoff from all sub-areas routed through the watershed reach system, by definition, is flow at the watershed outlet.

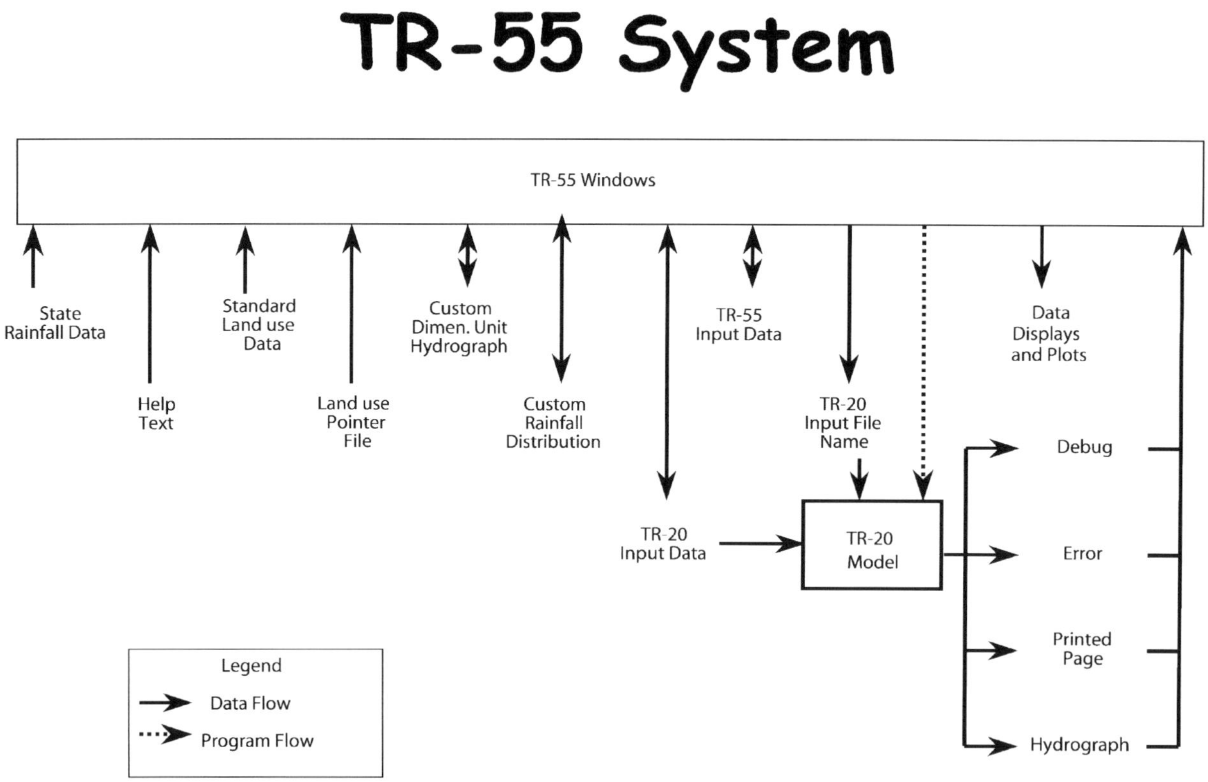

Figure 3.30. WinTR-55 system schematic (NRCS 2002a).

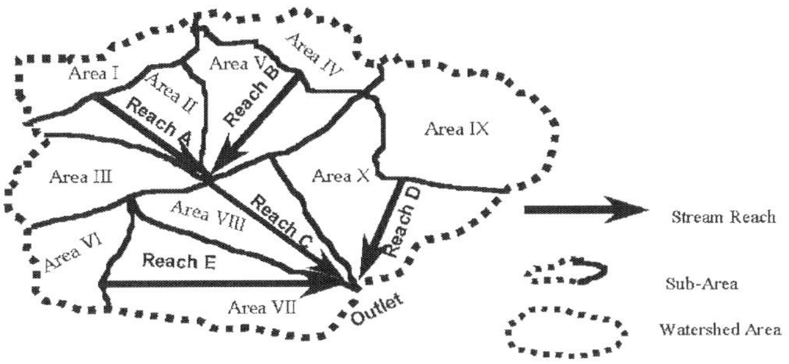

Figure 3.31. Sample Watershed Schematic (NRCS 2002a).

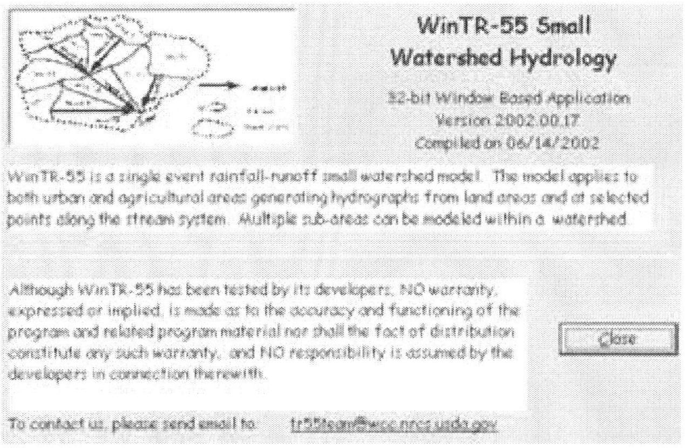

Figure 3.32. WinTR-55 opening screen.

Processes

WinTR-55 relies on the TR-20 model for all hydrograph processes, including hydrograph generation, combining hydrographs, channel routing, and structure routing. The program now uses a Muskingum-Cunge method of channel routing (Chow, *et al.* 1988; Maidment, 1993; Ponce, 1989). The storage-indication method (NRCS NEH Part 630, Chapter 17) is used to route structure hydrographs.

Example: WinTR-55 Setup and Operation

An application using WinTR-55 and the previously presented urban watershed example, is shown on Figures 3.32 through 3.41. Figures 3.42 and 3.43 are other screens available in WinTR-55 that can be used to aid in the calculation of some of the site data, while Figure 3.44 is used for detention facilities (structures).

This WinTR-55 example resulted in a peak flow for the 2-yr storm of about 730 cfs, compared to the previously calculated value of 910 cfs. This difference is due to the different routing procedure used, plus the more precise hydrograph development procedure in the updated WinTR-55 version compared to the tabular hydrograph method.

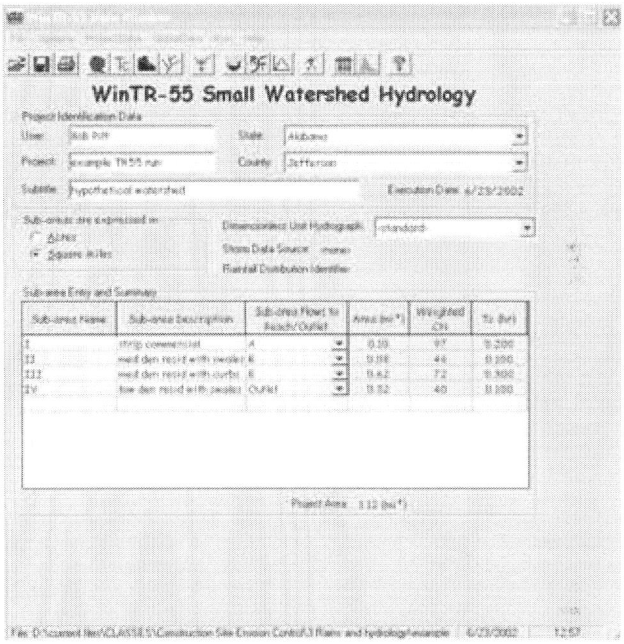

Figure 3.33. WinTR-55 small watershed basic information screen.

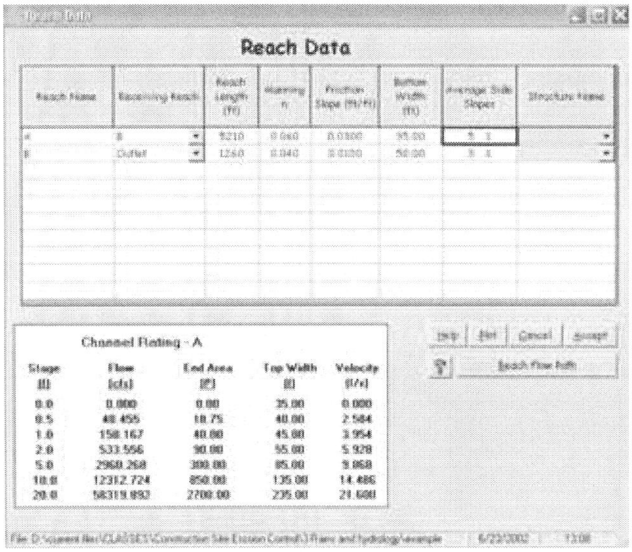

Figure 3.34. WinTR-55 reach data screen.

TABLE 3.29. Sample Watershed Flows (NRCS 2002a).

Subarea	Flows into Upstream End of	Reach	Flows into
Area I	Reach A	Reach A	Reach C
Area II	Reach C	Reach B	Reach C
Area III	Reach C	Reach C	OUTLET
Area IV	Reach B	Reach D	OUTLET
Area V	Reach C	Reach E	OUTLET
Area VI	Reach E		
Area VII	OUTLET		
Area VIII	OUTLET		
Area IX	Reach D		
Area X	OUTLET		

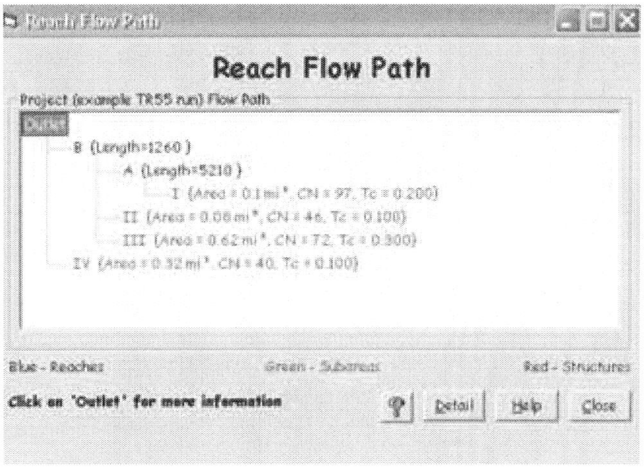

Figure 3.35. WinTR-55 reach flow path screen.

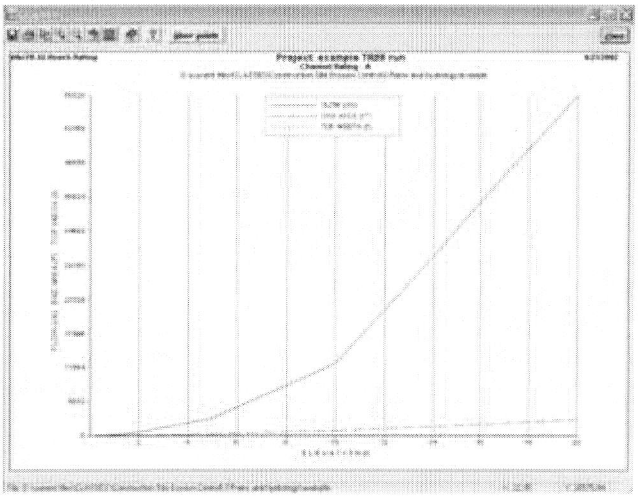

Figure 3.36. WinTR-55 reach routing screen.

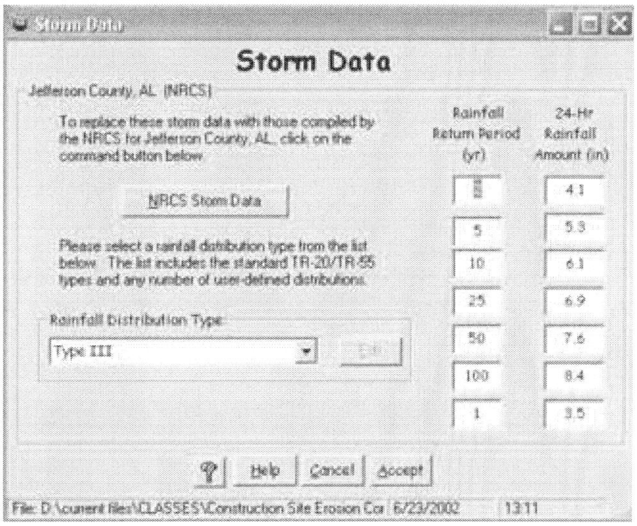

Figure 3.37. WinTR-55 storm data screen (information automatically determined by WinTR-55 based on location).

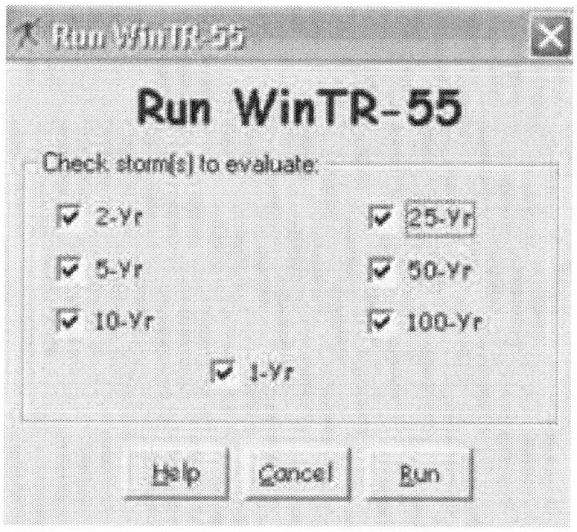

Figure 3.38. WinTR-55 event selection/run screen.

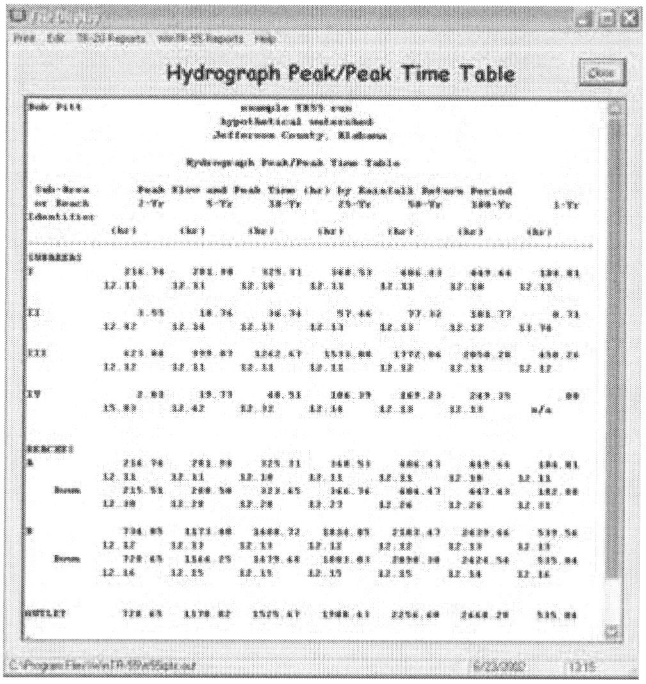

Figure 3.39. WinTR-55 calculated hydrograph summary screen.

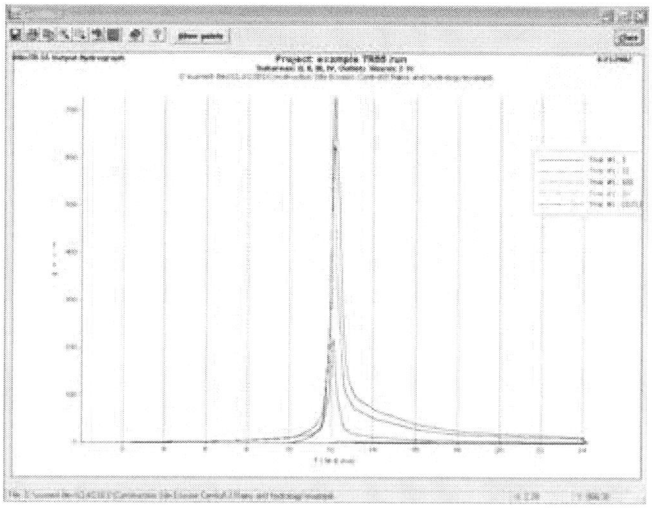

Figure 3.40. WinTR-55 hydrograph plot screen.

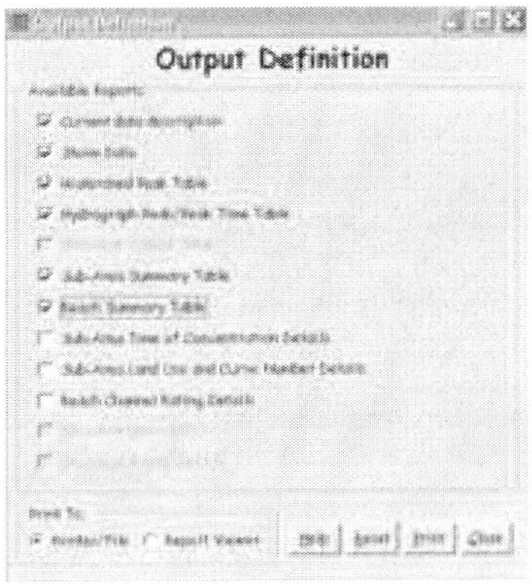

Figure 3.41. WinTR-55 report generation screen.

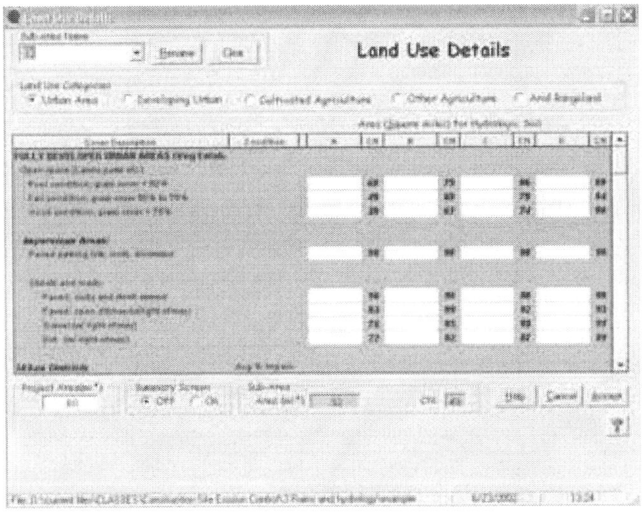

Figure 3.42. WinTR-55 land use details screen (if data not directly entered).

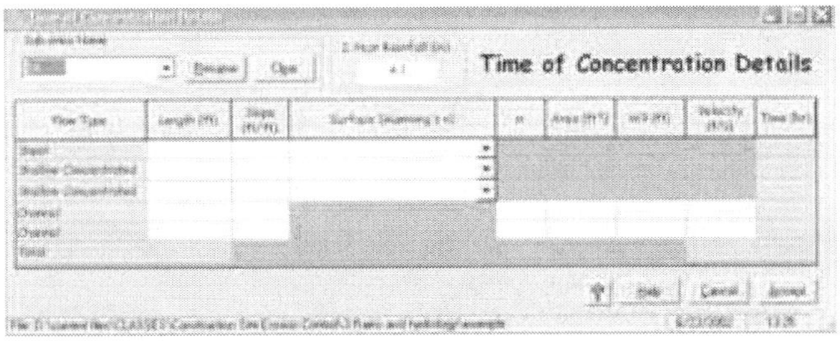

Figure 3.43. WinTR-55 time of concentration details screen/calculator (if data not directly entered).

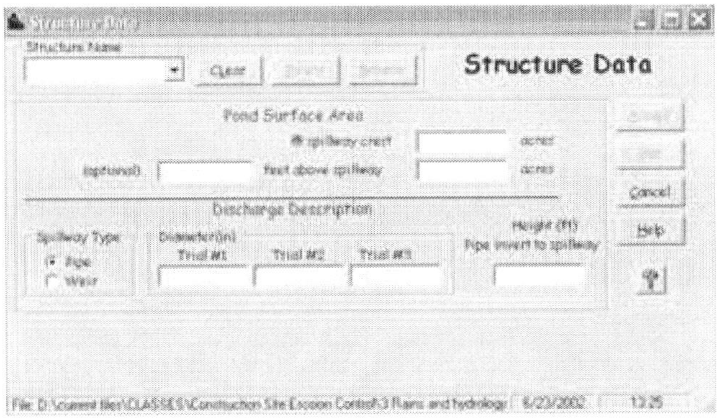

Figure 3.44. WinTR-55 structure data screen for detention facilities.

Example: WINTR-55 Applications to Construction Sites

As indicated previously, there are a number of situations where WinTR-55 (or TR-55) can be used to advantage when evaluating construction sites, including the design of erosion and sediment controls. These may include:

- Determination of flows leaving the site that may affect downstream areas. Downstream erosion controls may include filter fencing along the project perimeter, or sediment ponds, depending on flow conditions. These controls must be completed before any on-site construction is started.
- Determination of upland flows coming towards the disturbed areas. These flows must be diverted by swales or dikes, or safely carried through the construction sites. Channel design will be based on the expected flow conditions. These controls must be completed after the downstream controls, and before any on-site controls are started.
- Determination of on-site flows on slopes going towards filter fencing, sediment ponds, or other controls. These flows also will be needed to evaluate shear stress on channels and on slopes.

Figure 3.45 is an example map (base map: a portion of a USGS quadrangle sheet with 20 ft contours) showing a construction site, and the associated upland and downslope

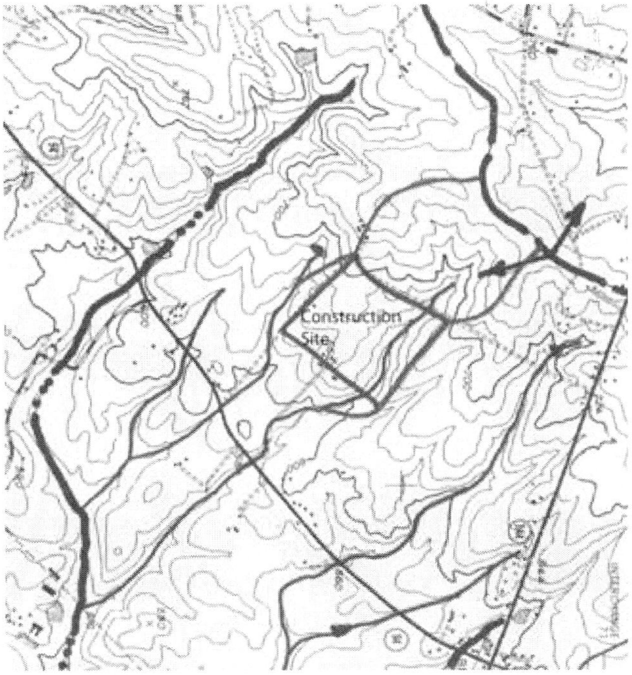

Figure 3.45. Determination of general upslope and downslope drainage areas from construction site.

drainages. This chapter has illustrated how it is possible to easily calculate the runoff characteristics affecting the site and downslope areas for different rain conditions. In addition, detailed site and rainfall conditions for different project phases can be evaluated and incorporated in the design of appropriate erosion and sediment controls.

Figure 3.46 shows subdrainages for the upslope, downslope, and on-site areas for this example construction site. Table 3.30 summarizes the characteristics of these areas, along with the hydrologic information needs for each area. Most of the site will be cleared and graded, except for the two small areas near the downslope edge. The upslope diversions (for U2 and U3) will carry the upslope water to the main channel. As an example, the diversion length for U2 is 900 ft long and the elevation drop is 70 ft. The channel slope for this diversion is therefore 70/900 = 0.08, or 8%.The runoff from the O1 and O2 on-site areas will be controlled by slope mulches and filter fences, before the runoff drains to the on-site main channel. A sediment pond will be constructed at the downslope property boundary before this main channel leaves the site, receiving runoff from U1, U2, U3, O1, and O2. This table shows 2 different rain depths for some conditions, based on the following discussion.

Table 3.31 and Figure 3.47 is an example using WinTR55 for this site. This example is for a sediment pond at the downslope boundary. Subareas O3, O4, O5, O6, and O7 are all very small and do not drain to this pond site, but drain towards the perimeter filter fabric fences. The reach data

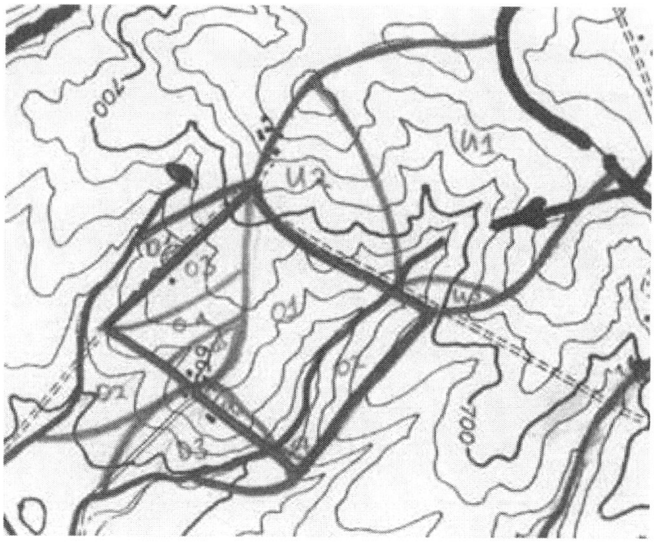

Figure 3.46. Subdrainage areas on and near construction site.

TABLE 3.30. Upslope and On-Site Subdrainage Area Characteristics for Construction Site and TR-55 Calculations.

Area Notation	Location	Objective	Area (acres)	Area (A_m, mi^2)	Cover n	Average Flow Path Slope	*CN* (all "C" soils)	*Ia* (in.)	Rain depth, *P* (in.)
U1	Upslope—direct to on site stream	Hydrograph (to be combined with U2 and U3)	37.4	0.058	0.4	8%	73	0.74	5.5
U2	Upslope—diversion to on site stream	Peak flow rate and hydrograph (to be combined with U1 and U3)	14.6	0.023	0.4	11.5	73	0.74	5.5
U3	Upslope—diversion to on site stream	Peak flow rate and hydrograph (to be combined with U1 and U2)	2.4	0.0038	0.4	12.7	73	0.74	5.5
O1	On site—drainage to sediment pond and main site stream (also slope protection needed)	Peak flow rate and hydrograph	12.6	0.020	0.011	10	91	0.198	6.6 8.4
O2	On site—drainage to filter fence and main site stream (also slope protection needed)	Peak flow rate and hydrograph	7.1	0.011	0.011	10.5	91	0.198	4.0 6.0
O3	On site—towards perimeter filter fence (also slope protection needed)	Peak flow rate and hydrograph	6.1	0.0095	0.011	5	91	0.198	4.0 6.0
O4	On site—towards perimeter filter fence (also slope protection needed)	Peak flow rate and hydrograph	3.1	0.0048	0.011	6.7	91	0.198	4.0 6.0
O5	On site—towards perimeter filter fence (also slope protection needed)	Peak flow rate and hydrograph	1.8	0.0028	0.011	11.3	91	0.198	4.0 6.0
O6	On site—nothing (will remain undisturbed)	na	1.3	0.0020	0.24	6.7	na	na	na
O7	On site—nothing (will remain undisturbed)	na	0.3	0.00047	0.24	10	na	na	na

(continued)

TABLE 3.30 (continued). Upslope and On-Site Subdrainage Area Characteristics for Construction Site and TR-55 Calculations.

Area Notation	Location	Objective	Ia/P	T_c (min)	T_c (hr)	Direct Runoff, Q (inches)	area-depth (A_mQ), (mi^2-inches)	Peak Unit Area Flow Rate (csm/in)	Peak Discharge (cfs)
U1	Upslope—direct to on site stream	Hydrograph (to be combined with U2 and U3)	0.13	29	0.48	2.8	0.16	411	66
U2	Upslope—diversion to on site stream	Peak flow rate and hydrograph (to be combined with U1 and U3)	0.13	25	0.42	2.8	0.064	449	29
U3	Upslope—diversion to on site stream	Peak flow rate and hydrograph (to be combined with U1 and U2)	0.13	20.7	0.35	2.8	0.011	449	4.9
O1	On site—drainage to sediment pond and main site stream (also slope protection needed)	Peak flow rate and hydrograph	0.03 0.02	3.5	0.06	5.4 7.3	0.11 0.15	662	73 99
O2	On site—drainage to filter fence and main site stream (also slope protection needed)	Peak flow rate and hydrograph	0.05 0.03	1.6	0.03	3.0 5.0	0.033 0.055	662	22 36
O3	On site—towards perimeter filter fence (also slope protection needed)	Peak flow rate and hydrograph	0.05 0.03	4.1	0.07	3.0 5.0	0.029 0.048	662	19 32
O4	On site—towards perimeter filter fence (also slope protection needed)	Peak flow rate and hydrograph	0.05 0.03	3.3	0.06	3.0 5.0	0.014 0.024	662	9.3 16
O5	On site—towards perimeter filter fence (also slope protection needed)	Peak flow rate and hydrograph	0.05 0.03	1.5	0.03	3.0 5.0	0.0084 0.014	662	5.6 9.3
O6	On site—nothing (will remain undisturbed)	na	na	na	na	na	na	na	na
O7	On site—nothing (will remain undisturbed)	na	na	na	na	na	na	na	na

assumed for reach A (the main channel to the outlet) is as follows: 1240 ft. long at 0.04 (4%) slope, n = 0.08, and bottom width = 10 ft. The channel side slopes are 1 to 3. Table 3.31 shows subareas O1 and O2 draining into reach A, but they actually drain directly to the outlet (the pond).

- Filter fences will be located along the side and bottom edges of the site, affected by O3, O4, O5, O6, and O7 subdrainage areas.
- Upslope channel diversions will be located along the upper edge of the site; subdrainage areas U2 and U3 will drain towards the site and drain into the on-site channel.
- All upslope areas, U1, U2, and U3, will be directed to the on-site drainage channel. The O1 and O2 on-site subdrainage areas will also drain to this on-site channel.
- A sediment pond will be located at the downslope edge of the property on the on-site drainage channel and collects the water from U1, U2, U3, O1, and O2.

Table 3.30 summarizes the subarea hydrologic site features, including the T_c values. This table also shows the calculated peak discharge rate for each of these areas. The following WinTR-55 example shows the calculations for the hydrograph entering the sediment pond (using Tuscaloosa, AL, rain conditions).

Design Storms for Different Site Controls

All of the information needed to calculate the expected flows from these upslope and on-site areas is shown on Table 3.32. The area has a SCS type III rain distribution and the construction period is assumed to be one year. The different site features will require different design storms due to the different levels of protection that are appropriate. Table 3.32 lists the features and the (assumed) acceptable failure rates during this one-year period, along with the corresponding design storm frequency and associated 24-hr rain total appropriate for the area. The design storms range from 4.0 to 8.4 inches in depth and the times of concentration range from 1.5 to 30 minutes. The design rain intensities could be very large for some of these design elements.

TABLE 3.31. WinTR55 Example for Sediment Pond (10-year rain event).

WinTR-55 Current Data Description

--- Identification Data ---

User:	Bob Pitt	Date:	7/21/2003
Project:	site diversions	Units:	English
SubTitle:	construction site example	Areal Units:	Acres
State:	Alabama		
County:	Tuscaloosa		

Filename: C:\Documents and Settings\rpitt.000\Application Data\WinTR-55\example erosion file.w55

--- Sub-Area Data ---

Name	Description	Reach	Area(ac)	RCN	Tc
U1	upslope to site stream	A	37.4	73	0.480
U2	upslope to diversion	A	14.6	73	0.420
U3	upslope to diversion	A	2.4	73	0.350
O1	on site to pond	A	12.6	91	0.100
O2	on site to pond	A	7.1	91	0.100

Total area: 74.10 (ac)

--- Storm Data --

Rainfall Depth by Rainfall Return Period

2-Yr (in)	5-Yr (in)	10-Yr (in)	25-Yr (in)	50-Yr (in)	100-Yr (in)	1-Yr (in)
4.2	5.4	6.3	7.1	7.8	8.6	3.6

Storm Data Source:	Tuscaloosa County, AL (NRCS)
Rainfall Distribution Type:	Type III
Dimensionless Unit Hydrograph:	<standard>

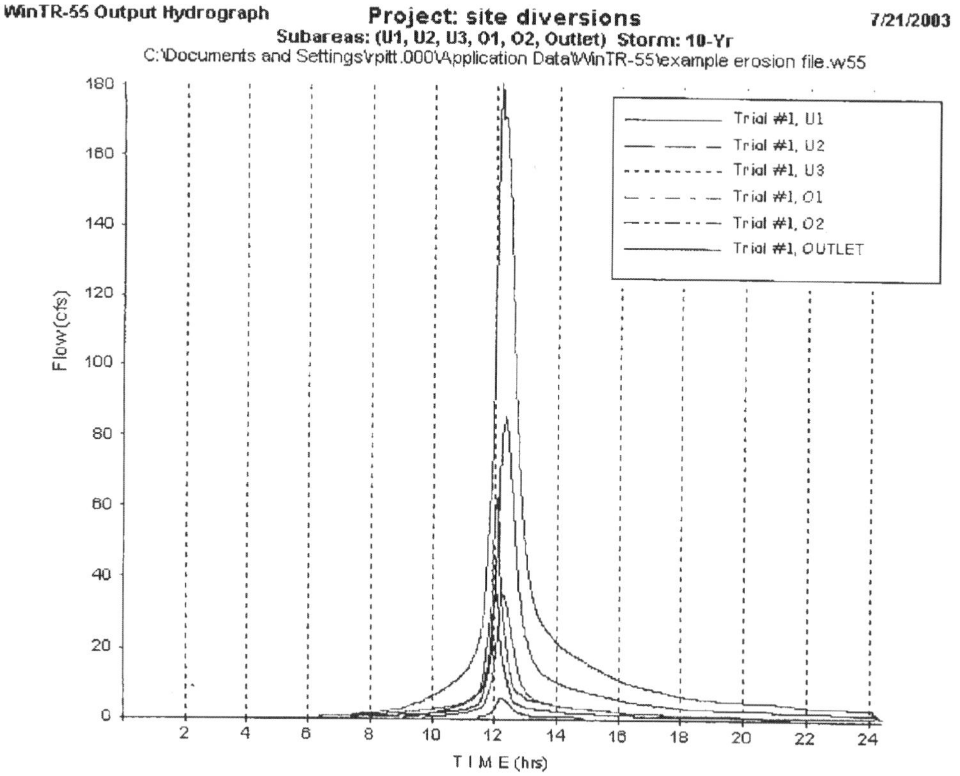

Figure 3.47. Subcatchment and outfall hydrographs for sediment pond location, WinTR55 example.

TABLE 3.32. Acceptable Levels of Protection for Different Site Activities.

Site Construction Control	Acceptable Failure Rate during 1-year Site Construction Activities	Design Storm Return Period (years)	24-hr Rain Depth Associated with this Design Storm Return Period
Diversion channels	25%	4	5.5
Main site channel	5%	20	6.6
Site slopes	10%	10	6.0
Site filter fences	50%	2	4.0
Sediment pond	5% and 1%	20 and 100	6.6 and 8.4
Downslope perimeter filter fences	10%	10	6.0

Runoff Water Depth

In some construction erosion control designs (such as those that use the shear stress calculations in Chapter 5), the water depth is needed for sheetflow conditions. The following equation can be used to calculate the estimated water depth for sheetflow, based on the Manning's equation (R, the hydraulic radius is equal to the flow depth for sheetflow):

$$y = \left(\frac{qn}{1.49s^{0.5}}\right)^{3/5}$$

where,

y = the flow depth (in feet),
q = the unit width flow rate (Q/W, the total flow rate, in ft^3/sec, divided by the slope width, in ft, ft^2/sec)
n = the sheet flow roughness coefficient, and
s = the slope (as a fraction)

Figure 3.48 contains plots of calculated flow depths for different slope conditions, using Birmingham, AL, rain conditions. These data are used later in Chapter 5 for calculating slope stability and needed reinforcements. These calculations used the Rational formula for the rain falling directly on the slopes, with the time of concentrations equal to the travel time of runoff down the slopes (as shown earlier in Figure 3.22). The Rational coefficients were varied depending on the slopes, according to typical values given for lawns in good condition: $C = 0.11$ for slopes <2%, $C = 0.16$ for slopes between 2 and 7%, and $C = 0.24$ for slopes >7%. These coefficients are averaged for sandy and heavy soil conditions. The calculations were made for several

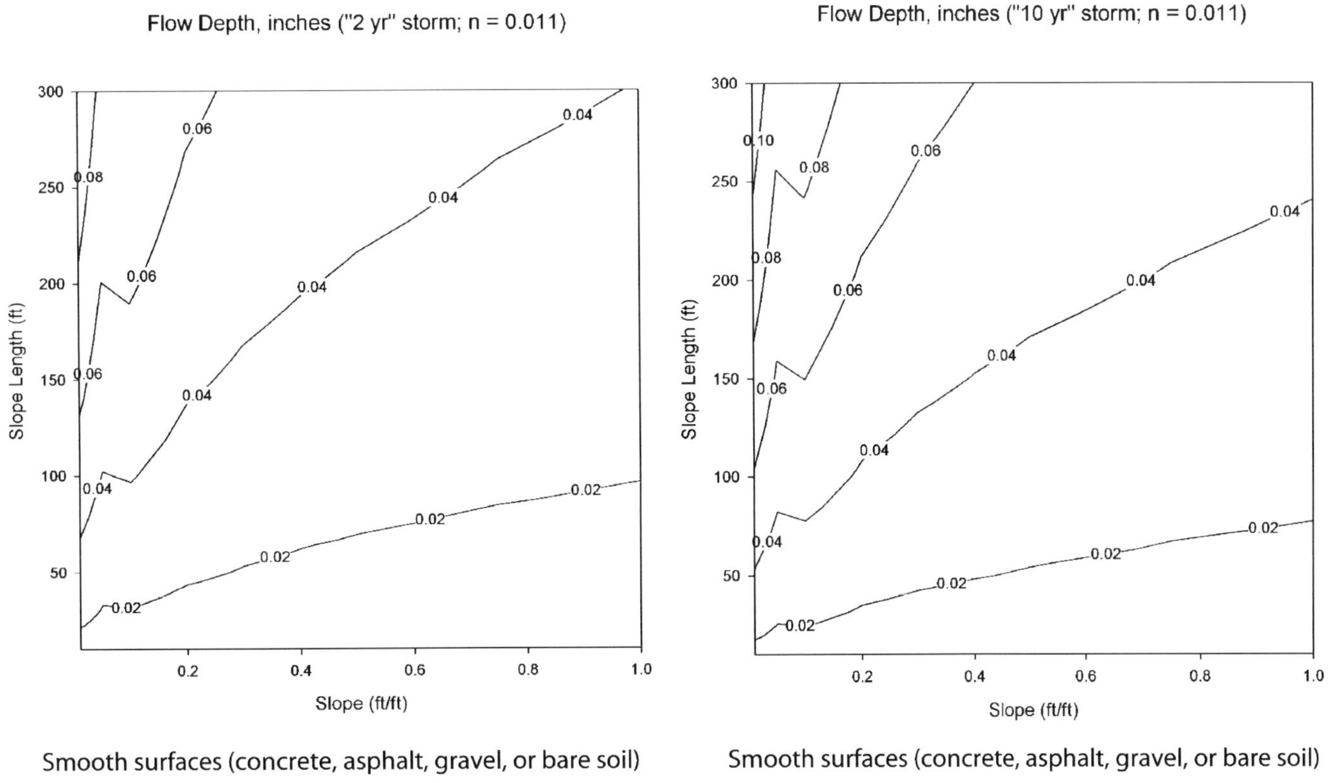

Figure 3.48. Calculated flow depths for different slope conditions.

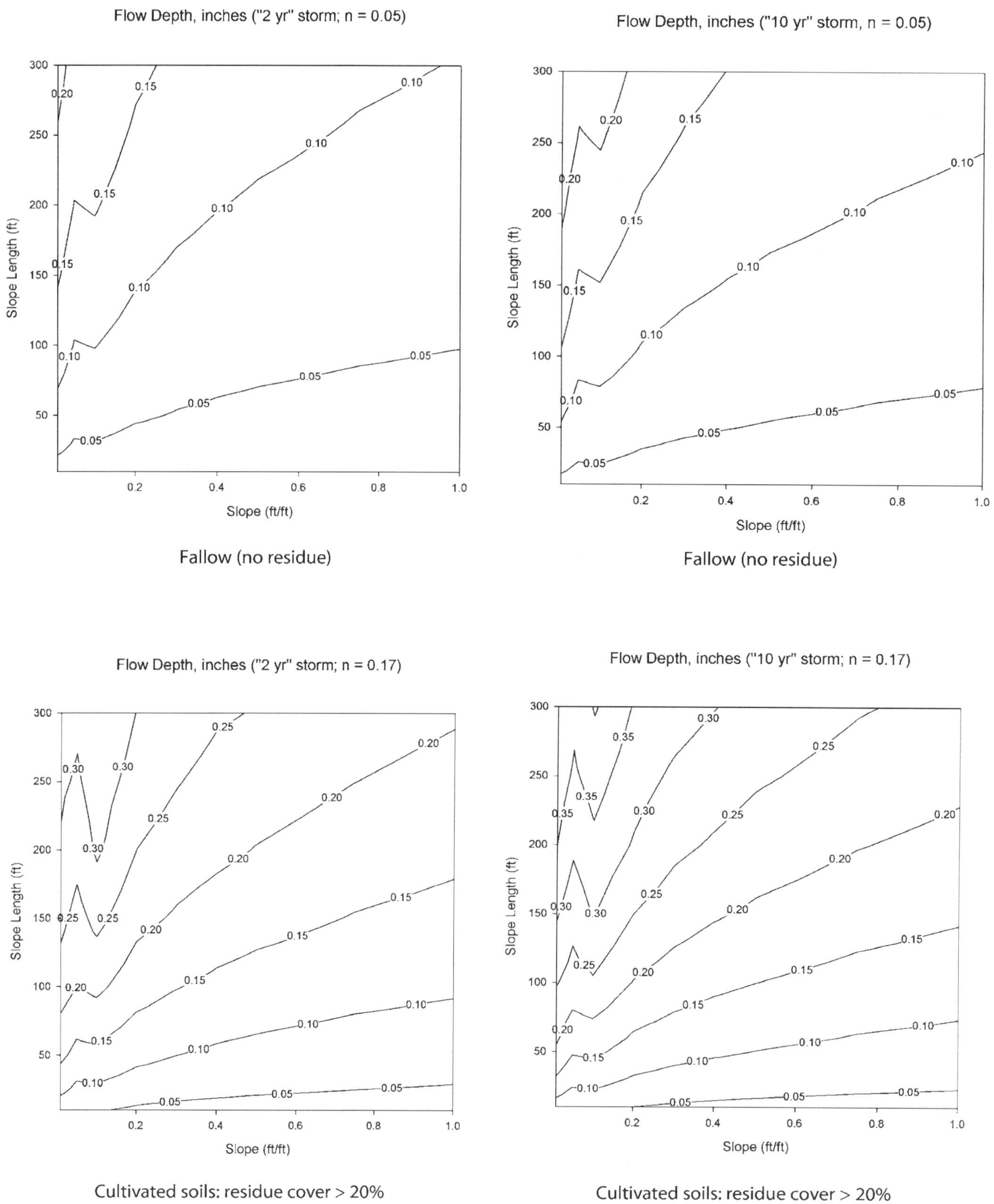

Figure 3.48 (continued). Calculated flow depths for different slope conditions.

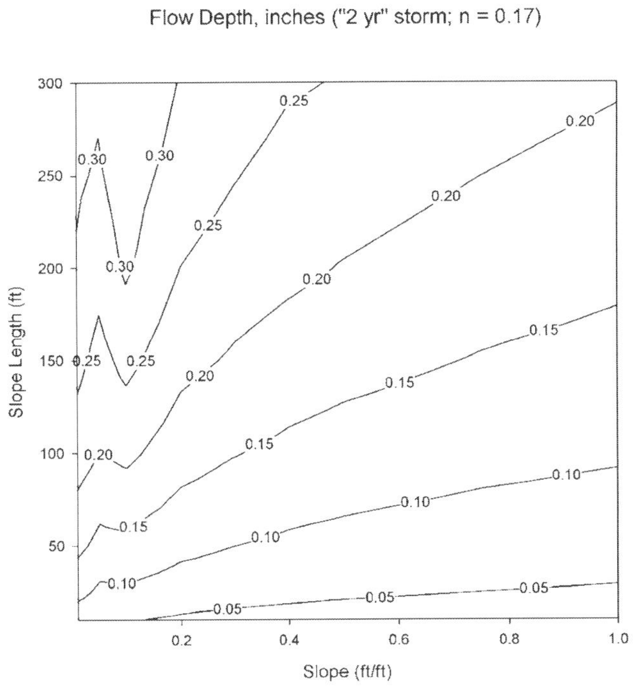

Cultivated soils: residue cover > 20%

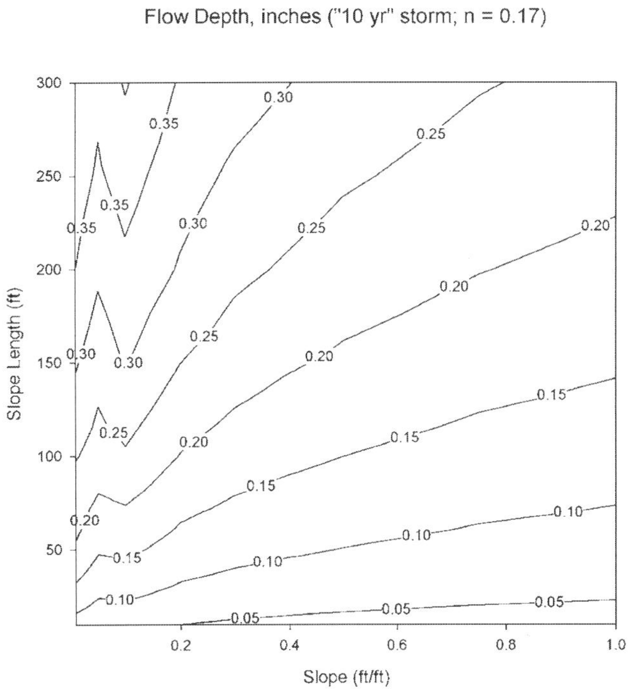

Cultivated soils: residue cover > 20%

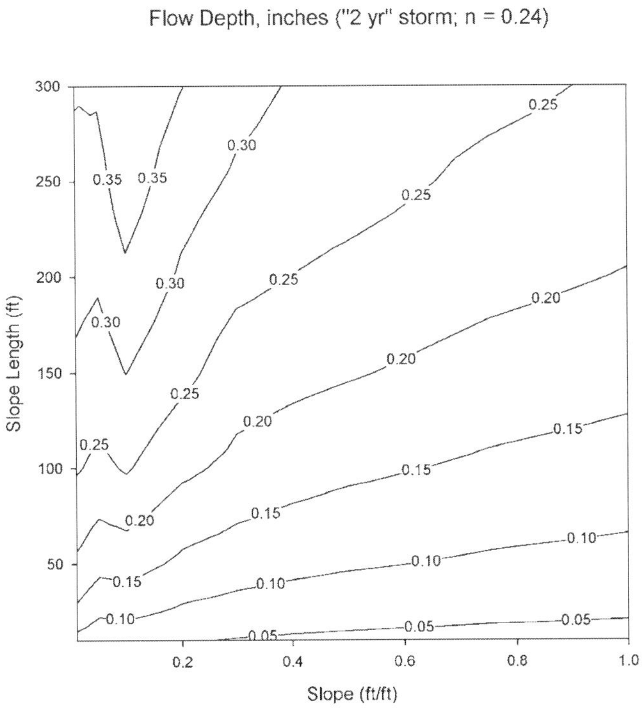

Grass, dense (weeping lovegrass, bluegrass, buffalo grass, blue gama grass, and native grass mixtures

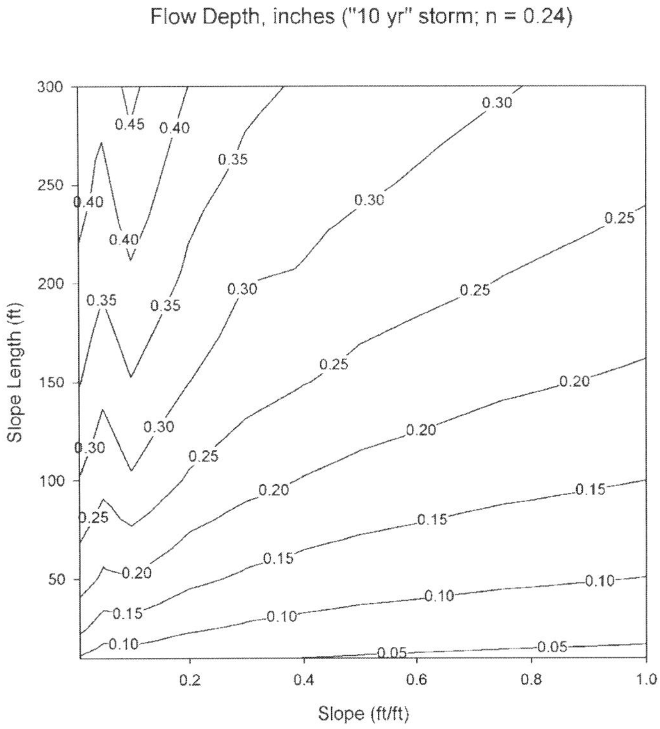

Grass, dense (weeping lovegrass, bluegrass, buffalo grass, blue gama grass, and native grass mixtures

Figure 3.48 (continued). Calculated flow depths for different slope conditions.

surface roughness conditions representing a range of slope surfaces at construction sites, including smooth surfaces (bare soil), fallow, cultivated soils, dense grass, and light underbrush. The slopes ranged from 1 to 100 percent and the slope lengths were as long as 300 ft, the generally maximum accepted slope length for silt fences, or for terrace spacing.

The Birmingham, AL, IDF curves for 2 and 10 year frequency storms (events having a 50 and 10% chance of occurring in any one year), are example design storms for erosion controls on construction site slopes. These IDF curves (shown earlier in Figure 3.9) are for NRCS type III rainfall distributions and have 24-hr total rain depths of 6 inches for the 10-yr event and 4.2 inches for the 2-yr event. The IDF curves assume the same rain intensities for all times of concentrations less than 5 minutes. That, plus changes in the Rational runoff coefficient for different slopes, cause the discontinuity on these plots at about 10 percent slopes.

The deepest water depths were for the flattest, but longest slopes, conditions that maximize the catchment area, increase the likelihood of substantial friction effects, and hinder drainage. Typical maximum water depths on the slopes are about 0.25 to 0.5 inches when the slopes have some residue, or growing grasses. If bare, the maximum depths can be much less. The slope length appears to be about twice as important as the slope angle in determining the water depth.

SUMMARY

This chapter reviewed rain conditions that affect erosion at construction sites. In many cases, a relatively few of the annual rains are responsible for the vast majority of the erosion potential. The much more common small rains likely contribute a very small fraction of the annual erosion losses from construction sites. The larger rains result in the greatest erosion and translate into much more substantial and costly sediment controls than if the focus could be only on the smaller rains. As frequently noted in this book, preventative erosion control strategies are much more cost effective than many of the treatment options.

This chapter also examines several approaches for calculating runoff conditions at construction sites. For some design objectives, peak flow rates are needed, while complete hydrographs may be necessary to meet other objectives. WinTR-55 is emphasized as a suitable and simple method for obtaining design flows and hydrographs for construction site erosion control design and for site evaluation. Long-term continuous simulations would be preferred for site evaluations, but a comprehensive model that considers construction site features and potential controls is not readily available.

The chapter ends with a comprehensive example for determining site hydrographic and hydrologic conditions at construction sites. This chapter is a fundamental component of a complete approach for evaluating and solving construction site erosion problems. These tools will be referenced frequently in the other book chapters.

IMPORTANT INTERNET LINKS

Maps and Aerial Photographs:
http://Virtualearth.msn.com

http://maps.google.com

http://Seamless.usgs.gov

Alabama Rainfall Atlas:
http://bama.ua.edu/~rain/

WinTR-55 computer program (new windows version, ver. 1.0.08, Jan 2005):
http://www.wcc.nrcs.usda.gov/hydro/hydro-tools-models-wintr55.html

TR-55 1986 documentation and early version of TR55 program:
http://www.wcc.nrcs.usda.gov/hydro/hydro-tools-models-tr55.html

TR-20 computer program (new windows January 2005 version):
http://www.wcc.nrcs.usda.gov/hydro/hydro-tools-models-wintr20.html

National Engineering Handbook, Part 630 HYDROLOGY
http://www.wcc.nrcs.usda.gov/hydro/hydro-techref-neh-630.html

U.S. Army Corps of Engineers, Hydrologic Management System User Guide (replacement for HEC-1) and River Analysis System User Guide for water surface profile calculations (replacement for HEC-2):
http://www.hec.usace.army.mil/

REFERENCES

Aldridge, B.N., and Garrett, J.M. *Roughness Coefficients for Stream Channels in Arizona.* U.S. Geological Survey Open-File Report, 87 p. 1973.

Arcement, Jr., G.J.and V.R. Schneider. *Guide for Selecting Manning's Roughness Coefficients for Natural Channels and Flood Plains.* United States Geological Survey Water-supply Paper 2339 and FHWA-TS-84-204. undated. (http://www.fhwa.dot.gov/bridge/wsp2339.pdf).

Barnes, H.H., Jr. *Roughness Characteristics of Natural Channels.* U.S. Geological Survey Water-Supply Paper 1849, 213 p. 1967.

Benson, M.A., and Dalrymple, Tate. *General Field and Office Procedures for Indirect Discharge Measurements.* U.S. Geological Survey Techniques of Water-Resources Investigations book 3, chap. Al, 30 p. 1967.

Chow, V.T. *Open-Channel Hydraulics.* New York, McGraw- Hill Book Co., 680 p. 1959.

Chow, V.T., Maidment, D. R, and Mays, L. W., *Applied Hydrology,* McGraw-Hill, 586 pages. 1988.

Cowan, W.L.. *Estimating Hydraulic Roughness Coefficients.* Agricultural Engineering, v. 37, no. 7, p. 473–475. 1956.

HEC. *HEC-RAS User's Manual,* Version 2.0. U.S. Army Corps of Engineers, Hydrologic Engineering Center, April 1997.

Illinois. *Illinois Procedures and Standards for Urban Soil Erosion and Sedimentation Control.* Association of Illinois Soil and Water Conservation Districts, Springfield, IL 62703. 1989.

Maidment, D. R. (ed.), *Handbook of Hydrology,* McGraw-Hill, 1422 pages. 1993.

McGee, T.J. *Water Supply and Sewerage.* McGraw-Hill, Inc., New York. 1991. NRCS. National Engineering Handbook, Part 630 HYDROLOGY, downloaded June 23, 2002 at: http://www.wcc.nrcs.usda.gov/water/quality/common/neh630/4content.html.

NRCS. *SITES Water Resource Site Analysis Computer Program User's Guide.* United States Department of Agriculture, Natural Resources Conservation Service. 469 pp. 2001.

NRCS. *TR-20 System: User Documentation.* United States Department of Agriculture, Natural Resources Conservation Service. 105 pp. 2002b (draft).

NRCS. *WinTR-55 User Manual.* U.S. Dept. of Agriculture, Natural Resources Conservation Service. Downloaded on June 23, 2002 from: http://www.wcc.nrcs.usda.gov/water/quality/common/tr55/tr55-beta.html Version dated April 23, 2002a.

Pitt, R. *Small Storm Urban Flow and Particulate Washoff Contributions to Outfall Discharges,* Ph.D. Dissertation, Civil and Environmental Engineering Department, University of Wisconsin, Madison, WI, November 1987.

Pitt, R. and S.R. Durrans. *Drainage of Water from Pavement Structures.* Alabama Dept. of Transportation. 253 pgs. September 1995.

Pitt, R., J. Lantrip, R. Harrison, C. Henry, and D. Hue. *Infiltration through Disturbed Urban Soils and Compost-Amended Soil Effects on Runoff Quality and Quantity.* U.S. Environmental Protection Agency, Water Supply and Water Resources Division, National Risk Management Research Laboratory. EPA 600/R-00/016. Cincinnati, Ohio. 231 pgs. December 1999.

Pitt, R., M. Lilburn, S. Nix, S.R. Durrans, S. Burian, J. Voorhees, and J. Martinson *Guidance Manual for Integrated Wet Weather Flow (WWF) Collection and Treatment Systems for Newly Urbanized Areas (New WWF Systems).* U.S. Environmental Protection Agency. 612 pgs. 1999.

Pitt, R, S. Chen, and S. Clark. "Compacted urban soils effects on infiltration and bioretention stormwater control designs." *Global Solutions for Urban Drainage; 9IUCD.* CD-ROM Proceedings of the 9th International Urban Drainage Conference, edited by E.W. Strecker and W.C. Huber., Sept 8–13, 2002, Portland, OR. Sponsored by the ASCE, Reston, VA, and the International Water Association, London. 2002.

Ponce, V.M., *Engineering Hydrology,* Prentice Hall, 640 pages. 1989.

SCS (now NRCS). *Urban Hydrology for Small Watersheds.* Technical Release 55, U.S. Department of Agriculture, Soil Conservation Service. 91 pp. 1975.

SCS. *Urban Hydrology for Small Watersheds.* Technical Release 55. U.S. Dept. of Agric., Soil Conservation Service. 156 pgs. 1986.

SCS. *Time of Concentration,* Hydrology Technical Note No. N4. United States Department of Agriculture, Soil Conservation Service, Northeast National Technical Center. 12 pp. 1986.

Thronson, R.E. *Comparative Costs of Erosion and Sediment Control, Construction Activities.* U.S. Environmental Protection Agency. EPA430/9-73.016. Washington, D.C. 1973.

Welle, P.I., Woodward, D. E., Fox Moody, H., *A Dimensionless Unit Hydrograph for the Delmarva Peninsula,* Paper No. 80-2013, ASAE 1980 Summer Meeting, 18 pp. 1980.

PROBLEMS

1. A rectangular, forested 10-acre parcel of property has been purchased by a developer for conversion to a three-shop strip mall. Can this plot of land be considered a watershed? Why or why not? What factors support your decision?

2. Delineate the watershed that is draining to the specified outlet in the maps given below. Compute the watershed slope, the channel length and the channel slope. The interval contours are 20 ft. Describe the site soils and determine the areas for each soil type in the watershed.

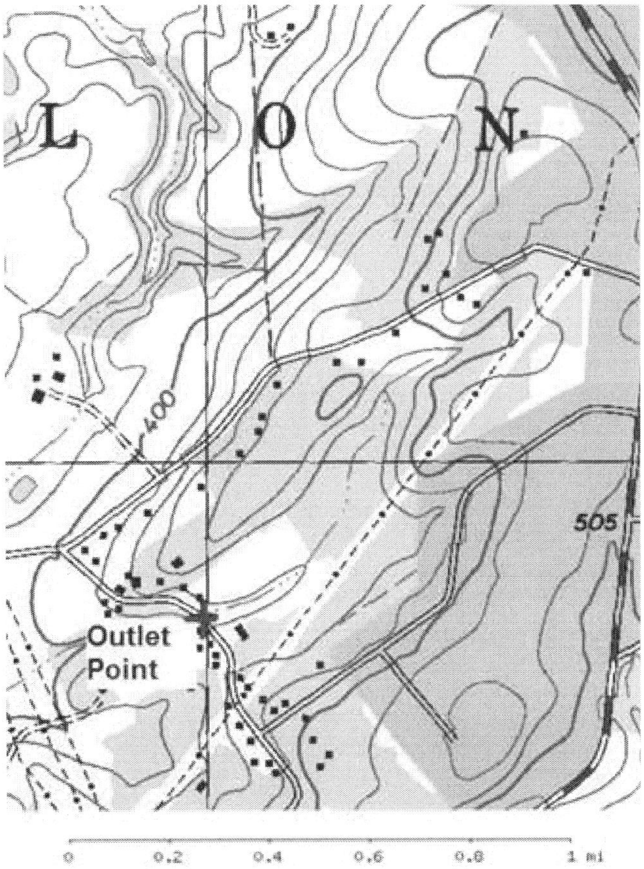

Note: The above map has 20 ft contour lines

3. Sources, and the resultant effects, of uncertainty are

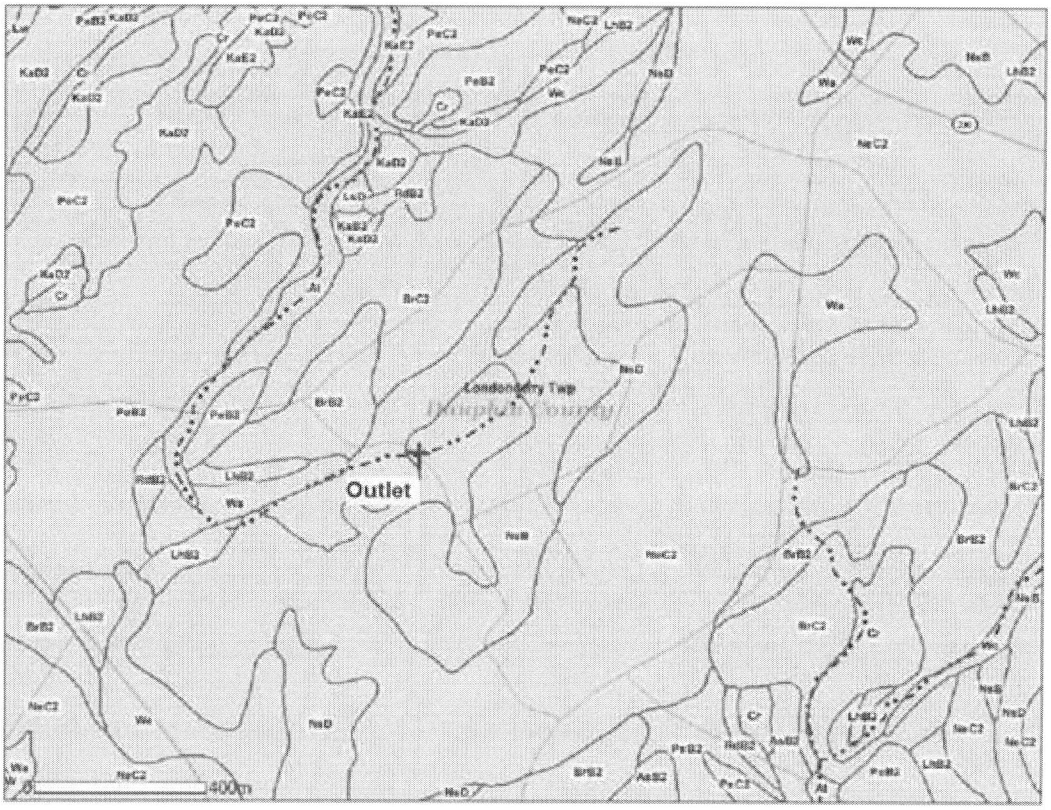

Soil map (*source:* Dauphin County, PA USDA, obtained from http://soilmap.psu.edu).

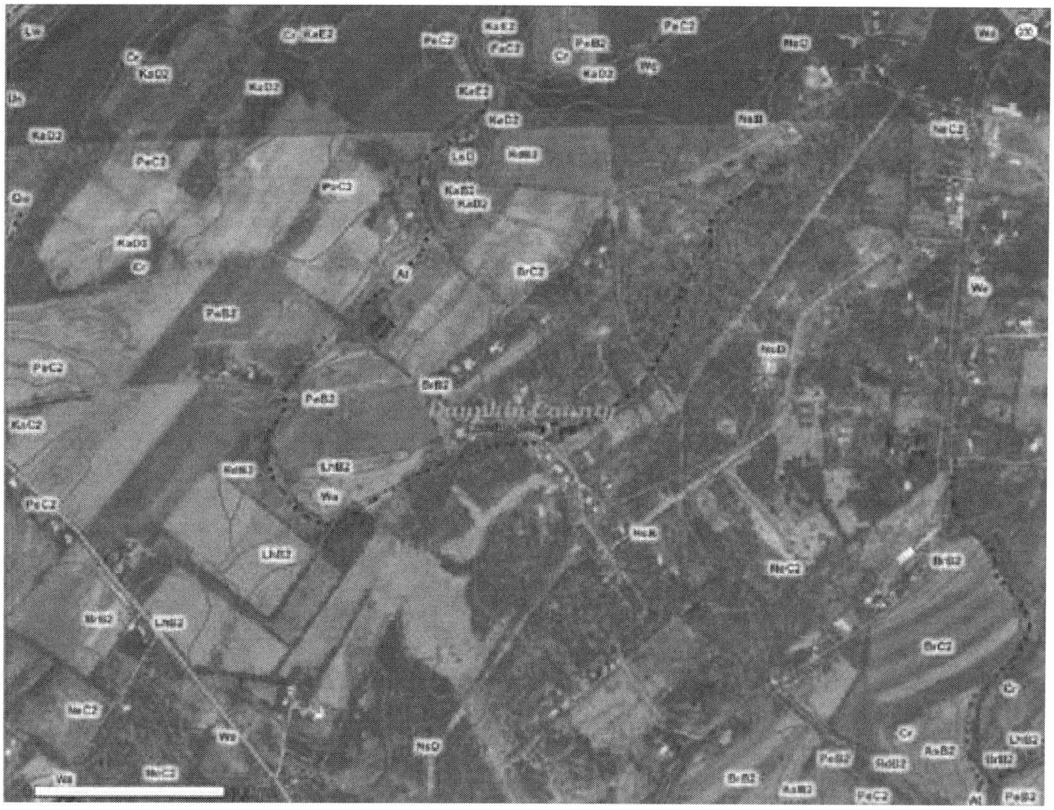

Soil map with aerial photograph (*source:* Dauphin County, PA USDA, obtained from http://soilmap.psu.edu).

always a concern when making hydrologic calculations. Compute the channel slope between sections 1 and 4 and for each of the three reaches. Average the computed slopes for the reaches. Is the slope calculated based on averaging the reach slopes similar to the overall watershed slope? Why or why not? How will this affect design decisions for the site assuming the entire watershed is developed?

4. For the reaches described in Problem 3, calculate the average velocity in these channels assuming that the channel is concrete lined, has side slopes of 3:1 (*h*:*v*), and the depth of flow is 0.2 ft. The base width is 4 ft. How does the velocity change if the channel is grass-lined? Does slope have a greater effect on the velocity for concrete- or grass-lined channels?

Survey Section	Elevation (ft)	Distance from Outlet
1	82	0
2	92	10,300
3	103	13,600
4	105	15,800

5. On a construction site, it is necessary to construct a grass-lined diversion channel. The cross section has a width of 10 ft, a depth of 1.5 ft, and side slopes of 4:1 (h:v), find the velocity assuming a slope of 0.002 ft/ft and an earthen surface with short grass (<6 inches high).

6. A small forested watershed (light understory brush) has an elevation drop of 15 ft and a principal flow path of 1000 ft. Compute the travel time along this flow path using the NRCS Time of Concentration method, assuming that the slope is consistent along this flow path and no channel flow occurs. Compare the results for sheetflow lengths of 100 ft and 300ft. Use the 2-year storm for your local area.

7. A graded, but unpaved, highway section under construction has a concrete gutter, with a longitudinal slope of 4% and a length of 10,800 ft. Determine the travel time using the NRCS Time of Concentration method. The sheetflow path will be across the lane section that is 40 ft wide with 0.5 lateral slope. Use the 2-year storm for your local area. Assume the gutter flow is shallow-concentrated flow.

8. The critical flow path for the time of concentration consists of the following sections. Estimate the time of concentration using the NRCS Time of Concentration method.

Section	Slope (%)	Length (ft)	Land Use
1	5.5	160	Forest (light underbrush)
2	3.1	690	Short gras
3	2.4	370	Bare ground
4	1.1	520	Riprap-lined waterway

9. Using the T_c from Problem 8, estimate the peak flow rate on a 3,280-ft section of asphalt roadway that is 60 ft wide using the NRCS tabular hydrograph method. Assume a 10-yr design frequency and your local IDF curve and rain type.

10. Calculate the time of concentration for the watershed shown in Problem 2, assuming the natural channel is 5 ft wide at the bottom and had 5:1 (*h*:*v*) side slopes. Assume good wood cover for the watershed and make reasonable assumptions as needed.

11. For the watershed delineated in Problem 2 and the T_c calculated in Problem 10, calculate the peak runoff rate for the 25-year storm (assuming B soils and good wood cover and making other reasonable assumptions as needed) using your local IDF curve.

12. The newest construction site in the watershed shown in Problem 1 has been delineated (the limits are outlined in black on the map copied below). Delineate the watershed that will drain the entire construction site to the creek.

Note: The above map has 20 ft contour lines

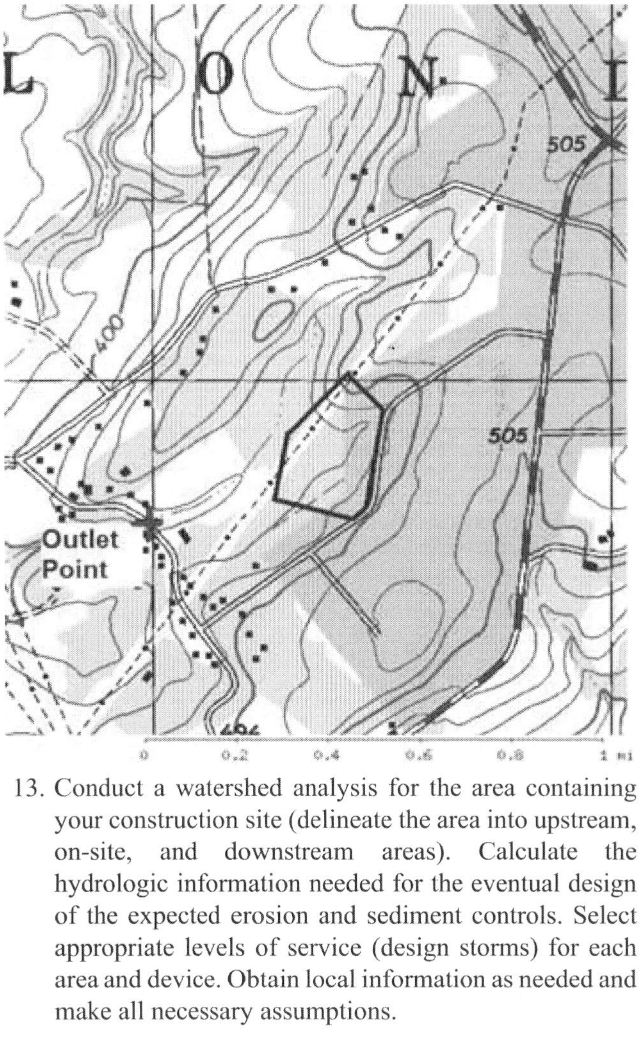

13. Conduct a watershed analysis for the area containing your construction site (delineate the area into upstream, on-site, and downstream areas). Calculate the hydrologic information needed for the eventual design of the expected erosion and sediment controls. Select appropriate levels of service (design storms) for each area and device. Obtain local information as needed and make all necessary assumptions.

APPENDIX 3.A. TABULAR HYDROGRAPH UNIT DISCHARGES (FROM TR-55, SCS 1986)

Exhibit 5-I: Tabular hydrograph unit discharges (csm/in) for type I rainfall distribution

HYDROGRAPH TIME (HOURS)

IA/P = 0.10 * * * TC = 0.1 HR * * * IA/P = 0.10

TRVL TIME (hr)	9.0	9.3	9.6	9.9	10.0	10.1	10.2	10.3	10.4	10.5	10.6	10.7	10.8	11.0	11.2	11.4	11.6	11.8	12.0	12.3	12.6	13.0	13.5	14.0	14.5	15.0	15.5	16.0	17.0	18.0	20.0	24.0
0.0	30	40	56	183	337	504	326	155	122	107	93	81	73	66	60	56	54	52	49	46	44	40	36	32	30	29	28	27	26	24	20	13
.10	26	35	48	93	153	276	428	360	223	156	123	103	83	72	65	59	56	54	51	47	45	42	37	33	30	29	28	28	26	24	21	13
.20	23	30	41	60	82	129	227	361	360	269	194	147	118	85	71	63	58	55	53	49	46	43	39	35	31	29	29	28	26	24	21	14
.30	22	29	39	56	73	111	188	303	341	293	227	173	136	94	75	65	60	56	54	50	47	43	39	35	31	29	29	28	26	25	21	14
.40	18	25	34	46	53	66	96	157	255	312	300	251	199	126	90	73	64	59	56	52	48	44	40	36	32	30	29	28	26	25	21	14
.50	18	24	32	44	50	61	84	133	214	280	293	265	221	144	99	77	66	60	56	52	49	45	41	37	32	30	29	28	27	25	21	14
.75	14	19	25	34	38	43	49	62	88	134	190	234	252	221	162	115	87	71	63	56	52	47	43	39	35	31	29	29	27	25	22	15
1.0	11	14	19	26	28	31	34	38	44	52	68	98	141	222	238	191	139	101	79	63	56	51	45	41	37	33	30	29	27	26	22	15
1.5	9	10	13	17	18	20	22	25	27	30	34	38	44	74	132	191	211	190	151	101	73	58	50	45	41	37	33	30	28	27	23	16
2.0	6	7	9	11	12	13	14	15	16	18	20	22	24	29	38	58	97	148	193	193	141	89	61	51	46	41	37	33	29	27	24	17
2.5	4	5	7	8	9	9	10	11	11	12	13	14	16	19	23	29	39	58	93	154	181	147	87	61	51	45	41	37	30	28	25	18
3.0	2	3	5	6	6	7	7	8	9	9	10	10	11	13	15	19	23	28	39	72	124	170	138	86	61	50	45	40	33	29	26	19

IA/P = 0.30 * * * TC = 0.1 HR * * * IA/P = 0.30

TRVL TIME (hr)	9.0	9.3	9.6	9.9	10.0	10.1	10.2	10.3	10.4	10.5	10.6	10.7	10.8	11.0	11.2	11.4	11.6	11.8	12.0	12.3	12.6	13.0	13.5	14.0	14.5	15.0	15.5	16.0	17.0	18.0	20.0	24.0
0.0	0	0	0	61	195	343	232	129	113	103	91	81	76	71	66	64	62	61	59	56	54	51	47	43	40	40	39	39	37	36	31	21
.10	0	0	0	12	45	145	277	247	169	131	112	98	87	76	70	65	64	62	60	57	55	53	49	44	40	40	39	39	38	36	32	21
.20	0	0	0	9	33	107	220	238	192	151	125	107	94	79	71	66	64	62	61	58	56	53	49	45	41	40	40	39	38	36	32	21
.30	0	0	0	1	6	24	79	173	216	200	168	139	118	90	77	70	66	64	62	59	57	54	50	46	41	40	40	39	38	36	32	22
.40	0	0	0	1	4	17	59	135	189	196	177	152	129	97	81	72	67	64	63	60	57	54	51	46	42	40	40	39	38	36	32	22
.50	0	0	0	0	0	3	12	43	104	161	185	180	161	121	93	79	71	66	64	61	58	55	52	48	43	40	40	39	38	37	33	22
.75	0	0	0	0	0	1	5	18	49	92	130	153	159	142	114	92	79	71	66	63	60	56	53	49	44	41	40	40	38	37	33	23
1.0	0	0	0	0	0	0	0	0	2	9	27	56	92	144	152	128	103	86	75	67	63	59	55	51	47	43	40	40	39	37	34	24
1.5	0	0	0	0	0	0	0	0	0	0	0	2	6	32	79	121	136	127	109	85	72	64	58	55	51	47	43	40	39	38	34	25
2.0	0	0	0	0	0	0	0	0	0	0	0	0	0	1	9	33	70	105	126	118	97	75	63	58	54	50	46	42	40	39	35	26
2.5	0	0	0	0	0	0	0	0	0	0	0	0	0	0	0	2	11	32	63	105	118	100	75	63	58	54	50	46	40	39	36	27
3.0	0	0	0	0	0	0	0	0	0	0	0	0	0	0	0	0	1	3	12	43	84	112	96	74	63	58	54	50	42	40	37	28

IA/P = 0.50 * * * TC = 0.1 HR * * * IA/P = 0.50

TRVL TIME (hr)	9.0	9.3	9.6	9.9	10.0	10.1	10.2	10.3	10.4	10.5	10.6	10.7	10.8	11.0	11.2	11.4	11.6	11.8	12.0	12.3	12.6	13.0	13.5	14.0	14.5	15.0	15.5	16.0	17.0	18.0	20.0	24.0
0.0	0	0	0	0	0	3	13	28	39	44	45	46	46	48	49	51	53	54	54	55	56	55	53	50	50	49	49	49	49	48	45	32
.10	0	0	0	0	0	2	9	22	33	41	44	45	46	48	49	50	52	54	54	54	56	55	53	50	50	49	49	49	49	48	45	32
.20	0	0	0	0	0	0	1	7	17	28	36	41	44	46	48	49	51	53	54	54	56	56	54	51	50	50	49	49	49	48	45	32
.30	0	0	0	0	0	0	0	1	5	13	23	32	38	44	47	48	50	51	53	54	54	55	54	52	50	50	49	49	49	48	46	33
.40	0	0	0	0	0	0	0	0	3	10	18	27	34	43	46	48	49	51	53	54	54	55	54	52	50	50	49	49	49	48	46	33
.50	0	0	0	0	0	0	0	0	0	2	7	15	23	36	43	46	48	50	51	53	54	55	55	53	50	50	49	49	49	48	46	34
.75	0	0	0	0	0	0	0	0	0	1	3	7	13	27	37	43	46	48	50	52	54	55	55	54	51	50	50	49	49	48	46	34
1.0	0	0	0	0	0	0	0	0	0	0	0	0	1	8	20	32	40	45	47	50	52	54	55	54	52	50	50	49	49	49	47	36
1.5	0	0	0	0	0	0	0	0	0	0	0	0	0	0	1	5	13	23	33	42	47	51	54	55	54	53	50	50	49	49	48	37
2.0	0	0	0	0	0	0	0	0	0	0	0	0	0	0	0	0	1	5	12	25	37	46	51	54	55	54	52	50	49	49	48	39
2.5	0	0	0	0	0	0	0	0	0	0	0	0	0	0	0	0	0	0	1	5	15	31	45	51	53	55	54	52	50	49	48	41
3.0	0	0	0	0	0	0	0	0	0	0	0	0	0	0	0	0	0	0	0	0	3	14	32	45	51	53	54	54	51	49	48	42

RAINFALL TYPE = I * * * TC = 0.1 HR * * * SHEET 1 OF 10

Exhibit 5-I: Tabular hydrograph unit discharges (csm/in) for type I rainfall distribution—continued

HYDROGRAPH TIME (HOURS)

IA/P = 0.10 * * * TC = 0.2 HR * * * IA/P = 0.10

TRVL TIME (hr)	9.0	9.3	9.6	9.9	10.0	10.1	10.2	10.3	10.4	10.5	10.6	10.7	10.8	11.0	11.2	11.4	11.6	11.8	12.0	12.3	12.6	13.0	13.5	14.0	14.5	15.0	15.5	16.0	17.0	18.0	20.0	24.0
0.0	28	37	52	126	220	379	405	267	168	129	108	92	80	69	62	58	55	53	50	47	44	41	37	33	30	29	28	27	26	24	21	13
.10	24	32	44	71	108	182	313	375	303	213	157	124	103	78	68	61	57	55	52	49	46	42	38	34	30	29	29	28	26	24	21	14
.20	21	28	38	53	65	94	153	260	336	314	247	187	145	97	77	67	60	57	54	50	47	44	39	35	31	30	29	28	26	25	21	14
.30	20	27	36	50	60	82	129	216	296	308	267	214	168	110	82	69	62	58	55	51	47	44	40	36	32	30	29	28	26	25	21	14
.40	17	23	31	42	47	56	74	111	181	258	291	275	234	152	103	79	68	61	57	53	49	45	41	37	33	30	29	28	27	25	22	14
.50	16	22	30	40	45	52	66	96	153	223	269	273	247	171	115	86	71	63	58	54	50	46	42	37	33	30	29	28	27	25	22	14
.75	13	17	24	32	35	39	44	52	68	99	145	194	229	240	183	132	97	77	66	58	53	48	44	39	35	31	30	29	27	26	22	15
1.0	11	13	17	24	26	29	32	35	39	45	56	75	107	189	229	206	158	115	88	67	58	52	46	42	38	34	31	29	28	26	23	15
1.5	8	10	12	15	17	19	21	23	25	28	31	35	39	60	106	166	204	197	165	112	79	60	51	46	42	37	33	30	28	27	23	16
2.0	5	7	9	10	11	12	13	14	15	17	18	20	22	27	34	49	79	126	169	188	152	97	64	52	46	42	38	34	29	28	24	17
2.5	3	5	6	8	8	9	10	10	11	12	13	14	15	18	22	26	34	49	77	136	176	155	95	64	52	46	42	37	31	28	25	18
3.0	2	3	4	5	6	6	7	7	8	8	9	9	10	11	13	16	19	23	29	49	91	154	167	102	67	53	47	42	34	29	26	19

IA/P = 0.30 * * * TC = 0.2 HR * * * IA/P = 0.30

TRVL TIME (hr)	9.0	9.3	9.6	9.9	10.0	10.1	10.2	10.3	10.4	10.5	10.6	10.7	10.8	11.0	11.2	11.4	11.6	11.8	12.0	12.3	12.6	13.0	13.5	14.0	14.5	15.0	15.5	16.0	17.0	18.0	20.0	24.0
0.0	0	0	0	22	76	206	258	207	144	119	104	92	82	74	68	65	63	62	60	57	55	52	48	44	40	40	39	39	38	36	32	21
.10	0	0	0	3	16	56	156	224	213	167	135	114	99	81	73	67	65	63	61	58	56	53	50	45	41	40	40	39	38	36	32	22
.20	0	0	0	2	11	41	116	189	205	179	150	126	108	85	75	69	65	63	62	59	56	54	50	46	41	40	40	39	38	36	32	22
.30	0	0	0	0	2	8	30	58	155	188	182	161	138	103	83	74	68	65	63	60	57	55	51	47	42	40	40	39	38	36	32	22
.40	0	0	0	0	1	6	22	66	126	167	177	166	147	111	88	76	69	66	63	61	58	55	51	47	43	40	40	39	38	37	33	22
.50	0	0	0	0	0	1	4	16	50	100	145	167	166	136	105	85	75	69	65	62	59	56	53	49	44	41	40	40	38	37	33	23
.75	0	0	0	0	0	0	2	7	22	50	87	119	140	148	124	101	85	74	69	64	61	57	54	50	45	42	40	40	38	37	33	23
1.0	0	0	0	0	0	0	0	0	1	3	11	28	55	114	142	135	113	93	80	69	64	60	56	52	48	44	41	40	39	37	34	24
1.5	0	0	0	0	0	0	0	0	0	0	0	1	2	17	52	97	124	129	115	91	75	65	59	55	52	48	43	41	39	38	35	25
2.0	0	0	0	0	0	0	0	0	0	0	0	0	0	0	4	19	49	84	111	120	102	79	65	59	55	51	47	43	40	39	36	26
2.5	0	0	0	0	0	0	0	0	0	0	0	0	0	0	0	1	6	20	45	89	113	104	78	65	59	55	51	47	41	39	36	28
3.0	0	0	0	0	0	0	0	0	0	0	0	0	0	0	0	0	0	2	7	30	68	108	99	77	65	58	54	50	43	40	37	29

IA/P = 0.50 * * * TC = 0.2 HR * * * IA/P = 0.50

TRVL TIME (hr)	9.0	9.3	9.6	9.9	10.0	10.1	10.2	10.3	10.4	10.5	10.6	10.7	10.8	11.0	11.2	11.4	11.6	11.8	12.0	12.3	12.6	13.0	13.5	14.0	14.5	15.0	15.5	16.0	17.0	18.0	20.0	24.0
0.0	0	0	0	0	0	0	4	15	28	38	43	45	45	48	49	50	52	53	53	53	53	53	53	50	50	50	49	49	49	48	45	32
.10	0	0	0	0	0	0	3	11	22	32	40	43	45	47	48	50	52	53	53	53	53	53	53	50	50	50	49	49	49	48	45	32
.20	0	0	0	0	0	0	2	8	17	27	36	41	43	46	48	49	51	53	53	53	53	53	53	51	50	50	49	49	49	48	45	33
.30	0	0	0	0	0	0	1	6	13	23	31	38	41	45	48	49	51	52	53	53	53	53	53	51	50	50	49	49	49	48	46	33
.40	0	0	0	0	0	0	1	4	10	19	27	34	39	44	47	49	50	52	53	53	53	53	53	51	50	50	49	49	49	49	46	33
.50	0	0	0	0	0	0	1	3	8	15	23	30	36	43	46	48	50	51	53	53	53	53	53	52	50	50	49	49	49	48	46	33
.75	0	0	0	0	0	0	0	0	3	8	14	20	27	37	43	46	48	50	52	53	53	53	53	52	50	50	50	49	49	48	46	34
1.0	0	0	0	0	0	0	0	0	0	0	0	2	4	14	26	36	42	46	48	51	52	53	53	53	52	50	50	49	49	49	47	35
1.5	0	0	0	0	0	0	0	0	0	0	0	0	0	1	5	13	23	33	40	46	49	52	53	53	53	51	50	50	49	49	48	37
2.0	0	0	0	0	0	0	0	0	0	0	0	0	0	0	1	3	8	16	26	37	44	50	52	53	53	53	51	50	49	49	48	38
2.5	0	0	0	0	0	0	0	0	0	0	0	0	0	0	0	0	1	3	8	19	31	43	50	52	53	53	52	51	50	49	48	39
3.0	0	0	0	0	0	0	0	0	0	0	0	0	0	0	0	0	0	0	0	3	11	25	41	49	52	53	53	52	50	49	48	41

RAINFALL TYPE = I * * * TC = 0.2 HR * * * SHEET 2 OF 10

Exhibit 5-I: Tabular hydrograph unit discharges (csm/in) for type I rainfall distribution—continued

IA/P = 0.10 * * * TC = 0.3 HR * * * IA/P = 0.10

TRVL TIME (hr)	9.0	9.3	9.6	9.9	10.0	10.1	10.2	10.3	10.4	10.5	10.6	10.7	10.8	11.0	11.2	11.4	11.6	11.8	12.0	12.3	12.6	13.0	13.5	14.0	14.5	15.0	15.5	16.0	17.0	18.0	20.0	24.0
0.0	25	34	46	87	138	242	349	346	256	182	141	115	97	76	66	60	57	54	52	48	45	42	38	34	30	29	28	28	26	24	21	13
.10	24	32	44	77	118	200	299	330	281	215	166	132	109	82	69	62	58	55	52	49	46	42	38	34	30	29	29	28	26	24	21	14
.20	21	28	38	54	69	102	167	255	305	289	240	190	152	103	79	67	61	57	54	50	47	44	40	35	31	30	29	28	26	25	21	14
.30	20	27	36	51	63	89	141	217	276	285	255	212	172	115	85	71	63	58	55	51	48	44	40	36	32	30	29	28	26	25	21	14
.40	17	23	31	42	48	58	79	120	185	246	272	260	228	156	108	82	69	62	57	53	49	45	41	37	33	30	29	28	27	25	22	14
.50	16	22	30	40	45	54	70	104	158	216	253	258	238	173	120	89	72	63	58	54	50	46	42	37	33	30	29	28	27	25	22	14
.75	13	17	24	32	35	39	45	54	72	104	147	189	219	231	182	135	100	79	67	58	53	48	44	40	35	32	30	29	27	26	22	15
1.0	11	13	17	24	26	29	32	35	40	46	58	79	110	184	221	202	158	118	90	68	58	52	46	42	38	34	31	29	28	26	23	15
1.5	8	10	12	15	17	19	21	23	25	28	31	35	40	62	107	163	199	193	165	114	81	61	52	46	42	37	33	31	28	27	23	16
2.0	5	7	9	10	11	12	13	14	15	17	18	20	22	27	35	50	80	125	165	184	152	99	64	53	47	42	38	34	29	28	24	17
2.5	3	5	6	8	8	9	10	10	11	12	13	14	15	18	22	27	35	50	78	134	173	154	96	64	52	46	42	37	31	28	25	18
3.0	2	3	4	6	6	7	7	8	8	9	9	10	11	12	14	17	21	26	34	61	107	164	145	94	64	52	46	41	33	29	26	19

IA/P = 0.30 * * * TC = 0.3 HR * * * IA/P = 0.30

TRVL TIME (hr)	9.0	9.3	9.6	9.9	10.0	10.1	10.2	10.3	10.4	10.5	10.6	10.7	10.8	11.0	11.2	11.4	11.6	11.8	12.0	12.3	12.6	13.0	13.5	14.0	14.5	15.0	15.5	16.0	17.0	18.0	20.0	24.0
0.0	0	0	0	8	32	98	198	217	192	148	125	108	95	79	72	67	64	63	61	58	56	53	49	45	41	40	40	39	38	36	32	22
.10	0	0	0	1	6	23	73	156	197	194	164	138	118	91	77	70	66	64	62	59	57	54	50	46	41	40	40	39	38	36	32	22
.20	0	0	0	1	4	17	54	122	172	187	171	149	128	98	81	72	67	64	63	60	57	54	51	46	42	40	40	39	38	36	32	22
.30	0	0	0	0	0	3	12	40	95	146	173	172	157	120	34	79	71	67	64	61	58	55	52	48	43	40	40	39	38	37	33	23
.40	0	0	0	0	0	2	9	30	73	122	156	167	160	128	100	83	73	68	65	62	59	56	52	48	44	41	40	39	38	37	33	23
.50	0	0	0	0	0	1	6	22	56	100	137	157	159	135	107	87	76	69	65	62	59	56	53	49	44	41	40	40	38	37	33	23
.75	0	0	0	0	0	0	1	3	9	26	53	86	115	143	134	113	93	80	72	65	62	58	54	51	46	42	40	40	39	37	33	23
1.0	0	0	0	0	0	0	0	0	1	5	14	32	57	111	138	132	113	94	81	69	64	60	56	52	48	44	41	40	39	37	34	24
1.5	0	0	0	0	0	0	0	0	0	0	0	0	1	8	34	74	111	126	121	99	80	67	61	56	53	49	44	41	39	38	35	26
2.0	0	0	0	0	0	0	0	0	0	0	0	0	0	0	2	11	33	66	97	118	108	84	67	60	56	52	48	44	40	39	36	27
2.5	0	0	0	0	0	0	0	0	0	0	0	0	0	0	0	0	3	12	32	74	106	112	83	67	60	56	52	48	41	39	37	28
3.0	0	0	0	0	0	0	0	0	0	0	0	0	0	0	0	0	0	1	4	21	54	98	107	81	67	60	55	51	43	40	37	29

IA/P = 0.50 * * * TC = 0.3 HR * * * IA/P = 0.50

TRVL TIME (hr)	9.0	9.3	9.6	9.9	10.0	10.1	10.2	10.3	10.4	10.5	10.6	10.7	10.8	11.0	11.2	11.4	11.6	11.8	12.0	12.3	12.6	13.0	13.5	14.0	14.5	15.0	15.5	16.0	17.0	18.0	20.0	24.0
0.0	0	0	0	0	0	0	1	6	16	27	36	41	44	46	48	49	51	52	52	52	52	52	52	51	50	50	49	49	49	48	46	33
.10	0	0	0	0	0	0	1	4	12	22	31	38	42	45	48	49	51	52	52	52	52	52	52	51	50	50	49	49	49	48	46	33
.20	0	0	0	0	0	0	0	3	9	18	27	34	39	44	47	49	50	51	52	52	52	52	52	51	50	50	49	49	49	48	46	33
.30	0	0	0	0	0	0	0	2	7	14	23	30	36	43	46	48	50	51	52	52	52	52	52	51	50	50	49	49	49	48	46	33
.40	0	0	0	0	0	0	0	1	5	11	19	26	33	41	46	48	49	51	52	52	52	52	52	52	50	50	49	49	49	48	46	33
.50	0	0	0	0	0	0	0	1	4	9	15	23	30	39	44	47	49	50	51	52	52	52	52	52	50	50	49	49	49	48	46	33
.75	0	0	0	0	0	0	0	0	2	4	8	14	20	31	40	44	47	49	51	52	52	52	52	52	50	50	50	49	49	48	46	34
1.0	0	0	0	0	0	0	0	0	0	0	0	1	2	9	20	31	39	44	47	50	51	52	52	52	52	50	50	50	49	49	47	36
1.5	0	0	0	0	0	0	0	0	0	0	0	0	0	0	3	9	18	28	36	44	48	51	52	52	52	51	50	50	49	49	48	37
2.0	0	0	0	0	0	0	0	0	0	0	0	0	0	0	0	2	5	12	21	33	42	48	51	52	52	52	51	50	49	49	48	38
2.5	0	0	0	0	0	0	0	0	0	0	0	0	0	0	0	0	0	2	5	15	27	40	48	51	52	52	52	51	50	49	48	40
3.0	0	0	0	0	0	0	0	0	0	0	0	0	0	0	0	0	0	0	1	3	11	25	41	48	51	52	52	51	50	49	48	41

RAINFALL TYPE = I * * * TC = 0.3 HR * * * SHEET 3 OF 10

Exhibit 5-I: Tabular hydrograph unit discharges (csm/in) for type I rainfall distribution—continued

IA/P = 0.10 * * * TC = 0.4 HR * * * IA/P = 0.10

TRVL TIME (hr)	9.0	9.3	9.6	9.9	10.0	10.1	10.2	10.3	10.4	10.5	10.6	10.7	10.8	11.0	11.2	11.4	11.6	11.8	12.0	12.3	12.6	13.0	13.5	14.0	14.5	15.0	15.5	16.0	17.0	18.0	20.0	24.0
0.0	23	31	42	66	96	157	250	310	304	244	186	149	122	89	73	64	59	56	53	49	46	43	39	35	31	29	29	28	26	25	21	14
.10	22	29	40	61	84	133	211	277	295	261	211	170	138	98	78	66	60	56	54	50	47	43	39	35	31	29	29	28	26	25	21	14
.20	19	26	34	47	56	75	114	178	244	278	267	230	190	128	93	75	65	59	56	52	48	44	40	36	32	30	29	28	27	25	21	14
.30	18	24	33	45	53	67	98	152	213	257	263	241	207	143	102	80	68	61	57	52	49	45	41	37	32	30	29	28	27	25	21	14
.40	15	21	28	38	43	49	61	86	130	185	233	253	245	188	132	97	77	66	60	55	51	46	42	38	34	30	29	28	27	25	22	14
.50	15	20	27	36	41	46	56	76	112	161	209	238	243	201	146	106	82	69	61	55	51	47	42	38	34	31	29	29	27	25	22	14
.75	12	16	21	29	32	35	39	46	57	77	108	147	184	220	200	157	118	91	74	61	55	50	45	40	36	32	30	29	27	26	22	15
1.0	10	12	16	21	24	26	29	32	36	40	48	61	83	147	202	212	178	138	105	75	62	54	47	43	39	35	31	29	28	26	23	16
1.5	8	9	11	14	16	17	19	21	23	25	28	31	35	50	83	134	179	193	177	131	92	65	53	47	42	38	34	31	29	27	24	16
2.0	5	6	8	10	10	11	12	13	14	15	17	18	20	25	31	42	64	102	144	179	163	111	70	55	48	43	39	35	30	28	25	17
2.5	3	4	6	7	8	8	9	9	10	11	12	13	14	16	20	24	30	42	63	114	160	170	107	70	54	47	43	38	31	29	25	18
3.0	2	3	4	5	6	6	7	7	8	8	9	9	10	11	13	16	19	23	30	51	90	148	161	104	69	54	47	42	34	29	26	19

IA/P = 0.30 * * * TC = 0.4 HR * * * IA/P = 0.30

TRVL TIME (hr)	9.0	9.3	9.6	9.9	10.0	10.1	10.2	10.3	10.4	10.5	10.6	10.7	10.8	11.0	11.2	11.4	11.6	11.8	12.0	12.3	12.6	13.0	13.5	14.0	14.5	15.0	15.5	16.0	17.0	18.0	20.0	24.0
0.0	0	0	0	3	14	48	115	184	192	178	148	127	111	88	77	70	66	64	62	59	56	54	50	46	41	40	40	39	38	36	32	22
.10	0	0	0	0	2	10	35	89	152	179	178	158	137	105	85	75	69	65	63	60	58	55	51	47	42	40	40	39	38	36	32	22
.20	0	0	0	0	2	7	26	68	124	161	172	163	146	113	90	78	70	66	64	61	58	55	52	47	43	40	40	39	38	37	33	22
.30	0	0	0	0	1	5	19	52	100	140	162	162	151	120	96	81	72	67	64	61	59	55	52	48	43	41	40	39	38	37	33	23
.40	0	0	0	0	0	1	4	14	39	80	120	148	158	142	114	92	79	71	67	63	60	57	53	49	45	41	40	40	38	37	33	23
.50	0	0	0	0	0	1	3	10	29	63	101	132	152	145	120	97	82	73	68	63	60	57	53	50	45	41	40	40	38	37	33	23
.75	0	0	0	0	0	0	0	1	4	13	31	58	87	130	138	123	103	87	77	68	63	59	55	51	47	43	41	40	39	37	34	24
1.0	0	0	0	0	0	0	0	0	0	2	7	18	36	86	125	134	122	104	88	73	66	61	57	53	49	45	41	40	39	38	34	24
1.5	0	0	0	0	0	0	0	0	0	0	0	0	1	10	36	75	109	124	120	100	81	68	61	56	53	49	44	41	40	38	35	26
2.0	0	0	0	0	0	0	0	0	0	0	0	0	0	0	1	6	22	50	83	116	116	91	70	61	57	53	49	45	40	39	36	27
2.5	0	0	0	0	0	0	0	0	0	0	0	0	0	0	0	0	2	8	22	60	97	111	88	70	61	56	53	48	42	39	37	28
3.0	0	0	0	0	0	0	0	0	0	0	0	0	0	0	0	0	0	0	2	14	43	89	106	86	69	61	56	52	44	40	37	29

IA/P = 0.50 * * * TC = 0.4 HR * * * IA/P = 0.50

TRVL TIME (hr)	9.0	9.3	9.6	9.9	10.0	10.1	10.2	10.3	10.4	10.5	10.6	10.7	10.8	11.0	11.2	11.4	11.6	11.8	12.0	12.3	12.6	13.0	13.5	14.0	14.5	15.0	15.5	16.0	17.0	18.0	20.0	24.0
0.0	0	0	0	0	0	0	0	2	8	17	27	34	40	44	47	49	50	51	51	51	51	51	51	51	50	50	49	49	49	48	46	33
.10	0	0	0	0	0	0	0	0	2	6	13	22	30	40	45	47	49	50	51	51	51	51	51	51	50	50	50	49	49	48	46	34
.20	0	0	0	0	0	0	0	0	0	1	4	10	18	33	41	45	48	49	51	51	51	51	51	51	50	50	50	49	49	49	46	34
.30	0	0	0	0	0	0	0	0	0	0	1	3	8	22	35	42	46	48	50	51	51	51	51	51	51	50	50	49	49	49	47	35
.40	0	0	0	0	0	0	0	0	0	0	1	2	6	19	32	40	45	47	49	51	51	51	51	51	51	50	50	49	49	49	47	35
.50	0	0	0	0	0	0	0	0	0	0	0	0	2	9	23	34	41	45	48	50	51	51	51	51	51	50	50	49	49	49	47	35
.75	0	0	0	0	0	0	0	0	0	0	0	0	1	5	14	26	35	42	46	49	50	51	51	51	51	50	50	50	49	49	47	36
1.0	0	0	0	0	0	0	0	0	0	0	0	0	0	0	1	5	14	25	35	44	48	50	51	51	51	51	50	50	49	49	48	37
1.5	0	0	0	0	0	0	0	0	0	0	0	0	0	0	0	0	2	6	13	27	39	47	50	51	51	51	51	50	49	49	48	39
2.0	0	0	0	0	0	0	0	0	0	0	0	0	0	0	0	0	0	0	2	9	21	36	47	50	51	51	51	51	50	49	48	40
2.5	0	0	0	0	0	0	0	0	0	0	0	0	0	0	0	0	0	0	0	1	6	19	37	47	50	51	51	51	50	40	48	42
3.0	0	0	0	0	0	0	0	0	0	0	0	0	0	0	0	0	0	0	0	0	1	8	25	40	48	50	51	51	50	50	49	43

RAINFALL TYPE = I * * * TC = 0.4 HR * * * SHEET 4 OF 10

Exhibit 5-I: Tabular hydrograph unit discharges (csm/in) for type I rainfall distribution—continued

TRVL TIME (hr)	9.0	9.3	9.6	9.9	10.0	10.1	10.2	10.3	10.4	10.5	10.6	10.7	10.8	11.0	11.2	11.4	11.6	11.8	12.0	12.3	12.6	13.0	13.5	14.0	14.5	15.0	15.5	16.0	17.0	18.0	20.0	24.0
IA/P = 0.10 * * * TC = 0.5 HR * * * IA/P = 0.10																																
0.0	21	23	38	56	73	111	175	247	282	271	229	186	152	108	84	71	63	58	55	51	47	44	40	35	31	30	29	28	26	25	21	14
.10	20	27	36	52	66	96	149	214	259	267	242	204	169	119	90	74	65	59	55	51	48	44	40	36	32	30	29	28	26	25	21	14
.20	17	23	31	42	49	60	84	127	185	234	256	247	219	156	112	86	72	63	58	53	49	45	41	37	33	30	29	28	27	25	22	14
.30	16	22	30	40	46	55	74	110	160	210	241	245	227	171	123	93	76	66	59	54	50	46	42	38	33	30	29	28	27	25	22	14
.40	14	19	26	35	38	43	51	67	95	138	186	222	237	210	157	115	89	73	64	57	52	47	43	39	35	31	29	29	27	25	22	15
.50	14	18	25	33	37	41	48	61	84	120	164	203	229	216	170	126	96	77	66	58	53	48	43	39	35	31	30	29	27	25	22	15
.75	11	14	19	26	29	32	35	40	48	61	83	114	148	202	210	176	137	106	84	66	57	51	46	41	37	33	30	29	28	26	22	15
1.0	10	11	15	19	21	24	26	29	32	36	41	50	65	116	176	203	190	156	122	85	67	56	49	44	40	35	32	30	28	26	23	16
1.5	7	9	11	13	14	16	17	19	21	23	25	28	31	42	67	110	157	187	183	146	104	72	55	48	43	39	35	32	29	27	24	17
2.0	4	6	7	9	10	10	11	12	13	14	15	17	18	22	27	36	53	83	123	175	175	124	77	57	49	44	40	35	30	28	25	18
2.5	3	4	5	7	7	8	8	9	9	10	11	12	13	15	18	22	27	36	53	97	146	166	119	76	57	49	44	39	32	29	25	18
3.0	1	2	3	5	5	6	6	6	7	8	8	9	9	11	12	15	17	21	27	43	76	135	158	114	75	57	48	43	35	30	26	19
IA/P = 0.30 * * * TC = 0.5 HR * * * IA/P = 0.30																																
0.0	0	0	0	1	6	22	64	123	168	174	165	143	126	100	84	74	68	65	63	60	57	54	51	47	42	40	40	39	38	36	32	22
.10	0	0	0	1	5	16	48	98	144	164	165	150	134	106	87	77	70	66	63	60	58	55	51	47	42	40	40	39	38	36	32	22
.20	0	0	0	0	1	3	12	36	77	122	150	160	154	126	101	85	75	69	65	62	59	56	52	48	44	41	40	39	38	37	33	23
.30	0	0	0	0	0	2	9	27	60	101	134	151	153	132	107	89	77	70	66	63	60	56	53	49	44	41	40	40	38	37	33	23
.40	0	0	0	0	0	0	2	6	20	47	83	117	140	148	125	102	86	76	69	64	61	57	54	50	46	42	40	40	39	37	33	23
.50	0	0	0	0	0	0	1	5	15	36	68	101	127	144	130	108	90	78	71	65	62	58	54	50	46	42	40	40	39	37	33	23
.75	0	0	0	0	0	0	0	2	6	17	36	60	87	125	133	121	103	88	78	68	63	59	55	52	47	43	41	40	39	37	34	24
1.0	0	0	0	0	0	0	0	0	0	1	3	9	21	62	106	130	126	112	96	78	69	62	57	54	50	46	42	40	39	38	34	25
1,5	0	0	0	0	0	0	0	0	0	0	0	0	1	6	23	56	92	115	120	106	87	71	62	57	53	49	45	42	40	38	35	26
2.0	0	0	0	0	0	0	0	0	0	0	0	0	0	0	0	3	14	36	67	104	113	96	74	63	58	54	50	46	40	39	36	27
2.5	0	0	0	0	0	0	0	0	0	0	0	0	0	0	0	0	1	5	15	47	85	109	93	73	63	57	53	49	42	40	37	28
3.0	0	0	0	0	0	0	0	0	0	0	0	0	0	0	0	0	0	0	1	9	32	77	105	90	72	62	57	53	45	40	38	30
IA/P = 0.50 * * * TC = 0.5 HR * * * IA/P = 0.50																																
0.0	0	0	0	0	0	0	0	1	4	10	18	26	33	41	45	47	49	50	50	50	50	50	50	50	50	50	50	49	49	48	46	33
.10	0	0	0	0	0	0	0	1	3	8	15	22	30	39	44	47	49	50	50	50	50	50	50	50	50	50	50	49	49	48	46	34
.20	0	0	0	0	0	0	0	0	2	6	12	19	26	37	43	46	48	50	50	50	50	50	50	50	50	50	50	49	49	48	46	34
.30	0	0	0	0	0	0	0	0	1	4	9	16	23	34	41	45	48	49	50	50	50	50	50	50	50	50	50	49	49	49	46	34
.40	0	0	0	0	0	0	0	0	1	3	7	13	19	31	39	44	47	49	50	50	50	50	50	50	50	50	50	49	49	49	46	34
.50	0	0	0	0	0	0	0	0	1	2	6	10	16	28	37	43	46	48	50	50	50	50	50	50	50	50	50	49	49	49	46	34
.75	0	0	0	0	0	0	0	0	0	1	3	6	10	20	30	38	43	46	48	50	50	50	50	50	50	50	50	49	49	49	47	35
1.0	0	0	0	0	0	0	0	0	0	0	0	0	0	3	10	20	30	38	43	47	49	50	50	50	50	50	50	50	49	49	47	36
1.5	0	0	0	0	0	0	0	0	0	0	0	0	0	0	2	6	14	23	31	41	46	49	50	50	50	50	50	50	49	49	48	38
2.0	0	0	0	0	0	0	0	0	0	0	0	0	0	0	0	0	2	6	13	25	36	45	49	50	50	50	50	50	49	49	48	39
2.5	0	0	0	0	0	0	0	0	0	0	0	0	0	0	0	0	0	1	4	12	23	37	46	49	50	50	50	50	50	49	48	40
3.0	0	0	0	0	0	0	0	0	0	0	0	0	0	0	0	0	0	0	0	2	8	21	37	46	49	50	50	50	50	49	48	42

RAINFALL TYPE = I * * * TC = 0.5 HR * * * SHEET 5 OF 10

Exhibit 5-I: Tabular hydrograph unit discharges (csm/in) for type I rainfall distribution—continued

TRVL TIME (hr)	9.0	9.3	9.6	9.9	10.0	10.1	10.2	10.3	10.4	10.5	10.6	10.7	10.8	11.0	11.2	11.4	11.6	11.8	12.0	12.3	12.6	13.0	13.5	14.0	14.5	15.0	15.5	16.0	17.0	18.0	20.0	24.0
IA/P = 0.10 * * * TC = 0.75 HR * * * IA/P = 0.10																																
0.0	16	22	30	41	47	59	82	119	167	209	229	233	212	161	122	97	80	69	62	56	51	46	42	38	33	31	29	28	27	25	22	14
.10	14	19	26	35	39	44	54	72	104	146	188	215	227	197	150	116	93	77	68	59	53	48	43	39	35	31	30	29	27	25	22	15
.20	13	16	22	30	33	37	42	50	65	91	128	168	200	218	183	142	110	89	75	63	56	49	44	40	36	32	30	29	27	26	22	15
.30	12	16	21	28	31	35	39	46	59	80	112	149	183	214	191	151	118	94	79	65	57	50	45	41	36	32	30	29	27	26	22	15
.40	12	15	20	27	30	33	37	43	54	71	98	132	166	207	197	161	126	100	83	67	58	51	45	41	37	33	30	29	27	26	22	15
.50	11	13	18	24	26	29	32	36	41	49	64	87	117	178	204	186	152	120	96	74	62	53	47	42	38	34	31	29	28	26	23	15
.75	10	12	15	21	23	25	28	31	35	40	49	63	83	136	180	190	171	142	115	85	69	57	49	44	39	35	32	30	28	26	23	16
1.0	8	10	12	16	17	19	21	23	25	28	31	35	42	66	110	158	186	180	157	116	86	65	53	46	42	38	34	31	28	27	23	16
1.5	5	7	8	10	11	12	13	14	15	17	18	20	22	27	35	52	83	124	159	174	147	102	69	54	47	42	38	34	30	28	24	17
2.0	3	5	6	8	8	9	9	10	11	12	13	14	15	18	22	27	36	52	80	131	165	148	99	68	54	47	42	38	31	28	25	18
2.5	2	3	4	5	6	6	7	7	8	9	9	10	11	12	14	17	21	26	35	63	107	157	141	96	67	54	46	42	34	29	26	19
3.0	1	1	2	3	4	4	4	5	5	6	6	7	7	8	10	11	13	15	19	26	42	86	151	141	101	71	55	47	38	31	27	20
IA/P = 0.30 * * * TC = 0.75 HR * * * IA/P = 0.30																																
0.0	0	0	0	0	1	4	14	35	69	105	131	143	146	124	104	89	79	72	68	64	60	56	53	49	44	41	40	40	38	37	33	23
.10	0	0	0	0	0	1	3	10	27	55	88	117	134	142	118	100	87	77	71	66	62	58	54	50	46	42	40	40	39	37	33	23
.20	0	0	0	0	0	0	2	7	20	43	73	102	123	137	123	105	90	80	73	66	62	58	54	50	46	42	40	40	39	37	33	23
.30	0	0	0	0	0	0	0	1	5	15	34	60	88	128	134	118	101	88	78	69	64	60	55	52	48	43	41	40	39	37	34	24
.40	0	0	0	0	0	0	0	1	4	12	27	49	75	118	131	121	105	91	80	71	65	60	56	52	48	44	41	40	39	37	34	24
.50	0	0	0	0	0	0	0	0	1	3	9	21	39	87	122	128	116	101	88	75	68	62	57	53	49	45	42	40	39	38	34	24
.75	0	0	0	0	0	0	0	0	0	1	4	10	21	56	95	118	121	110	98	82	72	64	58	54	50	46	42	41	39	38	34	25
1.0	0	0	0	0	0	0	0	0	0	0	0	1	2	12	39	76	106	119	115	98	82	69	61	57	53	49	44	42	40	38	35	26
1.5	0	0	0	0	0	0	0	0	0	0	0	0	0	0	3	14	37	67	95	112	105	85	69	61	56	52	48	44	40	39	36	27
2.0	0	0	0	0	0	0	0	0	0	0	0	0	0	0	0	1	4	15	35	74	102	107	84	69	61	56	52	48	41	39	37	28
2.5	0	0	0	0	0	0	0	0	0	0	0	0	0	0	0	0	0	1	5	23	55	94	103	82	68	60	56	52	44	40	37	29
3.0	0	0	0	0	0	0	0	0	0	0	0	0	0	0	0	0	0	0	0	3	15	50	93	100	81	68	60	55	47	41	38	30
IA/P = 0.50 * * * TC = 0.75 HR * * * IA/P = 0.50																																
0.0	0	0	0	0	0	0	0	0	1	2	5	10	17	29	37	42	45	48	49	49	49	49	49	49	49	49	49	49	49	49	47	34
.10	0	0	0	0	0	0	0	0	0	2	4	8	14	26	35	41	45	47	49	49	49	49	49	49	49	49	49	49	49	49	47	35
.20	0	0	0	0	0	0	0	0	0	1	3	7	11	23	33	39	44	47	48	49	49	49	49	49	49	49	49	49	49	49	47	35
.30	0	0	0	0	0	0	0	0	0	1	2	5	9	20	30	38	43	46	48	49	49	49	49	49	49	49	49	49	49	49	47	35
.40	0	0	0	0	0	0	0	0	0	1	2	4	8	17	28	36	41	45	47	49	49	49	49	49	49	49	49	49	49	49	47	35
.50	0	0	0	0	0	0	0	0	0	0	1	3	6	15	25	34	40	44	47	48	49	49	49	49	49	49	49	49	49	49	47	35
.75	o	0	0	0	0	0	0	0	0	0	1	1	3	9	18	27	35	40	44	47	48	49	49	49	49	49	49	49	49	49	47	36
1.0	0	0	0	0	0	0	0	0	0	0	0	0	1	4	10	18	27	35	40	45	48	49	49	49	49	49	49	49	49	49	48	37
1.5	0	0	0	0	0	0	0	0	0	0	0	0	0	0	2	6	13	21	29	38	44	48	49	49	49	49	49	49	49	49	48	38
2.0	0	0	0	0	0	0	0	0	0	0	0	0	0	0	0	1	4	9	15	26	36	44	48	49	49	49	49	49	49	49	48	39
2.5	0	0	0	0	0	0	0	0	0	0	0	0	0	0	0	0	0	1	4	11	21	34	44	48	48	48	48	48	48	48	48	40
3.0	0	0	0	0	0	0	0	0	0	0	0	0	0	0	0	0	0	0	0	2	8	19	35	44	48	48	48	48	48	48	48	42

RAINFALL TYPE = I * * * TC = 0.75 HR * * * SHEET 6 OF 10

Exhibit 5-I: Tabular hydrograph unit discharges (csm/in) for type I rainfall distribution—continued

TRVL TIME (hr)	9.0	9.3	9.6	9.9	10.0	10.1	10.2	10.3	10.4	10.5	10.6	10.7	10.8	11.0	11.2	11.4	11.6	11.8	12.0	12.3	12.6	13.0	13.5	14.0	14.5	15.0	15.5	16.0	17.0	18.0	20.0	24.0
IA/P = 0.10													* * * TC = 1.0 HR * * *																IA/P = 0.10			
0.0	14	19	25	34	38	46	58	79	106	136	165	186	202	185	153	122	101	85	74	63	56	50	44	40	35	32	30	29	27	26	22	15
.10	13	18	24	32	36	42	53	70	94	122	151	174	194	188	160	130	106	89	77	65	57	50	44	40	36	32	30	29	27	26	22	15
.20	12	15	21	28	31	34	40	49	63	83	109	137	162	190	180	152	124	102	86	70	61	53	46	41	37	33	31	29	27	26	22	15
.30	12	15	20	26	29	33	37	45	57	75	98	124	149	187	183	159	131	107	90	73	62	53	47	42	37	33	31	29	28	26	23	15
.40	10	13	17	23	25	28	31	35	42	52	67	88	112	160	183	176	151	125	103	81	67	56	48	43	39	35	31	30	28	26	23	15
.50	10	13	16	22	24	27	30	33	39	48	61	79	101	148	181	181	157	131	108	84	69	58	49	43	39	35	32	30	28	26	23	16
.75	9	11	14	19	21	23	26	29	33	38	47	59	75	115	153	172	167	148	125	96	77	62	51	45	40	36	33	30	28	27	23	16
1.0	8	9	11	15	16	17	19	21	23	26	29	33	40	61	95	134	162	169	157	126	97	72	57	49	43	39	35	32	29	27	24	16
1.5	5	6	8	10	10	11	12	13	14	16	17	19	21	25	33	48	74	107	138	160	148	111	76	59	50	44	39	35	30	28	25	17
2.0	3	4	6	7	8	8	9	9	10	11	12	13	14	17	20	25	33	48	71	115	153	153	108	75	58	49	43	39	32	29	25	18
2.5	2	3	4	5	5	6	6	7	7	8	9	9	10	12	14	16	20	25	33	56	94	139	147	104	74	58	49	43	34	30	26	19
3.0	1	1	2	3	3	4	4	4	5	5	6	6	7	8	9	10	12	14	17	24	38	76	131	143	108	78	60	50	39	32	27	20
IA/P = 0.30													* * * TC = 1.0 HR * * *																IA/P = 0.30			
0.0	0	0	0	0	1	2	7	17	34	55	79	99	114	128	114	100	89	80	74	68	63	59	54	51	46	43	41	40	39	37	33	23
.10	0	0	0	0	0	0	2	5	13	27	46	68	89	116	124	110	98	87	79	71	65	60	56	52	48	43	41	40	39	37	34	24
.20	0	0	0	0	0	0	1	4	10	21	37	58	78	109	121	113	101	90	81	72	66	61	56	52	48	44	41	40	39	37	34	24
.30	0	0	0	0	0	0	0	1	3	8	17	31	49	87	113	118	109	98	87	76	69	63	57	53	49	45	42	40	39	38	34	24
.40	0	0	0	0	0	0	0	1	2	6	13	25	41	78	107	117	111	101	90	78	70	63	58	54	50	45	42	41	39	38	34	25
.50	0	0	0	0	0	0	0	0	2	5	10	20	34	69	100	115	113	103	93	80	71	64	58	54	50	46	42	41	39	38	34	25
.75	0	0	0	0	0	0	0	0	0	1	2	5	10	31	61	90	107	110	104	90	79	69	61	56	52	48	44	41	40	38	35	25
1.0	0	0	0	0	0	0	0	0	0	0	0	1	3	12	33	61	89	105	109	99	86	73	64	58	54	50	45	42	40	38	35	26
1.5	0	0	0	0	0	0	0	0	0	0	0	0	0	1	4	13	31	55	80	104	104	88	73	63	58	53	49	45	41	39	36	27
2.0	0	0	0	0	0	0	0	0	0	0	0	0	0	0	0	0	2	8	20	51	82	100	90	74	64	58	54	50	42	40	37	28
2.5	0	0	0	0	0	0	0	0	0	0	0	0	0	0	0	0	0	1	3	13	37	76	98	88	73	64	58	53	45	41	38	30
3,0	0	0	0	0	0	0	0	0	0	0	0	0	0	0	0	0	0	0	0	2	9	34	77	95	86	73	63	58	49	42	38	31
IA/P = 0.50													* * * TC = 1.0 HR * * *																IA/P = 0.50			
0.0	0	0	0	0	0	0	0	0	0	1	3	5	9	18	28	35	40	44	47	48	48	48	48	48	48	48	48	48	48	48	47	35
.10	0	0	0	0	0	0	0	0	0	1	2	4	7	16	25	33	39	43	46	48	48	48	48	48	48	48	48	48	48	48	47	35
.20	0	0	0	0	0	0	0	0	0	1	2	3	6	14	23	31	38	42	45	47	48	48	48	48	48	48	48	48	48	48	47	36
.30	0	0	0	0	0	0	0	0	0	0	1	3	5	12	21	29	36	41	44	47	48	48	48	48	48	48	48	48	48	48	47	36
.40	0	0	0	0	0	0	0	0	0	0	1	2	4	10	19	27	34	40	44	47	48	48	48	48	48	48	48	48	48	48	47	36
.50	0	0	0	0	0	0	0	0	0	0	1	2	3	8	16	25	33	38	43	46	48	48	48	48	4&	48	48	48	48	48	47	36
.75	0	0	0	0	0	0	0	0	0	0	1	2	3	8	15	23	30	36	41	45	47	48	48	48	48	48	48	48	48	48	47	36
1.0	0	0	0	0	0	0	0	0	0	0	0	0	0	2	5	11	19	27	33	41	45	47	48	48	48	48	48	48	48	48	48	37
1.5	0	0	0	0	0	0	0	0	0	0	0	0	0	0	1	2	5	11	17	28	37	44	47	48	48	48	48	48	48	48	48	39
2.0	0	0	0	0	0	0	0	0	0	0	0	0	0	0	0	0	1	3	7	16	26	38	45	47	48	48	48	48	48	48	48	40
2.5	0	0	0	0	0	0	0	0	0	0	0	0	0	0	0	0	0	0	1	5	12	24	38	45	47	47	47	47	47	47	47	41
3.0	0	0	0	0	0	0	0	0	0	0	0	0	0	0	0	0	0	0	0	1	5	14	29	40	46	47	47	47	47	47	47	42
RAINFALL TYPE = I													* * * TC = 1.0 HR * * *																SHEET 7 OF 10			

Exhibit 5-I: Tabular hydrograph unit discharges (csm/in) for type I rainfall distribution—continued

TRVL TIME (hr)	9.0	9.3	9.6	9.9	10.0	10.1	10.2	10.3	10.4	10.5	10.6	10.7	10.8	11.0	11.2	11.4	11.6	11.8	12.0	12.3	12.6	13.0	13.5	14.0	14.5	15.0	15.5	16.0	17.0	18.0	20.0	24.0
IA/P = 0.10													* * * TC = 1.25 HR * * *																IA/P = 0.10			
0.0	12	16	22	29	33	38	47	61	80	103	127	149	164	180	163	138	115	98	85	71	62	53	47	41	37	33	31	29	27	26	22	15
.10	11	14	19	25	28	31	36	44	55	72	92	116	138	167	175	156	132	111	95	77	66	56	48	43	38	34	31	30	28	26	23	15
.20	11	14	18	24	27	30	34	40	50	65	83	105	127	160	172	160	138	116	99	80	68	57	49	43	39	34	31	30	28	26	23	15
.30	10	12	16	21	23	25	28	32	38	46	58	75	95	136	164	169	154	132	112	89	74	61	51	45	40	36	32	30	28	26	23	16
.40	9	12	15	20	22	24	27	30	35	42	53	68	86	126	157	167	157	137	117	92	76	62	52	46	40	36	33	30	28	26	23	16
.50	9	11	13	17	19	21	23	26	29	33	39	49	61	97	134	160	165	152	132	104	84	67	55	47	42	37	33	31	28	27	23	16
.75	7	9	11	14	15	17	18	20	22	25	28	32	38	59	90	124	150	159	152	126	101	77	60	51	45	40	35	32	29	27	24	17
1.0	6	8	10	12	13	14	15	17	19	20	23	25	29	40	60	91	123	148	157	144	118	88	66	54	47	41	37	33	29	27	24	17
1.5	4	6	7	9	9	10	11	12	13	14	15	16	18	22	28	40	59	85	114	145	150	121	86	65	54	46	41	37	31	28	25	13
2.0	2	3	5	6	6	7	7	8	9	9	10	11	11	13	16	19	24	33	47	80	119	144	124	90	68	55	47	42	33	29	26	19
2.5	1	2	3	4	4	5	5	6	6	7	7	8	8	10	11	13	16	19	24	38	65	111	140	120	88	67	54	47	37	31	27	20
3.0	1	1	2	2	3	3	3	4	4	5	5	5	6	7	8	9	11	13	15	21	32	62	114	136	116	86	66	54	41	33	27	20
IA/P = 0.30													* * * TC = 1.25 HR * * *																IA/P = 0.30			
0.0	0	0	0	0	0	1	4	10	21	35	53	71	86	106	115	104	94	86	79	72	66	61	56	52	48	44	41	40	39	37	34	24
.10	0	0	0	0	0	0	1	3	8	16	29	45	62	92	107	113	101	92	84	75	69	63	58	53	49	45	42	41	39	38	34	24
.20	0	0	0	0	0	0	1	2	6	13	23	37	54	84	104	110	104	94	86	77	70	64	58	54	49	45	42	41	39	38	34	24
.30	0	0	0	0	0	0	0	0	2	4	10	19	31	62	90	106	109	101	92	81	73	66	60	55	51	46	43	41	39	38	34	25
.40	0	0	0	0	0	0	0	0	1	3	8	15	26	55	83	102	107	103	94	83	74	67	60	56	51	47	43	41	39	38	35	25
.50	0	0	0	0	0	0	0	0	0	1	3	6	12	33	62	38	103	106	100	88	78	69	62	57	53	48	44	42	40	38	35	25
.75	0	0	0	0	0	0	0	0	0	0	0	1	3	12	31	56	80	97	103	97	87	75	65	59	55	50	46	43	40	39	35	26
1.0	0	0	0	0	0	0	0	0	0	0	0	0	1	4	13	32	56	79	95	101	94	80	69	61	56	52	48	44	40	39	36	27
1.5	0	0	0	0	0	0	0	0	0	0	0	0	0	0	1	4	13	30	51	81	98	94	79	68	61	56	52	47	42	39	36	28
2.0	0	0	0	0	0	0	0	0	0	0	0	0	0	0	0	0	1	5	13	37	66	95	95	78	68	61	56	51	44	40	37	29
2.5	0	0	0	0	0	0	0	0	0	0	0	0	0	0	0	0	0	0	2	9	26	61	93	89	77	67	60	55	47	41	38	30
3.0	0	0	0	0	0	0	0	0	0	0	0	0	0	0	0	0	0	0	0	1	6	24	64	91	87	76	66	60	50	43	39	31
IA/P = 0.50													* * * TC = 1.25 HR * * *																IA/P = 0.50			
0.0	0	0	0	0	0	0	0	0	0	1	2	3	6	13	21	29	35	40	43	47	47	47	47	47	47	47	47	47	47	47	47	36
.10	0	0	0	0	0	0	0	0	0	0	1	3	5	11	19	27	34	39	42	47	47	47	47	47	47	47	47	47	47	47	47	36
.20	0	0	0	0	0	0	0	0	0	0	1	2	4	9	17	25	32	37	41	46	47	47	47	47	47	47	47	47	47	47	47	36
.30	0	0	0	0	0	0	0	0	0	0	1	2	3	8	15	23	30	36	40	45	47	47	47	47	47	47	47	47	47	47	47	36
.40	0	0	0	0	0	0	0	0	0	0	0	1	2	7	13	21	28	35	39	45	47	47	47	47	47	47	47	47	47	47	47	37
.50	0	0	0	0	0	0	0	0	0	0	0	1	2	6	12	19	27	33	38	44	46	47	47	47	47	47	47	47	47	47	47	37
.75	0	0	0	0	0	0	0	0	0	0	0	0	1	3	8	14	21	28	34	41	45	47	47	47	47	47	47	47	47	47	47	37
1.0	0	0	0	0	0	0	0	0	0	0	0	0	0	1	2	5	11	17	24	34	41	46	47	47	47	47	47	47	47	47	47	38
1.5	0	0	0	0	0	0	0	0	0	0	0	0	0	0	0	1	2	5	10	19	29	39	45	47	47	47	47	47	47	47	47	40
2.0	0	0	0	0	0	0	0	0	0	0	0	0	0	0	0	0	0	1	3	9	18	30	41	46	47	47	47	47	47	47	47	41
2.5	0	0	0	0	0	0	0	0	0	0	0	0	0	0	0	0	0	0	0	2	7	17	31	41	46	47	47	47	47	47	47	42
3.0	0	0	0	0	0	0	0	0	0	0	0	0	0	0	0	0	0	0	0	0	1	6	18	32	42	46	47	47	47	47	47	43
RAINFALL TYPE = I													* * * TC = 1.25 HR * * *																SHEET 8 OF 10			

Exhibit 5-I: Tabular hydrograph unit discharges (csm/in) for type I rainfall distribution—continued

HYDROGRAPH TIME (HOURS) — * * * TC = 1.5 HR * * *

TRVL TIME (hr)	9.0	9.3	9.6	9.9	10.0	10.1	10.2	10.3	10.4	10.5	10.6	10.7	10.8	11.0	11.2	11.4	11.6	11.8	12.0	12.3	12.6	13.0	13.5	14.0	14.5	15.0	15.5	16.0	17.0	18.0	20.0	24.0
IA/P = 0.10																																
0.0	11	14	19	25	28	32	39	48	60	76	94	112	129	151	163	147	130	112	97	81	69	59	50	44	39	35	32	30	28	26	23	15
.10	10	12	16	22	24	27	30	36	44	55	69	86	103	135	152	159	143	125	108	88	75	62	52	46	41	36	32	30	28	26	23	16
.20	9	11	14	19	21	23	25	29	34	41	50	63	78	112	140	155	152	138	121	98	81	67	55	43	42	37	33	31	28	27	23	16
.30	9	11	14	18	20	22	24	27	31	38	46	57	71	104	133	151	154	141	125	101	84	68	56	48	43	38	34	31	29	27	23	16
.40	8	10	13	17	19	21	23	26	30	35	42	52	65	96	126	147	152	144	129	105	87	70	57	49	43	38	34	31	29	27	23	16
.50	8	9	12	15	17	18	20	22	25	28	33	39	48	73	104	131	148	151	140	117	95	75	60	51	45	40	35	32	29	27	24	17
.75	7	8	11	13	15	16	18	19	21	24	27	32	38	56	82	110	133	146	144	127	106	83	65	54	47	41	36	33	29	27	24	17
1.0	5	7	9	11	11	12	13	15	16	18	19	22	24	32	47	69	96	122	139	145	127	99	74	60	51	44	39	35	30	28	24	17
1.5	4	5	6	8	8	9	10	10	11	12	13	14	16	19	24	32	46	66	90	124	139	128	97	73	59	50	44	39	32	29	25	18
2.0	2	3	4	5	5	6	6	7	7	8	9	9	10	12	14	17	21	27	38	63	96	135	129	100	76	61	51	44	35	30	26	19
2.5	1	2	2	3	4	4	4	5	5	6	6	7	7	8	10	11	14	16	20	31	51	91	132	124	98	75	60	51	39	32	27	20
3.0	0	1	1	2	2	2	2	3	3	3	4	4	5	6	7	8	9	10	12	16	23	42	86	123	128	101	77	62	45	35	28	21
IA/P = 0.30																																
0.0	0	0	0	0	0	1	3	7	13	21	33	46	60	83	97	106	97	90	84	76	70	64	59	54	50	45	43	41	39	38	34	25
.10	0	0	0	0	0	0	1	2	5	10	18	28	40	66	86	98	103	95	88	80	73	66	60	55	51	47	43	41	39	38	35	25
.20	0	0	0	0	0	0	0	1	4	8	14	23	34	60	81	95	101	97	90	81	74	67	61	56	52	47	44	41	40	38	35	25
.30	0	0	0	0	0	0	0	0	1	3	6	12	19	41	65	85	98	100	95	85	77	69	62	57	53	48	44	42	40	38	35	25
.40	0	0	0	0	0	0	0	0	1	2	5	9	16	36	59	80	94	99	96	87	79	70	63	58	53	49	45	42	40	38	35	26
.50	0	0	0	0	0	0	0	0	1	2	4	8	13	31	54	75	91	98	97	88	80	71	64	58	54	49	45	42	40	38	35	26
.75	0	0	0	0	0	0	0	0	0	0	1	2	4	13	28	49	69	85	95	95	87	76	67	61	56	51	47	44	40	39	36	26
1.0	0	0	0	0	0	0	0	0	0	0	0	0	1	5	13	29	49	69	84	95	92	81	71	63	58	53	49	45	41	39	36	27
1.5	0	0	0	0	0	0	0	0	0	0	0	0	0	0	1	3	9	20	35	63	84	92	83	72	64	58	54	49	43	40	37	28
2.0	0	0	0	0	0	0	0	0	0	0	0	0	0	0	0	0	1	3	9	25	50	79	90	81	71	63	58	53	45	41	38	29
2.s	0	0	0	0	0	0	0	0	0	0	0	0	0	0	0	0	0	0	1	6	18	46	79	88	80	70	63	57	49	42	38	31
3.0	0	0	0	0	0	0	0	0	0	0	0	0	0	0	0	0	0	0	0	1	4	17	50	79	87	79	70	62	52	45	39	32
IA/P = 0.50																																
0.0	0	0	0	0	0	0	0	0	0	0	1	2	3	8	15	22	29	34	39	44	46	46	46	46	46	46	46	46	46	46	46	37
.10	0	0	0	0	0	0	0	0	0	0	0	1	2	5	10	17	24	30	35	41	46	46	46	46	46	46	46	46	46	46	46	37
.20	0	0	0	0	0	0	0	0	0	0	0	1	1	4	9	15	22	28	34	40	45	46	46	46	46	46	46	46	46	46	46	37
.30	0	0	0	0	0	0	0	0	0	0	0	0	1	3	7	13	20	27	33	39	44	46	46	46	46	46	46	46	46	46	46	38
.40	0	0	0	0	0	0	0	0	0	0	0	0	1	3	6	12	18	25	31	38	44	46	46	46	46	46	46	46	46	46	46	38
.50	0	0	0	0	0	0	0	0	0	0	0	0	1	2	5	10	17	23	30	37	43	46	46	46	46	46	46	46	46	46	46	38
.75	0	0	0	0	0	0	0	0	0	0	0	0	0	1	3	7	12	19	25	33	40	45	46	46	46	46	46	46	46	46	46	38
1.0	0	0	0	0	0	0	0	0	0	0	0	0	0	0	1	2	5	10	15	25	33	42	46	46	46	46	46	46	46	46	46	39
1.5	0	0	0	0	0	0	0	0	0	0	0	0	0	0	0	0	1	3	7	14	23	34	43	46	46	46	46	46	46	46	46	40
2.0	0	0	0	0	0	0	0	0	0	0	0	0	0	0	0	0	0	0	1	5	11	22	35	43	46	46	46	46	46	46	46	42
2.0	0	0	0	0	0	0	0	0	0	0	0	0	0	0	0	0	0	0	0	1	3	10	23	36	43	46	46	46	46	46	46	43
3.0	0	n	0	0	0	0	0	0	0	0	0	0	0	0	0	0	0	0	0	0	1	4	14	27	38	44	46	46	46	46	46	44

RAINFALL TYPE = I * * * TC = 1.5 HR * * * SHEET 9 OF 10

Exhibit 5-I: Tabular hydrograph unit discharges (csm/in) for type I rainfall distribution—continued

HYDROGRAPH TIME (HOURS) — * * * TC = 2.0 HR * * *

TRVL TIME (hr)	9.0	9.3	9.6	9.9	10.0	10.1	10.2	10.3	10.4	10.5	10.6	10.7	10.8	11.0	11.2	11.4	11.6	11.8	12.0	12.3	12.6	13.0	13.5	14.0	14.5	15.0	15.5	16.0	17.0	18.0	20.0	24.0
IA/P = 0.10																																
0.0	9	11	14	19	21	24	28	33	40	49	59	70	82	106	123	138	138	125	115	97	83	70	59	50	44	39	35	32	29	27	23	16
.10	8	10	13	17	18	20	23	26	31	37	45	54	65	89	110	125	135	130	123	106	90	75	62	53	46	40	36	33	29	27	24	16
.20	8	10	12	16	17	19	22	25	29	34	41	50	60	83	105	121	132	131	125	109	92	77	63	53	46	41	36	33	29	27	24	17
.30	7	9	11	14	15	17	18	21	23	27	32	38	46	66	89	109	123	131	129	117	100	82	66	56	48	42	38	34	30	28	24	17
.40	7	8	10	13	15	16	18	20	22	25	30	35	43	61	83	104	120	131	131	119	103	84	68	57	49	43	38	34	30	28	24	17
.50	6	8	10	13	14	15	17	19	21	24	28	33	39	56	78	99	116	127	130	121	106	86	69	58	50	44	39	35	30	28	24	17
.75	5	7	8	11	12	13	14	15	16	18	20	23	27	38	53	73	93	111	123	127	118	98	77	63	54	47	41	37	31	28	25	17
1.0	4	5	7	8	9	10	11	12	13	14	15	17	18	24	32	45	63	83	102	122	126	113	89	71	59	51	44	39	32	29	25	18
1.5	3	4	5	6	7	7	8	8	9	10	10	11	12	15	18	23	32	44	60	88	111	123	110	87	70	58	50	44	35	30	26	19
2.0	1	2	3	4	5	5	5	6	6	7	7	8	9	10	12	14	18	23	31	49	74	106	121	107	86	69	58	49	38	32	27	20
2.5	1	1	2	2	3	3	3	4	4	4	5	5	6	7	8	9	11	13	16	22	35	62	101	118	108	88	71	59	44	35	28	21
3.0	0	0	1	1	1	2	2	2	3	3	3	3	4	5	5	6	8	9	10	14	19	34	67	103	116	105	86	70	50	39	29	21
IA/P = 0.30																																
0.0	0	0	0	0	0	0	1	3	6	10	16	23	31	49	65	77	84	92	86	80	75	69	63	58	53	49	45	43	40	38	35	26
.10	0	0	0	0	0	0	0	1	2	5	9	13	19	35	53	68	79	85	90	83	77	71	64	59	55	50	46	43	40	39	35	26
.20	0	0	0	0	0	0	0	0	1	2	4	7	11	24	40	57	71	81	89	89	80	73	66	61	56	51	47	44	41	39	35	26
.30	0	0	0	0	0	0	0	0	1	1	3	6	9	20	36	53	68	78	84	88	81	74	67	61	56	52	48	44	41	39	36	26
.40	0	0	0	0	0	0	0	0	0	0	1	2	5	12	24	40	57	70	80	87	84	76	69	63	58	53	49	45	41	39	36	27
.50	0	0	0	0	0	0	0	0	0	0	1	2	4	10	21	36	53	67	77	87	85	77	69	63	58	54	49	46	41	39	36	27
.75	0	0	0	0	0	0	0	0	0	0	0	1	2	6	14	26	41	56	69	82	85	80	72	65	60	55	51	47	42	40	36	27
1.0	0	0	0	0	0	0	0	0	0	0	0	0	0	1	4	10	20	34	49	68	81	85	77	69	63	58	53	49	43	40	37	28
1.5	0	0	0	0	0	0	0	0	0	0	0	0	0	0	0	1	4	10	19	38	58	78	83	76	68	62	57	53	45	41	37	29
2.0	0	0	0	0	0	0	0	0	0	0	0	0	0	0	0	0	0	2	4	13	29	55	77	82	75	68	62	57	48	43	38	30
2.5	0	0	0	0	0	0	0	0	0	0	0	0	0	0	0	0	0	0	0	2	6	22	51	75	81	75	68	62	53	45	39	32
3.0	0	0	0	0	0	0	0	0	0	0	0	0	0	0	0	0	0	0	0	0	1	6	25	54	75	80	75	68	57	48	43	33
IA/P = 0.50																																
0.0	0	0	0	0	0	0	0	0	0	0	0	1	2	4	8	13	18	24	29	36	41	46	46	46	46	46	46	46	46	46	46	38
.10	0	0	0	0	0	0	0	0	0	0	0	1	1	3	7	12	17	22	28	35	40	45	46	46	46	46	46	46	46	46	46	38
.20	0	0	0	0	0	0	0	0	0	0	0	0	1	2	4	8	13	18	24	31	38	44	46	46	46	46	46	46	46	46	46	39
.30	0	0	0	0	0	0	0	0	0	0	0	0	0	1	4	7	12	17	22	30	37	43	45	46	46	46	46	46	46	46	46	39
.40	0	0	0	0	0	0	0	0	0	0	0	0	0	1	3	6	11	16	21	29	36	42	45	46	46	46	46	46	46	46	46	39
.50	0	0	0	0	0	0	0	0	0	0	0	0	0	1	3	5	9	14	20	28	35	42	45	45	45	45	45	45	45	45	45	39
.75	0	0	0	0	0	0	0	0	0	0	0	0	0	1	2	5	9	13	18	26	33	41	45	45	45	45	45	45	45	45	45	39
1.0	0	0	0	0	0	0	0	0	0	0	0	0	0	0	0	1	3	5	9	16	24	33	42	45	45	45	45	45	45	45	45	40
1.5	0	0	0	0	0	0	0	0	0	0	0	0	0	0	0	0	1	2	4	8	15	25	36	43	45	45	45	45	45	45	45	42
2.0	0	0	0	0	0	0	0	0	0	0	0	0	0	0	0	0	0	0	1	2	6	14	26	37	43	45	45	45	45	45	45	43
2.5	0	0	0	0	0	0	0	0	0	0	0	0	0	0	0	0	0	0	0	0	2	6	15	27	37	43	45	45	45	45	45	44
3.0	0	0	0	0	0	0	0	0	0	0	0	0	0	0	0	0	0	0	0	0	0	2	3	19	30	39	44	45	45	45	45	45

RAINFALL TYPE = I * * * TC = 2.0 HR * * * SHEET 10 OF 10

Exhibit 5-IA: Tabular hydrograph unit discharges (csm/in) for type IA rainfall distribution

TRVL TIME (hr)	7.0	7.3	7.6	7.9	8.0	8.1	8.2	8.3	8.4	8.5	8.6	8.7	8.8	9.0	9.2	9.4	9.6	9.8	10.0	10.3	10.6	11.0	11.5	12.0	12.5	13.0	13.5	14.0	15.0	16.0	18.0	22.0
IA/P = 0.10 * * * TC = 0.1 HR * * * IA/P = 0.10																																
0.0	28	36	50	143	154	163	140	103	87	76	68	67	65	61	54	49	45	44	44	41	40	39	36	33	32	32	31	30	30	29	26	21
.10	27	32	43	104	130	146	157	145	117	97	83	73	69	65	59	53	48	45	45	42	40	39	37	34	33	32	32	31	30	29	27	22
.20	26	29	37	59	89	116	136	150	147	127	107	91	79	68	63	57	52	47	45	44	41	39	37	35	33	32	32	31	30	29	27	22
.30	25	28	35	53	77	103	125	142	145	133	116	99	86	71	65	59	53	48	46	44	41	39	38	35	33	32	32	31	30	29	27	22
.40	24	26	31	41	49	68	91	114	132	141	136	122	107	82	70	63	57	52	48	45	43	40	38	36	33	32	32	31	30	29	27	22
.50	24	26	30	39	46	60	81	103	123	135	135	127	114	88	73	65	59	53	49	45	43	40	39	36	33	33	32	31	30	29	27	22
.75	20	24	27	32	35	39	47	60	76	94	111	122	125	114	94	79	68	61	55	48	45	42	39	37	35	33	32	32	30	30	27	22
1.0	16	20	24	27	28	30	32	36	41	49	62	77	94	118	122	104	86	74	65	55	49	44	41	39	36	34	33	32	31	30	28	23
1.5	12	15	18	22	23	24	25	27	28	30	32	36	42	61	86	107	112	104	91	73	60	50	44	41	38	36	34	33	31	30	28	23
2.0	7	10	12	15	16	18	19	20	21	22	24	25	26	30	36	50	69	90	106	103	87	67	52	45	41	39	36	34	32	31	29	24
2.5	4	6	8	11	12	13	14	15	16	17	18	19	21	23	26	30	36	49	66	91	101	89	66	51	44	41	38	36	33	31	29	25
3.0	2	3	5	7	8	9	10	11	11	12	13	14	16	18	20	23	25	29	36	55	79	98	86	65	51	44	41	38	34	32	30	25
IA/P = 0.30 * * * TC = 0.1 HR * * * IA/P = 0.30																																
0.0	0	0	1	45	65	89	78	64	59	54	54	53	51	50	48	45	44	43	43	43	42	42	41	40	40	40	40	40	39	39	38	33
.10	0	0	0	18	36	55	78	78	69	62	57	55	53	51	50	47	44	44	43	43	42	42	41	40	40	40	40	40	39	39	38	34
.20	0	0	0	13	29	46	67	74	71	65	60	56	54	52	50	48	45	44	43	43	43	42	41	41	40	40	40	40	39	39	38	34
.30	0	0	0	10	22	38	58	69	70	67	62	58	56	52	51	48	46	44	44	43	43	42	41	41	40	40	40	40	39	39	38	34
,40	0	0	0	2	7	17	31	49	62	67	67	64	60	55	52	50	48	45	44	43	43	42	42	41	40	40	40	40	39	39	38	34
,50	0	0	0	1	5	13	25	41	55	63	66	64	61	56	53	51	48	46	44	43	43	42	42	41	40	40	40	40	39	39	38	34
.75	0	0	0	0	0	2	6	13	24	36	47	55	61	61	57	54	51	49	47	44	43	43	42	41	41	40	40	40	40	39	38	34
1.0	0	0	0	0	0	1	3	8	15	25	36	46	53	60	59	55	53	50	48	45	44	43	42	42	41	40	40	40	40	39	38	35
1.5	0	0	0	0	0	0	0	0	1	2	5	9	15	31	47	55	57	56	53	50	46	44	43	42	41	41	40	40	40	39	39	35
2.0	0	0	0	0	0	0	0	0	0	0	0	0	1	5	15	28	42	51	55	54	51	47	44	43	42	41	41	40	40	39	39	36
2.5	0	0	0	0	0	0	0	0	0	0	0	0	0	0	2	6	14	26	38	50	54	52	47	44	43	42	41	40	40	40	39	36
3.0	0	0	0	0	0	0	0	0	0	0	0	0	0	0	0	0	2	6	13	29	43	53	51	47	44	43	42	41	40	40	39	37
IA/P = 0.50 * * * TC = 0.1 HR * * * IA/P = 0.50																																
0,0	0	0	0	0	0	0	0	0	0	0	0	0	0	2	7	12	14	18	21	23	26	30	32	33	37	38	41	42	46	48	53	49
.10	0	0	0	0	0	0	0	0	0	0	0	0	0	0	4	9	12	15	19	22	25	29	32	32	35	39	41	42	45	48	53	49
.20	0	0	0	0	0	0	0	0	0	0	0	0	0	0	3	7	11	14	18	22	24	28	32	32	35	39	41	42	45	48	53	49
.30	0	0	0	0	0	0	0	0	0	0	0	0	0	0	0	4	8	12	15	20	23	27	31	31	34	39	40	41	45	48	53	49
.40	0	0	0	0	0	0	0	0	0	0	0	0	0	0	0	1	5	9	13	18	22	25	30	31	33	37	39	41	44	47	53	49
.50	0	0	0	0	0	0	0	0	0	0	0	0	0	0	0	0	2	6	10	tS	20	24	29	31	32	36	39	41	44	47	53	53
.75	0	0	0	0	0	0	0	0	0	0	0	0	0	0	0	0	0	2	6	11	16	21	26	30	31	34	38	40	43	46	53	53
1.0	0	0	0	0	0	0	0	0	0	0	0	0	0	0	0	0	0	0	1	6	11	17	23	28	31	32	35	39	42	45	53	53
1.5	0	0	0	0	0	0	0	0	0	0	0	0	0	0	0	0	0	0	0	0	2	8	17	23	28	31	32	35	40	43	49	52
2.0	0	0	0	0	0	0	0	0	0	0	0	0	0	0	0	0	0	0	0	0	0	2	9	17	23	28	31	32	38	42	48	52
2.5	0	0	0	0	0	0	0	0	0	0	0	0	0	0	0	0	0	0	0	0	0	0	3	10	18	24	28	31	36	40	47	52
3.0	0	0	0	0	0	0	0	0	0	0	0	0	0	0	0	0	0	0	0	0	0	0	0	4	11	18	24	28	33	39	45	52
RAINFALL TYPE = IA * * * TC = 0.1 HR * * * SHEET 1 OF 10																																

Exhibit 5-IA: Tabular hydrograph unit discharges (csm/in) for type IA rainfall distribution—continued

TRVL TIME (hr)	7.0	7.3	7.6	7.9	8.0	8.1	8.2	8.3	8.4	8.5	8.6	8.7	8.8	9.0	9.2	9.4	9.6	9.8	10.0	10.3	10.6	11.0	11.5	12.0	12.5	13.0	13.5	14.0	15.0	16.0	18.0	22.0
IA/P = 0.10 * * * TC = 0.2 HR * * * IA/P = 0.10																																
0.0	27	34	46	124	143	153	153	127	103	87	76	68	67	63	57	51	47	45	44	42	40	39	36	33	33	32	32	31	30	29	27	21
.10	27	33	44	108	131	146	151	135	113	96	83	73	69	65	59	53	48	45	45	42	40	39	37	33	33	32	32	31	30	29	27	22
.20	26	29	38	67	95	119	137	146	139	122	105	90	79	69	63	57	52	47	45	44	41	39	37	34	33	32	32	31	30	29	27	22
.30	25	27	33	45	60	83	107	127	140	139	128	112	97	76	68	62	56	50	47	45	42	40	38	35	33	33	32	31	30	29	27	22
.40	24	27	32	43	54	73	96	117	132	137	131	118	104	81	70	63	57	52	48	45	43	40	38	36	33	33	32	32	30	29	27	22
.50	23	25	29	37	41	50	65	86	106	124	132	131	123	98	79	69	62	56	51	46	44	41	39	37	34	33	32	32	30	29	27	22
.75	20	24	27	33	36	41	50	63	80	97	112	121	123	111	93	78	68	61	55	48	45	42	39	37	34	33	32	32	30	30	27	22
1.0	16	20	24	27	29	31	33	37	43	52	65	81	96	120	120	102	85	73	64	55	49	44	41	39	36	34	33	32	31	30	28	23
1.5	12	15	19	22	23	24	26	27	28	30	33	37	44	63	88	107	111	102	89	72	60	50	44	41	38	36	34	33	32	30	28	23
2.0	7	10	12	16	17	18	19	20	22	23	24	25	27	30	38	52	71	91	105	101	86	66	52	45	41	39	36	34	32	31	29	24
2.5	4	6	9	11	12	13	14	15	16	17	19	20	21	23	26	30	38	50	68	92	101	88	65	51	44	41	38	36	33	31	29	25
3.0	2	3	5	7	7	8	9	10	11	12	13	14	15	17	19	22	24	28	33	49	72	97	89	68	53	45	41	39	34	32	30	25
IA/P = 0.30 * * * TC = 0.2 HR * * * IA/P = 0.30																																
0.0	0	0	0	26	45	64	76	74	65	59	55	53	52	51	49	46	44	44	43	43	42	42	41	40	40	40	40	40	39	39	38	33
.10	0	0	0	20	37	55	69	72	67	62	57	54	53	52	50	47	44	44	43	43	42	42	41	40	40	40	40	40	39	39	38	34
.20	0	0	0	5	15	30	47	61	68	68	64	59	56	53	52	49	46	44	44	43	43	42	42	41	40	40	40	40	39	39	38	34
.30	0	0	0	4	11	23	39	54	64	66	65	61	58	54	52	50	47	45	44	43	43	42	42	41	40	40	40	40	39	39	38	34
.40	0	0	0	3	8	18	32	47	58	64	64	62	59	55	52	50	48	45	44	43	43	42	42	41	40	40	40	40	40	39	38	34
.50	0	0	0	0	2	6	14	26	40	52	60	63	62	58	54	52	50	47	45	44	43	42	42	41	40	40	40	40	40	39	38	34
.75	0	0	0	0	1	3	7	14	24	35	45	53	57	59	56	54	51	49	47	44	43	43	42	41	41	40	40	40	40	39	38	34
1.0	0	0	0	0	0	0	1	4	8	15	25	35	44	56	58	56	54	51	49	46	44	43	42	42	41	40	40	40	40	39	39	35
1.5	0	0	0	0	0	0	0	0	0	1	2	5	9	23	38	50	55	56	54	51	47	44	43	42	42	41	40	40	40	39	39	35
2.0	0	0	0	0	0	0	0	0	0	0	0	1	1	6	15	27	40	49	54	54	51	47	44	43	42	41	41	40	40	40	39	36
2.5	0	0	0	0	0	0	0	0	0	0	0	0	0	0	2	6	14	25	36	49	53	51	47	44	43	42	41	41	40	40	39	36
3.0	0	0	0	0	0	0	0	0	0	0	0	0	0	0	0	1	2	6	13	28	42	52	51	47	44	43	42	41	40	40	39	37
IA/P = 0.50 * * * TC = 0.2 HR * * * IA/P = 0.50																																
0.0	0	0	0	0	0	0	0	0	0	0	0	0	0	0	5	10	13	16	19	23	25	29	32	33	36	39	41	42	46	48	50	49
.10	0	0	0	0	0	0	0	0	0	0	0	0	0	0	2	6	10	14	17	21	24	28	31	32	35	40	41	42	45	48	50	49
.20	0	0	0	0	0	0	0	0	0	0	0	0	0	0	0	3	7	11	15	19	23	26	31	31	34	39	40	41	45	48	50	49
.30	0	0	0	0	0	0	0	0	0	0	0	0	0	0	0	1	4	8	12	17	21	25	30	31	33	37	39	41	44	47	50	50
.40	0	0	0	0	0	0	0	0	0	0	0	0	0	0	0	0	2	5	9	15	19	23	28	31	32	35	39	41	44	47	50	50
.50	0	0	0	0	0	0	0	0	0	0	0	0	0	0	0	0	0	3	6	12	17	22	27	31	32	34	39	40	43	46	50	50
.75	0	0	0	0	0	0	0	0	0	0	0	0	0	0	0	0	0	1	4	9	14	20	25	30	31	33	37	39	42	46	50	50
1.0	0	0	0	0	0	0	0	0	0	0	0	0	0	0	0	0	0	0	1	4	9	16	22	27	31	32	35	38	42	45	49	50
1.5	0	0	0	0	0	0	0	0	0	0	0	0	0	0	0	0	0	0	0	0	3	8	17	23	28	31	32	35	40	43	49	50
2.0	0	0	0	0	0	0	0	0	0	0	0	0	0	0	0	0	0	0	0	0	0	2	9	17	23	28	31	32	38	42	48	50
2.5	0	0	0	0	0	0	0	0	0	0	0	0	0	0	0	0	0	0	0	0	0	0	2	9	16	22	27	30	35	40	46	50
3.0	0	0	0	0	0	0	0	0	0	0	0	0	0	0	0	0	0	0	0	0	0	0	0	3	9	17	23	27	33	38	45	50
RAINFALL TYPE = IA * * * TC = 0.2 HR * * * SHEET 2 OF 10																																

Exhibit 5-IA: Tabular hydrograph unit discharges (csm/in) for type IA rainfall distribution—continued

HYDROGRAPH TIME (HOURS)

IA/P = 0.10 * * * TC = 0.3 HR * * * IA/P = 0.10

TRVL TIME (hr)	7.0	7.3	7.6	7.9	8.0	8.1	8.2	8.3	8.4	8.5	8.6	8.7	8.8	9.0	9.2	9.4	9.6	9.8	10.0	10.3	10.6	11.0	11.5	12.0	12.5	13.0	13.5	14.0	15.0	16.0	18.0	22.0
0.0	26	31	41	92	120	138	145	144	125	105	90	78	71	66	60	54	49	46	45	43	40	39	37	34	33	32	32	31	30	29	27	22
.10	26	30	39	80	106	128	139	142	131	113	98	85	76	68	62	56	51	47	45	43	41	39	37	34	33	32	32	31	30	29	27	22
.20	25	28	34	51	70	94	117	131	139	133	120	105	92	74	67	60	54	49	46	44	42	40	38	35	33	33	32	31	30	29	27	22
.30	24	26	31	40	47	62	84	106	123	133	133	124	112	87	73	65	59	53	49	45	43	40	39	36	33	33	32	32	30	29	27	22
.40	23	26	30	38	44	56	75	95	114	127	131	127	117	93	76	67	60	55	50	46	43	40	39	36	34	33	32	32	30	29	27	22
.50	21	25	28	34	37	42	51	67	86	104	119	127	127	110	88	74	66	59	53	47	45	41	39	37	34	33	32	32	30	30	27	22
.75	19	23	26	31	33	36	42	51	64	80	96	109	119	116	102	85	73	65	58	50	46	43	40	38	35	33	33	32	31	30	28	23
1.0	15	19	23	26	27	29	31	33	37	44	53	66	80	106	116	109	93	79	69	58	51	45	41	39	37	34	33	32	31	30	28	23
1.5	11	14	17	21	22	23	24	26	27	29	31	34	38	53	75	97	109	106	96	78	64	52	45	41	39	36	34	33	32	30	28	24
2.0	6	9	11	14	16	17	18	19	20	21	23	24	25	28	34	44	61	81	97	103	91	71	54	46	42	39	37	34	32	31	29	24
2.5	4	5	8	10	11	12	13	14	15	16	17	19	20	22	25	28	34	44	58	84	59	92	70	54	46	41	39	36	33	32	29	25
3.0	2	3	5	7	7	8	9	10	11	12	13	14	15	17	19	22	24	28	33	49	71	96	89	68	53	45	41	39	34	32	30	25

IA/P = 0.30 * * * TC = 0.3 HR * * * IA/P = 0.30

TRVL TIME (hr)	7.0	7.3	7.6	7.9	8.0	8.1	8.2	8.3	8.4	8.5	8.6	8.7	8.8	9.0	9.2	9.4	9.6	9.8	10.0	10.3	10.6	11.0	11.5	12.0	12.5	13.0	13.5	14.0	15.0	16.0	18.0	22.0
0.0	0	0	0	13	28	45	63	69	69	65	60	55	53	53	51	48	45	44	43	43	42	42	42	40	40	40	40	40	39	39	38	34
.10	0	0	0	9	21	37	54	64	67	65	62	57	54	53	51	49	46	44	43	43	42	42	42	41	40	40	40	40	40	39	38	34
.20	0	0	0	7	17	30	46	58	64	65	63	59	56	53	52	49	46	44	44	43	42	42	42	41	40	40	40	40	40	39	38	34
.30	0	0	0	1	5	13	25	39	52	60	63	63	60	55	53	51	49	46	44	43	43	42	42	41	40	40	40	40	40	39	38	34
.40	0	0	0	1	4	10	20	33	45	55	61	62	61	56	54	52	49	46	45	43	43	42	42	41	40	40	40	40	40	39	38	34
.50	0	0	0	1	3	7	15	27	39	50	57	60	61	57	54	52	50	47	45	44	43	42	42	41	40	40	40	40	40	39	38	34
.75	0	0	0	0	1	3	8	15	24	34	43	51	55	58	56	54	52	49	47	44	43	43	42	41	41	40	40	40	40	39	38	34
1.0	0	0	0	0	0	0	0	2	4	9	16	25	34	49	56	57	55	53	50	47	44	43	42	42	41	40	40	40	40	39	39	35
1.5	0	0	0	0	0	0	0	0	0	1	3	5	10	23	37	49	54	55	54	51	47	44	43	42	42	41	40	40	40	39	39	35
2.0	0	0	0	0	0	0	0	0	0	0	0	0	1	3	10	21	33	44	51	54	52	48	44	43	42	41	41	40	40	40	39	36
2.5	0	0	0	0	0	0	0	0	0	0	0	0	0	0	1	4	10	19	30	45	53	53	48	44	43	42	41	41	40	40	39	37
3.0	0	0	0	0	0	0	0	0	0	0	0	0	0	0	0	0	1	4	9	23	37	50	52	47	44	43	42	41	40	40	39	37

IA/P = 0.50 * * * TC = 0.3 HR * * * IA/P = 0.50

TRVL TIME (hr)	7.0	7.3	7.6	7.9	8.0	8.1	8.2	8.3	8.4	8.5	8.6	8.7	8.8	9.0	9.2	9.4	9.6	9.8	10.0	10.3	10.6	11.0	11.5	12.0	12.5	13.0	13.5	14.0	15.0	16.0	18.0	22.0
0.0	0	0	0	0	0	0	0	0	0	0	0	0	0	0	3	7	11	14	18	22	24	28	32	32	35	40	41	42	45	48	48	48
.10	0	0	0	0	0	0	0	0	0	0	0	0	0	0	1	4	8	12	15	20	23	27	31	31	34	39	40	41	45	48	48	48
.20	0	0	0	0	0	0	0	0	0	0	0	0	0	0	0	1	5	9	13	18	21	25	30	31	33	37	39	41	44	47	48	48
.30	0	0	0	0	0	0	0	0	0	0	0	0	0	0	0	1	4	8	12	17	21	25	30	31	33	37	39	41	44	47	48	48
.40	0	0	0	0	0	0	0	0	0	0	0	0	0	0	0	0	2	5	9	14	19	23	28	31	32	35	39	41	44	47	48	48
,50	0	0	0	0	0	0	0	0	0	0	0	0	0	0	0	0	0	1	4	10	15	21	26	30	31	33	38	40	43	46	48	48
.75	0	0	0	0	0	0	0	0	0	0	0	0	0	0	0	0	0	1	3	7	13	19	24	29	31	33	36	39	42	46	48	48
1.0	0	0	0	0	0	0	0	0	0	0	0	0	0	0	0	0	0	0	1	4	9	16	22	27	31	32	35	38	42	45	48	48
1.5	0	0	0	0	0	0	0	0	0	0	0	0	0	0	0	0	0	0	0	0	3	8	16	23	27	31	32	35	40	43	48	48
2.0	0	0	0	0	0	0	0	0	0	0	0	0	0	0	0	0	0	0	0	0	0	2	8	16	22	27	30	32	38	42	47	48
2.5	0	0	0	0	0	0	0	0	0	0	0	0	0	0	0	0	0	0	0	0	0	0	2	9	16	22	27	30	35	40	46	48
3.0	0	0	0	0	0	0	0	0	0	0	0	0	0	0	0	0	0	0	0	0	0	0	0	3	9	17	23	27	33	38	45	48

RAINFALL TYPE = IA * * * TC = 0.3 HR * * * SHEET 3 OF 10

Exhibit 5-IA: Tabular hydrograph unit discharges (csm/in) for type IA rainfall distribution—continued

HYDROGRAPH TIME (HOURS)

IA/P = 0.10 * * * TC = 0.4 HR * * * IA/P = 0.10

TRVL TIME (hr)	7.0	7.3	7.6	7.9	8.0	8.1	8.2	8.3	8.4	8.5	8.6	8.7	8.8	9.0	9.2	9.4	9.6	9.8	10.0	10.3	10.6	11.0	11.5	12.0	12.5	13.0	13.5	14.0	15.0	16.0	18.0	22.0
0.0	26	29	37	67	93	117	134	137	136	121	105	91	81	70	64	58	52	48	46	44	41	40	37	34	33	33	32	31	30	29	27	22
.10	25	29	36	60	82	105	124	133	135	126	112	98	87	73	66	59	53	49	46	44	41	40	38	34	33	33	32	31	30	29	27	22
.20	24	27	32	43	55	73	94	114	127	132	128	117	104	83	71	64	58	52	48	45	43	40	38	35	33	33	32	32	30	29	27	22
.30	24	26	31	41	50	65	85	104	119	128	128	121	110	88	74	66	59	54	49	46	43	40	39	35	34	33	32	32	30	29	27	22
.40	22	25	28	35	39	46	59	76	95	111	122	126	122	104	85	72	64	58	52	47	44	41	39	37	34	33	32	32	30	29	27	22
.50	20	23	26	31	34	37	43	54	69	86	103	116	123	117	98	82	71	63	56	49	46	42	40	38	34	33	33	32	30	30	27	22
.75	16	20	24	27	29	31	33	37	43	53	65	80	94	116	116	101	86	74	65	56	49	44	41	39	36	34	33	32	31	30	28	23
1.0	14	18	22	25	26	27	29	31	34	38	45	54	67	93	111	113	101	87	75	62	53	47	42	40	37	34	33	33	31	30	28	23
1.5	10	13	16	20	21	22	23	24	26	27	29	31	34	45	64	86	102	107	101	84	69	55	46	42	39	37	34	33	32	30	28	24
2.0	6	8	11	13	14	16	17	18	19	20	21	23	24	27	31	39	52	71	88	102	96	76	57	47	42	40	37	35	33	31	29	24
2.5	3	5	7	9	10	11	12	13	14	15	16	17	18	21	23	26	31	38	51	75	94	98	74	57	47	42	39	37	33	32	30	25
3.0	1	2	4	6	7	7	8	9	10	11	12	13	14	16	18	20	23	26	30	43	64	89	95	72	56	47	42	39	34	33	30	25

IA/P = 0.30 * * * TC = 0.4 HR * * * IA/P = 0.30

TRVL TIME (hr)	7.0	7.3	7.6	7.9	8.0	8.1	8.2	8.3	8.4	8.5	8.6	8.7	8.8	9.0	9.2	9.4	9.6	9.8	10.0	10.3	10.6	11.0	11.5	12.0	12.5	13.0	13.5	14.0	15.0	16.0	18.0	22.0
0.0	0	0	0	6	15	29	46	60	65	65	63	59	56	53	52	49	47	44	44	43	42	42	42	41	40	40	40	40	40	39	38	34
.10	0	0	0	4	12	23	38	53	61	63	63	60	57	54	53	50	47	45	44	43	42	42	42	41	40	40	40	40	40	39	38	34
.20	0	0	0	3	9	19	32	46	56	61	62	61	59	55	53	51	48	45	44	43	43	42	42	41	40	40	40	40	40	39	38	34
.30	0	0	0	2	7	15	26	39	50	57	61	61	59	56	53	51	49	46	44	44	43	42	42	41	40	40	40	40	40	39	38	34
.40	0	0	0	0	2	5	11	21	33	44	53	58	60	58	55	53	51	48	45	44	43	42	42	41	40	40	40	40	40	39	38	34
.50	0	0	0	0	1	4	9	17	28	39	48	55	59	59	56	53	51	49	46	44	43	42	42	41	40	40	40	40	40	39	38	34
.75	0	0	0	0	0	2	4	9	16	25	34	43	49	57	57	55	53	50	48	45	44	43	42	41	41	40	40	40	40	39	39	35
1.0	0	0	0	0	0	0	1	2	5	10	17	25	34	48	55	56	55	53	50	47	45	43	42	42	41	40	40	40	40	39	39	35
1.5	0	0	0	0	0	0	0	0	0	0	1	3	6	16	30	43	51	55	55	52	49	45	43	42	42	41	40	40	40	40	39	36
2.0	3	0	0	0	0	0	0	0	0	0	0	0	0	2	6	15	27	39	47	53	53	49	45	43	42	42	41	40	40	40	39	36
2.5	0	0	0	0	0	0	0	0	0	0	0	0	0	0	1	2	7	14	25	40	50	52	49	45	43	42	41	41	40	40	39	37
3.0	0	0	0	0	0	0	0	0	0	0	0	0	0	0	0	0	1	3	7	18	32	47	52	48	45	43	42	41	40	40	39	37

IA/P = 0.50 * * * TC = 0.4 HR * * * IA/P = 0.50

TRVL TIME (hr)	7.0	7.3	7.6	7.9	8.0	8.1	8.2	8.3	8.4	8.5	8.6	8.7	8.8	9.0	9.2	9.4	9.6	9.8	10.0	10.3	10.6	11.0	11.5	12.0	12.5	13.0	13.5	14.0	15.0	16.0	18.0	22.0
0.0	0	0	0	0	0	0	0	0	0	0	0	0	0	0	1	5	9	13	16	20	23	27	31	32	34	39	40	42	45	47	47	47
.10	0	0	0	0	0	0	0	0	0	0	0	0	0	0	0	2	6	10	13	18	22	26	30	31	33	38	39	41	45	47	47	47
.20	0	0	0	0	0	0	0	0	0	0	0	0	0	0	0	1	3	7	11	16	20	24	29	31	33	36	39	41	44	47	47	47
.30	0	0	0	0	0	0	0	0	0	0	0	0	0	0	0	0	1	4	8	13	18	23	28	31	32	35	39	41	43	47	47	47
.40	0	0	0	0	0	0	0	0	0	0	0	0	0	0	0	0	0	2	5	11	16	21	26	30	32	34	38	40	43	46	47	47
.50	0	0	0	0	0	0	0	0	0	0	0	0	0	0	0	0	0	1	3	8	13	19	25	29	31	33	37	39	42	46	47	47
.75	0	0	0	0	0	0	0	0	0	0	0	0	0	0	0	0	0	0	2	6	11	17	23	28	31	32	35	39	42	45	47	47
1.0	0	0	0	0	0	0	0	0	0	0	0	0	0	0	0	0	0	0	0	2	6	12	20	25	29	31	33	37	41	44	47	47
1.5	0	0	0	0	0	0	0	0	0	0	0	0	0	0	0	0	0	0	0	0	1	5	13	20	26	30	31	34	40	43	47	47
2.0	0	0	0	0	0	0	0	0	0	0	0	0	0	0	0	0	0	0	0	0	0	1	6	14	21	26	30	32	37	41	47	47
2.5	0	0	0	0	0	0	0	0	0	0	0	0	0	0	0	0	0	0	0	0	0	0	2	7	15	21	26	30	34	40	46	47
3.0	0	0	0	0	0	0	0	0	0	0	0	0	0	0	0	0	0	0	0	0	0	0	0	2	8	15	22	27	32	38	45	47

RAINFALL TYPE = IA * * * TC = 0.4 HR * * * SHEET 4 OF 10

Exhibit 5-IA: Tabular hydrograph unit discharges (csm/in) for type IA rainfall distribution—continued

* * * TC = 0.5 HR * * *

TRVL TIME (hr)	7.0	7.3	7.6	7.9	8.0	8.1	8.2	8.3	8.4	8.5	8.6	8.7	8.8	9.0	9.2	9.4	9.6	9.8	10.0	10.3	10.6	11.0	11.5	12.0	12.5	13.0	13.5	14.0	15.0	16.0	18.0	22.0
IA/P = 0.10																																
0.0	25	28	34	53	72	94	115	129	130	129	117	104	92	76	68	61	55	50	47	44	42	40	38	34	33	33	32	31	30	29	27	22
.10	24	27	33	49	64	84	105	121	127	128	121	109	98	80	70	63	57	51	48	45	42	40	38	34	33	33	32	32	30	29	27	22
.20	23	26	30	39	46	58	75	95	112	122	126	123	114	93	78	69	61	55	50	46	44	41	39	36	34	33	32	32	30	29	27	22
.30	21	24	27	34	37	43	53	68	86	103	116	123	123	108	89	76	67	60	54	48	45	42	39	37	34	33	33	32	30	30	27	22
.40	20	24	27	32	35	40	49	61	78	95	109	118	121	111	93	79	69	62	55	49	45	42	40	37	34	33	33	32	30	30	27	22
.50	18	22	25	29	31	34	38	45	56	70	87	101	113	118	106	89	77	67	60	52	47	43	40	38	35	33	33	32	31	30	28	23
.75	15	19	23	26	27	29	31	34	38	45	54	67	80	104	112	106	93	80	70	59	51	46	41	39	36	34	33	33	31	30	28	23
1.0	13	16	20	24	25	26	27	29	31	34	39	46	56	80	102	110	105	93	81	66	56	48	43	40	38	35	33	33	31	30	28	23
1.5	9	12	15	18	20	21	22	23	24	26	27	29	31	39	55	75	94	104	103	89	74	58	48	43	40	37	35	33	32	31	29	24
2.0	5	7	10	12	13	14	15	17	18	19	20	21	23	25	29	35	45	62	79	100	100	81	61	49	43	40	38	35	33	31	29	25
2.5	3	4	6	8	9	10	11	12	13	14	15	16	17	20	22	25	28	34	45	67	88	96	79	60	49	43	40	37	33	32	30	25
3.0	1	2	3	5	6	7	7	8	9	10	11	12	13	15	17	19	21	24	28	38	56	83	93	77	59	48	43	40	35	33	30	26
IA/P = 0.30																																
0.0	0	0	0	2	8	18	31	46	57	61	61	61	58	54	53	51	48	45	44	44	43	42	42	41	40	40	40	40	40	39	38	34
.10	0	0	0	2	6	14	25	39	51	58	60	61	59	55	53	51	49	46	44	44	43	42	42	41	40	40	40	40	40	39	38	34
.20	0	0	0	0	1	4	11	21	33	45	54	58	60	58	55	53	51	48	46	44	43	42	42	41	40	40	40	40	40	39	38	34
.30	0	0	0	0	1	3	8	16	27	39	49	55	59	59	55	53	51	49	46	44	43	42	42	41	40	40	40	40	40	39	39	34
.40	0	0	0	0	1	2	6	13	23	34	44	51	56	58	56	54	52	49	47	44	43	42	42	41	40	40	40	40	40	39	39	34
.50	0	0	0	0	0	0	2	5	10	18	29	39	47	56	57	55	53	51	49	45	44	43	42	41	41	40	40	40	40	39	39	35
.75	0	0	0	0	0	0	1	2	5	10	17	25	34	48	54	56	55	53	50	47	45	43	42	42	41	40	40	40	40	39	39	35
1.0	0	0	0	0	0	0	0	0	0	1	3	6	11	26	41	51	55	55	54	50	47	44	43	42	41	41	40	40	40	40	39	35
1.5	0	0	0	0	0	0	0	0	0	0	0	0	1	4	11	23	36	47	52	54	52	48	44	43	42	41	40	40	40	40	39	36
2.0	0	0	0	0	0	0	0	0	0	0	0	0	0	0	2	7	16	27	38	49	53	51	47	44	42	42	41	40	40	40	39	36
2.5	0	0	0	0	0	0	0	0	0	0	0	0	0	0	0	1	3	7	15	30	43	52	50	46	44	42	42	41	40	40	39	37
3.0	0	0	0	0	0	0	0	0	0	0	0	0	0	0	0	0	0	1	3	10	23	40	51	50	46	44	42	42	40	40	39	37
IA/P = 0.50																																
0.0	0	0	0	0	0	0	0	0	0	0	0	0	0	0	1	3	7	11	14	19	22	26	30	31	34	38	40	41	45	46	46	46
.10	0	0	0	0	0	0	0	0	0	0	0	0	0	0	0	1	4	8	12	17	21	25	29	31	33	37	39	41	44	46	46	46
.20	0	0	0	0	0	0	0	0	0	0	0	0	0	0	0	0	2	5	9	14	19	23	28	31	32	35	39	41	43	46	46	46
.30	0	0	0	0	0	0	0	0	0	0	0	0	0	0	0	0	1	3	6	12	17	22	27	31	32	34	38	40	43	46	46	46
.40	0	0	0	0	0	0	0	0	0	0	0	0	0	0	0	0	0	2	5	11	16	21	26	30	32	34	38	40	43	46	46	46
.50	0	0	0	0	0	0	0	0	0	0	0	0	0	0	0	0	0	1	3	8	13	19	25	29	31	33	37	39	42	46	46	46
.75	0	0	0	0	0	0	0	0	0	0	0	0	0	0	0	0	0	0	2	6	11	17	23	28	31	32	35	39	42	45	46	46
1.0	0	0	0	0	0	0	0	0	0	0	0	0	0	0	0	0	0	0	0	1	4	11	18	24	29	31	33	36	41	44	46	46
1.5	0	0	0	0	0	0	0	0	0	0	0	0	0	0	0	0	0	0	0	0	1	4	11	19	24	29	31	33	39	42	46	46
2.0	0	0	0	0	0	0	0	0	0	0	0	0	0	0	0	0	0	0	0	0	0	1	5	12	19	25	29	31	37	41	46	46
2.5	0	0	0	0	0	0	0	0	0	0	0	0	0	0	0	0	0	0	0	0	0	0	1	6	13	20	25	29	34	39	45	46
3.0	0	0	0	0	0	0	0	0	0	0	0	0	0	0	0	0	0	0	0	0	0	0	0	2	7	14	20	26	32	37	44	46

RAINFALL TYPE = IA   * * * TC = 0.5 HR * * *   SHEET 5 OF 10

Exhibit 5-IA: Tabular hydrograph unit discharges (csm/in) for type IA rainfall distribution—continued

* * * TC = 0.75 HR * * *

TRVL TIME (hr)	7.0	7.3	7.6	7.9	8.0	8.1	8.2	8.3	8.4	8.5	8.6	8.7	8.8	9.0	9.2	9.4	9.6	9.8	10.0	10.3	10.6	11.0	11.5	12.0	12.5	13.0	13.5	14.0	15.0	16.0	18.0	22.0
IA/P = 0.10																																
0.0	22	25	29	38	45	56	71	87	103	114	117	117	111	95	81	72	65	58	53	48	45	41	39	36	33	33	33	32	30	29	27	22
.10	21	25	28	36	42	51	64	79	95	108	114	116	113	99	85	74	66	60	54	48	45	42	39	37	34	33	33	32	30	30	27	22
.20	19	23	26	32	35	39	47	58	72	87	101	110	114	109	95	82	72	65	58	51	47	43	40	38	34	33	33	32	31	30	27	23
.30	19	22	26	31	33	37	44	54	66	80	94	104	112	110	98	85	75	67	60	52	47	43	40	38	34	33	33	32	31	30	28	23
.40	17	21	24	28	30	32	36	41	49	61	74	87	99	111	106	94	82	73	65	56	49	45	41	39	35	33	33	33	31	30	28	23
.50	16	20	23	27	29	31	34	39	46	56	68	81	93	109	107	97	85	75	67	57	50	45	41	39	36	33	33	33	31	30	28	23
.75	15	18	22	25	27	28	31	34	38	45	54	64	75	95	104	102	93	83	74	62	54	47	42	40	37	34	33	33	31	30	28	23
1.0	12	15	18	22	23	24	25	27	29	31	34	39	46	65	85	100	103	98	88	74	62	52	45	41	39	35	33	33	32	30	28	23
1.5	7	10	12	15	17	18	19	20	21	22	24	25	26	31	39	53	71	87	99	99	85	68	53	46	41	39	36	34	33	31	29	24
2.0	4	6	8	11	12	13	14	15	16	17	18	19	21	23	26	31	39	51	67	88	95	86	67	53	45	41	38	36	33	32	29	25
2.5	2	3	5	7	8	9	10	11	11	12	13	14	16	18	20	23	26	30	38	56	77	93	84	66	52	45	41	38	34	33	30	25
3.0	0	1	2	4	4	5	6	6	7	8	8	9	10	12	14	16	18	21	24	30	42	67	90	84	68	54	46	42	36	33	30	26
IA/P = 0.30																																
0.0	0	0	0	0	2	4	10	19	29	39	48	53	56	56	54	53	51	49	46	44	44	42	42	40	40	40	40	40	40	39	39	34
.10	0	0	0	0	1	3	8	15	24	34	43	50	54	56	54	53	52	49	47	45	44	42	42	41	40	40	40	40	40	39	39	34
.20	0	0	0	0	0	1	2	6	12	20	30	39	46	54	55	54	53	51	49	46	44	43	42	41	40	40	40	40	40	39	39	35
.30	0	0	0	0	0	1	2	5	10	17	25	34	42	52	55	54	53	51	49	46	44	43	42	42	40	40	40	40	40	39	39	35
.40	0	0	0	0	0	0	1	4	8	14	21	30	38	50	55	55	53	52	50	47	45	43	42	42	40	40	40	40	40	39	39	35
.50	0	0	0	0	0	0	1	3	6	11	18	26	34	47	53	54	54	52	50	47	45	43	42	42	40	40	40	40	40	39	39	35
.75	0	0	0	0	0	0	0	1	3	6	10	16	23	37	47	52	53	53	52	49	46	44	42	42	41	40	40	40	40	40	39	35
1.0	0	0	0	0	0	0	0	0	0	1	2	4	7	17	30	42	50	53	53	51	49	45	43	42	41	40	40	40	40	40	39	36
1.5	0	0	0	0	0	0	0	0	0	0	0	0	1	4	11	21	33	43	49	52	51	48	45	43	42	41	40	40	40	40	39	36
2.0	0	0	0	0	0	0	0	0	0	0	0	0	0	0	1	4	11	20	30	43	50	51	48	45	43	42	41	40	40	40	39	37
2.5	0	0	0	0	0	0	0	0	0	0	0	0	0	0	0	0	2	5	10	23	36	48	51	47	44	43	42	41	40	40	39	37
3.0	0	0	0	0	0	0	0	0	0	0	0	0	0	0	0	0	0	0	2	7	17	34	48	50	47	44	43	42	40	40	39	38
IA/P = 0.50																																
0.0	0	0	0	0	0	0	0	0	0	0	0	0	0	0	0	1	3	6	10	15	19	23	28	31	32	36	39	41	44	44	44	44
.10	0	0	0	0	0	0	0	0	0	0	0	0	0	0	0	0	1	4	7	12	17	22	27	30	32	34	38	40	43	44	44	44
.20	0	0	0	0	0	0	0	0	0	0	0	0	0	0	0	0	1	2	5	10	15	20	25	30	31	33	37	39	43	44	44	44
.30	0	0	0	0	0	0	0	0	0	0	0	0	0	0	0	0	0	1	3	7	12	18	24	29	31	33	36	39	42	44	44	44
.40	0	0	0	0	0	0	0	0	0	0	0	0	0	0	0	0	0	0	1	5	10	16	22	27	30	32	35	38	42	44	44	44
.50	0	0	0	0	0	0	0	0	0	0	0	0	0	0	0	0	0	0	1	3	7	14	21	26	30	31	34	38	41	44	44	44
.75	0	0	0	0	0	0	0	0	0	0	0	0	0	0	0	0	0	0	0	2	5	11	19	24	29	31	33	36	41	44	44	44
1.0	0	0	0	0	0	0	0	0	0	0	0	0	0	0	0	0	0	0	0	1	3	8	16	22	27	30	32	35	40	43	44	44
1.5	0	0	0	0	0	0	0	0	0	0	0	0	0	0	0	0	0	0	0	0	0	3	9	16	23	27	30	32	38	42	44	44
2.0	0	0	0	0	0	0	0	0	0	0	0	0	0	0	0	0	0	0	0	0	0	0	3	10	17	23	28	30	36	40	44	44
2.5	0	0	0	0	0	0	0	0	0	0	0	0	0	0	0	0	0	0	0	0	0	0	1	4	11	18	23	28	33	39	44	44
3.0	0	0	0	0	0	0	0	0	0	0	0	0	0	0	0	0	0	0	0	0	0	0	0	1	4	10	17	23	30	36	43	44

RAINFALL TYPE = IA   * * * TC = 0.75 HR * * *   SHEET 6 OF 10

Exhibit 5-IA: Tabular hydrograph unit discharges (csm/in) for type IA rainfall distribution—continued

TRVL TIME (hr)	7.0	7.3	7.6	7.9	8.0	8.1	8.2	8.3	8.4	8.5	8.6	8.7	8.8	9.0	9.2	9.4	9.6	9.8	10.0	10.3	10.6	11.0	11.5	12.0	12.5	13.0	13.5	14.0	15.0	16.0	18.0	22.0
IA/P = 0.10														* * * TC = 1.0 HR * * *															IA/P = 0.10			
0.0	19	23	26	33	37	43	52	63	76	88	97	104	108	102	91	81	72	65	59	52	48	43	40	38	34	33	33	32	31	30	28	23
.10	19	22	26	32	35	41	48	58	70	82	92	100	105	103	94	83	75	67	61	53	48	44	40	38	35	33	33	32	31	30	28	23
.20	17	21	24	28	30	34	38	45	54	65	76	87	96	104	100	91	81	73	66	57	51	45	41	39	36	33	33	33	31	30	28	23
.30	16	20	23	27	29	32	36	42	50	60	71	81	91	103	101	93	84	75	67	58	52	46	42	39	36	34	33	33	31	30	28	23
.40	15	18	22	25	27	29	31	35	40	46	55	65	76	94	102	99	91	81	73	63	55	48	43	40	37	34	33	33	31	30	28	23
.50	14	18	21	25	26	28	30	33	37	43	51	61	71	90	101	101	93	84	75	64	56	49	43	40	37	34	33	33	31	30	28	23
.75	13	16	19	23	24	25	27	29	32	37	42	49	58	76	91	98	96	89	81	70	60	51	45	41	38	35	34	33	32	30	28	23
1.0	10	13	16	19	20	22	23	24	26	27	30	33	37	50	67	83	94	97	93	81	70	58	48	43	40	37	34	33	32	31	29	24
1.5	6	8	10	13	14	15	16	18	19	20	21	22	24	27	33	42	56	71	84	93	89	75	59	49	44	40	37	35	33	31	29	24
2.0	3	5	7	9	10	11	12	13	14	15	16	17	18	21	23	27	32	41	53	74	88	91	74	59	49	43	40	37	34	32	30	25
2.5	1	2	4	6	7	7	8	9	10	11	11	12	13	16	18	20	23	26	32	45	64	84	89	72	58	48	43	40	34	33	30	26
3.0	0	1	2	3	3	4	4	5	6	6	7	8	9	10	12	14	16	18	21	26	35	55	81	87	73	59	50	44	37	34	31	26
IA/P = 0.30														* * * TC = 1.0 HR * * *															IA/P = 0.30			
0.0	0	0	0	0	1	2	5	10	16	23	31	38	44	51	53	52	52	50	48	46	44	43	42	41	40	40	40	40	40	40	39	35
.10	0	0	0	0	0	1	2	4	8	13	20	27	35	46	51	53	52	51	50	47	45	44	42	42	40	40	40	40	40	40	39	35
.20	0	0	0	0	0	0	1	3	6	11	17	24	31	43	50	52	52	51	50	47	45	44	42	42	40	40	40	40	40	40	39	35
.30	0	0	0	0	0	0	1	2	5	9	14	21	28	40	48	52	52	52	50	48	46	44	43	42	40	40	40	40	40	40	39	35
.40	0	0	0	0	0	0	0	1	2	4	7	12	18	31	42	49	52	52	51	49	47	45	43	42	41	40	40	40	40	40	39	35
.50	0	0	0	0	0	0	0	1	1	3	6	10	15	28	39	47	51	52	51	49	47	45	43	42	41	40	40	40	40	40	39	35
.75	0	0	0	0	0	0	0	0	1	1	3	6	9	19	31	41	47	50	51	50	48	46	43	42	41	40	40	40	40	40	39	36
1.0	0	0	0	0	0	0	0	0	0	0	0	1	2	6	14	25	36	44	49	51	50	47	45	43	42	41	40	40	40	40	39	36
1.5	0	0	0	0	0	0	0	0	0	0	0	0	0	1	2	6	13	23	33	44	49	50	47	44	43	42	41	40	40	40	39	37
2.0	0	0	0	0	0	0	0	0	0	0	0	0	0	0	0	1	2	6	13	25	38	48	50	47	44	43	42	41	40	40	39	37
2.5	0	0	0	0	0	0	0	0	0	0	0	0	0	0	0	0	0	1	3	9	19	36	48	49	47	44	43	42	40	40	39	38
3.0	0	0	0	0	0	0	0	0	0	0	0	0	0	0	0	0	0	0	0	2	6	18	37	47	49	46	44	43	41	40	40	38
IA/P = 0.50														* * * TC = 1.0 HR * * *															IA/P = 0.50			
0.0	0	0	0	0	0	0	0	0	0	0	0	0	0	0	0	0	2	4	7	11	16	21	26	30	32	34	38	40	43	43	43	43
.10	0	0	0	0	0	0	0	0	0	0	0	0	0	0	0	0	1	2	5	9	14	19	25	29	31	33	37	39	42	43	43	43
.20	0	0	0	0	0	0	0	0	0	0	0	0	0	0	0	0	0	1	3	7	11	17	23	28	30	32	36	39	42	43	43	43
.30	0	0	0	0	0	0	0	0	0	0	0	0	0	0	0	0	0	0	1	5	9	15	21	27	30	32	35	38	41	43	43	43
.40	0	0	0	0	0	0	0	0	0	0	0	0	0	0	0	0	0	0	1	3	7	13	20	25	29	31	34	37	41	43	43	43
.50	0	0	0	0	0	0	0	0	0	0	0	0	0	0	0	0	0	0	0	2	5	11	18	24	28	31	33	36	41	43	43	43
.75	0	0	0	0	0	0	0	0	0	0	0	0	0	0	0	0	0	0	0	1	3	8	16	22	27	30	32	35	40	43	43	43
1.0	0	0	0	0	0	0	0	0	0	0	0	0	0	0	0	0	0	0	0	0	2	6	13	20	25	29	31	34	39	42	43	43
1.5	0	0	0	0	0	0	0	0	0	0	0	0	0	0	0	0	0	0	0	0	0	2	7	14	20	25	29	31	37	41	43	43
2.0	0	0	0	0	0	0	0	0	0	0	0	0	0	0	0	0	0	0	0	0	0	0	2	7	14	20	26	29	34	39	43	43
2.5	0	0	0	0	0	0	0	0	0	0	0	0	0	0	0	0	0	0	0	0	0	0	0	3	8	15	21	26	32	37	43	43
3.0	0	0	0	0	0	0	0	0	0	0	0	0	0	0	0	0	0	0	0	0	0	0	0	0	3	8	14	20	29	34	42	43
RAINFALL TYPE = IA														* * * TC = 1.0 HR * * *															SHEET 7 OF 10			

Exhibit 5-IA: Tabular hydrograph unit discharges (csm/in) for type IA rainfall distribution—continued

TRVL TIME (hr)	7.0	7.3	7.6	7.9	8.0	8.1	8.2	8.3	8.4	8.5	8.6	8.7	8.8	9.0	9.2	9.4	9.6	9.8	10.0	10.3	10.6	11.0	11.5	12.0	12.5	13.0	13.5	14.0	15.0	16.0	18.0	22.0
IA/P = 0.10														* * * TC = 1.25 HR * * *															IA/P = 0.10			
0.0	17	21	24	29	33	37	43	52	61	71	81	89	95	100	94	86	78	71	64	57	51	46	42	39	35	34	33	33	31	30	28	23
.10	16	19	23	27	29	31	35	41	48	57	66	76	85	96	99	92	84	76	69	60	54	47	43	40	36	34	34	33	31	30	28	23
.20	15	19	22	26	28	30	33	38	45	53	62	71	80	93	97	93	86	78	71	62	55	48	43	40	37	34	34	33	31	30	28	23
.30	14	17	20	24	25	27	29	32	36	42	49	58	67	84	95	96	91	84	76	66	58	50	44	41	38	35	34	33	31	30	28	23
.40	13	16	20	23	25	26	28	31	34	39	46	54	62	80	92	96	93	86	78	68	59	51	45	41	38	35	34	33	31	30	28	23
.50	12	15	18	21	23	24	25	27	29	33	37	43	50	67	83	93	95	91	84	73	63	54	46	42	39	36	34	33	32	30	28	23
.75	10	13	16	20	21	22	23	25	27	29	32	36	42	55	71	84	93	93	88	78	68	57	49	43	40	37	34	34	32	31	28	24
1.0	8	10	13	16	17	19	20	21	22	23	25	27	29	37	49	63	78	88	92	88	78	65	53	46	42	39	36	34	33	31	29	24
1.5	5	7	9	12	13	14	15	16	17	18	19	20	22	25	29	36	47	60	73	87	89	79	64	53	46	42	38	36	33	32	29	25
2.0	2	4	6	8	9	10	10	11	12	13	14	15	16	19	21	24	29	36	45	64	80	87	77	63	52	46	41	38	34	32	30	25
2.5	1	2	3	4	5	6	6	7	8	8	9	10	11	13	15	17	19	22	26	35	49	72	85	78	65	53	46	42	36	33	30	26
3.0	0	1	1	2	3	3	4	4	5	5	6	7	7	9	11	12	14	17	19	23	31	47	73	84	76	63	53	46	38	34	31	26
IA/P = 0.30														* * * TC = 1.25 HR * * *															IA/P = 0.30			
0.0	0	0	0	0	0	1	3	6	10	15	22	28	34	43	48	51	51	51	49	47	45	44	43	42	40	40	40	40	40	40	39	35
.10	0	0	0	0	0	0	1	2	5	8	13	19	25	36	45	49	51	50	50	48	46	44	43	42	40	40	40	40	40	40	39	35
.20	0	0	0	0	0	0	1	2	4	7	11	16	22	34	43	48	50	50	50	48	46	44	43	42	41	40	40	40	40	40	39	35
.30	0	0	0	0	0	0	1	1	3	5	9	14	19	31	40	46	49	50	50	48	46	45	43	42	41	40	40	40	40	40	39	35
.40	0	0	0	0	0	0	0	1	2	4	8	12	17	28	38	45	49	50	50	49	47	45	43	42	41	40	40	40	40	40	39	35
.50	0	0	0	0	0	0	0	1	2	4	6	10	14	25	36	43	48	50	50	49	47	45	43	42	41	40	40	40	40	40	39	35
.75	0	0	0	0	0	0	0	0	0	1	2	3	6	13	23	33	41	46	49	50	48	46	44	43	42	40	40	40	40	40	39	36
1.0	0	0	0	0	0	0	0	0	0	0	0	1	2	6	14	23	32	40	46	49	49	47	45	43	42	41	40	40	40	40	39	36
1.5	0	0	0	0	0	0	0	0	0	0	0	0	0	1	2	6	13	21	30	41	47	49	47	45	43	42	41	40	40	40	39	37
2.0	0	0	0	0	0	0	0	0	0	0	0	0	0	0	0	1	3	6	12	23	35	45	48	47	45	43	42	41	40	40	40	37
2.5	0	0	0	0	0	0	0	0	0	0	0	0	0	0	0	0	0	1	3	8	18	33	45	48	46	44	43	42	40	40	40	38
3.0	0	0	0	0	0	0	0	0	0	0	0	0	0	0	0	0	0	0	0	2	6	17	34	45	48	46	44	43	41	40	40	38
IA/P = 0.50														* * * TC = 1.25 HR * * *															IA/P = 0.50			
0.0	0	0	0	0	0	0	0	0	0	0	0	0	0	0	0	0	1	3	5	9	14	19	24	28	31	33	37	39	42	42	42	42
.10	0	0	0	0	0	0	0	0	0	0	0	0	0	0	0	0	0	1	3	7	11	17	23	27	30	32	36	38	42	42	42	42
.20	0	0	0	0	0	0	0	0	0	0	0	0	0	0	0	0	0	1	2	5	9	15	21	26	29	32	35	38	41	42	42	42
.30	0	0	0	0	0	0	0	0	0	0	0	0	0	0	0	0	0	0	1	3	7	13	19	25	29	31	34	37	41	42	42	42
.40	0	0	0	0	0	0	0	0	0	0	0	0	0	0	0	0	0	0	0	2	5	11	17	23	28	30	33	36	40	42	42	42
.50	0	0	0	0	0	0	0	0	0	0	0	0	0	0	0	0	0	0	0	1	3	8	15	22	26	30	32	35	40	42	42	42
.75	0	0	0	0	0	0	0	0	0	0	0	0	0	0	0	0	0	0	0	1	2	7	13	20	25	29	31	34	39	42	42	42
1.0	0	0	0	0	0	0	0	0	0	0	0	0	0	0	0	0	0	0	0	0	1	4	11	17	23	27	30	33	38	42	42	42
1.5	0	0	0	0	0	0	0	0	0	0	0	0	0	0	0	0	0	0	0	0	0	1	5	11	18	23	28	30	36	40	42	42
2.0	0	0	0	0	0	0	0	0	0	0	0	0	0	0	0	0	0	0	0	0	0	0	1	6	12	18	24	28	33	39	42	42
2.5	0	0	0	0	0	0	0	0	0	0	0	0	0	0	0	0	0	0	0	0	0	0	0	2	6	13	19	24	31	37	42	42
3.0	0	0	0	0	0	0	0	0	0	0	0	0	0	0	0	0	0	0	0	0	0	0	0	0	2	6	12	18	28	33	42	42
RAINFALL TYPE = IA														* * * TC = 1.25 HR * * *															SHEET 8 OF 10			

Exhibit 5-IA: Tabular hydrograph unit discharges (csm/in) for type IA rainfall distribution—continued

IA/P = 0.10 * * * TC = 1.5 HR * * *

TRVL TIME (hr)	7.0	7.3	7.6	7.9	8.0	8.1	8.2	8.3	8.4	8.5	8.6	8.7	8.8	9.0	9.2	9.4	9.6	9.8	10.0	10.3	10.6	11.0	11.5	12.0	12.5	13.0	13.5	14.0	15.0	16.0	18.0	22.0
0.0	15	19	22	27	29	32	37	42	49	57	65	73	80	89	94	89	83	76	70	62	55	49	44	40	37	35	34	33	31	30	28	23
.10	14	17	20	24	26	28	31	35	40	46	53	61	69	82	90	92	87	81	74	65	58	51	45	41	38	35	34	33	31	30	28	23
.20	13	16	20	23	25	27	29	33	37	43	50	57	65	79	88	91	88	82	76	67	59	52	46	42	38	35	34	33	32	30	28	23
.30	12	15	18	22	23	24	26	28	31	35	41	47	54	69	81	89	90	86	81	71	63	54	47	43	39	36	34	34	32	30	28	24
.40	11	14	18	21	22	24	25	27	30	34	38	44	51	65	78	87	90	87	82	73	64	55	48	43	40	36	34	34	32	30	28	24
.50	10	13	16	19	20	22	23	24	26	29	32	36	41	54	69	81	88	89	86	77	68	58	50	44	41	37	35	34	32	31	29	24
.75	9	11	14	18	19	20	21	22	24	26	28	31	35	46	58	71	82	88	88	81	73	62	52	46	42	38	35	34	32	31	29	24
1.0	7	9	11	14	15	16	18	19	20	21	22	24	26	32	40	52	65	76	84	87	81	70	58	49	44	40	37	35	33	31	29	24
1.5	4	6	8	10	11	12	13	14	15	16	17	18	19	22	26	31	39	50	62	77	85	81	69	57	49	44	40	37	34	32	30	25
2.0	2	3	5	7	7	8	9	10	11	12	12	13	14	17	19	22	25	31	38	53	69	83	80	68	56	48	43	40	35	33	30	25
2.5	1	1	2	3	4	5	5	6	6	7	8	9	9	11	13	15	17	20	23	30	41	61	82	82	69	58	49	44	37	34	31	26
3.0	0	0	1	2	2	2	3	3	4	4	5	5	6	8	9	11	13	15	17	21	27	40	63	80	78	68	57	49	40	35	31	27

IA/P = 0.30 * * * TC = 1.5 HR * * *

TRVL TIME (hr)	7.0	7.3	7.6	7.9	8.0	8.1	8.2	8.3	8.4	8.5	8.6	8.7	8.8	9.0	9.2	9.4	9.6	9.8	10.0	10.3	10.6	11.0	11.5	12.0	12.5	13.0	13.5	14.0	15.0	16.0	18.0	22.0
0.0	0	0	0	0	0	1	2	4	6	10	14	19	24	34	41	46	48	49	48	47	46	44	43	42	41	40	40	40	40	40	39	35
.10	0	0	0	0	0	0	1	1	3	5	8	12	17	27	36	42	46	48	49	48	46	45	43	42	41	40	40	40	40	40	39	36
.20	0	0	0	0	0	0	0	1	2	4	7	10	15	24	33	41	45	48	49	48	47	45	43	43	41	40	40	40	40	40	39	36
.30	0	0	0	0	0	0	0	1	2	3	6	9	13	22	31	39	44	47	48	48	47	45	44	43	41	40	40	40	40	40	39	36
.40	0	0	0	0	0	0	0	0	1	1	3	5	7	15	24	33	40	45	47	48	47	46	44	43	42	40	40	40	40	40	39	36
.50	0	0	0	0	0	0	0	0	0	1	2	4	6	13	22	31	38	44	47	48	48	46	44	43	42	40	40	40	40	40	39	36
.75	0	0	0	0	0	0	0	0	0	1	1	2	4	9	16	24	33	39	44	47	48	47	45	43	42	41	40	40	40	40	39	36
1.0	0	0	0	0	0	0	0	0	0	0	0	0	1	2	6	12	20	29	36	44	47	48	46	44	43	42	40	40	40	40	40	36
1.5	0	0	0	0	0	0	0	0	0	0	0	0	0	0	1	3	6	12	19	30	39	46	47	46	44	43	42	40	40	40	40	37
2.0	0	0	0	0	0	0	0	0	0	0	0	0	0	0	0	1	2	4	8	17	28	40	46	47	45	44	42	41	40	40	40	37
2.5	0	0	0	0	0	0	0	0	0	0	0	0	0	0	0	0	0	1	2	6	13	26	40	47	47	45	43	42	40	40	40	38
3.0	0	0	0	0	0	0	0	0	0	0	0	0	0	0	0	0	0	0	0	1	4	12	28	41	46	46	45	43	41	40	40	38

IA/P = 0.50 * * * TC = 1.5 HR * * *

TRVL TIME (hr)	7.0	7.3	7.6	7.9	8.0	8.1	8.2	8.3	8.4	8.5	8.6	8.7	8.8	9.0	9.2	9.4	9.6	9.8	10.0	10.3	10.6	11.0	11.5	12.0	12.5	13.0	13.5	14.0	15.0	16.0	18.0	22.0
0.0	0	0	0	0	0	0	0	0	0	0	0	0	0	0	0	0	1	2	3	7	11	16	22	27	30	32	35	38	42	42	42	42
.10	0	0	0	0	0	0	0	0	0	0	0	0	0	0	0	0	0	1	1	4	7	13	19	24	28	31	34	37	41	42	42	42
.20	0	0	0	0	0	0	0	0	0	0	0	0	0	0	0	0	0	0	1	3	7	12	19	24	28	31	33	36	41	42	42	42
.30	0	0	0	0	0	0	0	0	0	0	0	0	0	0	0	0	0	0	1	3	6	12	18	23	28	30	33	36	40	42	42	42
.40	0	0	0	0	0	0	0	0	0	0	0	0	0	0	0	0	0	0	1	3	6	11	17	23	27	30	33	36	40	42	42	42
.50	0	0	0	0	0	0	0	0	0	0	0	0	0	0	0	0	0	0	1	2	5	10	17	22	27	30	32	36	40	41	41	41
.75	0	0	0	0	0	0	0	0	0	0	0	0	0	0	0	0	0	0	1	2	5	10	16	22	26	29	32	35	40	41	41	41
1.0	0	0	0	0	0	0	0	0	0	0	0	0	0	0	0	0	0	0	0	0	2	5	11	17	23	27	30	33	38	41	41	41
1.5	0	0	0	0	0	0	0	0	0	0	0	0	0	0	0	0	0	0	0	0	0	1	5	12	18	23	27	30	36	40	41	41
2.0	0	0	0	0	0	0	0	0	0	0	0	0	0	0	0	0	0	0	0	0	0	0	2	6	12	18	24	28	33	39	41	41
2.5	0	0	0	0	0	0	0	0	0	0	0	0	0	0	0	0	0	0	0	0	0	0	0	2	7	13	19	24	31	36	41	41
3.0	0	0	0	0	0	0	0	0	0	0	0	0	0	0	0	0	0	0	0	0	0	0	0	1	3	7	13	19	28	34	41	41

RAINFALL TYPE = IA * * * TC = 1.5 HR * * * SHEET 9 OF 10

Exhibit 5-IA: Tabular hydrograph unit discharges (csm/in) for type IA rainfall distribution—continued

IA/P = 0.10 * * * TC = 2.0 HR * * *

TRVL TIME (hr)	7.0	7.3	7.6	7.9	8.0	8.1	8.2	8.3	8.4	8.5	8.6	8.7	8.8	9.0	9.2	9.4	9.6	9.8	10.0	10.3	10.6	11.0	11.5	12.0	12.5	13.0	13.5	14.0	15.0	16.0	18.0	22.0
0.0	12	15	18	22	24	26	28	32	36	41	46	52	58	68	76	80	84	79	76	68	62	55	48	44	40	37	35	34	32	31	28	24
.10	11	13	16	20	21	23	25	27	30	34	38	43	49	60	70	77	83	83	78	72	65	57	50	45	41	38	36	34	32	31	29	24
.20	9	12	15	18	19	21	22	24	26	29	32	36	41	52	63	72	78	82	81	75	68	60	52	46	42	39	36	35	32	31	29	24
.30	9	12	14	18	19	20	21	23	25	28	31	34	39	49	60	70	77	81	81	76	70	61	53	47	43	39	36	35	33	31	29	24
.40	9	11	14	17	18	19	21	22	24	26	29	33	37	47	57	67	75	80	81	77	71	62	53	47	43	39	37	35	33	31	29	24
.50	7	10	12	15	17	18	19	20	21	23	25	28	31	39	49	60	69	76	81	79	74	65	56	49	44	40	37	35	33	31	29	24
.75	6	9	11	14	15	16	17	18	20	21	23	25	27	34	42	52	62	70	76	79	76	68	59	51	46	41	38	36	33	31	29	24
1.0	5	6	9	11	12	13	14	15	16	17	18	20	21	25	30	38	47	57	66	76	79	75	64	55	49	44	40	37	34	32	29	25
1.5	3	4	6	8	8	9	10	11	12	13	14	15	16	18	21	24	30	37	45	59	71	78	73	63	55	48	43	40	35	33	30	25
2.0	1	2	3	4	5	5	6	7	7	8	9	10	10	12	14	16	19	22	26	36	48	65	77	73	64	56	49	44	37	34	31	26
2.5	0	1	1	2	3	3	4	4	5	5	6	6	7	8	10	12	14	16	18	23	31	46	66	76	72	63	55	48	40	35	31	27
3.0	0	0	1	1	1	2	2	2	3	3	3	4	4	5	7	8	10	11	13	17	21	30	49	67	75	71	63	54	43	37	32	27

IA/P = 0.30 * * * TC = 2.0 HR * * *

TRVL TIME (hr)	7.0	7.3	7.6	7.9	8.0	8.1	8.2	8.3	8.4	8.5	8.6	8.7	8.8	9.0	9.2	9.4	9.6	9.8	10.0	10.3	10.6	11.0	11.5	12.0	12.5	13.0	13.5	14.0	15.0	16.0	18.0	22.0
0.0	0	0	0	0	0	0	1	2	3	5	7	10	13	20	27	34	39	42	44	47	45	45	44	43	42	40	40	40	40	40	39	36
.10	0	0	0	0	0	0	0	1	1	2	4	6	8	15	22	29	35	40	43	45	46	45	44	43	42	41	40	40	40	40	40	36
.20	0	0	0	0	0	0	0	0	1	1	2	3	5	10	17	24	31	36	41	44	46	45	44	43	42	41	40	40	40	40	40	36
.30	0	0	0	0	0	0	0	0	0	1	2	3	4	9	15	22	29	35	39	44	46	45	44	43	42	41	40	40	40	40	40	36
.40	0	0	0	0	0	0	0	0	0	1	1	2	4	8	14	20	27	33	38	43	46	46	44	43	42	41	40	40	40	40	40	36
.50	0	0	0	0	0	0	0	0	0	0	1	1	2	5	9	15	22	29	35	41	44	45	45	44	43	42	40	40	40	40	40	36
.75	0	0	0	0	0	0	0	0	0	0	0	1	1	3	6	11	17	24	30	38	42	45	45	44	43	42	41	40	40	40	40	37
1.0	0	0	0	0	0	0	0	0	0	0	0	0	0	1	2	5	9	14	21	30	38	43	45	44	44	43	41	40	40	40	40	37
1.5	0	0	0	0	0	0	0	0	0	0	0	0	0	0	0	1	2	4	8	16	25	36	43	45	44	43	42	41	40	40	40	38
2.0	0	0	0	0	0	0	0	0	0	0	0	0	0	0	0	0	0	1	2	6	12	24	36	43	45	44	43	42	40	40	40	38
2.5	0	0	0	0	0	0	0	0	0	0	0	0	0	0	0	0	0	0	0	1	4	12	25	37	43	45	44	43	41	40	40	38
3.0	0	0	0	0	0	0	0	0	0	0	0	0	0	0	0	0	0	0	0	0	1	4	13	26	37	43	44	44	42	40	40	39

IA/P = 0.50 * * * TC = 2.0 HR * * *

TRVL TIME (hr)	7.0	7.3	7.6	7.9	8.0	8.1	8.2	8.3	8.4	8.5	8.6	8.7	8.8	9.0	9.2	9.4	9.6	9.8	10.0	10.3	10.6	11.0	11.5	12.0	12.5	13.0	13.5	14.0	15.0	16.0	18.0	22.0
0.0	0	0	0	0	0	0	0	0	0	0	0	0	0	0	0	0	0	1	2	4	7	12	18	23	27	30	33	36	40	40	40	40
.10	0	0	0	0	0	0	0	0	0	0	0	0	0	0	0	0	0	0	1	3	5	10	16	21	26	29	32	35	40	40	40	40
.20	0	0	0	0	0	0	0	0	0	0	0	0	0	0	0	0	0	0	1	2	4	8	14	20	25	28	31	34	39	40	40	40
.30	0	0	0	0	0	0	0	0	0	0	0	0	0	0	0	0	0	0	0	1	3	6	12	18	23	27	30	33	38	40	40	40
.40	0	0	0	0	0	0	0	0	0	0	0	0	0	0	0	0	0	0	0	1	2	5	11	16	22	26	29	32	38	40	40	40
.50	0	0	0	0	0	0	0	0	0	0	0	0	0	0	0	0	0	0	0	0	1	4	9	15	20	25	28	31	37	40	40	40
.75	0	0	0	0	0	0	0	0	0	0	0	0	0	0	0	0	0	0	0	0	1	3	7	13	19	24	27	30	36	40	40	40
1.0	0	0	0	0	0	0	0	0	0	0	0	0	0	0	0	0	0	0	0	0	0	1	3	8	14	20	24	28	34	39	40	40
1.5	0	0	0	0	0	0	0	0	0	0	0	0	0	0	0	0	0	0	0	0	0	0	1	5	10	16	21	25	32	37	40	40
2.0	0	0	0	0	0	0	0	0	0	0	0	0	0	0	0	0	0	0	0	0	0	0	0	2	5	11	16	21	29	35	40	40
2.5	0	0	0	0	0	0	0	0	0	0	0	0	0	0	0	0	0	0	0	0	0	0	0	0	2	6	11	17	26	32	40	40
3.0	0	0	0	0	0	0	0	0	0	0	0	0	0	0	0	0	0	0	0	0	0	0	0	0	0	2	6	11	21	29	39	40

RAINFALL TYPE = IA * * * TC = 2.0 HR * * * SHEET 10 OF 10

Exhibit 5-II: Tabular hydrograph unit discharges (csm/in) for type II rainfall distribution

HYDROGRAPH TIME (HOURS) — * * * TC = 0.1 HR * * *

TRVL TIME (hr)	11.0	11.3	11.6	11.9	12.0	12.1	12.2	12.3	12.4	12.5	12.6	12.7	12.8	13.0	13.2	13.4	13.6	13.8	14.0	14.3	14.6	15.0	15.5	16.0	16.5	17.0	17.5	18.0	19.0	20.0	22.0	26.0
IA/P = 0.10																																
0.0	24	34	53	334	647	1010	623	217	147	123	104	86	76	66	57	51	46	42	38	34	32	29	26	23	21	20	19	18	15	13	12	0
.10	21	29	43	134	267	520	847	701	378	224	157	122	98	75	64	56	50	45	41	36	33	30	27	24	21	20	19	18	16	13	12	0
.20	18	25	35	61	110	215	418	704	702	486	312	209	151	94	73	62	54	49	44	38	34	31	28	25	22	21	19	18	16	14	12	0
.30	17	23	33	56	92	174	337	582	662	545	389	269	190	109	79	65	56	50	45	39	35	32	29	25	22	21	20	18	16	14	12	0
.40	15	20	28	41	51	78	142	272	478	601	563	447	328	172	104	76	63	55	49	42	37	33	29	26	23	21	20	19	17	14	12	0
.50	14	19	26	39	47	68	117	220	392	531	553	482	380	209	121	84	67	57	51	43	38	33	30	27	23	21	20	19	17	14	12	0
.75	12	15	21	29	33	38	49	73	126	224	343	432	464	385	252	156	103	76	62	50	43	36	31	28	25	22	21	19	17	15	12	0
1.0	9	12	15	21	23	26	29	33	40	55	86	148	238	406	434	317	205	130	89	62	50	41	34	30	27	24	22	20	18	16	12	0
1.5	7	8	10	14	15	16	18	20	22	25	29	34	45	101	220	339	373	320	234	131	80	53	40	34	30	27	24	21	19	17	12	2
2.0	4	6	7	9	9	10	11	12	13	15	16	18	20	25	37	72	150	252	336	312	216	109	58	42	34	30	27	24	20	18	13	8
2.5	3	4	5	6	7	7	8	8	9	10	11	12	13	16	19	25	39	75	142	262	308	229	108	58	41	34	30	27	22	19	14	11
3.0	1	2	3	4	4	5	5	6	6	7	7	8	8	10	12	14	17	22	31	76	169	288	236	122	64	43	35	30	24	20	16	11
IA/P = 0.30																																
0.0	0	0	0	154	568	936	524	217	172	149	126	107	97	86	76	69	63	58	53	48	46	42	38	34	31	30	28	27	24	20	19	0
.10	0	0	0	19	109	415	762	603	346	230	176	143	119	96	84	74	68	62	57	50	47	44	40	35	32	30	29	27	24	21	19	0
.20	0	0	0	0	13	77	302	609	605	432	297	217	167	115	94	81	73	66	60	53	48	45	41	37	33	31	29	28	25	21	19	0
.30	0	0	0	0	9	54	219	479	563	476	357	263	199	129	99	85	75	68	62	54	49	45	41	37	33	31	29	28	25	21	19	0
.40	0	0	0	0	0	6	38	159	372	500	484	399	309	183	123	96	82	73	66	58	51	46	42	38	34	31	30	28	25	22	19	0
.50	0	0	0	0	0	4	27	115	287	429	465	421	346	213	138	103	86	76	68	59	52	47	43	39	34	32	30	29	25	22	19	0
.75	0	0	0	0	0	0	1	10	46	132	246	338	381	341	243	165	119	94	80	67	58	50	45	41	37	33	31	29	26	23	19	0
1.0	0	0	0	0	0	0	0	1	4	22	69	149	241	357	331	246	170	122	96	76	64	54	47	42	38	34	32	30	27	24	19	0
1.5	0	0	0	0	0	0	0	0	0	0	0	1	4	41	142	258	310	285	224	142	97	71	55	47	43	39	35	32	29	25	20	4
2.0	0	0	0	0	0	0	0	0	0	0	0	0	0	1	10	49	130	221	279	255	182	108	70	55	47	42	38	34	30	27	20	11
2.5	0	0	0	0	0	0	0	0	0	0	0	0	0	0	0	2	14	52	119	224	256	193	107	70	55	47	42	38	32	28	22	17
3.0	0	0	0	0	0	0	0	0	0	0	0	0	0	0	0	0	0	1	9	52	141	240	199	117	74	56	48	43	35	30	24	18
IA/P = 0.50																																
0.0	0	0	0	0	70	539	377	196	171	154	134	117	108	99	89	83	77	72	67	61	59	56	51	46	43	42	40	38	34	30	28	0
.10	0	0	0	0	47	375	376	256	199	169	146	126	114	102	92	85	79	73	68	62	59	56	52	47	43	42	40	38	34	30	28	0
.20	0	0	0	0	0	31	260	338	283	227	189	160	138	112	99	90	83	77	72	64	60	57	53	48	44	42	41	39	35	30	28	0
.30	0	0	0	0	0	0	21	180	285	284	246	208	176	131	110	97	88	82	76	68	62	59	54	50	45	43	41	39	36	31	28	0
.40	0	0	0	0	0	0	14	125	232	266	253	223	192	142	115	100	91	83	77	69	63	59	55	50	45	43	41	40	36	31	28	0
.50	0	0	0	0	0	0	9	86	183	239	248	231	205	154	122	104	93	85	79	71	64	59	55	51	46	43	41	40	36	32	28	0
.75	0	0	0	0	0	0	3	31	87	147	190	211	213	184	147	121	103	92	84	75	67	61	57	52	47	44	42	40	37	32	28	0
1.0	0	0	0	0	0	0	0	0	1	13	45	92	141	205	197	165	134	112	98	84	75	65	59	55	50	46	43	41	38	34	28	0
1.5	0	0	0	0	0	0	0	0	0	0	0	2	9	51	118	170	183	167	143	111	92	77	65	59	54	50	45	43	39	35	28	2
2.0	0	0	0	0	0	0	0	0	0	0	0	0	0	2	15	51	103	148	168	156	127	96	76	65	58	54	49	45	41	37	29	12
2.5	0	0	0	0	0	0	0	0	0	0	0	0	0	0	0	1	9	31	69	131	159	140	101	78	66	59	54	50	43	39	31	24
3.0	0	0	0	0	0	0	0	0	0	0	0	0	0	0	0	0	0	2	11	46	101	151	134	99	77	65	59	54	45	41	33	26

RAINFALL TYPE = II — * * * TC = 0.1 HR * * * — SHEET 1 OF 10

Exhibit 5-II: Tabular hydrograph unit discharges (csm/in) for type II rainfall distribution—continued

HYDROGRAPH TIME (HOURS) — * * * TC = 0.2 HR * * *

TRVL TIME (hr)	11.0	11.3	11.6	11.9	12.0	12.1	12.2	12.3	12.4	12.5	12.6	12.7	12.8	13.0	13.2	13.4	13.6	13.8	14.0	14.3	14.6	15.0	15.5	16.0	16.5	17.0	17.5	18.0	19.0	20.0	22.0	26.0
IA/P = 0.10																																
0.0	23	31	47	209	403	739	800	481	250	166	128	102	86	70	61	54	49	44	40	35	33	30	27	24	21	20	19	18	16	13	12	0
.10	19	26	39	86	168	325	601	733	565	355	229	161	122	83	69	59	53	47	43	37	34	31	28	25	22	21	19	18	16	14	12	0
.20	17	23	32	49	74	136	262	488	652	594	435	298	207	115	81	67	58	51	46	40	35	32	29	26	23	21	20	19	16	14	12	0
.30	16	22	30	46	64	112	212	396	566	585	485	360	258	139	90	71	60	53	48	41	36	32	29	26	23	21	20	19	16	14	12	0
.40	14	19	25	37	43	57	94	173	322	485	551	507	409	227	129	87	68	58	52	44	38	33	30	27	24	21	20	19	17	14	12	0
.50	13	18	24	35	40	52	80	142	262	410	504	506	441	269	153	98	73	61	53	45	39	34	30	27	24	22	20	19	17	15	12	0
.75	10	13	17	23	26	30	34	40	55	86	150	247	349	438	360	240	151	101	75	57	47	39	33	29	26	23	21	20	18	15	12	0
1.0	9	11	14	19	21	24	26	30	35	44	62	101	167	337	413	353	245	157	104	68	53	42	35	31	28	24	22	20	18	16	12	0
1.5	6	8	10	13	14	15	17	19	21	23	26	30	37	73	166	288	356	337	264	154	91	57	42	35	30	27	24	22	19	17	13	3
2.0	4	5	7	8	9	10	10	11	12	14	15	16	18	23	31	55	114	206	291	324	239	125	63	44	35	31	28	24	20	18	14	9
2.5	3	4	5	6	6	7	7	8	9	9	10	11	12	15	18	22	32	58	111	227	298	246	122	63	43	35	31	27	22	19	15	11
3.0	1	2	3	4	4	4	5	5	6	6	7	7	8	9	11	13	16	19	27	59	138	280	248	137	70	46	36	31	25	21	16	11
IA/P = 0.30																																
0.0	0	0	0	39	180	545	697	497	276	198	158	130	110	93	81	73	67	61	56	49	46	43	39	35	32	30	29	27	24	21	19	0
.10	0	0	0	2	27	129	407	600	532	361	252	190	150	108	90	79	71	65	59	52	48	44	41	36	32	31	29	28	25	21	19	0
.20	0	0	0	2	19	92	302	501	521	415	306	228	176	119	95	82	73	67	61	53	48	45	41	37	33	31	29	28	25	21	19	0
.30	0	0	0	0	1	13	66	223	408	484	438	350	269	163	114	93	80	72	65	57	51	46	42	38	34	31	30	28	25	22	19	0
.40	0	0	0	0	1	9	47	164	327	431	436	379	306	189	127	98	83	74	67	58	52	47	43	38	34	31	30	28	25	22	19	0
.50	0	0	0	0	0	0	6	33	120	258	374	415	391	271	173	121	95	81	72	62	55	48	44	40	35	32	30	29	26	22	19	0
.75	0	0	0	0	0	0	2	13	50	126	221	302	348	323	240	167	121	96	81	68	59	50	45	41	37	33	31	29	26	23	19	0
1.0	0	0	0	0	0	0	0	0	1	6	24	69	139	285	331	280	204	145	109	82	68	56	48	43	39	35	32	30	27	24	19	0
1.5	0	0	0	0	0	0	0	0	0	0	0	0	1	16	79	186	271	288	247	165	110	76	58	49	44	40	35	32	29	26	20	5
2.0	0	0	0	0	0	0	0	0	0	0	0	0	0	0	3	24	80	163	235	262	202	123	76	58	49	43	39	35	30	27	21	13
2.5	0	0	0	0	0	0	0	0	0	0	0	0	0	0	0	1	6	28	77	179	242	207	120	75	57	48	43	39	32	29	22	17
3.0	0	0	0	0	0	0	0	0	0	0	0	0	0	0	0	0	0	0	4	30	101	207	227	130	80	59	49	44	35	30	24	18
IA/P = 0.50																																
0.0	0	0	0	0	7	98	371	322	221	182	158	137	120	104	94	86	80	74	69	62	60	57	52	47	44	42	40	39	35	30	28	0
.10	0	0	0	0	4	67	270	305	249	204	174	149	130	108	97	88	82	76	71	64	60	57	53	48	44	42	41	39	35	30	28	0
.20	0	0	0	0	0	3	45	195	268	255	221	189	163	125	106	95	87	80	75	67	62	58	54	49	45	43	41	39	35	31	28	0
.30	0	0	0	0	0	2	31	140	226	245	229	203	176	134	111	98	89	82	76	68	62	59	55	50	45	43	41	39	36	31	28	0
.40	0	0	0	0	0	1	21	101	184	225	228	211	188	144	117	101	91	84	78	69	63	59	55	50	45	43	41	40	36	31	28	0
.50	0	0	0	0	0	0	1	14	72	146	199	218	213	175	137	113	99	89	82	73	66	60	56	52	47	43	42	40	36	32	28	0
.75	0	0	0	0	0	0	0	5	28	71	121	162	186	193	161	133	112	98	88	78	70	62	57	53	48	44	42	41	37	33	28	0
1.0	0	0	0	0	0	0	0	0	0	2	13	38	77	154	186	174	147	122	105	89	78	68	60	56	51	46	43	42	38	34	28	0
1.5	0	0	0	0	0	0	0	0	0	0	0	0	2	22	71	129	163	168	150	120	98	80	67	60	55	51	46	43	40	36	28	4
2.0	0	0	0	0	0	0	0	0	0	0	0	0	0	0	5	25	65	112	146	157	134	103	79	67	60	55	50	46	41	38	29	14
2.5	0	0	0	0	0	0	0	0	0	0	0	0	0	0	0	1	8	26	60	117	148	136	101	79	66	59	54	50	43	39	31	24
3,0	0	0	0	0	0	0	0	0	0	0	0	0	0	0	0	0	0	2	9	40	90	142	130	99	78	66	59	54	45	41	33	26

RAINFALL TYPE = II — * * * TC = 0.2 HR * * * — SHEET 2 OF 10

Exhibit 5-II: Tabular hydrograph unit discharges (csm/in) for type II rainfall distribution—continued

TRVL TIME (hr) / HYDROGRAPH TIME (HOURS)

IA/P = 0.10 * * * TC = 0.3 HR * * *

TRVL TIME (hr)	11.0	11.3	11.6	11.9	12.0	12.1	12.2	12.3	12.4	12.5	12.6	12.7	12.8	13.0	13.2	13.4	13.6	13.8	14.0	14.3	14.6	15.0	15.5	16.0	16.5	17.0	17.5	18.0	19.0	20.0	22.0	26.0
0.0	20	28	41	118	235	447	676	676	459	283	196	146	114	80	66	57	51	46	42	37	33	31	28	24	22	20	19	18	16	13	12	0
.10	19	26	39	99	189	361	571	641	520	362	251	181	136	89	70	60	53	48	43	37	34	31	28	25	22	21	19	18	16	14	12	0
.20	17	23	32	53	83	154	292	478	587	542	422	308	223	127	86	68	58	52	46	40	35	32	29	26	23	21	20	19	16	14	12	0
.30	16	22	30	49	72	127	237	398	524	536	460	359	268	151	97	73	61	53	48	41	36	32	29	26	23	21	20	19	16	14	12	0
.40	14	19	25	37	45	63	105	193	330	459	510	477	398	237	139	92	70	59	52	44	38	34	30	27	24	21	20	19	17	14	12	0
.50	13	18	24	35	42	56	89	158	272	397	472	475	424	274	163	104	76	62	54	46	39	34	30	27	24	22	20	19	17	15	12	0
.75	11	14	19	26	30	34	42	59	95	160	250	339	417	398	299	196	128	89	69	54	45	37	32	29	26	23	21	20	17	15	12	0
1.0	9	11	14	19	21	24	27	30	36	46	68	109	174	328	396	346	248	163	109	70	54	43	35	31	28	24	22	20	18	16	12	0
1.5	6	8	10	13	14	15	17	19	21	23	26	31	38	77	169	282	347	330	264	158	94	58	42	35	31	27	24	22	19	17	13	3
2.0	4	5	7	8	9	10	10	11	12	14	15	16	18	23	32	57	116	205	285	317	239	128	64	44	36	31	28	25	20	18	14	9
2.5	2	4	5	6	6	7	7	8	9	9	10	11	12	15	18	23	33	60	113	223	293	245	125	65	44	35	31	27	22	19	15	11
3.0	1	2	3	4	4	4	5	5	6	6	7	7	8	9	11	13	16	20	27	61	138	275	246	139	72	46	36	31	25	21	16	11

IA/P = 0.30 * * * TC = 0.3 HR * * *

TRVL TIME (hr)	11.0	11.3	11.6	11.9	12.0	12.1	12.2	12.3	12.4	12.5	12.6	12.7	12.8	13.0	13.2	13.4	13.6	13.8	14.0	14.3	14.6	15.0	15.5	16.0	16.5	17.0	17.5	18.0	19.0	20.0	22.0	26.0
0.0	0	0	0	11	64	251	525	574	454	303	221	173	140	104	88	77	70	64	58	51	47	44	40	36	32	31	29	28	24	21	19	0
.10	0	0	0	0	7	45	183	411	520	476	360	268	205	133	101	85	76	69	62	55	49	45	41	37	33	31	30	28	25	21	19	0
.20	0	0	0	0	5	32	132	318	452	468	396	310	240	151	109	90	78	70	64	56	50	46	42	38	33	31	30	28	25	22	19	0
.30	0	0	0	0	0	3	22	96	244	383	440	411	344	217	142	105	87	76	69	60	53	47	43	39	35	32	30	29	26	22	19	0
.40	0	0	0	0	0	2	16	69	186	317	399	407	365	246	160	115	92	79	71	61	54	48	43	39	35	32	30	29	26	22	19	0
,50	0	0	0	0	0	0	2	11	50	140	258	352	389	327	223	149	110	89	77	66	57	50	45	41	36	33	31	29	26	23	19	0
.75	0	0	0	0	0	0	1	4	20	63	135	219	290	335	281	205	146	110	89	72	62	52	46	42	38	34	31	30	27	23	19	0
1.0	0	0	0	0	0	0	0	0	0	2	9	32	78	216	320	306	243	176	128	90	72	59	49	44	40	36	33	31	28	24	19	1
1.5	0	0	0	0	0	0	0	0	0	0	0	0	2	20	84	185	264	281	246	168	112	77	58	49	44	40	36	32	29	26	20	5
2.0	0	0	0	0	0	0	0	0	0	0	0	0	0	0	1	12	50	121	200	257	224	141	83	61	50	44	40	36	31	28	21	14
2.5	0	0	0	0	0	0	0	0	0	0	0	0	0	0	0	0	3	16	51	145	239	223	137	82	60	50	44	40	33	29	22	17
3.0	0	0	0	0	0	0	0	0	0	0	0	0	0	0	0	0	0	0	2	19	74	184	224	146	89	63	51	45	36	31	24	18

IA/P = 0.50 * * * TC = 0.3 HR * * *

TRVL TIME (hr)	11.0	11.3	11.6	11.9	12.0	12.1	12.2	12.3	12.4	12.5	12.6	12.7	12.8	13.0	13.2	13.4	13.6	13.8	14.0	14.3	14.6	15.0	15.5	16.0	16.5	17.0	17.5	18.0	19.0	20.0	22.0	26.0
0.0	0	0	0	0	1	25	151	299	277	219	187	162	141	113	100	90	84	78	72	65	61	58	53	48	44	42	41	39	35	31	28	0
.10	0	0	0	0	1	17	106	235	263	234	202	175	152	120	104	93	85	79	73	66	61	58	54	49	44	42	41	39	35	31	28	0
.20	0	0	0	0	0	0	12	75	182	236	234	213	188	144	116	101	91	84	78	70	63	59	55	50	45	43	41	40	36	31	28	0
.30	0	0	0	0	0	0	8	52	138	203	224	217	197	154	123	105	94	86	79	71	64	59	55	51	46	43	42	40	36	32	28	0
.40	0	0	0	0	0	0	5	37	105	170	206	213	203	164	131	110	97	88	81	72	65	60	56	51	46	43	42	40	36	32	28	0
.50	0	0	0	0	0	0	0	4	26	78	140	184	203	191	155	126	107	95	86	76	69	62	57	53	48	44	42	41	37	33	28	0
.75	0	0	0	0	0	0	0	1	10	34	73	117	153	184	173	146	122	105	94	82	73	64	58	54	49	45	43	41	37	33	28	0
1.0	0	0	0	0	0	0	0	0	0	0	4	17	42	114	168	178	159	134	114	94	82	70	61	57	52	47	44	42	39	35	28	0
1.5	0	0	0	0	0	0	0	0	0	0	0	0	1	10	44	98	144	163	157	130	105	84	69	61	56	52	47	44	40	36	29	6
2.0	0	0	0	0	0	0	0	0	0	0	0	0	0	0	2	14	44	87	127	153	141	110	83	69	61	56	51	47	42	38	30	17
2.5	0	0	0	0	0	0	0	0	0	0	0	0	0	0	0	0	4	16	42	97	138	145	107	82	68	60	55	51	43	40	32	25
3.0	0	0	0	0	0	0	0	0	0	0	0	0	0	0	0	0	0	1	5	27	71	127	139	105	81	68	60	55	46	41	33	27

RAINFALL TYPE = II * * * TC = 0.3 HR * * * SHEET 3 OF 10

Exhibit 5-II: Tabular hydrograph unit discharges (csm/in) for type II rainfall distribution—continued

TRVL TIME (hr) / HYDROGRAPH TIME (HOURS)

IA/P = 0.10 * * * TC = 0.4 HR * * *

TRVL TIME (hr)	11.0	11.3	11.6	11.9	12.0	12.1	12.2	12.3	12.4	12.5	12.6	12.7	12.8	13.0	13.2	13.4	13.6	13.8	14.0	14.3	14.6	15.0	15.5	16.0	16.5	17.0	17.5	18.0	19.0	20.0	22.0	26.0
0.0	18	25	36	77	141	271	468	592	574	431	298	216	163	104	77	63	55	49	44	38	34	31	28	25	22	21	20	18	16	14	12	0
.10	18	24	34	67	116	219	385	523	557	473	357	263	196	119	84	67	57	51	46	39	35	32	29	25	22	21	20	19	16	14	12	0
.20	15	20	28	44	59	97	179	316	454	523	489	401	309	178	112	81	65	56	49	42	37	33	30	26	23	21	20	19	17	14	12	0
.30	15	20	27	41	53	82	147	260	389	478	486	429	349	210	129	89	69	58	51	43	38	33	30	27	24	21	20	19	17	14	12	0
.40	13	17	23	33	38	48	71	121	214	331	429	467	442	308	189	120	85	66	56	47	41	35	31	28	24	22	20	19	17	15	12	0
.50	12	16	22	31	36	44	62	102	176	279	379	438	440	339	218	137	94	71	59	49	42	35	31	28	25	22	21	19	17	15	12	0
.75	10	13	17	24	26	30	35	45	65	106	170	251	326	393	341	245	164	112	81	59	48	39	33	30	26	23	21	20	18	15	12	0
1.0	8	10	13	17	19	21	24	27	31	37	50	75	118	251	360	376	292	205	138	83	60	45	36	32	28	25	22	21	18	16	12	1
1.5	6	7	9	12	13	14	15	17	19	21	23	26	31	56	121	224	311	333	293	192	115	66	45	36	31	28	25	22	19	17	13	4
2.0	4	5	6	8	8	9	10	10	11	12	14	15	16	20	27	43	85	159	243	306	264	154	74	47	37	32	28	25	21	18	14	9
2.5	2	3	4	5	6	6	7	7	8	9	9	10	11	13	16	20	27	46	85	184	285	262	147	74	47	37	32	28	22	19	15	11
3.0	1	2	2	3	4	4	4	5	5	6	6	7	7	8	10	12	14	17	23	47	109	227	268	160	83	50	38	32	25	21	16	11

IA/P = 0.30 * * * TC = 0.4 HR * * *

TRVL TIME (hr)	11.0	11.3	11.6	11.9	12.0	12.1	12.2	12.3	12.4	12.5	12.6	12.7	12.8	13.0	13.2	13.4	13.6	13.8	14.0	14.3	14.6	15.0	15.5	16.0	16.5	17.0	17.5	18.0	19.0	20.0	22.0	26.0
0.0	0	0	0	4	26	113	296	480	495	413	306	234	186	127	100	84	74	67	61	54	49	45	41	37	33	31	29	28	25	21	19	0
.10	0	0	0	0	2	18	81	224	395	462	430	347	272	172	121	96	82	73	66	57	51	46	42	38	34	31	30	28	25	22	19	0
.20	0	0	0	0	2	13	59	169	320	414	424	373	305	196	134	103	85	75	67	59	52	47	43	39	34	32	30	29	25	22	19	0
.30	0	0	0	0	0	1	9	42	127	255	361	403	383	274	181	127	99	83	73	63	55	48	44	40	36	32	30	29	26	23	19	0
.40	0	0	0	0	0	1	6	30	94	202	308	372	379	298	203	141	106	87	76	65	56	49	44	40	36	32	31	29	26	23	19	0
.50	0	0	0	0	0	0	0	4	21	70	158	258	334	364	270	187	133	102	85	70	60	51	46	41	37	33	31	30	26	23	19	0
.75	0	0	0	0	0	0	0	2	8	30	76	145	219	321	305	241	177	130	102	78	65	55	47	43	38	34	32	30	27	24	19	0
1.0	0	0	0	0	0	0	0	0	0	1	4	15	42	150	267	308	272	209	154	103	79	62	51	45	41	37	33	31	28	25	19	1
1.5	0	0	0	0	0	0	0	0	0	0	0	0	1	10	51	136	226	274	263	195	131	85	62	51	45	41	36	33	29	26	20	6
2.0	0	0	0	0	0	0	0	0	0	0	0	0	0	0	1	6	31	86	162	252	239	162	93	64	52	45	41	37	31	28	21	15
2.5	0	0	0	0	0	0	0	0	0	0	0	0	0	0	0	0	1	9	33	112	202	235	155	92	64	52	45	41	33	29	23	18
3.0	0	0	0	0	0	0	0	0	0	0	0	0	0	0	0	0	0	0	2	21	76	182	221	148	90	63	51	45	36	31	24	18

IA/P = 0.50 * * * TC = 0.4 HR * * *

TRVL TIME (hr)	11.0	11.3	11.6	11.9	12.0	12.1	12.2	12.3	12.4	12.5	12.6	12.7	12.8	13.0	13.2	13.4	13.6	13.8	14.0	14.3	14.6	15.0	15.5	16.0	16.5	17.0	17.5	18.0	19.0	20.0	22.0	26.0
0.0	0	0	0	0	0	7	59	168	245	257	213	186	163	128	109	96	88	81	75	67	62	58	54	50	45	43	41	39	35	31	28	0
.10	0	0	0	0	0	0	5	41	125	205	240	222	198	154	123	106	94	86	79	71	64	60	56	51	46	43	42	40	36	32	28	0
.20	0	0	0	0	0	0	3	28	93	168	216	220	205	164	131	110	97	88	81	72	65	60	56	51	46	43	42	40	36	32	28	0
.30	0	0	0	0	0	0	0	2	20	69	135	189	209	192	155	126	107	95	86	77	69	62	57	53	48	44	42	41	37	33	28	0
.40	0	0	0	0	0	0	0	1	14	50	106	161	193	202	163	133	112	98	89	78	70	62	58	53	48	44	42	41	37	33	28	0
.50	0	0	0	0	0	0	0	1	9	37	83	135	174	194	171	140	117	102	91	80	71	63	58	54	49	45	43	41	37	33	28	0
.75	0	0	0	0	0	0	0	0	0	3	15	40	76	147	177	169	146	124	107	90	79	68	60	56	51	47	43	42	38	34	28	0
1.0	0	0	0	0	0	0	0	0	0	0	1	7	21	78	141	173	167	146	125	101	86	73	63	58	53	48	45	42	39	35	28	1
1.5	0	0	0	0	0	0	0	0	0	0	0	0	0	5	26	71	121	153	159	139	113	89	72	63	57	53	48	44	40	37	29	7
2.0	0	0	0	0	0	0	0	0	0	0	0	0	0	0	0	3	16	45	86	138	150	125	93	74	64	58	53	48	42	39	31	20
2.5	0	0	0	0	0	0	0	0	0	0	0	0	0	0	0	0	1	4	17	59	112	143	121	91	73	63	57	53	45	40	32	26
3.0	0	0	0	0	0	0	0	0	0	0	0	0	0	0	0	0	0	0	1	11	40	101	138	117	90	73	63	57	48	42	34	27

RAINFALL TYPE = II * * * TC = 0.4 HR * * * SHEET 4 OF 10

Exhibit 5-II: Tabular hydrograph unit discharges (csm/in) for type II rainfall distribution—continued

IA/P = 0.10 * * * TC = 0.5 HR * * * IA/P = 0.10

TRVL TIME (hr)	11.0	11.3	11.6	11.9	12.0	12.1	12.2	12.3	12.4	12.5	12.6	12.7	12.8	13.0	13.2	13.4	13.6	13.8	14.0	14.3	14.6	15.0	15.5	16.0	16.5	17.0	17.5	18.0	19.0	20.0	22.0	26.0
0.0	17	23	32	57	94	170	308	467	529	507	402	297	226	140	96	74	61	53	47	41	36	32	29	26	23	21	20	19	16	14	12	0
.10	16	22	30	51	80	140	252	395	484	499	434	343	265	162	108	80	65	55	49	42	36	33	29	26	23	21	20	19	16	14	12	0
.20	14	19	25	38	47	69	116	207	332	434	477	449	378	238	149	101	77	62	53	45	39	34	30	27	24	22	20	19	17	14	12	0
.30	13	18	24	35	43	60	97	170	278	382	446	448	401	270	171	114	83	66	56	46	40	34	31	27	24	22	20	19	17	15	12	0
.40	12	15	21	29	33	40	53	83	141	233	332	408	434	361	243	157	107	79	64	51	43	36	32	28	25	22	21	20	17	15	12	0
.50	11	15	20	28	31	37	48	71	118	194	286	367	412	378	271	178	119	86	68	53	44	37	32	29	25	23	21	20	17	15	12	0
.75	9	11	14	19	21	24	27	31	37	49	74	118	182	319	374	328	244	169	117	76	56	43	35	31	28	25	22	21	18	16	12	1
1.0	7	9	12	16	17	19	21	24	27	32	40	55	83	188	309	359	322	245	172	102	68	49	38	32	29	26	23	21	19	16	12	1
1.5	5	7	8	11	12	13	14	15	17	19	21	23	27	43	89	175	269	322	309	225	140	77	49	38	32	29	25	23	20	17	13	5
2.0	3	4	6	7	8	8	9	10	10	11	12	14	15	18	23	35	65	123	202	297	280	181	88	52	39	33	29	26	21	19	14	10
2.5	2	3	4	5	5	6	6	7	7	8	9	9	10	12	15	18	24	36	66	150	244	278	171	87	52	39	33	29	23	20	15	11
3.0	1	1	2	3	3	4	4	4	5	5	6	6	7	8	9	11	13	16	20	37	86	198	263	182	96	56	40	33	26	21	16	11

IA/P = 0.30 * * * TC = 0.5 HR * * * IA/P = 0.30

TRVL TIME (hr)	11.0	11.3	11.6	11.9	12.0	12.1	12.2	12.3	12.4	12.5	12.6	12.7	12.8	13.0	13.2	13.4	13.6	13.8	14.0	14.3	14.6	15.0	15.5	16.0	16.5	17.0	17.5	18.0	19.0	20.0	22.0	26.0
0.0	0	0	0	1	9	53	157	314	433	439	379	299	237	159	118	95	81	71	65	56	50	46	42	38	34	31	30	28	25	22	19	0
.10	0	0	0	0	1	6	37	117	248	372	416	391	330	218	150	113	92	79	70	60	53	47	43	39	35	32	30	29	26	22	19	0
.20	0	0	0	0	1	4	26	87	194	313	382	388	349	244	167	122	97	82	72	62	54	48	43	39	35	32	30	29	26	22	19	0
.30	0	0	0	0	0	0	3	19	64	151	259	341	372	316	223	156	117	94	80	67	58	50	45	41	36	33	31	29	26	23	19	0
.40	0	0	0	0	0	0	2	13	47	116	211	298	354	328	245	172	127	100	83	69	59	51	45	41	37	33	31	29	26	23	19	0
.50	0	0	0	0	0	0	0	1	9	34	89	170	255	341	303	225	161	120	96	76	64	54	47	42	38	34	31	30	27	24	19	0
.75	0	0	0	0	0	0	0	1	4	14	41	89	152	270	305	268	207	155	118	87	70	57	48	44	39	35	32	30	27	24	19	0
1.0	0	0	0	0	0	0	0	0	0	0	2	7	22	98	212	295	285	237	181	120	88	67	53	46	42	38	34	31	28	25	19	2
1.5	0	0	0	0	0	0	0	0	0	0	0	0	0	5	30	95	183	249	265	217	152	96	66	53	46	41	37	34	30	26	20	8
2.0	0	0	0	0	0	0	0	0	0	0	0	0	0	0	0	3	18	59	125	221	245	182	105	69	54	47	42	38	32	28	22	16
2.5	0	0	0	0	0	0	0	0	0	0	0	0	0	0	0	0	1	5	21	84	174	230	172	103	69	54	46	42	34	30	23	18
3.0	0	0	0	0	0	0	0	0	0	0	0	0	0	0	0	0	0	0	1	13	56	157	217	163	101	68	53	46	37	31	25	18

IA/P = 0.50 * * * TC = 0.5 HR * * * IA/P = 0.50

TRVL TIME (hr)	11.0	11.3	11.6	11.9	12.0	12.1	12.2	12.3	12.4	12.5	12.6	12.7	12.8	13.0	13.2	13.4	13.6	13.8	14.0	14.3	14.6	15.0	15.5	16.0	16.5	17.0	17.5	18.0	19.0	20.0	22.0	26.0
0.0	0	0	0	0	0	2	26	89	170	217	229	200	179	144	119	104	93	85	78	70	64	59	55	51	46	43	41	40	36	32	28	0
.10	0	0	0	0	0	0	1	18	65	135	190	216	205	170	137	115	101	91	83	74	67	61	56	52	47	44	42	40	36	32	28	0
.20	0	0	0	0	0	0	1	12	47	106	162	198	203	178	145	121	105	94	85	76	68	61	57	52	48	44	42	40	37	32	28	0
.30	0	0	0	0	0	0	0	1	8	34	82	135	177	194	168	139	117	102	92	80	71	63	58	54	49	45	43	41	37	33	28	0
.40	0	0	0	0	0	0	0	0	6	25	63	111	155	189	174	146	122	106	94	82	73	64	58	54	50	45	43	41	37	33	28	0
.50	0	0	0	0	0	0	0	0	4	18	48	90	133	184	177	152	128	110	97	84	74	65	59	55	50	45	43	41	38	33	28	0
.75	0	0	0	0	0	0	0	0	1	7	22	47	80	142	169	164	144	124	108	91	79	68	61	56	51	47	44	42	38	34	28	0
1.0	0	0	0	0	0	0	0	0	0	0	1	3	11	51	112	155	166	154	134	109	91	76	65	59	54	49	45	43	39	35	28	2
1.5	0	0	0	0	0	0	0	0	0	0	0	0	0	2	16	50	97	136	154	145	121	95	75	64	58	54	49	45	41	37	29	10
2.0	0	0	0	0	0	0	0	0	0	0	0	0	0	0	0	4	18	47	86	134	146	125	94	75	64	58	53	49	42	39	31	21
2.5	0	0	0	0	0	0	0	0	0	0	0	0	0	0	0	0	0	3	11	44	95	140	127	97	77	65	58	54	45	41	33	26
3.3	0	0	0	0	0	0	0	0	0	0	0	0	0	0	0	0	0	0	1	7	29	86	135	122	95	76	65	58	49	43	35	27

RAINFALL TYPE = II * * * TC = 0.5 HR * * * SHEET 5 OF 10

Exhibit 5-II: Tabular hydrograph unit discharges (csm/in) for type II rainfall distribution—continued

IA/P = 0.10 * * * TC = 0.75 HR * * * IA/P = 0.10

TRVL TIME (hr)	11.0	11.3	11.6	11.9	12.0	12.1	12.2	12.3	12.4	12.5	12.6	12.7	12.8	13.0	13.2	13.4	13.6	13.8	14.0	14.3	14.6	15.0	15.5	16.0	16.5	17.0	17.5	18.0	19.0	20.0	22.0	26.0
0.0	13	18	24	36	46	68	115	194	294	380	424	410	369	252	172	123	93	74	61	49	41	35	31	27	24	22	20	19	17	15	12	0
.10	13	17	23	34	42	59	97	162	250	337	395	405	381	279	191	135	100	79	65	51	42	36	31	28	25	22	21	19	17	15	12	0
.20	11	15	20	28	32	39	52	82	135	211	295	362	391	351	255	178	127	95	75	57	46	38	32	29	26	23	21	20	17	15	12	0
.30	11	14	19	26	30	36	47	70	113	179	256	326	379	360	277	196	140	103	80	60	48	38	33	29	26	23	21	20	18	15	12	0
.40	10	12	16	22	25	28	33	42	61	96	151	221	291	367	336	255	182	131	98	69	54	42	34	30	27	24	22	20	18	16	12	0
.50	9	12	16	21	24	27	31	39	53	82	128	190	258	358	343	274	200	144	106	74	56	43	35	30	27	24	22	20	18	16	12	0
.75	8	10	13	17	18	21	23	26	31	39	55	82	122	230	314	329	281	217	161	104	72	51	38	33	29	26	23	21	19	16	12	1
1.0	6	8	10	13	14	15	17	19	21	23	27	32	42	89	177	272	319	303	249	163	105	66	45	36	31	27	24	22	19	17	13	3
1.5	4	6	7	9	10	10	11	12	14	15	16	18	20	27	46	90	163	241	295	275	204	119	66	45	35	31	27	24	20	18	13	7
2.0	3	4	5	6	7	7	8	9	9	10	11	12	13	16	20	28	48	89	151	245	274	213	115	65	44	35	30	27	22	19	14	10
2.5	1	2	3	4	4	5	5	6	6	7	7	8	8	10	12	14	17	24	37	86	170	260	219	127	71	47	36	31	24	20	16	11
3.0	1	1	2	3	3	3	4	4	4	5	5	5	6	7	8	10	11	14	17	30	64	157	247	205	122	70	46	36	27	22	17	12

IA/P = 0.30 * * * TC = 0.75 HR * * * IA/P = 0.30

TRVL TIME (hr)	11.0	11.3	11.6	11.9	12.0	12.1	12.2	12.3	12.4	12.5	12.6	12.7	12.8	13.0	13.2	13.4	13.6	13.8	14.0	14.3	14.6	15.0	15.5	16.0	16.5	17.0	17.5	18.0	19.0	20.0	22.0	26.0
0.0	0	0	0	0	1	6	30	86	174	266	326	348	328	246	181	138	110	92	79	66	57	49	44	40	36	32	31	29	26	23	19	0
.10	0	0	0	0	0	1	4	22	65	137	223	292	329	303	228	170	131	106	89	73	61	52	46	41	37	33	31	29	26	23	19	0
.20	0	0	0	0	0	0	3	15	48	108	185	256	305	321	245	184	141	112	93	75	63	53	46	42	37	34	31	30	27	23	19	0
.30	0	0	0	0	0	0	2	11	36	84	151	221	277	308	260	199	152	120	98	78	65	54	47	42	38	34	31	30	27	23	19	0
.40	0	0	0	0	0	0	0	1	8	27	65	122	188	286	301	243	187	144	114	87	71	57	48	43	39	35	32	30	27	24	19	1
.50	0	0	0	0	0	0	0	1	6	20	50	98	158	263	292	254	200	155	122	91	74	59	49	44	40	35	32	30	27	24	19	1
.75	0	0	0	0	0	0	0	0	0	2	8	23	51	140	231	269	253	211	167	119	90	68	53	46	42	37	34	31	28	25	19	2
1.0	0	0	0	0	0	0	0	0	0	0	0	1	4	29	96	186	249	261	231	169	120	84	61	50	44	40	36	33	29	26	20	5
1.5	0	0	0	0	0	0	0	0	0	0	0	0	0	1	8	34	91	163	220	241	197	131	83	61	50	44	40	35	31	27	21	12
2.0	0	0	0	0	0	0	3	0	0	0	0	0	0	0	0	2	11	36	85	174	226	200	127	82	60	49	44	39	32	29	22	17
2.5	o	0	0	0	0	0	0	0	0	0	0	0	0	0	0	0	0	1	6	37	105	196	214	135	87	62	51	44	36	31	24	18
3.0	0	0	0	0	0	0	0	0	0	0	0	0	0	0	0	0	0	0	0	4	24	96	205	189	130	85	62	50	39	32	26	18

IA/P = 0.50 * * * TC = 0.75 HR * * * IA/P = 0.50

TRVL TIME (hr)	11.0	11.3	11.6	11.9	12.0	12.1	12.2	12.3	12.4	12.5	12.6	12.7	12.8	13.0	13.2	13.4	13.6	13.8	14.0	14.3	14.6	15.0	15.5	16.0	16.5	17.0	17.5	18.0	19.0	20.0	22.0	26.0
0.0	0	0	0	0	0	0	2	16	45	92	137	166	185	170	146	125	110	98	89	79	70	63	58	53	48	44	42	41	37	33	28	0
.10	0	0	0	0	0	0	0	1	11	34	73	115	149	180	163	141	122	107	96	84	74	65	59	54	50	45	43	41	38	33	28	0
.20	0	0	0	0	0	0	0	1	8	25	57	96	131	173	166	146	126	111	99	86	76	66	59	55	50	46	43	41	38	34	28	0
.30	0	0	0	0	0	0	0	0	1	5	18	44	79	143	170	160	141	122	108	92	81	69	61	56	52	47	44	42	38	34	28	1
.40	0	0	0	0	0	0	0	0	0	4	14	34	64	127	166	162	145	127	111	95	82	70	62	57	52	47	44	42	38	34	28	1
.50	0	0	0	0	0	0	0	0	0	0	2	10	26	82	138	162	157	140	123	103	88	75	64	58	53	49	45	43	39	35	28	2
.75	0	0	0	0	0	0	0	0	0	0	1	4	12	47	98	139	154	148	135	113	96	80	67	60	55	50	46	43	39	36	29	3
1.0	0	0	0	0	0	0	0	0	0	0	0	0	0	6	30	73	119	146	151	134	113	91	74	63	58	53	48	45	41	37	29	7
1.5	0	0	0	0	0	0	0	0	0	0	0	0	0	0	1	9	30	66	105	143	143	117	90	73	63	57	52	48	42	39	30	18
2.0	0	0	0	0	0	0	0	0	0	0	0	0	0	0	0	0	2	11	30	77	121	137	114	88	72	63	57	52	44	40	32	25
2.5	0	0	0	0	0	0	0	0	0	0	0	0	0	0	0	0	0	1	3	19	55	111	132	111	87	71	62	56	47	42	34	27
3.0	0	0	0	0	0	0	0	0	0	0	0	0	0	0	0	0	0	0	0	2	12	51	112	128	108	86	71	62	51	44	36	27

RAINFALL TYPE = II * * * TC = 0.75 HR * * * SHEET 6 OF 10

Exhibit 5-II: Tabular hydrograph unit discharges (csm/in) for type II rainfall distribution—continued

IA/P = 0.10 * * * TC = 1.0 HR * * * IA/P = 0.10

TRVL TIME (hr)	11.0	11.3	11.6	11.9	12.0	12.1	12.2	12.3	12.4	12.5	12.6	12.7	12.8	13.0	13.2	13.4	13.6	13.8	14.0	14.3	14.6	15.0	15.5	16.0	16.5	17.0	17.5	18.0	19.0	20.0	22.0	26.0
0.0	11	15	20	29	35	47	72	112	168	231	289	329	357	313	239	175	133	103	83	63	50	40	33	29	26	23	21	20	17	15	12	0
.10	10	13	17	24	27	33	42	62	95	144	202	260	306	340	293	222	165	126	98	72	56	43	35	30	27	24	22	20	18	15	12	0
.20	10	13	17	23	26	30	38	54	82	123	176	232	281	332	303	238	179	136	105	76	59	45	35	30	27	24	22	20	18	16	12	1
.30	9	12	16	22	24	28	35	48	70	105	152	205	256	323	310	254	193	146	113	81	61	46	36	31	27	24	22	20	18	16	12	1
.40	8	11	14	19	21	23	27	32	42	61	91	132	181	276	318	294	237	181	138	95	70	51	39	32	28	25	23	21	18	16	12	1
.50	8	10	13	18	20	22	25	30	38	53	78	114	159	253	311	300	251	195	149	102	74	53	40	33	29	25	23	21	18	16	12	1
.75	7	8	11	14	16	17	19	21	25	30	38	53	76	146	228	284	293	256	208	143	99	66	46	36	31	27	24	22	19	17	13	2
1.0	5	7	8	11	12	13	14	16	17	19	22	25	31	57	111	188	256	286	272	208	144	90	56	41	33	29	26	23	20	17	13	4
1.5	4	5	6	8	8	9	10	11	12	13	14	15	17	22	33	59	107	171	231	268	235	157	88	56	41	33	29	25	21	18	14	8
2.0	2	3	4	5	5	6	6	7	7	8	9	9	10	12	15	19	27	44	78	157	231	252	167	96	59	42	34	29	23	20	15	11
2.5	1	2	2	3	4	4	4	5	5	6	6	7	7	8	10	12	15	19	27	58	120	214	241	159	94	59	42	34	26	21	16	11
3.0	0	1	1	2	2	3	3	3	4	4	4	5	5	6	7	8	10	12	14	22	44	113	214	231	152	91	58	42	29	23	17	12

IA/P = 0.30 * * * TC = 1.0 HR * * * IA/P = 0.30

TRVL TIME (hr)	11.0	11.3	11.6	11.9	12.0	12.1	12.2	12.3	12.4	12.5	12.6	12.7	12.8	13.0	13.2	13.4	13.6	13.8	14.0	14.3	14.6	15.0	15.5	16.0	16.5	17.0	17.5	18.0	19.0	20.0	22.0	26.0
0.0	0	0	0	0	1	4	16	42	83	137	195	243	271	292	227	178	143	117	98	79	66	55	47	42	38	34	31	30	27	23	19	0
.10	0	0	0	0	0	0	3	12	32	66	113	168	218	279	260	213	169	136	113	88	72	59	49	43	39	35	32	30	27	24	19	1
.20	0	0	0	0	0	0	2	9	24	52	93	143	193	271	271	225	180	145	119	92	75	60	50	44	39	35	32	30	27	24	19	1
.30	0	0	0	0	0	0	1	6	18	41	75	120	169	246	264	234	191	153	125	96	78	62	51	44	40	36	33	31	27	24	19	1
.40	0	0	0	0	0	0	0	1	4	14	32	61	100	190	251	259	222	181	146	109	86	67	53	46	41	37	33	31	28	25	19	2
.50	0	0	0	0	0	0	0	1	3	10	24	49	83	168	237	254	230	191	155	115	90	69	54	47	42	37	34	31	28	25	19	2
.75	0	0	0	0	0	0	0	0	0	1	4	12	25	76	150	213	239	228	198	149	112	82	61	50	44	39	35	32	29	26	20	4
1.0	0	0	0	0	0	0	0	0	0	0	0	1	2	15	51	113	182	226	234	197	150	104	72	56	47	42	38	34	30	27	20	7
1.5	0	0	0	0	0	0	0	0	0	0	0	0	0	0	4	18	51	104	162	220	210	158	102	71	56	47	42	37	31	28	22	13
2.0	0	0	0	0	0	0	0	0	0	0	0	0	0	0	0	1	5	20	49	121	187	209	152	100	70	55	47	41	34	29	23	17
2.5	0	0	0	0	0	0	0	0	0	0	0	0	0	0	0	0	0	2	7	32	87	171	199	146	98	69	54	46	37	31	24	18
3.0	0	0	0	0	0	0	0	0	0	0	0	0	0	0	0	0	0	0	0	2	13	62	158	192	151	103	73	56	41	34	26	18

IA/P = 0.50 * * * TC = 1.0 HR * * * IA/P = 0.50

TRVL TIME (hr)	11.0	11.3	11.6	11.9	12.0	12.1	12.2	12.3	12.4	12.5	12.6	12.7	12.8	13.0	13.2	13.4	13.6	13.8	14.0	14.3	14.6	15.0	15.5	16.0	16.5	17.0	17.5	18.0	19.0	20.0	22.0	26.0
0.0	0	0	0	0	0	0	1	7	21	42	71	101	126	160	154	138	123	110	100	87	77	67	60	55	50	46	43	41	38	34	28	1
.10	0	0	0	0	0	0	0	1	5	15	33	58	87	134	156	149	134	120	108	93	82	71	62	57	52	47	44	42	38	34	28	1
.20	0	0	0	0	0	0	0	1	4	12	26	48	74	123	153	153	137	123	111	95	84	72	63	57	52	47	44	42	38	34	28	1
.30	0	0	0	0	0	0	0	0	3	9	20	38	62	111	143	150	140	127	114	98	86	73	63	58	53	48	45	42	39	35	28	1
.40	0	0	0	0	0	0	0	0	0	2	6	16	31	75	120	145	148	137	123	106	91	77	66	59	54	49	45	43	39	35	29	2
.50	0	0	0	0	0	0	0	0	0	1	5	12	25	64	109	139	146	139	127	108	94	79	67	60	55	50	46	43	39	36	29	3
.75	0	0	0	0	0	0	0	0	0	0	2	5	12	39	78	115	136	140	134	117	101	84	70	62	56	51	47	44	40	36	29	4
1.0	0	0	0	0	0	0	0	0	0	0	0	0	1	7	26	59	96	125	139	133	117	97	78	66	59	54	49	46	41	37	29	8
1.5	0	0	0	0	0	0	0	0	0	0	0	0	0	0	2	9	26	54	86	123	133	119	95	77	66	59	54	49	43	39	31	17
2.0	0	0	0	0	0	0	0	0	0	0	0	0	0	0	0	0	3	10	25	64	104	129	116	93	76	65	58	53	45	41	33	24
2.5	0	0	0	0	0	0	0	0	0	0	0	0	0	0	0	0	0	0	2	10	34	84	125	117	96	78	66	59	49	43	35	27
3.0	0	0	0	0	0	0	0	0	0	0	0	0	0	0	0	0	0	0	0	1	6	32	89	122	114	94	77	66	53	45	37	27

RAINFALL TYPE = II * * * TC = 1.0 HR * * * SHEET 7 OF 10

Exhibit 5-II: Tabular hydrograph unit discharges (csm/in) for type II rainfall distribution —continued

IA/P = 0.10 * * * TC = 1.25 HR * * * IA/P = 0.10

TRVL TIME (hr)	11.0	11.3	11.6	11.9	12.0	12.1	12.2	12.3	12.4	12.5	12.6	12.7	12.8	13.0	13.2	13.4	13.6	13.8	14.0	14.3	14.6	15.0	15.5	16.0	16.5	17.0	17.5	18.0	19.0	20.0	22.0	26.0
0.0	10	13	18	25	29	38	54	81	118	163	213	256	284	311	266	212	163	129	104	78	61	47	37	31	27	24	22	20	18	16	12	1
.10	10	13	17	23	27	34	47	69	102	143	189	234	267	297	274	226	175	138	111	82	64	48	38	31	27	24	22	20	18	16	12	1
.20	9	11	15	20	22	26	31	42	60	88	124	168	212	280	292	261	212	166	131	95	72	53	40	33	28	25	23	21	18	16	12	1
.30	8	11	14	19	21	24	29	38	53	76	108	148	190	263	288	268	224	177	140	101	76	55	41	34	29	25	23	21	18	16	12	2
.40	8	10	13	18	20	23	27	34	46	66	94	130	170	245	282	273	235	188	149	107	80	58	42	34	29	26	23	21	19	16	12	2
.50	7	9	12	16	17	19	22	25	31	41	58	82	114	190	256	279	262	222	178	127	93	65	46	36	31	27	24	22	19	17	13	2
.75	6	8	10	14	15	17	19	21	25	31	41	56	78	139	207	254	265	245	208	152	110	75	51	39	32	28	25	22	19	17	13	3
1.0	5	6	8	10	11	13	14	15	17	19	22	26	33	60	109	173	230	261	255	208	153	100	64	46	36	30	26	24	20	18	13	5
1.5	3	4	5	7	7	8	9	9	10	11	12	13	15	19	27	45	79	130	186	247	239	180	108	68	48	37	31	27	22	19	14	10
2.0	2	3	4	5	6	6	7	7	8	8	9	10	11	13	16	22	35	59	98	171	236	236	156	95	62	44	35	30	23	20	15	11
2.5	1	2	2	3	4	4	4	5	5	5	6	6	7	8	10	12	14	19	28	58	114	197	226	163	102	65	46	36	26	21	16	11
3.0	0	1	1	2	2	2	2	3	3	3	4	4	4	5	6	7	9	10	13	19	35	88	184	218	169	109	70	49	31	24	18	12

IA/P = 0.30 * * * TC = 1.25 HR * * * IA/P = 0.30

TRVL TIME (hr)	11.0	11.3	11.6	11.9	12.0	12.1	12.2	12.3	12.4	12.5	12.6	12.7	12.8	13.0	13.2	13.4	13.6	13.8	14.0	14.3	14.6	15.0	15.5	16.0	16.5	17.0	17.5	18.0	19.0	20.0	22.0	26.0
0.0	0	0	0	0	0	2	9	25	50	86	130	174	208	253	235	201	164	136	115	92	76	61	51	44	39	35	32	30	27	24	19	1
.10	0	0	0	0	0	0	1	6	19	40	71	110	153	217	247	227	191	157	131	103	84	66	53	46	41	36	33	31	28	24	19	2
.20	0	0	0	0	0	0	1	4	14	31	58	93	133	202	239	231	199	165	138	108	87	68	55	47	41	37	33	31	28	25	19	2
.30	0	0	0	0	0	0	0	1	3	10	24	46	77	152	210	236	222	190	158	122	97	74	58	49	43	38	34	32	28	25	20	3
.40	0	0	0	0	0	0	0	0	2	8	19	37	64	134	196	232	225	198	166	127	101	77	59	50	43	38	35	32	28	25	20	3
.50	0	0	0	0	0	0	0	0	0	2	6	14	30	82	151	206	228	217	189	146	113	85	64	52	45	40	36	33	29	26	20	5
.75	0	0	0	0	0	0	0	0	0	1	2	7	15	49	105	164	205	218	205	166	129	95	69	55	47	41	37	33	29	26	20	6
1.0	0	0	0	0	0	0	0	0	0	0	0	0	1	9	32	77	134	185	214	203	166	120	83	63	52	45	39	35	30	27	21	10
1.5	0	0	0	0	0	0	0	0	0	0	0	0	0	0	2	11	33	72	121	184	203	171	117	82	62	51	44	39	32	29	22	15
2.0	0	0	0	0	0	0	0	0	0	0	0	0	0	0	0	0	1	7	21	67	132	194	174	123	86	64	52	45	35	31	24	18
2.5	0	0	0	0	0	0	0	0	0	0	0	0	0	0	0	0	0	0	2	13	46	121	187	166	119	84	63	52	39	32	25	18
3.0	0	0	0	0	0	0	0	0	0	0	0	0	0	0	0	0	0	0	0	1	8	44	129	180	160	116	83	63	44	35	27	18

IA/P = 0.50 * * * TC = 1.25 HR * * * IA/P = 0.50

TRVL TIME (hr)	11.0	11.3	11.6	11.9	12.0	12.1	12.2	12.3	12.4	12.5	12.6	12.7	12.8	13.0	13.2	13.4	13.6	13.8	14.0	14.3	14.6	15.0	15.5	16.0	16.5	17.0	17.5	18.0	19.0	20.0	22.0	26.0
0.0	0	0	0	0	0	0	1	5	13	26	44	68	91	125	142	142	128	117	107	94	83	72	63	57	52	47	44	42	38	34	28	2
.10	0	0	0	0	0	0	0	0	3	10	20	36	57	100	129	140	136	125	114	100	88	76	65	59	54	49	45	43	39	35	29	3
.20	0	0	0	0	0	0	0	0	2	7	16	30	48	90	122	139	139	127	117	102	90	77	66	60	54	49	45	43	39	35	29	3
.30	0	0	0	0	0	0	0	0	0	2	5	12	24	59	98	126	137	134	125	109	96	82	69	61	56	51	46	44	40	36	29	4
.40	0	0	0	0	0	0	0	0	0	1	4	10	19	51	89	119	134	136	127	112	98	83	70	62	56	51	47	44	40	36	29	5
.50	0	0	0	0	0	0	0	0	0	1	3	7	15	43	79	112	131	135	129	114	100	85	71	63	57	52	47	44	40	36	29	6
.75	0	0	0	0	0	0	0	0	0	0	0	1	3	15	39	71	102	123	130	125	112	94	78	67	60	54	49	46	41	37	29	9
1.0	0	0	0	0	0	0	0	0	0	0	0	0	1	4	17	40	71	101	121	129	121	103	84	71	62	56	51	47	42	38	30	13
1.5	0	0	0	0	0	0	0	0	0	0	0	0	0	0	0	3	10	26	51	92	119	125	105	86	72	63	57	52	44	40	32	23
2.0	0	0	0	0	0	0	0	0	0	0	0	0	0	0	0	0	1	3	11	35	72	112	122	103	85	71	63	56	47	42	34	26
2.5	0	0	0	0	0	0	0	0	0	0	0	0	0	0	0	0	0	0	1	7	24	66	111	119	101	83	71	62	51	44	36	27
3.0	0	0	0	0	0	0	0	0	0	0	0	0	0	0	0	0	0	0	0	1	4	23	71	110	116	99	82	70	55	46	37	27

RAINFALL TYPE = II * * * TC = 1.25 HR * * * SHEET 8 OF 10

Exhibit 5-II: Tabular hydrograph unit discharges (csm/in) for type II rainfall distribution—continued

TRVL TIME (hr) — HYDROGRAPH TIME (HOURS)	11.0	11.3	11.6	11.9	12.0	12.1	12.2	12.3	12.4	12.5	12.6	12.7	12.8	13.0	13.2	13.4	13.6	13.8	14.0	14.3	14.6	15.0	15.5	16.0	16.5	17.0	17.5	18.0	19.0	20.0	22.0	26.0
IA/P = 0.10 — * * * TC = 1.5 HR * * * — IA/P = 0.10																																
0.0	9	11	15	21	25	31	41	58	82	112	147	184	216	255	275	236	198	159	129	98	76	57	43	35	30	25	23	21	18	16	12	1
.10	8	10	13	18	20	23	28	37	51	72	98	131	166	226	265	254	226	187	151	113	86	63	46	37	31	26	23	21	19	16	13	2
.20	8	10	13	17	19	22	26	33	45	63	87	116	149	212	259	259	233	197	160	119	90	66	48	38	32	27	24	22	19	16	13	2
.30	7	9	12	16	18	21	24	30	40	55	76	103	134	197	244	255	238	206	169	125	95	68	49	38	32	27	24	22	19	17	13	2
.40	7	8	11	14	15	17	19	23	28	36	49	67	91	151	208	247	252	230	196	146	109	77	54	41	34	29	25	22	19	17	13	3
.50	6	8	10	13	15	16	18	21	26	33	43	59	80	136	194	238	249	235	204	154	115	81	56	42	34	29	25	23	20	17	13	3
.75	5	7	8	11	12	13	14	16	18	21	25	32	42	76	125	179	222	240	233	193	148	102	67	48	38	32	27	24	20	18	13	5
1.0	4	5	7	8	9	10	11	12	13	14	16	18	22	34	59	101	152	201	236	230	193	135	86	59	44	35	30	26	21	18	14	7
1.5	3	4	5	6	6	7	8	8	9	10	11	12	13	16	22	34	58	95	141	203	226	197	131	84	58	43	35	29	23	20	15	10
2.0	1	2	3	4	4	5	5	6	6	7	7	8	9	10	12	16	22	34	56	110	172	218	187	126	82	57	43	34	25	21	16	11
2.5	1	1	2	2	3	3	3	4	4	4	5	5	6	7	8	9	11	14	18	34	69	141	210	190	133	87	60	44	30	23	17	12
3.0	0	0	1	1	2	2	2	2	3	3	3	3	4	5	5	6	8	9	11	16	27	66	149	204	181	128	85	58	35	25	18	12
IA/P = 0.30 — * * * TC = 1.5 HR * * * — IA/P = 0.30																																
0.0	0	0	0	0	0	1	6	15	31	53	80	112	144	193	225	208	186	157	134	108	89	70	56	48	42	37	34	31	28	25	20	2
.10	0	0	0	0	0	0	1	4	12	25	43	68	97	157	198	219	203	178	151	120	98	77	60	50	44	38	35	32	28	25	20	3
.20	0	0	0	0	0	0	0	1	3	9	19	35	57	114	168	201	213	196	171	135	108	84	64	53	46	40	36	33	29	26	20	4
.30	0	0	0	0	0	0	0	1	2	7	15	29	48	100	155	193	210	200	177	140	113	87	66	54	46	41	36	33	29	26	20	5
.40	0	0	0	0	0	0	0	0	2	5	12	23	39	87	141	184	207	202	182	146	117	89	68	55	47	41	36	33	29	26	20	5
,50	0	0	0	0	0	0	0	0	0	1	4	9	18	51	101	153	190	205	197	164	131	99	73	58	49	43	38	34	30	26	20	7
.75	0	0	0	0	0	0	0	0	0	0	2	4	9	30	68	116	160	189	197	179	147	110	80	62	52	45	39	35	30	27	21	8
1.0	0	0	0	0	0	0	0	0	0	0	0	0	1	5	20	49	92	138	175	195	178	137	97	72	57	48	42	37	31	28	21	12
1.5	0	0	0	0	0	0	0	0	0	0	0	0	0	0	1	7	21	47	85	145	187	178	133	95	71	57	48	42	34	29	23	16
2.0	0	0	0	0	0	0	0	0	0	0	0	0	0	0	0	0	1	4	13	45	97	162	180	138	99	74	58	49	38	32	25	18
2.5	0	0	0	0	0	0	0	0	0	0	0	0	0	0	0	0	0	0	1	8	31	89	161	174	133	97	72	58	42	34	26	18
3.0	0	0	0	0	0	0	0	0	0	0	0	0	0	0	0	0	0	0	0	1	5	29	98	160	169	129	95	71	48	37	28	19
IA/P = 0.50 — * * * TC = 1.5 HR * * * — IA/P = 0.50																																
.0	0	0	0	0	0	0	0	3	8	16	27	42	59	92	116	128	130	121	112	100	90	78	67	60	55	50	46	43	39	35	29	4
.10	0	0	0	0	0	0	0	2	6	12	22	35	51	84	110	125	128	123	114	102	91	79	68	61	55	50	46	43	39	35	29	4
.20	0	0	0	0	0	0	0	0	1	4	10	18	29	60	91	114	126	128	120	108	97	83	71	63	57	52	47	44	40	36	29	5
.30	0	0	0	0	0	0	0	0	1	3	8	14	24	52	83	108	123	126	122	110	98	85	72	63	57	52	48	44	40	36	29	6
.40	0	0	0	0	0	0	0	0	0	1	2	6	12	31	60	90	112	124	126	116	104	90	75	66	59	54	49	45	41	37	29	8
.50	0	0	0	0	0	0	0	0	0	0	2	4	9	26	53	83	106	121	125	118	106	91	77	67	60	54	49	46	41	37	29	8
.75	0	0	0	0	0	0	0	0	0	0	1	2	5	16	36	62	88	108	119	122	112	97	81	69	62	56	51	47	42	38	30	11
1.0	0	0	0	0	0	0	0	0	0	0	0	0	0	3	10	26	49	75	98	118	121	108	90	76	66	59	54	49	43	39	31	16
1.5	0	0	0	0	0	0	0	0	0	0	0	0	0	0	1	3	11	25	45	80	107	118	106	89	75	65	59	53	45	41	32	23
2.0	0	0	0	0	0	0	0	0	0	0	0	0	0	0	0	0	1	4	11	32	63	100	115	104	87	74	65	58	48	42	34	26
2.5	0	0	0	0	0	0	0	0	0	0	0	0	0	0	0	0	0	0	1	4	16	48	94	113	105	89	76	66	53	45	36	27
3.0	0	0	0	0	0	0	0	0	0	0	0	0	0	0	0	0	0	0	0	0	3	15	54	96	111	103	88	75	58	48	38	28
RAINFALL TYPE = II — * * * TC = 1.5 HR * * * — SHEET 9 OF 10																																

Exhibit 5-II: Tabular hydrograph unit discharges (csm/in) for type II rainfall distribution—continued

TRVL TIME (hr) — HYDROGRAPH TIME (HOURS)	11.0	11.3	11.6	11.9	12.0	12.1	12.2	12.3	12.4	12.5	12.6	12.7	12.8	13.0	13.2	13.4	13.6	13.8	14.0	14.3	14.6	15.0	15.5	16.0	16.5	17.0	17.5	18.0	19.0	20.0	22.0	26.0
IA/P = 0.10 — * * * TC = 2.0 HR * * * — IA/P = 0.10																																
0.0	7	9	12	16	18	21	27	36	49	64	82	104	127	171	201	226	208	193	171	132	105	79	58	45	36	30	26	23	20	17	13	3
.10	6	8	10	14	15	17	20	25	33	43	57	74	94	139	179	204	218	205	188	150	118	88	63	48	38	32	27	24	20	17	13	4
.20	6	8	10	13	14	16	19	23	29	39	51	66	84	128	169	198	213	207	192	157	123	91	65	49	39	33	28	24	20	17	13	4
.30	6	7	9	12	14	15	18	21	27	35	45	59	76	117	159	191	211	208	196	163	128	95	68	51	40	33	28	25	20	18	13	4
.40	5	6	8	11	12	13	15	17	20	24	31	41	53	87	128	167	197	209	205	180	145	106	75	55	43	35	30	26	21	18	14	5
.50	5	6	8	10	11	13	14	16	18	22	28	37	48	78	118	158	190	208	208	185	151	111	77	57	44	36	30	26	21	18	14	5
.75	4	6	7	9	10	11	12	13	15	18	22	27	35	58	91	129	164	191	202	194	167	125	87	63	48	38	32	27	22	18	14	6
1.0	3	4	6	7	8	8	9	10	11	12	14	16	18	28	46	74	110	147	178	201	193	156	108	76	56	43	35	30	23	19	14	8
1.5	2	3	3	5	5	5	6	6	7	8	8	9	10	12	16	23	36	57	86	137	178	195	160	113	79	58	45	36	26	21	16	11
2.0	1	2	2	3	3	4	4	4	5	5	6	6	7	8	10	12	16	23	35	67	112	169	190	154	110	78	57	44	30	23	17	11
2.5	0	1	1	2	2	2	3	3	3	4	4	4	5	6	7	8	9	12	16	28	52	105	170	185	149	107	76	56	35	26	18	12
3.0	0	0	1	1	1	1	1	2	2	2	2	3	3	3	4	5	6	7	8	12	18	41	99	161	180	152	112	80	45	30	19	12
IA/P = 0.30 — * * * TC = 2.0 HR * * * — IA/P = 0.30																																
0.0	0	0	0	0	0	1	3	8	15	25	38	54	74	115	148	168	185	170	159	131	110	89	70	57	49	42	38	34	29	26	20	5
.10	0	0	0	0	0	0	0	2	6	12	21	32	47	85	124	153	169	180	168	145	120	96	75	60	51	44	39	35	30	26	20	6
.20	0	0	0	0	0	0	0	2	4	10	17	27	41	75	114	146	165	175	170	149	124	99	76	62	52	45	39	35	30	27	21	6
.30	0	0	0	0	0	0	0	0	1	3	7	14	23	49	86	122	151	170	174	160	136	107	82	66	54	47	41	37	31	27	21	8
.40	0	0	0	0	0	0	0	0	1	2	6	11	19	43	77	113	144	165	173	163	140	111	85	67	55	47	41	37	31	27	21	8
.50	0	0	0	0	0	0	0	0	1	2	4	9	16	37	68	104	136	160	171	165	144	114	87	69	56	48	42	37	31	27	21	9
.75	0	0	0	0	0	0	0	0	0	0	1	2	5	15	34	62	96	127	152	167	160	132	100	77	62	52	45	40	32	28	22	11
1.0	0	0	0	0	0	0	0	0	0	0	0	0	0	3	10	24	48	79	111	150	166	153	118	90	71	58	49	43	34	29	23	14
1.5	0	0	0	0	0	0	0	0	0	0	0	0	0	0	1	3	10	24	45	88	130	161	148	115	88	70	57	48	37	31	24	17
2.0	0	0	0	0	0	0	0	0	0	0	0	0	0	0	0	0	1	4	10	32	68	122	157	143	113	87	68	56	42	34	26	18
2.5	0	0	0	0	0	0	0	0	0	0	0	0	0	0	0	0	0	0	1	4	16	51	114	153	144	116	89	70	49	38	27	19
3.0	0	0	0	0	0	0	0	0	0	0	0	0	0	0	0	0	0	0	0	0	3	15	59	118	150	140	113	88	57	42	29	19
IA/P = 0.50 — * * * TC = 2.0 HR * * * — IA/P = 0.50																																
0.0	0	0	0	0	0	0	0	1	4	8	13	20	28	51	73	92	104	111	112	106	97	86	75	66	60	54	49	46	41	37	30	7
.10	0	0	0	0	0	0	0	1	3	6	11	17	24	45	68	87	101	109	112	107	98	88	76	67	60	55	50	46	41	37	30	8
.20	0	0	0	0	0	0	0	1	2	5	9	14	21	40	62	82	98	107	111	108	100	89	77	68	61	55	50	47	41	37	30	8
.30	0	0	0	0	0	0	0	0	0	2	4	7	12	26	46	67	86	100	108	111	104	93	80	70	63	57	52	48	42	38	30	10
.40	0	0	0	0	0	0	0	0	0	1	3	6	10	22	41	62	81	96	106	110	105	94	81	71	63	57	52	48	42	38	30	11
.50	0	0	0	0	0	0	0	0	0	0	1	2	4	13	27	46	67	85	99	110	108	98	85	74	66	59	54	49	43	39	31	13
.75	0	0	0	0	0	0	0	0	0	0	0	1	2	7	18	33	52	71	88	104	108	102	89	77	68	61	55	50	44	39	31	15
1.0	0	0	0	0	0	0	0	0	0	0	0	0	0	1	5	13	25	43	62	87	103	108	97	84	73	65	59	53	45	41	32	20
1.5	0	0	0	0	0	0	0	0	0	0	0	0	0	0	0	2	5	12	24	48	74	99	106	95	83	72	64	58	48	43	34	25
2.0	0	0	0	0	0	0	0	0	0	0	0	0	0	0	0	0	0	2	5	17	37	69	99	104	94	82	72	64	52	45	36	27
2.5	0	0	0	0	0	0	0	0	0	0	0	0	0	0	0	0	0	0	0	2	8	27	65	95	102	95	83	73	58	49	38	28
3.0	0	0	0	0	0	0	0	0	0	0	0	0	0	0	0	0	0	0	0	0	1	8	32	68	95	101	93	82	64	52	40	28
RAINFALL TYPE = II — * * * TC = 2.0 HR * * * — SHEET 10 OF 10																																

Exhibit 5-III: Tabular hydrograph unit discharges (csm/in) for type III rainfall distribution

HYDROGRAPH TIME (HOURS)

IA/P = 0.10 — * * * TC = 0.1 HR * * *

TRVL TIME (hr)	11.0	11.3	11.6	11.9	12.0	12.1	12.2	12.3	12.4	12.5	12.6	12.7	12.8	13.0	13.2	13.4	13.6	13.8	14.0	14.3	14.6	15.0	15.5	16.0	16.5	17.0	17.5	18.0	19.0	20.0	22.0	26.0
0.0	29	38	57	172	241	425	662	531	345	265	191	130	101	83	68	62	58	54	50	44	41	37	32	27	23	21	19	16	14	13	11	0
.10	26	32	47	98	147	210	353	559	540	410	313	231	164	101	80	67	61	57	53	47	43	39	34	28	24	22	19	17	14	13	11	0
.20	25	31	44	86	127	182	296	471	517	446	357	273	200	117	86	70	63	58	54	48	44	39	34	29	24	22	20	17	14	13	11	0
.30	22	28	37	57	76	110	158	250	398	477	457	390	312	178	111	83	69	62	57	51	45	41	36	31	25	23	20	18	15	13	11	0
.40	21	27	35	53	68	96	137	213	336	430	448	410	345	210	128	90	72	64	58	52	46	41	36	31	26	23	20	18	15	13	11	0
.50	19	24	30	43	49	62	85	120	182	284	382	426	415	305	188	120	86	71	63	55	49	43	38	33	27	24	21	19	15	14	11	0
.75	17	22	27	37	41	49	62	84	120	181	258	327	375	353	264	177	120	88	72	59	52	45	39	34	29	25	22	20	15	14	11	0
1.0	13	17	22	27	30	33	37	43	52	66	91	131	190	315	358	307	220	149	104	72	60	50	43	37	32	27	23	21	16	14	12	0
1.5	9	11	14	18	19	21	23	25	27	29	33	37	44	70	134	229	304	318	269	172	106	68	52	44	38	33	28	24	19	15	12	2
2.0	6	8	10	13	14	15	16	17	19	20	22	24	26	32	45	73	130	207	271	292	216	121	68	51	43	37	32	27	21	16	13	6
2.5	3	4	6	8	9	10	10	11	12	13	14	16	17	20	23	29	38	57	97	189	271	244	136	75	53	44	38	33	24	19	14	9
3.0	1	2	4	5	6	6	7	8	8	9	10	11	12	14	16	19	23	28	38	74	146	256	226	131	74	53	44	37	27	21	14	10

IA/P = 0.30 — * * * TC = 0.1 HR * * *

TRVL TIME (hr)	11.0	11.3	11.6	11.9	12.0	12.1	12.2	12.3	12.4	12.5	12.6	12.7	12.8	13.0	13.2	13.4	13.6	13.8	14.0	14.3	14.6	15.0	15.5	16.0	16.5	17.0	17.5	18.0	19.0	20.0	22.0	26.0
0.0	0	0	0	48	106	296	597	496	368	300	221	155	125	106	89	83	79	74	69	62	59	54	47	40	35	32	28	25	22	20	17	0
.10	0	0	0	35	82	225	473	488	408	336	260	190	147	113	94	85	80	75	70	63	59	54	48	40	35	32	29	25	22	20	17	0
.20	0	0	0	7	26	64	171	372	449	422	365	295	225	142	109	92	84	79	74	66	61	56	50	43	36	33	30	26	22	20	17	0
.30	0	0	0	5	19	49	130	291	397	414	381	323	258	161	118	96	86	80	75	68	62	57	50	43	37	33	30	27	22	20	17	0
.40	0	0	0	0	3	14	37	99	227	340	389	384	343	229	152	113	94	85	79	71	65	59	52	46	38	34	31	28	23	21	17	0
.50	0	0	0	0	2	10	28	75	177	286	355	374	354	256	170	123	99	87	80	73	66	60	53	46	39	35	31	28	23	21	18	0
.75	0	0	0	0	0	1	4	13	35	86	161	238	296	325	266	194	141	110	93	80	71	63	56	50	43	37	33	30	24	21	18	0
1.0	0	0	0	0	0	0	0	2	6	19	48	99	165	282	311	264	197	144	112	88	77	67	59	52	45	39	34	31	24	22	18	0
1.5	0	0	0	0	0	0	0	0	0	0	0	1	4	29	99	197	265	277	236	162	113	84	69	60	53	46	39	35	28	23	19	2
2.0	0	0	0	0	0	0	0	0	0	0	0	0	0	1	8	35	94	172	233	253	196	124	83	68	59	52	45	39	31	25	20	8
2.5	0	0	0	0	0	0	0	0	0	0	0	0	0	0	0	2	11	37	88	184	235	201	122	83	67	59	52	45	34	27	21	13
3.0	0	0	0	0	0	0	0	0	0	0	0	0	0	0	0	0	0	1	7	38	110	222	202	131	88	69	60	52	39	31	22	15

IA/P = 0.50 — * * * TC = 0.1 HR * * *

TRVL TIME (hr)	11.0	11.3	11.6	11.9	12.0	12.1	12.2	12.3	12.4	12.5	12.6	12.7	12.8	13.0	13.2	13.4	13.6	13.8	14.0	14.3	14.6	15.0	15.5	16.0	16.5	17.0	17.5	18.0	19.0	20.0	22.0	26.0
0.0	0	0	0	0	0	107	226	282	258	209	155	130	123	107	97	95	91	87	82	78	74	69	61	52	47	43	39	35	32	29	25	0
.10	0	0	0	0	0	71	174	246	254	224	178	146	130	112	100	96	92	88	83	78	75	70	62	53	48	44	40	35	32	29	25	0
.20	0	0	0	0	0	0	48	132	208	239	229	195	162	127	109	99	95	91	87	80	77	72	65	56	49	45	41	36	32	30	25	0
.30	0	0	0	0	0	0	32	99	172	216	225	205	176	136	113	101	96	92	88	81	77	72	65	57	50	45	41	37	32	30	26	0
.40	0	0	0	0	0	0	0	21	73	139	191	213	208	164	131	111	100	95	91	85	79	74	68	60	51	47	43	38	33	30	26	0
.50	0	0	0	0	0	0	0	14	53	110	164	197	204	174	139	116	103	97	92	86	80	75	68	60	52	47	43	39	33	30	26	0
.75	0	0	0	0	0	0	0	5	22	54	96	137	166	180	159	134	115	103	96	89	83	77	70	63	54	48	44	40	33	31	26	0
1.0	0	0	0	0	0	0	0	0	0	2	10	29	60	132	175	169	146	124	109	97	89	81	74	67	59	52	47	43	34	31	27	0
1.5	0	0	0	0	0	0	0	0	0	0	0	0	2	17	58	112	150	159	148	122	104	91	81	74	67	59	51	46	38	32	28	1
2.0	0	0	0	0	0	0	0	0	0	0	0	0	0	0	4	20	54	98	133	149	133	108	90	80	73	66	58	51	42	34	29	7
2.5	0	0	0	0	0	0	0	0	0	0	0	0	0	0	0	0	2	12	35	87	131	141	111	92	81	74	66	59	46	38	30	18
3.0	0	0	0	0	0	0	0	0	0	0	0	0	0	0	0	0	0	1	4	22	63	120	136	110	91	81	73	66	51	42	31	22

RAINFALL TYPE = III — * * * TC = 0.1 HR * * * — SHEET 1 OF 10

Exhibit 5-III: Tabular hydrograph unit discharges (csm/in) for type III rainfall distribution—continued

HYDROGRAPH TIME (HOURS)

IA/P = 0.10 — * * * TC = 0.2 HR * * *

TRVL TIME (hr)	11.0	11.3	11.6	11.9	12.0	12.1	12.2	12.3	12.4	12.5	12.6	12.7	12.8	13.0	13.2	13.4	13.6	13.8	14.0	14.3	14.6	15.0	15.5	16.0	16.5	17.0	17.5	18.0	19.0	20.0	22.0	26.0
0.0	27	34	50	119	176	258	448	565	483	358	270	194	137	93	75	65	60	56	52	46	42	38	33	28	24	21	19	17	14	13	11	0
.10	24	30	42	70	103	151	222	372	501	489	402	314	234	128	91	73	64	59	55	49	44	40	35	30	25	22	20	17	15	13	11	0
.20	23	29	39	64	90	131	192	312	438	472	425	351	273	153	101	78	66	61	56	50	45	40	35	30	25	22	20	18	15	13	11	0
.30	21	26	33	48	59	79	114	166	263	380	441	431	378	238	141	96	75	65	60	53	47	42	37	32	26	23	21	18	15	13	11	0
.40	20	25	32	45	54	71	99	144	224	328	404	422	392	271	165	108	81	68	61	54	48	43	37	32	27	23	21	19	15	14	11	0
.50	18	22	28	38	43	51	64	88	125	191	282	363	402	356	241	151	102	78	66	57	51	44	39	34	28	24	22	19	15	14	11	0
.75	15	18	23	30	33	37	42	50	64	87	125	184	253	359	341	260	177	121	89	67	57	48	42	36	31	26	23	21	16	14	12	0
1.0	12	16	20	25	28	30	33	37	43	53	68	94	135	251	345	330	260	181	125	81	63	52	44	38	33	28	24	21	17	14	12	0
1.5	8	10	13	17	18	19	21	23	25	27	30	33	38	55	99	179	266	308	289	201	124	74	54	45	39	34	29	24	19	15	12	2
2.0	5	7	9	12	13	14	15	16	17	19	20	22	24	29	38	57	99	166	238	284	239	141	75	54	44	38	33	28	21	17	13	6
2.5	2	4	5	7	8	9	10	10	11	12	13	14	16	18	22	26	33	46	75	155	240	264	154	83	56	46	39	34	25	19	14	9
3.0	1	2	3	5	5	6	6	7	8	8	9	10	11	13	15	18	21	25	33	59	118	220	250	147	82	56	45	39	28	22	14	10

IA/P = 0.30 — * * * TC = 0.2 HR * * *

TRVL TIME (hr)	11.0	11.3	11.6	11.9	12.0	12.1	12.2	12.3	12.4	12.5	12.6	12.7	12.8	13.0	13.2	13.4	13.6	13.8	14.0	14.3	14.6	15.0	15.5	16.0	16.5	17.0	17.5	18.0	19.0	20.0	22.0	26.0
0.0	0	0	0	17	49	115	297	489	462	379	307	230	167	119	98	86	81	76	72	64	60	55	48	41	36	32	29	26	22	20	17	0
.10	0	0	0	12	37	89	228	402	442	400	338	266	200	131	104	89	82	78	73	65	61	56	49	42	36	33	29	26	22	20	17	0
.20	0	0	0	1	8	27	69	175	326	403	401	359	297	182	126	100	88	81	76	69	63	58	51	44	37	34	30	27	23	21	17	0
.30	0	0	0	1	6	20	52	134	262	356	386	368	321	208	140	107	91	83	78	70	64	58	52	45	38	34	31	27	23	21	17	0
.40	0	0	0	1	4	15	40	103	209	307	360	365	336	234	157	116	95	85	79	72	65	59	53	46	39	34	31	28	23	21	17	0
.50	0	0	0	0	0	3	11	30	78	165	260	327	352	305	213	147	111	93	84	75	68	61	55	48	41	35	32	29	23	21	18	0
.75	0	0	0	0	0	1	5	14	37	83	150	220	276	311	264	196	144	112	94	80	72	64	56	50	43	37	33	30	24	21	18	0
1.0	0	0	0	0	0	0	0	0	2	7	20	48	94	215	299	286	230	171	128	95	81	69	61	54	47	40	35	32	25	22	18	0
1.5	0	0	0	0	0	0	0	0	0	0	0	0	2	13	56	139	224	266	253	186	128	89	71	61	54	48	41	36	29	23	19	3
2.0	0	0	0	0	0	0	0	0	0	0	0	0	0	0	3	17	58	126	196	245	214	140	89	71	61	54	47	40	32	25	20	9
2.5	0	0	0	0	0	0	0	0	0	0	0	0	0	0	0	1	5	21	57	145	228	213	136	89	70	60	53	46	35	28	21	13
3.0	0	0	0	0	0	0	0	0	0	0	0	0	0	0	0	0	0	0	3	23	78	180	215	145	94	72	61	54	41	32	22	15

IA/P = 0.50 — * * * TC = 0.2 HR * * *

TRVL TIME (hr)	11.0	11.3	11.6	11.9	12.0	12.1	12.2	12.3	12.4	12.5	12.6	12.7	12.8	13.0	13.2	13.4	13.6	13.8	14.0	14.3	14.6	15.0	15.5	16.0	16.5	17.0	17.5	18.0	19.0	20.0	22.0	26.0
0.0	0	0	0	0	0	10	110	206	245	245	208	162	137	117	102	96	93	90	85	79	76	71	63	54	48	44	40	36	32	29	25	0
.10	0	0	0	0	0	7	76	163	217	235	217	181	152	122	106	98	94	90	86	80	76	71	64	55	49	45	41	36	32	29	25	0
.20	0	0	0	0	0	0	5	52	126	187	219	218	193	145	119	104	97	93	89	83	78	73	66	58	50	46	42	37	32	30	26	0
.30	0	0	0	0	0	0	3	36	96	156	198	211	199	155	125	108	99	94	90	84	79	74	67	59	51	46	42	38	32	30	26	0
.40	0	0	0	0	0	0	0	2	24	72	128	175	199	184	147	121	106	98	93	87	81	76	69	61	53	48	44	39	33	30	26	0
.50	0	0	0	0	0	0	0	1	17	54	103	151	183	194	156	127	109	100	95	88	82	76	70	62	54	48	44	40	33	30	26	0
.75	0	0	0	0	0	0	0	0	6	23	52	90	127	174	168	146	124	109	100	92	85	78	72	65	56	49	45	41	34	31	26	0
1.0	0	0	0	0	0	0	0	0	0	0	2	11	29	91	149	169	157	135	117	100	92	83	76	69	61	53	48	43	35	32	27	0
1.5	0	0	0	0	0	0	0	0	0	0	0	0	0	7	33	81	128	155	155	130	109	94	83	75	68	60	53	47	39	33	28	2
2.0	0	0	0	0	0	0	0	0	0	0	0	0	0	0	1	10	34	72	112	145	139	114	93	82	74	67	60	52	43	35	29	9
2.5	0	0	0	0	0	0	0	0	0	0	0	0	0	0	0	0	3	12	33	83	126	139	112	92	81	74	67	59	47	38	30	18
3.0	0	0	0	0	0	0	0	0	0	0	0	0	0	0	0	0	0	1	4	21	60	116	133	110	92	81	73	66	51	42	31	22

RAINFALL TYPE = III — * * * TC = 0.2 HR * * * — sheet 2 of 10

Exhibit 5-III: Tabular hydrograph unit discharges (csm/in) for type III rainfall distribution—continued

IA/P = 0.10 * * * TC = 0.3 HR * * * IA/P = 0.10

TRVL TIME (hr)	11.0	11.3	11.6	11.9	12.0	12.1	12.2	12.3	12.4	12.5	12.6	12.7	12.8	13.0	13.2	13.4	13.6	13.8	14.0	14.3	14.6	15.0	15.5	16.0	16.5	17.0	17.5	18.0	19.0	20.0	22.0	26.0
0.0	25	31	44	84	124	181	287	441	498	451	358	276	204	118	87	70	63	58	54	48	44	39	34	29	24	22	20	17	14	13	11	0
.10	22	28	37	56	74	108	156	244	375	457	453	389	314	180	113	83	69	62	57	51	46	41	36	31	26	23	20	18	15	13	11	0
.20	21	27	35	53	67	94	136	208	319	411	439	406	345	212	130	91	73	64	59	52	46	42	36	31	26	23	21	18	15	13	11	0
.30	19	24	30	42	49	61	83	118	178	272	365	414	409	305	190	121	87	71	63	55	49	43	38	33	27	24	21	19	15	14	11	0
.40	18	23	29	40	46	56	74	103	153	232	321	383	400	329	217	138	96	75	65	57	50	44	38	33	28	24	21	19	15	14	11	0
.50	16	21	26	34	38	43	52	67	91	132	199	280	349	383	297	196	128	91	73	60	53	46	40	35	30	25	22	20	16	14	11	0
.75	14	18	23	30	33	37	43	52	66	91	131	187	251	347	331	260	182	126	92	68	57	49	42	36	31	26	23	21	16	14	12	0
1.0	11	15	18	23	25	28	30	33	38	44	54	71	98	192	296	334	294	221	154	94	69	55	45	40	34	29	25	22	17	15	12	0
1.5	7	9	12	15	17	18	19	21	23	25	27	30	33	45	75	137	222	287	300	233	149	84	57	47	40	35	30	25	20	16	13	3
2.0	4	6	8	11	12	13	14	15	16	17	19	20	22	26	33	46	76	131	201	278	258	165	85	57	46	40	34	29	22	17	13	7
2.5	2	3	5	7	7	8	9	10	10	11	12	13	14	17	20	24	29	39	60	125	213	260	177	94	60	47	40	35	25	20	14	9
3.0	1	2	3	4	5	5	6	6	7	8	8	9	10	12	14	16	19	23	29	48	95	195	247	167	93	60	47	40	29	22	15	10

IA/P = 0.30 * * * TC = 0.3 HR * * * IA/P = 0.30

TRVL TIME (hr)	11.0	11.3	11.6	11.9	12.0	12.1	12.2	12.3	12.4	12.5	12.6	12.7	12.8	13.0	13.2	13.4	13.6	13.8	14.0	14.3	14.6	15.0	15.5	16.0	16.5	17.0	17.5	18.0	19.0	20.0	22.0	26.0
0.0	0	0	0	6	22	58	146	308	424	422	367	303	234	145	111	92	84	79	74	67	62	57	50	43	36	33	30	26	22	20	17	0
.10	0	0	0	4	16	44	112	243	364	402	379	328	266	166	120	97	86	80	75	68	62	57	51	44	37	33	30	27	22	21	17	0
.20	0	0	0	3	12	33	86	190	306	370	376	344	292	189	132	103	89	82	77	69	63	58	51	44	37	34	30	27	23	21	17	0
.30	0	0	0	0	2	8	25	65	149	254	331	361	350	261	175	126	100	88	31	73	66	60	53	46	39	35	31	28	23	21	18	0
.40	0	0	0	0	1	6	19	50	116	208	290	338	346	282	195	138	107	91	83	74	67	60	54	47	40	35	32	28	23	21	18	0
.50	0	0	0	0	0	1	4	14	38	90	168	250	308	333	256	180	131	103	89	78	71	63	56	49	42	36	33	29	23	21	18	0
.75	0	0	0	0	0	0	2	6	17	43	89	150	213	299	286	229	171	129	104	85	75	65	58	51	44	38	34	30	24	22	18	0
1.0	0	0	0	0	0	0	0	0	1	3	9	24	53	153	253	288	257	200	150	105	85	72	62	55	48	41	36	32	26	22	19	0
1.5	0	0	0	0	0	0	0	0	0	0	0	1	2	15	59	138	217	258	248	187	130	91	71	62	55	48	41	36	29	23	19	3
2.0	0	0	0	0	0	0	0	0	0	0	0	0	0	0	1	9	36	90	159	239	227	158	97	74	63	55	48	42	32	26	20	10
2.5	0	0	0	0	0	0	0	0	0	0	0	0	0	0	0	0	2	12	37	113	194	224	151	96	73	62	55	48	36	29	21	14
3.0	0	0	0	0	0	0	0	0	0	0	0	0	0	0	0	0	0	1	4	24	78	176	211	145	95	73	62	54	41	32	22	15

IA/P = 0.50 * * * TC = 0.3 HR * * * IA/P = 0.50

TRVL TIME (hr)	11.0	11.3	11.6	11.9	12.0	12.1	12.2	12.3	12.4	12.5	12.6	12.7	12.8	13.0	13.2	13.4	13.6	13.8	14.0	14.3	14.6	15.0	15.5	16.0	16.5	17.0	17.5	18.0	19.0	20.0	22.0	26.0
0.0	0	0	0	0	0	2	33	116	193	221	221	200	165	129	110	99	95	92	87	81	77	72	65	56	49	45	41	37	32	30	25	0
.10	0	0	0	0	0	1	23	85	157	200	214	205	178	138	115	102	96	92	88	82	77	73	66	57	50	46	42	37	32	30	26	0
.20	0	0	0	0	0	1	15	62	125	175	201	203	187	147	121	105	98	94	89	83	78	73	66	58	50	46	42	38	32	30	26	0
.30	0	0	0	0	0	0	0	10	45	99	149	184	197	174	140	117	104	97	93	86	80	75	69	61	52	47	43	39	33	30	26	0
.40	0	0	0	0	0	0	0	7	32	76	125	164	189	179	148	123	107	99	94	87	81	76	69	62	53	48	44	39	33	30	26	0
.50	0	0	0	0	0	0	0	0	5	23	59	103	144	183	169	141	119	105	98	90	84	78	71	64	55	49	45	41	33	31	26	0
.75	0	0	0	0	0	0	0	0	2	9	27	55	89	148	168	156	135	117	105	94	87	80	73	66	58	51	46	42	34	31	27	0
1.0	0	0	0	0	0	0	0	0	0	0	1	4	14	59	119	157	163	145	126	105	95	85	77	71	63	55	49	44	36	32	27	0
1,5	0	0	0	0	0	0	0	0	0	0	0	0	0	3	19	56	103	139	151	138	116	97	85	77	70	62	54	48	40	33	28	3
2.0	0	0	0	0	0	0	0	0	0	0	0	0	0	0	1	5	21	52	91	135	143	120	96	84	76	69	61	54	44	35	29	11
2.5	0	0	0	0	0	0	0	0	0	0	0	0	0	0	0	0	1	7	22	65	112	136	117	96	83	75	68	60	48	39	30	19
3.0	0	0	0	0	0	0	0	0	0	0	0	0	0	0	0	0	0	0	2	13	45	102	131	114	95	83	75	67	53	43	31	22

RAINFALL TYPE = III * * * TC = 0.3 HR * * * SHEET 3 OF 10

Exhibit 5-III: Tabular hydrograph unit discharges (csm/in) for type III rainfall distribution—continued

IA/P = 0.10 * * * TC = 0.4 HR * * * IA/P = 0.10

TRVL TIME (hr)	11.0	11.3	11.6	11.9	12.0	12.1	12.2	12.3	12.4	12.5	12.6	12.7	12.8	13.0	13.2	13.4	13.6	13.8	14.0	14.3	14.6	15.0	15.5	16.0	16.5	17.0	17.5	18.0	19.0	20.0	22.0	26.0
0.0	23	29	39	65	91	132	198	308	422	449	417	345	274	162	108	82	68	61	57	50	45	41	35	30	25	22	20	18	15	13	11	0
.10	20	26	33	48	60	80	114	170	262	368	422	418	370	242	149	102	79	67	60	53	47	42	37	32	27	23	21	18	15	13	11	0
.20	20	25	32	45	55	72	100	147	224	320	388	408	383	272	171	114	85	70	62	54	48	43	37	32	27	23	21	19	15	14	11	0
.30	17	22	28	38	43	51	65	88	127	191	277	351	389	349	244	157	108	81	68	58	51	45	39	34	29	24	22	19	15	14	11	0
.40	17	21	27	36	41	47	59	79	111	165	240	314	378	359	268	178	120	88	72	60	52	45	40	34	29	25	22	20	15	14	11	0
.50	15	19	24	31	34	38	44	54	71	98	142	207	278	364	332	243	163	113	84	65	56	48	41	36	31	26	23	20	16	14	12	0
.75	12	16	20	25	27	30	33	38	44	54	70	97	138	249	333	320	257	185	131	85	65	53	44	39	33	28	24	22	17	14	12	0
1.0	11	13	17	22	23	25	28	30	34	38	46	57	75	145	245	322	311	255	188	114	77	58	47	41	35	30	26	23	18	15	12	1
1.5	6	9	11	14	15	17	18	19	21	23	25	27	30	39	59	105	180	255	292	257	176	98	61	48	41	36	31	26	20	16	13	4
2.0	4	6	8	10	11	12	13	14	15	16	17	19	20	24	30	39	61	103	166	253	272	189	98	61	48	41	35	30	23	18	13	8
2.5	2	3	4	6	7	7	8	9	10	10	11	12	13	16	18	22	26	34	49	100	183	255	198	108	66	50	42	36	26	20	14	9
3.0	1	1	2	4	4	5	5	6	6	7	8	8	9	11	13	15	18	21	26	40	77	169	243	185	106	65	49	41	30	23	15	10

IA/P = 0.30 * * * TC = 0.4 HR * * * IA/P = 0.30

TRVL TIME (hr)	11.0	11.3	11.6	11.9	12.0	12.1	12.2	12.3	12.4	12.5	12.6	12.7	12.8	13.0	13.2	13.4	13.6	13.8	14.0	14.3	14.6	15.0	15.5	16.0	16.5	17.0	17.5	18.0	19.0	20.0	22.0	26.0
0.0	0	0	0	2	10	30	78	177	306	379	379	347	293	187	133	105	90	82	77	69	63	58	51	44	38	34	30	27	23	21	17	0
.10	0	0	0	2	7	22	59	138	250	336	365	353	313	212	146	112	94	84	78	71	64	59	52	45	38	34	31	27	23	21	17	0
.20	0	0	0	1	5	17	45	107	202	292	341	349	325	235	162	121	98	87	80	72	65	59	53	46	39	34	31	28	23	21	17	0
.30	0	0	0	0	1	4	12	34	83	162	249	310	336	298	215	152	116	96	85	76	68	61	55	48	41	36	32	29	23	21	18	0
.40	0	0	0	0	0	3	9	26	64	130	209	276	324	307	234	168	125	101	88	77	70	62	55	49	41	36	32	29	23	21	18	0
.50	0	0	0	0	0	0	2	7	19	49	103	173	242	313	285	216	157	119	98	82	73	65	57	51	44	37	33	30	24	22	18	0
.75	0	0	0	0	0	0	1	3	9	23	52	97	153	253	285	253	199	151	118	91	78	68	59	53	46	39	35	31	25	22	18	0
1.0	0	0	0	0	0	0	0	0	0	1	4	13	30	104	204	276	276	226	175	120	92	75	64	57	50	43	37	33	26	22	19	1
1.5	0	0	0	0	0	0	0	0	0	0	0	0	1	8	36	98	177	236	250	207	148	99	75	63	56	49	42	37	29	24	19	4
2.0	0	0	0	0	0	0	0	0	0	0	0	0	0	0	1	5	22	62	124	210	232	176	107	78	65	57	50	43	33	27	20	11
2.5	0	0	0	0	0	0	0	0	0	0	0	0	0	0	0	0	1	7	24	85	167	219	167	106	77	64	56	49	37	29	21	14
3.0	0	0	0	0	0	0	0	0	0	0	0	0	0	0	0	0	0	0	2	15	58	152	207	159	104	76	64	55	42	33	22	15

IA/P = 0.50 * * * TC = 0.4 HR * * * IA/P = 0.50

TRVL TIME (hr)	11.0	11.3	11.6	11.9	12.0	12.1	12.2	12.3	12.4	12.5	12.6	12.7	12.8	13.0	13.2	13.4	13.6	13.8	14.0	14.3	14.6	15.0	15.5	16.0	16.5	17.0	17.5	18.0	19.0	20.0	22.0	26.0
0.0	0	0	0	0	0	0	10	54	121	182	204	204	191	146	121	106	98	94	90	83	78	73	66	58	50	46	42	38	32	30	26	0
.10	0	0	0	0	0	0	7	38	94	153	187	198	193	155	128	110	100	95	91	84	79	74	67	59	51	46	42	38	32	30	26	0
.20	0	0	0	0	0	0	5	27	71	126	166	187	191	164	134	114	103	96	92	85	80	75	68	60	52	47	43	38	33	30	26	0
.30	0	0	0	0	0	0	0	3	19	54	102	145	173	185	155	129	111	101	95	88	82	76	70	62	54	48	44	40	33	31	26	0
.40	0	0	0	0	0	0	0	2	13	40	81	124	157	180	161	135	116	104	97	90	83	77	71	63	55	49	44	40	33	31	26	0
.50	0	0	0	0	0	0	0	0	1	9	30	64	104	163	174	154	130	113	102	93	86	79	73	66	57	50	46	42	34	31	27	0
.75	0	0	0	0	0	0	0	0	0	3	13	32	59	120	157	162	145	126	111	98	90	82	75	68	60	52	47	43	35	31	27	0
1.0	0	0	0	0	0	0	0	0	0	0	0	2	6	36	90	138	158	152	135	112	98	88	79	72	65	56	50	45	37	32	27	0
1.5	0	0	0	0	0	0	0	0	0	0	0	0	0	1	10	37	80	121	148	143	123	102	87	78	71	64	56	49	41	34	28	4
2.0	0	0	0	0	0	0	0	0	0	0	0	0	0	0	0	2	13	36	71	121	140	126	101	86	78	70	63	55	44	36	29	13
2.5	0	0	0	0	0	0	0	0	0	0	0	0	0	0	0	0	1	4	14	49	97	135	122	100	86	77	70	62	49	40	30	20
3.0	0	0	0	0	0	0	0	0	0	0	0	0	0	0	0	0	0	0	1	9	33	88	130	119	98	85	76	69	54	44	32	22

RAINFALL TYPE = III * * * TC = 0.4 HR * * * SHEET 4 OF 10

Exhibit 5-III: Tabular hydrograph unit discharges (csm/in) for type III rainfall distribution—continued

TRVL TIME (hr)	11.0	11.3	11.6	11.9	12.0	12.1	12.2	12.3	12.4	12.5	12.6	12.7	12.8	13.0	13.2	13.4	13.6	13.8	14.0	14.3	14.6	15.0	15.5	16.0	16.5	17.0	17.5	18.0	19.0	20.0	22.0	26.0
IA/P = 0.10																* * * TC = 0.5 HR * * *													IA/P = 0.10			
0.0	21	27	35	54	70	97	144	217	316	397	411	388	330	214	139	99	78	67	60	52	47	42	36	31	26	23	21	18	15	13	11	0
.10	19	24	30	43	50	64	86	125	186	273	355	392	390	296	194	129	94	75	65	56	49	43	38	33	28	24	21	19	15	14	11	0
.20	18	23	29	40	47	58	77	109	161	235	315	367	382	318	218	145	103	80	68	57	50	44	39	33	28	24	22	19	15	14	11	0
.30	16	21	26	34	38	44	53	69	95	139	203	278	337	367	289	199	135	98	77	62	54	46	40	35	30	25	22	20	16	14	11	0
.40	16	20	25	33	36	41	49	62	84	121	176	244	306	358	306	220	151	107	83	64	55	47	41	35	30	25	23	20	16	14	12	0
.50	14	18	22	28	31	35	39	46	57	75	106	152	213	323	346	282	202	140	102	73	59	50	42	37	32	27	23	21	16	14	12	0
.75	12	16	20	25	28	30	34	38	45	56	75	104	145	246	319	308	252	187	135	89	67	53	44	39	33	28	24	22	17	14	12	0
1.0	10	12	16	20	22	23	25	28	31	34	39	47	60	110	197	280	309	279	220	138	90	63	49	42	37	31	26	23	18	15	12	1
1.5	6	8	10	13	14	15	17	18	19	21	23	25	27	34	49	82	143	218	283	271	203	116	68	51	43	37	32	27	21	16	13	4
2.0	3	5	7	9	10	11	12	13	14	15	16	17	19	22	27	34	50	82	135	226	265	211	114	67	50	42	37	31	23	18	13	8
2.5	2	3	4	6	7	7	8	9	10	10	11	12	13	16	18	22	26	34	50	102	182	249	197	111	67	50	42	36	26	20	14	9
3.0	1	1	2	3	4	4	5	5	6	6	7	8	8	10	12	14	16	19	23	34	63	144	238	201	121	72	52	43	31	23	15	10
IA/P = 0.30																* * * TC = 0.5 HR * * *													IA/P = 0.30			
0.0	0	0	0	1	4	15	40	101	198	295	345	345	325	232	161	122	100	88	80	72	65	59	53	46	39	34	31	28	23	21	18	0
.10	0	0	0	1	3	11	30	77	158	249	313	335	329	253	178	132	106	91	82	73	66	60	53	47	40	35	31	28	23	21	18	0
.20	0	0	0	0	2	8	23	59	125	208	278	316	324	271	196	144	112	95	85	75	67	61	54	47	40	35	32	28	23	21	18	0
.30	0	0	0	0	0	2	6	17	45	98	171	242	291	313	249	182	136	108	92	80	71	63	56	49	42	36	33	29	24	21	18	0
.40	0	0	0	0	0	1	4	13	34	77	140	208	264	304	263	198	148	115	97	81	72	64	57	50	43	37	33	30	24	21	18	0
.50	0	0	0	0	0	0	1	3	10	26	60	113	177	276	295	244	185	140	111	88	77	67	59	52	45	39	34	31	24	22	18	0
.75	0	0	0	0	0	0	0	1	4	12	29	60	104	204	271	263	222	174	136	101	83	70	61	54	47	40	35	32	25	22	18	0
1.0	0	0	0	0	0	0	0	0	0	1	2	6	16	67	155	235	263	242	198	138	102	80	66	58	51	44	38	34	27	23	19	1
1.5	0	0	0	0	0	0	0	0	0	0	0	0	0	4	22	67	138	205	241	221	167	110	79	66	58	51	44	38	30	24	20	5
2.0	0	0	0	0	0	0	0	0	0	0	0	0	0	0	0	3	13	42	93	182	225	191	119	83	67	58	51	44	34	27	21	12
2.5	0	0	0	0	0	0	0	0	0	0	0	0	0	0	0	0	1	4	15	62	139	213	180	117	82	67	58	51	38	30	22	15
3.0	0	0	0	0	0	0	0	0	0	0	0	0	0	0	0	0	0	0	1	10	41	127	203	171	114	81	66	57	43	33	23	16
IA/P = 0.50																* * * TC = 0.5 HR * * *													IA/P = 0.50			
0.0	0	0	0	0	0	0	3	24	68	124	174	190	190	162	133	114	103	97	92	85	80	75	68	60	52	47	43	39	33	30	26	0
.10	0	0	0	0	0	0	2	17	51	100	149	177	186	169	140	119	106	99	93	86	81	75	69	61	52	47	43	39	33	30	26	0
.20	0	0	0	0	0	0	1	12	38	79	126	160	181	173	147	124	109	101	95	88	81	76	69	62	53	48	44	39	33	30	26	0
.30	0	0	0	0	0	0	0	1	8	28	62	105	141	176	165	141	120	107	99	91	84	78	71	64	56	49	45	41	33	31	26	0
.40	0	0	0	0	0	0	0	1	6	20	48	86	123	172	172	146	125	111	101	92	85	79	72	65	56	50	45	41	34	31	26	0
.50	0	0	0	0	0	0	0	0	4	15	37	70	105	157	167	151	130	114	104	94	87	79	73	66	57	50	46	42	34	31	27	0
.75	0	0	0	0	0	0	0	0	0	1	6	17	37	91	139	157	150	134	119	103	93	84	76	69	62	54	48	44	35	32	27	0
1.0	0	0	0	0	0	0	0	0	0	0	1	3	9	40	91	135	153	149	135	113	99	88	79	72	65	57	50	45	37	32	27	1
1.5	0	0	0	0	0	0	0	0	0	0	0	0	0	0	5	24	59	101	132	144	130	107	90	80	73	65	57	51	41	34	29	6
2.0	0	0	0	0	0	0	0	0	0	0	0	0	0	0	0	1	7	25	55	106	138	130	105	89	79	72	65	57	45	37	30	15
2.5	0	0	0	0	0	0	0	0	0	0	0	0	0	0	0	0	0	2	9	36	81	133	133	104	88	79	71	64	50	41	31	21
3.0	0	0	0	0	0	0	0	0	0	0	0	0	0	0	0	0	0	0	0	5	24	74	128	122	103	88	78	70	55	45	32	23

RAINFALL TYPE = III * * * TC = 0.5 HR * * * SHEET 5 OF 10

Exhibit 5-III: Tabular hydrograph unit discharges (csm/in) for type III rainfall distribution—continued

TRVL TIME (hr)	11.0	11.3	11.6	11.9	12.0	12.1	12.2	12.3	12.4	12.5	12.6	12.7	12.8	13.0	13.2	13.4	13.6	13.8	14.0	14.3	14.6	15.0	15.5	16.0	16.5	17.0	17.5	18.0	19.0	20.0	22.0	26.0
IA/P = 0.10																* * * TC = 0.75 HR * * *													IA/P = 0.10			
0.0	17	22	28	39	45	56	73	104	151	215	281	328	343	310	228	163	121	94	77	63	53	45	39	34	29	25	22	20	15	14	11	0
.10	17	21	27	37	42	51	66	91	131	187	250	302	336	319	247	179	131	101	82	65	55	46	40	35	29	25	22	20	16	14	11	0
.20	15	19	24	31	35	40	48	60	81	114	163	221	275	328	298	229	167	124	96	73	60	49	42	36	31	26	23	20	16	14	12	0
.30	14	18	23	30	33	38	44	55	72	100	142	194	248	320	305	245	182	135	103	76	62	50	42	37	31	27	23	21	16	14	12	0
.40	13	16	21	26	29	32	36	41	50	65	88	124	171	268	313	288	228	170	127	88	68	54	44	38	33	28	24	21	17	14	12	0
.50	12	16	20	25	28	30	34	39	46	59	78	109	150	244	306	294	242	184	138	94	71	56	45	39	34	28	24	22	17	14	12	0
.75	10	13	16	21	23	25	27	30	33	38	46	58	77	140	221	277	287	248	197	133	92	66	50	42	36	31	26	23	18	15	12	1
1.0	8	10	13	16	18	19	21	23	25	27	30	34	39	61	109	181	249	280	265	198	134	85	58	46	40	34	29	25	20	16	12	2
1.5	5	7	9	12	13	14	15	16	17	19	20	22	24	30	40	63	106	167	225	261	226	147	84	58	46	39	34	29	22	17	13	5
2.0	2	4	5	7	8	9	9	10	11	12	13	14	16	18	22	26	34	50	80	155	226	246	158	91	61	47	40	34	25	20	14	9
2.5	1	2	3	5	5	6	6	7	8	8	9	10	11	13	15	18	21	26	34	62	120	209	234	151	90	60	47	39	29	22	15	10
3.0	0	1	2	3	3	3	4	4	5	5	6	7	7	9	10	12	15	17	21	29	50	113	209	224	144	88	59	46	33	25	16	10
IA/P = 0.30																* * * TC = 0.75 HR * * *													IA/P = 0.30			
0.0	0	0	0	0	1	3	8	24	58	113	182	243	283	287	233	178	139	114	98	83	73	64	56	50	43	37	33	30	24	21	18	0
.10	0	0	0	0	0	2	6	18	45	91	151	212	259	284	245	191	149	120	102	85	75	65	57	50	43	37	33	30	24	21	18	0
.20	0	0	0	0	0	0	1	5	14	35	72	125	183	263	277	230	180	142	116	93	80	68	59	52	45	39	34	31	25	22	18	0
.30	0	0	0	0	0	0	1	3	10	26	57	102	156	245	270	240	192	151	122	96	82	69	60	53	46	39	35	31	25	22	18	0
.40	0	0	0	0	0	0	0	1	2	8	20	45	83	182	252	264	226	181	144	108	89	73	62	55	48	41	36	32	26	22	19	0
.50	0	0	0	0	0	0	0	0	2	6	15	35	67	158	235	259	235	192	153	113	92	75	63	56	49	42	36	33	26	22	19	1
.75	0	0	0	0	0	0	0	0	0	0	1	2	7	35	100	178	232	242	217	163	121	90	71	61	54	47	40	35	28	23	19	2
1.0	0	0	0	0	0	0	0	0	0	0	0	1	4	21	68	140	205	236	229	181	135	97	74	63	55	48	41	36	29	24	19	3
1.5	0	0	0	0	0	0	0	0	0	0	0	0	0	0	2	13	42	94	155	221	212	158	103	77	64	56	49	42	33	26	20	10
2.0	0	0	0	0	0	0	0	0	0	0	0	0	0	0	0	1	4	15	42	113	184	209	151	101	76	63	55	48	36	29	21	14
2.5	0	0	0	0	0	0	0	0	0	0	0	0	0	0	0	0	0	1	5	27	80	168	199	146	100	75	63	55	41	32	22	15
3.0	0	0	0	0	0	0	0	0	0	0	0	0	0	0	0	0	0	0	0	1	10	56	154	191	151	105	78	64	48	37	24	16
IA/P = 0.50																* * * TC = 0.75 HR * * *													IA/P = 0.50			
0.0	0	0	0	0	0	0	0	3	13	37	71	108	140	167	156	136	120	108	101	92	85	78	72	64	56	50	45	41	34	31	26	0
.10	0	0	0	0	0	0	0	2	10	28	57	91	124	163	158	140	124	111	103	94	86	79	72	65	57	50	46	41	34	31	27	0
.20	0	0	0	0	0	0	0	0	1	7	21	45	76	135	159	153	136	120	109	98	90	82	74	67	59	52	47	43	35	31	27	0
.30	0	0	0.	0	0	0	0	0	1	5	15	35	62	121	157	157	140	124	112	100	91	83	75	68	60	52	47	43	35	31	27	0
.40	0	0	0	0	0	0	0	0	1	3	11	27	50	107	146	154	143	128	115	102	93	84	76	69	61	53	48	43	35	32	27	0
.50	0	0	0	0	0	0	0	0	0	0	2	8	21	66	118	148	152	139	125	108	97	87	78	71	63	55	49	45	36	32	27	1
.75	0	0	0	0	0	0	0	0	0	0	1	4	10	38	82	122	142	144	133	116	103	91	80	73	66	58	51	46	37	32	28	1
1.0	0	0	0	0	0	0	0	0	0	0	0	0	0	5	24	60	102	132	142	133	116	99	86	77	70	62	55	49	40	33	28	3
1.5	0	0	0	0	0	0	0	0	0	0	0	0	0	0	1	7	25	55	91	128	136	119	98	85	77	69	62	54	44	36	30	11
2.0	0	0	0	0	0	0	0	0	0	0	0	0	0	0	0	0	2	9	25	66	109	131	116	97	85	76	69	61	48	39	30	18
2.5	0	0	0	0	0	0	0	0	0	0	0	0	0	0	0	0	0	0	3	16	47	100	127	114	96	84	75	68	53	43	31	22
3.0	0	0	0	0	0	0	0	0	0	0	0	0	0	0	0	0	0	0	0	2	10	44	102	124	112	95	83	75	59	47	33	23

RAINFALL TYPE = III * * * TC = 0.75 HR * * * SHEET 6 OF 10

Exhibit 5-III: Tabular hydrograph unit discharges (csm/in) for type III rainfall distribution—continued

TRVL TIME (hr)	11.0	11.3	11.6	11.9	12.0	12.1	12.2	12.3	12.4	12.5	12.6	12.7	12.8	13.0	13.2	13.4	13.6	13.8	14.0	14.3	14.6	15.0	15.5	16.0	16.5	17.0	17.5	18.0	19.0	20.0	22.0	26.0
IA/P = 0.10														* * * TC = 1.0 HR * * *															IA/P = 0.10			
0.0	15	19	24	32	37	44	54	71	98	136	181	227	264	297	270	215	164	128	103	78	64	52	43	36	31	26	23	21	16	14	12	0
.10	13	17	22	28	31	35	41	49	64	87	120	161	205	273	289	254	201	155	122	90	71	56	45	38	33	28	24	21	17	14	12	0
.20	13	16	21	27	29	33	38	46	58	77	105	142	184	257	285	263	214	167	130	95	74	57	46	39	33	28	24	22	17	14	12	0
.30	12	16	20	26	28	31	36	42	53	69	93	126	165	240	279	268	225	178	139	100	77	59	47	39	34	29	25	22	17	15	12	1
.40	11	14	18	23	25	27	30	34	40	48	62	83	112	185	251	276	256	213	168	118	87	65	50	41	35	30	26	23	18	15	12	1
.50	11	13	17	22	24	26	29	32	37	45	56	74	99	167	235	270	261	223	179	126	92	67	51	42	36	31	26	23	18	15	12	1
.75	8	10	13	17	18	19	21	23	25	28	31	36	44	72	122	186	239	258	243	189	136	90	62	48	40	34	29	25	20	16	12	2
1.0	6	9	11	14	15	17	18	20	21	23	25	28	32	46	75	124	185	234	253	226	170	110	71	53	43	37	31	27	21	16	13	4
1.5	4	6	8	10	11	12	13	14	15	16	17	19	21	25	32	46	74	118	170	230	239	179	108	70	52	43	36	31	23	18	13	7
2.0	2	3	4	6	7	7	8	9	10	10	11	12	13	16	18	22	28	38	58	111	179	228	185	116	75	54	44	37	27	21	14	9
2.5	1	1	2	4	4	5	5	6	6	7	8	8	9	11	13	15	18	22	28	46	87	167	219	176	113	73	54	43	31	23	15	10
3.0	0	0	1	2	2	2	3	3	3	4	4	5	5	7	8	10	12	14	16	21	32	68	156	210	179	120	78	56	37	27	16	11
IA/P = 0.30														* * * TC = 1.0 HR * * *															IA/P = 0.30			
0.0	0	0	0	0	0	1	5	13	30	57	95	141	186	243	249	213	174	142	119	97	83	70	60	53	46	39	35	31	25	22	18	0
.10	0	0	0	0	0	1	3	10	23	46	79	120	164	230	245	221	183	150	125	101	85	72	61	53	46	40	35	31	25	22	18	0
.20	0	0	0	0	0	0	1	3	7	18	36	65	102	183	233	241	210	174	144	112	92	76	64	56	48	42	36	33	26	22	19	1
.30	0	0	0	0	0	0	1	2	6	14	29	53	86	163	221	237	217	183	151	117	95	78	65	56	49	42	37	33	26	22	19	1
.40	0	0	0	0	0	0	0	0	1	4	11	23	43	107	180	225	233	207	175	133	105	84	68	59	51	44	38	34	27	23	19	1
.50	0	0	0	0	0	0	0	0	1	3	8	18	34	91	162	214	230	213	183	139	109	86	70	60	52	45	39	34	27	23	19	1
.75	0	0	0	0	0	0	0	0	0	0	1	4	9	33	82	145	196	218	211	174	135	101	77	64	56	48	42	37	29	24	19	3
1.0	0	0	0	0	0	0	0	0	0	0	0	1	2	11	37	85	144	192	214	199	160	116	85	69	59	51	44	38	30	25	20	5
1.5	0	0	0	0	0	0	0	0	0	0	0	0	0	0	1	7	23	56	104	174	203	177	123	89	71	60	52	45	35	27	21	11
2.0	0	0	0	0	0	0	0	0	0	0	0	0	0	0	0	0	2	8	24	73	139	194	169	120	87	70	59	51	39	30	22	14
2.5	0	0	0	0	0	0	0	0	0	0	0	0	0	0	0	0	0	1	3	15	51	127	186	162	117	86	69	59	44	34	23	16
3,0	0	0	0	0	0	0	0	0	0	0	0	0	0	0	0	0	0	0	0	1	6	35	117	180	164	122	90	71	51	39	25	16
IA/P = 0.50														* * * TC = 1.0 HR * * *															IA/P = 0.50			
0.0	0	0	0	0	0	0	0	2	6	17	34	57	83	127	151	142	130	118	109	99	91	83	75	68	59	52	47	43	35	31	27	1
.10	0	0	0	0	0	0	0	1	5	13	27	47	71	117	146	144	133	121	112	101	92	84	76	68	60	53	48	43	35	32	27	1
.20	0	0	0	0	0	0	0	1	3	10	21	38	60	106	138	143	135	124	114	102	94	85	76	69	61	54	48	44	35	32	27	1
.30	0	0	0	0	0	0	0	0	0	2	7	16	31	73	114	139	142	132	121	108	98	88	79	71	64	56	50	45	36	32	27	1
.40	0	0	0	0	0	0	0	0	0	2	5	13	25	62	104	133	140	134	124	110	99	89	80	72	64	56	50	45	37	32	27	1
.50	0	0	0	0	0	0	0	0	0	0	1	4	10	35	74	112	134	138	131	117	104	93	82	74	67	59	52	47	38	33	28	2
.75	0	0	0	0	0	0	0	0	0	0	0	2	5	19	43	84	115	131	134	123	110	97	85	77	69	61	54	48	39	33	28	3
1.0	0	0	0	0	0	0	0	0	0	0	0	0	1	6	21	49	84	113	129	132	119	103	89	80	72	64	56	50	41	34	28	5
1.5	0	0	0	0	0	0	0	0	o	0	0	0	0	0	1	7	21	46	75	113	128	120	102	88	79	71	63	56	45	37	29	12
2.0	0	0	0	0	0	0	0	0	0	0	0	0	0	0	0	0	1	4	14	42	82	119	124	104	90	80	72	64	50	41	31	20
2.5	0	0	0	0	0	0	0	0	0	0	0	0	0	0	0	0	0	0	1	9	29	75	117	121	102	89	79	71	56	45	32	23
3,0	0	0	0	0	0	0	0	0	0	0	0	0	0	0	0	0	0	0	0	1	5	28	80	118	118	101	88	79	63	49	34	24
RAINFALL TYPE = III														* * * TC = 1.0 HR * * *															SHEET 7 OF 10			

Exhibit 5-III: Tabular hydrograph unit discharges (csm/in) for type III rainfall distribution—continued

TRVL TIME (hr)	11.0	11.3	11.6	11.9	12.0	12.1	12.2	12.3	12.4	12.5	12.6	12.7	12.8	13.0	13.2	13.4	13.6	13.8	14.0	14.3	14.6	15.0	15.5	16.0	16.5	17.0	17.5	18.0	19.0	20.0	22.0	26.0
IA/P = 0.10														* * * TC = 1.25 HR * * *															IA/P = 0.10			
0.0	13	17	22	28	32	37	44	56	75	100	133	170	206	255	264	236	194	155	125	95	75	59	47	39	33	28	24	22	17	14	12	1
.10	13	16	21	27	30	34	41	51	67	89	118	152	188	243	261	243	204	164	133	100	79	61	48	40	34	29	25	22	17	15	12	1
.20	11	15	18	24	26	29	33	38	47	60	79	105	137	203	248	256	232	193	156	115	88	67	51	42	35	30	26	23	18	15	12	1
.30	11	14	18	23	25	27	31	36	43	54	71	94	122	187	237	254	238	203	165	121	92	69	53	43	36	30	26	23	18	15	12	1
.40	10	12	16	20	22	24	26	29	34	40	49	64	84	140	201	242	250	228	193	142	106	77	57	46	38	32	27	24	19	15	12	1
.50	9	12	15	19	21	23	25	28	32	37	45	58	75	126	186	232	248	233	201	150	111	80	58	47	39	33	28	24	19	15	12	2
.75	8	11	14	17	19	20	22	24	27	31	36	44	56	93	144	196	230	237	220	174	131	91	64	50	41	34	29	25	20	16	12	2
1.0	6	8	11	14	15	16	17	19	20	22	25	28	32	46	74	119	170	213	234	218	174	120	79	58	46	38	32	27	21	17	13	4
1.5	3	5	7	9	9	10	11	12	13	14	16	17	18	22	27	38	58	91	134	197	223	194	128	83	60	47	39	33	24	19	14	7
2.0	2	3	4	6	6	7	8	8	9	10	11	12	13	15	18	22	27	38	57	106	166	214	184	124	82	59	47	39	28	21	14	9
2.5	1	1	2	3	3	4	4	5	5	6	7	7	8	10	11	13	16	19	24	38	68	137	206	186	130	87	62	48	33	24	15	10
3.0	0	0	1	2	2	2	2	3	3	4	4	5	5	6	8	9	11	13	16	21	31	66	145	200	178	126	85	61	39	28	17	11
IA/P = 0.30														* * * TC = 1.25 HR * * *															IA/P = 0.30			
0.0	0	0	0	0	0	1	2	7	18	35	61	94	130	192	222	218	191	161	136	110	93	77	65	56	48	41	36	32	26	22	19	1
.10	0	0	0	0	0	1	2	5	14	28	50	79	113	177	214	219	197	168	142	115	96	79	66	57	49	42	37	33	26	22	19	1
.20	0	0	0	0	0	0	1	4	10	22	41	66	97	162	205	217	202	175	149	119	99	81	67	58	50	43	37	33	26	23	19	1
.30	0	0	0	0	0	0	0	1	3	8	17	33	55	115	174	208	214	195	168	133	108	87	70	60	52	45	39	34	27	23	19	2
.40	0	0	0	0	0	0	0	1	2	6	14	26	45	100	159	200	212	199	174	138	112	89	72	61	53	45	39	35	28	23	19	2
.50	0	0	0	0	0	0	0	0	1	2	5	11	21	60	116	170	203	209	192	156	124	97	76	64	55	47	41	36	29	24	19	3
.75	0	0	0	0	0	0	0	0	0	1	2	5	11	35	79	131	175	201	201	172	140	107	82	67	58	50	43	37	30	24	20	4
1.0	0	0	0	0	0	0	0	0	0	0	0	0	1	6	23	57	105	153	186	198	172	131	95	75	63	54	47	41	32	25	20	7
1.5	0	0	0	0	0	0	0	0	0	0	0	0	0	0	2	8	24	54	96	157	189	174	127	94	74	63	54	46	35	28	21	12
2.0	0	0	0	0	0	0	0	0	0	0	0	0	0	0	0	0	2	9	24	69	127	182	167	124	93	74	62	53	40	31	22	15
2.5	0	0	0	0	0	0	0	0	0	0	0	0	0	0	0	0	0	0	2	10	35	99	176	167	129	96	76	63	46	35	23	16
3.0	0	0	0	0	0	0	0	0	0	0	0	0	0	0	0	0	0	0	0	1	6	34	107	171	161	125	94	75	53	40	25	17
IA/P = 0.50														* * * TC = 1.25 HR * * *															IA/P = 0.50			
0.0	0	0	0	0	0	0	0	1	4	10	21	36	55	95	123	138	131	122	114	104	95	86	78	70	62	55	49	44	36	32	27	1
.10	0	0	0	0	0	0	0	0	1	3	8	16	30	66	102	125	134	128	120	108	99	89	80	73	65	57	51	46	37	32	28	2
.20	0	0	0	0	0	0	0	0	0	2	6	13	24	57	94	120	132	130	122	110	101	91	81	73	66	58	51	46	37	33	28	2
.30	0	0	0	0	0	0	0	0	0	1	4	10	19	49	85	113	131	131	124	112	102	92	82	74	66	58	52	47	38	33	28	2
.40	0	0	0	0	0	0	0	0	0	1	3	8	16	42	76	106	126	130	125	114	104	93	83	75	67	59	52	47	38	33	28	3
.50	0	0	0	0	0	0	0	0	0	0	1	2	6	22	51	84	112	127	129	120	109	97	86	77	70	61	54	49	40	33	28	4
.75	0	0	0	0	0	0	0	0	0	0	0	1	3	12	32	61	91	113	125	123	114	101	89	80	72	64	56	50	41	34	28	5
1.0	0	0	0	0	0	0	0	0	0	0	0	0	0	1	7	22	47	76	102	122	124	111	96	85	76	69	61	54	43	36	29	9
1.5	0	0	0	0	0	0	0	0	0	0	0	0	0	0	0	2	8	22	43	82	111	121	109	95	84	76	68	60	48	39	30	16
2.0	0	0	0	0	0	0	0	0	0	0	0	0	0	0	0	0	1	3	9	30	63	104	118	107	94	83	75	67	53	43	31	21
2.5	0	0	0	0	0	0	0	0	0	0	0	0	0	0	0	0	0	0	1	6	20	58	104	116	106	93	83	74	59	47	33	23
1.0	0	0	0	0	0	0	0	0	0	0	0	0	0	0	0	0	0	0	0	0	3	19	64	104	114	104	92	82	65	52	35	24
RAINFALL TYPE = III														* * * TC = 1.25 HR * * *															SHEET 8 OF 10			

Exhibit 5-III: Tabular hydrograph unit discharges (csm/in) for type III rainfall distribution—continued

IA/P = 0.10 * * * TC = 1.5 HR * * * IA/P = 0.10

TRVL TIME (hr)	11.0	11.3	11.6	11.9	12.0	12.1	12.2	12.3	12.4	12.5	12.6	12.7	12.8	13.0	13.2	13.4	13.6	13.8	14.0	14.3	14.6	15.0	15.5	16.0	16.5	17.0	17.5	18.0	19.0	20.0	22.0	26.0
0.0	12	15	19	25	27	31	37	45	57	75	97	122	151	203	231	238	213	182	150	115	91	70	54	44	37	30	26	23	18	15	12	1
.10	10	13	17	21	23	26	29	34	41	52	67	87	111	165	210	233	227	205	173	131	102	77	58	46	39	32	27	24	19	15	12	1
.20	10	13	16	21	23	25	28	32	38	47	61	78	100	152	199	228	231	210	181	138	107	80	59	47	39	33	28	24	19	15	12	2
.30	9	12	15	20	22	24	27	30	36	44	55	70	90	139	188	221	228	215	188	145	112	83	61	48	40	33	28	24	19	15	12	2
.40	8	11	14	18	19	21	23	25	29	33	40	50	64	103	152	196	223	226	208	166	127	92	66	52	42	35	30	25	20	16	13	2
.50	8	10	13	17	18	20	22	24	27	31	37	46	58	93	140	186	216	224	212	173	133	96	69	53	43	36	30	26	20	16	13	3
.75	6	8	11	14	15	16	18	19	21	23	26	31	36	55	87	130	173	205	217	202	165	119	81	60	48	39	33	28	21	17	13	4
1.0	5	6	8	11	12	13	14	15	16	18	19	21	24	31	46	71	109	151	189	214	200	153	102	72	55	44	37	31	23	18	13	6
1.5	3	4	5	7	8	9	10	11	11	12	14	15	16	19	24	31	45	69	103	159	207	207	147	99	71	54	44	36	26	20	14	8
2.0	1	2	3	5	5	6	6	7	8	9	9	10	11	13	16	19	23	31	45	81	132	189	199	142	97	69	53	43	30	22	15	10
2.5	0	1	1	2	3	3	4	4	4	5	6	6	7	8	10	12	14	17	20	31	53	107	179	194	147	102	73	55	36	26	16	10
3.0	0	0	1	1	1	2	2	2	3	3	3	4	4	5	7	8	10	11	13	18	26	51	116	178	188	142	100	71	43	30	18	11

IA/P = 0.30 * * * TC = 1.5 HR * * * IA/P = 0.30

TRVL TIME (hr)	11.0	11.3	11.6	11.9	12.0	12.1	12.2	12.3	12.4	12.5	12.6	12.7	12.8	13.0	13.2	13.4	13.6	13.8	14.0	14.3	14.6	15.0	15.5	16.0	16.5	17.0	17.5	18.0	19.0	20.0	22.0	26.0
0.0	0	0	0	0	0	1	2	5	11	22	38	59	84	138	180	200	195	176	154	125	105	86	71	60	52	44	39	34	27	23	19	2
.10	0	0	0	0	0	0	0	1	4	9	18	31	50	99	149	184	198	190	170	139	115	93	75	63	55	47	40	35	28	23	19	3
.20	0	0	0	0	0	0	0	1	3	7	14	25	41	86	137	176	195	192	175	144	119	95	76	64	55	47	41	36	28	24	19	3
.30	0	0	0	0	0	0	0	0	1	2	5	11	21	53	100	147	181	194	187	159	130	103	81	68	58	50	43	37	29	24	19	4
.40	0	0	0	0	0	0	0	0	0	2	4	9	17	45	88	136	172	192	189	164	135	106	83	69	59	51	43	38	30	24	20	4
,50	0	0	0	0	0	0	0	0	0	0	1	3	7	23	56	101	145	177	190	177	149	116	89	73	62	53	45	39	31	25	20	6
.75	0	0	0	0	0	0	0	0	0	0	0	1	3	13	35	71	113	151	176	184	162	128	97	77	65	55	48	41	32	26	20	7
1.0	0	0	0	0	0	0	0	0	0	0	0	0	0	2	8	24	53	92	132	174	182	153	114	88	72	61	52	45	34	27	21	10
1.5	0	0	0	0	0	0	0	0	0	0	0	0	0	0	0	3	9	24	50	103	152	175	148	112	87	71	60	51	38	30	22	14
2.0	0	0	0	0	0	0	0	0	0	0	0	0	0	0	0	0	0	2	6	24	62	127	170	151	116	89	73	61	45	34	23	16
2.5	0	0	0	0	0	0	0	0	0	0	0	0	0	0	0	0	0	0	0	4	16	58	131	165	146	113	88	72	52	39	25	16
3.0	0	0	0	0	0	0	0	0	0	0	0	0	0	0	0	0	0	0	0	0	2	16	67	134	161	142	111	87	60	44	27	17

IA/P = 0.50 * * * TC = 1.5 HR * * * IA/P = 0.50

TRVL TIME (hr)	11.0	11.3	11.6	11.9	12.0	12.1	12.2	12.3	12.4	12.5	12.6	12.7	12.8	13.0	13.2	13.4	13.6	13.8	14.0	14.3	14.6	15.0	15.5	16.0	16.5	17.0	17.5	18.0	19.0	20.0	22.0	26.0
0,0	0	0	0	0	0	0	0	1	2	6	13	22	34	65	94	114	129	122	117	108	100	91	81	74	66	58	52	46	38	33	28	2
.10	0	0	0	0	0	0	0	0	0	2	5	10	18	42	73	99	116	126	121	112	104	94	84	76	68	60	53	48	39	33	28	3
.20	0	0	0	0	0	0	0	0	0	1	4	8	15	36	65	93	112	123	122	113	105	95	85	77	69	61	54	48	39	34	28	4
.30	0	0	0	0	0	0	0	0	0	0	1	3	6	20	44	72	97	114	122	118	109	99	88	79	71	63	56	50	41	34	28	5
.40	0	0	0	0	0	0	0	0	0	0	1	2	5	16	38	65	91	110	121	119	110	100	89	80	72	64	57	51	41	34	29	5
.50	0	0	0	0	0	0	0	0	0	0	0	1	4	13	33	59	85	105	118	120	112	101	90	81	73	65	57	51	41	35	29	6
.75	0	0	0	0	0	0	0	0	0	0	0	1	2	7	20	41	66	89	107	118	115	105	93	83	75	67	60	53	43	35	29	8
1.0	0	0	0	0	0	0	0	0	0	0	0	0	0	1	4	14	31	53	78	106	117	113	100	89	80	72	64	57	46	37	30	12
1.5	0	0	0	0	0	0	0	0	0	0	0	0	0	0	0	1	5	14	29	60	91	115	111	99	88	79	71	63	50	41	31	18
2.0	0	0	0	0	0	0	0	0	0	0	0	0	0	0	0	0	0	2	5	20	45	85	113	109	98	87	78	70	56	45	32	22
2.5	0	0	0	0	0	0	0	0	0	0	0	0	0	0	0	0	0	0	0	3	13	42	87	111	107	97	86	78	62	49	34	24
3.0	0	0	0	0	0	0	0	0	0	0	0	0	0	0	0	0	0	0	0	0	1	9	39	82	109	109	98	87	70	56	37	25

RAINFALL TYPE = III * * * TC = 1.5 HR * * * SHEET 9 OF 10

Exhibit 5-III: Tabular hydrograph unit discharges (csm/in) for type III rainfall distribution—continued

IA/P = 0.10 * * * TC = 2.0 HR * * * IA/P = 0.10

TRVL TIME (hr)	11.0	11.3	11.6	11.9	12.0	12.1	12.2	12.3	12.4	12.5	12.6	12.7	12.8	13.0	13.2	13.4	13.6	13.8	14.0	14.3	14.6	15.0	15.5	16.0	16.5	17.0	17.5	18.0	19.0	20.0	22.0	26.0
0.0	9	11	15	19	21	23	27	31	39	48	60	75	91	129	164	187	200	191	178	147	119	92	69	55	45	37	31	26	20	16	13	3
.10	8	10	13	17	18	20	22	25	29	36	44	55	68	101	139	170	189	197	188	163	132	101	75	59	47	39	33	28	21	17	13	3
.20	7	10	13	16	18	19	21	24	28	33	40	50	62	93	129	162	185	195	190	168	137	104	77	60	48	40	33	28	21	17	13	4
.30	7	9	12	15	17	18	20	23	26	31	37	46	56	85	120	154	179	193	191	172	142	108	80	62	49	41	34	29	21	17	13	4
.40	6	8	11	14	15	16	18	19	22	25	29	34	42	64	94	129	161	183	192	183	157	120	87	66	53	43	36	30	22	17	13	5
.50	6	8	10	13	14	16	17	19	21	23	27	32	38	58	87	121	153	177	191	185	162	124	90	68	54	44	36	31	22	18	13	5
.75	5	6	8	11	12	13	14	15	16	18	20	23	26	37	55	81	113	144	169	186	179	147	106	78	61	49	40	33	24	19	14	6
1.0	4	5	7	9	10	11	12	13	14	15	17	18	20	27	38	56	82	113	143	176	185	164	121	88	67	53	43	36	26	20	14	7
1.5	2	3	4	5	6	7	7	8	9	10	10	11	12	15	18	23	31	45	65	106	148	180	166	126	92	69	55	44	31	23	15	9
2.0	1	1	2	3	4	4	5	5	6	6	7	8	8	10	12	14	18	23	31	52	87	139	176	160	122	90	68	54	36	26	16	10
2.5	0	0	1	2	2	2	2	3	3	4	4	5	5	6	7	9	11	13	16	22	36	71	132	172	161	126	94	71	45	31	18	11
3.0	0	0	0	1	1	1	1	1	2	2	2	3	3	4	5	6	7	9	10	14	19	35	78	136	168	156	123	92	54	36	20	11

IA/P = 0.30 * * * TC = 2.0 HR * * * IA/P = 0.30

TRVL TIME (hr)	11.0	11.3	11.6	11.9	12.0	12.1	12.2	12.3	12.4	12.5	12.6	12.7	12.8	13.0	13.2	13.4	13.6	13.8	14.0	14.3	14.6	15.0	15.5	16.0	16.5	17.0	17.5	18.0	19.0	20.0	22.0	26.0
0.0	0	0	0	0	0	0	1	2	6	11	18	29	41	75	111	140	159	170	163	145	124	103	84	70	60	51	44	39	30	25	20	4
.10	0	0	0	0	0	0	0	1	2	4	9	15	24	50	84	118	145	160	167	155	134	110	89	74	63	54	46	40	31	25	20	5
.20	0	0	0	0	0	0	0	0	1	3	7	12	20	43	76	110	138	157	165	157	138	113	91	75	64	55	47	41	32	26	20	5
.30	0	0	0	0	0	0	0	0	1	3	5	10	17	38	68	101	131	152	164	159	141	116	93	77	65	56	48	41	32	26	20	6
.40	0	0	0	0	0	0	0	0	0	1	2	4	8	22	45	76	109	137	155	163	151	125	99	81	68	58	50	43	33	26	20	7
.50	0	0	0	0	0	0	0	0	0	1	1	3	7	18	39	69	101	130	151	162	153	128	101	83	69	59	51	44	34	27	20	7
.75	0	0	0	0	0	0	0	0	0	0	0	1	2	6	17	36	63	93	122	151	158	143	114	92	76	64	55	47	36	28	21	9
1.0	0	0	0	0	0	0	0	0	0	0	0	0	0	2	7	18	37	63	93	132	157	153	125	100	82	68	58	50	38	29	21	11
1.5	0	0	0	0	0	0	0	0	0	0	0	0	0	0	0	1	4	12	26	59	100	142	154	128	102	83	70	59	44	34	23	15
2.0	0	0	0	0	0	0	0	0	0	0	0	0	0	0	0	0	0	2	5	18	43	93	142	150	125	101	82	69	50	38	24	16
2.5	0	0	0	0	0	0	0	0	0	0	0	0	0	0	0	0	0	0	0	3	12	41	98	141	147	122	99	81	58	43	26	17
3.0	0	0	0	0	0	0	0	0	0	0	0	0	0	0	0	0	0	0	0	0	1	8	38	92	135	144	124	102	69	50	30	18

IA/P = 0.50 * * * TC = 2.0 HR * * * IA/P = 0.50

TRVL TIME (hr)	11.0	11.3	11.6	11.9	12.0	12.1	12.2	12.3	12.4	12.5	12.6	12.7	12.8	13.0	13.2	13.4	13.6	13.8	14.0	14.3	14.6	15.0	15.5	16.0	16.5	17.0	17.5	18.0	19.0	20.0	22.0	26.0
0.0	0	0	0	0	0	0	0	0	1	3	6	11	16	33	54	75	91	102	114	114	103	96	87	79	72	64	57	51	41	35	29	6
.10	0	0	0	0	0	0	0	0	1	2	5	9	14	29	49	70	87	99	106	113	104	97	88	80	72	65	58	52	42	35	29	6
.20	0	0	0	0	0	0	0	0	0	1	2	4	7	18	34	54	74	90	101	112	107	99	90	82	75	67	60	53	43	36	29	8
.30	0	0	0	0	0	0	0	0	0	0	1	3	6	15	30	49	69	86	98	111	108	100	91	83	75	68	60	54	43	36	29	8
.40	0	0	0	0	0	0	0	0	0	0	0	1	2	8	19	35	55	73	89	105	110	103	94	86	78	70	62	56	45	37	30	10
.50	0	0	0	0	0	0	0	0	0	0	0	1	2	6	16	31	50	69	85	102	109	104	95	86	78	71	63	56	45	37	30	11
.75	0	0	0	0	0	0	0	0	0	0	0	0	1	4	10	21	37	55	73	93	107	107	97	89	81	73	65	58	47	38	30	13
1.0	0	0	0	0	0	0	0	0	0	0	0	0	0	0	2	7	15	29	46	72	93	107	103	94	86	78	70	62	50	40	31	16
1.5	0	0	0	0	0	0	0	0	0	0	0	0	0	0	0	1	2	7	15	34	59	89	105	101	93	85	77	69	55	44	32	21
2.0	0	0	0	0	0	0	0	0	0	0	0	0	0	0	0	0	0	1	3	10	25	55	90	104	100	92	84	76	61	49	34	23
2.5	0	0	0	0	0	0	0	0	0	0	0	0	0	0	0	0	0	0	0	2	7	24	59	91	103	99	91	83	68	54	36	25
3.0	0	0	0	0	0	0	0	0	0	0	0	0	0	0	0	0	0	0	0	0	1	7	28	62	91	101	98	90	75	60	39	26

RAINFALL TYPE = III * * * TC = 2.0 HR * * * SHEET 10 OF 10

Appendix 3.B. RAINFALL DISTRIBUTION FOR THE U.S. (FROM TR-55, SCS, AND TP-40)

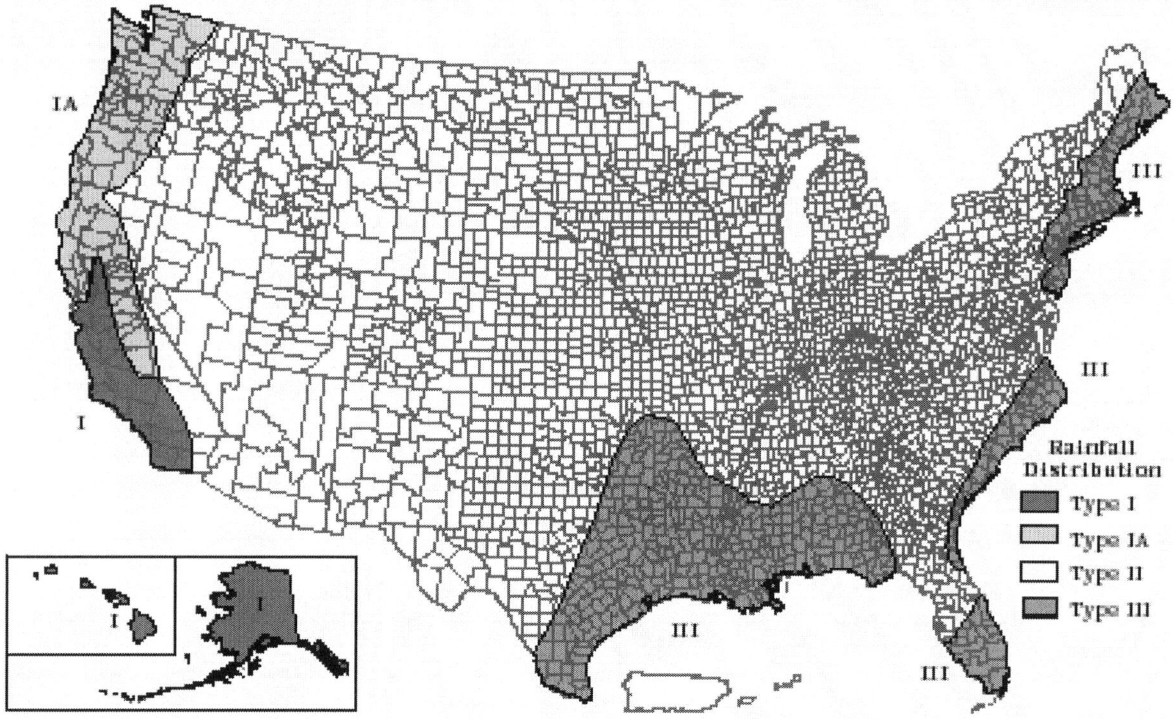

U.S. DEPARTMENT OF AGRICULTURE NATURAL RESOURCES CONSERVATION SERVICE

2-Year 24-Hour Rainfall (inches)

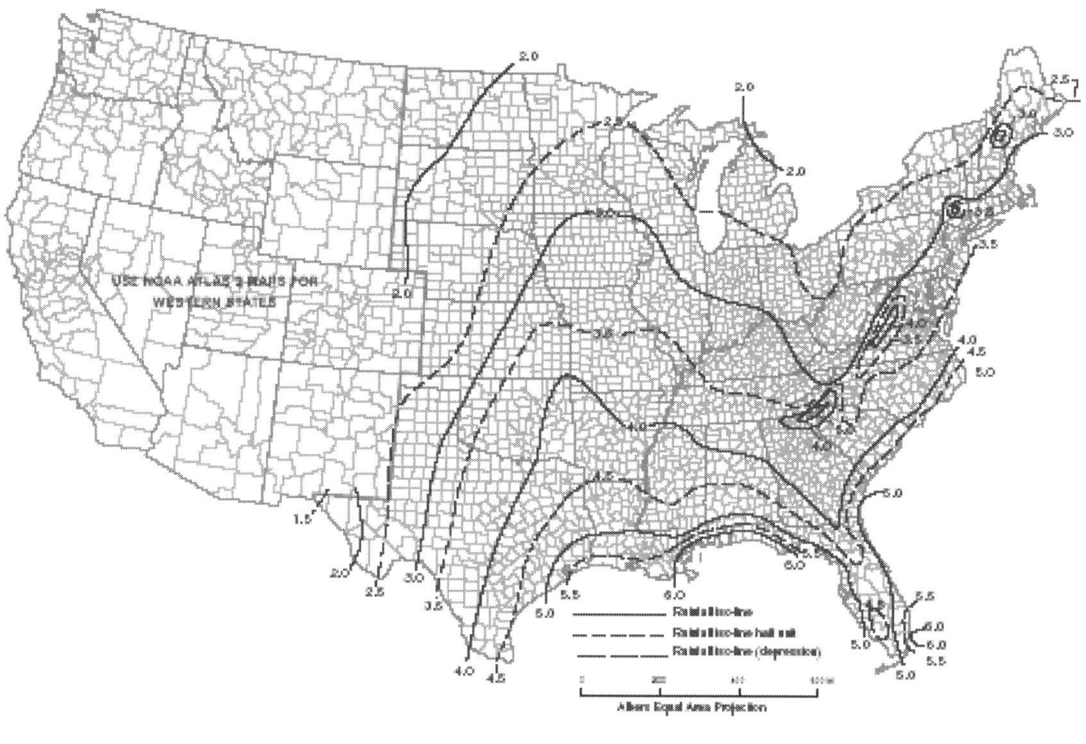

U.S. DEPARTMENT OF AGRICULTURE

NATURAL RESOURCES CONSERVATION SERVICE

5-Year 24-Hour Rainfall (inches)

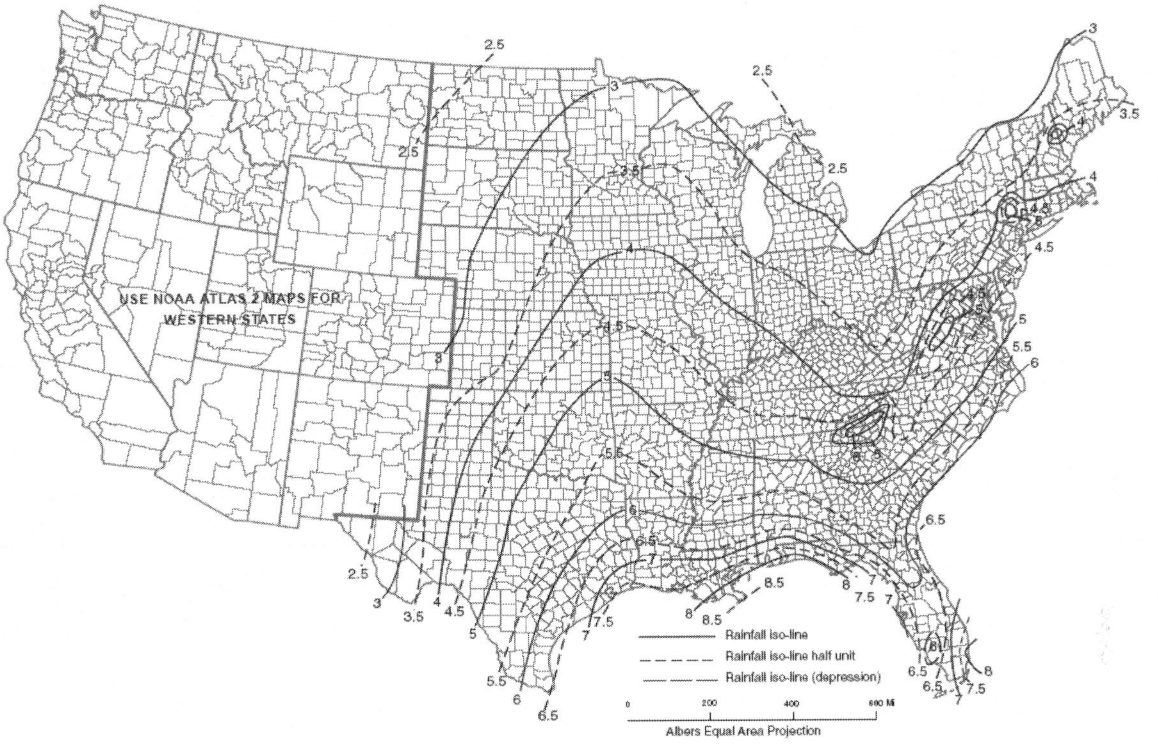

U.S. DEPARTMENT OF AGRICULTURE

NATURAL RESOURCES CONSERVATION SERVICE

10-Year 24-Hour Rainfall (inches)

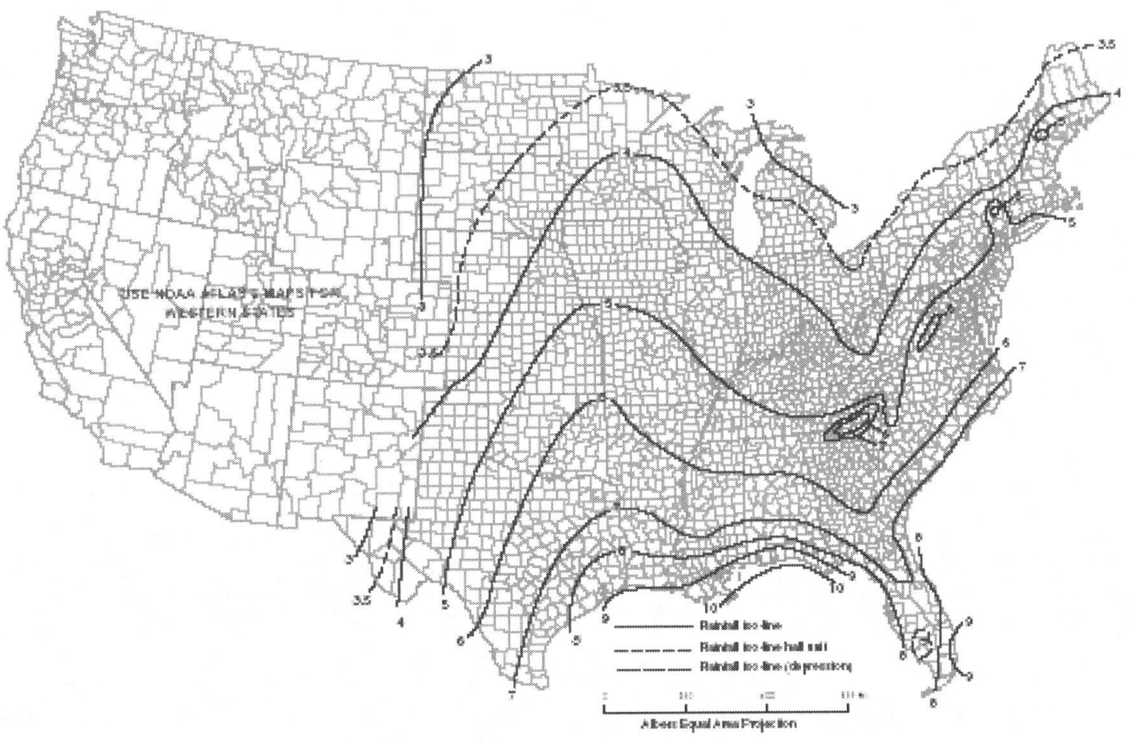

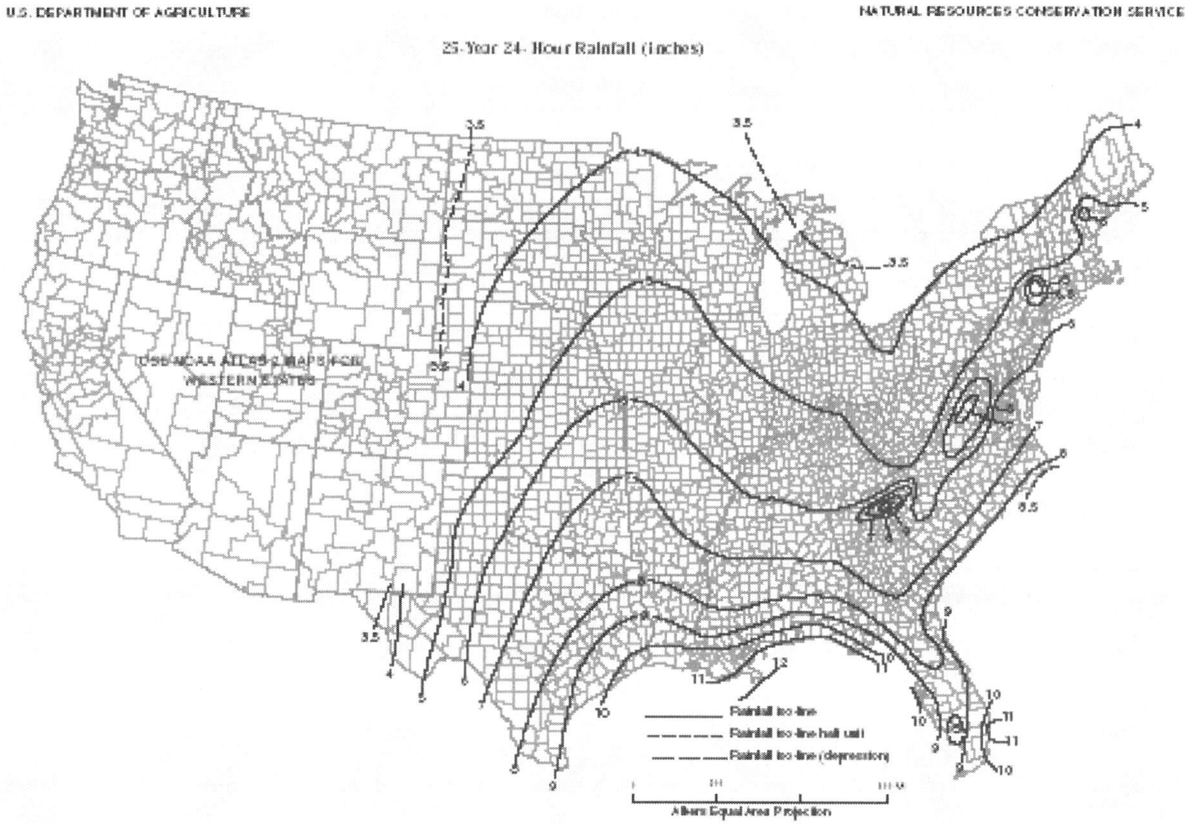
U.S. DEPARTMENT OF AGRICULTURE
NATURAL RESOURCES CONSERVATION SERVICE
25-Year 24-Hour Rainfall (inches)
USE NOAA ATLAS 2 MAPS FOR WESTERN STATES
Rainfall iso-line
Rainfall iso-line half unit
Rainfall iso-line (depression)
Albers Equal Area Projection

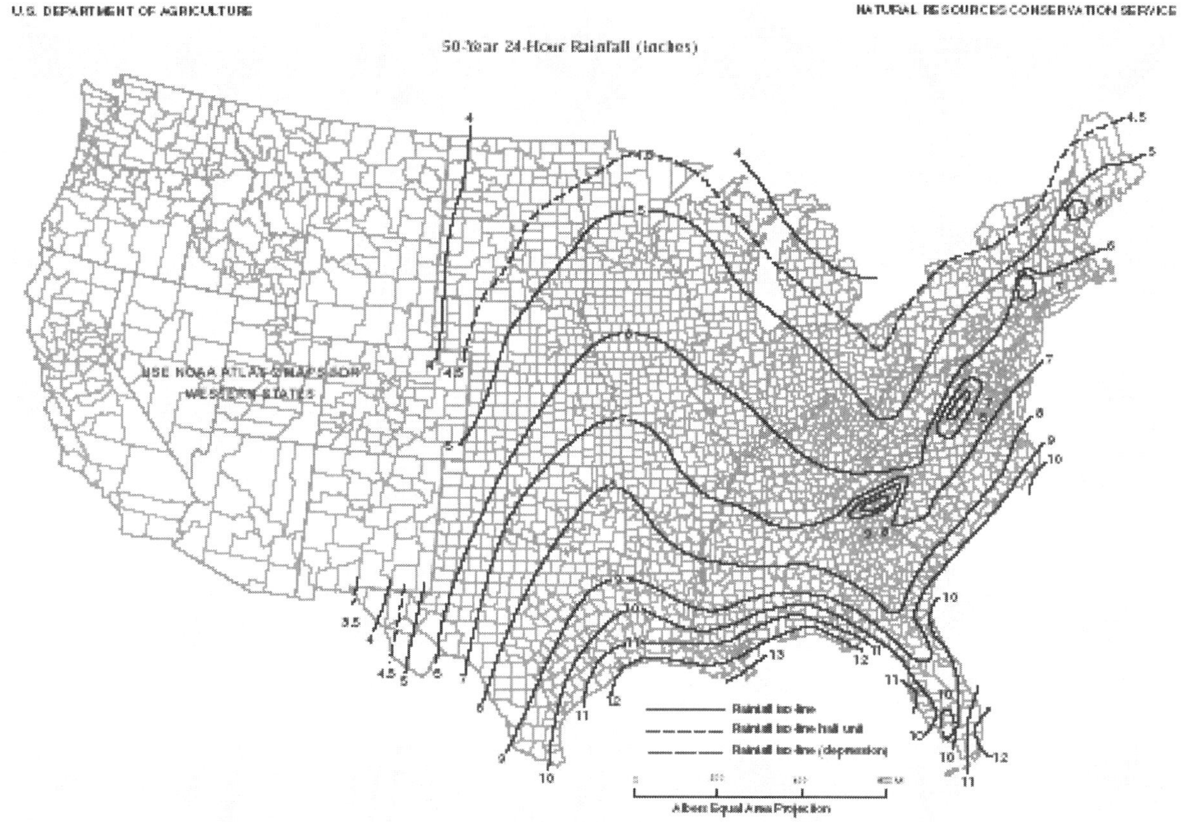
U.S. DEPARTMENT OF AGRICULTURE
NATURAL RESOURCES CONSERVATION SERVICE
50-Year 24-Hour Rainfall (inches)
USE NOAA ATLAS 2 MAPS FOR WESTERN STATES
Rainfall iso-line
Rainfall iso-line half unit
Rainfall iso-line (depression)
Albers Equal Area Projection

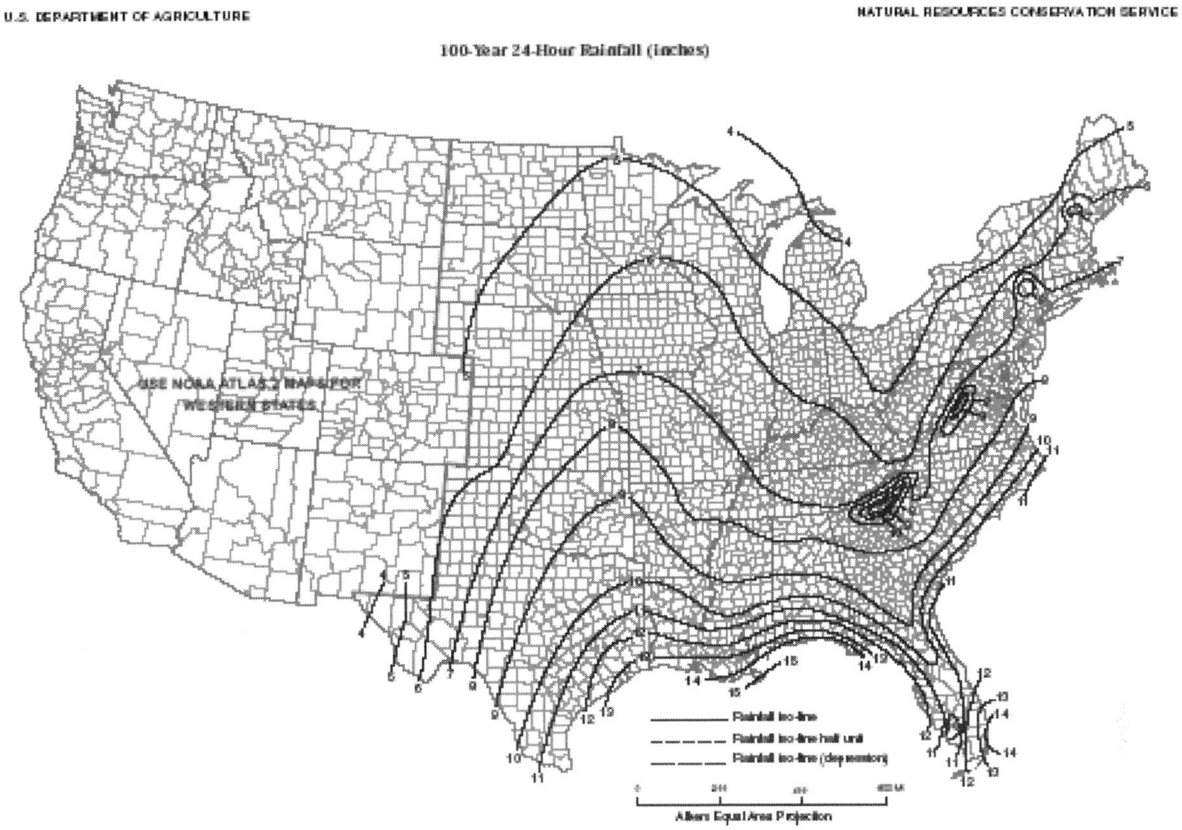
U.S. DEPARTMENT OF AGRICULTURE
NATURAL RESOURCES CONSERVATION SERVICE
100-Year 24-Hour Rainfall (inches)
USE NOAA ATLAS 2 MAPS FOR WESTERN STATES
Rainfall iso-line
Rainfall iso-line half unit
Rainfall iso-line (depression)
Albers Equal Area Projection

Erosion Mechanisms, the Revised Universal Soil Loss Equation (RUSLE), and Vegetation Erosion Controls

INTRODUCTION

KNOWLEDGE of the potential erosion problems on a construction site enables the site planner to better manage site development and erosion controls to minimize soil movement off the property. Prevention (erosion control) is much more effective than trying to improve the water quality of the runoff (sediment control). Information in this chapter enables a planner to understand basic erosion mechanisms and how they vary for different site conditions. Characteristics of construction site erosion material are highly dependent on site conditions and the local rainfall. This chapter describes how the Revised Universal Soil Loss Equation (RUSLE) can be used to predict the amount of erosion from a site, and it introduces some preventative practices to minimize site erosion. An introduction to RUSLE2 is also provided—an emerging powerful tool that should provide more powerful and accurate insights to construction site erosion problems, and their control—as it becomes more fully developed over the next several years.

BASIC EROSION MECHANISMS AND RAIN ENERGY

Soil erosion results when soil is exposed to the erosive powers of rainfall energy and flowing water (Barfield, *et al.* 1983). Rain (along with the shearing force of flowing water) acts to detach soil particles, while runoff transports the soil particles downslope. The most significant factor causing sheet erosion is raindrop impact, while the shearing force of flowing water is most important in rill and gully erosion.

Erosion Mechanisms

Soil detachment usually has been related to raindrop parameters and soil parameters (Huang, *et al.* 1982). The most important rain parameter is kinetic energy, while the most important soil parameter is shear strength. Soil detachment occurs when rain energy overcomes the soil's shear strength. This is why the use of surface mulches over bare soils can greatly decrease the transfer of energy to the soil, thereby lessening erosion losses.

When a raindrop strikes a surface, pressure acts to destabilize the particles. The raindrop impact loading function is very different from a uniform loading function (Huang, *et al.* 1982). The initial loading magnitudes are very high, but diminish very rapidly. These loadings are also not uniform and are concentrated at the edge of the contact area. When the drop strikes a surface, lateral jet streams impinge on adjacent irregular surfaces or dirt particles, as shown on Figure 4.1, further destabilizing the surrounding area (Springer, 1976). It is very difficult to model the specific drop impact forces due to these irregularities and simple approximations are usually used.

Kinnell (1981) defines two forms of raindrop kinetic energy, the rate of expenditure of energy per unit time (Err, in units of energy per area per time) and the amount of rainfall kinetic energy expended per unit quantity of rain (*Era*, in units of energy per area per rain depth). Based on typical drop sizes of about 1.5 mm, known drop populations (see Figures 4.2 and 4.3), and a terminal velocity of about 5.5 m/sec, it can be calculated that each drop contains about 3×10^{-4} joules of kinetic energy (Springer, 1976). A 3 mm per hour rain delivers about 11 joules per m^2 per minute (*Err*), while a 12 mm per hour rain delivers about 30 joules per m^2 per minute. *Err* and *Era* are related:

$$Era = Err\ (I)^{-1}$$

where I is the rain intensity. The Universal Soil Loss Equation (USLE) (Wischmeier and Smith, 1965) uses a similar equation to predict rain energy.

THE REVISED UNIVERSAL SOIL LOSS EQUATION (RUSLE) AND RELATING RAIN ENERGY TO EROSION YIELD

The Universal Soil Loss Equation (USLE) (Wischmeier and Smith, 1965) was based on many years of data from about 10,000 small test plots from throughout the U.S. Most

Various Erosion Mechanisms Found at Construction Sites

Small-scale sheet erosion on tops, rill erosion forms further downslope, and finally deposition zones, on a material stockpile at a construction site.

Sheet flows forming concentrated flows which will eventually form rill and possibly gully erosion.

Large-scale sheet and rill erosion and isolated gully erosion beginning to start at an inadequately protected construction site.

Extensive gully erosion on unprotected steeper slopes of detention pond.

Sheet and rill erosion on hillside.

(continued)

Various Erosion Mechanisms Found at Construction Sites *(continued)*

Several inches of material have been eroded by sheetflows at this construction site.

Gully erosion beginning to form where concentrated flows form after sheetflows.

Large gully from concentrated flow (Bill Morton photo).

Gully erosion where concentrated flows formed, and down gradient deposition area in seeded construction area.

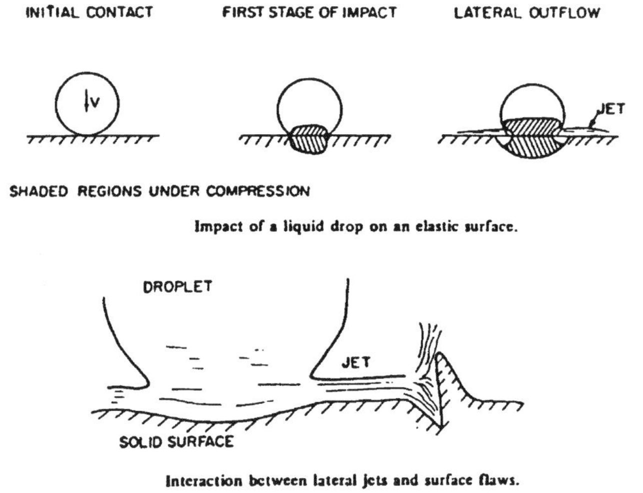

Figure 4.1. Raindrop impact with ground surface (from Springer, 1976. Adapted with permission. *Erosion by Liquid Impact.* 1976. ©V.H. Winston & Son. Inc., 360 South Ocean Boulevard, Palm Beach, FL 33480. All rights reserved).

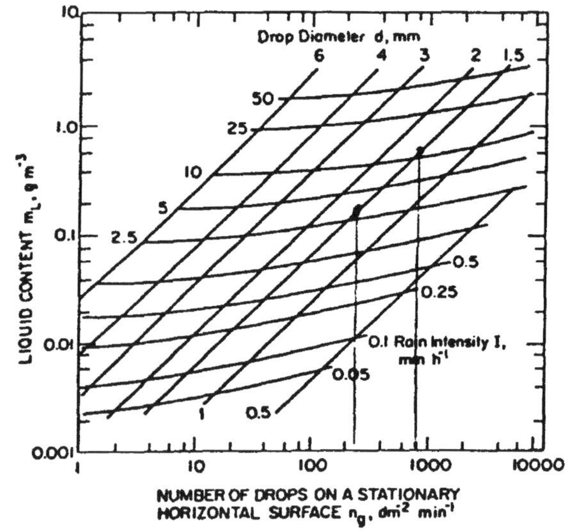

Figure 4.3. Characteristics of an idealized natural rain consisting of constant diameter spherical droplets distributed uniformly in air (from Springer, 1976. Adapted with permission. *Erosion by Liquid Impact.* 1976. ©V.H. Winston & Son. Inc., 360 South Ocean Boulevard, Palm Beach, FL 33480. All rights reserved).

test plots had approximately 22-m flow lengths at 9% slopes. All were operated in a similar manner, allowing the soil loss measurements to be combined into a predictive tool. The USLE has been extensively used for conservation planning in agricultural operations for decades. Many of the features, and the original database, also allow it to be used to predict erosion losses, and the benefits of some erosion controls, at construction sites. The RUSLE only predicts sheet and rill erosion; it does not predict the effects of concentrated runoff and gully formations.

The Revised Universal Soil Loss Equation (RUSLE) (Renard, *et al.* 1987) was developed to incorporate new research since the earlier USLE publication in 1978 (Wischmeier and Smith, 1978). The basic form of the equation has remained the same, but modifications in several of the factors have been made. There are many sources of information for the RUSLE, including the USDA's National Sedimentation Laboratory where extensive information can be obtained (http://www.sedlab.olemiss.edu/rusle/). The RUSLE document (Renard, *et al.* 1987) and the material on this referenced web site should be consulted for much greater detail on RUSLE than can be given in this chapter. This chapter focuses on construction site erosion issues and is greatly simplified compared to the complete RUSLE that stresses agricultural operations, but does periodically refer to construction site issues.

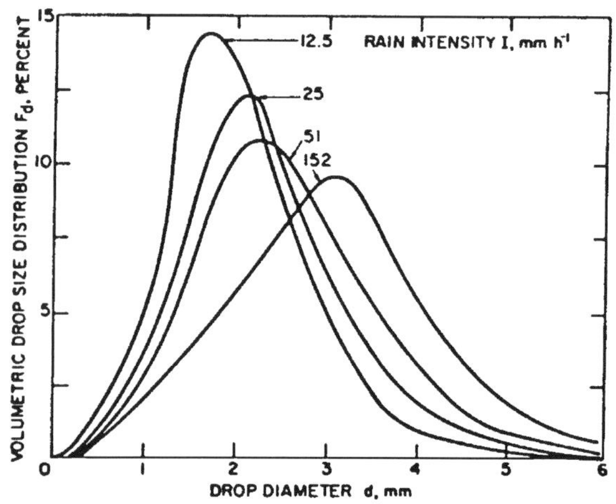

Figure 4.2. Typical rain drop size distribution (from Springer, 1976. Adapted with permission. *Erosion by Liquid Impact.* 1976. ©V.H. Winston & Son. Inc., 360 South Ocean Boulevard, Palm Beach, FL 33480. All rights reserved).

The underlying assumption in the RUSLE is that detachment and deposition are controlled by the sediment content of the flow. The erosion material is not source limited, but the erosion is limited by the carrying capacity of the flow for sediment. When the sediment load reaches the carrying capacity of the flow, no further sediment can be carried along by the flow. Sedimentation must also occur during the receding portion of the hydrograph as the flow rate decreases (Novotny and Chesters, 1981).

The RUSLE relates the rate of erosion per unit area (A) to the erosive power of the rain (R), the soil erodibility (K), the land slope and length (LS), the degree of soil cover (C), and conservation practices (P):

$$A = (R)(K)(LS)(C)(P)$$

The important aspect of this equation to note is the linear relationship between the equation parameters. As any parameter is changed, the resulting erosion yield is similarly changed. Also, the default values for LS, C, and P are all 1.0. They are changed by the planner as specific site and management conditions change. Many of these factors will change seasonally, especially those corresponding to plant

TABLE 4.1. Conversion Factors to Estimate Volume of Eroded Material.

Soil Texture Class	Conversion Factor to Convert tons to cubic yards
Sands, loamy sands, sand loam	0.70
Sand clay loam, silt loams, loams, and silty clay	0.87
Clay loams, sandy clays, and silty clays	1

growth and those affected by changes in rain and temperature characteristics. A modified version of RUSLE, RUSLE2, is currently being developed. It will incorporate many of these seasonal changes. Some of these (especially the seasonal variations in rainfall erosivity) can be considered in RUSLE (see later description of RUSLE2).

In this chapter, this equation is used to predict the amount of soil that may be eroded from construction sites. Specifically, it enables the most critical source areas to be identified, and allows predictions of the benefits of basic mulching and seedbed controls. Also, the erodibility of different slope and timing options can be compared for better preventive design. In addition, RUSLE can be used to predict the amount of sediment that may enter a sediment pond. Table 4.1 includes conversion factors that can be used to predict the volume of sediment given the weight of sediment generated, according to the RUSLE calculations. As an example, if a site is predicted to erode about 450 tons of silty-clay soil, the associated volume in cubic yards, is about 0.87 times this amount, or about 390 cubic yards of material.

Rainfall Energy (*R*)

The RUSLE implies that rain energy is directly related to erosion yield. Originally, the USLE was used with an annual R value to predict annual erosion yields, but Barfield, *et al.* (1983) summarizes several procedures and studies that have demonstrated relationships between individual storm energies and erosion yields. Therefore, the example rain energy calculations in the following subsections are used to directly relate the probabilities of individual rain events to approximate erosion yields.

Wischmeier (1959) found that the best predictor of R was:

$$R = \frac{1}{n}\sum_{j=1}^{n}\left[\sum_{k=1}^{m}(E)(I_{30})_k\right]$$

where E is the total storm kinetic energy in hundreds of ft-tons per acre, I_{30} is the maximum 30-minute rainfall intensity, j is the counter for each year used to produce the average, k is the counter for the number of storms in a year, m is the number of storms each year, and n is the number of years used to obtain the average R.

The calculated erosion potential for an individual storm is usually designated EI. The total annual R is therefore the sum of the individual EI values for each rain in the year.

Wischmeier also found that the rain kinetic energy (E) could be predicted by:

$$E = 916 + (331)\log_{10}(I), \text{ in ft-tons/acre per inch or rain}$$

where I is the average rain intensity. E is given in ft-tons per acre per inch of rain, if intensities in inches per hour are used (for up to 3 in/hr). Hence, the rain energy (and R parameter) is dependent only on rain intensities. Table 4.2 shows the calculated kinetic energy per inch of rain for different rain intensities (calculated using this equation). As an example, a rain having an average intensity of 0.37 in/hr would have a calculated kinetic energy of 773 ft-tons per acre of land per inch of rain. The maximum calculated kinetic energy using this equation is 1074 ft-tons/acre/in. It would be applied to rain intensities of 3.0 inches/hr and greater. This equation has been used to calculate the R values for the maps in RUSLE (Renard, *et al.* 1987). However, Renard, *et al.* (1987) recommend the following equation for all future R calculations:

$$E = 1099\,[1 - 0.72\exp(-1.27I)],$$
also in ft-tons/acre per inch of rain

TABLE 4.2. Rainfall Energy for Different Rain Intensities (ft-tons/acre-inch).

Intensity (in/hr)	0.00	0.01	002	003	0.04	0.05	0.06	0.07	0.08	0.09
0	0	254	354	412	453	485	512	534	553	570
0.1	585	599	611	623	633	643	653	661	669	677
0.2	685	692	698	705	711	717	722	728	733	738
0.3	743	748	752	757	761	765	769	773	777	781
0.4	784	788	791	795	798	801	804	807	810	814
0.5	816	819	822	825	827	830	833	835	838	840
0.6	843	845	847	850	852	854	856	858	861	863
0.7	865	867	869	871	873	875	877	878	880	882
0.8	884	886	887	889	891	893	894	896	898	899
0.9	901	902	904	906	907	909	910	912	913	915
1.0	916	930	942	954	964	974	984	992	1000	1008
2.0	1016	1023	1029	1036	1042	1048	1053	1059	1064	1069
3.0	1074*	1074	1074	1074	1074	1074	1074	1074	1074	1074

*1074 ft-tons/acre/inch is the maximum value and is applied for all intensities greater than 3.0 inches per hour of rain.

TABLE 4.3. Procedure for Calculating Kinetic Energy using a Rain Gage Record (Wischmeier and Smith, 1978).

Rain Gage Chart Readings		Storm Increments			Kinetic energy	
Time	Accumulative Depth (inches)	Duration (minutes)	Amount (inches)	Intensity (in/hr)	Per Inch (ft-tons per acre per inch of rain)	For Increment (ft-tons per acre)
4:00	0					
4:20	0.05	20	0.05	0.15	643	32
4:27	0.12	7	0.07	0.60	843	59
4:36	0.35	9	0.23	1.53	977	225
4:50	1.05	14	0.70	3.00	1074	752
4:57	1.20	7	0.15	1.29	953	143
5:05	1.25	8	0.05	0.38	777	39
5:15	1.25	10	0	0	0	0
5:30	1.30	15	0.05	0.20	685	34
Totals	1.30	90	1.30			1284

They found less than a 1% difference in *EI* for example storms calculated using the two methods. The largest difference was for less intense events where little erosion occurs.

Wischmeier and Smith (1978) present an example for calculating the rainfall kinetic energy from a rain gauge record, as illustrated in Table 4.3. In this example, the total kinetic energy of the storm equals 1284 ft-tons per acre, or 12.84 hundreds of ft-tons per acre. The maximum 30 minute rainfall during this 90-minute storm was 1.08 inches, occurring from 4:27 to 4:57. The corresponding I_{30} was therefore 2.16 inches per hour. The *EI* for this storm is calculated as (2.16)(12.84) = 27.7. (*Note:* If the storm duration is less than 30 minutes, the I_{30} used is twice the total rain depth, with a maximum used I_{30} value of 2.5 in/hr.).

Figures 4.4 through 4.7 (the isoerodent maps) presents values of *R* for the eastern U.S. and the western states. The USDA's National Sedimentation Laboratory (at http://www.sedlab. olemiss. edu/rusle/) contains extensive information on RUSLE and rainfall erosivity. The values shown in this figure were averaged from 20 to 25 years of data. The break between individual rains was defined as 6 hours, or more, having less than 0.5 inches of rain. Rains of less than 0.5 inches, separated from other showers by 6 hours, or more, were omitted from the calculation, unless the maximum 15-minute intensity was greater than 0.95 in/hr. Also, the maximum I_{30} value used in the calculations was 2.5 in/hr.

Locations in the southeast experience very high values of *R*, compared to other U.S. locations. As an example, the lowest values in Alabama are found in the northern part of the state, with *R* values of about 300. Most of the state has *R* values between 300 and 400, while values greater than 600 are shown for Mobile and Baldwin counties. Only the southern tip of Louisiana has a larger value of *R* in the continental U.S. (slightly more than 700).

Example: How Do the Rainfall Patterns Affect Erosion Control Strategies?

There can be large year-to-year variations in the annual *R* values and individual storms may be responsible for large fractions of the annual rain energy. Table 4.4 presents measured probabilities of the annual *R* values for three Alabama locations (Wischmeier and Smith, 1978). The 50 percent probability values are the values plotted in Figure 4.4. Table 4.5 shows the frequency of expected magnitudes of the calculated single-storm erosion index (*EI*) values. For example, there is a 5% chance that a single storm in any year could cause about half of the total annual erosion in the Birmingham and Montgomery areas (annual *R* values between 350 and 400), and about 30% of the total annual erosion in Mobile (annual *R* values between 600 and 650). The typical worst storm in any one year may cause about 15 to 20% of the total annual erosion in any of these cities.

As was discussed in Chapter 3, rainfall is distributed unevenly throughout the year in a single location, resulting in

TABLE 4.4. Probabilities of Annual *R* Values for the Calculation Period for Alabama Locations (Wischmeier and Smith, 1978).

	Observed 22-year Range	50 Percent Probability	20 Percent Probability	5 Percent Probability
Birmingham	179–601	354	461	592
Mobile	279–925	673	799	940
Montgomery	164–780	359	482	638

TABLE 4.5. Probabilities of Individual Storm Erosion Index (*EI*) Values for Alabama Locations (Wischmeier and Smith, 1978).

	Probability of Single Storm Exceeding EI Value in Any One Year:				
	100%	50%	20%	10%	5%
Birmingham	54	77	110	140	170
Mobile	97	122	151	172	194
Montgomery	62	86	118	145	172

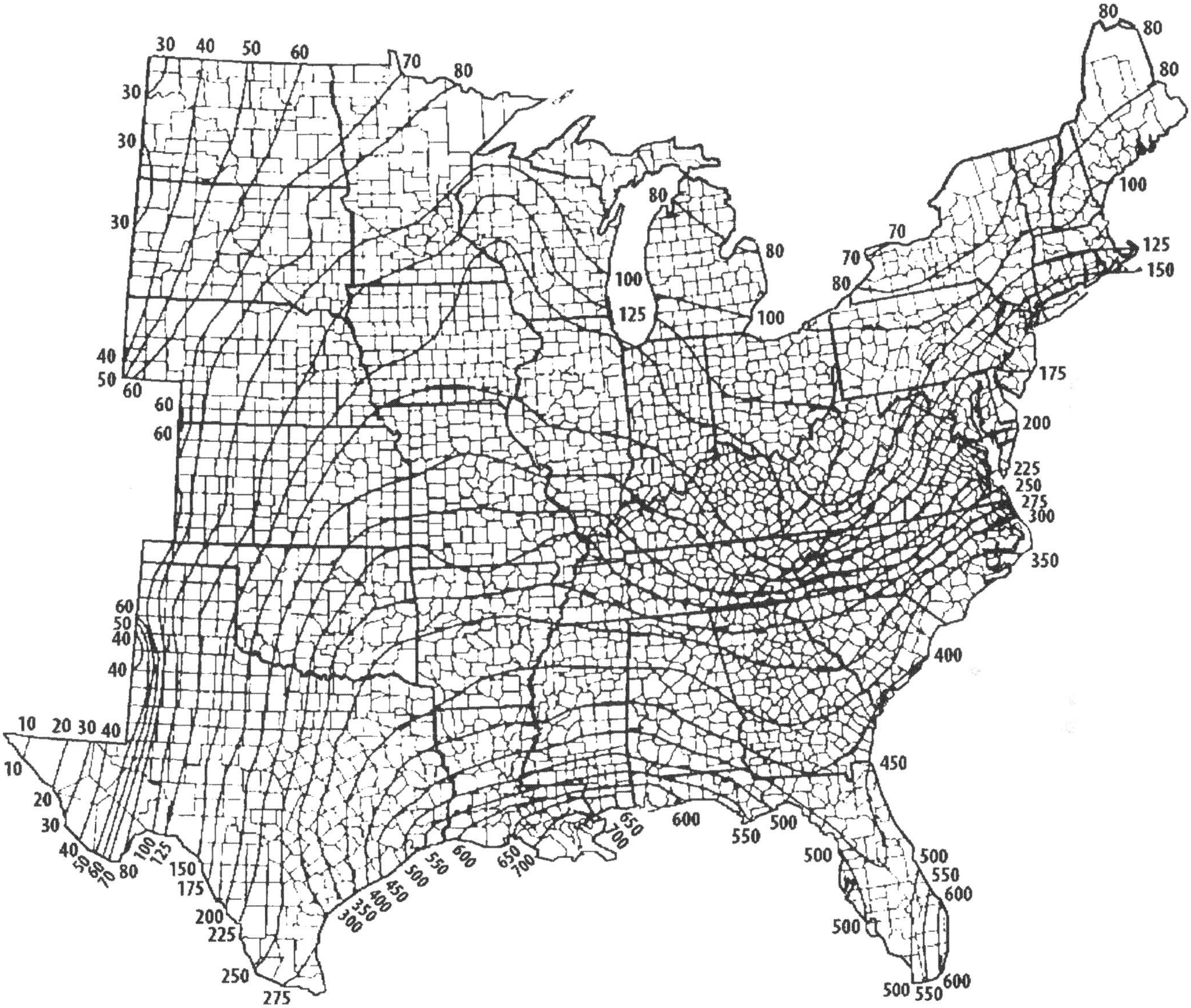

Figure 4.4. Isoerodent map of the Eastern U.S. (EPA, 2001).

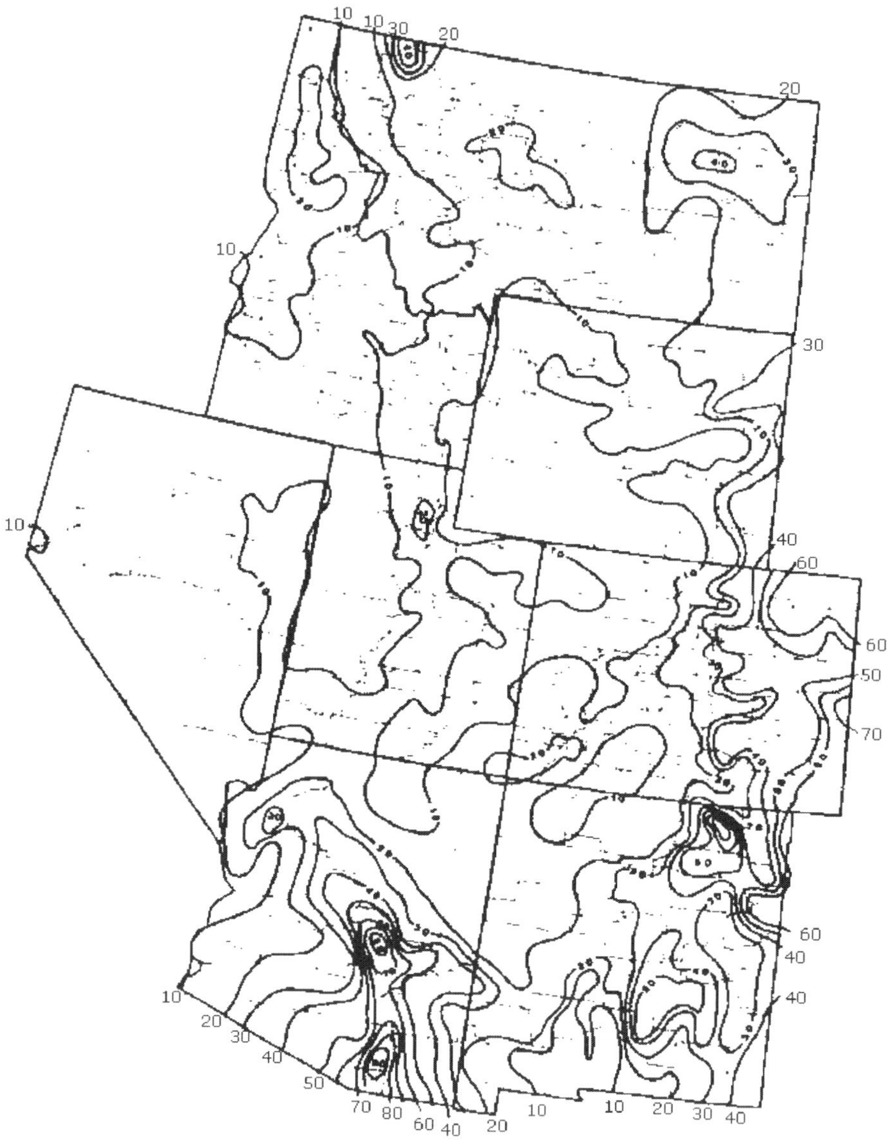

Figure 4.5. Isoerodent map of the Western U.S. (EPA, 2001).

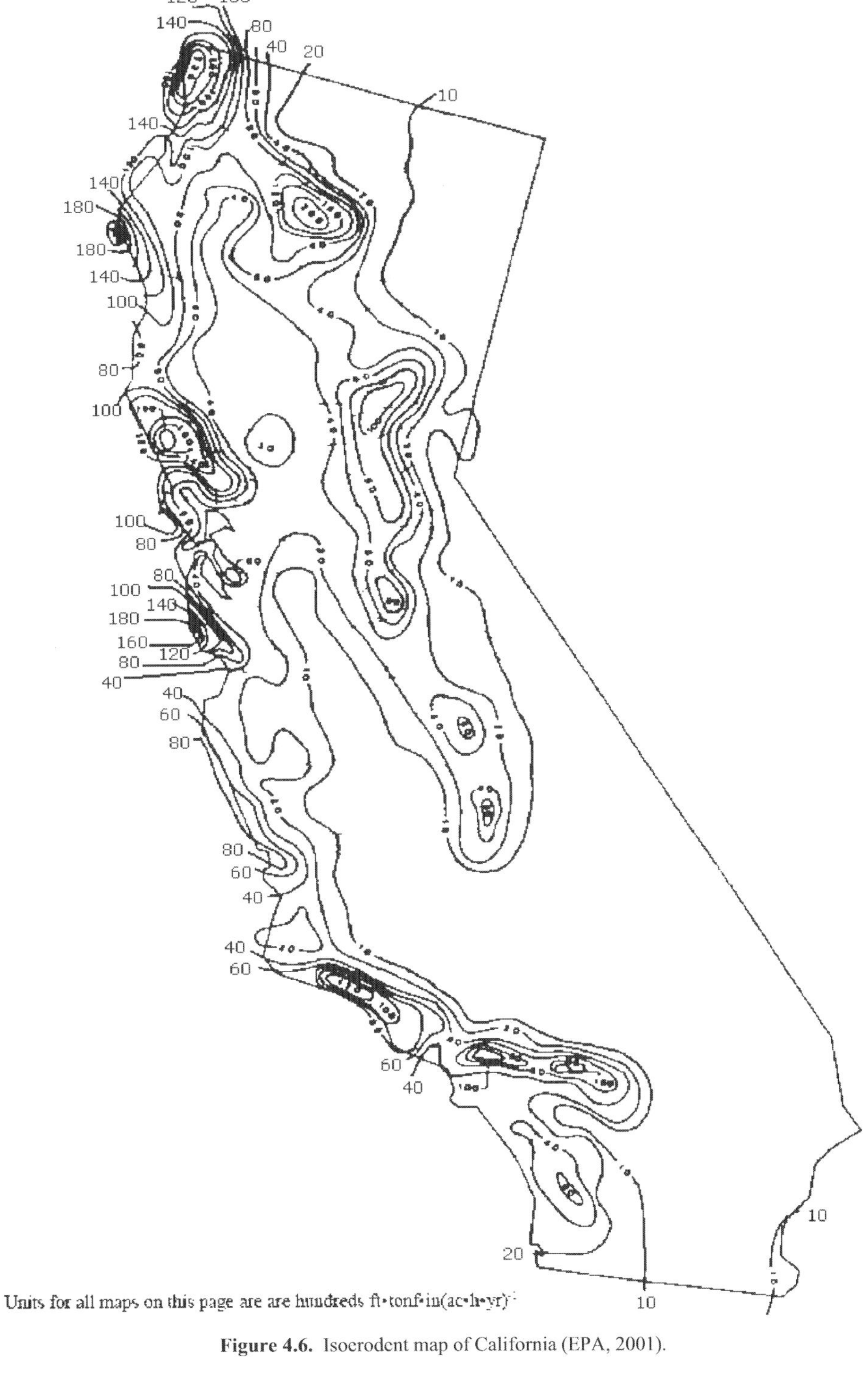

Figure 4.6. Isoerodent map of California (EPA, 2001).

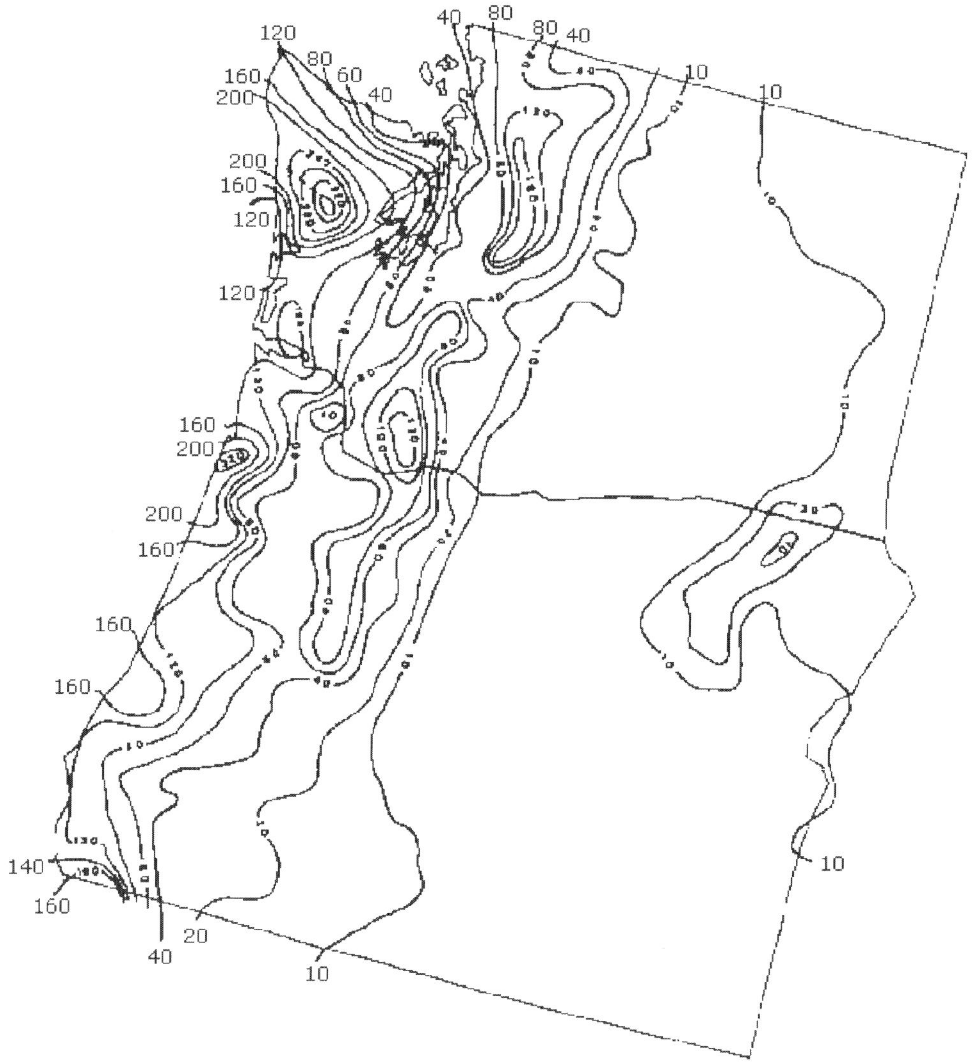

Note: Units for all maps on this page are are hundreds ft*tonf*in(ac*h*yr)^-1

Figure 4.7. Isoerodent map of Oregon and Washington U.S. (EPA, 2001).

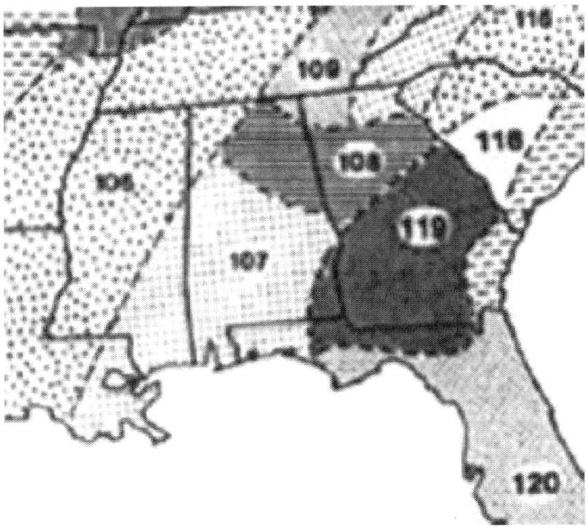

Figure 4.8. Rainfall erosion index zones for southeastern U.S. (Renard, *et al.* 1987).

TABLE 4.6. Distribution of the Erosivity Index Values for Different Time Periods throughout the Year for Index Zones in the Southeast.

Period	106	107	108	109	119
Jan 1–15	3	3	3	3	1
Jan 16–31	3	2	3	3	1
Feb 1–15	3	2	3	4	2
Feb 16–29	4	3	3	3	2
Mar 1–15	4	4	4	3	1
Mar 16–31	4	4	4	3	2
Apr 1–15	6	5	4	4	3
Apr 16–30	6	4	4	3	3
May 1–15	5	4	5	3	3
May 16–31	6	4	5	4	5
Jun 1–15	5	4	5	6	8
Jun 16–30	6	6	7	8	9
Jul 1–15	6	8	9	11	5
Jul 16–31	6	7	10	10	9
Aug 1–15	4	7	6	7	6
Aug 16–31	4	7	5	5	9
Sep 1–15	3	6	4	3	6
Sep 16–31	3	4	3	3	10
Oct 1–15	3	2	3	2	4
Oct 16–31	2	2	2	2	4
Nov 1–15	4	2	2	2	1
Nov 16–31	4	3	2	3	1
Dec 1–15	3	2	2	2	1
Dec 16–31	3	5	2	3	1

Source: EPA's Construction Rainfall Erosivity Waiver, Fact Sheet 3.1. EPA 833-F-00-014. Jan, 2001.

an uneven distribution of the rainfall energy. For that reason, the U.S. has been divided into rainfall erosion index zones. In each zone, the distribution of R in a year (i.e., the percentage of R that can be associated with any specific range of dates) is similar. Figure 4.8 shows the rainfall erosion index values for the southeast. Appendix 4A includes the erosion index map and associated tables for the entire country. The USDA's National Sedimentation Laboratory webpage (at http://www.sedlab.olemiss.edu/rusle/) also catalogs this data. In Alabama, there are five regions, although most of the state is in regions 107 or 108. In contrast, North Dakota has one zone. These regions are used to predict the fraction of the annual R that occurs in 2-week increments throughout the year. Incremental R information is useful for planning relatively rapid, but sensitive, construction practices, and to see if a potential project may be eligible for the possible "$R \leq 5$ total" exemption rule. Table 4.6 lists these distribution values for R for these areas in the state, while Appendix 4A includes the values for all regions of the U.S.

The values in Table 4.6 are the percentage of the total annual R values that occur in each 2 week period. If the R of ≤ 5 waiver will be available in Alabama and much of the southeast U.S., only the very rare construction activity may be eligible. Only small portions of region 119 may possibly qualify (if the annual $R<500$) and if the construction activity could be completed within a 2-week period during November, December, or January. The erosivity index values range from lows of 1% to a high of 11% per two week period. Periods greater than the average of 4.1% indicate periods when higher amounts of erosion than the overall average may occur. Depending on location, these periods are generally from the first of April through August, or September. Periods with the lowest erosion potentials are in the fall, winter and early spring. In contrast, construction in North Dakota could feasibly last for two-to-three months and still meet the "$R \leq 5$" waiver, depending on the selected construction period. The same construction plan would not meet the waiver requirement if construction occurred during the times when North Dakota typically receives its "heaviest" rains.

As indicated above, a relatively few rains can contribute much more of the annual rainfall energy than most, with the more intense rains contributing greater erosion losses per inch of runoff than the less intense rains. As an example, the most important single rain in the Birmingham area that may occur in any one year has an R value of about 54, and therefore contributes about 15 percent of the annual erosion losses. The most important single rain that may occur once every ten years has an R value of about 140 and may therefore contribute about 40 percent of the annual erosion losses for that year. This ten year rain would only contribute about four percent of the average ten year total erosion losses in any one year, however.

An analysis was conducted using the typical 1977 Birmingham rains to determine the distributions of erosion factors for individual rains and their recurrence intervals. This year was selected due to its similarity to the long-term average rain conditions (based on total annual rain depth and the distribution of the rains throughout the year). Most of the erosion is produced by a relatively few highly-erosive rains that may occur during any month. About 50 percent of the annual erosion yield is associated with only 11 individual rains (out of 96 that occurred in 1977). Approximately 40

TABLE 4.7. Probabilities of Highly Erosive Rains Occurring During Different Time Periods (Birmingham 1977 data).

Percentage of Annual Erosion Yield During Event	Estimated Erosion Yield During Single Event (with some site controls) (lb/acre)	Probability of Event Occurring at Least Once per:		
		7 days	14 days	30 days
7%	3,500	3%	6%	12%
5	3,000	8	16	31
3	1,800	17	31	55
2	1,200	29	50	77
1	600	45	70	92
Probable number of events per time period (out of 96):		2	4	8
Probable total erosion yield per time period (lb/acre):		1,200	2,300	5,000

percent of the individual rains were responsible for more than 90 percent of the annual erosion yield, and about 25 percent of the rains were responsible for about 75 percent of the annual erosion yield.

The probabilities of different highly erosive rains occurring during 7-, 14-, and 30-day periods for Birmingham 1977 conditions were calculated. Table 4.7 indicates these probabilities and the expected erosion yields for these time periods. Most erosion-protection regulations require disturbed areas inactive for more than 14 days to have suitable site erosion controls. During a 14-day period of time, more than a ton of sediment could be washed from each disturbed acre during four separate rain events. There is a 30 percent chance that the same amount of sediment could be washed from the site during a single event during this time period. If this time period was lengthened, the amount of sediment that could be lost and the probability of highly-erosive rains occurring would increase proportionately. Because of these potential significant sediment losses, most regulations also require appropriate downslope controls to capture any sediment that may move from uncontrolled disturbed areas on the site. However, downslope controls are not adequate by themselves in controlling all sediment during highly erosive rains. The onsite protection offered by mulching of inactive disturbed areas (in addition to the diversion of waters from upslope offsite areas) greatly lessens the burden on the downslope controls and allows them to remain useful during severe (but common) rains.

Soil Erodibility Factor (K)

Soil texture, and other soil characteristics, affect the soils susceptibility to erosion. The soil K factors were determined experimentally in test plots that were 73-ft (22-m) long and had a uniform slope of 9%. Normally, more than 10 years of runoff plot data was needed to determine these values in order to eliminate any effects from prior organic material and mulch, as well as effects associated with mechanical disturbance from constructing the plots. Figure 4.9 is the nomograph used to determine the K factor for a soil, based on its texture (% silt plus very fine sand, %sand, %organic matter), structure, and permeability. The NRCS county soil maps list the K factors for all soils in each county. However, significant disturbance and modifications of the soil obviously occurs at construction sites and care needs to be taken to ensure that the K factor used in the calculations is based on the actual surface soil conditions. As an example, the organic matter (decreases as the top soils are removed), permeability (decreases with compaction with heavy equipment), and soil structure (subsurface soils more massive than surface soils) could all likely change, causing the K factor to increase for a soil undergoing modification at a construction site.

Soil Classifications

The designation for a sand or clay is given in the *Unified Soil Classification System,* ASTM D 2487. Sandy soils, by definition, must have more than half of the material be larger than the No. 200 sieve, and more than half of that fraction must be smaller than the No. 4 sieve. Similarly, for clayey soils, more than half of the material is required to be smaller than the No. 200 sieve. Silt soils are intermediate between sands and clays in their size. Figure 4.10 is the standard soil texture triangle defining the different soil texture categories and Table 4.8 shows the standard USDA particle size ranges for the different soil texture categories.

Silt particles are barely visible to the naked eye and have many properties that fall between the values for sand and clay are intermediate in many properties between sand and clay. Silt is characterized by its plasticity and stickiness.

TABLE 4.8. USDA Particle Size Ranges for Different Soil Texture Categories.

Soil Particle	Size Range		
	micrometers	millimeters	inches
Cobble	150,000 to 300,000	150 to 300 mm	6 to 12 in.
Gravel	2,000 to 150,000	2 to 150	0.08 to 6
Sand*	50 to 2,000	0.05 to 2.00	0.002 to 0.08
Silt	2 to 50	0.002 to 0.05	0.00008 to 0.002
Clay	<2	<0.002	<0.00008

*"very fine sand" is in the 50 to 100 μm range

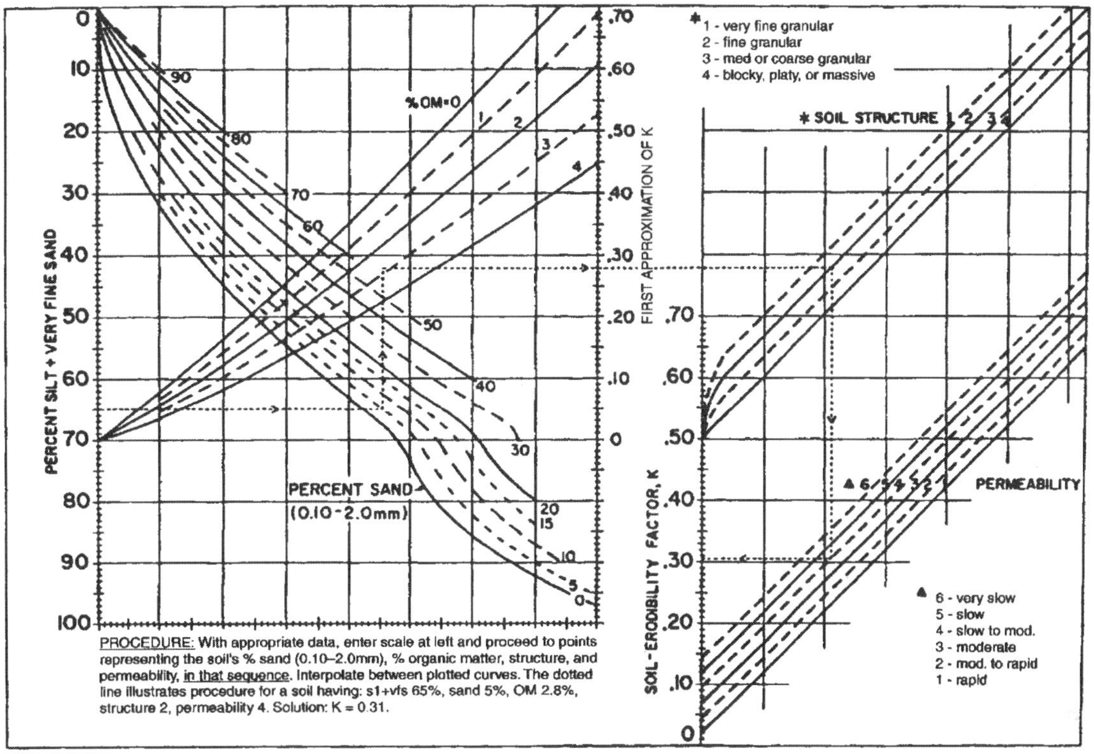

Figure 4.9. USDA nomograph used to calculate soil erodibility (*K*) factor.

According to the USDA (1993), the silt content is an important characteristic for determining erodibility because silt-sized particles are easily detached and transported in runoff. The small particle size also makes silt difficult to capture in sediment traps or basins. There are two major types of clays found in the natural environment—kaolinite and montmorillonite. Kaolinite is relatively inactive and fairly stable. Montmorillonite is a very active clay that shrinks when dry and swells when wet. These characteristics affect the permeability of soils and are very important to their use and management. Clayey soils retain water that should be available for plant growth, but these soils are often dense, hard, wet, airtight, acidic, and infertile. They can restrict root growth even though other factors are favorable.

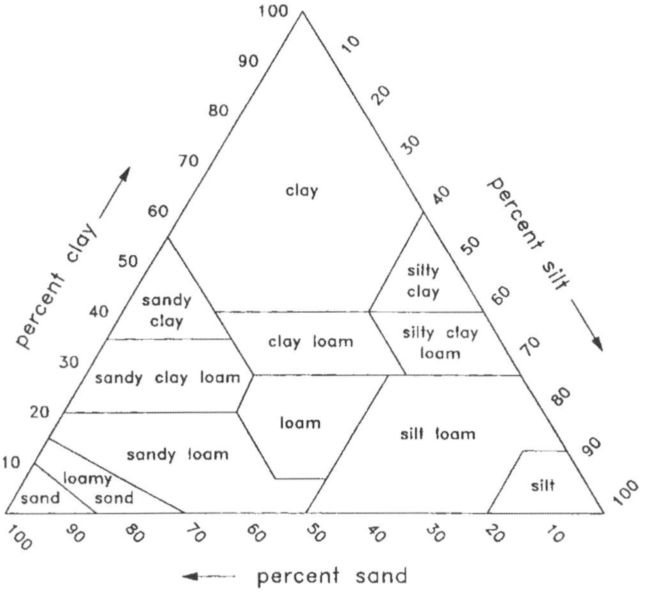

Figure 4.10. Standard USDA soil texture triangle.

The AASHTO system classifies soils according to the properties that affect roadway construction and maintenance. The fraction of a mineral soil that is less than 3 inches in diameter is classified in one of seven groups from A-1 through A-7 on the basis of grain-size distribution, liquid limit, and plasticity index. Soils in group A-1 are coarse grained and low in silt and clay. Soils in group A-7 are fine grained. Highly organic soils are in Group A-8 and are classified on the basis of visual inspection.

Problem: An Evaluation of Soil Conditions Affecting Construction Site Erosion Problems

The Alabama Soil and Water Conservation Committee produced the *Alabama Handbook for Erosion Control, Sediment Control, and Stormwater Management on Construction Sites and Urban Areas* in 1993 (USDA 1993), which was recently revised in 2003 in time for use with the Phase II stormwater regulations. This discussion is summarized from that manual.

Soil formation in Alabama has been influenced primarily by parent materials and relief. The Appalachian Plateau, Limestone Valleys and Uplands, and Piedmont Plateau of

Northern Alabama are all products of uplift and extended geologic erosion. The Coastal Plain and Blackland Prairie sections of the state represent the sedimentation and deposition products from millions of years of geologic erosion. As a result, soils differ among the major soil areas throughout the state.

Many characteristics of soils, including texture, organic matter, fertility, acidity, moisture retention, drainage, and slope, have an influence on the soils' vulnerability to erosion. Except for most of the Prairie area, most disturbed sites after grading end up with a surface layer of acid infertile subsoil materials. The soils of these sites can be toxic to many plants and may not be capable of supporting growth sufficient to prevent erosion. Construction activities further restrict plant growth by increasing compaction and altering the slopes and drainage patterns. To offset these problems, the original site topsoil should be removed, stockpiled, and reapplied to the disturbed area. Soil amendments (limestone and fertilizer) should also be applied based on a soil test of the area. In some areas, special seedbed preparation will also be necessary.

County soil surveys are available from local Natural Resources Conservation Service (NRCS, formally the Soil Conservation Service, or SCS) offices. These surveys include a tremendous amount of information about local soils, including special concerns about different land uses in those areas. The following information is summarized from the Jefferson County, AL, soil survey prepared by the SCS in 1981, and is presented as an example of the type of information available from the county soil surveys.

Soil information for the 10 most common Jefferson County, AL, soils are listed in Tables 4.9 and 4.10. These ten soils cover about 75% of the county. The urban soils currently comprise much more than the amounts shown on this table due to the urban development that has occurred during the past 25 years since these surveys were last updated. In many areas of the country, this information is available on the Internet. The NRCS will provide the updated soil surveys through their website (http://websoilsurvey.nrcs.usda.gov/app/). In Pennsylvania, the soils information compiled by NRCS is hosted by the Penn State server (http://soilmap.psu.edu). The type of graphical information provided by the Soilmap PA website is shown in Figure 4.11 Availability on the Internet is fortunate, as many of the older bound-paper surveys are out of print.

For this Jefferson County example, the K values range from 0.17 to 0.37. No K values are available for the urban soils, as they have been dramatically disturbed and no generic values could be assigned. For "urban soils," soil samples should be collected and analyzed and the nomograph in Figure 4.6 used to estimate appropriate K values. It is interesting to note that almost all of these most common soils are on moderately-steep to steep slopes. Also, the soil erodibility factors are given for several soil horizons for most soils. The K values may increase or decrease with depth for the different soils. The K factors for different soil horizons can be used to determine the erosion rates for a site for different stages of excavation as these lower soil horizons are exposed. In areas of fill, the characteristics of the "new" exposed soil must be considered, and Figure 4.9 must be used to determine an estimated K value based on the measured properties.

These generally clayey soils in northern Alabama have surface horizon K factors of 0.24 to 0.37, with the most common Jefferson County soils (Montevallo and Nauvoo)

TABLE 4.9. Ten Most Common Soils in Jefferson County, AL, in 1980.

	Area in Jefferson County:			
Soil Name	Map Symbol	Acres	%	Soil Type
Montevallo-Nauvoo association, steep	29	260,930	36.3	Montevallo Nauvoo
Nauvoo fine sandy loam, 8 to 15% slope	31	51,440	7.2	Nauvoo
Nauvoo-Montevallo association, steep	34	44,010	6.2	Nauvoo Montevallo
Palmerdale complex, steep	35	29,390	4.1	Palmerdale
Urban land	44	27,080	3.8	Urbaland
Townley-Nauvoo complex, 8 to 15% slope	40	25,870	3.6	Townley Nauvoo
Bodine-Birmingham association, steep	8	25,560	3.6	Bodine Birmingham
Fullerton-urban land complex, 8 to 15% slopes	18	21,990	3.1	Fullerton Urban land
Bodine-Fullerton association, steep	9	20,720	2.9	Bodine Fullerton
Sullivan-State complex, 0 to 2% slopes	39	19,600	2.7	Sullivan State

TABLE 4.10. Erodibility Factors, K, for the Most Common Soils in Jefferson County, AL.

Soil name	Soil Horizon Depth and Soil Erodibility K Factor		
Birmingham	0 to 5 inches (0.24)	5 to 29 inches (0.28)	
Bodine	0 to 72 inches (0.28)		
Fullerton	0 to 6 inches (0.28)	6 to 35 inches (0.24)	35 to 65 inches (0.20)
Montevallo	0 to 6 inches (0.37)	6 to 16 inches (0.32)	
Nauvoo	0 to 12 inches (0.28)	12 to 46 inches (0.32)	
Palmerdale	0 to 60 inches (0.24)		
State	0 to 40 inches (0.28)	40 to 60 inches (0.17)	
Sullivan	0 to 66 inches (0.32)		
Townley	0 to 4 inches (0.37)		
Urban land	No specific information		

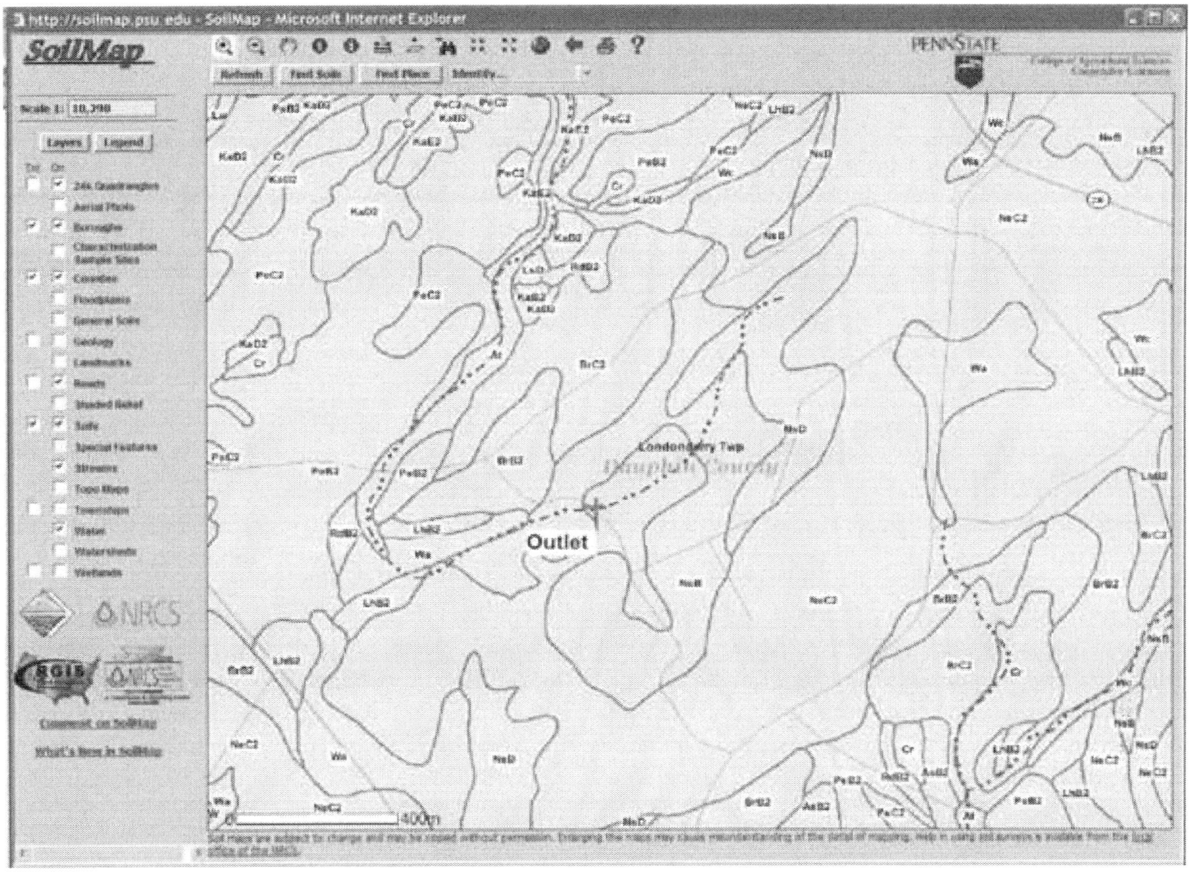

Figure 4.11. Soil map, Pennsylvania (URL: soilmap.psu.edu).

having 0.37 and 0.28 *K* values. Sandy soils with low organic content and with high permeability (similar to Gulf Coast soils) may have *K* values that are less than half of these values and could conceivably be as low as 0.05, although 0.10 is the more commonly seen minimum *K* value for Alabama soils. *K* values for soils having different textures are listed below (Dion, 2002):

Sandy, fine sand, loamy sand	0.10
Loamy sand, loamy fine sand, sandy loam, loamy, silty loam	0.15
Loamy, silty loam, sandy clay loam, fine sandy loam	0.24
Silty clay loam,silty clay, clay, clay loam, loamy	0.28

There is substantial overlap for the different soil textures, as there are other factors besides texture that are used to determine the *K* value, but this list does illustrate that *K* values generally increase as the soil particle sizes decrease. County soil surveys need to be consulted to determine the RUSLE *K* factors for the construction-site soils of interest.

Length-Slope Factor (*LS*)

The erosion of soil from a slope increases as the slope increases and lengthens. RUSLE defines a parameter called the length-slope (*LS*) factor that is used to calculate the impact of the interaction between these two parameters on erosion losses. The slope length, λ, is the horizontal distances from the start of the erosion area (typically a ridge, but not in all cases) to the start of the area where deposition of eroded sediment occurs (Figures 4.12 and 4.13). The slope length is used in RUSLE for calculating interrill (sheet) and rill erosion (but not gully erosion). Several example slope lengths are shown on Figure 4.13 (Renard, *et al.* 1987):

- *Slope A*—If undisturbed forest soil above the slope does not yield surface runoff, the top of the slope starts with the edge of the undisturbed forest soil and extends down slope to the windrow, if runoff is concentrated by the windrow.
- *Slope B*—Point of origin of runoff to the windrow, if the runoff is concentrated by the windrow.
- *Slope C*—From windrow to flow concentration point.
- *Slope D*—Point of origin of runoff to road that concentrates runoff.
- *Slope E*—From road to flood plain where deposition would occur.
- *Slope F*—On nose of hill, from point of origin of runoff to flood plain where deposition would occur.
- *Slope G*—Point of origin of runoff to slight depression where runoff would concentrate

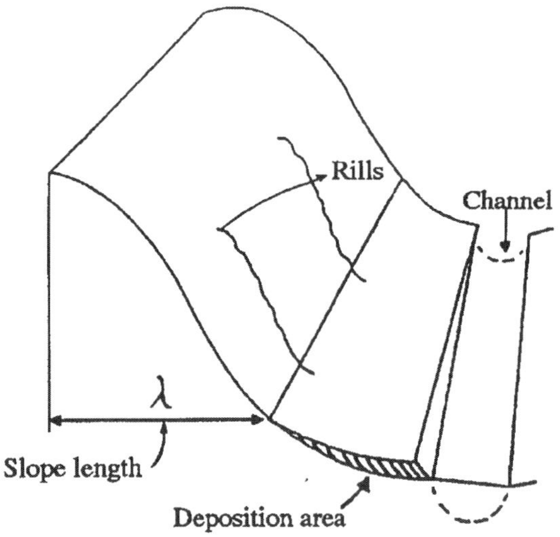

Figure 4.12. Definition of slope length as used in RUSLE (Renard, *et al.* 1987).

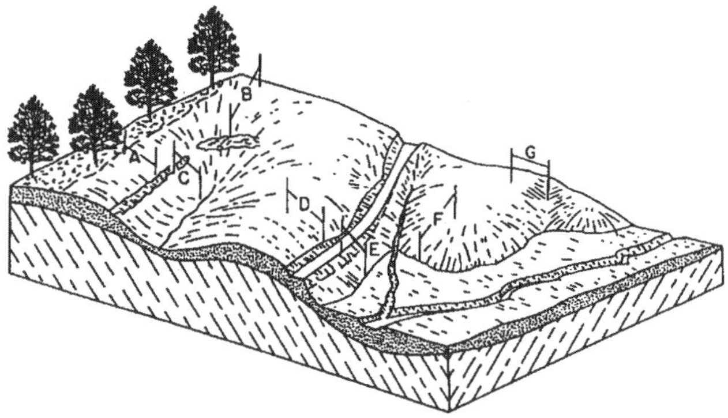

Figure 4.13. Examples of different slope length measurements (Renard, *et al.* 1987).

Once the slope length has been measured (such as from a detailed topographic map), RUSLE includes a table (Table 4.11) for selecting the length-slope factor, *LS*, according to these site characteristics. Values of 1.0 (the base condition) correspond to the standard condition of 9% slope and about 73 ft slope length (the dimensions and slopes of the erosion test plots). If the length of the slope is 300 ft., or less, the *LS* factor would be less than 0.10 for all slopes of 0.5%, or less. Roadway side cuts of 1:2 (50%) would have *LS* factors greater than 1.0 for all slope lengths of about 6 ft, or longer. Long and steep slopes, frequently occurring along roadway cuts in hilly terrain, can have extremely large LS factors. It is interesting to note that more than 80% of Jefferson County, AL, lands have slopes greater than 8% (1981 USDA Jefferson County Soil Survey). Land slopes are much less steep in Alabama below the fall line and approaching the gulf coast.

The RUSLE *LS* factors have been significantly changed compared to the original USLE *LS* values. There are now four separate *LS* tables, although Table 4.11 is the only one appropriate for construction sites (freshly prepared sites that are highly disturbed). The *LS* values have also been generally reduced compared to the original values, sometimes by as much as 50% for the largest values. *LS* values for slopes less than 20% are similar in both versions. Also, steepness and length are now more evenly sensitive to the *LS* factor, while previously, slope steepness was much more critical.

If ponding occurs on a site due to heavy rain intensities, low infiltration rates, and small slopes, the erosion loss will be substantially less than predicted using the above *LS*

TABLE 4.11. LS Values for Freshly Prepared Construction and other Highly Disturbed Soil, with Little, or No Cover (Renard, *et al.* 1987).

	Slope Length in Feet																
Slope	<3	6	9	12	15	25	50	75	100	150	200	250	300	400	600	800	1000
%																	
0.2	0.05	0.05	0.05	0.05	0.05	0.05	0.05	0.05	0.05	0.05	0.06	0.06	0.06	0.06	0.06	0.06	0.06
0.5	0.07	0.07	0.07	0.07	0.07	0.07	0.08	0.08	0.09	0.09	0.10	0.10	0.10	0.11	0.12	0.12	0.13
1.0	0.09	0.09	0.09	0.09	0.09	0.10	0.13	0.14	0.15	0.17	0.18	0.19	0.20	0.22	0.24	0.26	0.27
2.0	0.13	0.13	0.13	0.13	0.13	0.16	0.21	0.25	0.28	0.33	0.37	0.40	0.43	0.48	0.56	0.63	0.69
3.0	0.17	0.17	0.17	0.17	0.17	0.21	0.30	0.36	0.41	0.50	0.57	0.64	0.69	0.80	0.96	1.10	1.23
4.0	0.20	0.20	0.20	0.20	0.20	0.26	0.38	0.47	0.55	0.68	0.79	0.89	0.98	1.14	1.42	1.65	1.86
5.0	0.23	0.23	0.23	0.23	0.23	0.31	0.46	0.58	0.68	0.86	1.02	1.16	1.28	1.51	19.1	2.25	2.55
6.0	0.26	0.26	0.26	0.26	0.26	0.36	0.54	0.69	0.82	1.05	1.25	1.43	1.60	1.90	2.43	2.89	3.30
8.0	0.32	0.32	0.32	0.32	0.32	0.45	0.70	0.91	1.10	1.43	1.72	1.99	2.24	2.70	3.52	4.24	4.91
10.0	0.35	0.37	0.38	0.39	0.40	0.57	0.91	1.20	1.46	1.92	2.34	2.72	3.09	3.75	4.95	6.03	7.02
12.0	0.36	0.41	0.45	0.47	0.49	0.71	1.15	1.54	1.88	2.51	3.07	3.60	4.09	5.01	6.67	8.17	9.57
14.0	0.38	0.45	0.51	0.55	0.58	0.85	1.40	1.87	2.31	3.09	3.81	4.48	5.11	6.30	8.45	10.40	12.23
16.0	0.39	0.49	0.56	0.62	0.67	0.98	1.64	2.21	2.73	3.68	4.56	5.37	6.15	7.60	10.26	12.69	14.96
20.0	0.41	0.56	0.67	0.76	0.84	1.24	2.10	2.86	3.57	4.85	6.04	7.16	8.23	10.24	13.94	17.35	20.57
25.0	0.45	0.64	0.80	0.93	1.04	1.56	2.67	3.67	4.59	6.30	7.88	9.38	10.81	13.53	18.57	23.24	27.66
30.0	0.48	0.72	0.91	1.08	1.24	1.86	3.22	4.44	5.58	7.70	9.67	11.55	13.35	16.77	23.14	29.07	34.71
40.0	0.53	0.85	1.13	1.37	1.59	2.41	4.24	5.89	7.44	10.35	13.07	15.67	18.17	22.95	31.89	40.29	48.29
50.0	0.58	0.97	1.31	1.62	1.91	2.91	5.16	7.20	9.13	12.75	16.16	19.42	22.57	28.60	39.95	50.63	60.84
60.0	0.63	1.07	1.47	1.84	2.19	3.36	5.97	8.37	10.63	14.89	18.92	22.78	26.51	33.67	47.18	59.93	72.15

factors. The basic method to correct for this over-prediction is to estimate the land area subject to ponding and reduce the site area accordingly.

Cover Management Factor (*C*)

The methods used to protect the soil surface will affect the amount of soil erosion that may occur. Chapter 5 on channel and slope stability, and Chapter 7 on vegetation controls contain additional information pertaining to this factor, and to mulches in general. Wischmeier and Smith (1978) commented in their original USLE report regarding the model's applicability and use for construction sites. The following paragraphs are summarized from their prior discussion.

Site preparations that remove all vegetation and also the root zone of the soil not only leave the surface completely without protection, but also remove the residual effects of prior vegetation. This condition is comparable to the standard continuous fallow condition, and $C = 1$. Roots and residual effects of prior vegetation, and partial covers of mulch or vegetation, substantially reduce soil erosion. These reductions are reflected in the soil loss prediction by *C* values of less than 1.0.

Mechanical Mulches

Applied mulches immediately restore protective cover on denuded areas and drastically reduce the *C* values, and hence erosion. Where mulch effects are insignificant, these *C* values equal 1.0, the standard value. Straw or hay mulches applied on steep construction slopes and not tied to the soil by anchoring and tacking equipment are usually much less effective than equivalent mulch rates on relatively flat land.

Table 4.12 presents approximate *C* values for straw, crushed stone, and woodchip mulches on construction site slopes where no canopy cover exists. This table also shows the maximum slope lengths for which these values may be assumed to be applicable. These values are from the original (USLE, 1978) guidance and can now be better determined by making calculations based on specific site and rainfall conditions, as described in the chapters on hydrology (Chapter 3) and slope stability (Chapter 5). Also, currently available mulching products and erosion control blankets offer a much greater range of options for controlling erosion on construction site slopes. However, the values given here are suitable for calculating the effects of a basic mulch.

The percentage mulch cover is what generally determines the effectiveness of the mulch. This is the percentage of the soil surface that is covered by mulch laying on the surface. According to Wischmeier and Smith (1978), a simple method of estimating mulch cover is with a line at least 50 ft long that has 100 equally spaced markings. The line is stretched over the mulched surface and the marks that contact a piece of mulch are counted. The number of counted marks indicates the percentage coverage of mulch on the site. This is repeated randomly on the site to obtain an average value along with an indication of the variation. Table 4.13 shows the approximate percentage coverage for different mulching rates for straw, along with the range of erosion control (Wischmeier and Smith 1978).

TABLE 4.12. Construction Site Mulching *C* Factors and Length Limits for Different Slopes (Wischmeier and Smith, 1978).

Type of Mulch	Mulch Rate (tons per acre)	Land Slope (%)	Mulching *C* Factor	Length Limit (ft)*
None	0	all	1.0	n/a
Straw or hay, tied down by anchoring and tacking equipment	1.0	1–5	0.20	200
	1.0	6–10	0.20	100
	1.5	1–5	0.12	300
	1.5	6–10	0.12	150
	2.0	1–5	0.06	400
	2.0	6–10	0.06	200
	2.0	11–15	0.07	150
	2.0	16–20	0.11	100
	2.0	21–25	0.14	75
	2.0	26–33	0.17	50
	2.0	34–50	0.20	35
Crushed stone, 1/4 to 1-1/2 inch	135	<16	0.05	200
	135	16–20	0.05	150
	135	21–33	0.05	100
	135	34–50	0.05	75
	240	<21	0.02	300
	240	21–33	0.02	200
	240	34–50	0.02	150
Wood chips	7	<16	0.08	75
	7	16–20	0.08	50
	12	<16	0.05	150
	12	16–20	0.05	100
	12	21–33	0.05	75
	25	<16	0.02	200
	25	16–20	0.02	150
	25	21–33	0.02	100
	25	34–50	0.02	75

*Maximum slope lengths for which the specified mulch rate is considered effective. If these limits are exceeded, either a higher application rate or mechanical shortening of the effective slope length is required (such as with terracing).

Vegetative Covers

It is very important to establish vegetation on denuded areas as quickly as possible. A good sod has a *C* value of 0.01 or less, but such a low *C* value can be obtained quickly only by laying sod on the area at a substantial cost. When grass or small grain is started from seed, the probable soil loss for the period while cover is developing can be computed by the standard procedure for estimating crop stage-period soil losses. If the seeding is on topsoil without a mulch, the soil loss ratios given in Table 4.14 are appropriate for crop stage *C* values.

When the seedbed is protected by a mulch covering, the pertinent mulch factor from Table 4.12 is applicable until good canopy cover is attained. When grass is established in

TABLE 4.13. Straw Mulching Rates, Approximate Coverage and Corresponding Erosion Control (data from Wischmeier and Smith, 1978).

Straw Mulch Rate (tons per acre)	Percent Coverage	Erosion Control for Selected Coverages
0.10	10%	
0.25	30	
0.5	50	
1.0	70	80%
1.5	84	88%
2.0	92	80 to 94%
2.5	96	
3.0	97	

small grain as a nurse crop, it can usually be evaluated as "established meadow" about 2 months after the grain is harvested after which values in the following discussion can be used.

Table 4.15 (from the NRCS's National Engineering Handbook) lists cover management *C* factors for land covers with no trees. This table can be applied to construction sites having temporary or permanent vegetative covers, or mulches. It indicates the improved erosion control as the ground coverage increases. With good coverage (more than 80% ground cover), the erosion control could be 95%, or greater. These values assume that the vegetation or mulch is randomly distributed over the entire area. In areas with canopies where the rain drops have much less effective drop heights, and correspondingly less energy, the *C* factors are further decreased. A mechanically prepared site with no topsoil and no forest residue mixed in would have a *C* close to 1.0 if no cover was applied. With an 80% cover of mulch, this type of site (indicative of most construction sites) would have about 90% erosion control. In comparison, the C factor for a woodland with 100 percent duff cover (partly decayed organic matter on the forest floor) would be a low 0.0001 (99.99% erosion control), the lowest reported value.

Supporting Practices Factor (*P*)

The method of tillage and crop rotations all affect the soil erosion rate for an agricultural operation. This factor is rarely applicable for construction sites and is therefore given a value of 1.0 for this application, although some construction site erosion decision support models use the *P* factor when considering the effects of on-site controls (Dion, 2002). Other chapters in this book describe specific hydrologic and sediment transport functions that enable these effects to be directly calculated for specific site and design conditions.

RUSLE2 INFORMATION

The following description of RUSLE2 is based on information provided by the USDA. RUSLE2 is an upgrade of the text-based RUSLE DOS version 1 model. It is a computer model containing both empirical and process-based processes in a Windows environment. It predicts rill and interrill (sheet) erosion by rainfall and runoff. The USDA-Agricultural Research Service (ARS) is the lead agency for developing the RUSLE2 model, including developing the technical processes in the model and the model interface. The official NRCS RUSLE2 Internet site is at: http://fargo.nserl. purdue.edu/rusle2_dataweb/Tutorial.htm. The model can be downloaded from this site, along with supporting documents and other materials.

RUSLE2 has evolved from a series of previous erosion prediction methods. The USLE was entirely an empirically-based equation and was limited in its application to conditions where experimental data were available for deriving factor values. While RUSLE2 uses the USLE basic formulation of the unit plot, the calculations of RUSLE2 are based on daily predictions. The major visible change in RUSLE2 is its graphical user interface.

Development of RUSLE2 and its support is on-going. A current project being conducted by the University of Tennessee for the USDA is providing support to enhance the model and to further develop the User Guides. This project (*Development of Documentation Plans for RUSLE2 Science Component,* ARS project #6408-12130-012-12) is scheduled for completion in 2006.

TABLE 4.14. Cover Factor *C* Values for Different Growth Periods for Planted Cover Crops for Erosion Control at Construction Sites (data from Wischmeier and Smith 1978).

Type of Mulch	SB (seedbed preparation)	Period 1 (establishment)	Period 2 (development)	Period 3a (maturing crop)	Period 3b (maturing crop)	Period 3c (maturing crop)
Crop canopy*	0 to 10%	10 to 50%	50 to 75%	75 to 80%	75 to 90%	75 to 96%
Seeding is on topsoil, without a mulch	0.79	0.62	0.42	0.17	0.11	0.06
Seeding is on a desurfaced area, where residual effects of prior vegetation are no longer significant	1.0	0.75	0.50	0.17	0.11	0.06

*Percent canopy cover is the percentage of the land surface that would not be hit by directly falling rain drops because the drops would be intercepted by the plant. It is the portion of the soil surface that would be covered by shadows if the sun were directly overhead.

TABLE 4.15. Cover Factor *C* Values for Established Plants (data from NEH chapter 3 and Wischmeier and Smith, 1978).

	Percent Cover[1]	Plant Type	Percentage of Surface Covered by Residue in Contact with the Soil 0%	20	40	60	80	95+
C factor for grass, grasslike plants, or decaying compacted plant litter.	0	Grass	0.45	0.20	0.10	0.042	0.013	0.003
C factor for broadleaf herbaceous plants (including most weeds with little lateral root networks), or undecayed residues.	0	Weeds	0.45	0.24	0.15	0.091	0.043	0.011
Tall weeds or short brush with average drop height[2] of ≥20 inches	25	Grass	0.36	0.17	0.09	0.038	0.013	0.003
		Weeds	0.36	0.20	0.13	0.083	0.041	0.011
	50	Grass	0.26	0.13	0.07	0.035	0.012	0.003
		Weeds	0.26	0.16	0.11	0.076	0.039	0.011
	75	Grass	0.17	0.10	0.06	0.032	0.011	0.003
		Weeds	0.17	0.12	0.09	0.068	0.038	0.011
Mechanically prepared sites, with no live vegitation and no top soil, and no litter mixed in	0	None	0.94	0.44	0.30	0.20	0.10	Not given

[1]Percent cover is the portion of the total area surface that would be hidden from view by canopy if looking straight downward.
[2]Drop height is the average fall height of water drops falling from the canopy to the ground.

The following overview is based on information on RUSLE2 provided by the USDA.

Although mostly intended for agricultural erosion prevention and farm operation planning, RUSLE2 can be applied to other erosion problems including construction sites. Earlier sections of this chapter discussed the major components of RUSLE, which are generally applicable to RUSLE2. RUSLE2 is very easy to use; with the exception of the site topography, the RUSLE2 model user describes the site-specific field conditions by selecting the appropriate values and control practices from menus. When a menu selection is made, RUSLE2 "pulls" values stored in the RUSLE2 database and uses them as input values to compute the expected erosion rates. The user enters site-specific values for slope length and steepness to represent site topography.

The following are several important enhancements available in RUSLE2 that will aid the erosion control planner. With the development of expanded User Guides, model enhancements, and model templates in the next several years, RUSLE2 should become the preferred tool to predict erosion rates for construction sites.

- Although RUSLE can calculate erosion rates for 2-week increments (through the use of the detailed seasonal rainfall erosivity values for all parts of the U.S.), RUSLE2 extends the resolution to daily erosion predictions. RUSLE2 also uses seasonal temperature information, along with rainfall, to predict the longevity of applied mulches for erosion control. Simply selecting the location of the study site automatically uses the correct daily erosivity, precipitation, and temperature values in the model. The following figure shows plots of the erosivity variations throughout the year for sites in California, Tennessee, North Dakota, and Maryland (USDA 2003):

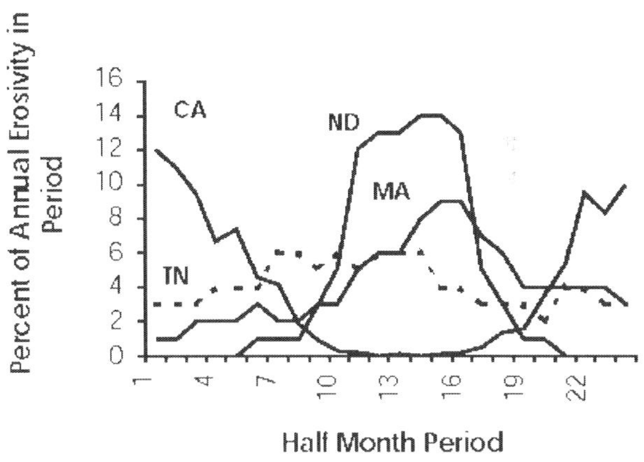

In the above example, erosivity is nearly uniform at Memphis, Tennessee, while 80 percent of the erosivity occurs in the months of May, June and July in North Dakota (the months having most of the annual rainfall). Soil erodibility also varies during the year. Erosion is greatest when peak soil erodibility, rain erosivity, and vulnerability of cover-management all occur simultaneously.

- Another important enhancement of RUSLE2 is its ability to vary the soil erodibility by season. The RUSLE2 user typically selects a soil by soil-map unit name from a list of soils in the RUSLE2 database. Soil erodibility, *K*, varies by season. It tends to be high early in the spring during and immediately following thawing and other periods when the soil is wet. The value entered for *K* is a base value. RUSLE2 uses monthly precipitation and temperature to compute monthly *K* values that vary about the base *K* value. The monthly values are then disaggregated into daily values. Example variations of *K* computed by RUSLE2 for St. Paul, MN, Birmingham, AL, and Tombstone, AZ, are shown below (USDA, 2003).

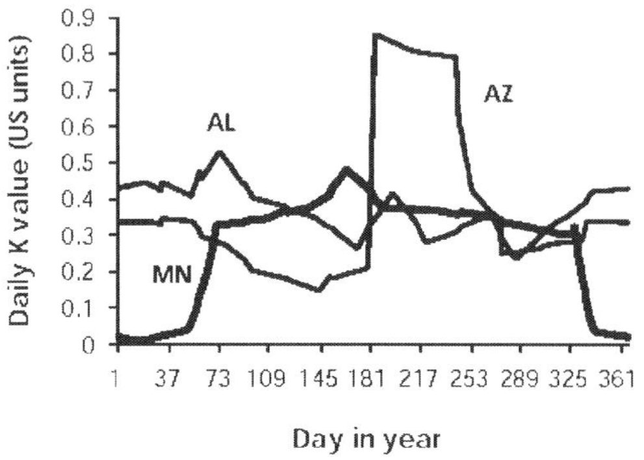

The low values for St. Paul during the winter months represent frozen soil that is nonerodible. RUSLE2 does not fully represent the thawing period in early spring in St. Paul, primarily because observed data are too few to determine a relationship for this period. The peak for Birmingham in March results from rainfall rather than from temperature. The main influence of temperature on temporally varying K values is in late summer when increased temperature increases soil evaporation and reduces runoff and erosion. The peak erodibility during the summer for Tombstone is because most of the annual rainfall at the location occurs during this period.

RUSLE2 assumes that soil erodibility is 2.2 times higher immediately after a mechanical disturbance than after the soil has become fully consolidated. Therefore, erosion decreases with time as the soil becomes more consolidated. Care must be taken with this factor for construction sites, where soil compaction (and associated soil density) increases during construction operations is very common. RUSLE assumes a decrease in soil density—contrary to what actually occurs on a site during construction.

- *Topography*—Slope length, steepness, and shape are the topographic characteristics that most affect rill and interrill erosion. Site-specific values are entered for these variables. The following examples are from the Technology User's Guide (USDA, 2003) and describe some important RUSLE2 topographic features for construction sites.

On a complex slope, the sediment yield is reduced by deposition on a downslope concave slope section:

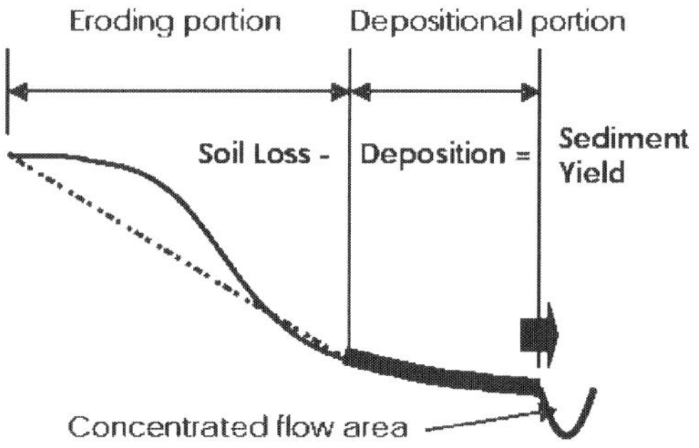

On uniform or convex slopes, the sediment yield is equal to the soil loss, because there is no depositional area:

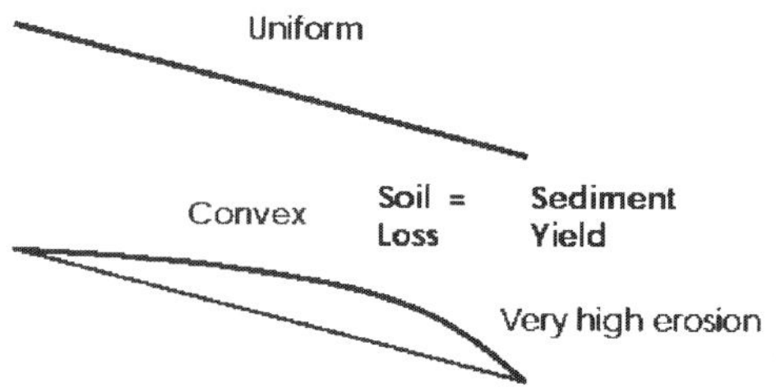

An important complex hillslope shape is shown below where a concave section occurs in the middle of the hillslope. This example is for a cut slope (e.g., road-fill) slope that is common in hilly terrain. Deposition can occur on the mid-section of the hillslope where the road is located. Soil loss occurs on the cut slope and on the fill slope where overland flow continues across the road onto the cut slope. Although the steepness and length of the fill slope is the same as that for the upper cut slope, soil loss is likely to be much greater on the cut slope than on the fill slope because of the increased amounts of overland flow of water. Although USLE and RUSLE cannot easily describe this hillslope, it can be easily described in RUSLE2, which also determines the appropriate overland flow slope lengths, and computes soil loss on the two eroding portions of the hillslope, deposition on the depositional portion of the hillslope, and final sediment yield from the hillslope. (*Note:* The slope-length used in RUSLE2 does not end where deposition begins for the hillslope profile, as it does in earlier model versions).

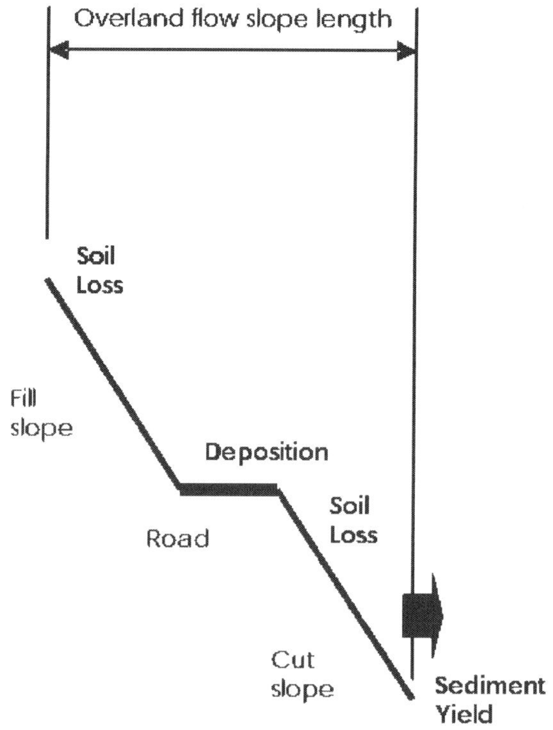

- Cover Management Practice: Important features on a construction site include whether or not the land is bare, the soil material is a cut or fill, mulch is applied, or the slope is recently reseeded. The description of any cover-management practice is created, named, and stored in the RUSLE2 database. When RUSLE2 is run, the cover-management practice that fits the site-specific field condition is selected from the menu of choices.
- Support practices, include contouring, vegetative strips and buffer strips, silt fences, terraces, diversions, and sediment basins, all reduce eroded soil discharges primarily by reducing the erosivity of surface runoff and by causing deposition. Support practices are selected from a list of these practices in the RUSLE2 database. Site-specific information, such as the location of a diversion on the hillslope, is entered as required for each practice.

If the control segment is sufficiently long (the grass strip is sufficiently wide) and the increase in transport capacity with distance is less than the detachment quantity, deposition ends within the segment, as illustrated below (USDA 2003). Erosion may occur further down gradient where transport capacity is available. In this case, the sediment load exceeds the transport capacity at the upper end of the grass strip, while both sediment and transport capacity increase within the strip segment. RUSLE2 computes the location where deposition ends and sediment load equals transport capacity, as well as the additional erosion.

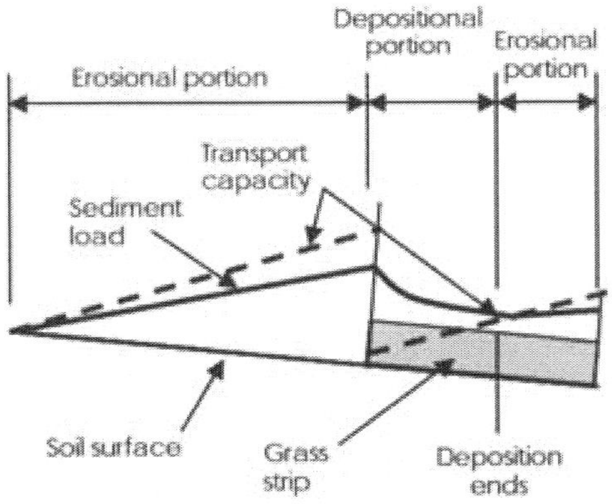

The following list (USDA, 2003) shows the various RUSLE2 database components that comprise the different parts of the model. The input information is organized using these components, allowing excellent organization and sensitivity analyses:

Worksheet—Computes soil loss for alternative management practices, alternative profiles, and average soil loss for an area.

Profile—Computes soil loss for a single hillslope profile, the basic computational unit in RUSLE2.

Climate—Contains data on average annual erosivity, EI30, rainfall amount and temperature.

Storm erosivity—Contains data on the distribution of erosivity during the year.

Soil—Contains soil data, including erodibility, texture, hydrologic soil group, time to consolidation, sediment characteristics, and soil erodibility nomographs.

Management—Contains descriptions of cover-management systems. Includes dates, operations, vegetation, type and amount of applied materials.

Operation—Contains data on operations (events that affect soil), vegetation and residue. Includes the sequence of processes used to describe each operation, such as for an operation placing residue in the soil: values for flattening, burial and resurfacing ratios; ridge heights; and initial soil roughness.

Vegetation—Contains data on vegetation, like values for residue type, yield, above-ground biomass at maximum canopy, senescence, flow retardance, root biomass, canopy cover, fall height, and live ground cover.

Residue—Contains data that describes the residue assigned to each vegetation. Includes values for decomposition, mass-cover relationship, and how residue responds to tillage.

Contouring–Contains values for row grade used to describe degree of contouring.

Strips/barriers—Contains data that describes filter strips, buffer strips and rotational strip cropping. Includes cover-management in strips, width of strips, number of strips across slope length, whether or not a strip is at the end of the slope, and offset of rotation by strip.

Hydraulic system—Identifies the hydraulic elements and their sequence (e.g., describing the hydraulic systems of diversions, terraces and impoundments). Includes numbers across slope length, and whether or not a system is at the end of the slope or specific locations on the slope length.

Hydraulic element—Contains data on the grade of the named channel for terraces and diversions.

Subsurface drainage system—Contains data on the percent of the area covered by optimum drainage.

BASIC PREDICTIONS OF SOIL LOSSES FROM A CONSTRUCTION SITE

Construction site evaluations have several dimensions: different construction phases lasting for different time periods, different soils on different locations and at different times reflecting cut and fill operations, changes in the gradients and lengths of slopes, and varying cover conditions. Therefore, in order to conduct a site evaluation, these different dimensions need to be clearly organized.

Construction Phases

The most basic dimension is understanding the construction phasing, beginning with site clearing and grubbing to final contouring. The basic time phases of interest for erosion evaluation and control may include the following activities on the site:

1. Install downslope sediment controls (filter fencing and sediment ponds)
2. Install upslope diversions and protect on-site channels that will remain (diversion berms and swales, channel lining, establish buffers, and filter fencing)
3. Clear and grub first area (minimize area exposed and phase-completion time)
4. Do final contouring of first area (stabilize exposed areas before moving on to next area)
5. Repeat above 2 steps for all other areas, dividing the whole planned disturbed construction site into areas as small as possible
6. Establish roadways and parking areas and install utilities (leaving road bed base, or preliminary pavement, protect inlets, etc.)
7. Erect buildings (provide adequate storage for materials and for construction vehicle parking, practice good housekeeping, etc.)
8. Do final landscaping (remove temporary controls, replace with permanent stormwater facilities, irrigate vegetation until established)

Site Information

Site layouts and erosion control plans are needed for each major phase that alters the construction site contours and soil cover. Specifically, RUSLE should be applied for (1) the initial clearing and grubbing operation, (2) the site reflecting the final contouring, and (3) the final phases during roadway and utility construction and building erection. As indicated above, it is hoped that the site can be divided into small units where the clearing to final contouring operations can be completed as rapidly as possible, and temporary soil protection can then be applied before moving to the next area. Obviously, small areas, and sites where massive grading is needed simultaneously over most of the site, will prevent this type of phasing. In this situation, the objective will be to complete the grading quickly, and, hopefully, to schedule it during periods when the erosion potential is reduced.

During each phase, the following site information will be needed to use RUSLE:

1. Expected start and finish dates, and corresponding "partial" R based on monthly rain variations
2. Surface soil K values
3. Various slopes and slope lengths over the site for calculating the LS factor
4. Type of mulch or vegetated cover

The LS factor may be the most confusing for a developing site. Basically, the site will need to be divided into separate sections for each slope, from the ridges to the toe of the slopes. The R factor will be uniformly applied to the whole site for each phase period, and the soil maps will help indicate the appropriate K factors. Therefore, RUSLE erosion yields will need to be computed for each separate slope, with the results summed to create a total-site erosion yield. The complete site will need to be represented, even for undisturbed areas (using natural cover conditions).

Example: Quantifying Site Erosion for Different Construction Phases

An example site may be represented by the conditions shown in Tables 4.16 through 4.18. Once the conditions for each site area are fully described and a map prepared showing the site areas, the resulting factors can be determined, and calculated soil losses can be displayed in tables such as these. This type of analysis also has the advantage of high-lighting areas responsible for most of the site erosion, possibly leading to further modifications in the erosion control plan.

The following example construction area in Birmingham, AL, is on a moderately-sloped site, with most slopes of 10 and 12%. About 22 of the 27 site acres will be graded, with about 5 acres left undisturbed. Approximately 18 acres will be used as parking, on-site roads, and commercial buildings, with about 4.5 acres used for relatively-steep embankments and road cuts.

Table 4.16 shows the erosion predictions for the first construction phase, the initial grubbing of existing vegetation. The erosion control plan calls for temporary mulching on the newly cleared land and limiting the active construction area to 5 acres. The 5 acres is being graded to the final site contours. When completed, that area will be stabilized with appropriate erosion controls and then another area will be graded. During this 3 month period, about 1600 tons of sediment may be eroded from the site, the vast majority from the active area that has no preventative erosion control measures. Sediment control measures (as described later in Chapter 6) will be used to provide further reductions in sediment losses from the site.

The new site contours will result in milder slopes so the calculations for this phase likely represent worst case conditions. The next phase represents the end of the grading operations when more established controls are in place, but still there will be areas of active construction.

Table 4.17 represents site conditions at the end of the rough grading operations. All site contours are basically in place, and erosion controls have been newly established. There is still the last 5 acres of active construction that is unprotected, but at it is a much less severe slope. It is seen that once re-graded and properly protected, the site's

TABLE 4.16. Example RUSLE Calculations for Initial Grubbing Phase (same site contours as pre-development, but stripped cover and with temporary mulch).

Site Areas	Area Description	Land Areas (acres)	*R* for Phase Period (June 16 to Sept 15)[1]	*K* Soil Factor[2]	*LS* Slope Length Factor	*C* Cover Factor[3]	Calculated Unit Area Soil Loss (tons/acre/period)	Calculated Total Area Soil Loss (tons/period)
A	Undisturbed area (L=50 ft; S=3%)	5.23	144	0.15	0.30	0.001	0.01	0.03
B	Future development, temp. mulch (L=350 ft; S=10%)	5.81	144	0.37	1.46	0.2	15.6	90
C	Future development, temp. mulch (L=600 ft; S=12%)	11.03	144	0.28	1.88	0.2	13.5	150
D	Future development, active construction (L=600 ft; S=12%)	5.0	144	0.28	6.67	1.0	269	1300
Total Site		27.07						1600 tons over 3 months

[1]41% of annual *R*; annual *R* is 350, so project phase partial R is: (0.41)(350) = 144.
[2]From county soil map and anticipated surface soils during this phase.
[3]C factors based on native good cover for undisturbed areas, grubbing debris and 1 ton/ac of straw tacked on newly denuded areas having temporaryberms to limit slope length to 100 ft., and nothing on active construction area (5 acres maximum is allowed to be under active construction at any time).

TABLE 4.17. Example RUSLE Calculations for Rough Grading Phase (final site contours, but still working on final grades).

Site Areas	Area Description	Land Areas (acres)	*R* for Phase Period (Sept 16 to Feb 28)[1]	*K* Soil Factor[2]	*LS* Slope Length Factor	*C* Cover Factor[3]	Calculated Unit Area Soil Loss (tons/acre/period)	Calculated Total Area Soil Loss (tons/period)
1a	Undisturbed area (L=50 ft; S=3%)	1.51	105	0.15	0.30	0.001	0.01	0.01
1b	Undisturbed area (L=100 ft; S=5%)	3.72	105	0.17	0.68	0.005	0.06	0.2
2	Road cut (L=50 ft; S=25%)	0.54	105	0.28	2.67	0.02	1.6	0.9
3	Road cut (L=100 ft; S=25%)	1.37	105	0.37	4.59	0.02	3.6	4.9
4a	Main embankment (L=15 ft; S=10%)	0.84	105	0.28	0.40	0.55	6.5	5.4
4b	Main embankment (L=200 ft; S=16%)	0.33	105	0.37	4.56	0.17	30.1	9.9
4c	Main embankment (L=300 ft; S=10%)	1.15	105	0.17	3.09	0.07	3.9	4.4
5	Parking area (L=500 ft; S=0.2%)	5.5	105	0.28	0.06	0.02	0.03	0.2
5a	Parking area (L=500 ft; S=0.2%) Active construction	5	105	0.28	0.06	1	1.8	8.8
6	Building areas (L=250 ft; S=0.2%)	5.53	105	0.35	0.06	0.02	0.04	0.2
7a	Road segment (L=200 ft; S=3%)	0.26	105	0.17	0.57	0.02	0.2	0.1
7b	Road segment (L=400 ft; S=1%)	0.95	105	0.28	0.22	0.02	0.1	0.1
7c	Road segment (L=250 ft; S=0.5%)	0.37	105	0.28	0.10	0.02	0.1	0.02
Total Site		27.07						35 tons over 5.5 months

[1]30% of annual *R*; annual *R* is 350, so project phase partial R is: (0.30)(350) = 105.
[2]From county soil map and anticipated surface soils during this phase.
[3]*C* factors based on native good cover for undisturbed areas, erosion control mats for road cuts, planted vegetation or tacked mulches on embankments, and gravel pads for parking, building, and road areas. The vegetation *C* factor was calculated based on plant growth stages during this construction phase.

sediment losses are significantly reduced. However, failure of erosion controls on any of the steep slopes can have important consequences.

Table 4.18 illustrates the same site for the final phase, when building finishing is occurring and all grading and final erosion controls are in place and well established. The calculated erosion rate for this site for this last construction phase is also quite low, being only about 2 tons per acre for this 5 month period. Obviously, this rate represents the established values due to the low *C* factors and assuming careful maintenance of the soil-protecting mulches.

This is an example of a phase-specific erosion control plan that is possible using modern techniques. If these eroding soils are mostly clay loams, the total volume of sediment eroded from this site during the total construction period would be about 1700 cubic yards, with almost all occurring during the initial grubbing and clearing operation and before the site is contoured to its final topography. This amount of material would be an important consideration when designing a sediment pond downstream of the eroding areas. This amount of sediment would require about 2 or 3 feet of sacrificial volume in a well-functioning and properly- designed sediment pond (see Chapter 6). However, it is likely that excessive erosion associated with failure of the erosion control materials on the steeper slopes may occur. As an example, more than 50 tons per acre could be lost for every month that one of the 10% slopes was in disrepair.

If this site had no erosion controls, an expected 3900 tons of sediment could be eroded over the 13.5 months of construction. This is about 130 tons per acre per year, typical for locally-monitored construction sites. These erosion controls are expected to reduce these losses to about 1600 tons, or a reduction of approximately 60%. Most of the sediment losses are expected to occur during the initial clearing and grubbing operations when the slopes have not been reduced. The percentage reductions of sediment losses during the final grading operations may be about 90%. Effective sediment controls, as described in Chapter 6, also will be needed for further reductions, especially for the grubbing operations, and in case of periodic slope-cover failures.

Evaluating Timing Options for Construction Operations

Timing of specific construction operations may have an important effect on the estimated soil erosion rate. As an example, the distribution of rainfall energy in Alabama

TABLE 4.18. Example RUSLE Calculations for Final Grading Phase.

Site Areas	Area Description	Land Areas (acres)	*R* for Phase Period (March 1 to July 31)[1]	*K* Soil Factor[2]	*LS* Slope Length Factor	*C* Cover Factor[3]	Calculated Unit Area Soil Loss (tons/acre/period)	Calculated Total Area Soil Loss (tons/period)
1a	Undisturbed area (L=50 ft; S=3%)	1.51	196	0.15	0.30	0.001	0.01	0.01
1b	Undisturbed area (L=100 ft; S=5%)	3.72	196	0.17	0.68	0.005	0.11	0.4
2	Road cut (L=50 ft; S=25%)	0.54	196	0.28	2.67	0.02	2.93	1.6
3	Road cut (L=100 ft; S=25%)	1.37	196	0.37	4.59	0.02	6.66	9.1
4a	Main embankment (L=15 ft; S=10%)	0.84	196	0.28	0.40	0.55	12.07	10
4b	Main embankment (L=200 ft; S=16%)	0.33	196	0.37	4.56	0.17	56.22	19
4c	Main embankment (L=300 ft; S=10%)	1.15	196	0.17	3.09	0.07	7.21	8.3
5	Parking area (L=500 ft; S=0.2%)	10.5	196	0.28	0.06	0.02	0.07	0.7
6	Building areas (L=250 ft; S=0.2%)	5.53	196	0.35	0.06	0.02	0.08	0.5
7a	Road segment (L=200 ft; S=3%)	0.26	196	0.17	0.57	0.02	0.38	0.1
7b	Road segment (L=400 ft; S=1%)	0.95	196	0.28	0.22	0.02	0.24	0.2
7c	Road segment (L=250 ft; S=0.5%)	0.37	196	0.28	0.10	0.02	0.11	0.04
Total Site		27.07						50 tons over 5 months

[1]56% of annual *R*; annual *R* is 350, so final project phase partial R is: (0.56)(350) = 196.
[2]From county soil map and anticipated surface soils during this phase.
[3]*C* factors based on native good cover for undisturbed areas, erosion control mats for road cuts, planted vegetation or tacked mulches on embankments, and gravel pads for parking, building, and road areas. The vegetation *C* factor was calculated based on plant growth stages during this construction phase.

TABLE 4.18. Alternative Slope Configurations and Corresponding Reductions in Erosion.

Original Slope			Alternative Terrace 1 (1 mid-slope bench)				Alternative Terrace 2 (5 benches)			
Slope	Length	*LS* Factor	New Slope	Length (and terrace width)	Approx. New *LS* Factor	Estimated Erosion Reduction	New Slope	Length (and Terrace width)	Approx. New *LS* Factor	Estimated Erosion Reduction
0.5%	300 ft.	0.10	0.54%	150 (10) ft.	0.095	5%	0.56%	50 (5) ft.	0.09	10%
3.0	300	0.69	3.2	150 (10)	0.51	26	3.3	50 (5)	0.29	58
10	300	3.09	10.7	150 (10)	1.9	39	11.1	50 (5)	1.0	68
25	300	10.81	26.8	150 (10)	6.0	44	27.8	50 (5)	2.8	74
50	300	22.57	53.6	150 (10)	10.6	53	55.6	50 (5)	5.0	78

(Table 4.6) indicates that for most of the state, June through September is the period having the highest erosion potential. These 4 months have about half of the total annual erosion-rainfall-related energy. October through February are usually the driest Alabama months, with only about 30% of the annual rainfall related energy occurring during these 5 months. Therefore, if possible, construction activities near sensitive waters could beneficially be scheduled during these drier months, but highly erosive rains may still occur during any period of the year.

Planning for vegetative covers also must consider the growing season and the need for supplemental irrigation. Table 4.14 showed how the *C* cover factors dramatically change for different growth stages. Obviously, plants that rapidly germinate, become established, and mature early, are important for erosion control. Mature crops with extensive canopies are also desired. Local NRCS and agricultural extension services can provide suitable lists of plants with these attributes for a local site. If using erosion control mats or sod, differences in cover *C* factors with time are not very large, and excellent control is available as soon as these are installed. This is especially important for channel linings. If relying on seeded plantings, several weeks to months may pass before the *C* factor reduces to less than 0.25 for slopes, and much more time is needed to establish a strong root system to withstand flowing waters. However, because of the high costs of erosion control mats, they are usually only used in the most critical areas, with less expensive mulches used over prepared seed beds whenever possible. Information presented in other chapters allow site hydrologic conditions and associated shear stresses to be calculated for specific site conditions, ensuring the most efficient use of the different cover products.

Comparing Different Slope Design Options

The information presented in Table 4.11 enables the erodibility of different slope conditions to be evaluated. In most cases, these conditions cannot be changed easily, as they were established for the most cost-effective development options. However, it is obvious that very steep slopes are not a good idea. Erosion on slopes greater than 15% can dominate the total erosion from a construction site. Similarly, efforts should be made to terrace long slopes, shortening the flow paths down their embankments. Chapter 5 will outline the procedures for evaluating specific erodibility and erosion-control solutions for slopes.

Terracing can be considered as a control option with relatively little effect on the use of the land. Long slopes can be divided into separate sections with great benefit. The terraces can be built as diversion swales to carry the accumulated water to a collection point. A reinforced drop chute then can be used to minimize the water flowing across downslope areas. Table 4.19 illustrates some options for modifying slopes with terracing. The slope angles will increase as slope length is decreased by the width of the terrace/diversion, which would somewhat offset the decrease in slope length, if no additional land was used for the slope. This table shows that significant reductions in expected erosion can occur with terracing, even with the slightly increased slopes. The largest benefits are associated with steeper initial slopes. Of course, almost all slopes will need to be stabilized with erosion control mats (especially required if steep), or at least tacked mulches (if less steep and relatively short). These slope protection calculations are presented in Chapter 5. They will show that terracing also decreases the cost of this needed slope protection.

Erosion and Construction Scheduling

The following is excerpted from a homework assignment prepared by Heather Hill, a student at the University of Alabama at Birmingham, as part of the Construction Site Erosion and Sediment Control Class taken during the summer of 2005.

The Assignment

1. Describe the different construction phases for your site (initial grubbing and clearing, using pre-development contours; final grading contours during active construction activities, at least). Describe site soils and land cover. Describe the timing of the construction site erosion and sediment controls for your site.

The project site is at "the mound of dirt on U.S. 280," a large pile of previously excavated dirt placed on this 16 acre site near Birmingham, AL, many years ago in anticipation of construction which was delayed for many years. This area has been under constant scrutiny by the residents of the area and the city of Mountain Brook. This area is finally under construction. The site contractor is working 6 days a week, 12 hours a day, to get this project going. The site engineer donated the site plans, a detailed site topographic map, and their erosion control plan for assistance for this project.

The first phase of construction was the construction of a 15′ diameter corrugated metal pipe to channel the flow of the tributary of Little Shades Creek that runs through the site. In order to do this, a series of pump systems had to be installed for water diversions. One pump was installed at the beginning of the creek to collect the waters before entering the active area of the construction site. This was done by placing riprap in the creek bed and lining the upstream side with plastic. The water was then collected and pumped approximately 1000 feet downstream and released in a basin and allowed to settle a little before releasing into the original creek bed. The other pump was installed at the catchbasin which collects the site runoff water and then routes it to the holding pond. The water was then released into a set of baffles for sedimentation control in the pond and then released back to the creek.

A culvert pipe was also placed in the stream to allow access to the construction site on the other side of the creek. A road was cut from Green Valley Road to access the stream and install the pump. Another road was also cut around the side of the mound to access the area for the holding pond and a laydown area for the fabrication of the 15′ diameter culvert. At the same time, Highway 280 was given a facelift to create turn lanes for access to this new commercial area. Curbs were installed and then the road was paved and the median and the edge of the property were grassed and had excelsior blankets placed over them.

Currently, phase II has started which includes the major site grading. The mound of dirt is being excavated and sieved to acquire good backfill for the site. The final site grading will consist of covering the 15′ culvert with approximately 25′ of backfill and taking the mound down to near the original site grade (approximately elevation 750′). The entire site will be fairly flat and consist mainly of parking lots, roads and buildings. Green Valley Road will be rerouted to come down the center of the site and the current Green Valley Road will become part of the commercial development.

The soils on the site are described as silty clay and clayey sand by the project manager. The county soil survey describes the soils as a silty loam. The northern portion of the site is a sandstone and shale ridge. The mound consists of all kinds of soils, including rock and debris.

The site is densely vegetated in areas along the creek and the ridge with underbrush and mature trees and weeds. The mound and the access road for the site had sparse vegetation and mainly weeds and little grass.

Construction schedule for the site work is as follows:

Task	Start	Finish
Culvert Procurement	Jun. 13, 2005	Aug. 8, 2005
Culvert Preparation Work	Jun. 27, 2005	Jul. 25, 2005
Grading/Undercut for grocery store	Jul. 5, 2005	Oct. 17, 2005
Culvert Installation	Jul. 26, 2005	Sep. 19, 2005
Culvert Backfill	Aug. 9, 2005	Oct. 3, 2005
Grading North of Green Valley	Nov. 1, 2005	Nov. 21, 2005
Retail/Residential Grading	Mar. 22, 2006	May 16, 2006
Parking Lot Construction	May 17, 2006	Sep. 22, 2006

Silt fences have been installed in some areas and excelsior blankets have been placed on the flat seeded areas that have been disrupted. The pumps are working and the holding basin is collecting water and working well. Silt fence has been installed around the creek channel to divert water to the catchbasin. Approximately 6 acres of the site is undergoing active construction. Final plans for the site cover consist of asphalt parking lots, landscaping and sod at the entrance and around the parking lot. ALDOT seed mix will be used on the cut/fill areas.

The Assignment Continued

2. Apply RUSLE for each of the phases (Table 4.20).
3. Select the appropriate temporary and permanent plants to be used for construction site erosion control at your site, and describe planting and mulching conditions, etc. Consider the likely dates for the plantings (Table 4.21).

Temporary cover for the holding pond and the areas disrupted during the installation of the turn lanes can be millet and ryegrass for this time of year. Millet is suggested for use in Central Alabama for April 1 to August 15 and Ryegrass for September 1 to October 15. Most of the areas would need to be covered with straw or a temporary erosion control blanket. Most of the area to be seeded and mulched is flat with less than 2% grade. The holding basin area has approximately 30% slopes. Permanent plantings would mainly be sod. The area for sod would be relatively flat with about a 3% slope. When the site is ready for sodding, probably in August or September, Bermuda grass or fescue would be appropriate. As with any landscaped area in the Mountain Brook area, the entrance to the site will be planted with trees and shrubs and perennial flowers that will be changed with the seasons. For the most part, there will be little of the final 16 acre high intensity commercial site that will be vegetated.

TABLE 4.20. Initial Grubbing, predevelopment contours

Site Areas	Area Description	Land Areas	*R* for Phase Period (6/27–9/19)[1]	*K* Soil Factor[2]	*LS* Slope Length Factor	*C* Cover Factor[3]	Calculated Unit Area Soil Loss (tons/acre/period)	Calculated Total Area Soil Loss (tons/period)
A	Undisturbed (*L*=120 ft, *S*=25%)	9	143.5	0.1	5.1	0.003	0.22	2.0
B	Future Development Mulch /Straw (*L*=20 ft, *S*=0.2%)	1	143.5	0.24	0.05	0.2	0.344	0.344
C	Active Construction Mound (*L*=70 ft, *S*=28%)	4	143.5	0.15	3.67	1	79	316
D	Active Construction Roads (*L*=1000 ft, *S*=3%)	2	143.5	0.15	1.23	1	26	53
Total Site		16						371 tons over 3 months

[1]41% of Annual *R*, annual *R* is 350, so project phase partial *R* is (350)(0.41)=143.5.
[2]From Jefferson County soils map and anticipated surface soils during this phase.
[3]*C* factors are based on native good cover for undisturbed areas, grubbing debris and 1 ton/ac of straw tacked on newly denuded areas, and nothing on active construction areas Final Grading contours (after active construction, all land covered).

TABLE 4.21. Final Grading contours (after active construction, all land covered).

Site Areas	Area Description	Land Areas	*R* for Phase Period (9/20–11/21)[1]	*K* Soil Factor[2]	*LS* Slope Length Factor	*C* Cover Factor[3]	Calculated Unit Area Soil Loss (tons/acre/period)	Calculated Total Area Soil Loss (tons/period)
A1	Road (*L*=500 ft, *S*=5%)	1.1	35	0.28	1.71	0.02	0.34	0.37
A2	Road (*L*=450 ft, *S*=2%)	1.3	35	0.17	0.5	0.02	0.060	0.0775
B1	Parking Lot (*L*=300 ft, *S*=2%)	3.7	35	0.15	0.43	0.02	0.045	0.17
B2	Parking Lot (*L*=150 ft, *S*=2%)	3.1	35	0.28	0.33	0.02	0.065	0.20
C	Area to be Landscaped (*L*=75 ft, *S*=10%)	1.3	35	0.28	1.2	0.2	2.4	3.1
D	Runoff Pond (*L*=25 ft, *S*=30%)	0.5	35	0.15	1.86	0.2	2.0	0.98
E	Undisturbed Area (*L*=100 ft, *S*=50%)	1	35	0.15	9.13	0.003	0.14	0.14
F1	Building 1 (*L*=120 ft, *S*=0.2%)	1.8	35	0.28	0.05	0.02	0.0098	0.018
F2	Building 2 (*L*=350 ft, *S*=0.2%)	1.7	35	0.28	0.06	0.02	0.012	0.020
G	Slopes (*L*=50 ft, *S*=25%)	0.5	35	0.15	2.67	0.2	2.8	1.4
Total Site		16	35					6.4 tons over 2 months

[1]10% of Annual *R*, annual *R* is 350, so project phase partial *R* is (350)(0.1)=35.
[2]From county soils map and anticipated surface soils during this phase.
[3]*C* factors are based on native good cover for undisturbed areas, gravel pads for roads, buildings, and parking lots, and mulch in areas and slopes to be landscaped and the runoff pond.

Predicting the Benefits of Alternative Mulches

The USLE (and now the RUSLE) has long been used to estimate the benefits of different management systems on reducing erosion rates from construction sites. This has mostly been done by estimating *C* and *P* values for different control strategies. Mulches have been directly studied at many erosion test plots, enabling some basic *C* factors to be determined, as shown in Table 4.12. These earlier measured *C* factors did not include the modern erosion control mats. Many of the mat producers have sponsored independent evaluations of *C* factors and tolerable shear stress conditions for their mats to enable the developer to select suitable selection of different materials. Chapter 6 will present this additional information.

Use and Selection of Vegetation at Construction Sites

As is obvious from the preceding discussions, erosion prevention at construction sites is critical. The following chapters will show that sediment control to remove particulates and other pollutants from the water flowing from a construction site is generally much more costly and less effective than preventing the erosion from occurring in the first place. The use of vegetation to protect disturbed areas soon after clearing and grading is one of the most important erosion preventive practices. The following information in this chapter presents additional information on "vegetation controls" that can be used to meet these local needs, mostly summarized from the *Alabama Handbook for Erosion Control, Sediment Control, and Stormwater Management on Construction Sites and Urban Areas* (USDA, 2003)—an example of the type of guidance information usually available from regional construction site "handbooks." Obviously, other guidance documents are usually available for other local areas and should be used whenever available.

As stated in the *Alabama Handbook,* a dense, vigorous growing vegetative cover protects the soil surface from raindrop impacts, a major force in causing erosion losses. Also, vegetation will shield the soil surface from the scouring effects of overland flows and decrease the erosive capacity of the flowing water by reducing its velocity. The shielding effect of a plant canopy is augmented by roots and rhizomes that hold the soil together, improve its physical condition, and increase the rate of infiltration, further decreasing runoff. Plants also reduce the moisture content of the soil through transpiration, thus increasing its capacity to absorb water. Suitable vegetative cover therefore offers excellent erosion protection. It is also essential to the design and stabilization of many structural erosion control practices. Vegetative cover is relatively inexpensive to achieve and maintain. Also, it is often the only practical, long-term solution to stabilization and erosion control on many disturbed sites. Planning from the start for vegetative establishment reduces its cost, minimizes maintenance and repair, and makes structural erosion control measures more effective and less costly to maintain.

Plant selection should be considered early in the process of preparing the erosion and sedimentation control plan. A wide diversity of plant species can be grown in Alabama due to the variation in both soils and climate. However, for practical, economical stabilization and long-term protection of disturbed sites, plant selection should be made with care. Many plants are inappropriate for soil stabilization because they do not protect the soil effectively, or they can not be established quick and easy. Some plants may be very effective for soil stabilization, but are not aesthetically acceptable on some sites. Some plants may even become troublesome pests.

The Story of Kudzu

Excerpted from the "Amazing Story of Kudzu" (http://www.cptr.ua.edu/kudzu/):

> In Georgia, the legend says
> That you must close your windows
> At night to keep it out of the house.
> The glass is tinged with green, even so . . .
>
> From the poem, Kudzu, by James Dickey

Kudzu was introduced to the United States in 1876 at the Centennial Exposition in Philadelphia, Pennsylvania. Countries were invited to build exhibits to celebrate the 100th birthday of the U.S. The Japanese government constructed a beautiful garden filled with plants from their country. The large leaves and sweet-smelling blooms of kudzu captured the imagination of American gardeners who used the plant for ornamental purposes.

Florida nursery operators, Charles and Lillie Pleas, discovered that animals would eat the plant and promoted its use for forage in the 1920s. Their Glen Arden Nursery in Chipley sold kudzu plants through the mail. A historical marker there proudly proclaims "Kudzu Developed Here." During the Great Depression of the 1930s, the Soil Conservation Service promoted kudzu for erosion control. Hundreds of young men were given work planting kudzu through the Civilian Conservation Corps. Farmers were paid as much as eight dollars an acre as incentive to plant fields of the vines in the 1940s.

The problem is that it just grows too well! The climate of the Southeastern U.S. is perfect for kudzu. The vines grow as much as a foot per day during summer months, climbing trees, power poles, and anything else they contact. Under ideal conditions kudzu vines can grow sixty feet each year.

While they help prevent erosion, the vines can also destroy valuable forests by preventing trees from getting sunlight.

Kudzu (rumored to have been imported into the U.S. for land conservation purposes) can readily take over and kill the existing vegetation.

Kudzu covering a pasture in Alabama.

Kudzu covering trees in Alabama.

This problem led Dr. James H. Miller of the U.S. Forest Service in Auburn, Alabama, to research methods for killing kudzu. In eighteen years of research, he has found that one herbicide actually makes kudzu grow better while many have little effect. Miller recommends repeated herbicide treatments for at least four years, but some kudzu plants may take as long as ten years to kill, even with the most effective herbicides."

Currently, kudzu covers about seven million acres of the south. The USDA declared it a weed in 1972.

Plant Hardiness Zones

The U.S. Department of Agriculture has produced plant hardiness zone maps. They are normally used to help determine the suitability of different plants for an area. These maps are based on the annual average low temperatures and are therefore most appropriate for permanent vegetation. Short-term vegetation use does not necessarily have to following the same selection guidelines needed for permanent vegetation. In all cases, it is important to contact the local NRCS office, or other erosion control specialists, for the most suitable vegetation to consider for a specific site. Figure 4.14 and Table 4.22 shows the current USDA hardiness zone map and the annual average minimum temperatures for selected cities.

It is possible to simplify this map into fewer zones for some vegetation types. As an example, the Patten Seed company (http://www.pattenseed.com/info-chsel.html) simplified the map into five zones for the purpose of selecting permanent turfgrasses. This was possible because these grasses are generally adaptable to a broader range of temperatures than other plants, such as flowers, shrubs and trees. The following lists their recommendations for turfgrasses in each of these consolidated areas. Not all of these turfgrasses are suitable for erosion control applications, but this list does illustrate a simplified approach:

Area 1—This area includes lower coastal North Carolina, coastal South Carolina, coastal and south Georgia, all of Florida, and lower and coastal sections of Alabama, Mississippi, Louisiana, and Texas. This area should use the Hot Climate Grasses which include Bermuda, Bahia, Centipede, Carpet, St. Augustine, and Zoysia.

Area 2—This zone is north of Area 1 and includes north coastal North Carolina, much of central South Carolina, central Georgia, north and central Alabama, northern Louisiana, south west Tennessee, all except the most northern part of Arkansas, most of central Texas, and the southern portion of Oklahoma. This area should use a limited set of the Hot Climate Grasses including Bermuda, Centipede, and Zoysia.

Area 3—This area covers much of the middle U.S. including parts of New Jersey, Maryland, Delaware, Virginia, western North Carolina, western Tennessee, western Kentucky, southern Indiana, southern Illinois, southern Missouri, southern Kansas, northern Oklahoma, northern Texas, most of New Mexico, southern Arizona, and most of coastal California. This area should use Cool Season Grasses including Tifway Bermuda, Meyer Zoysia, and Zenith Zoysia.

Area 4—This area covers a band of the upper central U.S., including parts of Rhode Island and Connecticut, a small portion of southern New York, northern New Jersey, eastern Pennsylvania, eastern West Virginia, northern Virginia, east Tennessee, central Kentucky, most western Ohio, northern Indiana, southern Michigan, northern Illinois, southern Iowa, northern Missouri, southern Nebraska, northern Kansas, central Colorado, northwest New Mexico, northern Arizona, southeast Utah, the southern tip of Nevada, much of central California, coastal Oregon, and south coastal Washington. This zone should use Cool Season Grasses including Meyer Zoysia, and Zenith Zoysia.

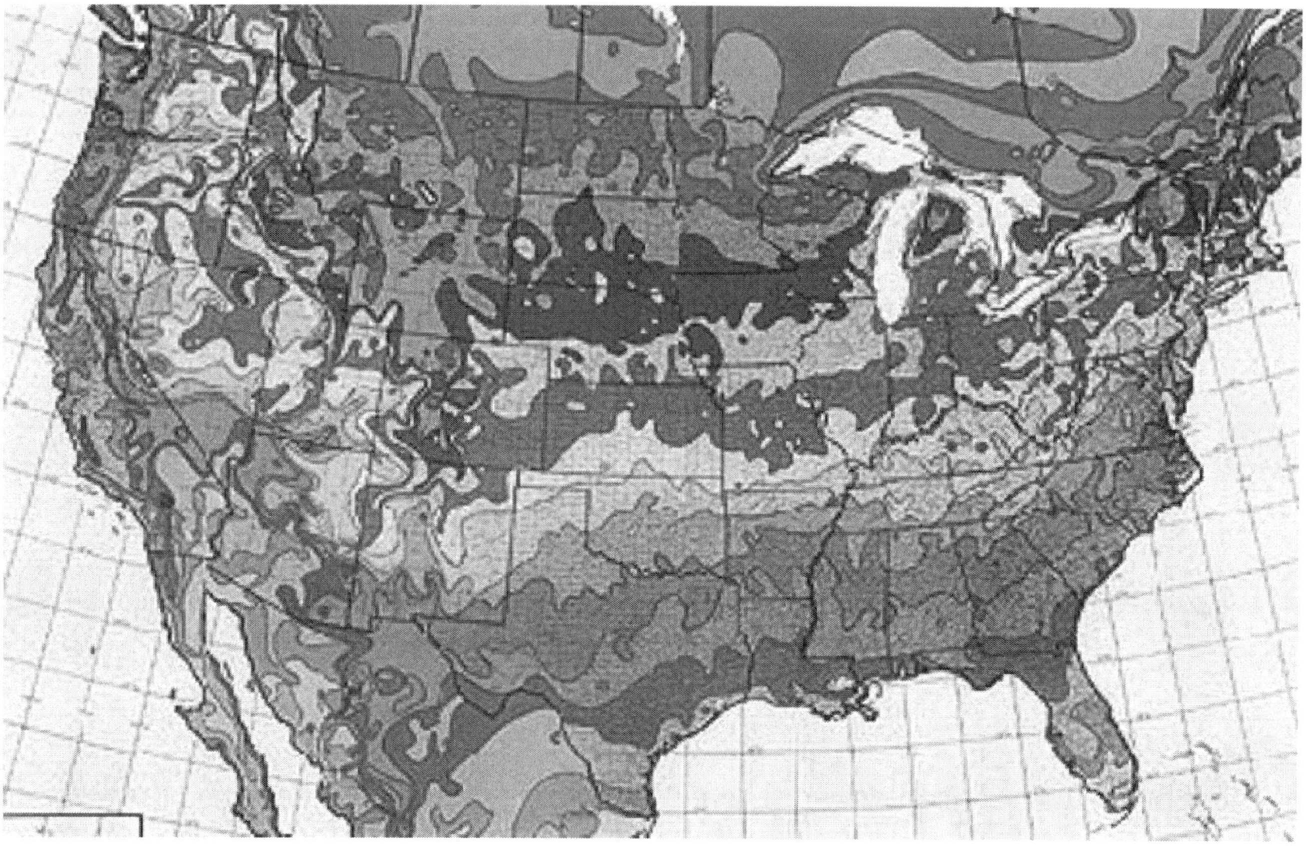

Figure 4.14. USDA Plant Hardiness Zone Map (Source: Agricultural Research Service, USDA).

Area 5—This area covers the upper U.S., north of Area 4 and should use Cool Season Grasses.

Cebeco International Seeds (http://www.intlseed.com/index.html) provides an example of seed selection guidance for erosion control. This information is specifically for the Pacific Northwest, but many of these grass types are used in other areas of the country. The following is a description of introduced grass species commonly used for erosion-control seed mixtures, excerpted from a summary paper by Craig Edminster of Cebeco International Seeds. The

TABLE 4.22. Annual Average Minimum Temperatures for Selected Cities.

Fahrenheit	Celsius	Example Cities
Below –50°F	Below –45.6°C	Fairbanks, Alaska; Resolute, Northwest Territories (Canada)
–50 to –45°F	–42.8 to –45.5°C	Prudhoe Bay, Alaska; Flin Flon, Manitoba (Canada)
–45 to –40°F	–40.0 to –42.7°C	Unalakleet, Alaska; Pinecreek, Minnesota
–40 to –35°F	–37.3 to –39.9°C	International Falls, Minnesota; St. Michael, Alaska
–35 to –30°F	–34.5 to –37.2°C	Tomahawk, Wisconsin; Sidney, Montana
–30 to –25°F	–31.7 to –34.4°C	Minneapolis/St.Paul, Minnesota; Lewistown, Montana
–25 to –20°F	–28.9 to –31.6°C	Northwood, Iowa; Nebraska
–20 to –15°F	–26.2 to –28.8°C	Des Moines, Iowa; Illinois
–15 to –10°F	–23.4 to –26.1°C	Columbia, Missouri; Mansfield, Pennsylvania
–10 to –5°F	–20.6 to –23.3°C	St. Louis, Missouri; Lebanon, Pennsylvania
–5 to 0°F	–17.8 to –20.5°C	McMinnville, Tennessee; Branson, Missouri
0 to 5°F	–15.0 to –17.7°C	Oklahoma City, Oklahoma; South Boston, Virginia
5 to 10°F	–12.3 to –14.9°C	Little Rock, Arkansas; Griffin, Georgia
10 to 15°F	–9.5 to –12.2°C	Tifton, Georgia; Dallas, Texas
15 to 20°F	–6.7 to –9.4°C	Austin, Texas; Gainesville, Florida
20 to 25°F	–3.9 to –6.6°C	Houston, Texas; St. Augustine, Florida
25 to 30°F	–1.2 to –3.8°C	Brownsville, Texas; Fort Pierce, Florida
30 to 35°F	1.6 to –1.1°C	Naples, Florida; Victorville, California
35 to 40°F	4.4 to 1.7°C	Miami, Florida; Coral Gables, Florida
above 40°F	above 4.5°C	Honolulu, Hawaii; Mazatlan, Mexico

following excerpt from this paper illustrates the importance of proper seed selection and the assistance of an expert:

Ryegrass has been used extensively as a short-lived component in erosion control mixtures. Their key attribute in erosion control is rapid seedling establishment, tolerance to slightly acidic soils and excellent spring, and fall forage growth when rainfall is abundant in the Pacific Northwest. In addition, they serve as an excellent nurse crop in low input plantings. Ryegrass is intolerant of droughty, nutrient-deficient soils, and therefore may senescence and die during the early establishment period, which provides an excellent growing environment for long lived, grass species. *Lolium perenne* (Perennial ryegrass) tetraploid and diploid sources are commonly used in erosion control plantings, the diploid being more tolerant of grazing pressure (mowing) and more persistent than the larger leafed, more robust and less cold tolerant tetraploid. The use of very late maturing diploid perennials, such as Elka and Essence®, has been recommended to reduce reseeding potential and enhance long lived species establishment.

Annual ryegrass (*Lolium multiflorum*) is the most commonly used cool-season grass in conservation and erosion control in the Pacific Northwest. Annual ryegrass has the best seedling vigor and lowest cost per pound of all the cool season grass species. At low planting rates it can provide good to fair nurse or companion crop attributes. At extremely high seeding rates it can provide living mulch attributes. Annual ryegrass has excellent reseeding capability and seeds can remain dormant in soil for up to five years. Therefore, its use is often discouraged where mixed-species longevity is desired. Westerwold ryegrass, and genetic mixtures containing high percentages of Westerwold germplasm, are readily available in the Pacific Northwest (cv Gulf, Oregon Common). Westerwold ryegrass requires a very short floral induction period for plant vernalization and results in reseeding potential. Under these circumstances, annual ryegrass can become a weedy grass in erosion-control mixtures. True Italian ryegrass cultivars (cv Sultan, Total), developed in Europe, that require significantly more floral induction to induce seed production should be considered as an alternative if annual ryegrass is used.

There are six species of fine fescue recognized for their use in turf and forage production systems in the Pacific Northwest. They include, but are not limited to, chewings fescue *F. rubra* L. subsp. commutata, hard fescue *F. longifolia*, and sheeps or blue fescue *F. ovina*; and the rhizomatous type: slender creeping red fescue F. *rubra* L. subsp. *tricholphylla* and strong creeping red fescue *Festuca rubra* L. subsp. *rubra*. Strong creeping red fescue has been used extensively in conservation and erosion control mixtures primarily because of excellent seedling vigor, tolerance to acidic soils, good shade tolerance (understory), and rhizomatous growth habit. Strong creeping red fescue requires very little supplemental fertilization once established, and grows well on shallow- and rocky-cut bank riparian and upland sites. Strong creeping red fescue is a moderately tall plant species and is highly compatible with many other tall and short serial species of introduced grass.

Timothy (*Phleum pratense*) has been used as a minor component in mixtures for wetland, bottomland and stream bank restoration where imperfect soil drainage may be a limiting factor. It is poorly adapted for erosion control mixtures because of its lack of seedling vigor. Therefore, mixtures containing rapid establishing species as a nurse crop are advised. Timothy is also intolerant of drought soils so its establishment on well-drained, sloped areas in riparian and upland sites is not recommended.

Orchardgrass (*Dactylis glomerata*) is a bunchgrass that has been used extensively in erosion control mixtures in West Coast Mountain Region. It has good seedling vigor, early spring forage growth, but requires well drained soil sites to persist. It is tolerant of mild soil acidity, and moderately shade tolerant, but requires supplemental fertilizer for proper growth. Orchardgrass cultivars are segregated into different maturity groups (early, medium and late) for their relative feed value when used in legume-based forage production systems. Early-maturing short-statured varieties such as Paiute, Palestine are often recommended because they enter dormancy during the summer when soil moisture is depleted in the Pacific Northwest. Upon dehydration in the fall, they regrow and persist.

Tall fescue (*Festuca arundinacea*) has been used on occasions in conservation and erosion control with mixed results. Tall fescue has poor seedling vigor, but exhibits good shade tolerance. Once established, it is a very dominate forage producer and may require aggressive management to constrain growth (mowing, burning). Tall fescue is tolerant of acidic, poorly-drained, shallow-soil sites, but prefers well-drained sandy loam soil sites. In contrast to other cool-season grasses, tall fescue may not enter into summer-induced dormancy or rest period. Its deep, extensive root system facilitates deep soil-profile water uptake during the summer, and tall fescue can dominate a riparian, upland or wetland site.

Kentucky bluegrass (*Poa pratensis*) has been used to a limited extent in the Pacific Northwest. Its most redeeming characteristic is the presence of rhizomes, which provides good soil and plant interface to reduce soil erosion potential. Its most limiting factors are that it has the poorest seedling vigor of all cool-season grasses and is intolerant of slightly acidic to acidic soils. To persist, it must be established in soils with excellent internal drainage. It also requires moderate to high soil nutrition and does best in a diurnal environment where summers are hot and winters cold.

Creeping bentgrass (*Agrostis palustris*), "the golf course greens grass," has been used to a very limited extent for erosion control in the Pacific Northwest. Bentgrass is very tolerant of acidic, poorly drained soils and exhibits fair to poor seedling vigor. If hydrated throughout the season, it can dominate a planting site because of its short, aggressive stoloniferous growth habit. It is therefore incompatible in grass seed mixtures. Established stands of creeping bentgrass will require burning or very short mowing to enhance persistence.

Highland bentgrass (*Agrostis castellana*) is very tolerant of acidic, poorly-drained, or shallow-soil sites and exhibits good to fair seedling vigor. It also exhibits better summer drought tolerance than creeping bentgrass. Highland bentgrass has larger, more robust stolons than creeping bentgrass, and provides more forage for grazing animals and wildlife. Similar to creeping bentgrass, it can dominate a planting site because of its aggressive stoloniferous growth habit and is therefore considered incompatible in grass seed mixtures.

Little colonial bentgrass (*Agrostis tenuis*) has been used in conservation and erosion control projects in the Pacific Northwest. This is more the result of short seed supplies than a lack of its adaptation in conservation program. Colonial bentgrass is the only Agrostis species that is compatible in mixture with other cool-season grass species. This short, acid-tolerant, fine-leaved species has short prolific stolons that grow more upright than prostate. It exhibits excellent drought tolerance, requires only modest soil fertility and has good to fair seedling germination.

Selecting the Right Grasses and/or Legumes

The *Alabama Handbook* states that single-species plantings are desired in some cases, but most of the time a mixture is more desirable. Mixtures can be selected that may provide protective cover more quickly and can be more enduring than a single species. Mixtures need not be elaborate. The addition of a quick-growing annual or short-lived perennial provides early protection and facilitates establishment of a slower-growing and longer-living perennial. It is important to evaluate the merits and weakness of each species in selecting the mixtures for the specific site to be treated. The addition of a companion or "nurse" crop (quick-growing annual or weak perennial added to permanent mixtures) is a good practice on difficult sites, when late seeding, or in situations where the development of permanent cover is likely to be slow. The companion crop germinates and grows rapidly, holding the soil until the perennial species becomes established. Seeding rate of the companion crop must be limited to avoid crowding, especially under optimum growing conditions.

Detailed information on plant species adapted for soil stabilization use in Alabama is contained in the following discussions and from the Internet sources listed at the end of this chapter. Most of these commercial suppliers of seeds and sod will help select the most appropriate species for local site conditions. Local USDA Agricultural Extension offices may also be able to provide updated guidance. Using this information makes plant selection more straightforward for most situations. Specific seeding rates and planting instructions are presented in specifications for local conditions. They often are provided by regulatory agencies.

Annual plants grow rapidly, mature, and die in one growing season. They are useful for quick, temporary cover or as a companion crop for slower growing perennials. Rye (cereal) is usually superior to other small grains (wheat, oats, or barley) for temporary cover. It has more cold hardiness than other annuals and will germinate and grow at lower temperatures. It will provide more fall and early winter growth and matures earlier than other small grains. Rye germinates quickly and is tolerant of poor soils. Including rye in fall-seeded perennial mixtures is particularly helpful on difficult soils and erodible slopes or when seeding is late. However, seeding rates of rye should be limited to the suggested rates because a thick stand will suppress the growth of the desired perennial seedlings. No more than 60 lb/acre should be planted when rye is used as a companion crop. Rye does grow fairly tall in the spring which may be undesirable. If this is a problem, some of the shorter growing varieties of wheat may be used. Annual ryegrass is not recommended for use as a companion crop in perennial mixtures in Alabama. It is highly competitive and, if included in mixtures, crowds out most other species before it matures in late spring or early summer, leaving little or no lasting cover. It will provide dense cover rapidly, so it can be effective as a temporary seeding, but if allowed to mature, the seed volunteers and can seriously interfere with subsequent efforts to establish permanent cover.

Millets (Browntop, Foxtail) are warm-season annuals, useful for temporary seeding or as a nurse crop. Browntop millet has early rapid growth, growing two to three feet in height. It is adapted to fine and medium textured soils of moderate productivity. Foxtail is a fine stemmed plant growing to a height of four to five feet. The leaves are broad and flat. Foxtail millets do best under fairly abundant moisture conditions. German millet is a type of foxtail millet.

Sudangrass and sorghum-sudangrass hybrids, like the millets, are warm-season annuals which are useful for temporary vegetation. They are better adapted to medium- to heavy-textured soils. The small-stemmed, shorter-growing varieties are more satisfactory for temporary vegetation than the tall coarse-stemmed varieties.

Annual lespedeza is a warm-season, reseeding annual legume growing to a height of six to twelve inches. It is tolerant of low fertility and is adapted to the climate and most soils throughout Alabama. It is not adapted to alkaline soils of the Black Belt or to deep sands. It is a good companion crop for spring-planted sericea lespedeza, filling in weak or spotty stands the first season without suppressing the sericea. Annual lespedeza can heal damaged areas in the perennial cover for several years after initial establishment. Two species of annual lespedeza are grown in Alabama. "Common" annual lespedeza volunteers in many parts of Alabama and is sold under the variety name Kobe. Korean lespedeza is a slightly larger, coarser and earlier-maturing plant sold under several variety names. Kobe is superior on sandy soils and generally preferable in south Alabama. Korean is better in north Alabama as the seeds mature earlier. The preferred seeding dates for annual lespedeza are in the late winter to early spring. It can be mixed with fall seeding, in which case some seeds remain dormant over the winter and germinate the following spring.

Perennial plants, once established, will live for more than one year. They may die back during a dormant period, but will grow back from their underground tubers or rhizomes in succeeding years. Stands of perennials will persist for a number of years under proper management and environmental conditions. They are the principal components of permanent vegetative covers. Cool-season perennials produce most of their growth during the spring and fall and are more cold-hardy than most warm-season species.

Tall fescue is the only cool-season perennial grass recommended for vegetating disturbed soils in Alabama. Tall fescue, a cool-season grass, is the most widely-used species in north Alabama for erosion control. It is well adapted to all of north Alabama and all but the most droughty soils of central Alabama. Also it can be grown on the Black Belt soils of south Alabama. It thrives in full sun to partial shade and is fairly easy to establish. It will provide stabilization the year of establishment. Because tall fescue has a bunch-growth habit, it is slow to fill in areas with poor stands. Therefore, some maintenance will be required on washed-out areas or areas of spotty stands to prevent further damage. A number of new varieties of tall fescue are becoming available for lawn and other turf use and several offer definite improvements. However, their higher cost over the standard, Kentucky 31, is seldom justified solely for purposes of stabilization and erosion control. Also, fescue seed infected with a fungal endophyte are preferred since endophyte-infected plants are more hardy, resulting in longer-lasting stands. Tall fescue is a fall-planted grass. Liberal fertilization and proper liming are also essential for prompt establishment, but once established, it can tolerate minimal maintenance almost indefinitely. White clover is sometimes planted with tall fescue.

Warm-season perennials initiate growth later in the spring than cool-season species and experience their greatest growth during the hot summer months. Most species of warm-season perennials do better in the southern one-half of Alabama, but there are species or varieties that will grow in north Alabama. The following grasses have proven the most useful for soil stabilization:

- Bahiagrass is a warm-season perennial grass particularly well adapted for growing on sandy soils in the southern half of Alabama. It will tolerate acid and low fertility soils, grow in full sun to light shade, and persist almost indefinitely with little or no maintenance after it is established. However, bahiagrass seedlings are small and lack the vigor some species of warm-season grasses possess; it usually takes two years to establish a good sod. Bahiagrass is established with seed. Bahiagrass does produce a fairly dense sod suitable for low maintenance areas. It has a high resistance to wear and recovers fairly fast from wear. It produces rhizomes and will fill in small bare spots fairly fast. Bahiagrass will produce seedheads about one to two feet in height throughout the growing season and, where this is not a problem, it is probably the best choice for stabilizing soil in the southern one half of the state. Pensacola is the better variety of bahiagrass for soil stabilization. It is more tolerant to upland sites and is more cold tolerant than Argentine bahiagrass.
- Common Bermudagrass is a long-lived perennial that spreads by creeping stolons and rhizomes outward several feet in a growing season. It will survive extreme heat and drought. It is not shade tolerant. Bermudagrass is best adapted to well-drained fertile soils. It does poorly on extremely droughty sandy soils and will not grow on poorly-drained soils. It responds well to fertilizer and will establish a dense sod quickly from seed. Common bermudagrass will grow in all areas of the state. Bermudagrass requires more maintenance than bahiagrass and, if a regular maintenance fertility program is not used, it will tend to slowly decline. It has a high resistance to wear and a fast recovery from wear which makes it a good choice for heavy use areas.

There are two types of bermudagrass which are important in soil stabilization. Common bermudagrass, which can be established with seeds or sprigs, and turf-type bermudagrasses which must be established from vegetative material. Common bermudagrass has longer internodes and larger leaves than the turf-type hybrid bermudagrass. When common bermudagrass will be used for permanent vegetation, only seeds that are 98% pure common bermudagrass should be planted. Common bermudagrass seeds are often contaminated with giant-type bermudagrass seeds. Giant-type bermudagrass is very competitive and fast growing, but is not cold hardy in Alabama. So when common bermudagrass seed contains even a small percent of giant type bermudagrass seed, they will be choked out by the giant-type bermudagrass. Since the giant-type bermudagrass is killed by the cold, a good sod the year of establishment becomes destroyed the second year.

The turf-type hybrid bermudagrasses have fine leaves and short internodes which make them desirable for lawn, golf courses and other areas where a quality turf is desired. However, turf-type hybrid bermudagrasses are more costly to establish because they must be planted from sprigs, plugs, or solid sodded. Tifway 419 is the most commonly used turf-type hybrid bermudagrass. The agronomic varieties of hybrid bermudagrasses do not lend themselves to soil stabilization of construction areas. They too must be established with vegetative material which makes them costly to establish.

Sericea lespedeza or sericea is a deep-rooted, drought-resistant perennial legume, adapted to all but the poorly drained and deep sandy soils of the state. It is long lived, tolerant of low fertility soils, pest free, and will fix nitrogen. It can be a valuable component in most low-maintenance mixtures. Sericea is slow to establish and will not contribute much to prevention of erosion the first year; however, once established it persists indefinitely on suitable sites. Plantings that include sericea require mulch and should include a companion crop such as browntop millet, annual lespedeza, or common bermudagrass. Sericea should be planted as early as possible within the planting date range so as to reduce as much weed competition as possible. Also, sericea may be planted in the late fall and winter months because many of the seeds will lie dormant until germination the following spring. Sericea does not tolerate frequent mowing and may be considered unsightly because the old top growth breaks down slowly.

Crownvetch is a deep rooted, perennial legume adapted only to north exposures in the northern tier of counties in Alabama. It is useful on steep slopes and rocky areas that are likely to be left unmoved. It can be seeded in the spring or fall. Crownvetch requires a specific inoculant.

Summary: Selection of Erosion Control Grasses

This section was excerpted from material prepared by Jason Kirby (2003) as part of his MSCE thesis investigating the hydraulics of grass swales while at the Department of Civil and Environmental Engineering, the University of Alabama. All grasses are not the same for erosion control as they vary in their ability to protect and survive in a given environment. Ryegrass is moderately dark green with good density (measured by the number of blades of grass per square inch) and a fine texture. This species is known to establish quickly and produce a stable/hearty turf. In addition to its low maintenance requirements, ryegrass has good tolerance to sun, shade, drought, temperature, and wear. Bluegrass displays a dark green color with dense uniform coverage. Bluegrass requires moderate maintenance (watering, mowing, etc.) and is less tolerant of changes in temperature, shade, drought than rye grass. Bluegrass can withstand more abuse (foot traffic, wear) than other similar grasses. Finally, Fescue has deep green blades and is known for its rapid germination and establishment. Fescue is quite tolerant to changes in temperature, wear, shade, and drought. Fescue can be maintained with limited effort. Unfortunately, all of these above listed grasses are considered cool-season grasses and have limited application in the Southeast.

Bermuda, Centipede, and Zoysia share characteristics similar to the above listed grass, but are better suited to the hotter conditions in the Southeast. Commercial grass suppliers (S&S seeds, for example, at www.ssseeds.com) will recommend grass types/blends based on site location and other characteristics (slope, watering, etc). These recommendations will identify the appropriate species and the suggested method of application, such as by seed or sod.

The decision to use seed or sod to establish a specified grass type is a crucial one. While most grasses can be established either way, the initial costs and characteristics can be significantly different. The following table is a general comparison between seeding and sodding.

Sod, as a rule of thumb, cost about 20 times more than seeding to install; however, this cost is usually offset by sod's ability to be planted year round (in the southeast, at least), uniform establishment, and instant erosion protection. Sod is available throughout the country from various national and local sod farms. These farms carry numerous species with varying levels of quality. Rapid establishment in grass-lined drainage channels is a great benefit of sod over seeding, although the use of reinforcing turf mats (described in Chapter 5) enables the use of seed in channels with immediate benefit. In fact, the combination of reinforcing turf mats and grass seed may be superior to sod in a channel (but more expensive).

High quality sod is expensive (up to \$0.60 per ft^2) but will contain fewer weeds and have a better appearance. Lower quality sods have more weeds/pests but save money and will still establish a good ground cover. Laying sod can cost up to \$15,000 an acre, so while it has enhanced erosion control properties, it needs to be used as a permanent control or, if temporary, on a small scale to be cost effective.

Seeding an area is much less expensive than using sod (\$250 an acre) and can provide adequate erosion protection given time. Germination can take up to a month, and up to six months may be needed for grass establishment, depending on the grass type and planting conditions. Until full grass development, constant maintenance (watering, replanting, etc.) will be required. In addition to seeding a site for grass creation, annual species can be used to supplement established grasses that may go seasonally dormant. The extra attention seeding requires may make sod a more attractive option, depending on the site. The decision, in effect, comes down to a decision between excellent initial erosion protection at high cost, or low initial cost with less immediate erosion control.

Sod sizing will depend on the farm and grass type selected. Sod pieces can range from 1 ft × 2 ft (residential) to 8 ft × 32 ft (commercial applications, especially for golf courses). Staples may be required to anchor the sod into place until the root system is established.

Once grass has been established (seed or sod), its physical characteristics become indistinguishable (sod will have better erosion resistance initially, but once the seeds develop, the differences are minimal). Typically, grass can withstand a maximum permissible velocity of around 5 ft/s with an absolute maximum of 8 ft/s. Table 4.24 (USDA, 1954) lists the permissible velocities for several grasses.

TABLE 4.23. Comparison Between Seeding and Sodding.

	Seeding	Sod
Planting Season	Fall, and perhaps Spring	Anytime
Water Requirements	Very High for Germination/Establishment	Low (6″ initially then limited for next 3 weeks)
Soil Preparation	Tillage, fertilization, etc.	Same as for seeding
Weed control	Requires Herbicide	Minimal, if any
Uniformity	Varies based on weeds, washouts, etc.	99–100%
Usability (Traffic)	None for 2 months, then limited up to 6 months	Normal to high within 2 weeks
Erosion Control	None until established, rain will necessitate repair	Good control after installation
Cost	\$0.01 to \$0.04 per ft^2	\$0.14 to \$0.60 per ft^2

TABLE 4.24. Permissible Velocities for Several Grasses.

Cover	Slope Range	Erosion Resistant Soils Maximum Permissible Velocity (ft/s)	Easily Eroded Soils Maximum Permissible Velocity (ft/s)
	0–5	8	6
Bermudagrass	5–10	7	5
	over 10	6	4
	0–5	7	5
Kentucky Bluegrass	5–10	6	4
	over 10	5	3
Grass Mixture (Rye, Fescue)	0–5	5	4
	5–10	4	3
Crabgrass	0–5	3.5	2.5
Common Lespedeza	0–5	3.5	2.5

Handbook of Channel Design for Soil and Water Conservation. Technical Paper TP-61. 1954.

Temporary Vegitation—Seeding

The following is from the *Alabama Handbook* (USDA, 2003) and describes seedbed preparation guidance for temporary vegetation. Guidance such as this is usually presented in regional erosion control handbooks.

Definition

Planting rapid growing annual grasses, small grains, or legumes to provide initial, temporary cover for erosion control on disturbed areas.

Purpose

To temporarily stabilize bare areas that will not be brought to final grade for a period of more than 30 working days.

Temporary seedling controls runoff and erosion until permanent vegetation or other erosion control measures can be established. In addition, it provides residue for soil protection and seedbed preparation and reduces problems of mud and dust production from bare soil surfaces during construction.

Conditions Where Practice Applies

On any cleared, bare, or sparsely vegetated soil surface where vegetative cover is needed for less than one year. Application of this practice include diversions, dams, temporary sediment basins, temporary road banks, and soil stockpiles.

Planning Considerations

1. Temporary vegetative cover can provide short term protection before establishing perennial vegetation. It can control rills and excessive erosion on earthen sediment control structures such as diversions, dams, and sediment basins.
2. Temporary vegetation will reduce the amount of maintenance associated with sediment basins. The frequency of sediment basin cleanups will be reduced if watershed areas outside the active construction zone are stabilized.
3. Certain plant species used for temporary vegetation will produce large quantities of residue which can provide mulch for establishment of permanent vegetation.
4. Proper seedbed preparation and selection of appropriate species are important with this practice. Failure to follow establishment guidelines and recommendations carefully may result in an inadequate or short-lived stand of vegetation that will not control erosion.
5. Temporary vegetation is used to provide cover for no more than one year. Permanent vegetation should be established at the proper planting time for permanent vegetative cover.

Specifications

1. *Grading and shaping*—Minor grading and shaping may be needed to provide a surface on which equipment can safely and efficiently be used for seedbed preparation and seeding.
2. *Plant Selection*—Plant selection for temporary vegetation should be based on plant characteristics, site and soil conditions, time of year of planting, method of planting, and the needed use of the vegetative cover. Plant species commonly used for temporary cover are contained in Table 4.25.
3. *Soil Amendments*
   a. Apply lime according to soil test recommendations. If the pH of the soil in not known, use 2 tons of agricultural limestone or equivalent per acre on coarse textured soils and 3 tons per acre on fine textured soils. Do not apply lime to alkaline soils or to areas which have been limed during the preceding 2 years.
   b. Fertilizer application rates should be based on soil test

TABLE 4.25. Commonly Used Plants for Temporary Cover in Alabama.

		Seeding Dates		
Species	Seeding Rate/Ac	North Alabama	Central Alabama	South Alabama
Millet, Browntop or German	40 lbs	May 1–Aug 1	Apr 1–Aug 15	Apr 1–Aug 15
Rye	3 bu	Sep 1–Nov 15	Sep 15–Nov 15	Sep 15–Nov 15
Ryegrass	30 lbs	Aug 1–Sep 15	Sep 1–Oct 15	Sep 1–Oct 15
Sorghum-Sudan Hybrids	40 lbs	May 1–Aug 1	Apr 15–Aug 1	Apr 1–Aug 15
Sudangrass	40 lbs	May 1–Aug 1	Apr 15–Aug 1	Apr 1–Aug 15
Wheat	3 bu	Sep 1–Nov 1	Sep 15–Nov 15	Sep 15–Nov 15

results. When soil test are not possible, apply 500 to 700 pounds of 10-10-10 grade fertilizer.

4. *Seedbed Preparation*—Complete grading before preparing seedbeds and install all necessary erosion control practices, such as sediment basins. If soils become compacted during grading, loosen them to a depth of 6 to 8 inches using a ripper or chisel plow. Good seedbed preparation is essential to successful plant establishment. A good seedbed is well pulverized, loose, and smooth. Incorporate lime and fertilizer into the top 6 inches of soil during seedbed preparation. If rainfall has caused the surface to become sealed or crusted, loosen it just prior to seeding by disking, raking, harrowing, or other suitable methods. When hydroseeding methods are used, the surface should be left with a more irregular surface of clods.
5. *Planting*—Evenly apply seed using a cyclone seeder (broadcast), drill, cultipacker seeder, or hydroseeder. Use seeding rates given in Table 4-25. Broadcast seeding and hydroseeding are appropriate for steep slopes where equipment cannot operate safely.

   Small grains should be planted no more than 1 inch deep, and grasses and legumes no more than 1/2 inch deep. Broadcast seed must be covered by raking or chain dragging, and then lightly firmed with a roller or cultipacker, Hydroseeding mixtures should include a wood fiber mulch which is dyed an appropriate color to facilitate uniform application of the seed.
6. *Mulching*—The use of an appropriate mulch will help ensure establishment of vegetative cover under normal conditions and is essential to seeding success under harsh site conditions. Harsh site conditions include:

   - seeding in late fall for winter cover (wood fiber mulches are not considered adequate for this use),
   - slopes steeper than 3:1, and
   - adverse soils (shallow, rocky, or high in clay or sand).

   If the area to be mulched is subject to concentrated water flow, as in channels, anchor mulch with netting, or preferably use sod or an erosion control mat. See Chapter 5 for determining channel stability requirements.
7. *Irrigation*—Use irrigation when available and needed to insure establishment. Apply irrigation at a rate that will not cause runoff.
8. *Maintenance*—Reseed and mulch areas where seedlings emergence is poor, or where erosion occurs, as soon as possible. Do not mow. Protect from traffic as much as possible.

Permanent Seeding

The following is from the *Alabama Handbook* (USDA 2003) and describes seedbed preparation guidance for permanent vegetation. Similar guidance may be found in other regional erosion control handbooks.

Definition

Controlling runoff and erosion on disturbed areas by establishing perennial vegetative cover with seed.

Purpose

To reduce erosion and decrease sediment yield from disturbed areas, and to permanently stabilize such areas in a manner that is economical, adapts to site conditions, and allows selection of the most appropriate plant materials.

Conditions Where Practice Applies

Disturbed areas where permanent, long-lived vegetative cover is needed or the most effective method of stabilizing the soil. Permanent seeding may also be used on rough-graded areas that will not be brought to final grade for a year or more.

Planning Considerations

1. The most common and economical means of stabilizing disturbed soils is by seeding grasses and legumes. The advantages of seeding over other means of establishing plants include the smaller initial cost, lower labor input,

Permanent Seeding Along Highway Right-of-Way

and greater flexibility of method. Disadvantages of seeding include potential for erosion during the establishment stage, seasonal limitations on suitable seeding dates, and weather related problems such as droughts etc.

2. The probability of successful plant establishment can be maximized through good planning. The selection of plants for permanent vegetation must be site specific. Factors that should be considered are type of soils, climate, establishment rate, and management requirements of the vegetation. Other factors that may be important are wear, mowing tolerance, and salt tolerance of vegetation.
3. The use of irrigation (temporary or permanent) will greatly improve the success of vegetation establishment.
4. Endophyte-infected tall fescue appears to establish quicker and have better survival under adverse conditions than endophyte-free tall fescue.
5. The operation of equipment is restricted on slopes steeper than 3:1, severely limiting the quality of the seedbed that can be prepared. Provisions for establishment of vegetation on steep slopes can be made during final grading. In construction of fill slopes, for example, the last 4–6 inches might not be compacted. A loose, rough seedbed with irregularities that hold seeds and fertilizer is essential for hydroseeding. Cut slopes should be roughened.
6. Good mulching practices are critical to protect against erosion on steep slopes. When using straw, anchor with netting or asphalt. On slopes steeper than 2:1, jute, excelsior, or synthetic matting may be required to protect the slope.

Specifications

1. *Grading and shaping*—Minor grading and shaping may be needed to provide a surface on which equipment can safely and efficiently be used for seedbed preparation and seeding.
2. *Plant Selection*—Plant selection for permanent vegetation should be based on plant characteristics, site and soil conditions, time of year of planting, method of planting, and the intended use of the vegetated area. Climate factors can vary widely in Alabama and the three basic climatic zones were indicated previously.

   Plant selection may include companion plants to provide quick cover on difficult sites, late seedings, or in situations where the desired permanent cover may be slow to establishment. Annuals are usually used for companion plants. The plants used for temporary vegetation may be used for companion plants provided the seeding rate is reduced by one half. Ryegrass or other highly-competitive plants should not be used as a companion plant. Table 4.26 lists suitable perennial plants, along with the seeding rates and dates.
3. *Seedbed Requirements*—Establishment of vegetation should not be attempted on sites that are unsuitable due to inappropriate soil texture, poor drainage, concentrated overland flow, or steepness of slope, until measures have been taken to correct these problems.

   To maintain a good stand of vegetation, the soil must meet certain minimum requirements as a growth medium. A good growth medium should have these criteria:

   - Enough fine-grained (silt and clay) soil material to maintain adequate moisture and nutrient supply.
   - Sufficient pore space to permit root penetration.
   - Sufficient depth of soil to provide an adequate root zone. The depth to rock or impermeable layers such as hardpans should be 12 inches or more, except on slopes steeper than 2:1 where the addition of soil is not feasible.

TABLE 4.26. Perennial Grasses, Legumes and Mixtures; Seeding Rates; and Planting Dates for Disturbed Areas in Alabama.

		Seeding Dates and Adapted Area		
Species	Seeding Rate/Ac	North Alabama	Central Alabama	South Alabama
Bahiagrass, Pensacola	40 1bs	—	Mar l–July 1	Feb 1–Nov 1*
Bermudagrass, Common	10 lbs	Apr 1–July 1	Mar 15–July 15	Mar 1–July 15
Bahiagrass, Pensacola	30 lbs	—	Mar 1–July l	Mar 1–July 15
Common Bermudagrass	5 lbs	—	Mar 1–July l	Mar 1–July 15
Bermudagrass, Hybrid (Lawn Types)	Solid Sod	Anytime	Anytime	Anytime
Bermudagrass, Hybrid (Lawn Types)	Sprigs 1/sq ft	Mar 1–Aug 1	Mar 1–Aug 1	Feb 15–Sep 1
Fescue, Tall	40–50 lbs	Sep 1–Nov 1	Sep 1–Nov 1	—
Sericea	40–601bs	Mar 15–July 15	Mar 1–July 15	Feb 15–July 15
Sericea & Common Bermudagrass	40–60 lbs 10 lbs	Mar 15–July 15	Mar 1–July 15	Feb 15–July 15*

*Fall planting of bahia should contain 45 pounds of smallgrain to provide cover during winter months.

- A favorable pH range for plant growth, usually 6.0–6.5.
- Freedom from large roots, branches, stones, or large clods. Clods and stones may be left on slopes steeper than 3:1 if they are to be hydroseeded.

If any of the above criteria are not met—i.e., if the existing soil is too coarse, dense, shallow or acidic to foster vegetation—special amendments or topsoil should be used to improve soil conditions. The soil conditioners described below may be beneficial or, preferably, topsoil may be applied.

4. *Soil Conditioners*—In order to improve the structure or drainage characteristics of a soil, the following materials may be added. These amendments should only be necessary where soils have limitations that make them poor for plant growth or for turf establishment.
   a. *Peat*—Appropriate types are sphagnum moss peat, reed-sedge peat, or peat humus, all from freshwater sources. Peat should be shredded and conditioned in storage piles for at least 6 months after excavation.
   b. *Sand*—Clean and free of toxic materials.
   c. *Vermiculite*—Horticultural grade and free of toxic substances.
   d. *Rotted manure*—Stable or cattle manure not containing undue amounts of straw or other bedding materials.
   e. *Thoroughly rotted sawdust*—Free of stones and debris. All 6 lbs of nitrogen to each cubic yard.
5. *Soil Amendments*
   A. *Liming Materials*—Lime (Agricultural limestone) should have a neutralizing value of not less than 90 percent calcium carbonate equivalent and 90 percent will pass through a 10 mesh sieve and 50 percent will pass through a 60 mesh sieve. Selma chalk should have a neutralizing value of not less than 80 percent calcium carbonate equivalent and 90 percent will pass through a 10 mesh sieve.
   b. *Plant Nutrients*—Commercial-grade fertilizers that comply with current state fertilizer laws should be used to supply nutrients required to establish vegetation.
   c. *Rates of Soil Amendments*—Lime and fertilizer needs should be determined by soil tests. Soil testing can be performed by university soil testing laboratories. The local county Cooperative Extension Service can provide information on obtaining soil tests. Commercial laboratories that make recommendations based on soil analysis may be used.

   When soil tests are not available, use the following rates for application of soil amendments.

   Lime (Agricultural limestone or equivalent)

   - Light-textured, sandy soils: 2 tons/acre
   - Heavy-textured, clayey soils: 3 tons/acre (Do not apply lime to alkaline soils)

   Fertilizer

   - Grasses alone: 800 to 1200 lbs/acre of 10-10-10 or equivalent.
   - Grass-legume mixtures: 800 to 1200 lbs/acre of 5-10-10 or equivalent.
   - Legumes alone: 800 to 1200 lbs/acre of 0-10-10 or equivalent.

   d. *Application of Soil Amendments*—Apply lime and fertilizer evenly and incorporate into the top 6 inches of soil by disking, chiseling or other suitable means during seedbed preparation. Operate machinery on the contour.
6. *Seedbed Preparation*—Install necessary mechanical erosion and sedimentation control practices before seedbed preparation, and complete grading according to the approved plan.

Complete the seedbed preparation, which began with incorporation of soil amendments with tillage as a minimum, that will adequately loosen the soil to a depth of at least 6 inches. Break up large clods, alleviate compaction, and smooth and firm the soil into a uniform surface. Fill in or level depressions that can collect water.

7. Planting Methods
   a. *Seeding*—Use certified seed for permanent seeding whenever possible. All seed sold in Alabama is required by law to be tagged indicating it has been inspected, for example. Seed tags contain important information on seed purity, germination, and presence of weed seeds. Seed must meet State standards for content of noxious weeds. Do not accept seed containing prohibited noxious weed seed.

      Seeding dates are given in Table 4.26. Seeding properly carried out within the optimum dates have a higher probability of success. It is also possible to have satisfactory establishment when seeding outside these dates. However, if plantings are conducted outside of the optimum dates, the probability of failure increases rapidly. Seeding dates should be taken into account in scheduling land-disturbing activates.

      Inoculate legume seed with the *Rhizobium* bacteria appropriate to the species of legume.

      Plant seed uniformly with a cyclone seeder, drill, cultipacker seeder, or by hand on a fresh, firm, friable seedbed. If the seedbed has been sealed by rainfall, it should be disked so the seed will be sown in freshly prepared seedbed.

      When using broadcast-seeding methods, subdivide the area into workable sections and determine the amount of seed needed for each section. Apply one-half the seed while moving back and forth across the area, making a uniform pattern; then apply the second half in the same way, but moving at right angles to the first pass.

      Cover broadcast seed by raking or chain dragging; then firm the surface with a roller or cultipacker to provide good seed contact. Small grains should be planted no more than 1 inch deep and grasses and legume seed no more than 1/2 inch deep.

   b. *Hydroseeding*—Surface roughening is particularly important when hydroseeding, as roughened slopes will provide some natural coverage for lime, fertilizer, and seed. The surface should not be compacted or smooth. Fine seedbed preparation is not necessary for hydroseeding operations; large clods, stones, and irregularities provide cavities in which seeds can lodge.

      Mix seed, inoculant if required, and a seed carrier with water and apply as a slurry uniformly over the area to be treated. The seed carrier should be a cellulose fiber, natural wood fiber, or cane fiber mulch material which is dyed an appropriate color to facilitate uniform application of seed. Use the correct legume inoculant at four times the recommended rate when adding inoculant to a hydroseeder slurry. The mixture should be applied within one hour after mixing to reduce damage to seed.

      Fertilizer should be not be mixed with the seed inoculant mixture because fertilizer salts may damage seed and reduce germination and seedling vigor. Fertilizer may be applied with a hydroseeder as a separate operation after seedlings are established.

      Lime is not normally applied with a hydraulic seeder because it is abrasive, but if necessary, it can be added to the seed slurry and applied at seeding or it may be applied with the fertilizer mixture. Also lime can be blown onto steeper slopes in dry form.

   c. *Sprigging*—Hybrid bermudagrass cannot be grown from seed and must be planted vegetatively. Vegetative methods of establishing common and hybrid bermudagrass, centipedegrass, and zoysia include sodding, plugging and sprigging. Sprigs are fragments of horizontal stems which include at least one node (joint). They are normally sold by the bushel and can either be broadcast or planted in furrows using a tractor-drawn transplanter.

      Furrows should be 4–6 inches deep and 2 feet apart. Place sprigs about 2 feet apart in the row with one end at or above ground level.

      Broadcast sprigs at the specified rate. Press into the top 1/2 to 2 inches of soil with a cultipacker or with a disk set nearly straight so that the sprigs are not brought back to the surface. A mulch tacking machine may be used to press sprigs into the soil.

8. *Mulching*—The use of a mulch will help ensure establishment of vegetation under normal conditions and is essential to seeding success under harsh site conditions. Harsh site conditions include:

   - Seeding in late fall (wood fiber mulches are not adequate for this use),
   - Slopes steeper than 3:1, and
   - Adverse soils (shallow, rocky, or high in clay or sand),

9. *Irrigation*—Moisture is essential for seed germination and vegetation establishment. Supplemental irrigation can be very helpful in assuring adequate stands in dry seasons or to speed development of full cover. It is a requirement for establishment of vegetation from sprigs and should be used elsewhere when feasible. However, irrigation is rarely critical for low-maintenance vegetation planted at the appropriate time of the year. Water application rates must be carefully controlled to prevent runoff. Inadequate or excessive amounts of water can be more harmful than no supplemental water.

10. *Maintenance*—Generally, a stand of vegetation cannot be determined to be fully established until soil cover has been maintained for one full year from planting. Inspect vegetated areas for failure and make necessary repairs and vegetate as soon as possible.

If stand has inadequate cover, reevaluate choice of plant materials and quantities of lime and fertilizer. Re-establish the stand after seedbed preparation or over-seed the stand. Consider seeding temporary cover if the time of year is not appropriate for establishment of permanent vegetation.

If vegetation fails to grow, soil must be tested to determine if acidity or nutrient imbalance is responsible.

Fertilization—On the typical disturbed site, full establishment usually requires application of fertilizer in the second growing season. Turf grasses require annual maintenance fertilization. Use soil tests if possible or follow the guidelines given for the specific seeding mixtures.

Protect establishing vegetation from traffic that will be harmful. Use either temporary fences or barriers to protect areas that may be damaged by excessive traffic.

Sodding

The following is from the *Alabama Handbook* (USDA, 2003) and is similar to sodding guidance for temporary vegetation that is usually presented in regional erosion control handbooks.

Definition

Permanently stabilizing areas by laying a continuous cover of grass sod.

Sodding, or other reinforcement, is usually needed along concentrated flow pathways

Purpose

To prevent erosion and damage from sediment and runoff by stabilizing the soil surface with permanent vegetation; to provide immediate vegetative cover of critical areas; to stabilize disturbed areas with a suitable plant material that cannot be established by seed; and to stabilize drainage ways and channels and other areas of concentrated flow where flow velocities will not exceed that specified for a vegetated waterway.

Conditions Where Practice Applies

Disturbed areas which require immediate and permanent vegetative cover, or where sodding is preferred to other means of grass establishment such as waterways or sod flumes carrying intermittent flow at acceptable velocities, areas around drop inlets, and steep critical areas needing immediately cover.

Planning Considerations

1. Advantages of properly installed sod include immediate erosion control, nearly year-round establishment capability, less chance of failure than with seeding, and rapid stabilization of surfaces for traffic areas, channel linings, or critical areas.
2. Initially, it is more costly to install sod than to plant seed; however, the higher cost may be justified for specific situations where sod performs better than seed.
3. Sodding for soil stabilization eliminates the seeding and mulching operations, but the same site preparation is required. Sodding is a more reliable method of producing adequate cover and erosion control than seeding.
4. Sod can be laid during the times of the year when seeded grasses may fail, provided there is adequate water available for irrigation in the early establishment period. Irrigation is essential, at all times of the year, when installing sod.
5. In waterways and sod flumes that carry concentrated flow, properly pegged sod provides immediate protection and is preferable to seeding.
6. Sod placed around drop inlets can protect them from sediment and help maintain the necessary grade around the inlet.
7. The site should be prepared and ready for laying of sod when it is delivered. Leaving sod stacked or rolled can cause severe damage and loss of plant material.

Specifications

1. *Selection of appropriate types of sod*—The type of sod selected should be adapted to both the site and the

TABLE 4.27. Types of Sod Available in Alabama.

Warm-season Grasses	Varieties	Adaptable Region
Bermudagrass	Tifway, Tifgreen Tiflawn, common	North Alabama Central Alabama South Alabama
Bahiagrass	Pensacola	Central Alabama South Alabama
Centipede	No improved varieties	Central Alabama South Alabama
St. Augustine	Bitterblue, Raleigh, common	South Alabama
Zoysia	Emerald, Meyer	Central Alabama South Alabama
Cool-season Grasses:		
Tall Fescue	Kentucky 31	North Alabama

intended purpose. In Alabama, these are limited to bermuda, zoysia, centipede, St. Augustine, tall fescue, and bahiagrass. Tall fescue and bahiagrass are not readily available but can be obtained from some growers. Species selection is primarily determined by region, availability, and intended use (Tables 4.27 and 4.28).

2. *Sod Quality*—Sod should be machine cut at a uniform depth of 1/2 to 2 inches (excluding shoot growth and thatch). The sections of sod should be strong enough to support their own weight and retain their size and shape when lifted by one end. Sod should be placed within 36 hours of harvest.
3. *Site preparation*—Test soil to determine the exact requirements for lime and fertilizer. Soil test may be conducted by university soil testing laboratories (available through local agricultural extension offices for a nominal fee) or other laboratories that make recommendations based on soil analysis. When soil test recommendations are unavailable, the following soil amendments may be sufficient:
   - Agricultural limestone at a rate of 2 tons per acre (100 lbs per 1000 sq. ft.)
   - Fertilizer at a rate of 1000 lbs per acre (25 lbs per 1000 sq. ft.) of 10-10-10.

   Equivalent nutrients may be applied with other fertilizer formulations. The soil amendments should be spread evenly over the treatment area and incorporated into the top 6 inches of soil by disking, chiseling or other effective, means.

   Prior to laying sod, clear the soil surface of trash, debris, roots, branches, stones, and clods larger than 2 inches in diameter. Fill or level low spots in order to avoid standing water. Rake or harrow the site to achieve a smooth and level final grade.

   Complete soil preparation by rolling or cultipacking to firm the soil. Avoid using heavy equipment on the area, particularly when the soil is wet, as this may cause excessive compaction and make it difficult for the sod to take root.

4. *Sod installation*—A step-by-step procedure for installing sod is described below:
   a. Moistening the sod after it is unrolled helps maintain its viability. Store it in the shade during installation.
   b. Rake the soil surface to break the crust just before laying sod. During the summer, lightly irrigate the soil immediately before laying the sod to cool the soil and reduce root burning and dieback.
   c. Do not lay sod on gravel, frozen soils, or soils that have been recently sterilized or treated with herbicides.
   d. Lay the first row of sod in a straight line with subsequent rows placed parallel to and butting tightly against each other. Stagger strips in a brick-like pattern. Be sure that the sod is not stretched or overlapped and that all joints are butted tightly to prevent voids. Use a knife or sharp spade to trim and fit irregularly shaped areas.
   e. Install strips of sod with their longest dimension perpendicular to the slope. On slopes 3:1 or greater, or wherever erosion may be a problem, secure sod with pegs or staples.
   f. As sodding of clearly defined areas is completed, roll sod to provide firm contact between roots and soil.
   g. After rolling, irrigate until the soil is wet at least 6 inches below the sod.
   h. Keep sodded areas moist to a depth of 4 inches until the grass takes root. This can be determined by gently tugging on the sod. Resistance indicates that rooting has occurred.

TABLE 4.28. Characteristics of Grasses Used as Sod in Alabama (USDA 2003).

	Adaptation					Maintenance	
Grass	Shade	Heat	Cold	Drought	Wear	Mowing Height	Mowing Frequency
Bermudagrass	no	good	poor	excel.	excel.	1 in.	high
Bahiagrass	fair	good	poor	excel.	good	2–3 in.	high
Centipede	fair	good	poor	good	poor	1-1/2 in.	low
Tall fescue	good	fair	good	good	good	3 in.	high
St. Augustine	good	good	poor	poor	poor	2–3 in.	med.
Zoysia	fair	good	fair	excel.	good	1 in.	high

   i. Mowing should not be attempted until the sod is firmly rooted, usually 2 to 3 weeks.
5. *Sodded waterways*—Sod provides a resilient channel lining, providing immediate protection from concentrated flow and eliminating the need for installing mats or mulch. The following points apply to the use of sod in waterways:
   a. Prepare the soil as needed for good channel design. The sod type must be able to withstand the velocity of flow specified in the channel design.
   b. Lay sod strips perpendicular to the direction of flow, with the lateral joints staggered in a brick-like pattern. Edges should butt tightly together.
   c. After rolling or tamping to create a firm contact, peg or staple individual sod strips to resist washout during establishment. Jute or other netting material may be pegged over the sod for extra protection on critical areas.

Maintenance

1. After the first week, water as necessary to maintain adequate moisture in the root zone and prevent dormancy of the sod.
2. Do not remove more than one-third of the shoot during any one mowing. Grass height should be maintained between 2 and 3 inches, unless otherwise specified (see Table 4.28).
3. After the first growing season, established sod requires fertilization, and may also require lime.

ESTABLISHING VEGETATION

Site Preparation

The soil on a disturbed site must be modified to provide an optimum environment for germination and growth. Addition of topsoil, soil amendments, and tillage are used to prepare a good seedbed. At planting, the soil must be loose enough for water infiltration and root penetration, but firm enough to retain moisture for seedling growth. Tillage generally involves disking, harrowing, chiseling, or some similar method of land preparation. Tillage should be done on the contour where feasible to reduce runoff and erosion. Lime and fertilizer should be incorporated during the tillage. The following is from the *Alabama Handbook* (USDA 2003). Site preparation guidance for temporary vegetation usually is presented in regional erosion control handbooks.

Soil Amendments

Lime is almost always required on disturbed sites to decrease soil acidly. Lime raises the pH, reduces exchangeable aluminum, and supplies calcium and magnesium for vigorous plant growth. Only the alkaline soils of the Black Belt and north Alabama do not require lime. A soil test should be used to determine the need for liming materials.

Plant nutrients (fertilizers) will usually be required, even on the best soils. Plant nutrient application rates for a particular species of vegetative cover should be applied according to a soil test report. Soil amendments should be applied uniformly and well mixed with the top 6 inches of soil during seedbed preparation.

Planting Methods

Seeding is by far the fastest and most economical method that can be used with most grass species. However, some grasses do not produce seed and must be planted vegetatively. Seedbed preparation, liming, and fertilization are essentially the same regardless of the method chosen.

Uniform seed distribution is essential. This is best obtained using a cyclone seeder, conventional grain drill, cultipacker seeder, or hydraulic seeder. The grain drill and cultipacker seeder are pulled by a tractor and require a fairly clean, smooth seedbed. On steep slopes where equipment can not safely work, hydroseeding may be the most effective seeding method. A rough surface is particularly important when preparing slopes for hydroseeding. In contrast to other seeding methods, a rugged and even trashy seedbed gives the best results. Because uniform distribution is difficult to achieve with hand broadcasting, it should be considered only as a last resort. When hand broadcasting of seed is necessary, uneven distribution may be minimized by applying half the seed in one direction and the other half at right angles to the first. Small seed should be mixed with sand for better distribution. A sod seeder (no-till planter) can plant seed into an existing cover or mulch or be used to restore or repair a weak stand. It can be used on moderately uneven, rough surfaces. It is designed to penetrate the sod, open narrow slits, and deposit seed with a minimum of surface disturbance. Seeding rates recommended in the *Alabama Handbook* have taken into account the "insurance" effect of extra seed. Rates exceeding those given are not recommended because over-dense stands are more subject to drought and competitive interference.

Sprigging refers to planting stem fragments consisting of runners (stolons) or lateral, below-ground stems (rhizomes), which are sold by the bushel. Sprigs can be broadcast or planted in furrows using a transplanter. This method works well with bermudagrass. Also sprigs may be broadcast and covered with soil by light disking, or cultipacking. Broadcasting is easier but requires more planting material. Common and forage-type hybrid bermudagrass will cover over much more quickly than the lawn type bermudagrass. Plugging differs from sprigging only in the use of plugs cut from established sod, in place of sprigs. It requires more

Heavy mulch at a median grass swale area.

Hydroseeding, with mulch (SCS photo.)

Temporary hydroseeding for erosion control (SCS photo).

Newly established grass needs frequent watering.

Straw mulch application showing fair coverage.

Straw mulch applied over previously eroded ground.

Thin straw mulch application showing poor coverage.

Hydroseeded material, with mulch and tackifier to hold in place.

Straw mulch wind blown from adjacent area.

planting stock, but usually produces a complete cover more quickly than sprigging. It is usually used to introduce a superior grass into an old lawn.

In sodding, the soil surface is completely covered by laying cut sections of turf. It is limited primarily to lawns, steep slopes, and sod waterways in Alabama. Turf-type bermudas, centipede, and zoysia are usually the types of turf used for sodding. Plantings must be wet down immediately after planting, and kept well watered for a week or two thereafter. Sodding, though quite expensive, is warranted where immediate establishment is required, as in stabilizing drainageways and steep slopes, or in the establishment of high quality turf. If properly done, it is the most dependable method and the most flexible in seasonal requirements. Sodding can be done almost anytime of the year in Alabama.

Inoculation of Legumes

Legumes have a bacteria called rhizobia which invade the root hairs and form gall-like "nodules." The host plant supplies carbohydrates to the bacteria, which supply the plant with nitrogen compounds fixed from the atmosphere. A healthy stand of legumes, therefore, does not require nitrogen fertilizer. *Rhizobium* species are host specific in that a given species will inoculate some legumes but not others. Therefore, successful establishment of legumes requires the presence of specific strains of nodule-forming, nitrogen-fixing bacteria on their roots. In areas where a legume has been growing, sufficient bacteria may be present in the soil to inoculate seeded plants, but in other areas the natural *Rhizobium* population may be too low.

In acid subsoil material, if the specific *Rhizobium* is not already present, it must be supplied by mixing it with the seed at planting. Cultures for inoculating various legume seeds are usually available through seed dealers.

Among the legumes listed for use in the *Alabama Handbook*, crown vetch is the only one generally requiring inoculation. Lespedeza nodule bacteria are widely distributed in the soils of Alabama, unless the site has had all surface soil removed.

Irrigation

Irrigation, though not usually required, can extend seeding dates into the summer and insure seedling establishment. Damage can be caused by both under and over irrigating. If the amount of water applied penetrates only the first few inches of soil, plants may develop shallow root systems that are prone to desiccation during droughts. If supplementary water is used to get seedlings up, it must be continued until plants become completely established.

Mulching

Initial stabilization of most disturbed sites requires grasses and legumes that grow together without gaps. This is true even where part or the entire site is planted to trees or shrubs. In landscape plantings, disturbed soil between trees and shrubs must also be protected either by mulching or by permanent grass, legumes, or mixtures.

Mulch is essential to the vegetation of most disturbed sites, especially on difficult sites such as southern exposures, channels, and excessively dry soils. The steeper the slope and the poorer the soil, the more valuable mulch becomes. Mulch protects the site from erosion until the vegetation is established. In addition, mulch aids seed germination and seedling growth by reducing evaporation, preventing soil crusting, and insulating the soil against rapid temperature changes. Mulch may also protect surfaces that cannot be seeded. Mulch prevents erosion in the same manner as vegetation, by protecting the surface from raindrop impact and by reducing the velocity of overland flow.

Mulching is an application of a protective layer of straw (wheat, oats, barley or rye are the most widely used straw mulches), other plant residues, stone, or synthetic materials to the soil surface. Its purpose is to protect the soil surface from the forces of raindrop impact and overland flow. Mulch encourages the growth of vegetation, reduces evaporation, insulates the soil, and suppresses weed growth. If incorporated into the soil, mulch also improves many soil properties that have been altered by the construction activities.

Planning Considerations

1. A surface mulch is the most effective, practical means of controlling runoff and erosion on disturbed land prior to vegetation establishment. Mulch reduces evaporative moisture losses, prevents crusting and sealing of the soil surface, moderates soil temperatures, provides a suitable microclimate for seed germination, and increases the amount of infiltration into the soil.
2. Organic mulches, such as straw, wood chips, and shredded bark, have been found to be the most effective mulch materials. Materials containing weed and grass seeds which may compete with establishing vegetation should not be used. Also, decomposition of some wood products can tie up significant amounts of soil nitrogen, making it necessary to modify fertilization rates, or add fertilizer with the mulch.
3. A variety of erosion control blankets have been developed in recent years for use as mulches, particularly in critical areas such as waterways and channels. Various types of netting materials are also available to anchor organic mulches.
4. Chemical soil stabilizers, or soil binders, when used alone, are less effective than other types of mulches. These products are primarily useful for tacking wood fiber or straw mulches.

5. The choice of materials for mulching should be based on soil conditions, season, type of vegetation, and size of the area. Properly applied and tacked mulch is always beneficial. It is especially important when conditions of germination are not optimum, such as midsummer and early winter, and on difficult sites such as cut slopes and drought soils.
6. Organic Mulches
   a. Straw is the most commonly-used material in conjunction with seeding. Wheat straw is the most commonly-used straw, and can be spread by hand or with a mulch blower. If the site is susceptible to blowing wind, the straw needs to be tacked down to prevent loss.
   b. Wood chips are suitable for areas that will not be closely mowed, and around ornamental plantings. Chips do not require tacking. Because they decompose slowly they must be treated with 12 pounds of nitrogen per ton to prevent nutrient deficiency in plants. They can be an inexpensive mulch if the chips are obtained from trees cleared on the site.
   c. Wood fiber refers to short cellulose fibers applied as a slurry in hydroseeding operations. Wood fiber hydroseeder slurries may be used to tack straw mulch on steep slopes, critical areas, and where harsh climatic conditions exist. Wood fiber mulch does not provide sufficient erosion protection when used alone.
   d. Peanut hulls, cotton burs, and pine straw are organic materials that make excellent mulches but may only be available locally or seasonally. Creative use of these materials can reduce costs.
7. Erosion Control Blankets and Netting
   a. Jute mesh, or other types of netting, is very effective in holding mulch in place on waterways and slopes before grasses become established.
   b. Erosion control blankets promote seedling growth in the same way as organic mulches. They are very useful in establishing grass in channels and waterways. A wide variety of synthetic and organic materials are available such as wood excelsior, small grain straw, coconut fiber, or mixtures of these materials. When installing erosion control blankets, it is critical to obtain a firm, continuous contact between the material and the soil. Without such contact, the material is useless and erosion will occur underneath.

Specifications

1. Select a mulch material based on the site and practice requirements, availability of material, and availability of labor and equipment. Table 4.18 lists commonly used mulches and their application rates.

TABLE 4.18. Typical Mulching Materials and Application Rates.

Material	Rate Per Acre	Notes
Straw	1-1/2 to 2 tons	Spread by hand or machine; tack down when subject to blowing.
Wood chips	5 to 6 tons	Treat with 12 lbs nitrogen/ton.
Bark	35 cu yds	Can apply with mulch blower. Do not use asphalt tack.
Pine straw	1 to 2 tons	Spread by hand or machine; will not blow like straw.
Peanut hulls	10 to 20 tons	Will wash off slopes. Treat with 12 lbs nitrogen/ton.

2. Before mulching, complete the required grading, install sediment control practices, and prepare the seedbed. Also, plant and cover seed before mulching, except when seed is applied as part of a hydroseeder slurry containing wood fiber mulch.
3. Uniformly spread organic mulches by hand or with a mulch blower at a rate which provides about 75% ground cover. When spreading straw mulch by hand, divide the area to be mulched into sections of approximately 1,000 sq. ft. and place 70–90 pounds of straw (1-1/2 to 2 bales) in each section to facilitate uniform distribution. This will result in 1-1/2 to 2 tons of straw per acre. In hydroseeding operations, a green dye may be added to the slurry to assure a uniform application.
4. Anchoring Straw Mulch
   a. When straw mulch is subject to being blown away by wind, it must be anchored immediately after spreading. It can be anchored with a mulch-anchoring tool or a regular farm disk, with added weight and the disk set to run straight. The disk should not be sharp enough to cut the straw, but to punch it into the ground
   b. Liquid mulch binders can also be used to tack mulch subject to being blown away by wind. Applications of liquid mulch binders and tackifiers should be heaviest at the edges of areas and at crests of ridges and banks, to resist the wind. Binders should be applied uniformly to the rest of the area. Binders may be applied after mulch is spread or may be sprayed into the mulch as it is being blown onto the soil. Applying straw and binder together is the most effective method. Liquid binders include an array of commercially available synthetic binders. Asphalt-based binders have been used in the past, but runoff toxicity is a potential problem.
   c. Straw mulch may also be anchored with lightweight plastic, cotton, jute, wire or paper netting which are stapled over the mulch. The manufacturer's recommendations on stapling netting should be followed.

5. Installation of Erosion Control Blankets
   a. All smoothing, seedbed preparation, and vegetation operations must be completed prior to placing the erosion control blanket. Any rocks, clods, sticks, or other debris which would prevent the blanket from making close contact with the soil should be removed. The erosion control blanket should be placed immediately after planting seed. Some special erosion control blankets are also available with the seed incorporated in the blanket, allowing much more uniform seeding.
   b. Unroll the erosion control blanket from the top down, parallel to the direction of flow, in flumes and ditches and perpendicular to the direction of flow on slopes. Allow the blankets to lay loosely on the soil but without wrinkles, and do not stretch.
   c. To secure the blanket, bury the upslope end in a slot or trench no less than 6 inches deep, cover with soil, and tamp firmly. Staple the blanket every 12 inches across the top end and every 3 feet around the edges and bottom. Where erosion control blankets are laid side by side, the adjacent edges should be overlapped, with the uphill blanket on top, and stapled together. Each blanket should also be stapled down the center every 3 feet. Do not stretch the erosion control blanket when applying staples. Most manufactures provide specific installation and stapling instructions for their products and for specific situations. Manufactures of erosion control blankets also frequently specify a specific staple pattern that must be followed when using their products in order to obtain the specified level of performance.

Maintenance of Mulches

Inspect all mulches periodically, and after rainstorms, to check for rill erosion, dislocation, or failure. Where erosion is observed, apply additional mulch. If washout occurs, repair the slope grade, reseed, and reinstall mulch. Continue inspections until vegetation is firmly established.

Pitt, *et al.* (1999) described the effects of construction activities on soil structure and the use of compost soil amendments to improve soil characteristics. Land construction activities typically significantly compact soil, increasing the soil density with decreased rainwater infiltration and reduced plant viability. The use of organic amendments to the soil, and surface mulches, can be used to dramatically improve the sol texture, allowing better plant growth under these typically stressful conditions.

Maintenance of Vegetative Covers

Satisfactory stabilization and erosion control requires a complete vegetative cover. Even small breaches in vegetative cover can expand rapidly and, if not repaired, can result is excessive soil loss from an otherwise stable site. A single heavy rain will enlarge rills and bare spots and, the longer repairs are delayed, the more costly they become. Prompt action will keep soil loss, sedimentation damage, and repair cost down. New plantings should be inspected frequently and maintenance performed as needed. If rills and eroded areas develop, they must be repaired, seeded, and mulched as soon as possible.

Maintenance requirements extend beyond the seeding phase. Damage to vegetation from disease, insects, traffic, etc., can occur at any time. Pest control (weed or insect) may be needed at any time. Weak or damaged spots must be fertilized, seeded and mulched as promptly as possible.

Vegetation established on disturbed soils often requires additional fertilization. Frequency and amount of fertilizer to apply can best be determined through periodic soil testing. A fertilization program is required for the maintenance of turf and sod that is mowed frequently. Maintenance requirements should always be considered when selecting plant species for vegetation.

SUMMARY

This chapter introduced the Revised Universal Soil Loss Equation (RUSLE) and presented some specific information for using this model for construction sites. In addition, the application of vegetation controls that help prevent erosion from occurring were also outlined in this chapter. Several examples of how this information can be used to calculate the estimated soil erosion losses for construction sites were also presented.

IMPORTANT LINKS

The official NRCS RUSLE2 Internet site is at: http://fargo.nserl.purdue.edu/rusle2_dataweb/Tutorial.htm. The model can be downloaded from this site, along with supporting documents and other materials.

The Alabama Soil and Water Conservation Committee web site includes locations and contacts for local USDA/NRCS offices where soil information can be obtained. They also recently completed an updated version to the 1993 *Alabama Handbook for Erosion Control, Sediment Control, and Stormwater Management on Construction Sites and Urban Areas*. The updated handbook is available at: http://www.swcc.state.al.us/

Sources of Commercial Seeds and Plants

Sod
www.Gardnerturf.com
www.usaturf.com

Hydroseeding
www.htpa.org

Seed Suppliers
www.sylvanative.com
www.sroseed.com
www.turf-seed.com
www.seedland.com
www.erosionseed.com
www.seedswest.com
www.albrightseed.com

PROBLEMS

1. Explain the effects of the following factors on soil erosion and transport. (a) Climate; (b) Soil characteristics; (c) Topography; (d) Soil cover.
2. Explain the effect each of the following have on splash erosion. (a) Adhesive forces in the soil; (b) Kinetic energy of the raindrops; (c) The type of vegetal landcover; (d) Cohesive forces in soil.
3. Which one of the following is not a factor in determining shear stress when computing bedload with the tractive-force method? (a) The shape of the soil particles; (b) The specific weight of the fluid; (c) The specific weight of the soil particles; (d) The particle diameter; (e) All of the above are factors.
4. Use the appropriate equation to estimate the kinetic energy of a raindrop for exceedence frequencies of 2, 10, and 100 yr. Use your local IDF curve and a duration of 5 min.
5. Estimate the soil loss using the RUSLE for a square 0.8-acre plot at a 3% slope in the southwestern corner of Missouri. This soil loss is being estimated for the time period required to perform the grading—a time frame of four months starting on April 16. Assume the soil is 40% silt plus very fine sand, 10% sand ($0.1 < d < 2$ mm), no organic matter, fine granular soil structure, and moderate permeability. Assume bare ground with no cover practice.
6. For the conditions of Problem 5, show the variation of the soil loss as the percentage sand varies from 0 to 30%.
7. Assuming a void ratio of 34% and a specific weight of 135 lb/ft^3, estimate the depth of soil loss for the conditions of Problem 5.
8. A proposal is made to use a rainfall erosivity factor R of 225 in the state of Missouri. Show the spatial variation across the state of the error that results from this simplification.
9. A farmer has decided to sell his 120-acre farm to a developer, who plans to construct estate homes. The construction will be performed in three phases of approximately 40 acres each. The developer has two choices: grade each phase individually (allowing vegetation to establish between the end of one phase and the start of a new phase), or grading the entire site at one time. Assuming a construction schedule of three years (one year per construction phase), estimate the soil loss from the site using each of the development scenarios and the site information provided below. Assuming the soil is a loam, what is the difference is volume of soil material generated by erosion between the two scenarios? Site information: 120 acres, 1.5% slope; Farmland with wheat covering the entire acreage; Slope length of 400 ft (length from top of ridge to street level for lots with steepest grading problems); assume the soil is your local (non-urban) soils; use your local rain/erosion zone information.
10. Project Question:
    a. Describe the different construction phases for your site (initial grubbing and clearing, using pre-development contours; and final grading contours during active construction activities, at least). Describe site soils and land cover. Describe the timing of the construction site erosion and sediment controls for your site.
    b. Apply RUSLE for each of these phases (apply estimates for cover factors and durations of the phases; we will examine channels and slope protection during the next module, so this assignment will be a preliminary evaluation. However, consider different terracing options and other control choices described so far).
    c. Select the appropriate temporary and permanent plants to be used for construction site erosion control at your site, and describe planting and mulching conditions, etc. Consider the likely dates for the plantings).

REFERENCES

Barfield, B.J., R.C. Warner, and C.T. Haan. *Applied Hydrology and Sedimentology for Disturbed Areas.* Oklahoma Technical Press. Stillwater, Oklahoma. 1983.

Dion, T.R. *Land Development for Civil Engineers.* John Wiley & Sons, Inc. New York. 796 pgs. 2002.

Environmental Protection Agency (EPA). *Construction Rainfall Erosivity Waiver,* Fact Sheet 3.1. EPA 833-F-00-014. January 2001.

Huang, C., J.M. Bradford and J.H. Cushman. "A numerical study of raindrop impact phenomena: The elastic deformation case." *Soil Sci. Soc. Am. J.,* Vol. 47, pp. 855–861, 1983.

Kinnell, P.I.A. "Rainfall intensity-kinetic energy relationships for soil loss prediction." *Soil Sci. Soc. Am. J.,* Vol. 45, pp. 153–155, 1981.

Novotny, V. and G. Chesters. *Handbook of Nonpoint Pollution.* VanNostrand Reinhold, New York, 1981.

Pitt, R. and S.R. Durrans. *Drainage of Water from Pavement Structures.* Alabama Dept. of Transportation. 253 pgs. September 1995.

Renard, K.G., Foster, G.R., Weesies, G.A., McCool, D.K, and Yoder, D.C., coordinators. *Predicting Soil Erosion by Water: A Guide to Conservation Planning with the Revised Soil Loss Equation (RUSLE).* U.S. Dept. of Agriculture, Agric. Handbook No. 703, 404 pp. 1987.

Soil and Water Conservation Service (SCS). *Soil Survey of Jefferson County, Alabama.* USDA 1981–167-S/111.

Springer, G.S. *Erosion by Liquid Impact.* Scripta Pub. Co., John Wiley and Sons, New York, 1976.

U.S. Dept. of Agriculture (USDA), Soil Conservation Service. *National Engineering Handbook, Section 3 Sedimentation, Chapter 3 Erosion.* Undated (continuously updated).

USDA. *Alabama Handbook for Erosion Control, Sediment Control, and Stormwater Management on Construction Sites and Urban Areas.* Alabama Soil and Water Conservation Committee. Montgomery, AL. July 1993 and 2003.

USDA. *User's Guide (Technology), Revised Universal Soil Loss Equation Version 2 RUSLE2.* Agricultural Research Service, Washington, D.C. Draft, January 2003

Wischmeier, W.H. *A Rainfall Erosion Index for a Universal Soil Loss Equation.* Soil Science of America Proceedings. Vol. 23, pp.246–249. 1959.

Wischmeier, W.H. and D.D. Smith. *Rainfall Erosion Losses from Cropland East of the Rocky Mountains.* Agriculture Handbook No. 282, U.S. Dept. of Agric. Washington, D.C. 1965.

Wischmeier, W.H. and D.D. Smith. *Predicting Rainfall Erosion Losses—A Guide to Conservation Planning.* USDA. AH537. Washington, D.C. Dec 1978.

APPENDIX 4A. EROSIVITY INDICES BY LOCATION AND EROSION VARIATIONS BY SEASON

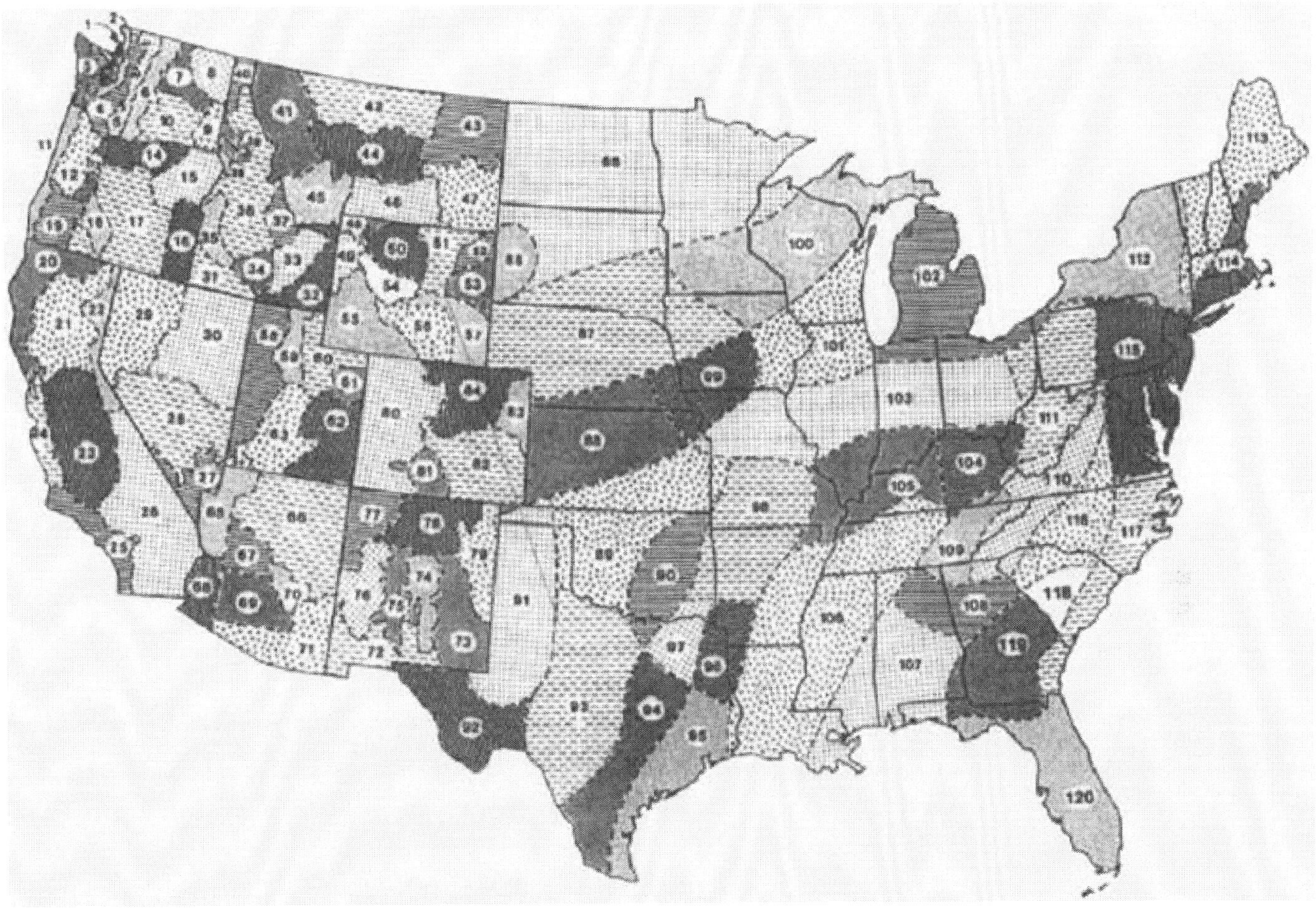

Figure 4A.1. Erosivity index zone map (EPA, 2001).

TABLE 4A.1. Erosivity Index Table (*EI* as a percentage of the annual average *R*, computed for geographical areas) (*Source:* EPA, 2001).

	Jan	Jan	Feb	Feb	Mar	Mar	Apr	Apr	May	May	Jun	Jun	Jul	Jul	Aug	Aug	Sep	Sep	Oct	Oct	Nov	Nov	Dec	Dec
EI#	1-15	16-31	1-15	16-29	1-15	16-31	1-15	16-30	1-15	16-31	1-15	16-30	1-15	16-31	1-15	16-31	1-15	16-31	1-15	16-31	1-15	16-31	1-15	16-31
1	0.0	4.3	8.3	12.8	17.3	21.6	25.1	28.0	30.9	34.9	39.1	42.6	45.4	48.2	50.8	53.0	56.0	60.8	66.8	71.0	75.7	82.0	89.1	95.2
2	0.0	4.3	8.3	12.8	17.3	21.6	25.1	28.0	30.9	34.9	39.1	42.6	45.4	48.2	50.8	53.0	56.0	60.8	66.8	71.0	75.7	82.0	89.1	95.2
3	0.0	7.4	13.8	20.9	26.5	31.8	35.3	38.5	40.2	41.6	42.5	43.6	44.5	45.1	45.7	46.4	47.7	49.4	52.8	57.0	64.5	73.1	83.3	92.3
4	0.0	3.9	7.9	12.6	17.4	21.6	25.2	28.7	31.9	35.1	38.2	42.0	44.9	46.7	48.2	50.1	53.1	56.6	62.2	67.9	75.2	83.5	90.5	96.0
5	0.0	2.3	3.6	4.7	6.0	7.7	10.7	13.9	17.8	21.2	24.5	28.1	31.1	33.1	35.3	38.2	43.2	48.7	57.3	67.8	77.9	86.0	91.3	96.9
6	0.0	0.0	0.0	0.5	2.0	4.1	8.1	12.6	17.6	21.6	25.5	29.6	34.5	40.0	45.7	50.7	55.6	60.2	66.5	75.5	85.6	95.9	99.5	99.9
7	0.0	0.0	0.0	0.0	0.0	1.2	4.9	8.5	13.9	19.0	26.1	35.4	43.9	48.8	53.9	64.5	73.4	77.5	80.4	84.8	89.9	96.6	99.2	99.7
8	0.0	0.0	0.0	0.0	0.0	0.9	3.6	7.8	15.0	20.2	27.4	38.1	49.8	57.9	65.0	75.6	82.7	86.8	89.4	93.4	96.3	99.1	100.0	100.0
9	0.0	0.8	3.1	4.7	7.4	11.7	17.8	22.5	27.0	31.4	36.0	41.6	46.4	50.1	53.4	57.4	61.7	64.9	69.7	79.0	89.6	97.4	100.0	100.0
10	0.0	0.3	0.5	0.9	2.0	4.3	9.2	13.1	18.0	22.7	29.2	39.5	46.3	48.8	51.1	57.2	64.4	67.7	71.1	77.2	85.1	92.5	96.5	99.0
11	0.0	5.4	11.3	18.8	26.3	33.2	37.4	40.7	42.5	44.3	45.4	46.5	47.1	47.4	47.8	48.3	49.4	50.7	53.6	57.5	65.5	76.2	87.4	94.8
12	0.0	3.5	7.8	14.0	21.1	27.4	31.5	35.0	37.3	39.8	41.9	44.3	45.6	46.3	46.8	47.9	50.0	52.9	57.9	62.3	69.3	81.3	91.5	96.7
13	0.0	0.0	0.0	1.8	7.2	11.9	16.7	19.7	24.0	31.2	42.4	55.0	60.0	60.8	61.2	62.6	65.3	67.6	71.6	76.1	83.1	93.3	96.2	99.6
14	0.0	0.7	1.8	3.3	6.9	16.5	26.6	29.9	32.0	35.4	40.2	45.1	51.9	61.1	67.5	70.7	72.8	75.4	78.6	81.9	86.4	93.6	97.7	99.3
15	0.0	0.0	0.0	0.5	2.0	4.4	8.7	12.0	16.6	21.4	29.7	44.5	56.0	60.8	63.9	69.1	74.5	79.1	83.1	87.0	90.9	96.6	99.1	99.8
16	0.0	0.0	0.0	0.5	2.0	5.5	12.3	16.2	20.9	26.4	35.2	48.1	58.1	63.1	66.5	71.9	77.0	81.6	85.1	88.4	91.5	96.3	98.7	99.6
17	0.0	0.0	0.0	0.7	2.8	6.1	10.7	12.9	16.1	21.9	32.8	45.9	55.5	60.3	64.0	71.2	77.2	80.3	83.1	87.7	92.6	97.2	99.1	99.8
18	0.0	0.0	0.0	0.6	2.5	6.2	12.4	16.4	20.2	23.9	29.3	37.7	45.6	49.8	53.3	58.4	64.3	69.0	75.0	86.6	93.9	96.6	98.0	100.0
19	0.0	1.0	2.6	7.4	16.4	23.5	28.0	31.0	33.5	37.0	41.7	48.1	51.1	52.0	52.5	53.6	55.7	57.6	61.1	65.8	74.7	88.0	95.8	98.7
20	0.0	9.8	18.5	25.4	30.2	35.6	38.9	41.5	42.9	44.0	45.2	48.2	50.8	51.7	52.5	54.6	57.4	58.5	60.1	63.2	69.6	76.7	85.4	92.4
21	0.0	7.5	13.6	18.1	21.1	24.4	27.0	29.4	31.7	34.6	37.3	39.6	41.6	43.4	45.4	48.1	51.3	53.3	56.6	62.4	72.4	81.3	88.9	94.7
22	0.0	1.2	1.6	1.6	1.6	1.6	1.6	2.2	3.9	4.6	6.4	14.2	32.8	47.2	58.8	69.1	76.0	82.0	87.1	96.7	99.9	99.9	99.9	99.9
23	0.0	7.9	15.0	20.9	25.7	31.1	35.7	40.2	43.2	46.2	47.7	48.8	49.4	49.9	50.7	51.8	54.1	57.7	62.8	65.9	70.1	77.3	86.8	93.5
24	0.0	12.2	23.6	33.0	39.7	47.1	51.7	55.9	57.7	58.6	58.9	59.1	59.1	59.2	59.2	59.3	59.5	60.0	61.4	63.0	66.5	71.8	81.3	89.6
25	0.0	9.8	20.8	30.2	37.6	45.8	50.6	54.4	56.0	56.8	57.1	57.1	57.2	57.6	58.5	59.8	62.2	65.3	67.5	68.2	69.4	74.8	86.6	93.0
26	0.0	2.0	5.4	9.8	15.6	21.5	24.7	26.6	27.4	28.0	28.7	29.8	32.5	36.6	44.9	55.4	65.7	72.6	77.8	84.4	89.5	93.9	96.5	98.4
27	0.0	0.0	0.0	1.0	4.0	5.9	8.0	11.1	13.0	14.0	14.6	15.3	17.0	23.2	39.1	60.0	76.3	86.1	89.7	90.4	90.9	93.1	96.6	99.1
28	0.0	0.0	0.0	0.0	0.2	0.5	1.5	3.3	7.2	11.9	17.7	21.4	27.0	37.1	51.4	62.3	70.6	78.8	84.6	90.6	94.4	97.9	99.3	100.0
29	0.0	0.6	0.7	0.7	0.7	1.5	3.9	6.0	10.5	17.9	28.8	36.6	43.8	51.5	59.3	68.0	74.8	80.3	84.3	88.8	92.7	98.0	99.8	99.9
30	0.0	0.0	0.0	0.0	0.0	0.2	0.8	2.8	7.9	14.2	24.7	35.6	45.4	52.2	58.7	68.5	77.6	84.5	88.9	93.7	96.2	97.6	98.3	99.6

(continued)

TABLE 4A.1 (continued). Erosivity Index Table (*EI* as a percentage of the annual average *R*, computed for geographical areas)

	Jan	Jan	Feb	Feb	Mar	Mar	Apr	Apr	May	May	Jun	Jun	Jul	Jul	Aug	Aug	Sep	Sep	Oct	Oct	Nov	Nov	Dec	Dec
EI#	1-15	16-31	1-15	16-29	1-15	16-31	1-15	16-30	1-15	16-31	1-15	16-30	1-15	16-31	1-15	16-31	1-15	16-31	1-15	16-31	1-15	16-31	1-15	16-31
31	0	0	0	0	0	0.2	1	3.5	9.9	15.7	26.4	47.2	61.4	65.9	69	77.2	86	91.6	94.8	98.7	100	100	100	100
32	0	0.1	0.1	0.1	0.1	0.6	2.2	4.3	9	14.2	23.3	34.6	46.3	54.2	61.7	72.9	82.5	89.6	93.7	98.2	99.7	99.9	99.9	99.9
33	0	0	0	0	0	0.6	2.3	4.2	8.8	16.1	30	46.9	57.9	62.8	66.2	72.1	79.1	85.9	91.1	97	98.9	98.9	98.9	98.9
34	0	0	0	0	0	1.8	7.3	10.7	15.5	22	29.9	35.9	42	48.5	56.9	67	76.9	85.8	91.2	95.7	97.8	99.6	100	100
35	0	0	0	0	0	2.5	10.2	15.9	22.2	27.9	34.7	43.9	51.9	56.9	61.3	67.3	73.9	80.1	85.1	89.6	93.2	98.2	99.8	99.8
36	0	0	0	0	0	0.9	3.4	6.7	12.7	18.5	26.6	36.3	46	53.5	60.2	68.3	75.8	82.6	88.3	96.3	99.3	99.9	100	100
37	0	0	0	0	0	0	0	1	3.9	9.1	19.1	26.7	36.3	47.9	61.4	75.1	84.5	92.3	96	99.1	100	100	100	100
38	0	0	0	1.1	4.3	7.2	11	13.9	17.9	22.3	30.3	43.1	55.1	61.3	65.7	72.1	77.9	82.6	86.3	90.3	93.8	98.4	100	100
39	0	0	0	0	0	1.6	6.5	11	17.8	24.7	33.1	42.8	50.3	54.9	59.7	68.9	78.1	83.6	87.5	93	96.5	99.2	100	100
40	0	0	0	0	0	1.5	6.2	10.1	16.3	23.3	32.5	42.2	50.1	55.6	60.5	67.5	74.3	79.4	84.1	91.1	95.8	99.1	100	100
41	0	0.1	0.2	0.2	0.2	0.2	0.2	0.4	1.1	6.8	22.9	40.1	54.9	63.8	70.7	81.5	89.8	96.3	98.7	99.2	99.3	99.4	99.4	99.7
42	0	0	0	0	0	0	0	0.2	0.9	5.2	17.3	33.8	53.2	66.5	75.9	87.6	93.7	97.5	99	99.7	100	100	100	100
43	0	0	0	0	0	0	0	0.1	0.4	2.7	9.5	21.9	42.7	58.6	71.1	84.6	91.9	97.1	99	99.8	100	100	100	100
44	0	1.7	2.3	2.4	2.4	2.4	2.4	2.7	3.5	7.6	18.5	34.3	52.5	64	72.3	83.3	90	95.1	97.3	98.5	98.9	98.9	98.9	99.2
45	0	0.2	0.2	0.3	0.3	0.4	0.6	0.8	1.4	3.7	10.2	22.6	41.8	54	64.5	78.7	88.4	96	98.7	99.4	99.7	99.7	99.8	99.9
46	0	0	0	0	0	0	0	0.6	2.6	7.5	19.6	32.9	48.9	63	73.5	83.3	89.5	95.6	98.3	99.6	100	100	100	100
47	0	0	0	0	0	0	0	0.4	1.6	5.8	17	33	52.5	66.4	75.7	85.5	91.3	96.5	98.8	100	100	100	100	100
48	0	0	0	0	0	0	0	0	0	2	8.1	15.4	27.8	40.7	52.6	61.1	69.3	82.6	92	98	100	100	100	100
49	0	0	0	0	0	0	0	0.7	2.7	8.3	20	27.5	35.6	44.6	56	70.2	81.3	89.2	93.6	98.5	100	100	100	100
50	0	0	0	0	0	0.1	0.4	2.4	8.2	13.7	23.8	38.8	55.1	66.1	73.6	81.8	87.7	93.8	97	99.4	100	100	100	100
51	0	0	0	0	0	0.3	1	3.1	8.7	18.8	35.8	49.6	60.4	70.2	77	84	88.8	93.8	96.6	99.1	100	100	100	100
52	0	0	0	0	0	0	0	0.6	2.5	6.8	17.5	29.8	46.1	60.5	72.7	86	92.8	96.8	98.4	99.7	100	100	100	100
53	0	0	0	0	0	0	0	0.8	3	9.5	24.2	35.3	48	63.1	76.1	87.7	93.5	97.2	98.6	99.5	99.8	99.9	100	100
54	0	0	0	0	0	0.2	0.7	2.4	7.2	14.7	27.2	37.2	47.3	58.8	67.6	74	79.2	86.7	92.6	97.9	99.8	99.9	100	100
55	0	0	0	0	0	0	0	1.3	5.4	13.3	25.5	31.6	38.8	52.5	66.8	75.5	81.2	87.9	92.8	98.3	100	100	100	100
56	0	0	0	0	0	0	0	1.3	5.1	11.4	22.3	29.5	38.5	51.1	65.2	77.8	85.6	91.7	95	98.7	100	100	100	100
57	0	0	0	0	0	0	0.1	1	3.5	9.2	21.5	31	43.5	60.4	75.1	86.1	91.6	96.2	98.1	99.4	99.9	99.9	100	100
58	0	0	0	0	0	0.2	0.9	2.9	8	13.2	21	29.1	38	45.9	54.5	65.4	74.8	82.1	87.5	95.4	98.8	99.7	100	100
59	0	0	0	0	0	0	0	2.2	8.9	15.6	24.2	31.1	38.3	46	54.9	64.2	73.2	81.9	88.5	95.7	98.6	99.4	99.7	99.7
60	0	0	0	0	0	0	0	0.4	1.5	4	9.5	13.3	20.5	33.6	52.8	66.5	76.7	88.1	94.2	98.6	100	100	100	100

(continued)

TABLE 4A.1 (continued). Erosivity Index Table (*EI* as a percentage of the annual average *R*, computed for geographical areas) (*Source:* EPA, 2001).

	Jan	Jan	Feb	Feb	Mar	Mar	Apr	Apr	May	May	Jun	Jun	Jul	Jul	Aug	Aug	Sep	Sep	Oct	Oct	Nov	Nov	Dec	Dec
EI#	1-15	16-31	1-15	16-29	1-15	16-31	1-15	16-30	1-15	16-31	1-15	16-30	1-15	16-31	1-15	16-31	1-15	16-31	1-15	16-31	1-15	16-31	1-15	16-31
61	0	0	0	0	0	0	0	1.3	5	8.5	15.5	29.8	41.8	46	49.2	56	65.1	71.6	78.6	91.1	97.3	99.3	100	100
62	0	0	0	0.1	0.3	0.8	2.1	3.6	6.5	9.7	13.7	16.5	20.8	27.3	40.1	56.9	72.6	83.4	89.4	95.5	98.1	99.6	100	100
63	0	0	0	0	0	0	0	0.9	3.7	7.8	13.3	15.8	19.9	29	46.8	64.7	78.3	88.8	93.9	98.5	100	100	100	100
64	0	0	0	0.7	2.8	7.4	12.4	14.4	15.6	17.3	19.4	21	24.4	32.3	48	61.4	72.1	81.9	87	90.1	92.4	98.1	100	100
65	0	3.6	7	9.6	11.4	13	14.4	16.3	17.7	18.4	19.3	20.5	23.6	32	50	66.2	77.2	85.4	88.8	90.4	91.3	92.7	94.8	97
66	0	0	0	0	0	0.1	0.5	1.1	2.2	3.6	6	7.6	11.1	19.8	38.9	59.7	74.4	83.2	88.1	94.6	97.7	99.4	100	100
67	0	0	0	0	0	0.1	0.4	0.9	1.6	1.9	2.4	5	12.1	24.8	48.3	73.6	86.5	92	94.3	96.6	97.9	99.5	100	100
68	0	2.3	4.5	7.8	10.4	12	13.3	16.3	17.7	18.1	18.2	18.3	18.4	19.9	24.5	35	54.4	69.4	78.6	85.7	89.2	91.9	93.9	97
69	0	2	3.7	5.7	7.8	10.5	12.4	13.7	14.3	14.7	15.1	15.7	17.1	22.7	36.7	50.4	63.6	75	81.8	87.8	90.8	93.2	94.9	97.5
70	0	0.5	0.7	1	1.3	1.7	2.2	2.8	3.4	3.9	4.7	5.4	7.4	15.7	36.5	55.8	70.3	80.9	86.4	90.9	93.4	96.4	98.1	99.4
71	0	0.7	1.2	1.6	2.1	2.8	3.3	3.6	4	4.5	5.6	6.5	9.1	18.5	40.6	59.7	74	86.3	91.7	94.7	96	96.7	97.3	98.8
72	0	0	0	0	0	0	0.1	0.2	0.7	0.8	1.3	3.5	9.9	24.7	51.4	71.5	83.6	93.8	97.7	99.2	99.8	99.9	99.9	100
73	0	0	0.1	0.1	0.2	0.2	0.3	0.6	1.3	4.1	11.5	18.1	28.3	40.2	54.1	67	77.2	87.7	93.3	97.5	99.1	99.6	99.8	100
74	0	0	0	0	0	0.1	0.2	0.5	1.2	2.7	6.4	10.2	18.4	31	50.7	68.7	81.2	91.6	96.1	98.4	99.2	99.8	100	100
75	0	0.1	0.1	0.1	0.2	0.5	1.3	1.9	3	4.1	6.6	10	17.6	28.3	44.7	59.4	71.6	83.9	90.3	94.7	96.7	98.8	99.6	99.9
76	0	0	0	0	0	0.1	0.2	0.6	1.3	2	3.5	4.9	8.4	17.4	37.3	57.5	72.9	83.7	89.5	95.8	98.4	99.6	100	100
77	0	0.2	0.3	0.3	0.4	0.8	1.5	2	2.8	3.9	5.9	7.2	10.3	21.5	46.5	66.3	78.3	86.5	90.8	96	98.2	99.1	99.5	99.8
78	0	0	0	0	0	0	0.2	0.5	1.6	3.8	8.9	13.2	21.8	35.8	56.6	75.4	86	92.9	95.9	98.2	99.2	99.8	100	100
79	0	0	0	0	0	0.2	0.7	1.3	2.7	5.8	12.7	18.8	28.8	41.6	58.4	75.7	86.5	94.2	97.3	98.9	99.5	99.9	100	100
80	0	0.6	1.2	1.6	2.1	2.5	3.3	4.5	6.9	10.1	15.5	19.7	26.6	36.4	51.7	67.5	79.4	88.8	93.2	96.1	97.3	98.2	98.7	99.3
81	0	0.1	0.1	0.2	0.4	0.5	0.8	0.9	1.5	3.9	9.9	12.8	18.2	30.7	54.1	77.1	89	94.9	97.2	98.7	99.3	99.6	99.7	99.9
82	0	0	0.1	0.1	0.2	0.2	0.5	1.2	3.1	6.7	14.4	20.1	29.8	44.5	64.2	83.1	92.2	96.4	98.1	99.3	99.7	99.8	99.8	99.9
83	0	0	0.1	0.1	0.1	0.3	0.9	1.6	3.5	8.3	19.4	30	44	59.2	72.4	84.6	91.2	96.5	98.6	99.5	99.8	99.9	100	100
84	0	0	0.1	0.1	0.2	0.3	0.6	1.7	4.9	9.9	19.5	27.2	38.3	52.8	68.8	83.9	91.6	96.4	98.2	99.2	99.6	99.8	99.8	99.9
85	0	0	0	0	0	0	1	2	3	6	11	23	36	49	63	77	90	95	98	99	100	100	100	100
86	0	0	0	0	0	0	1	2	3	6	11	23	36	49	63	77	90	95	98	99	100	100	100	100
87	0	0	0	0	1	1	2	3	6	10	17	29	43	55	67	77	85	91	96	98	99	100	100	100
88	0	0	0	0	1	1	2	3	6	13	23	37	51	61	69	78	85	91	94	96	98	99	99	100
89	0	0	1	1	2	3	4	7	12	18	27	38	48	55	62	69	76	83	90	94	97	98	99	100
90	0	1	2	3	4	6	8	13	21	29	37	46	54	60	65	69	74	81	87	92	95	97	98	99

(continued

TABLE 4A.1 (continued). Erosivity Index Table (*EI* as a percentage of the annual average *R*, computed for geographical areas) (*Source:* EPA, 2001).

	Jan	Jan	Feb	Feb	Mar	Mar	Apr	Apr	May	May	Jun	Jun	Jul	Jul	Aug	Aug	Sep	Sep	Oct	Oct	Nov	Nov	Dec	Dec
EI#	1-15	16-31	1-15	16-29	1-15	16-31	1-15	16-30	1-15	16-31	1-15	16-30	1-15	16-31	1-15	16-31	1-15	16-31	1-15	16-31	1-15	16-31	1-15	16-31
91	0	0	0	0	1	1	1	2	6	16	29	39	46	53	60	67	74	81	88	95	99	99	100	100
92	0	0	0	0	1	1	1	2	6	16	29	39	46	53	60	67	74	81	88	95	99	99	100	100
93	0	1	1	2	3	4	6	8	13	25	40	49	56	62	67	72	76	80	85	91	97	98	99	99
94	0	1	2	4	6	8	10	15	21	29	38	47	53	57	61	65	70	76	83	88	91	94	96	98
95	0	1	3	5	7	9	11	14	18	27	35	41	46	51	57	62	68	73	79	84	89	93	96	98
96	0	2	4	6	9	12	17	23	30	37	43	49	54	58	62	66	70	74	78	82	86	90	94	97
97	0	1	3	5	7	10	14	20	28	37	48	56	61	64	68	72	77	81	86	89	92	95	98	99
98	0	1	2	4	6	8	10	13	19	26	34	42	50	58	63	68	74	79	84	89	93	95	97	99
99	0	0	0	1	1	2	3	5	7	12	19	33	48	57	65	72	82	88	93	96	98	99	100	100
100	0	0	0	0	1	1	2	3	5	9	15	27	38	50	62	74	84	91	95	97	98	99	99	100
101	0	0	0	1	2	3	4	6	9	14	20	28	39	52	63	72	80	87	91	94	97	98	99	100
102	0	0	1	2	3	4	6	8	11	15	22	31	40	49	59	69	78	85	91	94	96	98	99	100
103	0	1	2	3	4	6	8	10	14	18	25	34	45	56	64	72	79	84	89	92	95	97	98	99
104	0	2	3	5	7	10	13	16	19	23	27	34	44	54	63	72	80	85	89	91	93	95	96	98
105	0	1	3	6	9	12	16	21	26	31	37	43	50	57	64	71	77	81	85	88	91	93	95	97
106	0	3	6	9	13	17	21	27	33	38	44	49	55	61	67	71	75	78	81	84	86	90	94	97
107	0	3	5	7	10	14	18	23	27	31	35	39	45	53	60	67	74	80	84	86	88	90	93	95
108	0	3	6	9	12	16	20	24	28	33	38	43	50	59	69	75	80	84	87	90	92	94	96	98
109	0	3	6	10	13	16	19	23	26	29	33	39	47	58	68	75	80	83	86	88	90	92	95	97
110	0	1	3	5	7	9	12	15	18	21	25	29	36	45	56	68	77	83	88	91	93	95	97	99
111	0	1	2	3	4	5	6	8	11	15	20	28	41	54	65	74	82	87	92	94	96	97	98	99
112	0	0	0	1	2	3	4	5	7	12	17	24	33	42	55	67	76	83	89	92	94	96	98	99
113	0	1	2	3	4	5	6	8	10	12	17	22	31	42	52	60	68	75	80	85	89	92	96	98
114	0	1	2	4	6	8	11	13	11	13	21	26	32	38	46	55	64	71	77	81	85	89	93	97
115	0	1	2	3	4	5	6	8	10	14	19	26	34	45	56	66	76	82	86	90	93	95	97	99
116	0	1	3	5	7	9	12	15	18	21	25	29	36	45	56	68	77	83	88	91	93	95	97	99
117	0	1	2	3	4	5	7	9	11	14	17	22	31	42	54	65	74	83	89	92	95	97	98	99
118	0	2	4	6	8	12	16	20	25	30	35	41	47	56	67	75	81	85	87	89	91	93	95	97
119	0	1	2	4	6	7	9	12	15	18	23	31	40	48	57	63	72	78	88	92	96	97	98	99
120	0	8	16	25	33	41	46	50	53	54	55	56	56.5	57	57.75	58	58.75	60	61	63	66.5	72	80	90

(continued

TABLE 4A.1 (continued). Erosivity Index Table (*EI* as a percentage of the annual average *R*, computed for geographical areas) (*Source:* EPA, 2001).

0	Jan	Jan	Feb	Feb	Mar	Mar	Apr	Apr	May	May	Jun	Jun	Jul	Jul	Aug	Aug	Sep	Sep	Oct	Oct	Nov	Nov	Dec	Dec
EI#	1-15	16-31	1-15	16-29	1-15	16-31	1-15	16-30	1-15	16-31	1-15	16-30	1-15	16-31	1-15	16-31	1-15	16-31	1-15	16-31	1-15	16-31	1-15	16-31
121	0	7	14	20	25.5	33.5	38	43	46	50	52.5	54.5	56	58	59	60	61.5	63	65	68	72	79	86	93
122	0	4	8	12	17	23	29	34	38	44	49	53	56	59	62	65	69	72	75	79	83	88	93	96
123	0	4	9	15	23	29	34	40	44	48	50	51	52	53	55	57	60	62	64	67	72	80	88	95
124	0	7	12	17	24	30	39	45	50	53	55	56	57	58	59	61	62	63	64	66	70	77	84	92
125	0	9	16	23	30	37	43	47	50	52	54	55	56	57	58	59	60	62	64	67	71	77	86	93
126	0	8	15	22	28	33	38	42	46	50	52	53	53	53	53	54	55	57	59	63	68	75	83	92
127	0	8	15	22	29	34	40	45	48	51	54	57	59	62	63	64	65	66	67	69	72	76	83	91
128	0	9	16	22	27	32	37	41	45	48	51	53	55	56	57	57	58	59	61	64	68	73	79	89
129	0	10	20	28	35	41	46	49	51	53	55	56	56	57	58	59	60	61	62	65	69	74	81	90
130	0	8	15	22	28	33	38	41	44	47	49	51	53	55	56	58	59	60	63	65	69	75	84	92
131	0	10	18	25	29	33	36	39	41	42	44	45	46	47	48	49	51	53	56	59	64	70	80	90
132	0	8	16	24	32	40	46	51	54	56	57	58	58	59	59	60	60	61	62	64	68	74	83	91
133	0	12	22	31	39	45	49	52	54	55	56	56	56	56	57	57	57	57	58	59	62	68	77	88
134	0	7	15	22	30	37	43	49	53	55	57	58	59	60	61	62	63	65	67	70	74	79	85	92
135	0	11	21	29	37	44	50	55	57	59	60	60	60	60	61	61	61	62	63	64	67	71	78	89
136	0	10	18	25	30	39	46	51	54	57	58	59	59	60	60	60	61	62	63	64	67	72	80	90
137	0	11	22	31	39	46	52	56	58	59	60	61	61	61	61	62	62	62	63	64	66	71	78	89
138	0	8	14	20	25	32	37	42	47	50	53	55	56	58	59	61	63	64	66	68	71	76	85	93
139	0	10.6	21.2	28.6	36	41.4	46.8	49.3	51.8	52.5	53.2	53.5	53.7	53.9	54	54.3	54.7	55.7	56.8	61.6	65.3	73.9	82.5	91.2
140	0	0.2	0.3	0.3	0.3	0.3	0.3	0.8	1.3	5.3	9.3	30.1	50.8	56.8	62.9	67.5	72.2	75.8	79.4	85.6	91.7	95.9	100	100
141	0	10.7	21.4	28.7	36	41.7	47.3	50.3	53.2	54.5	55.7	56.2	56.7	56.9	57	57.4	57.8	59	60.2	64.1	67.9	76.1	84.2	92.1
142	0	2.7	5.5	5.7	5.9	7.1	8.4	10	11.7	15.3	19	22.6	26.1	29	31.9	36.6	41.2	46	50.7	62.3	73.9	83.5	93.1	96.6
143	0	8.7	17.5	25.2	33	39.9	46.7	50.8	54.8	56.2	57.6	58	58.4	58.9	59.4	60.8	62.3	64.1	65.9	68.8	71.7	78.6	85.5	92.7
144	0	4.3	8.6	9.3	10.1	11.1	12	15.3	18.6	22.7	26.7	28.7	30.7	31.3	32	34	36	44.4	52.9	60.1	67.3	78.2	89.2	94.6
145	0	11.7	23.3	33.5	43.7	50.7	57.6	60.3	63	63.5	64.1	64.2	64.2	64.5	64.8	66.1	67.3	68.6	69.8	70.7	71.6	79.2	86.7	93.4
146	0	4.8	9.6	13.1	16.5	22.6	28.7	30.8	32.8	33.3	33.8	34	34.2	36.4	38.6	43	47.5	56	64.5	66.2	67.9	77.9	88	94
147	0	0	4.7	9.4	10.8	12.2	13.2	14.3	14.9	15.5	24.2	32.8	45.5	58.2	67.9	77.6	86.3	95.1	95.6	96.1	98	100	100	100
148	0	5.5	11	19.2	27.5	36.6	45.7	47.8	50	50.9	51.7	52.1	52.5	54.2	55.9	60.1	64.4	70.5	76.7	81.2	85.7	90.4	101	97.6
149	0	2.4	4.9	7.4	9.9	11.7	13.6	14.6	15.6	16.2	16.8	17.2	17.7	24.7	31.7	46.9	62.1	67	72	80.7	89.3	92.3	95.3	97.7

(continued

Channel and Slope Stability for Construction Site Erosion Control

INTRODUCTION

THIS chapter reviews the basic approaches and techniques available for the design of stable channels and slopes. Several alternatives that can be used are briefly described. Example problems are also presented. Specific issues associated with construction sites are stressed in this chapter, compared to the more general applications for which some of these techniques are usually applied. The information previously presented in Chapter 3 (Regional rainfall conditions and site hydrology for construction site erosion evaluations) and Chapter 4 (Erosion mechanisms and the Revised Universal Soil Loss Equation) is used in this chapter to design stable diversion, on-site, and downslope channels, plus to ensure stable slopes. These are some of the most critical erosion control practices on a construction site, as these are preventative measures, which are always more effective than sediment control (treatment) practices applied after erosion has occurred. The design approaches described in this chapter can be also modified to meet different criteria, based on allowable erosion yield objectives.

GENERAL CHANNEL STABILITY SHEAR STRESS RELATIONSHIP

An important reference on general shear stress relationships and channel bed movement is *Engineering and Design: Channel Stability Assessment for Flood Control Projects* (COE 1994; EM 1110-2-1418). Although this reference is specifically for large channels, many of the basic concepts are similar to what occurs at construction sites. These are specifically addressed in the following discussion. More extensive information on these topics is available in numerous textbooks and manuals on sediment transport and channel design.

Allowable Velocity Approach to Channel Design

Allowable velocity and allowable shear stress have been used to design stable channels that would have minimal channel erosion. Modifications of allowable velocity or shear stress to account for sediment transport have been proposed in a few references, but generally are not useful for construction site applications (see the discussion on the "regime" theory in McCuen 1998, for example).

The concept of allowable velocities for various soils and materials dates from the early days of hydraulics. An example of simple velocity criteria is given by Table 5.1 (COE undated, EM 1110-2-1601). Table 5.2 is a similar table from U.S. Bureau of Reclamation research (Fortier and Scobey, 1926; reprinted by McCuen, 1998) that also shows the corresponding allowable shear stresses and Manning's roughness values.

Figure 5.1 is additional guidance and is based on SCS data (USDA, 1977). This figure differentiates between "sediment-free" and "sediment-laden" flow, similar to the distinction made in the sediment quantity in the runoff water in Table 5.2.

Allowable Shear Stress Calculations

By the 1930's, boundary shear stress (sometimes called tractive force) was generally accepted as a more appropriate erosion criterion than allowable velocity. The average boundary shear stress in uniform flow (Figure 5.3) is calculated by

$$\tau_o = \gamma RS \text{ (lb/ft}^2\text{)}$$

where,

γ = specific weight of water (62.4 lbs/ft^3)
R = hydraulic radius (ft)
S = hydraulic slope (ft/ft)

Figure 5.2 (Chow 1959) shows a typical distribution of the shear stresses in a channel, indicating how, for straight channel reaches having constant depths, the maximum shear stress is applied along the center of the channel.

If the maximum shear stress is desired (typical for design conditions), then the flow depth is used instead of the hydraulic radius. For sheetflow conditions, the hydraulic radius (R) is very close to the depth of flow, and the above

Massive streambank failure after new development and a new outfall in a suburban area (WI DNR photo).

equation is modified, as shown in Figure 5.3, by using the depth of flow to replace the hydraulic radius.

Flow characteristics predicting the initiation of motion of sediment in noncohesive materials are usually presented in nondimensional form in the Shield's diagram (Figure 5.4). This diagram indicates the initial movement, or scour, of noncohesive uniformly graded sediments on a flat bed. The diagram plots the Shield's number (or mobility number), which combines shear stress with grain size and relative density, against a form of the Reynolds number that uses grain size as the length variable. The ASCE *Sedimentation Manual* (1975) uses a dimensionless parameter, shown on Figure 5.4, to select the dimensionless stress value. This stress value is calculated as follows:

$$\frac{d}{\nu}\left[0.1\left(\frac{\gamma_s}{\gamma}-1\right)gd\right]^{0.5}$$

where,

d = particle diameter (meters)
g = gravitational constant (9.81 m/sec^2)
ν = kinematic viscosity (1.306×10^{-6} m^2/sec for 10°C)
γ_s = specific gravity of the solid
γ = specific gravity of water

A series of parallel lines on Figure 5.4 represent these calculated values. The dimensionless shear stress value ($\tau *$) is selected where the appropriate line intersects the Shield's curve. The critical shear stress can then be calculated by:

$$\tau_c = \tau * (\gamma_s - \gamma)d$$

Post-Agnes Stream Stabilization

Sometimes desperate times require desperate measures. On June 21, 1972 Hurricane Agnes made its way up the east coast of the U.S. into the southern tier of New York and north central Pennsylvania. The resulting flooding and economic impact was dramatic and devastating. Downtown Elmira, New York recorded a flood depth of 17 feet above street level from the Chemung River. Across the valley from Elmira on a tributary to the river (Seeley Creek), people were trying to protect their property in any way that they could.

This photo, taken in July 1973, one year after the storm, shows a number of automobiles that were pushed over the creek bank to help prevent it from washing away. Although a gravel bar has deposited due to comparatively-reduced velocity and some "windshield vegetation" has been established, the effort is not in compliance with water quality standards. Many comprehensive streambank stabilization methods can be employed that both protect against erosion and provide aquatic habitat enhancements. These techniques are covered in many stream restoration and ecological engineering guidelines, handbooks, and textbooks.

Scrap Metal Stream Stabilization.

Bioengineered channel slopes (IECA photo).

Geogrids being filled with sand for bank protection (IECA photo).

Example:

The following example, presented by Chang (1988), illustrates the use of the Shield's diagram:

Determine the maximum depth of a wide canal for which scour of the bed material can just be prevented. The canal has rigid banks and an erodible bed; it is laid on a slope of 0.0005. The bed material has a median size of 2.5 mm and its specific gravity is 2.65. Assume a temperature of 10°C.

Therefore:

d = particle diameter (meters) = 2.5 mm = 0.0025 m
g = gravitational constant = 9.81 m/sec^2
ν = kinematic viscosity = 1.306×10^{-6} m^2/sec for 10°C
γ_s = specific gravity of the solid = 2.65
γ = specific gravity of water = 1

$$\frac{d}{\nu}\left[0.1\left(\frac{\gamma_s}{\gamma}-1\right)gd\right]^{0.5} =$$

$$\frac{0.0025}{1.306\times 10^{-6}}\left[0.1\left(\frac{2.65}{1}-1\right)(9.81)(0.0025)\right]^{0.5} = 121.8$$

This line intersects the Shield's curve at $\tau* = 0.043$. The critical shear stress is therefore:

$$\tau_c = \tau*(\gamma_s - \gamma)d = 0.043(2.65 - 1)0.0025 = 1.74 \text{ N/m}^2$$

Using the basic shear stress formula:

$$\tau_c = \gamma DS$$

TABLE 5.1. Example of Simple Allowable Velocity Objectives (From COE, undated, EM 1110-2-1601).

Channel Material	Mean Channel Velocity (ft/sec)
Fine Sand	2.0
Coarse Sand	4.0
Fine Gravel	6.0
Earth	
Sandy Silt	2.0
Silt clay	3.5
Clay	6.0
Grass-lined Earth (Slopes less than 5%)	
Bermuda Grass	
Sandy Silt	6.0
Silt Clay	8.0
Kentucky Blue Grass	
Sandy Silt	5.0
Silt Clay	7.0
Poor Rock (usually sedimentary)	10.0
Soft Sandstone	8.0
Soft Shale	3.5
Good Rock (usually igneous or hard metamorphic)	20.0

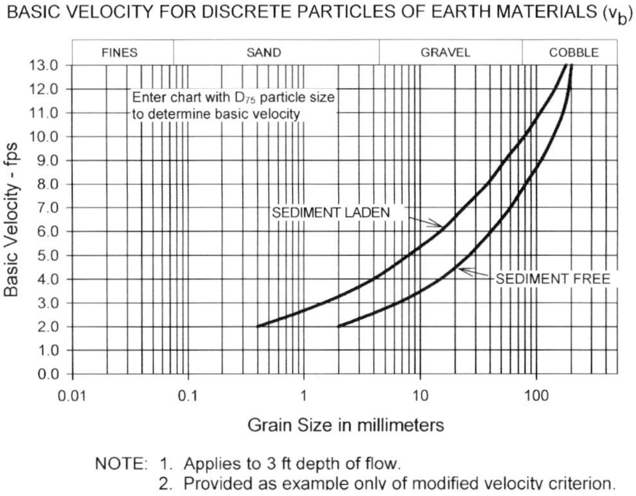

Figure 5.1. Example of allowable velocity data with provision for sediment transport (adapted from USDA, 1977).

TABLE 5.2. Maximum Permissible Velocities and Corresponding Unit Tractive Forces (Shear Stress) (U.S. Bureau of Reclamation research, Fortier and Scobey, 1926).

		Clear Water (diversion structures)		Water Transporting Colloidal Silts (on site and down slope)	
Material	*n*	*V* (ft/sec)	τ_o (lb/ft²)	*V* (ft/sec)	τ_o (lb/ft²)
Fine sand, colloidal	0.020	1.50	0.027	2.50	0.075
Sandy loam, noncolloidal	0.020	1.75	0.037	2.50	0.075
Silt loam, noncolloidal	0.020	2.00	0.048	3.00	0.11
Alluvial silts, noncolloidal	0.020	2.00	0.048	3.50	0.15
Ordinary firm loam	0.020	2.50	0.075	3.50	0.15
Volcanic ash	0.020	2.50	0.075	3.50	0.15
Stiff clay, very colloidal	0.025	3.75	0.26	5.00	0.46
Alluvial silts, colloidal	0.025	3.75	0.26	5.00	0.46
Shales and hardpans	0.025	6.00	0.67	6.00	0.67
Fine gravel	0.020	2.50	0.075	5.00	0.32
Graded loam to cobbles when noncolloidal	0.030	3.75	0.38	5.00	0.66
Graded silts to cobbles when noncolloidal	0.030	4.00	0.43	5.50	0.80
Coarse gravel, noncolloidal	0.025	4.00	0.30	6.00	0.67
Cobbles and shingles	0.035	5.00	0.91	5.50	1.10

Notes:

- An increase in velocity of 0.5 ft/sec can be added to these values when the depth of water is greater than 3 ft.
- A decrease in velocity of 0.5 ft/sec should be subtracted when the water contains very coarse suspended sediments.
- For high and infrequent discharges of short duration, up to 30% increases in velocity can be added

Rearranging gives (with the specific weight of water being 9.808 kN/m³, or 999.7 kg/m³ at 10°C):

$$D = \frac{\tau_c}{\gamma S} = \frac{1.74\ N/m^2}{(9{,}808\ N/m^3)(0.0005)} = 0.35\ m$$

The critical depth of flow (D) is therefore 0.35 meters.

For sediments in the gravel size range and larger, the Shield's number for beginning of bed movement is essentially independent of the Reynolds number. For wide channels, the relationship can then be expressed as:

$$\frac{dS}{(s-1)D} = \text{constant}$$

where,

S = channel slope
s = dry relative density of sediment
D = grain size
d = depth of flow

The constant is shown as 0.06 in Figure 5.4, but it is often taken as 0.045, or even as low as 0.03 if absolutely no movement is allowed. For widely graded bed materials, the median grain size by weight (D_{50}) is generally taken as the representative size, although some favor a smaller percentile, such as D_{35}.

An example evaluation is given by the COE (1994) in their assessment manual. In their example, the use of the Shield's diagram is shown to likely greatly over-predict the erodibility of the channel bottom material. The expected reason they give is that the Shield's diagram assumes a flat bottom channel and the total roughness is determined by the size of the granular bottom material. The actual Manning's roughness value is likely much larger because it is largely determined by bed forms, channel irregularities, and vegetation. They recommend, as a more realistic assessment,

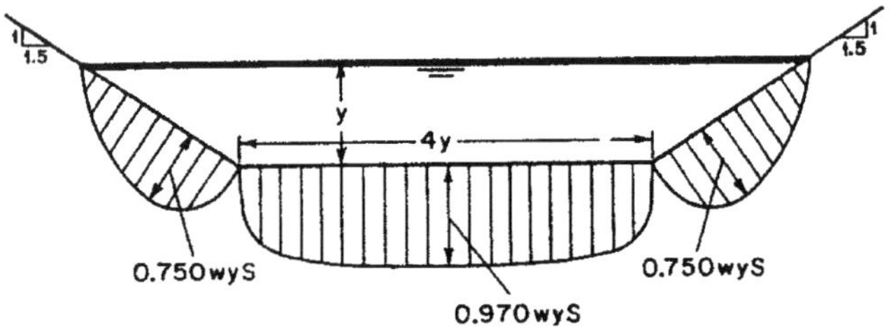

Figure 5.2. Typical shear stress distributions in a trapezoidal channel (Chow, 1959).

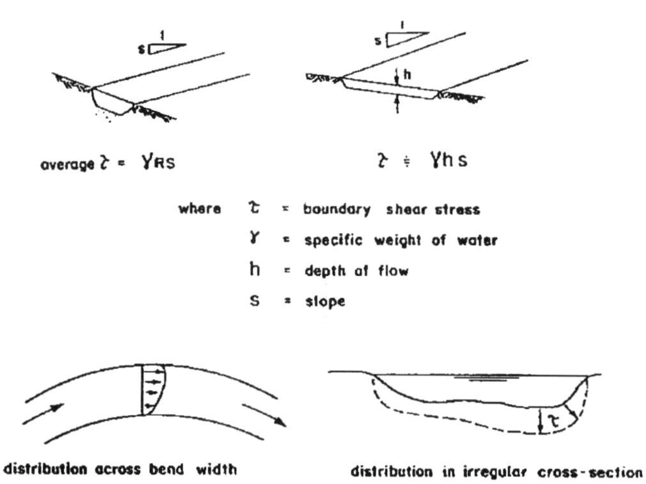

Figure 5.3. Boundary shear stress in uniform flow (COE, 1994).

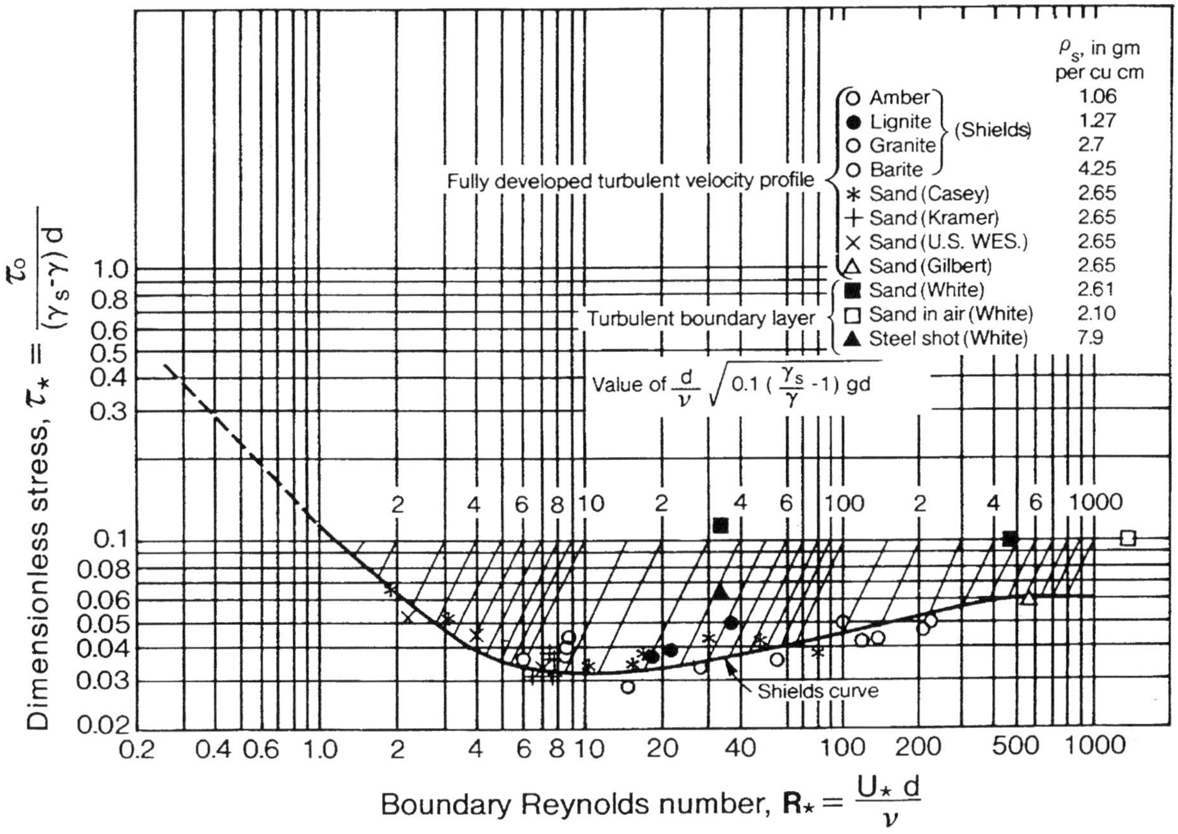

Figure 5.4. Shield's diagram for dimensionless critical shear stress in wide, flat-bottomed channels (COE, 1994).

that empirical data based on field observations be used. In the absence of local data, they present Figure 5.5 (from USDA, 1954) for applications to channels bedded in granular materials. This figure shows the permissible unit tractive force (shear stress) as a function of the average particle diameter and the fine sediment content of the flowing water. For construction-site diversion channels intercepting upland water from stable sites, the "clear" water curve is recommended. However, if the channel is on, or below, the construction site, the "high content" curve is more suitable.

The allowable shear stress concept has also been applied to semicohesive and noncohesive soils, but the values do not correlate well with standard geotechnical parameters because the resistance to erosion is affected by such factors as water chemistry, history of exposure to flows, and weathering (Raudkivi and Tan, 1984). Figure 5.6 gives an example of allowable shear stresses for a range of cohesive materials. Again, the COE recommends that local field observation or laboratory testing results be given preference.

Shear Stress in Channels having Bends

The basic shear stress formulas can be modified to account for the increased shear stress after bends in channels. Normally, the maximum shear stress is along the center part of a channel (usually the deepest area), but, after a change in direction, a hydrodynamic force is applied to the outside bend. Along the outside of the bend, increased water velocity and shear stress will increase the erosion potential, while sedimentation may occur along the inside of the bend where the water velocity slows. The basic shear stress formula is modified with a bend coefficient, as follows:

$$\tau_o = \frac{\gamma RS}{K_b}$$

where,

γ = specific weight of water (62.4 lbs/ft^3)
R = hydraulic radius (ft) (can be estimated by water depth, for relatively wide channels or sheetflows)
S = hydraulic slope (ft/ft)
K_b = bend coefficient

The bend coefficient can be estimated by (Croke, 2001):

$$K_b = \frac{R_c}{B}$$

where,

R_c = bend curvature (radius of the bend)
B = bottom width of the channel

As the bend curvature, R_c, increases, the effect of the bend decreases. These parameters are illustrated in Figure 5.7

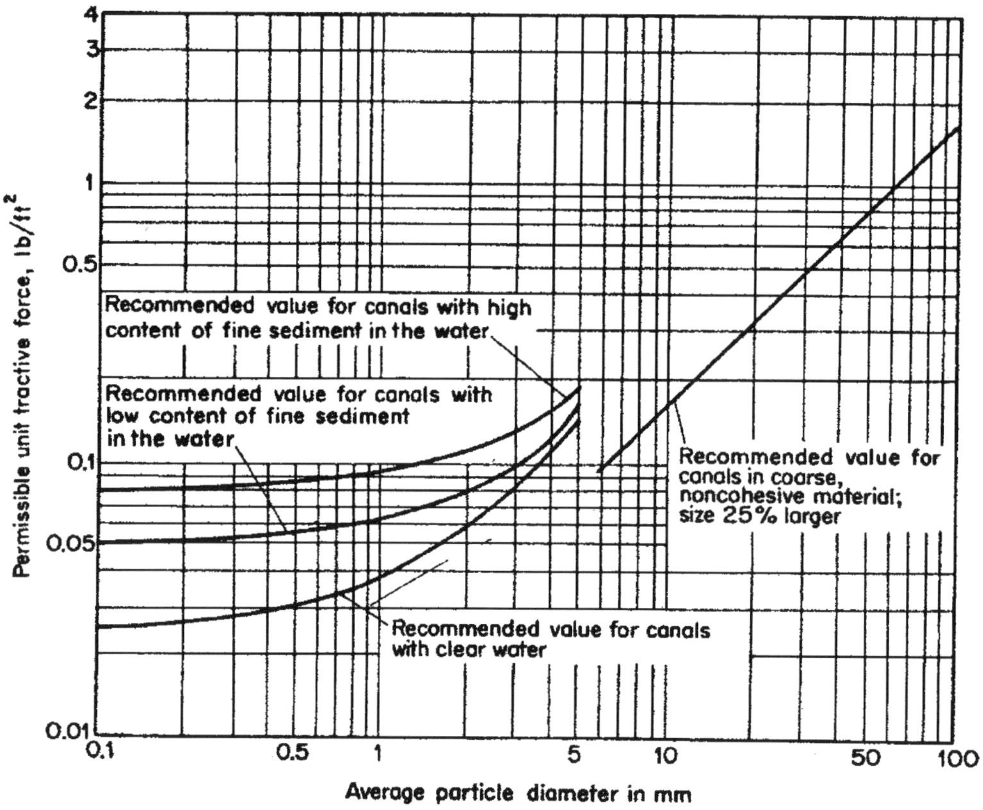

Figure 5.5. Allowable shear stresses (tractive forces) for canals in granular materials (U.S. Bureau of Reclamation, USDA, 1954).

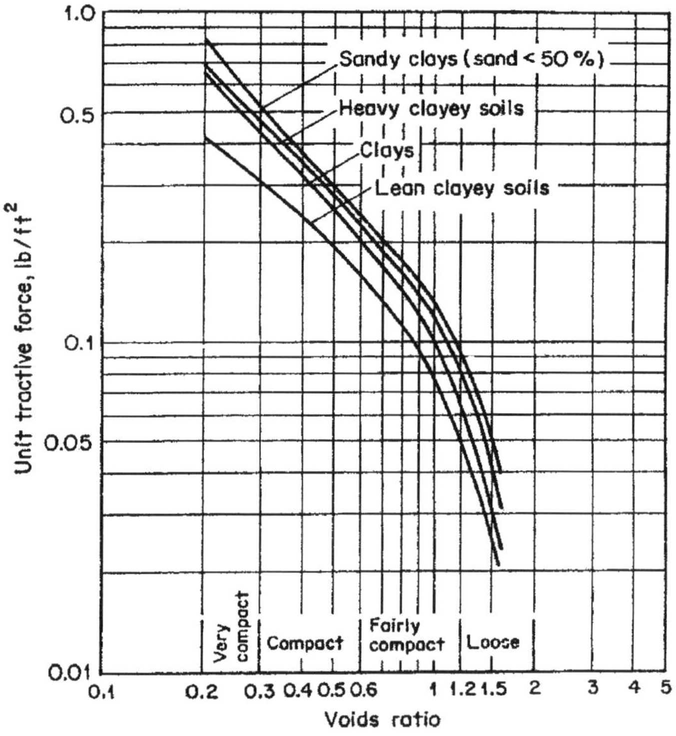

Figure 5.6. Example of allowable shear stresses (tractive forces) for cohesive materials (COE, 1994). *Note:* Leon clayey soils are hardpan soils where the soil grains become cemented together with bonding agents such as iron oxide or calcium carbonate, forming a hard, impervious mass.

(North American Green). This formula obviously cannot be used for a V-shaped channel, where the bottom width is zero.

The area being affected by the increased shear stress due to channel bends is usually assumed to begin immediately after the bend at the tangent to the downstream channel, as shown in Figure 5.7. The length of extra shear stress can be estimated by the following formula (after Croke, 2001):

$$L_p = \frac{0.604R^{1.17}}{n}$$

where,

L_p = length of extra protection needed due to increased shear stress on outside of bend (same units as R)
R = hydraulic radius = ratio of cross-sectional area of flow to wetted perimeter (A/P)
n = Manning's roughness coefficient for liner in the channel bend

As an example, assume the following conditions:

R = 3.0 ft
n = 0.042

then,

$$L_p = \frac{0.604(3)^{1.17}}{0.042} = 52 ft$$

In addition to the increased shear stress being exerted along the outside bend, water elevations also will rise due to centrifugal force This will create an additional channel depth needing protection along the outside bends.

Cautions Regarding Allowable Velocity or Shear Stress

The COE (1994) lists the following limitations of the allowable velocity and allowable shear stress approaches:

- For channels with substantial inflows of bed material, a minimum velocity or shear stress to avoid sediment deposition may be as important as a maximum value to avoid erosion. Such a value cannot be determined using allowable data for minimal erosion. [See the discussion of the "regime" theory in McCuen (1998)].
- In bends and meandering channels, bank erosion and migration may occur even if average velocities and boundary shear stresses are well below allowable values. Conversely, deposition may occur in local slack-water areas, even if average values are well above the values indicated for maximum deposition. Information on cross-sectional distributions of velocity and shear stress in bends is provided in COE (undated) (EM 1110-2-1601). *Authors' note:* There are design curves in many sediment transport books that allow the user to estimate if the flow will encourage scour or deposition, based on particle diameter, hydraulic radius and flow rate.
- The Shield's relationship (Figure 5.4) should be applied primarily to uniform flow over a flat bed. In sand-bed channels especially, the bed is normally covered with bed forms such as ripples or dunes, and shear stresses required for significant erosion may be much greater than indicated by the Shield's diagram. Bed forms and irregularities occur also in many channels with coarser beds. More complex approaches have been used that involve separating the total shear stress into two parts associated with the roughness of the sediment grains and of the bed forms. Then, only the first part contributes to erosion. In general, however, the Shield's approach is not very useful for the design of channels in fine-grained materials.

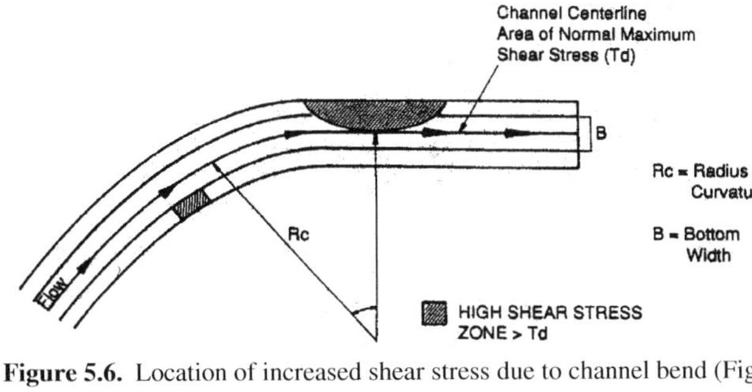

Figure 5.6. Location of increased shear stress due to channel bend (Figure reprinted with permission of North American Green).

Guidelines for Applications

The following guidelines are suggested by the COE (1994) for performing the computations and following the procedures listed for the allowable velocity and shear stress concepts:

- If cross sections and slope are reasonably uniform, computations can be based on an average section. Otherwise, divide the project length into reaches and calculate values for small, medium, and large sections.
- Determine the discharge that would cause the initiation of erosion from the stage-velocity or discharge-velocity curve, and determine its frequency from a flood-frequency or flow-duration curve. This may give some indication of the potential for instability. For example, if bed movement has a return period measured in years, which is the case with some cobble or boulder channels, the potential for extensive profile instability is likely to be negligible. On the other hand, if the bed is evidently active at relatively frequent flows, response to channel modifications may be rapid and extensive.

Design Steps for Maximum Permissible Velocity/Allowable Shear Stress Method

McCuen (1998) presents the following steps when designing a stable channel using the permissible velocity/allowable shear stress method:

1. For a given channel material, estimate the Manning's roughness coefficient (n), the channel slope (S), and the maximum permissible velocity (V) (such as from Tables 5.1 or 5.2).
2. Compute the hydraulic radius (R) using Manning's equation:

$$R = \left[\frac{Vn}{1.49\,S^{0.5}}\right]^{1.5}$$

where,

R = hydraulic radius, ft.
V = permissible velocity, ft/sec
S = channel slope, ft/ft
n = roughness of channel lining material, dimensionless

Some typical values for Manning's n for open channels (Chow, 1959) are as follows:

Very smooth surface (glass, plastic, machined metal)	0.010
Planed timber	0.011
Rough wood	0.012–0.015
Smooth concrete	0.012–0.013
Unfinished concrete	0.013–0.016
Brickwork	0.014
Rubble masonry	0.017
Earth channels, smooth no weeds	0.020
Firm gravel	0.020
Earth channel, with some stones and weeds	0.025
Earth channels in bad condition, winding natural streams	0.035
Mountain streams	0.040–0.050
Sand (flat bed), or gravel channels, d = median grain diameter, ft.	$0.034d^{1/6}$

Chow (1959) also provides an extensive list of n values, along with photographs. Most engineering hydrology and hydrologic texts (including McCuen, 1998) will also contain extensive guidance on the selection of Manning's n values for different channel conditions. A later section in this chapter presents the traditional trial-and-error method for determining Manning's n values for grass-lined channels, using measured *VR-n* relationships for different grass types.

3. Calculate the required cross-sectional area, using the continuity equation and the previously determined design storm peak flow rate (Q):

$$A = \frac{Q}{V}$$

where,

A = cross-sectional area of channel (wetted portion), ft^2
Q = peak discharge for design storm being considered, ft^3/sec
V = permissible velocity, ft/sec

4. Calculate the corresponding wetted perimeter (P):

$$P = \frac{A}{R}$$

where,

P = wetted perimeter, ft
A = cross-sectional area of channel (wetted portion), ft^2
R = hydraulic radius, ft.

5. Calculate an appropriate channel base width (b) and depth (y) corresponding to a specific channel geometry (usually a trapezoid channel, having a side slope of z:*1* side slopes [horizontal:vertical]).

Figure 5.8 (Chow, 1959) can be used to significantly shorten the calculation effort for the design of channels by skipping step 4 above and more effectively completing step 5. This figure is used to calculate the normal depth (y) of a channel based on the channel side slopes and known flow and channel characteristics. It requires using the Manning's equation in the following form:

$$AR^{2/3} = \frac{nQ}{1.49\,S^{0.5}}$$

Initial channel characteristics that must be know include the following: z (the side slope), and b (the channel bottom width, assuming a trapezoid or a rectangular cross-section). It is easy to examine several different channel options (varying z and b) by calculating the normal depth (y) for a given peak discharge rate, channel slope, and roughness. The most practical channel can then be selected from the alternatives.

Example:

Assume the following conditions:

Noncolloidal alluvial silts, channel lining material water transporting colloidal silts:

Manning's roughness coefficient (n) = 0.020

maximum permissible velocity (V) = 3.5 ft/sec
(the allowable shear stress is 0.15 lb/ft^2)

Peak discharge flow rate (Q) = 13 ft^3/sec

Channel slope = 1%, or 0.01 ft/ft

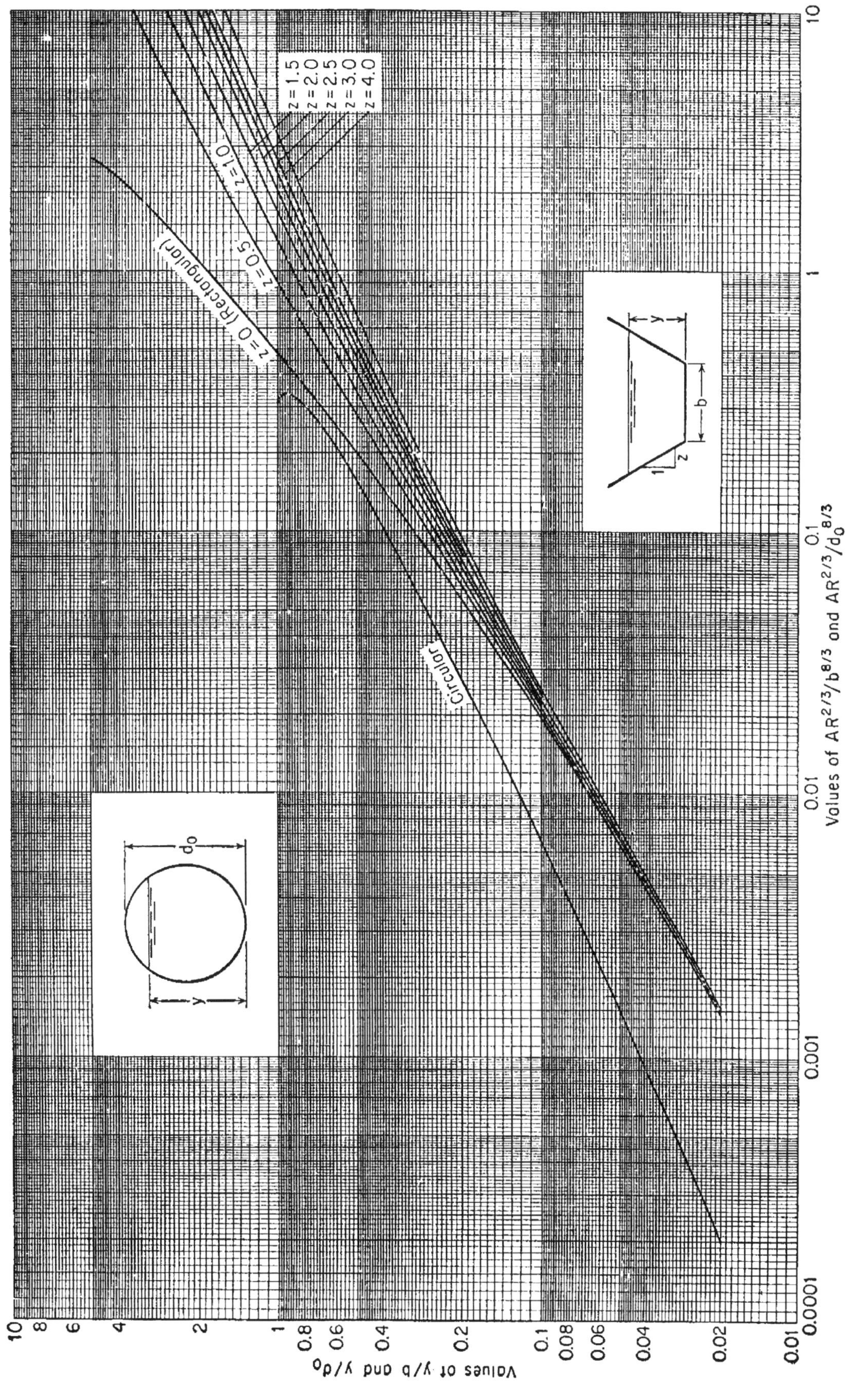

Figure 5.8. Curves for determining normal depth for various channel geometries (Chow, 1979, with permission).

Therefore,

The hydraulic radius (R) using Manning's equation:

$$R = \left[\frac{Vn}{1.49S^{0.5}}\right]^{1.5} = \left[\frac{3.5(0.020)}{1.49(0.01)^{0.5}}\right]^{1.5} = 0.32ft$$

The required cross-sectional area, using the continuity equation and the design storm peak flow rate (Q):

$$A = \frac{Q}{V} = \frac{13}{3.5} = 3.7ft^2$$

Therefore, $AR^{2/3}$ = (3.7)(0.32)2/3 = 1.7, and the wetted perimeter is A/R = 3.7/0.32 = 12 ft. Table 5.3 shows the calculated normal depth (y) for different channel options that all meet the allowable velocity criteria. Also shown on this table is the calculated maximum shear stress:

$$\gamma RS = (62.4\ \text{lb/ft}^3)\ (R\ \text{ft})\ (0.01\ \text{ft/ft}) = 0.62R$$

Since the allowable shear stress is 0.15 lb/ft^2, the hydraulic radius must be less than 0.24 ft (less than only about 3 inches deep). This will therefore require a relatively-wide channel, as the hydraulic radius approximates the depth of flow for wide and shallow channels. Also, the depth of flow can be used instead of the hydraulic radius as a conservative approach to calculate the maximum shear stress.

As the channel becomes wider, the side slopes have little effect on the normal depth and the calculated maximum shear stress, as expected. The safety factors are the ratios of the allowable shear stress (0.15 lb/ft^2) divided by the calculated maximum shear stress. None of these channels can satisfy the allowable shear stress with this natural material, unless the channel is very wide. A minimum channel width between 15 and 25 ft would result in a stable channel. However, a channel liner can be used to reinforce the channel, resulting in a larger allowable shear stress, which will enable a narrower channel to be used to safely transport the flow.

Table 5.3 shows both the shear stress calculated using the hydraulic radius, R, and the larger shear stress calculated using the normal depth, y. Also shown is the ratio of the hydraulic radius to the normal depth for different channel conditions. Figure 5.9 is a plot showing how the normal depth approaches the hydraulic depth, for this example, as the channel width to normal depth ratios increase. The maximum shear stress is therefore much larger when the normal depth is used instead of the hydraulic radius for relatively narrow channels, but the results are similar for wider channels.

A more direct approach is to use Figure 5.8 in reverse order. As shown previously, the maximum depth can be calculated based on the maximum allowable shear stress and the channel slope:

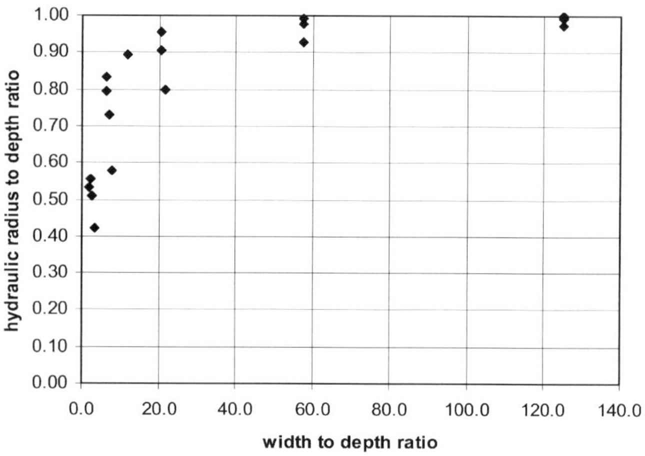

Figure 5.9. Relationship of hydraulic radius to normal depth for different channel width to depth conditions.

$$D = \frac{\tau_c}{\gamma S} = \frac{0.15lb/ft^2}{(62.4lb/ft^3)(0.01ft/ft)} = 0.24ft$$

With the known value for $AR^{2/3}$ ($3.7 \times 0.32^{2/3} = 1.7$), Table 5.4 shows the calculated maximum side slope for different channel bottom widths (b). All of these options will meet both the allowable velocity and shear stress criteria.

For this example, side slopes of about 5:1 and with a bottom width of 15 ft may be stable, or "any" side slope may be suitable for bottom widths of 20 ft, or wider. This example has shown that it may not be possible to design a stable channel only based on allowable maximum velocity. It is a good idea to also calculate the maximum shear stress, based on the normal depth. Without a channel liner, most stable channels in soils will need to be relatively wide. Because of the increased use of land needed for wide channels (see the calculated top width "T" in Table 5.3), it is usually necessary to consider channel liners, either grass-lined, or re-enforced with netting mats, as described in the following sections.

DESIGN OF GRASS-LINED CHANNELS

Temple, *et al.* (1987) in *Stability Design of Grass-Lined Open Channels,* USDA Agricultural Handbook # 667, shows that grass channel linings can be used to protect an erodible soil boundary and prevent channel degradation. They found that soil detachment begins at total shear stress levels small enough to be withstood by the vegetation without significant damage to the plants themselves, i.e., it is possible for the vegetation to be undercut and the weaker vegetation washed away. This vegetation loss decreases the density and uniformity of the vegetative cover, which in turn leads to greater stresses at the soil-water interface, resulting in an increased erosion rate. Supercritical channel flows cause a more severe problem compared to subcritical flows because small irregularities in the channel lining cause

TABLE 5.3. Alternative Channel Geometries Meeting Maximum Permissible Velocity (3.5 ft/sec) Compared to Allowable Shear Stress (0.15 lb/ft²).

Side Slope (z)	Bottom Width (b), ft	$b^{8/3}$	$AR^{2/3}/b^{8/3}$	y/b	Normal Depth (y), ft	Top Width (T), ft	Area (A), ft^2	Wetted Perimeter (P), ft	Hydraulic Radius (R), ft	b/y	R/y	Maximum Shear Stress Using y (τ), lb/ft^2	Safety Factor (allowable shear stress/ max. shear stress using y)	Maximum Shear Stress Using y (τ), lb/ft^2	Safety Factor (allowable shear stress/ max. shear stress using y)
4	2	6.4	0.27	0.32	0.62	7.0	2.8	10.6	0.26	3.2	0.42	0.32	0.39	0.16	0.92
4	4	41	0.041	0.13	0.52	8.2	3.2	10.5	0.30	7.7	0.58	0.32	0.47	0.19	0.80
4	8	260	0.0066	0.046	0.37	11.0	3.5	11.9	0.30	21.6	0.80	0.23	0.65	0.18	0.81
4	15	1400	0.0012*	0.017	0.26	17.1	4.2	17.3	0.24	57.7	0.93	0.16	0.94	0.15	0.99
4	25	5300	0.00032*	0.008	0.2	26.6	5.2	26.5	0.19	125.0	0.97	0.12	1.25	0.12	1.24
2	2	6.4	0.27	0.38	0.76	5.0	2.7	6.9	0.39	2.6	0.51	0.47	0.32	0.24	0.62
2	4	41	0.041	0.14	0.56	6.2	2.9	7.0	0.41	7.1	0.73	0.35	0.43	0.26	0.59
2	8	260	0.0066	0.049	0.39	9.6	3.4	9.7	0.35	20.5	0.91	0.24	0.63	0.22	0.68
2	15	1400	0.0012*	0.017	0.26	16.0	4.0	15.9	0.25	57.7	0.98	0.16	0.94	0.16	0.95
2	25	5300	0.00032*	0.008	0.2	25.8	5.1	25.6	0.20	125.0	0.99	0.12	1.25	0.12	1.21
1	2	6.4	0.27	0.44	0.88	3.8	2.5	5.2	0.49	2.3	0.55	0.55	0.27	0.30	0.49
1	4	41	0.41	0.16	0.64	5.3	3.0	5.8	0.51	6.3	0.79	0.40	0.38	0.32	0.47
1	8	260	0.0066	0.049	0.39	8.8	3.3	8.8	0.37	20.5	0.95	0.24	0.63	0.23	0.65
1	15	1400	0.0012*	0.017	0.26	15.5	4.0	15.4	0.26	57.7	0.99	0.16	0.94	0.16	0.93
1	25	5300	0.00032*	0.008	0.2	25.4	5.0	25.3	0.20	125.0	1.00	0.12	1.25	0.12	1.20
0.5	2	6.4	0.27	0.5	1	3.0	2.5	4.7	0.53	2.0	0.53	0.62	0.24	0.33	0.45
0.5	4	41	0.41	0.16	0.64	4.6	2.8	5.2	0.53	6.3	0.83	0.40	0.38	0.33	0.45
0.5	8	260	0.0066	0.049	0.69	8.7	5.8	9.4	0.62	11.6	0.89	0.24	0.63	0.38	0.39
0.5	15	1400	0.0012*	0.017	0.26	15.3	3.9	15.2	0.26	57.7	0.99	0.16	0.94	0.16	0.93
0.5	25	5300	0.00032*	0.008	0.2	25.2	5.0	25.1	0.20	125.0	1.00	0.12	1.25	0.12	1.20

*Estimated, as these values are under range from the plotted curves on Figure 5.8.

TABLE 5.4. Example Calculations for Required Side Slopes for Different Bottom Widths, Meeting Both the Allowable Velocity and Maximum Shear Stress Criteria.

b (ft)	y/b (with y = 0.24ft)	$AR^{2/3}/b^{8/3}$	Required Side Slope (z), or Longer
8	0.020	0.0066	>4
10	0.024	0.0036	>4
15	0.016	0.0012*	5 (?)
20	0.012	0.00057*	any (0.5 to 4)

*Estimated, as these values are under range from the plotted curves on Figure 5.8.

stress-concentration points to develop. For very erosion-resistant soils, the lining vegetation may sustain damage before the effective stress at the soil-water interface becomes large enough to detach soil material. Although the limiting condition in this case is the stress on the plants, failure progresses in a similar manner: damage to the plant cover results in an increase in effective stress on the soil boundary until conditions critical to erosion are exceeded. The resulting erosion further weakens the cover, and unraveling occurs. When plant failure occurs, it is a complex process involving removing young and weak plants, shredding and tearing of leaves, and fatigue weakening of stems.

Because of the many uncertainties and different methods of failure, the use of an approximate design approach is considered appropriate for most practical applications. Temple, *et al.* (1987) state that conservative design criteria are required, as the potential for rapid unraveling of a channel lining can occur once a weak point has developed, especially considering the variability of vegetative covers. Very dense and uniform covers will likely withstand stresses substantially greater than immature or spotty covers without significant damage. However, they recommend that poor maintenance should be assumed in conservative designs.

The design of a grass-lined open channel differs from the design of an unlined or structurally lined channel in that (1) the flow resistance is dependent on channel geometry and discharge, (2) a portion of the boundary stress is associated with drag on individual vegetation elements and is transmitted to the erodible boundary through the plant root system, and (3) the properties of the lining vary both randomly and periodically with time. Each of these differences requires special considerations in the design process. Temple, *et al.* (1987) presents detailed descriptions of the generalized step-by-step procedure for grass-lined channel design, including computer code.

Plant Species Selection for Vegetative-Lined Channels

The following is a general discussion and does not provide site-specific guidance for different climatic regions. However, it does describe the general problems associated with establishing plants in a channel environment. Local guidance (such as from the local USDA or University Extension services' offices) needs to be sought for specific recommendations for a particular location. Obviously, channels carrying water for long periods of the year may not be suitably lined with terrestrial vegetation. Extended wet periods will also affect plant selection. Again, local plant specialists need to be consulted for the proper selection of suitable plants for the anticipated growing conditions. The *Alabama Handbook for Erosion Control, Sediment Control, and Stormwater Management on Construction Sites and Urban Areas* (USDA 2003) contains further general guidance on plant selection for Alabama uses, for example.

Site Considerations

When a site will receive heavy use, such as a sports field, plant species that are wear resistant and have rapid wear recovery (such as bermudagrass) should be selected. Bermudagrass also has a fast establishment rate and is adapted to many geographical areas. Where a neat appearance is desired, plants that respond to frequent mowing should be used. Likely choices for quality turf in north Alabama are bermudagrass or tall fescue, while in central or south Alabama bermudagrass, centipede, or zoysia are good choices. At sites where low maintenance is desired, low fertility requirements and vegetation persistence are particularly important. Sericea lespedeza and tall fescue are good choices in north Alabama, while bahiagrass and centipede do well in central and south Alabama.

Seasonal Considerations

Planting guidance is available throughout the United States. For example, in Alabama, the most effective times for planting perennial grasses and legumes generally extend from March through May and from late August through October. Outside these dates, the probability of failure is greater. Growing seasons must be considered when selecting species. Grasses and legumes are usually classified as warm or cool-season in reference to their season of growth. Cool-season species produce most of their growth during the spring and fall and are relatively inactive or dormant during the hot summer months. Therefore, fall is the most dependable time to plant them. Warm-season plants grow most actively during the summer, and go dormant at the first frost in the fall. Spring and early summer are the preferred planting times for warm-season species.

Selecting the Right Grasses for Channel Lining

Information on plant species adapted for soil stabilization use is contained in most state erosion control manuals and from the Internet sources listed at the end of this chapter. Most of these commercial suppliers of seeds and sod will

Sod placed along a channel bottom, with grass seed along the edges.

Sod needing irrigation during dry period.

Seeding along median strip swale of highway project.

Mesh placed over seed and mulch.

help select the most appropriate species for local site conditions. Local USDA Agricultural Extension offices also may be able to provide updated guidance. Using this locally-generated information makes plant selection more straight forward for most situations. Specific seeding rates and planting instructions are presented in the specifications for local conditions by regulatory agencies.

According to Temple, *et al.* (1987), the selection of grass species for use in channels is based on important site-specific factors, including: (1) soil texture, (2) depth of underlying material, (3) management requirements of vegetation, (4) climate, (5) slope, and (6) type of structure or engineering design. The expected flow rates, availability of seed, ease of stand establishment, species or vegetative growth habit, plant cover, and persistence of established species are other factors that also should be considered in selecting appropriate grasses necessary for stable channel designs. Channel construction should be scheduled to allow establishment of the grass stand before subjecting the channel to excessive flows. The uses of modern channel lining systems, as discussed below, help alleviate this problem. The establishment of permanent covers involves liming and fertilizing, seed bed preparation, appropriate planting dates, seeding rates, and mulching.

Plants for Temporary Channel Linings

Based on flow tests on sandy clay channels, Temple, *et al.* (1987) recommends wheat (*Triticum aestivum L.*) for winter and sudangrass [*Sorghum sudanensis* (Piper) *Hitchc.*] for late-summer temporary covers. These temporary covers have been shown to rapidly increase the permissible discharge rate to five times that of an unprotected channel. Other recommended annual and short-lived perennials that can be used for temporary channel linings include:

- Barley (*Hordeum vulgare L.*), noted for its early fall growth;
- Oats [(*Avena*) *sativa L.*], in areas of mild winters;
- Mixtures of wheat, oats, barley, and rye (*Secale cereale L.*);
- Field bromegrass (*Bromus spp.*); and
- Ryegrasses (*Lolium spp.*).

Summer annuals, including German and foxtail millets (*Setaria* spp.), pearl millet [*Pennisetum americanurn (L.) Leeke*], and certain cultivated sorghums other than sudangrass, may also be used for temporary mid- to late-summer covers, according to Temple, *et al.* (1987). Since millets do not continue to grow as aggressively as sorghums after mowing, they may leave a more desirable, uniformly thin mulch for subsequent permanent seeding. Temporary seedings involve minimal cultural treatment, short-lived but quick germinating species, and little or no maintenance. The temporary covers should be close-drilled stands and not be allowed to go to seed. The protective cover provided by the temporary vegetation should provide stalks, roots, and litter into which permanent grass seeds can be drilled the following spring or fall.

Plants for Permanent Channel Linings

Many grasses can be used for permanent vegetative channel linings. Temple, *et al.* (1987) lists the following tight-sod-forming grasses as the most preferred warm- and cool-season grasses for channel linings: bermudagrass [*Cyodon dactylon var dactylon (L.) Pers.*], bahiagrass (*Paspalum notatum Fluggle*), buffalograss [*Buchloe dactyloides (Nutt.) Engelm.*], intermediate wheatgrass [*Agropyron intermedium (Host) Beauv.*], Kentucky bluegrass (*Poa ratensis L.*), reed canarygrass (*Phalaris arundinacea L.*), smooth bromegrass, (*Bromus inermis Leyss.*), vine mesquitegrass (*Panicum obtusum H.B.K.*), and Western wheatgrass (*Agropyron Smithii Rydb.*). These grasses are among the most widely used species for channel linings and grow well on a variety of soils. A grass mixture should include species adapted to the full range of soil moisture conditions anticipated to be encountered by the channel side slopes. The local NRCS and University Extension offices know the best soil-binding grass species for a particular area, as well as the associated planting and maintenance requirements. The most important characteristic of the selected grasses is its ability to survive and thrive in the channel environment.

Bermudagrass is probably the most widely-used grass in the southern region of the U.S. It will grow on many soil types, but it may require extra management. It forms a dense and persistent sod, if managed properly. Temple, *et al.* (1987) recommend that when bermudagrass is used, winter-hardy varieties should be obtained. Improved varieties, such as "Coastal," "Midland," "Greenfield," "Tifton," and "Hardie," do not produce seed, and must be established by sprigging. Where winters are mild, channels can be established quickly with seed of "Arizona Common" bermudagrass. "Seed of bermudagrass," a new seed-propagated variety with greater winter hardiness than Arizona Common, should be available now commercially. Bermudagrass is not shade tolerant and should not be used in mixtures containing tall grasses. However, the inclusion of winter annual legumes such as hairy vetch (*Vicia villosa Roth.*), narrowleaf vetch [*V. sativa L. subspecies nigra (L.) Ehrh.*], and/or a summer annual such as Korean lespedeza (*Lespedeza stipulacea Maxim.*), may be beneficial to stand maintenance.

The selection of grasses used in channels often depends on availability of seed or plant material. Chronic national seed shortages of some warm-season grasses, especially seed of native species, often have led to planting seed marginally suited to site situations. Lack of available seed of desired grass species and cultivars adapted to specific problem sites is a major constraint often delaying or frustrating seeding programs. In addition to the grass species or base mixture of grasses used for erosion control, carefully selected special-use plants may be added for a specific purpose or situation. Desirable wildlife food plants may be included in the mixture if they do not detrimentally compete with the base grasses used for erosion control. Locally-adapted legumes are often added if they are compatible with the grasses and noncompetitive. Additional information on establishment and maintenance of grass-lined channels is provided in Temple, *et al.* (1987).

Determination of Channel Design Parameters

The conditions governing the stability of a grass-lined open channel are the channel geometry and slope, the erodibility of the soil boundary, and the properties of the grass lining that relate to flow retardance potential and boundary protection.

Vegetation Parameters

The design of a stable grass-lined open channel needs to consider the effective stress imposed on the soil layer (Temple, *et al,* 1987). This requires the determination of two vegetation parameters: (1) the retardance curve index (C_I) which describes the potential of the vegetal cover to develop flow resistance, and (2) the vegetation cover factor (C_f) which describes the degree to which the vegetation cover prevents high velocities and stresses at the soil-water interface. These are described below.

Retardance Potential. The parameter describing the retardance potential of a vegetal cover is the retardance curve index, C_I. This parameter determines the limiting vegetation stress. Its relation to the measurable physical properties of the vegetal cover is given by:

$$C_I = 2.5(h\sqrt{m})^{1/3}$$

where,

h = the representative stem length
M = the stem density in stems per unit area.

When consistent units are used, the relation is dimensionless. This factor is commonly used in the following equation to estimate the maximum allowable stress on the vegetation (τ_{va}, in lb/ft^2):

$$\tau_{va} = 0.75C_I$$

The stem length usually will need to be estimated directly from knowledge of the vegetation conditions at the time of anticipated maximum flow. When two or more grasses with widely differing growth characteristics are involved, the representative stem length is determined as the root mean square of the individual stem lengths.

When this equation is used to estimate the retardance potential, an estimate of the stem density is also required. The reference stem densities shown in Table 5.5 may be used as a guide in estimating this parameter. Temple, *et al.* (1987) obtained the values of reference stem densities from a review of the available qualitative descriptions and stem counts reported by researchers studying channel resistance and stability.

Since cover conditions will vary from year to year and season to season, establishing an upper and a lower bound for the curve index (C_I) is often more realistic than selecting a single value. When this approach is taken, the lower value should be used in stability computations and the upper value should be used in determining channel capacity. Such an approach normally will result in satisfactory channel operation for lining conditions between the specified bounds. Whatever the approach used to obtain the flow retardance potential of the lining, the values selected should represent an average for the channel reach in question, since it will be used to infer an average energy loss per unit of boundary area for any given flow.

Vegetation Cover Factor

The vegetation cover factor, C_f, is used to describe the degree to which the vegetation cover prevents high velocities and stresses at the soil-water interface. Because the protective action described by this parameter is associated with the prevention of local erosion damage which may lead to channel unraveling, the cover factor should represent the weakest area in a reach, rather than an average for the cover type.

Observations of flow behavior and available data indicate that the cover factor is dominated by the density and uniformity of density in the immediate vicinity of the soil boundary. For relatively dense and uniform covers, uniformity of density is primarily dependent on the growth characteristics of the cover, which are in turn related to grass

TABLE 5.5. Properties of Grass Channel Linings (Temple, *et al.* 1987).

Cover Factor (C_f) (good uniform stands)	Covers Tested	Reference Stem Density (M), stem/ft^2
0.90	bermudagrass	500
0.90	centipedegrass	500
0.87	buffalograss	400
0.87	kentucky bluegrass	350
0.87	blue grama	350
0.75	grass mixture	200
0.50	weeping lovegrass	350
0.50	yellow bluestem	250
0.50	alfalfa	500
0.50	lespedeza sericea	300
0.50	common lespedeza	150
0.50	sudangrass	50

type. This relationship was used by Temple, *et al.* (1987) in the development of Table 5.5. This table does not account for such considerations as maintenance practices, or uniformity of soil fertility or moisture conditions.

Soil Parameters

Two soil parameters are required for the application of effective stress concepts to the design of stable lined or unlined channels having an erodible soil boundary: (1) soil grain roughness (n_s), and (2) allowable effective stress (τ_a). When the effective stress approach is used, the soil parameters are the same for both lined and unlined channels, satisfying sediment transport restrictions. The relations shown here were presented by Temple, *et al.* (1987) and were taken from the SCS (1977) channel stability criteria; the desired parameters, soil grain roughness and allowable stress, are determined from basic soil parameters. Ideally, the basic parameters should be determined from tests on representative soil samples from the site.

For effective stress design, soil grain roughness is defined as the roughness associated with particles or aggregates of a size that may be independently moved by the flow at incipient channel failure. Although this parameter is expressed in terms of a flow resistance coefficient (n_s), its primary importance in design of vegetated channels is its influence on effective stress, as shown below. Its contribution to the total flow resistance of a grass-lined channel is usually negligibly small.

The allowable stress is key to the effective stress design procedure. It is defined as the stress above which an unacceptable amount of particle or aggregate detachment would occur.

Noncohesive Soil

Noncohesive soils are defined as fine- or coarse-grained, based on whether d_{75} (the diameter for which 75 percent of the material is finer) is less than, or greater than, 0.05 in. For fine-grained soils, the soil grain roughness and allowable effective stress are constant, while for a coarse-grained soil, these parameters are a function of particle size. The allowable effective stress and roughness parameters for noncohesive soils are given in Figures 5.10 and 5.11, as a function of particle size.

Cohesive Soil

All cohesive soils are treated as fine-grained soils, having a constant soil grain roughness (about 0.0155, according to Figure 5.11). The allowable effective stresses presented here are taken directly from SCS (1977) permissible velocity

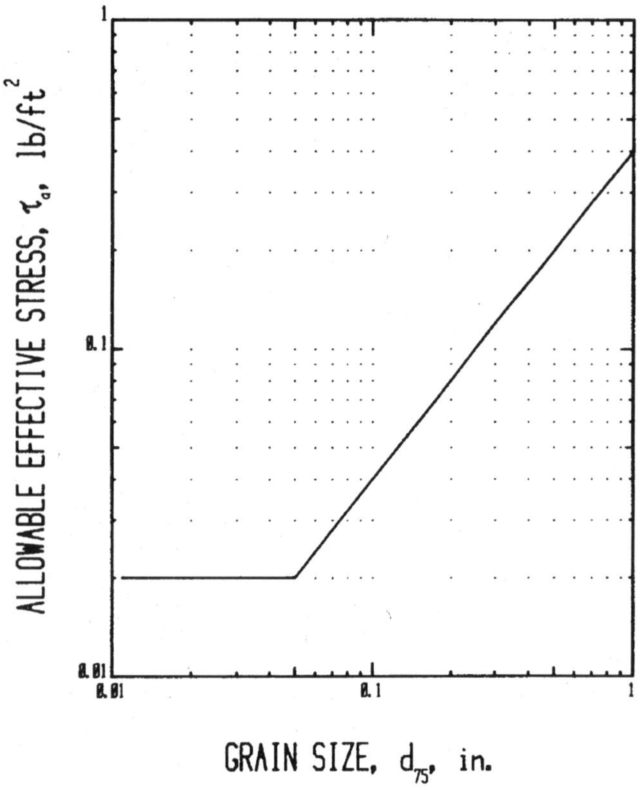

Figure 5.10. Allowable effective stress for noncohesive soils (Temple, *et al.* 1987).

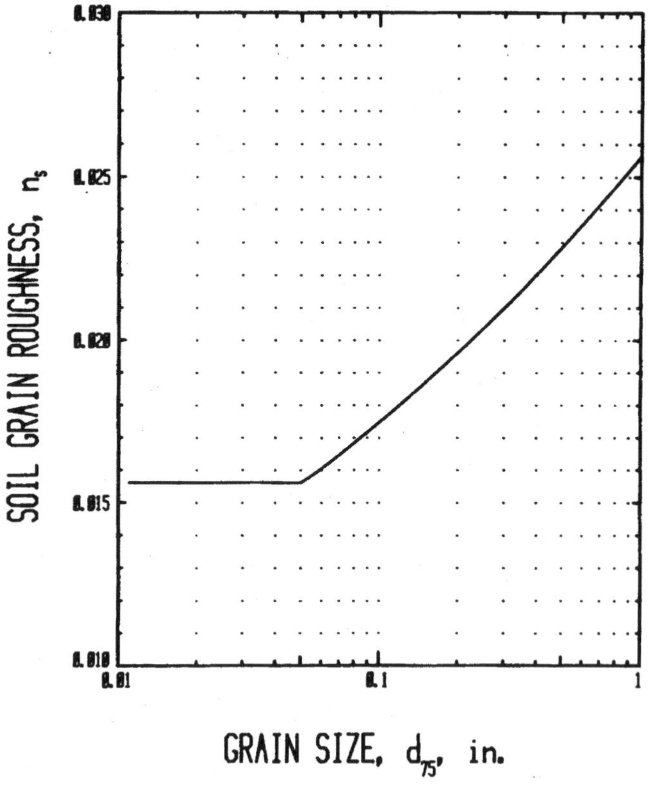

Figure 5.11. Soil grain roughness for noncohesive soils (Temple, *et al.* 1987).

design criteria. The soil properties required to determine the allowable effective stress are the soil's classification in the unified soil classification system, its plasticity index (I_w), and its void ratio (e). This calculation requires a basic allowable effective stress (τ_{ab}) that is determined from the soil classification and plasticity index. This basic value is then corrected for void ratio, according to the relation:

$$\tau_a = \tau_{ab} C_e^2$$

The basic allowable shear stress (τ_{ab}) is given in Figure 5.12, while the void ratio correction factor (C_e) is given in Figure 5.13. The soil classification information (plasticity index, I_w, and void ratio, e) are readily available for cohesive soils in standard soils references, and in Temple, *et al.* (1987). The previously presented Figure 5.6 (COE, 1994) is a simplified figure for determining allowable shear stress for cohesive soils, if these detailed soil characteristics are not available.

Selection of Roughness Factor for Grass Lined Channels

The value of Manning's n for grass-lined channel is a function of grass type and the product of velocity and hydraulic radius (VR). Grasses are divided into retardance classes based on their physical characteristics (height, width, density, etc.). Most sod forming grasses are classified as type C. These grasses can have n values ranging from 0.03–0.3 depending on VR, with a typical value of 0.03 in open channels. Figure 5.14 is an example of a VR-n curve based on data from the Stillwater, OK, USDA field tests. It was extended to cover smaller VR ranges appropriate for small drainage flows during extensive field and lab tests by Kirby (2003). The following example shows how the correct n

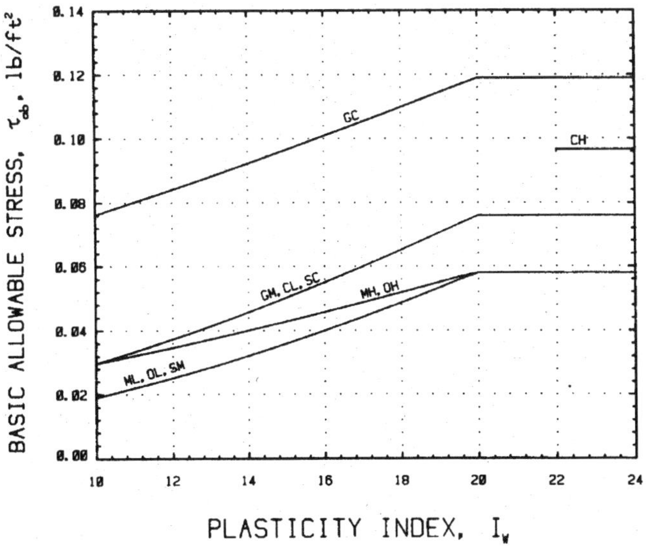

Figure 5.12. Basic allowable effective stress for cohesive soils (Temple, *et al.* 1987 and SCS, 1977).

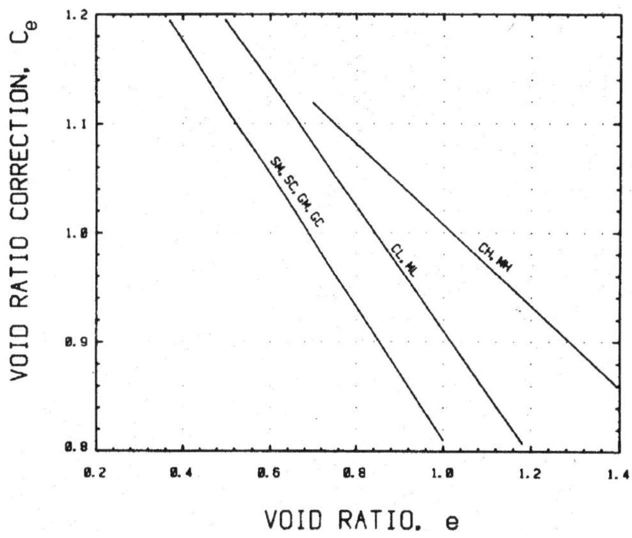

Figure 5.13. Void ratio correction factor for cohesive soils (Temple, *et al.* 1987 and SCS, 1977).

value is selected through a trial-and-error method, depending on the product of the velocity (V) and hydraulic radius (R).

Example: Selection of Roughness for Grass-Lined Channels

The appropriate Manning's n to use varies on the time frame: (1) bare soil retention and vegetation establishment (short-term), and (2) fully-grassed conditions (long-term) (Chow, 1959). Bare soil conditions can be examined using the procedures presented earlier. Mature grass-lined channel roughness values can be determined using typical procedures as illustrated in the following example, which shows how VR-n curves can be used for the proper selection of a roughness value for a grass-lined channel:

> Determine the roughness value for a 10-year design storm of 70 ft^3/sec (2 m^3/sec) in a grass-lined drainage channel having a slope of 0.05 ft/ft and a 4-foot (1.2-m) bottom width and 1:1 side slopes. The grass cover is expected to be in retardance group D.

Long-term Design, Based on Vegetated Channel Stability

- use $Q_{peak} = Q_{10year} = 70$ ft^3/s (2 m^3/s)
- initially assume that $n_{vegetated} = 0.05$

Determine the normal depth of flow, using Figure 5.8 (from Chow, 1959):

$$AR^{2/3} = \frac{nQ}{1.49 S^{0.5}} = \frac{0.05(70 \text{ cfs})}{1.49(0.05)^{0.5}} = 10.51$$

and

$$b^{8/3} = (4 \text{ ft})^{8/3} = 40.32$$

therefore,

$$AR^{2/3}/b^{8/3} = 10.51/40.32 = 0.26$$

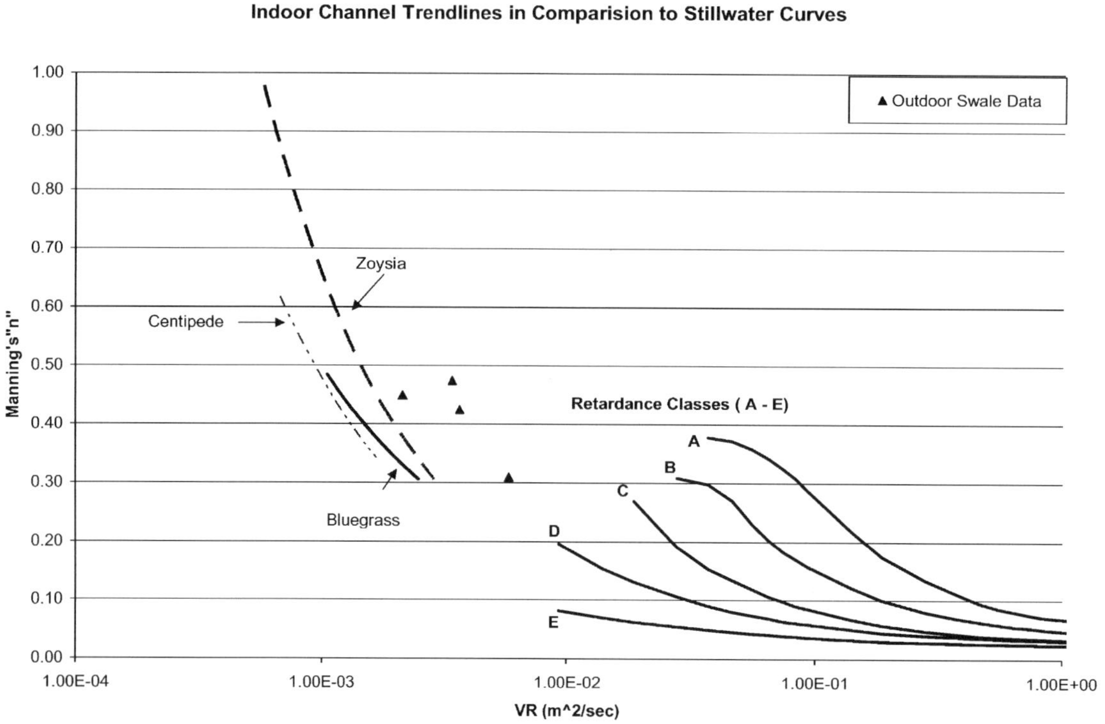

Figure 5.14. *VR-n* curve for different grasses, showing results for shallow flows (Kirby 2003). (Multiply ft^2/sec by 0.092 to obtain m^2/sec units).

With a 1:1 side slope trapezoidal channel, the ratio of y/b from Figure 5.8 is 0.43, and the depth is therefore: 4(0.43) = 1.7 ft.

The cross-sectional area is therefore 9.7 ft^2, the velocity is (70 ft^3/sec)/(9.7 ft^2) = 7.2 ft/sec, P is 8.8 ft, and R is 9.7/8.8 = 1.1 ft. VR is therefore (7.2 ft/sec)(1.1 ft) = 7.9 ft^2/sec (=0.73 m^2/sec). From Figure 5.14, the estimated new value for n is therefore 0.032, using a retardance class of D. Figure 5.14 can also be used if the VR product is very small, such as for small flows in small swales, which is common for many urban applications. The VR product is converted from ft^2/sec to m^2/sec. The depth must therefore be recalculated, using this new value for n:

$$AR^{2/3} = \frac{nQ}{1.49\,S^{0.5}} = \frac{0.032(70\text{ cfs})}{1.49(0.05)^{0.5}} = 6.72$$

and

$$b^{8/3} = (4\text{ ft})^{8/3} = 40.32$$

therefore,

$$AR^{2/3}/b^{8/3} = 6.72/40.32 = 0.17$$

With a 1:1 side slope trapezoidal channel, the ratio of y/b from Figure 5.8 is 0.34, and the depth is therefore: 4(0.34) = 1.4 ft.

The area is therefore 7.6 ft^2, the velocity is 70/7.6 = 9.2 ft/sec, P is 8.0 ft, and R is 7.6/8.0 = 0.95 ft. The revised VR is therefore (9.2 ft/sec)(0.95 ft) = 8.7 ft^2/sec (0.80 m^2/sec). Figure 5.14 shows that the revised value of n is still close to 0.032.

The maximum shear stress (using normal depth instead of hydraulic radius) is therefore:

$$\gamma DS = (62.4\text{ lb/ft}^3)\ (1.4\text{ ft})\ 0.05\text{ ft/ft}) = 4.4\text{ lb/ft}^2$$

Hence, this channel would be stable if the acceptable value is greater than this rather high value. A following discussion presents additional guidance on the selection and evaluation of turf-reinforcing mats that would likely be needed for this high shear-stress condition. The use of channel-lining mats protecting immature vegetation allows immediate protection of the sensitive soil boundary layer, as described in the following discussions. Also, free computer programs, such as supplied by North American Green (http://www.nagreen.com/), greatly help in the design of the most appropriate channel cross section and liner system.

DRAINAGE DESIGN USING TURF-REINFORCING MATS

Current practice is to design channel linings based on shear stress and less frequently on allowable velocity. Shear stress considers the weight of the water above the lining and therefore does a better job of predicting liner stability, compared to only using velocity. However, allowable

Examples of Channels Lined with Vegetation and other Materials

Installation of reinforced liner along thalweg of channel, with other material along sides (VA photo).

Concrete lined channel, with reinforced matting along overflow area.

Large rocks for channel reinforcement.

Close-up of rock reinforced channel, showing sediment accumulation between rocks.

(continued)

Examples of Channels Lined with Vegetation and other Materials (continued)

Reinforced liner along channel thalweg.

Plastic tarp used as a temporary liner.

Plastic tarp, with coir logs, for a temporary liner.

Close-up of temporary plastic tarp liner, showing edge staples.

velocity and the flow regime (if the flow is supercritical or subcritical) still should be examined to minimize unusual conditions.

If a channel will have intermittent flows, it is common to use turf-reinforcing mats as liners to increase the channel stability. However, if the channel will have perennial (or long-term) flows, grass will not be successful and mechanical liners must be used.

According to Croke (2000), drainage channel design using turf-reinforcement mats must consider three phases: (1) the original channel in an unvegetated state to determine if the matting alone will provide the needed protection before the vegetation is established, (2) the channel in a partially vegetated state, usually at 50% plant density, and (3) the permanent channel condition with vegetation fully established and reinforced by the matting's permanent net structure. The basic shear stress equation can be modified to predict the shear stress applied to the soil beneath a channel mat (Temple, *et al.* 1987):

$$\tau_{\varepsilon} = \gamma DS(1-C_f)\left(\frac{n_s}{n}\right)^2$$

where,

τ_e = effective shear stress exerted on soil beneath vegetation
γ = specific weight of water (62.4 lbs/ft^3)
D = the maximum flow depth in the cross section (ft)
S = hydraulic slope (ft/ft)
C_f = vegetation cover factor (this factor is 0 for an unvegetated channel)
n_s = roughness coefficient of underlying soil
n = roughness coefficient of vegetation and/or erosion control blanket (if vegetated, or not)

The flow depth, rather than the hydraulic radius, is used in this equation because this will result in the maximum shear stress developed, rather than the average stress (Temple, *et al.* 1987). In addition, the depth value is very close to the hydraulic radius for most channels, especially as sheetflow conditions are approached. The cover factor is a function of the grass and stem density, as previously described, while the roughness coefficients are standard Manning's roughness values for channels. The permissible shear stress for a liner mat should be available from manufacture's specifications, but it will vary for different growth phases, if vegetated. Obviously, the liner matting significantly reduces the shear stress exerted on the soil. Tables 5.6 through 5.14 summarize some typical values for a selection of these equation parameters for turf-reinforcing mats (products supplied by North American Green [from www.nagreen.com presented here as an example of the information sometimes available from product suppliers and manufactures. The mention of these materials should not be considered an endorsement from the authors or publishers]). Included on these tables are the conservation factor, C, values used in RUSLE for slope protection, along with roughness coefficients and maximum permissible shear stress values used in channel lining analyses. Only the P300 and C350 mats shown here are permanent liners and therefore have different values for different plant growth stages.

Values of C_f, the grass cover factor, were given in Table 5.5 (Temple, *et al.* 1987). They recommend multiplying the stem densities given by 1/3, 2/3, 1, 4/3, and 5/3, for poor, fair, good, very good, and excellent covers, respectively. C_f values for untested covers may be estimated by recognizing that the cover factor is dominated by density and uniformity of cover near the soil surface; the sod-forming grasses near the top of the table have higher C_f values than the bunch grasses and annuals near the bottom. For the legumes tested (alfalfa and *Lespedeza sericea*), the effective stem count for resistance (given in Table 5.5) is approximately five times the actual stem count very close to the bed. Similar adjustments may be needed for other unusually large-stemmed, branching, and/ or woody vegetation.

Example: Channel Lining

Consider the following conditions for a mature buffalograss on a channel liner mat:

$$\tau_o = \gamma DS = 2.83 \text{ lb/ft}^2$$

(previously calculated), requiring a NAG P300 permanent mat, for example:

n_s = the soil is 0.016
n = the vegetated mat is 0.042
C_f = the vegetated mat is 0.87

The permissible shear stress for the underlying soil is 0.08 lb/ft^2.

Therefore,

$$\tau_e = 2.83(1-0.87)\left(\frac{0.016}{0.042}\right)^2 = 0.053 \text{ lb / ft}^2$$

The calculated shear stress being exerted on the soil beneath the liner mat must be less than the permissible shear stress for the soil. In this example, the safety factor (permissible shear stress/allowable shear stress) is 0.08/0.053 = 1.5 and the channel lining system is expected to be stable.

Example: Permanent Channel Lining Design

An example of a permanent channel design and the selection of an appropriate reinforced liner is given below. The following example is for a channel that collects runoff from 14.6 acres. This channel is 900 ft. long and has an 8% slope. The peak discharge was previously calculated to be 29 ft^3/sec.

TABLE 5.6. S75 Straw Erosion Control Blanket (12 month life; 314 g/m² mass per unit area).

RUSLE Conservation Coefficients (*C*)		Channel Roughness Coefficients (*n*)	
	Slope Gradient (*S*)	Flow Depth	Manning's *n* (unvegetated)
Slope length (*L*)	All ≤ 3:1 slope:	≤0.50 ft (0.15 m)	0.055
≤20 ft (6 m)	0.029	0.50–2.00 ft	0.055–0.021
20 to 50 ft	0.110	≥2.00 ft (0.60 m)	0.021
≥50 ft (15 m)	0.190	Max. permissible shear stress: 1.55 lbs/ft² (74.4 Pa)	

TABLE 5.7. S150 Straw Erosion Control Blanket (12 month life; 323 g/m² mass per unit area).

RUSLE Conservation Coefficients (*C*)			Channel Roughness Coefficients (*n*)	
	Slope Gradient (*S*)		Flow Depth	Manning's *n* (unvegetated)
Slope length (*L*)	≤ 3:1	3:1 to 2:1	≤0.50 ft (0.15 m)	0.055
≤20 ft (6 m)	0.004	0.106	0.50–2.00 ft	0.055–0.021
20 to 50 ft	0.062	0.118	≥2.00 ft (0.60 m)	0.021
≥50 ft (15 m)	0.120	0.180	Max. permissible shear stress: 1.75 lbs/ft² (84.0 Pa)	

TABLE 5.8. S150BN Straw Erosion Control Blanket (10 month life; 352 g/m² mass per unit area).

RUSLE Conservation Coefficients (*C*)			Channel Roughness Coefficients (*n*)	
	Slope Gradient (*S*)		Flow Depth	Manning's *n* (unvegetated)
Slope length (*L*)	≤ 3:1	3:1 to 2:1	≤0.50 ft (0.15 m)	0.055
≤20 ft (6 m)	0.00014	0.039	0.50–2.00 ft	0.055–0.021
20 to 50 ft	0.010	0.070	≥2.00 ft (0.60 m)	0.021
≥50 ft (15 m)	0.020	0.100	Max. permissible shear stress: 1.85 lbs/ft² (88.0 Pa)	

TABLE 5.9. SC150 Straw Erosion Control Blanket (24 month life; 424 g/m² mass per unit area).

RUSLE Conservation Coefficients (*C*)				Channel Roughness Coefficients (*n*)	
	Slope Gradient (*S*)			Flow Depth	Manning's *n* (unvegetated)
Slope length (*L*)	≤ 3:1	3:1 to 2:1	≥2:1	≤0.50 ft (0.15 m)	0.050
≤20 ft (6 m)	0.001	0.048	0.100	0.50–2.00 ft	0.050–0.018
20 to 50 ft	0.051	0.079	0.145	≥2.00 ft (0.60 m)	0.018
≥50 ft (15 m)	0.100	0.110	0.190	Max. permissible shear stress: 2.00 lbs/ft² (96.0 Pa)	

TABLE 5.10. SC150BN Straw Erosion Control Blanket (18 month life; 424 g/m² mass per unit area).

RUSLE Conservation Coefficients (*C*)				Channel Roughness Coefficients (*n*)	
	Slope Gradient (*S*)			Flow Depth	Manning's n (unvegetated)
Slope length (*L*)	≤ 3:1	3:1 to 2:1	≥2:1	≤0.50 ft (0.15 m)	0.050
≤20 ft (6 m)	0.00009	0.029	0.063	0.50–2.00 ft	0.050–0.018
20 to 50 ft	0.005	0.055	0.092	≥2.00 ft (0.60 m)	0.018
≥50 ft (15 m)	0.010	0.080	0.120	Max. permissible shear stress: 2.10 lbs/ft² (100 Pa)	

TABLE 5.11. C125 Coconut Fiber Erosion Control Blanket (36 month life; 274 g/m² mass per unit area).

RUSLE Conservation Coefficients (*C*):				Channel Roughness Coefficients (*n*)	
	Slope Gradient (*S*)			Flow Depth	Manning's *n* (unvegetated)
Slope length (L)	≤ 3:1	3:1 to 2:1	≥2:1	≤0.50 ft (0.15 m)	0.022
≤20 ft (6 m)	0.001	0.029	0.082	0.50–2.00 ft	0.022–0.014
20 to 50 ft	0.036	0.060	0.096	≥2.00 ft (0.60 m)	0.014
≥50 ft (15 m)	0.070	0.090	0.110	Max. permissible shear stress: 2.25 lbs/ft² (108 Pa)	

TABLE 5.12. C125BN Coconut Fiber Erosion Control Blanket (24 month life; 360 g/m² mass per unit area).

RUSLE Conservation coefficients (*C*)				Channel Roughness Coefficients (*n*)	
	Slope Gradient (*S*)			Flow Depth	Manning's *n* (unvegetated)
Slope length (*L*)	≤ 3:1	3:1 to 2:1	≥2:1	≤0.50 ft (0.15 m)	0.022
≤20 ft (6 m)	0.00009	0.018	0.050	0.50–2.00 ft	0.022–0.014
20 to 50 ft	0.003	0.040	0.060	≥2.00 ft (0.60 m)	0.014
≥50 ft (15 m)	0.007	0.070	0.070	Max. permissible shear stress: 2.35 lbs/ft² (112 Pa)	

TABLE 5.13. P300 Polypropylene Fiber Erosion Control Blanket (permanent use; 456 g/m² mass per unit area).

RUSLE Conservation coefficients (*C*)	Slope Gradient (*S*)			Channel Roughness Coefficients (*n*)			
Slope length (*L*)	≤ 3:1	3:1 to 2:1	≥2:1	Flow depth	Manning's *n* (unvegetated)	Maximum Permissible Shear Stress	
≤20 ft (6 m)	0.001	0.29	0.082	≤0.50 ft (0.15 m)	0.049–0.034	Unvegetated	3.00 lb/ft² (144 Pa)
20 to 50 ft	0.036	0.060	0.096	0.50–2.00 ft	0.034–0.020	Partially vegetated	5.50 lb/ft² (264 Pa)
≥50 ft (15 m)	0.070	0.090	0.110	≥2.00 ft (0.60 m)	0.020	Fully vegetated	8.00 lb/ft² (383 Pa)

TABLE 5.14. Additional Permissible Shear Stress Information for Vegetated North American Green Products (permanent liners).

	Manning's Roughness Coefficient (*n*) for Flow Depths			Maximum Permissible Shear Stress	
Vegetated Blanket Type*	0 to 0.5 ft	0.5 to 2 ft	>2 ft.	Short Duration (<2 hours peak flow)	Long Duration (>2 hours peak flow)
C350 Phase 2	0.044	0.044	0.044	6.00 lb/ft² (288 Pa)	4.50 lb/ft² (216 Pa)
P300 Phase 2	0.044	0.044	0.044	5.50 lb/ft² (264 Pa)	4.00 lb/ft² (192 Pa)
C350 Phase 3	0.049	0.049	0.049	8.00 lb/ft² (384 Pa)	8.00 lb/ft² (384 Pa)
P300 Phase 3	0.049	0.049	0.049	8.00 lb/ft² (384 Pa)	8.00 lb/ft² (384 Pa)

*Phase 2 is 50% stand maturity, usually at 6 months, while Phase 3 is mature growth

Using the Manning's equation and the Chow (1959) shortcut on channel geometry (Figure 5.8):

$$AR^{2/3} = \frac{nQ}{1.49S^{0.5}}$$

where,

$n = 0.02$
$Q = 29$ CFS
$S = 8\%$ (0.08)

$$AR^{2/3} = \frac{(0.02)(29)}{1.49(0.08)^{0.5}} = 1.38$$

The following drawing shows the channel dimensions for this basic analysis:

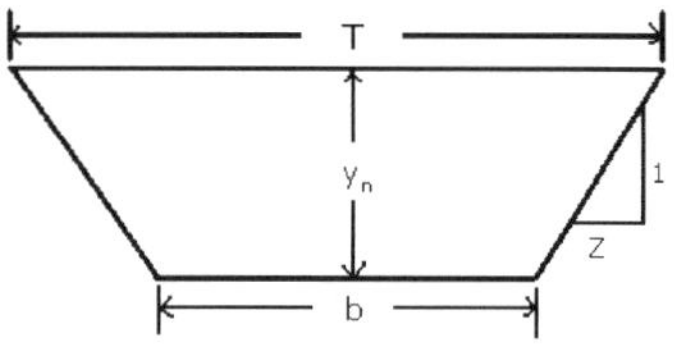

Figure 5.8 can be used to determine the normal depth (y_n) for many combinations of bottom width (b), and side slope (z). As an example, assume that the bottom width is 5 ft. and the side-slope parameter, z, is 3. The calculated $AR^{2/3}$ value (1.38) needs to be divided by $b^{8/3}$ ($5^{8/3} = 73.14$) for the shape factor used in Figure 5.8. This value is therefore: 1.38/73.14 = 0.018. For a side slope of $z = 3$, the figure indicates that the ratio of the depth to the bottom width (y/b) is 0.088. In this example, the bottom width was 5 ft, so the normal depth is: y_n = 0.088 (5 ft.) = 0.44 ft., which is only 5.3 inches. The following shows these dimensions on the channel cross-section:

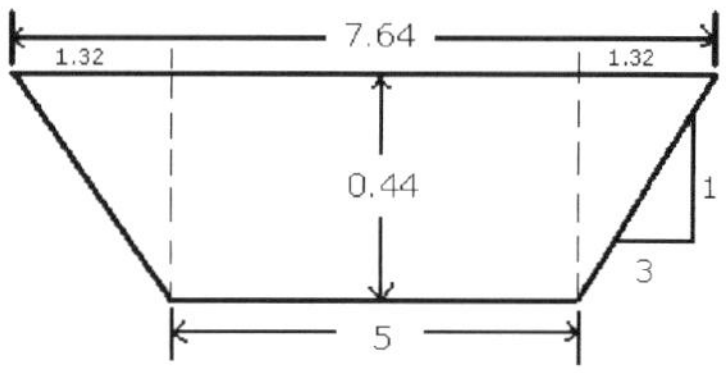

It is now possible to calculate the velocity and shear stress associated with this set of channel conditions:

$$A = [(7.64+5)/2]\ (0.44) = 2.78\ \text{ft}^2$$

$$V = Q/A = 29\ \text{ft}^3/\text{sec}/2.78\ \text{ft}^2 = 10.4\ \text{ft/sec}$$

$R = A/P$, and $P = 5 + 2(3.16)(0.44) = 7.78$ ft.

$$R = A/P = 2.78\ \text{ft}^2/7.78\ \text{ft.} = 0.36\ \text{ft.}$$

and

$$\tau = \gamma RS = (62.4\text{lb/ft}^3)(0.36\ \text{ft.})(0.08) = 1.8\ \text{lb/ft}^2$$

With a velocity of 10.4 ft/sec and a shear stress of 1.8 lb/ft^2, it is obvious that some type of channel reinforcement will be needed (refer to Table 5.2), or another design option will have to be considered. Using Figure 5.8, plus liner information (such as listed previously), it is possible to create a simple spreadsheet with multiple-cross section and liner alternatives, as shown in Table 5.15. This table shows the unvegetated conditions and calculations, along with the Phase 2 and Phase 3 vegetation conditions, for several channel cross-sections, considering both NAG P300 and C350 permanent channel liner mats. The shear stress values are calculated using the normal depth of flow, assuming worst-case design conditions, instead of using the hydraulic radius.

Example: Calculations for Permanent C350 Liner, 5 ft Bottom Width, z = 3 Side Slope, and Phase 3 Vegetation Plant Stage (mature)

$$AR^{2/3} = \frac{nQ}{1.49S^{0.5}} \frac{(0.049)(29)}{1.49(0.08)^{0.5}} = 3.38$$

$$b^{8/3} = 5^{8/3} = 73.1$$

$$AR^{2/3}/b^{8/3} = 3.38/73.1 = 0.046$$

with,

$$z = 3,\ y/b = 0.14$$

therefore,

$$y_n = 0.14\ (5) = 0.7\ \text{ft}$$

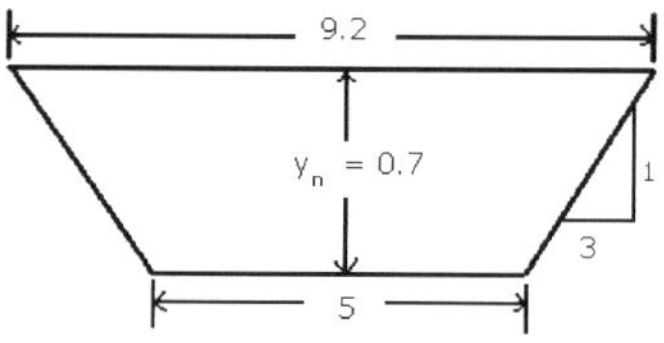

$A = [(5+9.2)/2]\ (0.7) = 4.97\ \text{ft}^2$
$P = 5 + 2(1.21) = 7.42$ ft
$R = A/P = 4.97/7.42 = 0.67$
$\tau = \gamma RS = (62.4\text{lb/ft}^3)(0.67\ \text{ft.})(0.08) = 3.34\ \text{lb/ft}^2$ (analysis case using hydraulic radius)
$\tau = \gamma DS = (62.4\text{lb/ft}^3)(0.70\ \text{ft.})(0.08) = 3.49\ \text{lb/ft}^2$ (design case using normal depth)
$V = Q/A = 29\ \text{ft}^3/\text{sec}/4.97\ \text{ft}^2 = 5.8\ \text{ft/sec}$

$$\tau_e = \gamma DS(1 - C_f)\left(\frac{n_s}{n}\right)^2 = 3.49\ \text{lb/ft}^2(1-0.87)\left(\frac{0.016}{0.049}\right)^2 = 0.048\ \text{lb/ft}^2$$

$n_s = 0.016$;
$C_f = 0.87$ phase 3

TABLE 5.15. Characteristics for Alternative Designs for Drainage Channel (Q = 29 ft^3/sec and S = 8%).

		Unvegetated NAG P300, n = 0.02 (allowable shear stress = 3.0 lb/ft^2) [data not given for C350, assumed to be similar to P300 for this example]					Channel with Reinforced Liner and Vegetation						
Bottom Width (b), ft	Side Slope (z)	Normal Depth (y_n), ft	Top Width (T), ft	Hydraulic Radius (R), ft	Shear Stress (τ), lb/ft^2 (using depth)	Velocity (V), ft/sec	Assumed NAG Material and Growing Conditions	Manning's Roughness (n)	Normal Depth (y_n), ft	Shear Stress (τ), lb/ft^2 (using depth and peak Q)	Peak Velocity (V), ft/sec	Allowable Shear Stress for NAG Product (short and long exposures), lb/ft^2	Effective Soil Shear Stress (τ_e), n_s = 0.016; C_f = 0.50 phase 2 C_f = 0.87 phase 3
3	1	0.63	4.3	0.48	3.1	12.7	P300 phase 2	0.044	0.80	4.0	9.5	5.5/4.0	0.26
							P300 phase 3	0.049	0.89	4.4	8.4	8.0/8.0	0.06
6	4	0.31	8.5	0.26	1.5	12.9	P300 phase 2	0.044	0.57	2.8	6.1	5.5/4.0	0.19
							P300 phase 3	0.049	0.65	3.2	5.2	8.0/8.0	0.04
8	4	0.30	10.4	0.14	1.5	11.0	P300 phase 2	0.044	0.54	2.7	5.3	5.5/4.0	0.18
							P300 phase 3	0.049	0.88	4.4	3.4	8.0/8.0	0.06
5	3	0.44	7.6	0.36	2.2	10.4	C350 phase 2	0.044	0.66	3.3	6.3	6.0/4.5	0.22
							C350 phase 3	0.049	0.70*	3.5*	5.8*	8.0/8.0	0.05*
6	1.5	0.43	7.3	0.38	2.1	10.1	C350 phase 2	0.044	0.68	3.4	6.1	6.0/4.5	0.22
							C350 phase 3	0.049	0.72	3.6	5.7	8.0/8.0	0.05
10	3	0.26	11.6	0.26	1.3	10.4	C350 phase 2	0.044	0.49	2.4	5.2	6.0/4.5	0.16
							C350 phase 3	0.049	0.52	2.6	4.8	8.0/8.0	0.04

*Calculation example given in text.

Based on these calculations, either the P300 or the C350 liner will be suitable for most conditions outlined in this example. When newly placed, with no vegetation growth, the Manning's *n* roughness is 0.02 for these liners. The maximum calculated maximum shear stress is 3.1 lb/ft^2 for the narrowest cross section examined, slightly greater than the maximum allowable value of 3.0 lb/ft^2. The calculated shear stresses are less than this allowable maximum value for the other cross-sections. Therefore, one of the wider channels should be used. Unfortunately, the velocities are all very high, ranging from 10.1 to 12.9 ft/sec before the establishment of vegetation. The use of check dams is therefore highly recommended for this channel. These can range from coir logs to rock check dams.

The calculations after vegetative growth show that either liner is acceptable. A range of conditions were examined for Phase 2 (50% stand maturity) and Phase 3 (mature growth), with Manning's roughness values of 0.044 and 0.049. The smallest (and steepest side sloped) channel resulted in the highest shear stress of 4.4 lb/ft^2, less than the maximum acceptable values. The short exposure critical values are for peak flows of <2 hours duration. After mature plant establishment in the channel, the maximum allowable shear stress increases to 8.0 lb/ft^2 for all conditions. The effective soil shear stress also is shown, which would be applicable to evaluate temporary channel liners. During the Phase 2 plant growth stage (50% plant growth), the resulting values are larger than soil tolerance conditions, while they are acceptable during the Phase 3 growth stage (mature plant growth). This emphasizes the need for a permanent liner in this case where the additional protection provided by the vegetation is not necessary. The steep slope (8% in this case) results in these relatively extreme solutions. If the slope for this example was about 2%, or less, temporary liners may be suitable (assuming that suitable growth conditions exist).

CHANNEL DESIGN USING CONCRETE AND RIPRAP LINER MATERIALS

For certain conditions when "soft-liner" materials are not suitable, it is common to use concrete or rocks (riprap). New advances in soft liners have produced some materials capable of withstanding large shear stresses, but the more common hard materials still are used frequently in demanding situations.

Historical practice has been to rely on concrete-lined channels for the most demanding applications. However, problems have occurred when water flows beneath the concrete structure, causing massive failure, as indicated in the following photograph. Flexible liners that can conform to instability of the soils may be a better choice. If moisture underneath the liner is permissible and likely, porous flexible liners, as previously described, may outperform rigid concrete.

Failed concrete channel liner due to undercutting (photo by Mark Burford).

The *Alabama Handbook* (USDA, 2003) includes the following guidance for hard-lined channels:

Lined Swale (*LS*)

Practice Description

A lined swale is a constructed channel with a permanent lining designed to carry concentrated runoff to a stable outlet. This practice applies where grass swales are unsuitable because of conditions such as steep channel grades, prolonged flow areas, soils that are too erodible or not suitable to support vegetation or insufficient space and where riprap-lined swales are not desired. The purpose of a lined swale is to conduct stormwater runoff without causing erosion problems in the area of channel flow. The material that provides the permanent lining may be concrete, a specialized type of erosion control blanket or manufactured concrete products.

Planning Considerations

A lined swale is used to convey concentrated runoff to a stable outlet in situations where a grass swale is inadequate. A lined swale can be lined with concrete, manufactured concrete products or manufactured erosion control products. Concrete lined swales are covered only in this practice. The practice standard for Erosion Control Blanket should be referenced for criteria on this type of swale lining. Product manufacturers and qualified design professional should be consulted for design requirements for manufactured concrete linings. Concrete lined swales are generally used in areas where riprap-lined swales are not desired due to aesthetics, safety, or maintenance concerns. Concrete lined swales allow easy maintenance of surrounding vegetation with normal lawn care equipment. The concrete generally provides a more visually pleasing structure than the riprap linings. Concrete lined swales are especially desirable in areas accessed by small children. In areas where stormwater infiltration is a concern, riprap and manufactured products should be considered rather than the concrete lining.

Design Criteria

Capacity

Lined swales should be capable of passing the peak flow expected from a 10-year, 24-hour duration storm.

Adjustments should be made for release rates from structures and other drainage facilities. Swales shall also be designed to comply with local stormwater ordinances, and should be designed for greater capacity whenever there is danger of flooding or out-of-bank flow cannot be tolerated. Peak rates of runoff values used to determine the capacity requirements should be calculated using accepted engineering methods. Some accepted methods are:

- Natural Resources Conservation Service, *Engineering Field Manual for Conservation Practices,* Chapter 2 Estimating Runoff.
- Natural Resources Conservation Service formerly Soil Conservation Service, Technical Release 55, *Urban Hydrology for Small Watersheds.*
- Natural Resources Conservation Service, *Alabama Engineering Field (Design) Manual,* Chapter 2 on Estimating Runoff.
- Other comparable methods.

Slope

This practice only applies to paved flumes that are installed on slopes of 25% or less. Slopes steeper than this should be designed by a qualified design professional. The slope in feet per 100 feet of length can be determined from a topographic map of the site or from a detailed survey of the planned lined swale location.

Cross Section

With peak flow (capacity) and slope known, the paved flume cross section can be determined by using Figures 5.16 through 5.18.

Concrete Flumes

Concrete flumes should be constructed of concrete with a minimum 28 day compressive strength of 3,000 psi. Flumes shall have a minimum concrete thickness of 4 inches.

Cutoff Walls

Cutoff walls shall be constructed at the beginning and end

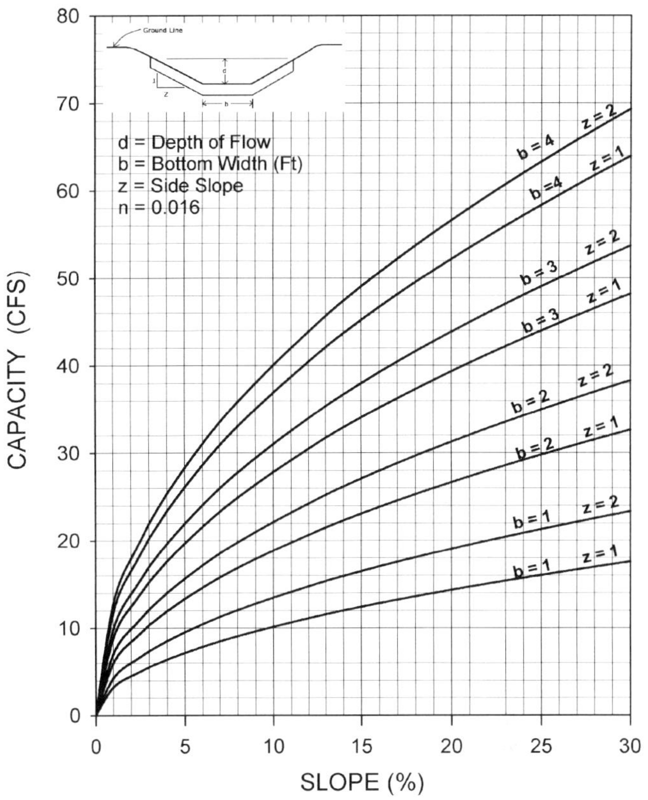

Figure 5.16. Capacity graph for concrete flumes, depth of flow = 0.50 feet (USDA 2003).

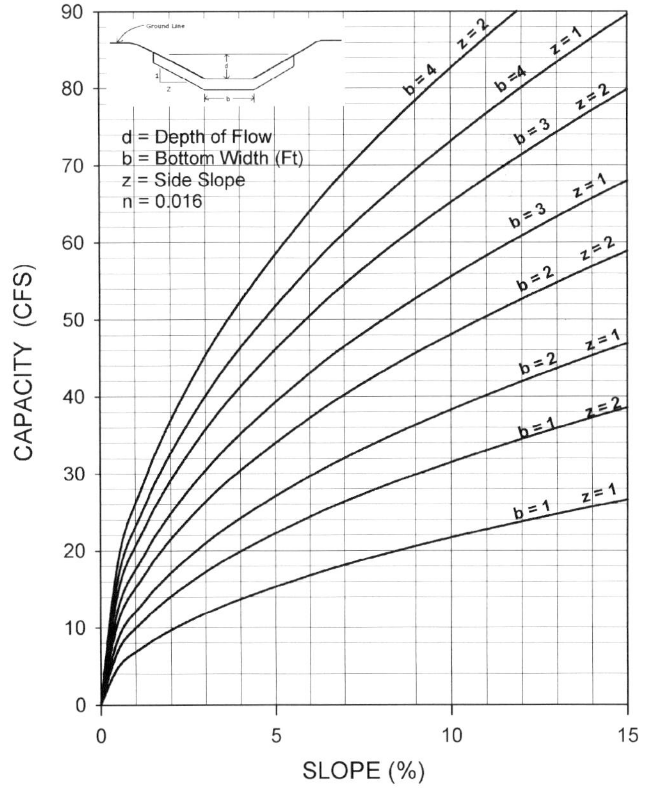

Figure 5.17. Capacity graph for concrete flumes, depth of flow = 0.75 feet (USDA 2003).

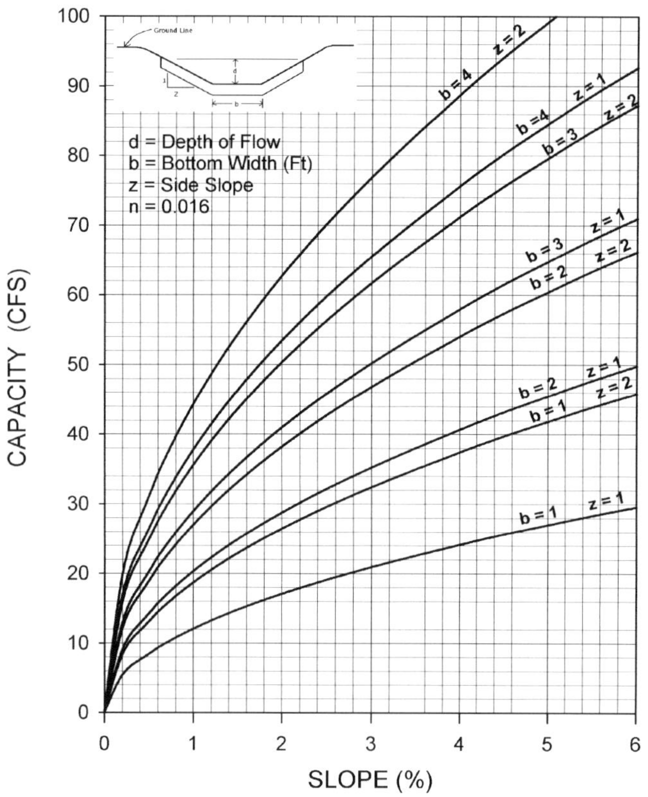

Figure 5.18. Capacity graph for concrete flumes, depth of flow = 1.00 feet (USDA 2003).

of every flume except where the flume connects with a catch basin or inlet.

Alignment

Keep paved flumes as straight as possible because they often carry supercritical flow velocities.

Inlet Section

The inlet section to the paved flume should be at least 6 feet long and have a bottom width equal to twice the bottom width of the flume itself. The bottom width should transition from twice the flume bottom width to the flume bottom width over the 6 feet length.

Outlet

Outlets of paved flumes shall be protected from erosion. The standard for Outlet Protection can be used to provide this protection. A method to dissipate the energy of low flows is to bury the last section of the flume in the ground. This will usually force the development of a "scour hole" which will stabilize and serve as a plunge basin. For the design of large capacity flumes it may be necessary to design a larger energy dissipator at the outlet.

***Riprap-lined Swale* (RS)**

Practice Description

A riprap-lined swale is a natural or constructed channel with an erosion-resistant rock lining designed to carry concentrated runoff to a stable outlet. This practice applies where grass swales are unsuitable because of conditions such as steep channel grades, prolonged flow areas, soils that are too erodible or not suitable to support vegetation or insufficient space.

Planning Considerations

Swales should be carefully built to the design cross section, shape and dimensions. Swales are hydraulic structures and as such depend upon the hydraulic parameters to serve satisfactorily. Swales may be used to:

- Serve as outlets for diversions and sediment control basins and stormwater detention basins.
- Convey water collected by road ditches or discharged through culverts.
- Rehabilitate natural draws and gullies carrying concentrations of runoff.

The design of a swale cross section and lining is based primarily upon the volume and velocity of flow expected in the swale. Riprap-lined swales should be used where velocities are in the range of 5 to 10 ft/sec. Besides the primary design considerations of capacity and velocity, a number of other important factors should be taken into account when selecting a cross section. These factors include land availability, compatibility with land use and surrounding environment, safety, maintenance requirements and outlet conditions, etc.

Riprap lined swales are trapezoidal in shape. Trapezoidal swales are often used where the quantity of water to be carried is large and conditions require that it be carried at a relatively high velocity.

Outlet conditions for all swales should be considered. This is particularly important for the transition from the riprap lining to a vegetative lining. Appropriate measures must be taken to dissipate the energy of the flow to prevent scour of the receiving swale.

Design Criteria

Capacity

Lined swales shall be designed to convey the peak rate of runoff from a 10-year, 24-hour rainfall event. Adjustments should be made for release rates from structures and other drainage facilities. Swales should also be designed to comply with local stormwater ordinances.

Swales should be designed for greater capacity whenever

there is danger of flooding or out-of-bank flow cannot be tolerated. The maximum capacity of the swale flowing at design depth should be 200 cubic ft/sec.

Peak rates of runoff values used to determine the capacity requirements should be calculated using accepted engineering methods. Some accepted methods are:

- Natural Resources Conservation Service, National Engineering Handbook Series, Part 650, *Engineering Field Handbook,* Chapter 2, Estimating Runoff.
- Natural Resources Conservation Service formerly Soil Conservation Service, Technical Release 55, *Urban Hydrology for Small Watersheds.*
- Natural Resources Conservation Service, *Alabama Engineering Field (Design) Manual,* Chapter 2 on Estimating Runoff.
- Other comparable methods.

Cross Section

The swale cross section should be trapezoidal in shape. The steepest permissible side slope of the swale should be 2:1. A bottom width should be selected based on area available for installation of the swale and available rock sizes. The bottom width will be used in determining stable rock size and flow depth.

Depth

Design flow depth should be determined by the following formula:

$$z = [n(q\ )/1.486(S)^{0.50}]^{3/5}$$

S = Bed slope, ft./ft.
z = Flow depth, ft.
q = Unit discharge, ft^3/s/ft (Total dischargeBottom width)
n = Manning's coefficient of roughness (see formula under velocities)

The design water surface elevation of a swale receiving water from other tributary sources should be equal to or less than the design water surface elevation of the contributing source. The design water surface elevation of contributing and receiving waters should be the same, whenever practical. A minimum depth may be necessary to provide adequate outlets for subsurface drains and tributary swales.

Freeboard

The minimum freeboard is 0.25 feet. Freeboard is not required on swales with less than 1% slope and where out-of-bank flow will not be damaging and can be tolerated from an operational point of view.

Stable Rock Size

Stable rock sizes, for rock lined swales having gradients between 2 percent and 40 percent, can be determined using the following formulas from "Design of Rock Chutes" by Robinson et al. 1998.

For swale slopes between 2% and 10%:

$$d_{50} = [q\ (S)^{1.5}/4.75(10)^{-3}]^{1/1.89}$$

For swale slopes between 10% and 40%:

$$d_{50} = [q\ (S)^{0.58}/3.93(10)^{-2}]^{1/1.89}$$

d_{50} = Particle size for which 50% of the sample is finer, inch
S = Bed slope, ft/ft
q = Unit discharge, ft^3/s/ft (Total discharge ÷ Bottom width)

After the stable median stone size is determined, the gradation of rock to be used should be specified using Tables 5.16 and 5.18. Table 5.16 is used to determine the weight of the median stone size ($d5_0$). Using this median weight, a gradation can be selected from Table 5.17, which shows the commercially available riprap gradations as classified by the Alabama Department of Transportation.

Velocities

Velocities should be computed by using Manning's Formula with a coefficient of roughness, n, as follows:

$$n = 0.047(d_{50} \times S)0.147$$

Applies on slopes between 2 and 40% with a rock mantle thickness of $2 \times d_{50}$ where:

d_{50} = median rock diameter (inch),
S = lined section slope (ft/ft) ($0.02 < S < 0.4$)

TABLE 5.16. Size of Riprap Stones (USDA 2003).

Weight (lbs)	Mean Spherical Diameter (feet)	Rectangular Shape Length Width (feet)	Rectangular Shape Length Height (feet)
50	0.8	1.4	0.5
100	1.1	1.75	0.6
150	1.3	2.0	0.67
300	1.6	2.6	0.9
500	1.9	3.0	1.0
1000	2.2	3.7	1.25
1500	2.6	4.7	1.5
2000	2.75	5.4	1.8
4000	3.6	6.0	2.0
6000	4.0	6.9	2.3
8000	4.5	7.6	2.5
12000	6.1	10.0	3.3

TABLE 5.17. Graded Riprap (USDA 2003).

	Weight (lbs.)					
Class	d_{10}	d_{15}	d_{25}	d_{50}	d_{75}	d_{90}
1	10	—	—	50	—	100
2	10	—	—	80	—	200
3	—	25	—	200	—	500
4	—	—	50	500	1000	—
5	—	—	200	1000	—	2000

Velocities exceeding critical velocity should be restricted to straight reaches.

Waterways or outlets with velocities exceeding critical velocity should discharge into an outlet protection structure to reduce discharge velocity to less than critical (see Outlet Protection practice).

Lining Thickness

The minimum lining thickness should be equal to the maximum stone size of the specified riprap gradation plus the thickness of any required filter or bedding.

Lining Durability

Stone for riprap should consist of field stone or rough unhewn quarry stone of approximately rectangular shape. The stone should be hard and angular and of such quality that it will not disintegrate on exposure to water or weathering and it should be suitable in all other respects for the purpose intended. The specific gravity of the individual stones should be at least 2.5.

Geotextiles

Geotextiles should be used where appropriate as a separator between rock and soil to prevent migration of soil particles from the subgrade through the lining material. Geotextiles should be Class I material as selected from Table 5.18.

TABLE 5.19. Selected U.S. Standard Sieve Sizes.

U.S. Standard Sieve Sizes	Sieve Screen Opening (μm, unless otherwise noted)	Sieve Screen Opening (inch)
4 in	100 mm	4.0
3 in	75 mm	3.0
2 in	50 mm	2.0
1 in	25 mm	1.0
3/4 in	19.0 mm	0.75
5/8 in	16.0 mm	0.63
3/8 in	9.5 mm	0.38
1/2 in	12.5 mm	0.500
1/4 in	6.3 mm	0.250
No. 4	4.75 mm	0.187
No. 6	3.35 mm	0.132
No. 8	2.36 mm	0.0929
No. 10	2.00 mm	0.0787
No. 12	1.70 mm	0.0669
No. 14	1.40 mm	0.0555
No. 16	1.18 mm	0.0465
No. 18	1.00 mm	0.0394
No. 20	850	0.0335
No. 25	710	0.0278
No. 35	500	0.0197
No. 40	425	0.0167
No. 45	355	0.0139
No. 50	300	0.0118
No. 60	250	0.0098
No. 70	212	0.0083
No. 80	180	0.0070
No. 100	150	0.0059
No. 120	125	0.0049
No. 140	106	0.0041
No. 170	90	0.0035
No. 200	75	0.0029
No. 230	63	0.0017
No. 270	53	0.0021
No. 325	45	0.0017
No. 400	38	0.0015
No. 450	32	0.0013
No. 500	25	0.0010
No. 635	20	0.0008

Filters or Bedding

Filters or bedding should be used where needed to prevent piping. Filters should be designed according to the requirements contained in the Subsurface Drain practice.

TABLE 5.18. Requirements for Nonwoven Geotextile (USDA 2003).

Property	Test Method	Class I	Class II	Class III	Class IV[1]
Tensile strength (lb)[2]	ASTM D 4632 grab test	180 minimum	120 minimum	90 minimum	115 minimum
Elongation at failure (%)[2]	ASTM D 4632	≥50	≥50	≥50	≥50
Puncture (pounds)	ASTM D 4833	80 minimum	60 minimum	40 minimum	40 minimum
Ultraviolet light (% residual tensile strength)	ASTM D 4355 150-hr exposure	70 minimum	70 minimum	70 minimum	70 minimum
Apparent opening size (AOS)	ASTM D 4751	As specified max. no. 40[3]	As specified max. no. 40[3]	As specified max. no. 40[3]	As specified max. no. 40[3]
Permittivity (sec^{-1})	ASTM D 4491	0.70 minimum	0.70 minimum	0.70 minimum	0.10 minimum

Table copied from NRCS Material Specification 592.

[1]Heat-bonded or resin-bonded geotextile may be used for classes III and IV. They are particularly well suited to class IV. Needle-punched geotextile are required for all other classes.

[2]Minimum average roll value (weakest principal direction).

[3]U.S. standard sieve size"

The minimum thickness of a filter or bedding should be 6″.

Check Dam (CD)

This section is excerpted from the *Alabama Handbook* (USDA, 1993) and describes check-dam use for erosion controls:

Definition

Small barriers or dams constructed across a swale, drainage ditch or areas of concentrated flow.

Purpose

To prevent or reduce erosion by lessening the gradient of the flow channel which reduces the velocity of storm water flows. Some sediment will be trapped upstream from the check dams, but its volume will be insignificant and should not be considered in off-site sediment reduction.

Conditions Where Practice Applies

This measure is limited to use in small open channels and drainage ways which drain 10 acres or less. It should not be used in a live stream. Specific applications include:

1. Temporary ditches or water courses which, because of their short length of service, cannot establish a nonerodible lining but still need some protection against erosion.
2. Permanent ditches or water courses which for some reason cannot establish an enduring non-erodible lining.
3. Either temporary or permanent ditches or water courses which need protection during the establishment of protective linings.

Planning Considerations

Check dams may be constructed of rock, logs, hay bales or other suitable material. Most check dams would be constructed of rock. Rock may not be acceptable in some installations because of aesthetics and hay bales or logs may need to be considered.

Rock check dams (Figures 5.19 and 5.20) are easier to install with backhoes or other suitable equipment. The rock is usually purchased and would increase cost. Some locations may not have rock readily available. Rock should be handled carefully in areas to be mowed. Some rock may be washed downstream and should be removed before each mowing operation.

Check dams should be planned to be compatible with the other features such as streets, walks, trails, sediment basins and rights-of-way or property lines. Check dams may be constructed in series and the dams should be located at a normal interval from other grade controls such as culverts or sediment basins. Needed hydraulic conveyance must also be confirmed.

Check dams constructed of hay bales (Figure CD-3) have the shortest life of the materials discussed. Hay bale check dams should not be used where permanent water course protection is needed. They should not be used where the drainage areas exceeds 5 acres.

Design Criteria

Formal design is not required. The following limiting factors shall be adhered to when designing check dams:

- Drainage Area:
  —10 acres or less (Rock)
  —5 acres or less (Hay bale check dam)

Check dam constructed of sand bags (Photograph by New York State Department of Transportation).

Typical check dams made of riprap.

Filter fabrics rarely made adequate check dams (Photograph by D. Lake).

Flows commonly erode around ends of filter fabric check dams.

Check dams of rock and filter fabric.

Series of riprap check dams spaced to cause ponding between dams (SCS photo).

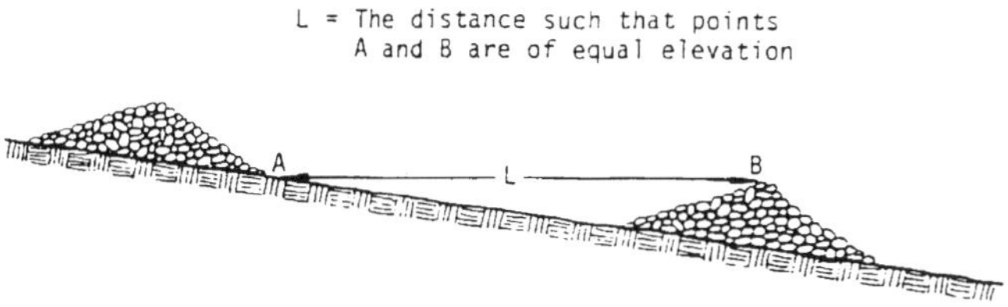

Figure 5.19. Spacing of typical rock check dams (USDA 2003).

- Maximum Height:
  —2 feet when drainage area is less than 5 acres
  —3 feet when drainage area is 5 to 10 acres
- Depth of Flow:
  —6 inches when drainage area is less than 5 acres
  —12 inches when drainage area is 5 to 10 acres
- Side Slopes:
  —2:1 or flatter
- Max. Spacing between dams:
  —Elevation of toe of upstream dam is at or below elevation of crest of downstream dam.

Top of dam, perpendicular to flow, should be parabolic. The center of the dam must be lower than the ends. The dam shall be constructed well into the abutment so that water cannot run around the dam.

Rock check dams should be constructed of durable rock riprap. Riprap gradation shall conform to the requirements of Class I Riprap, Alabama Highway Department, *Standard Specification for Highway Construction,* or equivalent.

Maintenance

Check dams may be removed when their useful life has been completed. Whenever check dams are removed, care shall be taken to minimize disturbance to the remainder of the watercourse.

The area where check dams are removed shall be shaped and smoothed to water course dimensions and seeded and mulched immediately. On rock check dams, care shall be taken to remove all rock if the area is to be mowed.

Periodic inspection is necessary on check dams. Repair should be done as soon as need is noted to minimize damage and expense of repair."

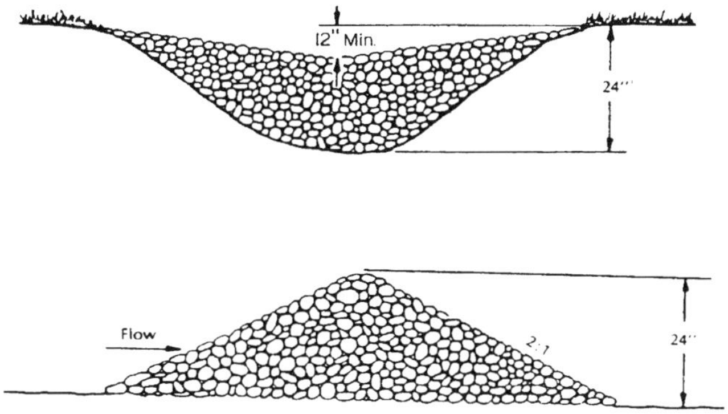

Figure 5.20. Cross sections of rock check dam (USDA 2003).

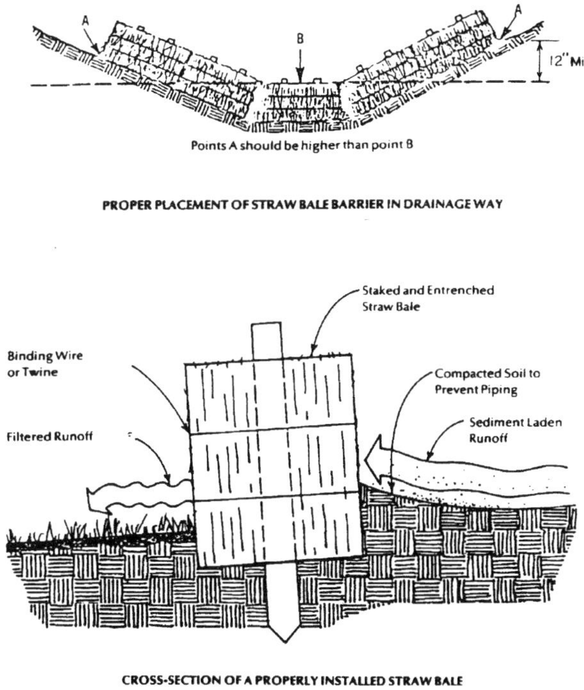

Figure 5.21. Typical hay bale check dam (USDA 2003).

Flow Rates through Rock Check Dams

The flow through a rock check dam can be calculated using the following equation:

$$Q = \frac{h^{2/3}W}{[(L/D) + 2.5 + L^2]^{0.5}}$$

where,

Q = Outflow through the rock check dam (cfs)
h = Ponding depth behind the check dam (ft)
W = Width of the check dam (ft), not to be confused with the horizontal flow path length through the check dam
L = Horizontal flow path length through the check dam (ft)
D = Average rock diameter in the check dam (ft)

This equation is from *Analysis of Flow through Porous Media as Applied to Gabion Dams Regarding the Storage and Release of Storm Water Runoff*, NAHB/NRC Designated Housing Research Center at Penn State, Report No. 10, August 1992. This equation was developed to calculate the flow through a gabion dam, which is usually a vertical walled structure composed of large stones confined in wire baskets. In order to apply it to rock check dams that have sloped faces, WinDETPOND (www.WinSLAMM.com), a continuous hydraulic and particle routing model, calculates the flow by dividing the rock check dam into horizontal slices. The flow through each slice is calculated and then the flows from all slices are summed to determine the total flow for a given depth.

Example: Calculation of Flows through Check Dams

The following examples assume uniform thickness (such as would apply in a gabion dam):

h = 3 ft
W = 15 ft
L = 3 ft
D = 9 inches = 0.75 ft

$$Q = \frac{h^{2/3}W}{[(L/D) + 2.5 + L^2]^{0.5}}$$

$$= \frac{(3\text{ ft})^{2/3}(15\text{ ft})}{[(3\text{ ft}/0.75\text{ ft}) + 2.5 + (3\text{ ft})^2]^{0.5}} = 7.9\text{ ft}^3/\text{sec}$$

The same rock check dam, but with only a 1 ft depth of water:

h = 1 ft
W = 15 ft
L = 3 ft
D = 9 inches = 0.75 ft

$$Q = \frac{h^{2/3}W}{[(L/D) + 2.5 + L^2]^{0.5}}$$

$$= \frac{(1\text{ ft})^{2/3}(15\text{ ft})}{[(3\text{ ft}/0.75\text{ ft}) + 2.5 + (3\text{ ft})^2]^{0.5}} = 3.8\text{ ft}^3/\text{sec}$$

This process can be repeated for an installation for different depths of water to create a stage-discharge curve for a rock check dam.

Example: Design for Reinforced Grass-Lined Channels with Check Dams and Level Spreader Pads

A new industrial site in Huntsville, AL, has several 2-acre individual building sites. Each of the sites will be served with a grass-lined channel that will carry site water to a larger swale system. The slopes of the channels vary from about 1 to 6.5%. The calculated peak flow from each construction site was calculated to be 16 ft³/sec (corresponding to the Huntsville, AL, 25-yr design storm of 6.3 inches for 24 hours). A grass-lined channel is to be designed for each site. The bare seed bed is assumed to have a hydraulic roughness of about 0.016. The channels are to be built to be 10 ft wide on the bottom and have 3 to 1 (*h:v*) side slopes. Table 5.20 summarizes the results of these calculations.

TABLE 5.20. Summary of Calculated Results.

Slope	Bare Seed Bed Shear Stress (lb/ft²)	Unvegetated Mat Shear Stress, Effect on Soil (lb/ft²)	Safety Factor (allowable shear stress of 0.05 lb/ft²)	Maximum Velocity with Mature Vegetation (ft/sec)
1%	0.14	0.012	4.2	3.1
3%	0.28	0.023	2.2	4.8
5%	0.42	0.035	1.4	5.5
6.5%	0.46	0.039	1.3	6.4

The seed bed has an allowable shear stress of about 0.05 lb/ft². The calculated values for unprotected conditions are all much larger. These values would be exceeded even with a much smaller 2-yr design storm (3.9 inches, 24 hrs). Therefore, an erosion control mat is needed to protect the seedbed until the grass can become established. A North American Green S75 mat was selected, having an allowable shear stress of 1.55 lb/ft² and a life of 12 months. The unvegetated mat has a roughness factor n of 0.055. The shear stress under the mat is calculated as follows (for the 6.5% slope condition), assuming a C_f = 0 for the unvegetated condition and a bare soil n of about 0.016.

$$\tau_e = 0.46(1-0)\left(\frac{0.016}{0.055}\right)^2 = 0.039 ft^2/\text{sec}$$

The unvegetated mat is seen to be suitable protection of the seed bed, with safety factors ranging from 4.2 (for the 1% slopes) to 1.3 (for the 6.5% slopes).

The allowable velocity for mature bermudagrass for slopes <5% is 6 ft/sec for sandy silt soils to 8 ft/sec for silty clay soils. The 6-ft/sec maximum is expected to be most applicable for the site soils. The swales greater than 5% slopes may be a problem, as they exceed the maximum slope criterion for bermudagrass. In addition, the 6.5% slopes have maximum velocities of about 6.4 ft/sec, slightly greater than the maximum permissible velocity. Although these maximum flows are very infrequent (associated with the 25.year storms), rock check dams were specified for the 5 and 6.5% slopes to provide a suitable safety factor.

The check dams, in the 3-ft deep channels, are assumed to be 2 ft high to the maximum over-topping elevation. In channels with 5% slopes, the check dams would have to be about 40 ft apart (or less) to ensure that the toes of the upstream check dams were at the same, or lower, elevations as the overflows of the downstream dams. Similarly, the check dams in the 6.5% sloped channels would have to be no more than 30 ft apart. The *Alabama Handbook* specifies ALDOT class 1 riprap for check dams. This rock has the following size:

d_{10} = 10 lbs
d_{100} = 100 lbs (1.1 ft in diameter)
d_{50} = 50 lbs (0.8 ft in diameter)

The roughness coefficients for class 1 riprap is dependent on channel slope:

S of 1%, $n = 0.033$
S of 3%, $n = 0.0393$
S of 5%, $n = 0.0423$
S of 6.5%, $n = 0.044$

The flow rate through the check dam (2-ft high, 10-ft wide at the base, and 4-ft average flow length) in the grass channels is estimated to be about 5 ft^3/sec. This would leave about 11 ft^3/sec to overtop the check dams during peak flows. There is adequate capacity in the channels to accommodate this overtopping. The velocity through the check dams is reduced to about 0.5 ft/sec. The ponding between the check dams and the low velocity through the check dams will substantially reduce the peak flows to values well below the critical values for the grass-lined channels.

The last check dams for each channel will be located at the end of the channels and will discharge onto level-spreader pads. These will be located at the end of all of the channels, even those at 1 and 3% slopes. These will dissipate the energy of the flow from each building area and produce sheetflow until the flows collect in the large main swales. The level spreader pads will each handle a maximum flow of 16 ft^3/sec. The *Alabama Handbook* provides design guidance for these pads. The pads will need to be at least 15 feet long, and spread flow out to a width of 25 ft from the 10-ft wide channel. For this example, the rock size for the pad needs to be 0.7 ft in diameter (d_{50}) and be spread at least 15 inches thick.

SLOPE STABILITY APPLIED TO CONSTRUCTION SITE EROSION CONTROL DESIGN

Much of the above information on channel stability can also be applied to slope stability evaluations. Of course, this discussion assumes that the slopes have been designed by geotechnical engineers to prevent slippage, as this discussion only addresses sheet and rill erosion.

As indicated in Chapter 4, it is possible to modify the Manning's formula to calculate the flow depth for the sheetflow conditions used for slope analyses:

$$y = \left(\frac{qn}{1.49s^{0.5}}\right)^{3/5}$$

where:

y = the flow depth (in feet),
q = the unit width flow rate (Q/W, the total flow rate, in ft^3/sec, divided by the slope width, in ft.)
n = the sheet flow roughness coefficient for the slope surface, and
s = the slope (as a fraction)

The basic shear stress equation can be used to calculate the maximum shear stress expected on a slope:

$$\tau_o = \gamma y S \ \ (\text{lb/ft}^2)$$

where,

γ = specific weight of water (62.4 lbs/ft^3)
y = flow depth (ft)
S = slope (ft/ft)

Information in Chapter 3 can be used to calculate the unit width flow rate (q) for the slope in question. Assume the following conditions:

Design storm peak flow rate (Q) = 2.2 ft^3/sec (from Chapter 3 procedures)

Slope width (W) = 200 ft

Slope roughness for sheetflow (n) = 0.24 (vegetated with dense grass stand; would be only about 0.055 for an erosion control mat before vegetation, however, using the vegetated mat results in deeper water and a worst case shear stress condition)

Steepness of slope (s) = 25% = 0.25

Therefore, the unit width peak flow = Q/W = 2.2 ft^3/sec/200 ft = 0.011ft^2/sec and the flow depth is:

$$y = \left(\frac{(0.011)0.24}{1.49(0.25)^{0.5}}\right)^{3/5} = 0.033 \text{ ft}$$

This depth corresponds to a flow depth of about 0.4 inches. The corresponding maximum shear stress would be:

$$\tau_o = (62.4)(0.033)(0.25) = 0.51 \text{ lb/ft}^2$$

For an ordinary firm loam soil, the Manning's roughness is 0.020 and the allowable shear stress is 0.15 lb/ft^2 (from Table 5.2). Without a protective mat, the calculated maximum shear stress is substantially greater than the allowable shear stress for the soil. The effective shear stress impacting the soil underneath an erosion control mat can be calculated (using the previously calculated maximum shear stress). The following calculation indicates the effective shear stress underneath the mat:

$$\tau_e = \tau_o(1 - C_f)\left(\frac{n_s}{n}\right)^2$$

where,

τ_e = effective shear stress exerted on soil beneath mat on slope
τ_o = maximum shear stress from the flowing water = 0.51 lb/ft^2
C_f = vegetal cover factor (this factor is 0 for an unvegetated channel) = 0 for critical unvegetated slope

n_s = roughness coefficient of underlying soil = 0.020
n = roughness coefficient of mat = assume 0.055 as a typical value for unvegetated mat on slope

therefore,

$$\tau_e = \tau_o(1-C_f)\left(\frac{n_s}{n}\right)^2 = 0.51(1-0)\left(\frac{0.020}{0.055}\right)^2 = 0.067 \text{ lb/ft}^2$$

The safety factor using these values is about 2.2 (0.15/0.067), so the slope should be adequately protected when an adequate mat is selected. In fact, any erosion mat with a Manning's roughness larger than 0.037 should be adequate for this example, shown by setting the effective shear stress equal to the allowable shear stress (0.15 lb/ft^2):

$$n = n_s\sqrt{\frac{\tau_o(1-C_f)}{\tau_e}} = (0.020)\sqrt{\frac{(0.51)(1-0)}{0.15}} = 0.37$$

Final mat selection is usually based on calculating the erosion yield from the slope, using the RUSLE [A = RK(LS)CP]. Table 5.21 lists some representative conservation factors (C) for different North American Green erosion control mats, and the following lists other needed RUSLE values, for example:

R = 350 (Birmingham, AL conditions)
K = 0.28
LS = 10.81 for length of 300 ft and slope of 25%(from Chapter 3)

This is a 200 ft by 300 ft area, or about 1.4 acres.

Soil loss for bare slope (C = 1):

Soil loss = (350)(0.28)(10.81)(1) = 1,060 tons/acre/yr

Soil loss for protected slope (example using S75, with an n = 0.055, and C = 0.19)

Soil loss = (350)(0.28)(10.81)(0.19) = 201 tons/acre/yr

The conversion factor for calculating the uniform inches lost per year from tons/acre is 0.00595 (for a typical loam soil). Therefore, for the unprotected bare slope, the soil loss would be about 6.3 inches, while it would be about 1.2 inches for the protected slope. Both these values are excessive. The USDA (1987) states that a reasonable tolerance limit to allow agricultural activity (plants to survive) would be about 0.5 inches. North American Green recommends that a tolerable soil loss value of 0.25 inches (at the bottom 10% of the slope length) be used for temporary erosion control blankets and new growth vegetation. This corresponds to a maximum soil loss of about 42 tons/acre/year (still about 10 times greater than the tolerable, T, value given for many soils in the USDA county soil maps for sustainable agriculture). Since these are for temporary controls while the vegetation is immature, it is expected that the soil losses would decrease substantially with time, as the plants on the slope mature. The tolerable soil loss for permanent slope protection is given as 0.03 inches/year, or about 5 tons/acre/year (close to the USDA tolerable, T, values in the soil maps).

Therefore, an erosion control mat having a smaller C factor for these slope conditions is needed. The target maximum C value can be estimated by the ratio of the maximum allowable to the bare soil conditions:

$$C_{\text{maximum}} = 42/1060 = 0.039$$
$$\text{Manning's } n_{\text{minimum}} = 0.037$$

For the long slope length of 300 ft and the 25% (4:1) slope, the suitable erosion control mats shown on Table 5.21 would be: S150BN (C = 0.020 and n = 0.055) or SC150BN (C = 0.010 and n = 0.050) both much more substantial than the initial selection of S75 based on shear stress alone. If the slope length was shorter, the lower rated mats would be suitable (such as S75 for slope lengths less than 20 ft, and C125 if the slope lengths are up to 50 ft). None of these erosion control mats would provide the soil loss protection on this long slope example for permanent installations (5/1060 = 0.0047) without vegetation. Both the C350 or

TABLE 5.21. North American Green Conservation Factors (C) for Different Erosion Control Mats, for Different Slopes and Slope Lengths.

Slope Length and Gradient	S75	S150	SC150	C125	S75BN	S150BN	CS150BN	C1250BN	C350	P300
Length ≤ 20 ft (6 m)										
S ≤ 3:1	0.029	0.004	0.001	0.001	0.029	0.00014	0.0009	0.00009	0.0005	0.001
S between 3:1 to 2:1	0.11	0.106	0.048	0.029	0.11	0.039	0.02	0.018	0.015	0.029
S ≥ 2:1	0.23	0.13	0.10	0.082	0.23	0.086	0.063	0.05	0.043	0.082
Length between 20 and 50 ft (6 to 15 m)										
S ≤ 3:1	0.11	0.062	0.51	0.036	0.11	0.010	0.005	0.003	0.018	0.036
S between 3:1 to 2:1	0.21	0.118	0.79	0.060	0.21	0.07	0.055	0.04	0.031	0.06
S ≥ 2:1	0.45	0.17	0.145	0.096	0.45	0.118	0.092	0.06	0.050	0.096
Length ≥ 50 ft (15 m)										
S ≤ 3:1	0.19	0.12	0.10	0.07	0.19	0.02	0.01	0.007	0.035	0.07
S between 3:1 to 2:1	0.30	0.18	0.11	0.09	0.30	0.10	0.08	0.07	0.047	0.09
S ≥ 2:1	0.66	0.22	0.19	0.11	0.66	0.15	0.12	0.07	0.057	0.11

Soil Roughening to Protect Slope

Sheep's foot compactor.

Roughened slope after compaction by Sheep's foot compactor.

Roughened and compacted slope.

Asphaltic Slope Protection in Heavy Shade.

Slopes Protected by Erosion Control Netting, Vegetation, Coir Logs and Soil Adhesives

Coir (cocoanut fiber) rolls/logs and soil adhesives for slope stabilization project at Newport Beach, CA, wildlife refuge.

Coir log slope protection and check dams in steep drainage channel.

(continued)

Slopes Protected by Erosion Control Netting, Vegetation, Coir Logs and Soil Adhesives (continued)

Reinforced slope netting and established vegetation at Birmingham, AL, highway project.

Slope being protected with netting after failure.

Netting used in areas to repair bank failure.

Netting used in areas to repair bank failure.

Installation of Erosion Control Matting (SCS photo).

Stockpile of Erosion Control Mats at Construction Site.

Netting over Mulch allowing Grass to Grow (Bill Morton photo).

Failing Mulch on Slope above Rip Rap.

Installation of Wire Netting over Mulch

Installation of Netting over Mulch (Photograph by Synthetic Industries).

Various Slope Protection Treatments and Tree Conservation (Photrgraph by D. Lake).

P300 mats would likely be suitable permanent solutions, with partial to full vegetation.

It is possible to design slope protection having different erosion control mats at different sections on the slope. The free software available from North American Green, for example, can recommend composite slope protection schemes using different mats for different areas on a slope. However, there are many other factors involved in selecting the most appropriate erosion control mat, and the manufacturers' information must be reviewed for proper selection.

RUSLE Cover Factors (*C*) for Grasses

Table 5.22 lists reported RUSLE cover (*C*) factors for different grass-covered slopes, having varying mulch rates, and for different growing periods. The use of erosion control mats and blankets significantly increases the immediate protection available (compared to seeding) under most conditions. The use of mulch rates of 2 tons per acre for slopes less than 20% may result in comparable initial performance and only slightly less protection for longer periods, assuming the mulch is securely anchored. With mulches or protective mats or blankets, grasses can provide 85 to 98% erosion control during the initial year, increasing to 99+% control the second year. Without mulches or other protection, the level of erosion control is much less before establishment. In fact, in many cases, grasses planted on slopes and without protection would likely be so severely damaged that successful grass stands would never occur.

Hydroseeding and Mulching

Hydroseeding and mulching provide a method of planting on moderate to steep slopes, but require large amounts of water. Mulches include:

1. Long-stem wheat straw (preferred), clean prairie hay, and so forth. Straw or hay mulches are either broadcast and "punched" in (4 to 5 inches deep) on moderate slopes with a straight disk, or broadcast along with an adhesive or tacking agent on steep slopes. About 1.0 to 1.5 tons/acre of straw is desired. Mulches conserve surface moisture and reduce summer soil surface temperatures and crusting. The disadvantages of hay and straw mulches are that they can be a source of weed seed, and too much surface mulch, regardless of the type, can cause seedling disease problems. Commercial wood fiber mulch materials are available for relatively level areas.
2. Soil retention blankets, or mats, made of various interlocking fabrics and plastic webbing can be used on moderate to steep slopes in areas with a high potential for runoff. These erosion blankets prevent seeds from being washed out by rain, and at the same time mulch and enhance germination and establishment.

Example: Slope Stability Calculation

The following simple example shows how it is possible to select the most appropriate erosion protection for a slope, based on allowable erosion rates. Assume a 0.25 acre hillside, as shown below, having a slope width of 104 ft and a slope length of 104 ft. The total critical flow rate off this hillside was previously calculated to be 1.2 ft^3/sec, and the steepest slope is 15%. The Manning's n of the soil (n_s) is 0.05.

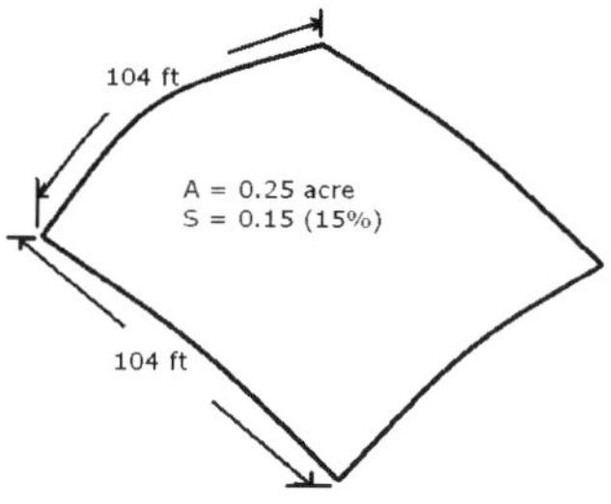

TABLE 5.22. RUSLE Cover *C*-Factors for Different Grass Growing Periods and Mulch Rates (Sprague, 1999).

Treatment	Mulch Rate (tons/acre)	Slope (%)	*C*-Factor for Growing Period for Humid Climates: <6 Weeks	1.5 to 6 Months	6–12 Months	First Year Weighted Total *C* Factor	Second Year Grass and Fully Vegetated Mats
Length ≤ 20 ft (6 m)	—	all	1.00	1.00	1.00	1.00	1.00
	none	all	0.70	0.10	0.05	0.15	0.01
	1	<10	0.20	0.07	0.03	0.07	0.01
	1.5	<10	0.12	0.05	0.02	0.05	0.01
	2	<10	0.06	0.05	0.02	0.04	0.01
Seeded grass	2	11–15	0.07	0.05	0.02	0.04	0.01
	2	16–20	0.11	0.05	0.02	0.04	0.01
	2	22–25	0.14	0.05	0.02	0.05	0.01
	2	26–33	0.17	0.05	0.02	0.05	0.010
	2	34–50	0.20	0.05	0.02	0.05	0.01
Organic and synthetic blankets and composite mats	—	all	0.07	0.07	0.005	0.02	0.005
Synthetic mats	—	all	0.14	0.14	0.005	0.03	0.005

The sheetflow depth can be calculated using Manning's equation:

$$y = \left(\frac{qn}{1.49s^{0.5}}\right)^{3/5}$$

Where q is the unit width flow = 1.2 ft³/sec/104 ft = 0.012 ft²/sec. Therefore:

$$y = \left(\frac{(0.012)(0.05)}{1.49(0.15)^{0.5}}\right)^{3/5} = 0.02 \text{ ft} = 0.21 \text{ in}$$

$$\tau_o = \gamma y S = (62.4 \text{ lb/ft}^2)(0.02 \text{ ft})(0.15) = 0.18 \text{ lb/ft}^2$$

The allowable shear stress for the soils on this hillside is only 0.11 lb/ft², and a vegetated mat will therefore be needed. Assuming the n for the mat to be 0.055, it is possible to calculate the resulting shear stress:

$$\tau_e = \tau_o(1-C)\left(\frac{n_s}{n}\right)^2 = 0.18(1-0)\left(\frac{0.05}{0.055}\right)^2 = 0.15 \text{ lb / ft}^2$$

This mat selection is therefore not adequate. The mat needs to have an n of at least:

$$0.18(1-0)\left(\frac{0.05}{n}\right)^2 = 0.11 \text{ lb / ft}^2$$

Solving for $n = 0.067$

The mat also needs a C factor to meet the maximum allowable erosion rate on the slope (0.25 inches, or less). Using RUSLE:

$R = 350/\text{yr}$
$k = 0.28$
$LS = 2.5$ (for 104 ft slope at 15%)

The base (unprotected) erosion rate is therefore: (350)(0.28)(2.5) = 245 tons/acre/year

This corresponds to 245 (0.00595) = 1.45 inches per year. With a maximum allowable erosion loss of 0.25 inches per year, the C factor for the mat must be: 0.25/1.45 = 0.17. Table 5.6 shows many mats that have this C factor for this slope condition (all except S75). In this example, the selection of a mat having an n of 0.067 or greater will be difficult. Most mats are in the range of 0.022 to 0.055. It will therefore be necessary to use filter fences, coir logs, or other methods to provide additional flow resistance on this slope. Alternatively, the slope length can be shortened with a bench and diversion.

USE OF NEWLY DEVELOPED EROSION CONTROLS

The following presents some brief information concerning new products for controlling soil erosion at construction sites. This is a rapidly expanding area, but this product category (chemicals to bind soil particles) seems to have the most products being developed and marketed. Few of these products have been evaluated in comprehensive field tests, but it is hoped they will offer additional tools to the erosion-control professional. The following are only examples of a few of these alternatives. Many more exist and this listing is not intended to be comprehensive, or an endorsement.

Chemical Treatment of Exposed Soils

There are a number of new products being developed and sold for the control of erosion and sediment at construction sites. One emerging area is the use of chemical polymers and coagulant agents. Older chemical products were mostly soil binding agents, including light asphalts. These newly developed materials act by chemically combining small soil particles into larger discrete particles that are more effective in settling in ponds and in channels. Polyacrylamide (PAM) is the most common chemical being sold now. The following information is from the Internet sites of several distributors or manufacturers of some of these chemicals. This list is very short and is not intended to include all products.

JRM Chemicals, Inc.
(http://www.soilmoist.com/agerosion.html)

Products

FI-1000 Soil Erosion Polymer:—FI-1000 is an anionic high molecular weight polymer designed to reduce soil loss and silt loss in furrow irrigation applications. FI-1000 will increase water infiltration and reduce fertilizer and other chemical runoff. The anionic polymer bonds the suspended particles in the water and they fall to the bottom of the water. Its application rate is one pound per acre (into 12,000 gallons of water).

FI-2000—FI-2000 is an anionic high molecular weight water-soluble polymer designed to reduce soil loss and silt loss in all aspects of agricultural irrigation. FI-2000 is an emulsion that can be applied to furrow, gated pipe, sprinkler and pivotal irrigation systems. Its application rate is 30 ppm.

Polyacrylamide (PAM)

The University of Nebraska Cooperative Extension service (http://www.ianr.unl.edu/pubs/water/g1356.htm) provides the following information on PAM:

> Polyacrylamide (PAM) is a long-chain synthetic polymer that acts as a strengthening agent, binding soil particles together. It is harder for water to move these larger, heavier particles of soil. USDA researchers in Kimberley, Idaho began working with PAM in the early 1990s as a method to reduce erosion in furrow irrigation. Their tests indicated PAM applied in the irrigation water reduced soil

erosion in furrows by over 95 percent, when compared to irrigation without the polymer.

Polyacrylamide used for erosion control should have a negative (anionic) molecular charge. Historically, similar compounds have been used in other industries like potable water treatment, food processing, paper manufacturing and wastewater treatment. Research conducted in Idaho showed that less than 5 percent of PAM applied during an irrigation left fields in the runoff water. This research also showed that after leaving the field, the PAM concentration in the runoff quickly fell below detectable limits (>1,500 yards). There is no indication of any adverse impact on soil, plant or aquatic systems when anionic PAM is used to control soil erosion. Because PAM limits soil erosion, using it can prevent attached pollutants from also leaving the area.

Many companies distribute PAM. HYDROSORB (1390 N. Manzanita St., Orange, CA 92867) presents the following information for their products. SOILFLOC™ is a water-soluble, linear polyacrylamide (PAM) polymer that was designed to be used for erosion control, soil structure improvement and dust abatement. SOILFLOC™ works by aggregating soil particles, increasing pore space and infiltration capacity, resulting in soils that are less susceptible to raindrop and scour erosion. SOILFLOC™ is environmentally safe and non-toxic. A variety of PAM products have been approved by NSF International for potable water clarification. They will naturally degrade with UV light and are consumed by microbiological attack. This product is compatible with almost all irrigation systems. PAM products are now registered throughout the western United States. MSDS available upon request. SOILFLOC™ is available in a dry granule form, liquid emulsion, and tablets.

HydroGrass Technologies
(http://www.hydrograsstech.com/cleansing.php)
also supplies PAM. The following describes their products:

APS 600 Series Silt Stop®
Polyacrylamide Erosion Control Emulsion
A soil specific tailored polyacrylamide copolymer liquid emulsion for erosion control. It reduces and prevents erosion of fine particles and colloidal clays from water. Applied with a water truck of hydroseeder or other spraying devices at a rate of 1 1/2 gallons per acre.

APS 700 Series Silt Stop®
Polyacrylamide Erosion Control Powder
A soil specific tailored polyacrylamide copolymer powder for erosion control. Used to reduce and prevent erosion of fine particles. Settles out suspended particles of sediment and colloidal clays from water. Applied with a hand spreader, mechanical disc or can be mixed with water and applied with a spraying device at a rate of approximately 10 pounds per acre.

APS Floc Log®
Polyacrylamide Semi-hydrated Gel Block
A soil and water chemistry tailored gel block, that when placed within stormwater or construction site damages will remove fine colloidal particles and reduce NTU values. Floc Logs are staked in place in a location close to active earth moving activities and can also be used in drop inlets, storm drains, retrofits and slope drains. The APS Floc Log will treat a flow rate of 60 to 75 gallons per minute.

The *Alabama Handbook* (USDA, 2003) provides the following guidance for use of PAM:

Chemical Stabilization (CHS)

Practice Description

Chemical erosion control on construction sites in the Southeast usually involves a water-soluble anionic polyacrylamide product referred to as PAM. It is used to minimize soil erosion caused by water and wind. PAM is typically applied with temporary seeding and or mulching on areas where the timely establishment of temporary erosion control is so critical that seedings and mulching need additional reinforcement. It may be used alone on sites where no disturbances will occur until site work is continued and channel erosion is not a significant potential problem. Only PAM is currently included in this practice.

Planning Considerations

Anionic PAM is available in emulsions, powders, and gel bars or logs. Anionic PAM should be used in combination with other Best Management Practices. The use of seed and mulch should be considered for providing erosion protection beyond the life of the anionic PAM. If the area where PAM is applied is disturbed after the application, the application will need to be repeated. Following are additional considerations to enhance the use of or avoid problems with the use of anionic PAM:

- Use setbacks when applying anionic PAM near natural water bodies.
- Decreased performance by the PAM can be expected if the PAM is exposed to ultraviolet light or if there is a delay between mixing the PAM with water and applying it to the exposed soil.
- When used in flow concentration channels, PAM's effectiveness for stabilization is decreased.
- If seed is applied with the anionic PAM, mulch should be used to protect the seed.
- Never add water to PAM, add PAM slowly to water. If water is added to PAM, the PAM tends to clot and form "globs" which can clog dispensers. This will result in an increased risk of under-application of the product.
- Only use anionic PAM. Not all polymers are PAM.
- Requests to use other products on permitted sites should be made to the state environmental agency.

Design Criteria

Application rates shall conform to manufacturer's guidelines for application. The following specific criteria shall be followed:

- Only the anionic form of PAM shall be used. Cationic PAM is toxic and shall NOT be used.
- PAM and PAM mixtures shall be environmentally benign, harmless to fish, wildlife, and plants. PAM and PAM mixtures shall be non-combustible.
- Anionic PAM, in pure form, shall have less than or equal to 0.05% acrylamide monomer by weight, as established by the Food and Drug Administration and the Environmental Protection Agency.
- To maintain less than or equal to 0.05% of acrylamide monomer, the maximum application rate of PAM, in pure form, shall not exceed 200 pounds/acre/year. Do not over apply PAM. Excessive application of PAM can lower infiltration rate or suspend solids in water, rather than promoting settling.
- Users of anionic PAM shall obtain and follow all Material Safety Data Sheet requirements and manufacturer's recommendations.
- Additives such as fertilizers, solubility promoters or inhibitors, etc. to PAM shall be non-toxic.
- The manufacturer or supplier shall provide written application methods for PAM and PAM mixtures. The application method shall ensure uniform coverage to the target and avoid drift to non-target areas including waters of the state. The manufacturer or supplier shall also provide written instructions to ensure proper safety, storage, and mixing of the product.
- Gel bars or logs of anionic PAM mixtures may be used in ditch systems. This application shall meet the same testing requirements as anionic PAM emulsions and powders.
- To prevent exceeding the acrylamide monomer limit in the event of a spill, the anionic PAM in pure form shall not exceed 200 pounds/batch at 0.05% acrylamide monomer or 400 pounds/batch at 0.025% acrylamide monomer.

SUMMARY

This chapter reviewed several techniques for designing stable channels and slopes at construction sites. The shear stress method was shown to be generally necessary for channel design, compared to only using an allowable velocity approach. However, liner vegetation in erosion resistant soils may still fail due to vegetation damage, thus requiring careful plant selection. For slopes, tolerable soil loss calculations may also be needed to verify the selection of slope protection solutions; the use of shear stress alone may not be suitable, especially in highly erosive locations.

It is critical that a construction site use suitable procedures to prevent erosion on site, instead of relying on sediment removal from the flowing water after erosion occurs. These techniques must be used, in conjunction with good construction planning, to minimize the amount of land exposed to erosion, and to decrease the amount of sediment erosion produced. The next chapter describes sediment control measures, and their design, for construction sites.

INTERNET SOURCES

U.S. Army Corps of Engineers Channel Stability Assessment Method Report: Engineering and Design—*Channel Stability Assessment for Flood Control Projects,* 1994:
http://www.usace.army.mil/inet/usace-docs/eng-manuals/em1110-2-1418/toc.htm
USDA Agricultural Handbook 667: *Stability Design of Grass-Lined Open Channels,* 1987:
http://www.info.usda.gov/CED/ftp/CED/AH-667.pdf

North America Green downloadable program for slope and channel protection:
http://www.nagreen.com/

Example Project Assignment on Slope and Swale Design

The following is excerpted from a homework assignment prepared by Heather Hill, a student at the University of Alabama at Birmingham, as part of the Construction Site Erosion and Sediment Control Class taken during the summer of 2005. The assignment was given as follows:

1. Identify several different slope categories on your construction evaluation site and propose suitable control practices for each type. Justify your selections with appropriate calculations.

The following is an example output screen from the North American Green software to assist in the selection of turf reinforcement mats for slopes.

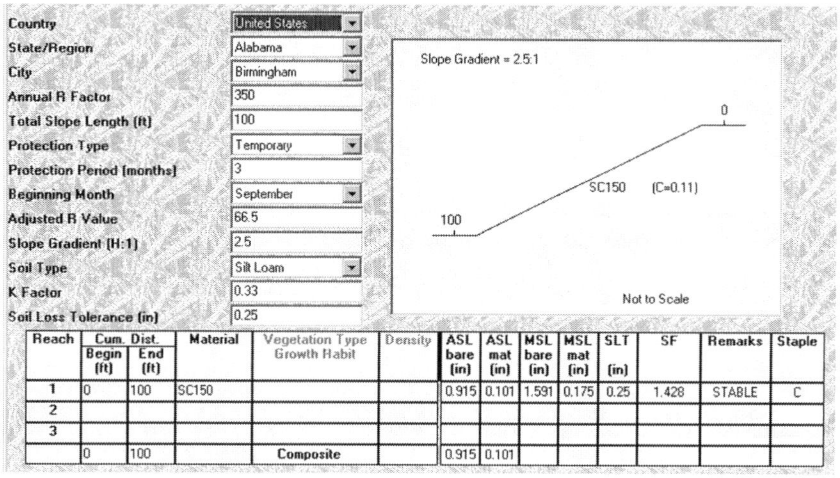

NAG SC150 Temporary Cover 100′ slope (Figure reprinted with the permission of North American Green).

2. Design an appropriate diversion swale for your evaluation site. Using the previously calculated flow rates, select a suitable channel lining, including the consideration of check dams. Justify your selections with appropriate calculations.

There is not a diversion swale at this construction site. The creek that runs through the site will actually be rereouted through a 15′ culvert pipe and covered to level out the site. In order to reroute the stream and install the culvert, an impermeable diversion dam was installed where the stream entered the site. A bypass pump was set up at this location to pump the water to the end of the site where it naturally releases. The average daily flow of this stream is approximately 2600 gpm. The following is an example of applying the North American Green software to evaluate a channel lining material to this channel, assuming that an open channel was an optional method.

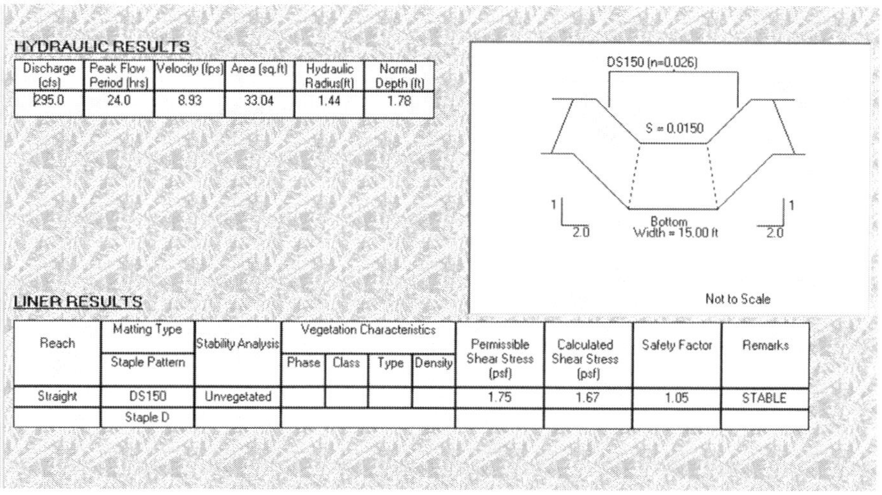

(Figure reprinted with the permission of North American Green).

Example Project Assignment on Slope and Swale Design

The following is modified from a homework assignment prepared by Regan Johnson, a student at the University of Alabama, as part of the Construction Site Erosion and Sediment Control Class taken during the summer of 2005.

1. Identify several different slope categories on your construction evaluation site and propose suitable control practices for each type. Justify your selections with appropriate calculations.

The construction site described here is located along Kicker Road in Tuscaloosa, AL. The City of Tuscaloosa has granted permission to develop St. Charles Place, a townhouse residential development located. The following is the detailed site topographic map donated by the site developer.

This site was separated into the following slope categories: 0–2%, 2–5%, 5–20%, and 20% and greater. One critical slope is examined in the following calculations for slope stability requirements, having the following characteristics:

Slope = 0.5 = 50%
Width of slope = 175ft
Manning's $n = 0.02$

The modified Manning's equation was used to calculate the nominal depth for sheetflow on this particular slope, assuming a previously calculated peak flow rate of 6.4 ft^3/sec.

$$y = \left(\frac{qn}{1.49s^{0.5}}\right)^{3/5}$$

where,

$y =$ the flow depth (in feet)
$q =$ the unit width flow rate (Q/W)
$n =$ the sheet flow roughness coefficient for the slope surface
$s =$ the slope (as a fraction)
$q = Q/W = 6.39/175 = 0.0365$ cfs
$n = 0.02$ for sandy loam soils
$s = 0.50$ ft/ft

y= ((0.0365)(0.02) /(1.49/ 0.500.5))3/5 = 0.051 ft,
or about a half an inch

Therefore, the basic shear stress equation can be used to calculate the maximum shear stress expected on a slope:

$$\tau_o = \gamma y S \ \ (\text{lb/ft}^2)$$

where:

$\gamma =$ specific weight of water (62.4 lbs/ft^3)
$y =$ flow depth (ft)
$S =$ slope (ft/ft)

Thus, $\tau_o = (62.4)(0.051)(0.50) = 1.58$ lb/ft^2

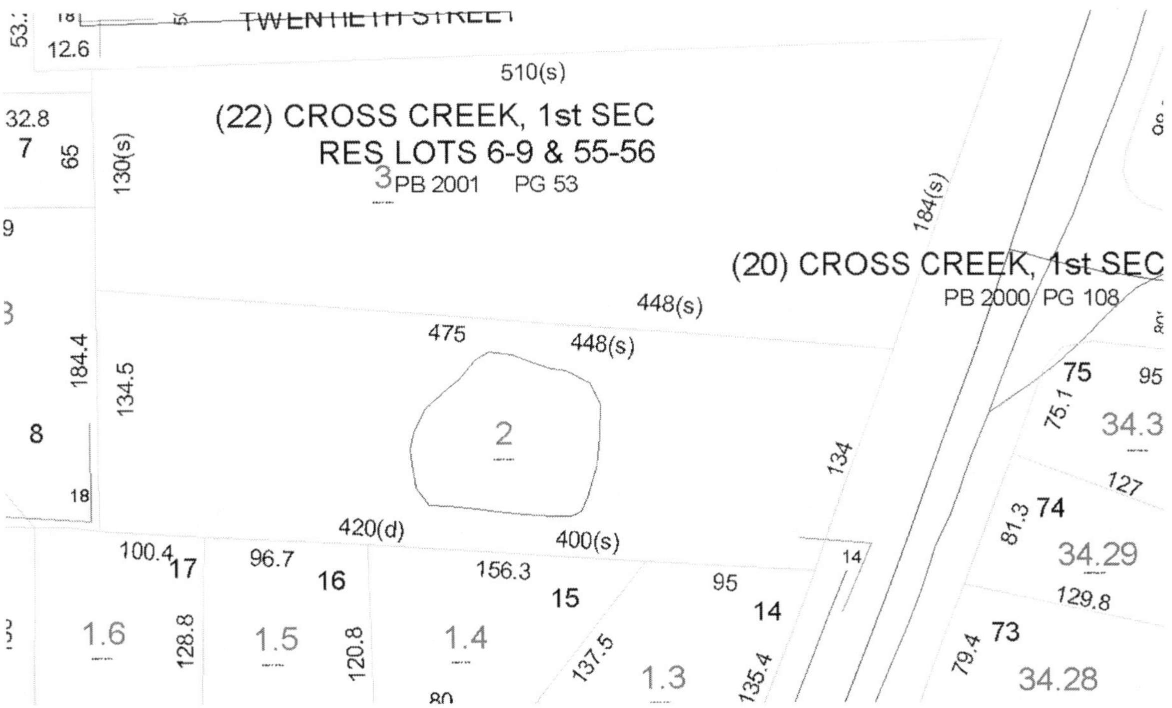

Since the allowable shear stress for the soils on this hillside is only 0.15 lb/ft^2, a vegetated mat will therefore be needed. The next step is to check the shear stress under the mat.

We can solve for the needed roughness factor n of the mat to find a mat that will work given the following equation:

$$\tau_e = \tau_o(1 - C_f)\left(\frac{n_s}{n}\right)^2$$

$\tau_e = 0.15 = 1.58(1–0)[0.02/n_{mat}]^2 \quad n_{mat} = 0.065$

Therefore, a mat is needed having an "n" value of at least 0.065 to provide proper soil protection. Additionally, the mat also needs a C factor to meet the maximum allowable erosion rate on the slope (0.25 inches, or less). Using RUSLE:

R = 350/yr (Tuscaloosa)
k = 0.21
LS = 12.75 (for 150 ft slope at 50%)

The base (unprotected) erosion rate is therefore: (350)(0.21)(12.75) = 643 tons/acre/year

This corresponds to 643(0.00595) = 3.8 inches per year. With a maximum allowable erosion loss of 0.25 to 0.5 inches per year, the C factor for the mat should therefore be: 0.5 to 3.8 = 0.13; or 0.25/3.8 = 0.065, or less

A NAG P300 mat has a C of 0.09 (intermediate in the above range) and an n of 0.02 for this slope and unvegetated condition.

The shear stress is too large, as the mat n is only 0.02, and not the desired 0.065. Therefore, the only reasonable solution for this steep and long slope is to use terraces to divide the slope into several segments, and to use diversion down-slope drains to collect the water from each terrace bench and safely carry it to the bottom of the slope.

If the slope was divided into 50 ft lengths, 1/3 of the original slope length, the Q would also be 1/3, or 2.1 ft^3/sec, and the q would be 0.012 cfs. The resulting flow depth would therefore be:

y= ((0.012)(0.02) /(1.49/ 0.500.5))3/5
= 0.0125 ft, or 0.15 inch

The resulting shear stress is therefore:

τ_o = (62.4)(0.0125)(0.50)= 0.39 lb/ft^2

The needed value for n (unvegetated) is therefore:

$\tau_e = 0.15 = 0.39(1–0)[0.02/n_{mat}]^2 \quad n_{mat} = 0.032$ at least

The NAG P300 still is not "rough" enough.

If the slope was divided into 25 ft lengths, 1/6 of the original slope length, the Q would also be 1/6, or 1.1 ft^3/sec, and the q would be 0.006 cfs. The resulting flow depth would therefore be:

y = ((0.006)(0.02) /(1.49/ 0.500.5))3/5
= 0.0043 ft, or 0.052 inch

The resulting shear stress is therefore:

τ_o = (62.4)(0.0043)(0.50)= 0.13 lb/ft^2

The needed value for n (unvegetated) is therefore:

$\tau_e = 0.15 = 0.13(1–0)[0.02/n_{mat}]2 \quad n_{mat} = 0.019$ at least

Therefore, this slope length is suitable, as the n for the mat is 0.02.

As an alternative, it may be suitable to re-examine the slope itself and consider reducing it from 50% to 40%, and with terraces at 50 ft spacing:

y = ((0.012)(0.02) /(1.49/ 0.400.5))3/5
= 0.007 ft, or 0.08 inch

The resulting shear stress is therefore:

τ_o = (62.4)(0.007)(040)= 0.17 lb/ft^2

The needed value for n (unvegetated) is therefore:

$\tau_e = 0.15 = 0.17(1–0)[0.02/n_{mat}]2 \quad n_{mat} = 0.021$ at least

This is close to the available n of 0.2 and is also a likely suitable solution. Either of these solutions to modify the slope would also reduce the resulting erosion rate.

2. Design an appropriate diversion swale for your evaluation site. Using the previously calculated flow rates, select a suitable channel lining, including the consideration of check dams. Justify your selections with appropriate calculations.

This site consists of one channel that diverts water from the upper portion of the watershed. This channel is located in the back of the site and it will be designed to handle the flow rates that were calculated through an earlier analysis. The cross section of the channel will be a trapezoidal in shape.

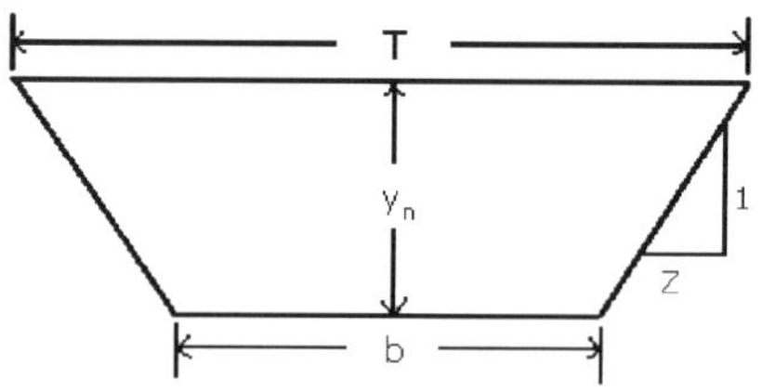

Other factors such as slope, Manning's n, and soil type also affect the channel's performance as well. There are two important parameters involved when designing a diversion swale (1) Allowable velocity (V_o) and (2) Allowable shear stress (τ_o). The first step of the design is to determine the applicable values associated with site specific soil conditions. The site soil is sandy loam. The following parameters should therefore be met for this design:

Maximum permissible velocity (V_o): 2.5 ft/sec

Allowable shear stress (τo): 0.075 lb/ft^2

For this particular swale design, the Manning's equation for open channel flow will be used with the Chow shape factor:

$$AR^{2/3} = \frac{nQ}{1.49 S^{0.5}}$$

where,

Q = 16.3 cfs
S = 0.055 ft/ft
n = 0.02

It is therefore possible to calculate the nominal depth of channel flow within the swale for different swale cross sections, using an Excel spreadsheet. The spreadsheet allowed an examination of various base widths (b) and side slopes (z) of the channel. The selected alternative for the channel dimension is one with a 3ft base and 2:1 side slope. The resulting shear stress and channel velocity are also shown on this table.

Since the shear stress is higher than permissible, the channel will be fitted with a liner or vegetation mat. Installing a channel liner will cause the effective shear stress to decrease, thus, reducing the potential of excessive sediment erosion. Moreover, the vegetation mat will provide adequate support for the channel's exposed sediment surface.

The North American Green website provides a list of potential mats to be used for erosion control for construction sites. For this channel, a P300 polypropylene fiber erosion control blanket was selected. The following calculations show that this liner meets the permissible shear stress criteria.

$$\tau_e = \gamma DS(1 - C_f)\left(\frac{n_s}{n}\right)^2$$

where,

τ_e = effective shear stress exerted on soil beneath vegetation
γ = 62.4 lbs/ft^3
D = the maximum flow depth in the cross section=1.05 ft
S = hydraulic slope = 0.055 ft/ft
C_f = vegetation cover factor =0.90 (Bermuda grass)
n_s = roughness coefficient of underlying soil = 0.02
n = roughness coefficient of erosion control blanket = 0.44

$\tau_e = 62.4 \times 1.05' \times 0.055(1\text{–}0.90)[0.02/0.044]2 = 0.074$

Therefore, $\tau_e < \tau_o$ and the NAG P300 mat will be an acceptable solution for this channel.

Channel Design Option

b (ft)	z (ft)	Top (ft)	$AR^{2/3}$	$b^{8/3}$	$AR^{2/3}/b^{8/3}$	y_n/b (From Chow's figure)	y_n	τ (lb/ft^2)	R	V (ft/s)
3	2	15	8.01	18.72	0.43	0.35	1.05	3.60	0.70	2.50
4	2	20	8.01	40.32	0.20	0.3	1.20	4.12	0.83	1.94
5	2	25	8.01	73.10	0.11	0.27	1.35	4.63	0.96	1.48
6	2	30	8.01	118.87	0.07	0.153	0.92	3.15	0.71	1.91
3	3	21	8.01	18.72	0.43	0.32	0.96	3.29	0.62	2.43
4	3	28	8.01	40.32	0.20	0.27	1.08	3.71	0.73	1.85
5	3	35	8.01	73.10	0.11	0.24	1.20	4.12	0.84	1.46
6	3	42	8.01	118.87	0.07	0.142	0.85	2.92	0.63	1.83

Note: Highlighted values indicate the selected channel for example.

PROBLEMS

1. Explain the influence of each of the following on the tractive force or shear stress along a channel bottom. (a) The shape of the soil particles; (b) The specific weight of the fluid; (c) The specific weight of the soil particles; (d) The particle diameter.
2. Using the allowable shear stress method, design an upslope diversion channel to carry a discharge of 10 ft^3/sec, a maximum velocity of 2 ft/sec, a channel slope of 0.5%, and that is located on loam soil. Is this channel stable if no protective mat or liner is installed?
3. An existing trapezoidal canal has a slope of 0.01 ft/ft, a base of 12 ft, a Manning's roughness coefficient of 0.035, and side slopes of 3.5H:1V. Determine the permissible velocity.
4. A new roadway is being cut through your area. The side slope (1H:1V) of 500 ft in width and 50 ft in slope length needs to be stabilized. Assuming the design storm for side-slope stabilization is the 25.yr storm (to prevent washout of roadway support). Design a slope stabilization scheme. Does this slope require protection above mulching while awaiting seed cover?
5. Project Questions:
   a. Identify several different slope categories on your construction evaluation site and propose suitable control practices for each type. Justify your selections with appropriate calculations.
   b. Design an appropriate diversion swale, or a main drainage swale for your evaluation site. Using the previously calculated flow rates, select a suitable channel lining, including the consideration of check dams. Justify your selections with appropriate calculations.

REFERENCES

ASCE. *Sedimentation Engineering.* Edited by Vito A. Vanoni. Manuals and Reports on Engineering Practice, No. 54. 1975.

Chang, H.H. *Fluvial Processes in River Engineering.* John Wiley & Sons. 1988.

Chow, Ven Te. *Open Channel Hydraulics.* McGraw-Hill. 1959.

Croke, T. "Reliable channel design using turf reinforcement mats." Erosion Discussion. Vol. 5, No. 1. North American Green (availabe from their web page: www.nagreen.com). June 2000.

Croke, T. "Accounting for bends in channel design." *Erosion Discussion.* Vol. 6, No. 1. North American Green (availabe from their web page: www.nagreen.com). Summer 2001.

Fortier, S. and F.C. Scobey. "Permissible canal velocities." *Trans. ASCE,* Vol. 89, paper No. 1588, pp. 940–984. 1926.

McCuen, R.H. *Hydrologic Analysis and Design,* 2nd Edition. Prentice Hall. 1998.

Penn State. *Analysis of Flow through Porous Media as Applied to Gabion Dams Regarding the Storage and Release of Storm Water Runoff,* NAHB/NRC Designated Housing Research Center, Report No. 10, August 1992.

Raudkivi, A. J., and Tan, S. K. "Erosion of cohesive soils," *Journal of Hydraulic Research,* Vol 22, No. 4, pp 21X–233. 1984.

Ree, W.O. and Palmer, V.J. *Flow in Channels Protected by Vegetative Linings.* Technical Bulletin No. 967, USDA: 1–22, 44–48. 1949.

Ree, W.O. "Hydraulic Characteristics of Vegetation for Vegetated Waterways." *Agricultural Engineering,* April: 184–189. 1949.

Ree, W.O. *Friction Factors for Vegetation Covered, Light Slope Waterways.* Research Report No. 335, USDA: 1–34. 1960.

Robinson, K.H., C.E. Rice, and K.C. Kadavy. Design of rock chutes, *Transactions of the American Society of Agricultural Engineers,* 41(3) pp. 621–626. 1982.

SCS (Soil Conservation Service). *Handbook for Channel Design for Soil and Water Conservation.* SCS-TP-61, rev. 1954.

SCS (Soil Conservation Service). *Design of Open Channels.* TR-25. 1977.

Sprague, C.J. "Green Engineering; Design principles and applications using rolled erosion control products." *CE News On-Line,* March 1999, at: http://www.cenews.com/edenviroeng0399.html

Temple, D.M. "Flow Retardance of Submerged Grass Channel Linings." *Transactions of the ASAE:* 1300–1303. 1982.

Temple, D.M. *Erosionally Effective Soil Stress in Grass Lined Open Channels.* ASAE Paper No SWR84-102. 1984.

U.S. Army Corps of Engineers (COE). *Hydraulic Design of Flood Control Channels.* EM 1110-2-1601. U.S. Army Engineer Waterways Experiment Station, Vicksburg, MS. undated.

U.S. Department of Agriculture (USDA). *Design of Open Channels.* Technical Release No. 25, Soil Conservation Service, Washington, DC. 1977.

U.S. Army Corps of Engineers (COE). *Engineering and Design: Channel Stability Assessment for Flood Control Projects.* EM 1110-2-1418. U.S. Army Engineer Waterways Experiment Station, Vicksburg, MS. 1994.

USDA. *Alabama Handbook for Erosion Control, Sediment Control, and Stormwater Management on Construction Sites and Urban Areas,* prepared by the Alabama Soil and Water Conservation Committee, Montgomery, AL. 2003.

USDA. *Handbook of Channel Design for Soil and Water Conservation.* Technical Paper TP-61, USDA: 1–34. 1954.

USDA. *Stability Design of Grass-Lined Open Channels.* Agricultural Handbook 667. 1987.

Williams, D. T., and Julien, P. Y. "Examination of stage-discharge relationships of compound/composite channels," *Proceedings of the International Conference on Channel Flow and Catchment Runoff.* B.C. Yen, ed., University of Virginia, Charlottesville, 22–26 May 1989, pp 478–488. 1989.

APPENDIX 5A: COMMERCIAL SOURCES FOR CHANNEL LINERS

The following lists various commercial sources for channel liners and turf-reinforcing mats for protecting channels. Obviously, this is not a comprehensive listing and their inclusion here does not imply any endorsement. These are included as examples of the types of products, and supporting information, currently available.

Rock Baskets (Gabions)

Maccaferri
www.maccaferri-usa.com

Terra Aqua
www.terraaqua.com

Concrete Flexible Grids (Revetment)

Armortec
www.armortec.com

Hydropace
info@hydropve.com

Plastic Grids

Invisible Structures, Inc, (Slope Tamer^2)
www.invisiblestructures.com

Presto (Geoweb) 1-800-548-3424
http://www.prestogeo.com/

Fabric Blankets and Channel Mat Liners

Synthetic Industries (Landlok erosion control blankets and Pyramat)
www.fixsoil.com

Construction Products, Inc. (Contech) Middleton, Ohio

Amoco Fabrics and Fibers Co. (Super Gro)
1-800-445.7732

Akzo Nobel (Enkamat)
1-800-365.7391

North American Green
www.nagreen.com

RoLanks Erosion Control Blankets
1-800-760-3215

Turf Grass

Gardner Turfgrass, Inc. (sod farms, also Stay Turf live matting)
www.Gardnerturf.com

Hydroseeding and Chemicals

Conwed fibers (mulch and blankets)
www.conwedfibers.com
1-800-366-1180

Soil Guard (bonded fiber matrix mulch)
www.soilguard.com

Soil Moist (soil erosion polymer)
www.soilmoist.com

Terra Mulch
www.terra-mulch.com

Applied Polymer Systems
lwinskis@aol.com

Gabion Wall Channel Protection

IECA photo.

APPENDIX 5B: KANSAS DEPARTMENT OF TRANSPORTATION BUREAU OF MATERIALS AND RESEARCH

The following is an example list containing the pre-approved products listed by the Kansas Department of Transportation, Bureau of Materials and Research, for different types of construction site erosion control applications.

This list is an example of pre-approved erosion control products and shows the variety of products available for slope protection. Other states' have similar lists, but may have differing standards and testing procedures. This is included here as an example of what is being developed. Obviously, it is important to select materials that will meet local, site-specific conditions, and that are also approved by the local regulatory agency.

Approved Erosion Control Products for Slope Protection

Class 1 "Slope Protection"

Approved Products for Flexible Channel Liners

Type A–Slopes 1:3 or Flatter–Clay Soils

Airtrol	Greenstreak Pec-Mat
Anti-Wash/Geojute	Landlok® BonTerra® EcoNetT™ ENS2
BioD-Mesh 60	Landlok® BonTerra® EcoNet™ ENCS2
Carthage Mills Veg Net	Landlok® BonTerra S 1
C-Jute	Landlok® BonTerra S2
Contech Standard	Landlok BonTerra CS2
Contech Standard Plus	Landlok TRM 435
Contech Straw/Coconut Fiber Mat w/Kraft Net	Maccaferri MX287
Contech C-35	Miramat TM8
Conwed 3000	North American Green S150
Curlex I	North American Green S75
Curlex™-LT	North American Green SC150
Earth Bound	North American Green® S150 BN
EcoAegis™	Pennzsuppress®
Econo Jute	Poplar Erosion Blanket
ECS Excelsior Blanket Standard	Soil Guard
ECS High Velocity Straw Mat	Soil Saver
ECS Standard Straw	SuperGro
EnviroGuard Plus	TerraControl®
Formula 480 Liquid Clay	TerraJute
Futerra™	Verdyol Ero-Mat
GeoTech TechMat™ SCKN	Verdyol Excelsior High Velocity
Grass Mat	Verdyol Excelsior Standard
Greenfix WS072	Webtec Terraguard 44P
Green Triangle Regular	Xcel Regular
Green Triangle Superior	Xcel Superior

Type B–Slopes 1:3 or Flatter–Sandy Soils

C-Jute	Landlok® BonTerra CS2
Carthage Mills Veg Net	Landlok® BonTerra® EcoNet™ ENCS2™
Contech Standard	Landlok® BonTerra® EcoNet™ ENS2
Contech Standard Plus	Landlok TRM 435
Contech Straw/Coconut Fiber Mat w/Kraft Net	Maccaferri MX287
Contech C-35	Miramat 1000
Curlex LT	Miramat TM8
Earth Bound	North American Green S75
ECS Standard Straw	North American Green® S75 BN
ECS Excelsior Blanket Standard	North American Green S150
ECS High Velocity Straw Mat	North American Green SC150
EcoAegiS™	North American Green® S150 BN
EnviroGuard Plus	Poplar Erosion Blanket
Futerra®	Soil Guard
Greenfix WS072	Terra-Control®
Geojute Plus 1	TerraJute
GeoTech TechMat™ SCKN	Verdyol Ero-Mat
Green Triangle Regular	Verdyol Excelsior Standard
Green Triangle Superior	Webtec Terraguard 44P
Landlok® BonTerra S1	Xcel Regular
Landlok® BonTerra S2	Xcel Superior

Type C–Slopes Steeper than 1:3–Clay Soils

Airtrol	Landlok® BonTerra® EcoNet™ ENCS2™
Anti-Wash/Geojute	Landlok® BonTerra S2
Carthage Mills Veg Net	Landlok BonTerra CS2
C-Jute	Landlok TRM 435
Contech Standard Plus	Maccaferri MX287
Contech Straw/Coconut Fiber Mat w/Kraft Net	Miramat TM8
Contech C-35	North American Green S150
Conwed 3000	North American Green S75
Curlex I	North American Green SC150
Earth Bound	North American Green® S150BN
Econo Jute	Pennzsuppress®
ECS High Velocity Straw Mat	Poplar Erosion Blanket
ECS Standard Straw	Soil Guard
EnviroGuard Plus	Soil Saver
Formula 480 Liquid Clay	SuperGro
Futerra®	TerraJute
Greenfix WS072	Verdyol Excelsior High Velocity
Green Triangle Superior	Webtec Terr
GeoTech TechMat™ SCKN	Xcel Superior
Greenstreak Pec-Mat	

Type D–Slopes Steeper than 1:3–Sandy Soils

C-Jute	Landlok® BonTerra S2
Carghage Mills Veg Net	Landlok® BonTerra CS2
Contech Standard Plus	Landlok® BonTerra® EcoNet™ ENCS2™
Contech Straw/Coconut Fiber Mat w/Kraft Net	Landlok TRM 435
Contech C-35	Maccaferri MX287
Curlex I	Miramat 1000
ECS High Velocity Straw Mat	Miramat TM8
ECS Standard Straw	North American Green S150
EnviroGuard Plus	North American Green SC150
Futerra®	North American Green® S150BN
Greenfix WS072	Soil Guard
Geojute Plus 1	TerraJute
GeoTech TechMat™ SCKN	Webtec Terraguard 44P
Green Triangle Superior	Xcel Superior

Type E–Shear Stress Range 0–96 Pascal (0–2 lb/ft^2)

Contech TRM C-45
Contech C-35
Contech C50
Contech Coconut/Poly Fiber Mat
Contech Coconut Mat w/Kraft Net
Curlex« Channel Enforcer 1
Curlex« Channel Enforcer II
Curlex« II Stitched
Curlex« III Stitched
Earth-Lock
Earth-Lock II
ECS High Impact Excelsior
ECS Standard Excelsior
ECS High Velocity Straw Mat
Enkamat 7018
Enkamat 7020
Enkamat Composite 30
Enviromat
Geotech TechMatTM CP 3-D
Geotech TechMatTM CKN
Greenfix CFO 72RR
Greenstreak Pec-Mat
Koirmat® 700
Landlok® BonTerra® C2
Landlok TRM 435
Landlok TRM 450
Landlok TRM 1050
Landlok TRM 1060
Maccaferri MX287
Miramat TM8
Multimat 100
North American Green C125 BN
North American Green C350 Three Phase
North American Green SC 150 BN
North American Green S350
North American Green® P350
North American Green S 150
Pyramat®
Webtec Terraguard 44P
Webtec Terraguard 45P
Xcel PP-5

Type F–Shear Stress Range 0–192 Pascal (0–4 lb/ft^2)

Curlexg II Stitched
Curlex® III Stitched
Curlex® Channel Enforcer 1
Curlex® Channel Enforcer 11
Contech C50
Contech TRM C-45
Contech C-35
Contech Coconut/Poly Fiber Mat
Contech Coconut Mat w/Kraft Net
Earth-Lock
Earth-Lock II
ECS High Impact Excelsior
ECS High Velocity Straw Mat
ECS Standard Excelsior
Enkamat 7018
Enkamat Composite 30
Enviromat
Geotech TechMatTM CP 3-D
Geotech TechMatTM CKN
Greenfix CFO 72RR
Greenstreak Pec-Mat
Koirmat® 700
Landlok® BonTerra® C2
Landlok® BonTerra® EcoNet™ ENC2
Landlok TRM 435
Landlok TRM 450
Landlok TRM 1050
Maccaferri MX287
Miramat TM8
Multimat 100
North American Green C125 BN
North American Green C350 Three Phase
North American Green SC 150 BN
North American Green S350
North American Green® P350
North American Green S 150
Pyramat®
Webtec Terraguard 44P
Webtec Terraguard 45P
Xcel PP-5

Type G–Shear Stress Range 0–287 Pascal (0–6 lb/ft^2)

Contech TRM C-45
Contech C-35
Contech C50
Contech Coconut/Poly Fiber Mat
Curlex« III Stitched
Curlex« Channel Enforcer 11
Earth-Lock
Earth-Lock II
Enkamat 7018
Enkamat Composite 30
Geotech TechMatTM CP 3-D
Greenstreak Pec-Mat
Koirmat« 700
Landlok TRM 1050
Landloc TRM 1060
Landlok TRM 435
Landlok TRM 450
North American Green C350 Three Phase
North American Green S350
North American Green« P350
Pyramat«
Webtec Terraguard 44P
Webtec Terraguard 45P

Type H–Shear Stress Range 0–383 Pascal (0–8 lb/ft^2)

Contech TRM C-45	Landlok TRM 1050
Contech C-35	Landlok TRM 1060
Contech C50	North American Green C350 Three Phase
Contech Coconut/Poly Fiber Mat	North American Green S350
Curlex® III Stitched	North American Green® P350
Geotech TechMat™ CP 3-D	Pyramat®
Landlok TRM 435	Webtec Terraguard 44P
Landlok TRM 450	Webtec Terraguard 45P

Temporary Ponds and Filter Fabric Barriers for Construction Site Sediment Control

INTRODUCTION

THE use of temporary ponds for sediment control is a common practice at many construction sites. In some cases, these ponds are re-built after the construction period and used as permanent ponds for stormwater control. However, often they are filled in and their area used as part of the land development. Because sediment ponds have relatively short lives, their design criteria and construction methods differ from more permanent stormwater control ponds. The particle trapping mechanisms are the same for both types of ponds, but the influent hydrology and particle size distributions can be substantially different. The following discussion therefore stresses the special features of temporary sediment control ponds for construction sites. Also discussed are filter fences for two reasons: (1) small drainage areas are usually controlled using filter fences, while large areas require sediment ponds (they are therefore complementary practices with similar objectives), and (2) filter fences remove sediment from the flowing water in much the same way as sediment ponds, by sedimentation (not "filtration," as their name implies).

Temporary construction site sediment ponds have sediment loads that are very large while the particulates in that load may be very small. Sizeable accumulations of sediment may occur in short periods of time. Due to the lack of protection from scour, dry detention ponds have much smaller removal benefits than wet ponds (which have at lest 3 ft. of standing water). If well designed and properly maintained, suspended solids removals of 70 to 90% can be obtained in wet ponds, while dry ponds seldom provide more than 30% suspended solids reductions.

There are a number of basic design guidelines needed to maximize sediment removal and to minimize potential problems in ponds, including the following:

- At least three feet of permanent standing water is needed over most of the pond to protect sediments from scouring. Additional depth is also needed for sediment storage between cleanout operations.
- Ideally, the pond length should be about three to five times the width for maximum detention efficiency.
- The inlets and outlets need to be widely spaced to minimize short-circuiting.
- Correct pond side slopes are very important for safety reasons and to minimize mosquito problems. An underwater shelf near the pond edge needs to be planted with rooted aquatic plants to hinder access to deep water, if the pond will be in place for several years. The temporary ponds commonly used at construction sites receive large sediment loads and their time is so short that vegetation cannot become established. Temporary ponds in urban areas may need fencing to prevent access by neighborhood children.
- Outlet structures should be designed for low outflows during low pond depths to maximize particulate retention. Place underwater dams or deeper sediment trapping forebays near pond inlets to decrease required dredging areas.
- Protect the inlet and outlet areas from scour erosion and cover the inlets and outlets with appropriate safety gratings. Provide an adequate emergency spillway.

Basic pond design guidelines must also be followed to provide the expected level of sediment removal. The following list is a typical example of these guidelines for proper design, installation and operation.

- Engineering design guidelines (covering such things as foundations, fill materials, embankments, gratings, anti-seep collars, and emergency spillway construction), such as published by the U.S. Natural Resources Conservation Service and the Corps of Engineers (SCS, 1982).
- Pond size is dictated mostly by desired particle control and water outflow rate. For construction sites, the pond water surface should be about 1.5% of the watershed area draining to the pond in order to achieve approximately 90% suspended solids reductions. If the pond area is only about 0.5% of the drainage area, the resulting removal would be reduced to about 65% (or

less) of suspended solids. The use of chemicals can increase the removal of sediment in ponds. In an early example, Colston (1974) used alum to increase suspended solids and turbidity removals up to about 85 to 97%. More recent examples show similar removal benefits when using chemical-assisted sedimentation.

Ponds can be classified according to their size and design objectives. Table 6.1 from the *Alabama Handbook* (USDA, 2003) is one way to classify ponds based on their size and spillway designs. The maximum water surfaces shown here are all very large for temporary ponds at construction sites, compared to ponds installed at other locations with different objectives than construction erosion control.

Safety of Wet Detention Ponds

The most important wet detention pond design guidelines are those that maintain public safety. The following discussion briefly summarizes common suggestions to maintain and improve safety at wet detention facilities. Death by drowning is the most common safety concern associated with wet detention ponds. Marcy and Flack (1981) state that drownings, in general, most often occur because of slips and falls into water, unexpected depths, cold water temperatures, and fast currents. Four methods to minimize these problems include the following: (1) eliminating or minimizing the hazard; (2) keeping people away; (3) making the onset of the hazard gradual; and (4) providing escape routes. Many of the design suggestions and specifications contained in this discussion are intended to accomplish these methods.

Jones and Jones (1982) consider safety and landscaping together because landscaping can be an effective safety element. They feel that appropriate slope grading and landscaping can provide a more desirable approach than wide-spread fencing around a wet detention pond. Unfortunately, landscaping is not very effective for temporary pond installations, so pond side slopes are most critical. Fences are expensive to install and maintain and usually produce unsightly pond edges. They collect trash and litter, challenge some individuals who like to defy barriers, and impede emergency access if needed. Marcy and Flack (1981) state that limited fencing may be appropriate in special areas. When the pond side slopes cannot be made gradual (such as when against a railroad right-of-way or close to a roadway), steep sides having submerged retaining walls may be needed. A chain link fence located directly on the top of the retaining wall very close to the water's edge would be needed (to prevent human occupancy of the narrow ledge on the water side of the fence). Another area where fencing may be needed is at the inlet or outlet structures. However, fencing usually gives a false sense of security, as most fences can be crossed easily (Eccher 1991). Temporary sediment ponds in urban areas may need fencing as neighborhood children are likely to be attracted to the pool and the temporary nature of construction site sediment ponds likely precludes the vegetative barriers recommended for permanent wet ponds.

TABLE 6.1. Stormwater Detention Basin Classification (*Alabama Handbook,* USDA 2003).

Type	Maximum Water Surface Area (acre)	Maximum Dam Height[2] (feet)	Emergency Spillway Design Storm Frequency[3]	Freeboard[4] (feet)
1[1]	20	7	10-yr 24-hr	0.5
2	20	10	10-yr 24-hr	0.5
3	50	15	25-yr 24-hr	1.0

[1]Type 1 basins may be used where site conditions prevent the construction of an emergency spillway on residual earth.
[2]Height is measured from the top of the dam to the low point on the original centerline survey of the dam.
[3]Runoff should be determined by NRCS methods or other methods accepted by local ordinances. Soil and cover conditions used should be based on those expected during the construction period.
[4]Vertical distance between basin water surface at maximum design stage and top of dam.

Gradual slopes near the water edge and a submerged ledge close to shore are usually together the best solution to maximize safety. Aquatic plants on the ledge decreases the chance of continued movement to deeper water and thick vegetation on-shore near the water's edge would discourage access to the water edge and decrease the possibility of falling into the water accidentally. Pathways should not be located close to the water's edge, or turn abruptly near the water.

Marcy and Flack (1981) also encourage the placement of escape routes in the water whenever possible. These could be floats on cables, ladders, hand-holds, safety nets, or ramps. They should not be placed to encourage entrance into the water.

The use of inlet and outlet trash racks and antivortex baffles is also needed to prevent access to locations having dangerous water velocities. Several types are recommended by the NRCS (SCS, 1982), as shown in Figures 6.1 and 6.2. Racks need to have openings smaller than about 6 inches to prevent people from passing through them and need to be placed where water velocities are less than three feet per second to allow people to escape (Marcy and Flack, 1981). Besides maintaining safe conditions, racks also help to keep trash from interfering with the outlet structures operation.

Eccher (1991) lists the following pond attributes to ensure maximum safety:

1. There should be no major abrupt changes in water depth in areas of uncontrolled access,
2. Slopes should be controlled to insure good footing,
3. All sloped areas should be designed and constructed to prevent or restrict weed and insect growth (generally requiring some form of hardened surface on the slopes), and
4. Shoreline erosion needs to be controlled.

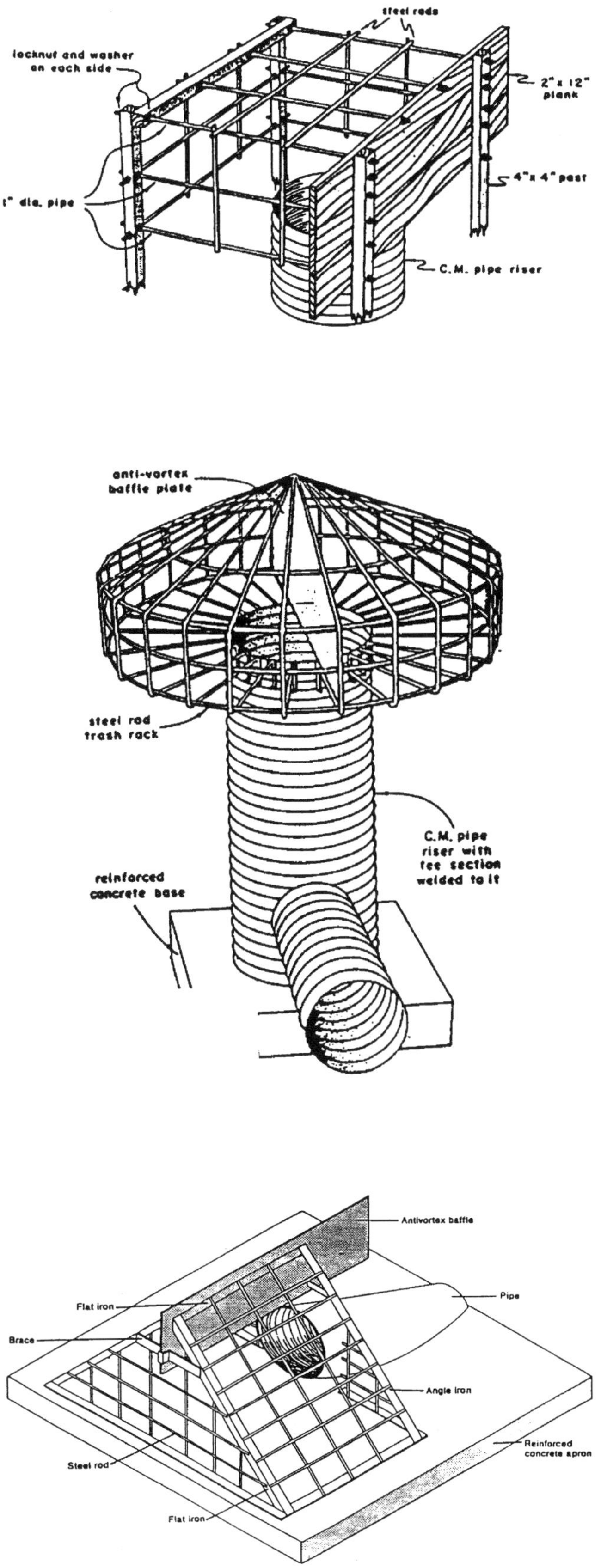

Figure 6.1. Various trash racks and baffles used by the SCS (NRCS). (SCS, 1982).

Maintenance Requirements of Wet Detention Ponds

The most important maintenance for temporary construction-site erosion ponds is to conduct periodic inspections and to make sure that the sediment accumulation is not excessive and prematurely filling the pond.

Temporary sediment ponds need to be inspected after each major storm. The inspection should include checking the pond embankments for subsidence and erosion. The conditions of the emergency spillway and inlets and outlets also need to be determined during the inspection. The adequacy of any channel erosion protection measures near the pond also should be investigated. Sediment accumulation in the pond (especially near, and in, the inlets and outlets) also needs to be examined.

Large sediment accumulations in detention ponds can have significantly adverse affects on pond performance. Bedner and Fluke (1980) reported on the long term effects of detention ponds that received little maintenance. Lack of dredging actually caused the silted-in ponds to become a major sediment source to downstream areas. Poorly-maintained ponds only delayed the eventual delivery of the sediment downstream; they did not prevent it.

During major storms, construction-site erosion ponds can literally fill up during a single storm. Most of the sedimentation would occur near the inlet and the resulting sediment accumulation would be very uneven throughout the pond. Normally, sediment removal in a permanent wet pond may be needed about every five to ten years, but it may be needed every few months at construction sites. It is therefore necessary to plan for required maintenance during the design and construction of sediment ponds. Ease of access of heavy equipment and the possible paving of a sediment trap near the inlet would ease maintenance problems. Dredged sediment is usually placed directly onto trucks, or on the pond banks for dewatering before hauling to the disposal location. One common practice is to keep an area adjacent to the detention pond available for on-site sediment disposal. Small mounds can be created of the dried sediment and covered with top soil and planted.

Poertner (1974) reviewed various sediment removal procedures. An underwater scoop can be pulled across the pond bottom and returned to the opposite side with guiding cables. If drains and underwater roads were built during the initial pond construction, the pond can be drained and front-end-loaders, draglines, and trucks can directly enter the pond area. Small hydraulic dredges can also be towed on trailers to ponds. The dredge pumps sediment through a floating line to the shore where the sediment then is dewatered and loaded into trucks or piled on site. A sediment trap (forebay) also can be constructed near the inlet of the pond. The pond entrances are widened and submerged dams are used to retain the heavier materials in a restricted area

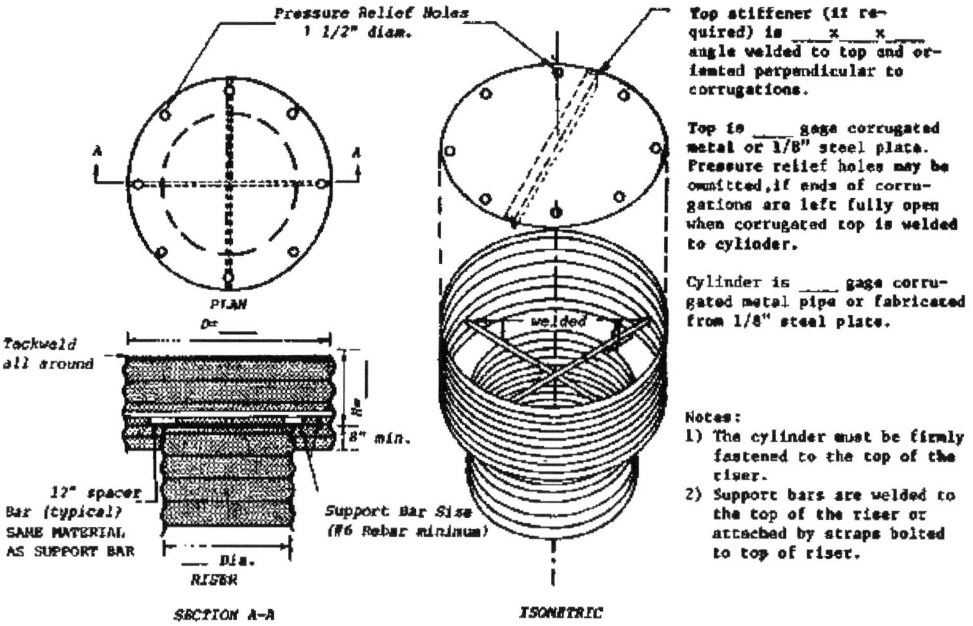

Figure 6.2. Anti-vortex design on riser outlet (*Alabama Handbook,* USDA, 2003).

near the inlets. This smaller area can then be cleaned much easier and with less expense than the complete pond.

Guidelines to Enhance Pond Performance

The NRCS (SCS, 1982) has prepared a design manual that addresses specific requirements for such things as anti-seep collars around outlet pipes, embankment widths, types of fill required, foundations, emergency spillways, etc., for a variety of wet detention pond sizes and locations. That manual must be followed for detailed engineering requirements. The *Alabama Handbook for Erosion Control* (2003) describes the construction and maintenance of sediment basins, and many other sediment and slope-control practices.

Pond Surface Area and Shape

Surface area is one of the most important design considerations for particle removal. Hittman (1976) reports that pond length-to-width ratios of about five have produced maximum pond efficiencies (decreased short-circuiting) during dye tests. If a long and narrow pond cannot be constructed, Schueler (1986) suggests that baffles or gabions be placed within the pond to lengthen the flow path between the inlets and outlets. Bondurat, *et al.* (1975) has also suggested that the idealized pond shape would be triangular: narrow near the inlet and wider near the outlet. This triangular configuration would allow more efficient particle settling by having a continually decreasing forward velocity. Very irregular pond shapes may decrease circulation and cause localized nuisance problems. The pond shape should be irregular for aesthetic considerations, but with minimal opportunities for water stagnation. Short-circuiting in adequately-sized ponds has little detrimental effect on pond performance. However, it can be serious in under-sized ponds. Stagnation is a much more serious problem degrading pond water quality.

Pond Water Depth

A storage volume above the permanent pool elevation of the pond affects the pond's ability to absorb excess flows for flood control. Harrington (1986) found that increasing the wet pool depth increases sedimentation efficiency (due to flocculation), but that surface area increases were much more effective in enhancing the water quality performance of wet ponds. A minimum wet pool depth is very critical in wet ponds to decrease scour losses of previously-settled material. Without an adequate permanent pool depth, very little water quality benefits can be expected from wet ponds.

Extra pond depth needs to be considered for sediment storage between removal operations (Schimmenti 1980). Wiegand, *et al.* (1986) state that it costs about five times as much to removal sediment during pond dredging operations (about \$14 per cubic yard) as it does to provide extra sediment storage capacity (sacrificial volume) during initial pond construction (about \$3 per cubic yard). This sacrificial storage should be provided as deeper forebays near the pond inlets (Driscoll, 1986). These forebays, or the use of underwater dams, need to be designed as pre-sedimentation traps to encourage the deposition of sediment in a relatively restricted area. This would result in more frequent sediment removal operations, but in a smaller area and at a much lower cost than dredging the entire pond.

Sufficient water depth (at least three feet over the

maximum deposited sediment thickness) is also needed to decrease the potential of sediment scour caused by increased flows during large storms (EPA, 1983). Hey and Schaefer (1983) found that a depth of five feet was sufficient to protect the unconsolidated sediment from resuspension in Lake Ellyn.

Pond Side Slopes

Reported recommended side slopes of detention ponds have ranged from 1:4 (one vertical unit to four horizontal units) to 1:10. Steeper slopes will cause problems with grass cutting and may erode. Steep slopes are not as aesthetically pleasing and are more dangerous than gentle slopes (Chambers and Tottle, 1980). Schueler (1986) also recommends a minimum slope of 1:20 for land near the pond to provide for adequate drainage.

The slope near the waterline, and for about one foot below, should be relatively steep (1:4) to provide relatively fast pond drawdown after common storms. However, a flat underwater shelf several feet wide and about one foot below the normal pond surface is needed as a safety measure to make it easier for anyone who accidentally falls into the pond to regain their footing and climb out. This shelf should also be planted with native rooted aquatic plants (macrophytes) to create a barrier making unauthorized access to deep water difficult for permanent ponds. If the installation is a temporary pond, a mild slope, without the planted safety ledge, is more common.

Outlet Structures

Most of the effort given to alternative outlet structure designs has been for dry detention ponds. Wet ponds usually only have a surface weir, outlet pipe, or other simple overflow device to allow the passage of displaced pond water during rains. With the use of a more sophisticated outlet device (such as a floating weir), located at the normal wet pond surface elevation, more efficient particulate removals and flood control benefits may occur.

Hittman (1976) recommends that wide outflow (and inflow) channels be used to decrease erosion. If wide flow channels are not possible, then energy dissipaters to reduce the water velocity should be used. The NRCS (SCS, 1982) has prepared design guidelines for wet-pond outlet structures. These guidelines include a turf-covered embankment having a trapezoidal cross section, a pipe with a metal riser and passing through the embankment as the major outlet, an upstream trash rack at the outlet, and an emergency spillway.

Controlled emptying of a detention pond at low outlet flow rates is desirable for effective sediment removal and flood control. A small diameter outlet pipe, or a small orifice on a plate, typically is used to achieve low outflows. The rate of discharge varies for these outlets because the elevation above the orifice controls the outflow rate. High flow rates occur with higher water levels, and the outlet flow rates decrease with falling water levels. Selecting an appropriate outlet structure has significant effects on pond performance. To have a constant pond performance for all events (if desired), the shape of the outlet must allow a constant upflow velocity (pond outflow rate divided by pond surface area for all pond stages). The following discussion is from the *Alabama Handbook* (USDA, 2003) and is an example of the guidance provided for outlet devices in many regional erosion control guidance documents.

Pond Outlet Protection

Definition

Structurally lined aprons of riprap, concrete or other acceptable energy-dissipating devices placed at the outlets of pipes or paved channel sections.

Purpose

To prevent scour at stormwater outlets and to minimize the potential for downstream erosion by reducing the velocity of concentrated stormwater flows.

Conditions Where Practice Applies

Applicable to the outlets of all pipes and paved channel sections where the velocity of flow at design capacity of the outlet will exceed the permissible velocity of the receiving channel or area. To prevent scour at stormwater outlets, a flow transition structure is needed which will absorb the initial impact of the flow and reduce the flow velocity to a level which will not erode the receiving channel or area.

Planning Considerations

The outlets of pipes and structurally-lined channels are points of critical erosion potential. Stormwater which is transported through man-made conveyance systems at design capacity generally reaches a velocity which exceeds the ability of the receiving channel or area to resist erosion. To prevent scour at stormwater outlets, a flow-transition structure is required which will absorb the initial impact of the flow and reduce the flow velocity to a level which will not erode the receiving channel or area.

The most commonly-used structure for outlet protection is an erosion-resistant lined apron. These aprons are generally lined with loose rock riprap, grouted riprap or concrete. They are constructed at zero grade for a distance which is related to the outlet flow rate and the tailwater level. Criteria for designing these structures are contained in this standard.

Where the flow is excessive for the economical use of an

apron, excavated stilling basins may be used. Acceptable designs for stilling basins may be found in the following documents available from the U.S. Government Printing Office.

Hydraulic Design of Energy Dissipators for Culverts and Channels, Hydraulics Engineering Circular No. 14, U.S. Department of Transportation, Federal Highway Administration.

Hydraulic Design of Stilling Basins and Energy Dissipators, Engineering monograph No. 25 U.S. Department of Interior—Bureau of Reclamation.

Design Criteria

Structurally lined aprons at the outlets of pipes and paved channel sections shall be designed according to the criteria shown in Figure 6.3.

Tailwater—The depth of tailwater immediately below the pipe outlet must be determined for the design capacity of the pipe. Manning's Equation may be used to determine tailwater depth. If the tailwater depth is less than half the diameter of the outlet pipe, it shall be classified as a Minimum Tailwater Condition. If the tailwater depth is greater than half the pipe diameter, it shall be classified as a Maximum Tailwater Condition. Pipes which outlet to flat areas with no defined channel may be assumed to have a Minimum Tailwater Condition.

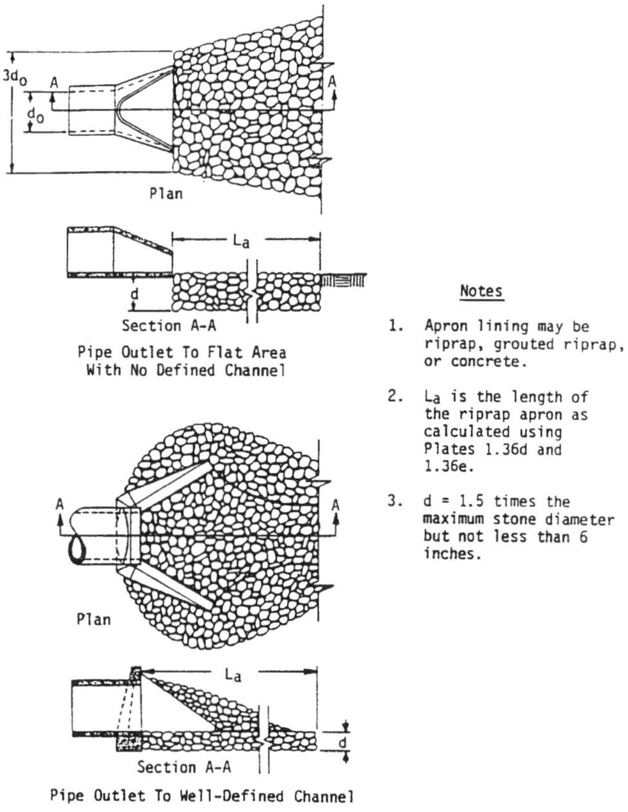

Figure 6.3. Pipe outlet conditions (USDA 2003).

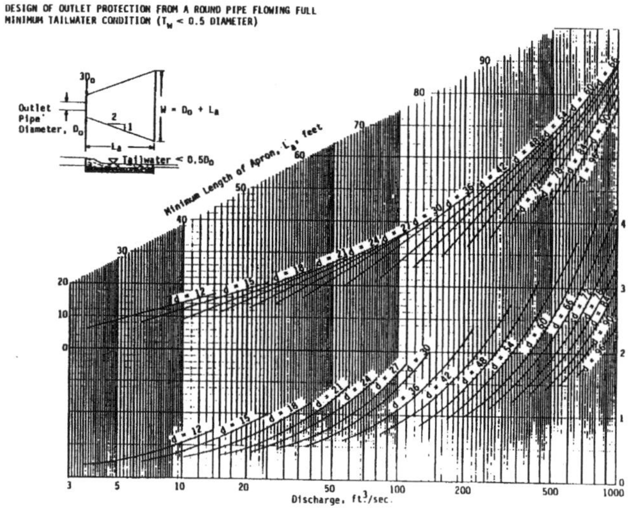

Figure 6.4. Outlet protection design for tailwater <0.5 diameter (USDA 2003).

Apron Length—The apron length shall be determined from Figure 6.4 or 6.5 according to the tailwater condition.

Apron Width—If the pipe discharges directly into a well-defined channel, the apron shall extend across the channel bottom and up the channel banks to an elevation one foot above the maximum tailwater depth or to the top of the bank (whichever is less).

If the pipe discharges onto a flat area with no defined channel, the width of the apron shall be determined as follows:

a. The upstream end of the apron, adjacent to the pipe, shall have a width three times the diameter of the outlet pipe.
b. For a Minimum Tailwater Condition, the downstream end of the apron shall have a width equal to the pipe diameter plus the length of the apron obtained from the figures.

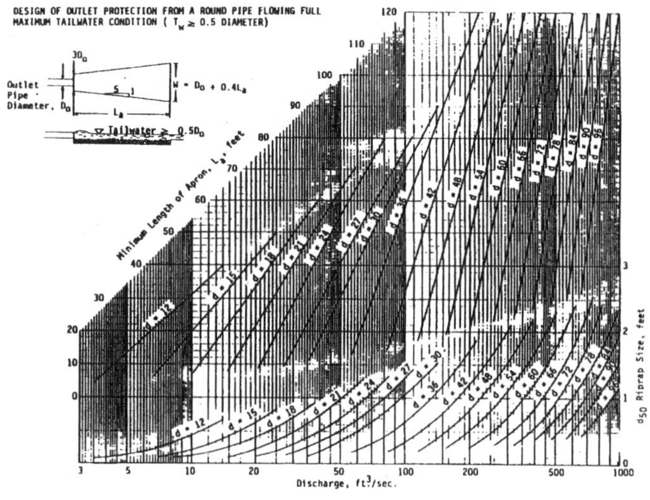

Figure 6.5. Outlet protection design for tailwater ≥0.5 diameter (USDA 2003).

c. For a Maximum Tailwater Condition, the downstream end shall have a width equal to the pipe diameter plus 0.4 times the length of the apron from Figures 6.4 or 6.5.

Bottom Grade—The apron shall be constructed with no slope along its length (0.0% grade). The invert elevation of the downstream end of the apron shall be equal to the elevation of the invert of the receiving channel. There shall be no overfall at the end of the apron.

Side Slope—If the pipe discharges into a well-defined channel, the side slopes of the channel shall not be steeper than 2:1 (Horizontal:Vertical).

Alignment—The apron shall be located so that there are no bends in the horizontal alignment.

Materials—The apron may be lined with loose rock riprap, grouted riprap, or concrete. The median sized stone for riprap shall be determined from the curves on Figure 6.4 and 6.5 according to the tailwater condition. The gradation, quality, and placement of riprap shall conform to Standard and Specification for Riprap presented earlier.

1. The flow velocity at the outlet of paved channels flowing at design capacity must not exceed the permissible velocity of the receiving channel.
2. The end of the paved channel shall merge smoothly with the receiving channel section. There shall be no overfall at the end of the paved section. Where the bottom width of the paved channel is narrower than the bottom width of the receiving channel, a transition section shall be provided. The maximum side divergence of the transition shall be 1 in 3F where:

F = Froude number = V/gD
V = Velocity at beginning of transition (ft/sec)
D = depth of flow at beginning of transition (ft)
g = 32.2 ft/sec^2

3. Bends or curves in the horizontal alignment of the transition are not allowed unless the Froude number (F) is 0.8 or less (implying supercritical flow), or the section is specifically designed for turbulent flow.

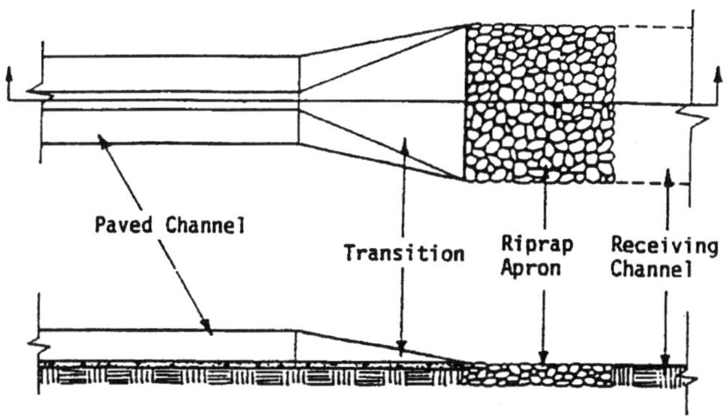

Figure 6.6. Paved channel outlet (USDA 2003).

Maintenance

Inspect riprap outlet after heavy rains to see if any erosion around or below the riprap has taken place or if stones have been dislodged. Immediately make all needed repairs to prevent further damage.

Construction Specification for Outlet Protection

Subgrade Preparation—Brush, trees, stumps, and other objetionable material shall be removed. The subgrade for the riprap or filter shall be prepared to the required lines and grades. Any fill required in the subgrade shall be compacted to a density approximating that of the surrounding undisturbed material.

Bedding or Filter Cloth—Filter bedding shall be placed to the depth, line and grade and in the manner specified. Geotextile shall be installed where specified and laid on the subgrade with sufficient slack so that it will not suffer extreme tension during placement of riprap or other linings.

Stone Placement—Placement of riprap should follow immediately after subgrade preparation. The riprap should be placed so that it produces a dense, well-graded mass of stone with a minimum of voids. The desired distribution of stones throughout the mass may be obtained by selective loading at the quarry, controlled dumping of successive loads during final placing, or by a combination of these methods.

The riprap should be placed to its full thickness in one operation, not in layers. The riprap should not be placed by dumping into chutes or similar methods which are likely to cause segregation of the various stone sizes. Care should be taken not to dislodge the underlying material when placing the stones. Some hand placing may be necessary to achieve the required grades and a good distribution of stone sizes.

Emergency Spillways

All detention ponds also must be equipped with emergency spillways. Mason (1982) states that the preferred location of an emergency spillway is on undisturbed ground, rather than over a prepared embankment, to reduce the erosion potential. Detention ponds treating runoff from small contributing areas can safely handle overflows as sheetflows through well-designed swales.

The NRCS guidelines for designing runoff control measures must be followed when designing emergency spillways for sediment ponds. In addition, if the detention pond is large, special regulations of the state and the Army Corps of Engineers must be followed.

SEDIMENT POND DESIGN FUNDAMENTALS

The basic design approaches for sediment ponds consider

Sediment pond at landfill.

Permanent pond acting as sediment trap during final construction.

Series of small sediment ponds at a complex construction site (Atlanta, GA).

Temporary pond at highway construction site in area where hauling trucks are washed prior to re-entering roads (WI).

the behavior of the water passing through the pond to be either plug flow or completely-mixed flow. Martin (1989) reviewed these flow regimes and conducted five tracer studies in a wet detention pond/wetland in Orlando, FL, to determine the actual flow patterns under several storm conditions. Completely-mixed flow conditions assume that the influent is completely and instantaneously mixed with the contents of the pond. The concentrations are therefore uniform throughout the pond. Under plug-flow conditions, the flow proceeds through the pond in an orderly manner, following streamlines and with equal velocity, i.e., the flow enters at a single time and travels through the pond to the outlet as a batch, displacing a slug of previously-captured water. The concentrations vary in the direction of flow and are uniform in cross section. The steady-state resident time for both flow patterns is the same, namely the pond volume divided by the discharge rate. Historically, wet detention ponds have been designed using the plug-flow concept, probably because it had been used in conventional clarifier designs for water and wastewater treatment. In reality, detention ponds exhibit a combination flow pattern that Martin terms moderately-mixed flow. He found that the type of mixing that actually occurs is dependent on the ratio of the storm volume to the pond storage volume (the flushing ratio). If the ratio is less than one, plug flow likely predominates. If the ratio is greater than one, the flow type is not as obvious. With faster moving water in the pond, short-circuiting may reduce the available pond storage volume (and therefore the resident time), resulting in less effective treatment.

Upflow Velocity

Linsley and Franzini (1964) stated that in order to get a fairly high percentage removal of particulates, it is necessary that a sedimentation pond be properly designed. In an ideal system, particles that do not settle below the bottom of the outlet will pass through the sedimentation pond, while particles that do settle below/before the outlet will be retained. The path of any particle is the vector sum of the water velocity (V) passing through the pond and the particle settling velocity (v).

Therefore, if the water velocity is slow (slower than the settling rate of the particles by gravity), slowly-falling particles can be retained, assuming the residence time is sufficiently long for the particle to settle below the outlet structure's drainage point. If the water velocity is fast, then only the heaviest (fastest-falling) particles are likely to be retained. The critical ratio of water velocity to particle settling velocity must therefore be equal to the ratio of the sedimentation pond length (L) to depth to the bottom of the outlet (D):

$$\frac{V}{v} = \frac{L}{D}$$

as shown on Figure 6.7.

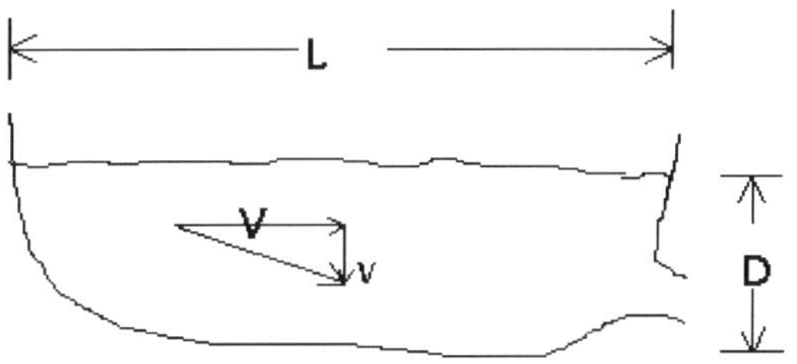

Figure 6.7. Critical Velocity and Pond Dimensions.

The water velocity is equal to the water volume rate (Q, often expressed in cubic feet per second) divided by the pond cross-sectional area (A_{cs}, or pond depth multiplied by pond width: DW):

$$V = \frac{Q}{A_{cs}}$$

or

$$V = \frac{Q}{DW}$$

The pond outflow rate equals the pond inflow rate under steady state conditions. The critical time period for steady-state conditions is the time of travel from the inlet to the outlet. During critical portions of a storm, the inflow rate (Q_{in}) will be greater than the outflow rate (Q_{out}) due to freeboard storage. Therefore, the outflow rate controls the water velocity through the pond. By substituting this definition of water velocity into the critical ratio:

$$\frac{Q_{out}}{WDv} = \frac{L}{D}$$

The water depth to the outlet bottom (D) cancels out, leaving:

$$\frac{Q_{out}}{Wv} = L$$

or

$$\frac{Q_{out}}{v} = LW$$

However, pond length (L) multiplied by pond width (W) equals pond surface area (A). Substituting leaves:

$$\frac{Q_{out}}{v} = A$$

and the definition of upflow velocity:

$$v = \frac{Q_{out}}{A}$$

where,

Q_{out} = pond outflow rate (cubic feet per second),
A = pond surface area (square feet: pond length times pond width), and
v = upflow velocity, or critical particle settling velocity (feet per second).

Therefore, for an ideal sedimentation pond, particles having settling velocities less than this upflow velocity will be removed. Only increasing the surface area, or decreasing the pond outflow rate, will increase pond settling efficiency. Increasing the pond depth does lessen the possibility of bottom scour, decreases the amount of attached aquatic plants, and decreases the chance of winter kill of fish. Deeper ponds may also be needed to provide sacrificial storage volumes for sediment between dredging operations. For construction-site sediment ponds, it should be assumed that inlet zones are restricted to the pond surface and that the outlet zones are full depth, providing a worst-case situation (as verified during field tests, such as during the Nationwide Urban Runoff Program, NURP, EPA 1983, where many ponds were monitored for several years).

For continuous flow conditions (such as for water or wastewater treatment), the following relationships can be shown:

$$t = \frac{Volume}{Flow\ rate}$$

and

$$Flow\ rate\ (Q_{out}) = \frac{Volume}{t}$$

where, t = detention (residence) time. With

$$v = \frac{Q_{out}}{A}$$

and substituting:

$$v = \frac{Volume}{(t)(A)}$$

but

$$Volume = (A)(depth)$$

therefore,

$$v = \frac{(A)(depth)}{(t)(A)}$$

leaving:

$$v = \frac{depth}{t}$$

It is seen that the surface overflow rate (Q/A) is equivalent to the ratio of pond depth to detention time. Therefore, it is not possible to predict pond performance by only specifying detention time. If the pond depth was also specified (or kept within a typical and narrow range), then the detention time could be used as a performance specification for a continuous or plug flow condition. However, it is not possible to hold all of the water in a detention pond for the specified detention time. Outlet devices typically release water at a high rate of flow when the pond stage is elevated (resulting in minimal detention times during peak flow conditions) and lower flow rates at lower stages, after most of the detained water has already been released. The average detention time is therefore difficult to determine and is likely very short for most of the water entering the pond during a moderate-to-large storm. For variable-flow stormwater conditions, it is much easier to design and predict pond performance using the surface overflow rate relationships.

The surface overflow rate (the ratio of outflow rate to pond surface area) can be kept constant (or less than a critical value) for all pond stages. This results in a substantially more direct method of designing or evaluating pond performance. Pond performance curves therefore can be easily prepared, where surface overflow rate (and therefore critical particle control) are related for all stages at a pond site.

Effects of Short-Circuiting on Particulate Removals in Wet Detention Ponds

Under dynamic conditions, particle trapping can be predicted using the basic Hazen theory presented by Fair and Geyer (1954) that considers short-circuiting effects:

$$\frac{y}{y_o} = 1 - \left[1 + \frac{v_o}{n(Q/A)}\right]$$

where,

- y_o = initial quantity of solids having settling velocity of v_o
- y = quantity of these particles removed
- y/y_o = proportion of particles removed having settling velocity of v_o
- Q = wet pond discharge
- A = wet pond surface area
- n = short-circuiting factor (number of hypothetical basins in series)

This equation is closely related to the basic upflow velocity equation (or surface overflow rate) developed previously. The short-circuiting factor is typically given a value of 1 for very poor conditions, 3 for good conditions, and 8 for very good conditions. Short-circuiting allows some large particles to be discharged that theoretically would be completely trapped in the pond. However, field monitoring of particle size distributions of detention pond effluent shows that this has a very small detrimental effect on the suspended solids (and pollutant) removal rate of a pond. Figure 6.8 shows the effects of different n values on the removal of particles having different settling rates (v) compared to the critical settling rate (Q/A). For a particle having a settling rate equal to the critical values (v = Q/A), the ideal settling indicates 100% removal, while for "best performance" ($n = \infty$), the actual removal would be only about 65%. If the pond had an n of 1 (very poor performance), the removal of this critical particle would be only 50%.

The degradation of performance is much worse for

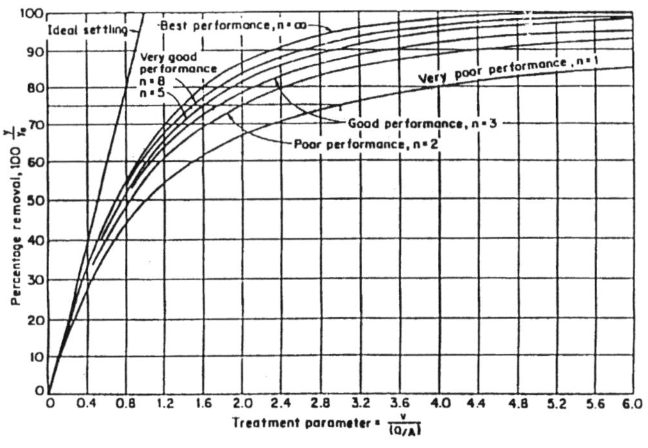

Figure 6.8. Performance curves for settling basins of varying effectiveness (AWWA, 1971, with permission).

particles having settling rates much larger than the critical rate. However, most wet detention ponds are greatly over-sized according to their ability to remove large particles, so this degraded performance has minimal effect on the overall suspended solids removal. The suggested detention pond design presented in this discussion only operates at the "design" stage (where the critical particle size is being removed) a few times a year. At all other times, the smallest particles being removed in the ponds are much smaller than the critical size used in the pond design. Almost all of the larger particles are effectively trapped because they are much larger than the design particle size (the pond is over-sized for these large particles), even if they are not being removed at their highest possible rate. In most cases, a few relatively large particles (much larger than the critical design particle size) will be observed in the pond effluent, but they have little effect on the overall SS removal.

Figure 6.9 shows example particle settling distributions

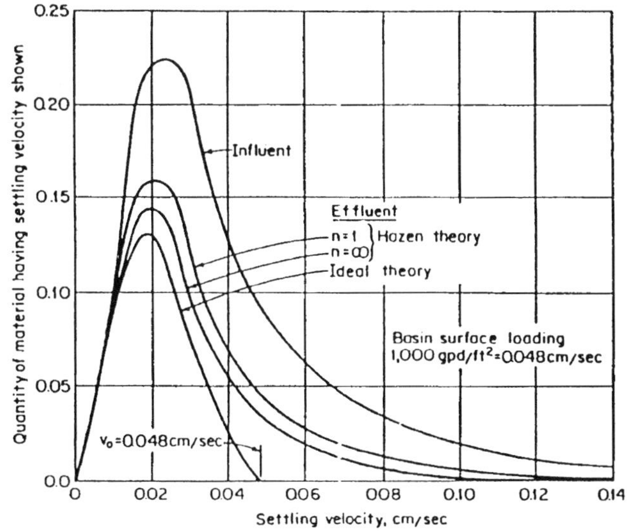

Figure 6.9. Influent and effluent particle settling rate distributions for settling basins of varying effectiveness (AWWA, 1971, with permission).

for a pond, comparing effluent conditions using the short-circuiting effects of Hazen's theory. The most common particle size (the mode of the distribution) changes very little for the different effluent conditions. However, there are more larger-sized particles present in the effluent using Hazen's theory compared to the ideal theory, and the median size obviously increases as the value for n decreases.

Very little degraded performance was observed at a pond monitored during NURP (EPA, 1983) in Lansing, MI, that was expected to have significant short-circuiting. A golf course pond located across the street from a commercial strip was converted into a stormwater pond, but the inlets and outlets were adjacent to each other in order to reduce construction costs. It was assumed that severe short-circuiting would occur because of the close proximity of the inlet and outlet, but the pond produced suspended-solids removals close to what was theoretically predicted, and similar to other ponds having similar pond-area-to-watershed-area ratios. Actually, the close inlet and outlet may have resulted in less short-circuiting because the momentum of the inflowing waters may have forced the water to travel in a general circular pattern around the pond, instead of directly flowing across the pond (and "missing" some edge area) as would be expected if the outlet was located at the opposite side of the pond.

Forty-seven events were studied at the Madison, WI, Monroe St, wet detention pond to find the short-circuiting *n* factors using observed and predicted particle size distributions in effluent water. Particle size distributions were measured using the Sedigraph method at the USGS Denver laboratory. This technique measures settling rates of different-size suspended-solid particulates down to 2 μm. The value of n is calculated using the concentrations of large particles that are found in the effluent. In ideal settling, no particles greater than the theoretical critical size (about 5 μm for Monroe St.) should appear in the effluent. However, there are always a small number of larger particles in the effluent. Generally, it is assumed that short-circuiting is responsible for these large particles. The measured values for n were one, or less, indicating a high degree of shortcircuiting in the pond. However, these observations were possibly affected by scour of bottom deposits near the subsurface effluent pipes. The maximum effect of short-circuiting on pond performance is shown in the following table, which shows the average reduction in suspended solids removals for different n values compared to the best performance (*n* value equal to 8):

n Value	% SS Removal (average)	Reduction in % SS Removal Compared to *n* = 8
8	85	
3	84	1
1	80.7	4.3
0.5	78.5	6.5
0.2	59	26

The calculated values of n (based on matching measured effluent particle size distributions with distributions calculated using different values of n) ranged from about 0.2 to 1, indicating "very poor performance", or worse. The median value of n observed was about 0.35, indicating degradation in the annual average suspended-solids capture efficiency of no more than about 10 percent. The effects of this shortcircuiting, even with the extremely-low values of n for the Monroe St. pond, only has a minimal effect on the suspended-solids percentage removals. The Monroe St. pond provided an average suspended solids reduction of 87%, compared to the design goal of 90% according to extensive monitoring results. These values are quite close and the short-circuiting has a negligible effect on actual performance, as the pond surface is relatively large (0.6% of the drainage area) and the outlets were efficiently modified during the retrofitting activities.

Although the pond is producing very good suspended-solids removals as designed, the particle size distributions of the effluent indicate some short-circuiting (some large particles are escaping from the pond). The short-circuiting has not significantly reduced the effectiveness of the pond (measured as the percentage of suspended solids captured). Therefore, care should be taken in locating and shaping ponds to minimize short-circuiting problems, but not at the expense of other more important factors (especially size, or constructing the pond at all). Poor pond shapes probably cause greater problems by producing stagnant areas where severe aesthetic and nuisance problems originate.

Residence Time and Extended Detention Ponds

Residence time is defined as the ratio of volume to average flow rate (volume divided by volume per time). It can be assumed to be the average length of time any parcel of water remains in the pond. As in any pond performance measure or design criteria, residence time values are very dependent on good pond configurations. Harrington (1986) stresses the need to subtract pond "dead zones" from pond volume when calculating residence times. Dead zones (and associated short-circuiting) can significantly reduce pond effectiveness.

Designing a wet pond for the treatment of runoff based on residence time alone is usually not recommended. Barfield (1986) states that residence (detention) time is not a good criteria for pond performance, but the ratio of peak discharge rate to pond surface area (the peak upflow velocity) is a good criteria of performance. The state of Maryland uses a residence time standard as part of their design criteria for "extended detention" ponds. These ponds are normally dry between events, or have a small and shallow wet pond area near the outlet, and greatly extend in surface area during storms. For these types of ponds, Harrington (1986) found, through computer modeling studies, that a residence time of about nine days is needed to achieve a 70 percent reduction of particulate residue. Nine days is longer than the interevent period for most rains in the midwest and the southeast (about three to five days). These types of ponds therefore are not expected to be very useful for locations where the interevent periods of rains is short, or the drain-down time of the pond is rapid.

Unfortunately, dry ponds usually do not allow permanent retention of the settled particles. Subsequent storms usually scour the fine particles previously settled to the pond bottom. As stated previously, dry detention ponds have not been shown to be consistently effective in water quality control. The use of a small permanently-wet detention pond or wetland at the downstream end of a dry detention pond could help recapture some of these scoured particles. A wet detention pond located immediately upstream of a dry pond is usually a much better solution, as the wet pond would act as a pre-treatment pond, keeping particles and debris out of the dry pond which should be designed for peak flow rate reductions.

The previous discussion on upflow velocity as a design criteria illustrated the relationship between particle settling rates and upflow velocity, while this discussion showed the relationship between particle settling rates and residence times. A relationship therefore must exist between residence time and upflow velocity. Residence time is dependent on pond volume and outlet rate, while upflow velocity is dependent on pond surface area and outflow rate. The relationship between residence time and upflow velocity therefore is equal to the relationship between pond volume and pond surface area, or the pond depth. When a pond depth of five feet is used, the residence times of ponds designed using the upflow velocity method are generally the same residence times needed for similar control levels using the residence time criteria. Even though the two procedures result in the same basic design, it is still recommended that the upflow procedure be used for evaluating wet detention ponds during storm events. The depth and configuration design criteria are very critical for the other pond uses (aquatic life, aesthetics, and safety, besides scour prevention) and they should not be varied as part of the major design elements.

Runoff Particle Size Distributions

Knowing the settling velocity characteristics associated with stormwater particulates is necessary when designing wet detention ponds. Particle size is directly related to settling velocity (using Stokes law, for example, and using appropriate shape factors, specific gravity and viscosity values), and settling velocity usually is used in the design of detention facilities. Particle size also can be more rapidly measured in the laboratory than settling velocities. Settling tests for stormwater particulates need to be conducted for about three days in order to quantify the smallest particles

Different Methods to Characterize Particle Size and Settling Rates

Cascading sieves (with total solids analyses after each sieve).

Andreseen pipette (miniature settling column).

Coulter Counter Multi-Sizer IIe.

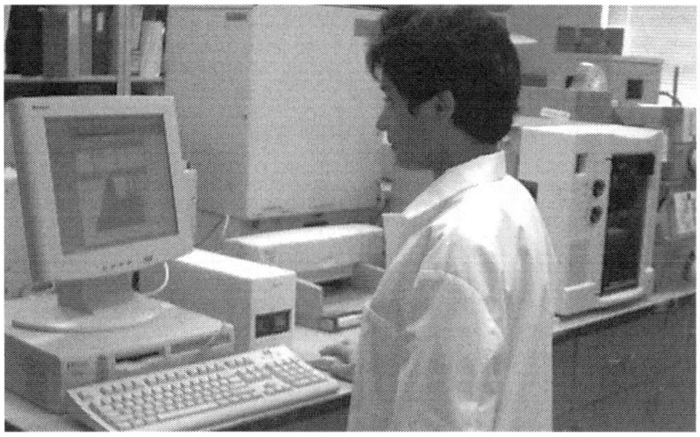

Coulter Counter Multi-Sizer 3.

Multi-Sizer 3 aperture tube and stirrer.

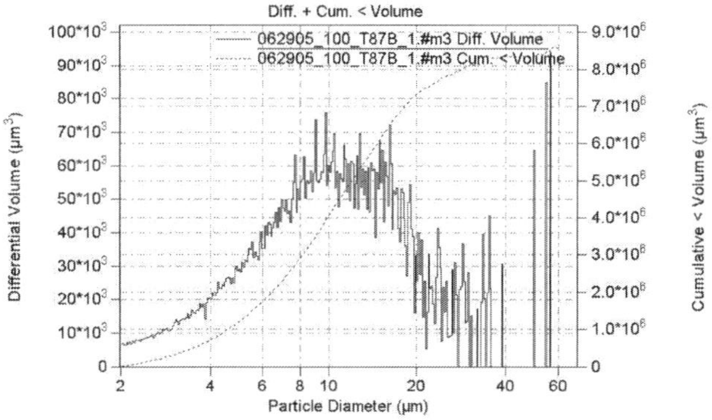

Multi-Sizer 3 computer display of particle size distribution.

(continued)

Different Methods to Characterize Particle Size and Settling Rates ***(continued)***

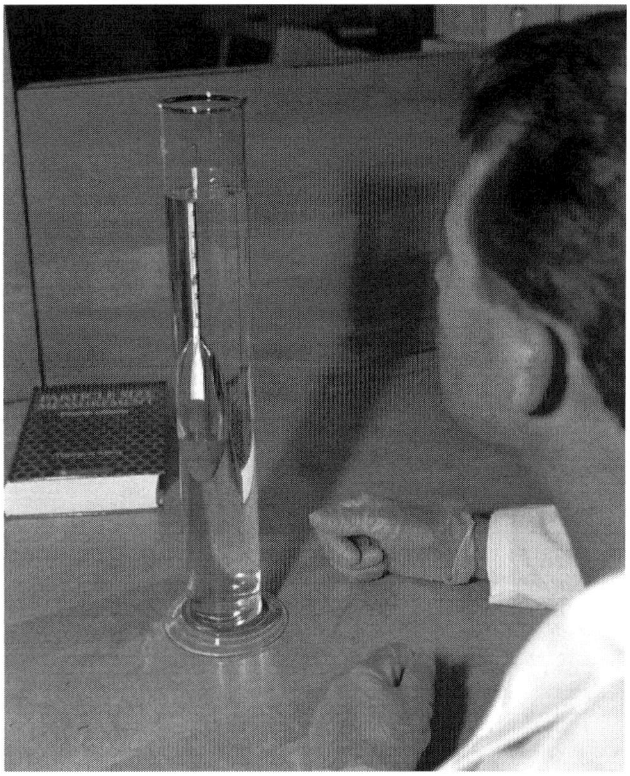

Pipette for high solids loadings.

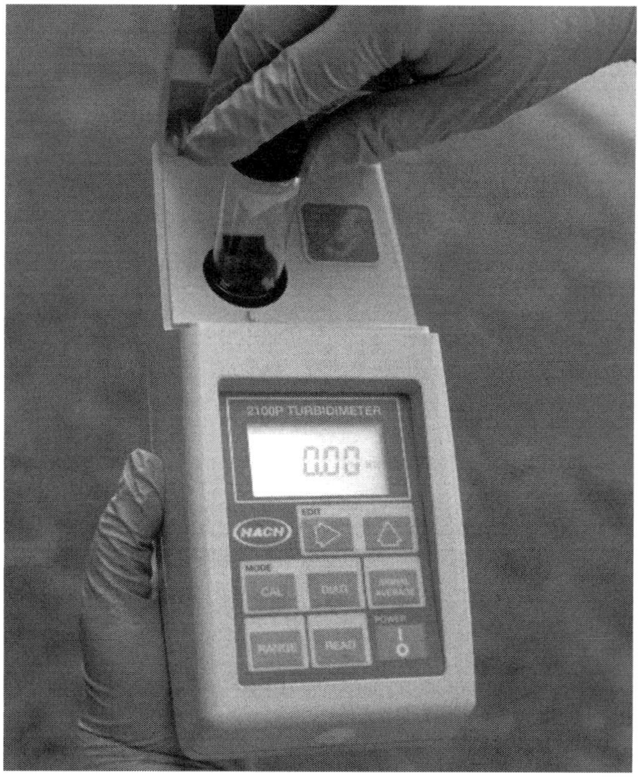

Field turbidimeter .

Teflon™ settling column.

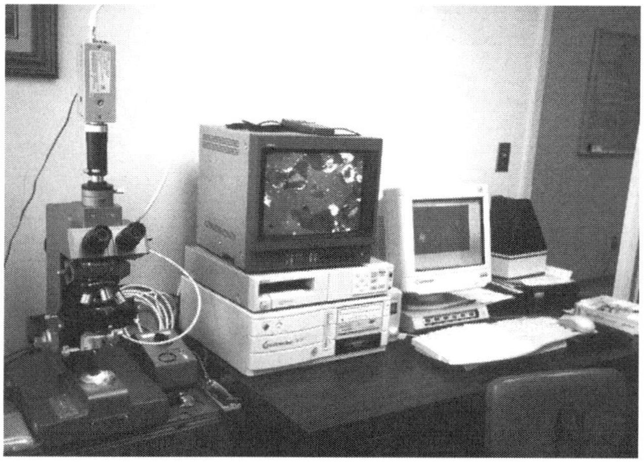

Research light microscope with automatic video analyses of particles.

that are of interest in the design of wet detention ponds. Probably the earliest description of conventional particle settling tests for stormwater samples was made by Whipple and Hunter (1981).

Whipple and Hunter (1981) contradict the assumption sometimes used in modeling detention pond performance that pollutants generally settle out in proportion to their concentrations (first-order rate equations). However, Grizzard and Randall (1986) have shown a relationship between particulate concentrations and particle size distributions. High particulate concentrations were found to be associated with particle-size distributions that had relatively high quantities of larger particulates, in contrast to waters having low particulate concentrations. The high-particulate-concentration water therefore would have increased particulate removals in detention ponds. This relationship is expected to be applicable for pollutants found mostly in particulate forms (such as suspended solids and most heavy metals), but the relationship between concentration and settling would be much poorer for pollutants that are mostly in soluble forms (such as filterable residue, chlorides and most nutrients). Therefore, the partitioning of specific pollutants between the "particulate" and "dissolved" forms, and eventually for different particulate size fractions, is needed.

Smith (1982) also states that settleability characteristics of the pollutants, especially their particle size distribution, are needed before detention pond analyses can be made. Kamedulski and McCuen (1979) report that as the fraction of larger particles increase, the fraction of the pollutant load that settles also increases. Randall, *et al.* (1982), in settleability tests of urban runoff, found that non-filterable residue (suspended solids) behaves like a mixture of discrete and flocculant particles. The discrete particles settled out rapidly, while the flocculant particles were very slow to settle out. Therefore, simple particle size information may not be sufficient when flocculant particles are also present. Particle size analyses could be supplemented with microscopic examinations to examine the extent of potential flocculation. Flocs can be readily distinguished from discrete particles due their nebulous characteristic in contrast to discrete grains.

Approximate stormwater particle size distributions derived from several upper Midwest and Ontario analyses, from all of the NURP data (Driscoll, 1986), and for several eastern U.S. sites that reflect various suspended solids (residue) concentrations are shown in Figure 6.10 (Grizzard and Randall, 1986). Pitt and McLean (1986) microscopically measured the particles in selected stormwater samples collected during the Humber River Pilot Watershed Study in Toronto. The upper Midwest data sources were from two NURP projects: Terstriep, *et al.* (1982), in Champaign/Urbana Ill. and Akeley (1980) in Washtenaw County, Michigan.

Tests have also been conducted to examine the routing of particles through the Monroe St. detention pond in Madison, Wisconsin (Roger Bannerman, Wisconsin Department of Natural Resources, personal communication). This detention pond serves an area that is mostly comprised of medium-density residential land uses, with some strip commercial areas. This joint project of the Wisconsin Department of Natural Resources and the U.S. Geological Survey has obtained a number of inlet and outlet particle-size distributions for a wide variety of storms, and included inlet bedload contributions. The observed median particle sizes for the inlet samples ranged from about 2 to 26 μm, with an average of 9 μm. Table 6.2 shows the average particle sizes corresponding to various distribution percentages for the Monroe St. outfall for inlet samples:

TABLE 6.2. Average Particle Sizes Corresponding to Various Distribution Percentages.

Percent Larger than Sze	Particle Size (μm)
10%	450
25	97
50	9.1
75	2.3
90	0.8

These distributions included bedload material that was also sampled and analyzed during these tests. Figure 6.11 shows the particle size distribution for the inflow events, including bedload, for a series of about 50 runoff events at the Monroe St. detention pond in Madison, WI. The median size is about 8 μm, but it ranges from about 2 to 30 μm. About 10% of the particles may be larger than 400 μm. The largest particle size observed was larger than 2 mm. The bedload material added about 10% of the mass of these particulates and was associated with the largest sizes. The settling velocities of discrete particles can be predicted using Stoke's and Newton's settling equations. Typically more than 90% of all stormwater particulates (by volume and mass) are in the 1 to 100 μm range, corresponding to low

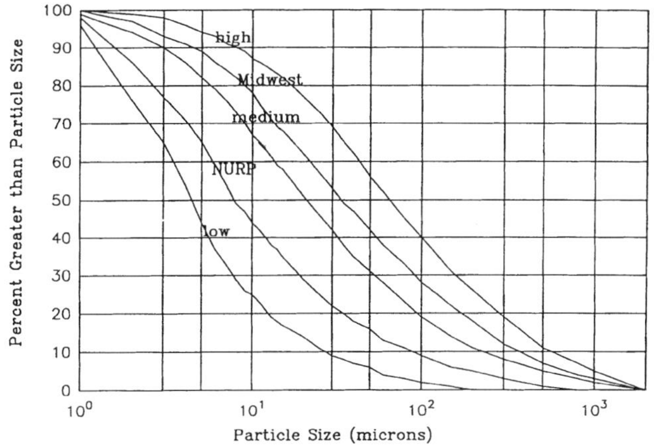

Figure 6.10. Particle size distributions for various stormwater sample groups.

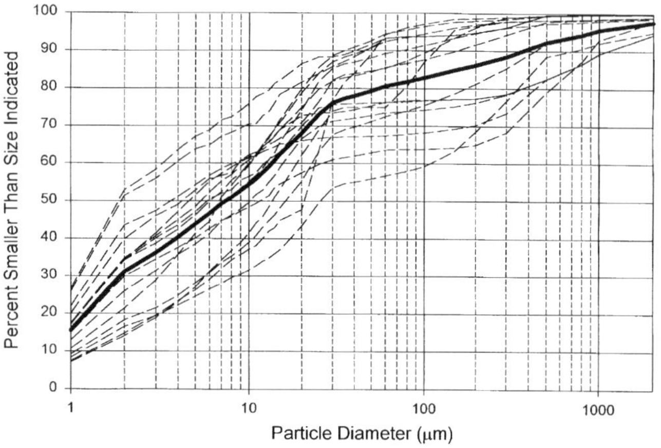

Figure 6.11. Inlet particle size distributions observed at the Monroe St. wet detention pond.

Reynolds's numbers and laminar flow conditions, as required for Stoke's law settling rate calculations. In most cases, stormwater particulates have specific gravities in the range of 1.5 to 2.5 (determined by conducting settling column, sieving, and microscopic evaluations of the samples, in addition to the particle counting), corresponding to a relatively narrow range of settling rates for a specific particle size.

Some data are also available concerning the particle-size distribution of erosion runoff from construction sites. Hittman (1976) reported erosion runoff having about 70 percent of the particles (by weight) in the clay fraction (less than 2 μm), while the exposed soil that is being eroded only had about 15 to 25 percent of the particles (by weight) in the clay fraction. When the available data is examined, it is apparent that many factors affect construction-site erosion runoff particle sizes. Rain characteristics, soil type, and on-site erosion controls are all important. The typical distribution is generally comparable to the "all NURP" particle size distribution presented previously. Table 6.3 lists the critical particle sizes corresponding to the 50 and 90 percent control values are as follows for the different data groups:

As dicussed in previous chapters, many Alabama and southeastern U.S. areas experience severe erosion problems. For example, in addition to high rain energy, many Alabama soils also are highly erosive and result in construction site runoff that is very difficult to control. About 70 construction site erosion samples were collected in the Birmingham area by Nelson (1996) and Pitt (1998). The characteristics of this runoff include the following:

- Measured suspended solids concentrations ranged from 100 to more than 25,000 mg/L (overall median about 4,000 mg/L).
- Turbidity ranged from about 300 to >50,000 NTU, with an average of about 4,000 NTU
- Particle sizes: 90% were smaller than about 20 μm (0.02 mm) in diameter and median size was about 5 μm (0.005 mm).
- Measured Birmingham construction site erosion discharges range from about 100 to 300 tons/acre/year

TABLE 6.3. Critical Particle Sizes Corresponding to the 50 and 90 Percent Control Values.

	90%	50%
Monroe St.	0.8 μm	9.1 μm
All NURP	1	8
Midwest	3.2	34
Low solids conc.	1.4	4.4
Medium solids conc.	3.1	21
High solids conc.	8	66

TABLE 6.4. Relationships between Rain Conditions and the Observed Runoff Quantity.

Measured Conditions:	Low Intensity Rains (<0.25 in/hr)	Moderate Intensity Rains (about 0.25 in/hr)	High Intensity Rains (>1 in/hr)
Suspended solids, mg/L	400	2,000	25,000
Particle size (median), μm	3.5	5	8.5

Nelson, 1996 and Pitt, 1998.

There were obvious relationships between rain conditions and the observed runoff quality during these local Birmingham studies, Table 6.4.

These construction site data would therefore correspond to the "low," or "all NURP" particle size distributions. The particle size distribution of material leaving construction sites is therefore quite small and hard to control. Small particle sizes are much more difficult to remove by most erosion control strategies, which usually employ sedimentation (sediment ponds and "filter" fences) without chemical addition. Particle sizes or associated settling velocities are used with the desired outflow rate to determine the required surface area for a sediment pond.

These data show that construction site runoff likely has smaller particle-size distributions than most stormwater; construction-site runoff has median sizes generally in the range of 3 to 8 μm, while stormwater at many locations contains larger particles, with median sizes from about 8 to 65 μm.

Particle Settling Velocities

The settling velocities of discrete particles are shown in Figure 6.12, based on Stoke's and Newton's settling relationships. It is likely that more than 90% of all runoff particulates are in the 1 to 100 μm range, corresponding to particles that will settle with low Reynolds's numbers, and hence laminar flow conditions, and the settling rates can therefore be calculated using Stoke's law. This figure also

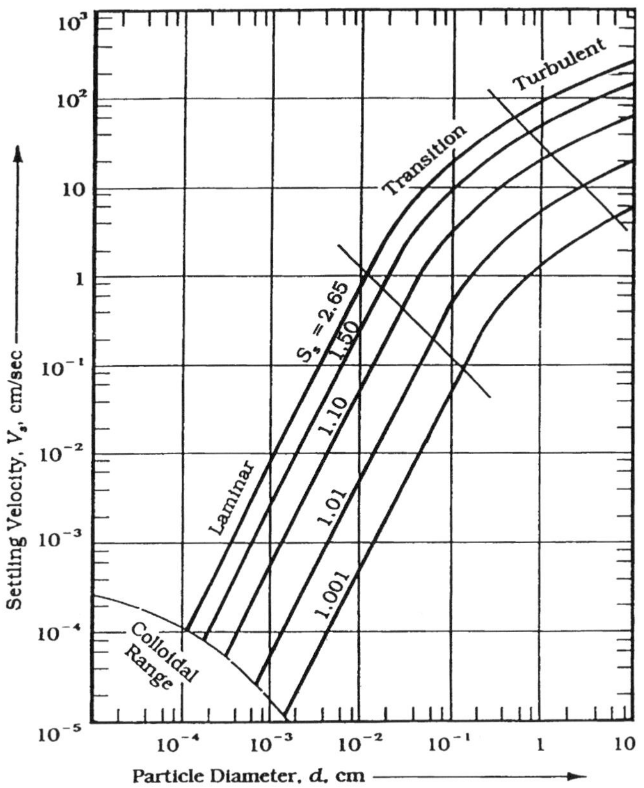

Figure 6.12. Type 1 (discrete) settling of spheres in water at 10°C.

illustrates the effects of different specific gravities on the settling rates. In most cases, stormwater particulates have specific gravities in the range of 1.5 to 2.5, while construction site runoff particles would be closer to 2.5. This corresponds to a relatively narrow range of settling rates for a specific particle size. Particle size is much easier to measure than settling rates. It is generally recommended to measure particle sizes using automated particle sizing equipment (such as a Coulter Counter Multi-Sizer III) and to conduct periodic settling column tests to determine the corresponding specific gravities. If the particle counting equipment is not available, then small-scale settling column tests (using 50 cm diameter Teflon™ columns about 0.7 m long) can be used. Sieve measurements, another method for creating particle size distributions, are limited to sizes greater than about 20 μm, although precision membrane filters can be used for much smaller sizes.

Particle settling observations in actual detention ponds have generally confirmed the ability of well-designed and well-operated detention ponds to capture the "design" particles. Gietz (1983) found that particles smaller than 20 μm predominated (comprised between 50 to 70 percent of the sediment) at the outlet end of a "long" monitored pond, while they only made up about ten to 15 percent of the sediment at the inlet end. Particles between 20 and 40 μm were generally uniformly distributed throughout the pond length, and particles greater than 40 μm were only found in the upper (inlet) areas of the pond. As a side note to pond performance, smaller particles also were resuspended during certain events.

Design Based on NURP Detention Pond Monitoring Results

The EPA (1983) determined that long-term detention pond performance could be estimated based on geographical location and the ratio of the pond surface area to contributing source area. Driscoll (1989) and EPA (1986) presented a basic methodology for the design and analysis of wet detention ponds. A pond operates under dynamic conditions when the storage of the pond is increasing with runoff entering the pond and with the stage rising, and when the storage is decreasing when the pond stage is lowering. Quiescent settling occurs during the dry period between storms when storage is relatively constant and when the previous flows are trapped in the pond, before being partially or completely displaced by the next storm. The relative importance of the two settling periods depends on the size of the pond, the volume of each runoff event, and the inter-event time between the rains.

Driscoll (1989) produced a summary curve (Figure 6.13) that relates wet-pond performance to the ratio of the pond surface area to the drainage area, based on the numerous NURP wet detention pond observations. The NURP ponds were in predominately residential areas and were drained with conventional curbs and gutters. This figure indicates that wet ponds from about 0.3 to 0.8 percent of the drainage area should produce about 90% reductions in suspended solids. Southeastern ponds need to be larger than ponds in the Rocky Mountain region because of the substantially larger quantities of rain and the increased size of the individual events in the southeast. Also, wet ponds designed to remove 90% of the suspended solids need to be about twice as large as ponds with only a 75% suspended solids removal objective.

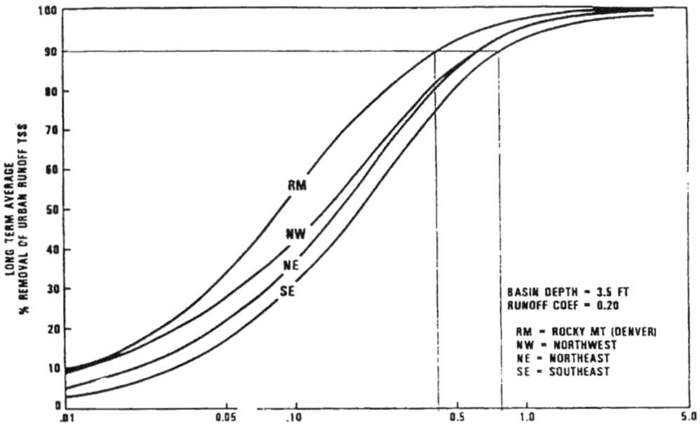

Figure 6.13. Regional differences in detention pond performance (EPA, 1983).

Introduction to the Storage-Indication Method

The discharged water from a detention pond is simply displaced pond water. In some cases, observed outlet water characteristics during a specific storm cannot be related to the inlet water characteristics. If the storm is small, the volume of water coming into the pond can be substantially less than the resident water in the pond. In these cases, the outlet water is mostly "left-over" water from a previous event or from relatively low volume (but long duration) baseflows that had previously entered the pond since the last storm. However, if the storm is large, then the water being discharged from the pond is mostly related to the specific event. Therefore, analyses of detention pond behavior must consider the relative displacement of pond water. Long-term continuous analyses comparing many adjacent storms resulting in seasonal inlet and outlet flows of pollutants may be more appropriate than monitoring simple paired samples of inlet and effluent flows during random events spread over time.

The following discussion on routing includes a fairly simple procedure to examine these pond water displacement considerations and their effects on particulate trapping. The Source Loading and Management Model (WinSLAMM) and the Detention Pond Analysis model (WinDETPOND) include a computerized version of the storage-indication method (www.WinSLAMM.com). The pond routing calculation procedure presented in the remainder of this section is based on the NRCS Technical Release-20 (TR-20) procedures (SCS, 1982), as presented by McCuen (1982). The reservoir routing subroutine in TR-20 (RESVOR) is based on the storage equation:

$$I - O = \frac{\Delta S}{\Delta T}$$

where I is the pond inflow and O is the pond outflow. The difference between the inflow and outflow must be equal to the change in pond storage per unit of time ($\Delta S/\Delta T$). McCuen presents a series of equations and their solutions that require the preparation of a "storage-indication" curve to produce the pond outflow hydrograph. The storage-indication curve is a plot of pond outflow (O) against the corresponding pond storage at that outflow (S) plus 1/2 of the outflow multiplied by the time increment. When the pond outflow hydrograph is developed, the upflow velocity procedure described earlier can be used to estimate pond pollutant removal and peak flow rate reduction performance.

The relationship between the pond stage and the surface area for the pond under study is also needed in order to calculate the storage volume available for specific pond stages. Figure 6.14 is an example stage-area curve developed from topographic maps of the Monroe Street detention pond in Madison, Wisconsin. The normal pond wet surface is at 13 feet (arbitrary datum) and the emergency spillway is located at 16 feet, for a resultant useable stage range of three feet.

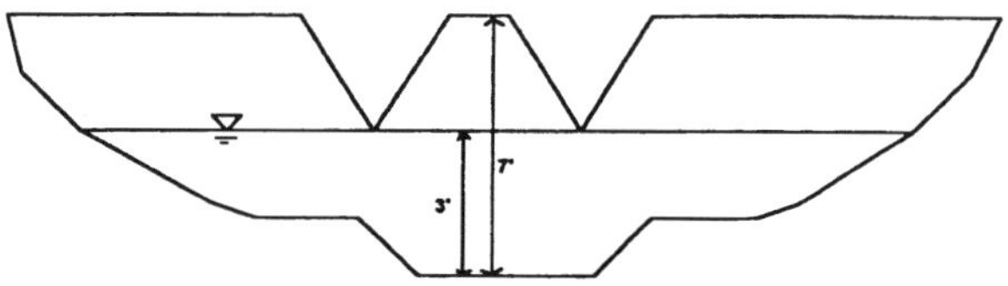

Diagram of Example Pond with Two 90 Degree V-Notch Weirs

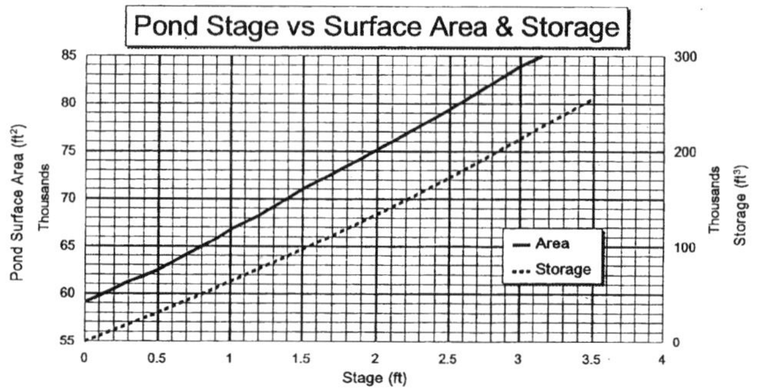

Figure 6.14. Pond-stage surface area relationship for example problem.

Table 6.5 shows the calculations used to produce the storage-indication figure (Figure 6.15) for the Monroe St. pond. This example reflects some pond modifications that were made to enhance pond performance: two 90° V-notch weirs, which increased the maximum stage range to 3.5 feet available before the emergency spillway is activated. The storage calculations assume an initial storage value of zero at the bottom of the V-notch weirs (13.0 feet). The time increment used in these calculations is ten minutes, or 600 seconds. The storage-indication curve shown as Figure 6.15 is therefore a plot of pond outflow (cfs) verses pond storage plus 300 (1/2 of 600 seconds) times the outflow rate. The storage-indication figure must also include the stage versus outflow and storage versus outflow curves (also from Table 6.5).

Design of Wet Detention Ponds for the Control of Construction Site Sediment

A wet detention pond performance specification for water quality control has two objectives: (1) to result in a consistent level of protection for a variety of conditions, and (2) to allow a site engineer a range of options to best fit the needs of the site. The pond design also must be easily evaluated by the reviewing agency and be capable of being integrated into the complete stormwater management program for the watershed. It should have minimal effects on the hydraulic routing of stormwater flows, unless a watershed-wide hydraulic analysis is available that specifies the specific hydraulic effects needed at the specific location.

TABLE 6.5. Calculation of Storage-Indication Relationships for Example Pond and 1.5.Inch, 3-Hour Rain.

Datum Stage (*H*) (ft)	Discharge Rate[1] (*O*) (ft^3/sec)	Surface Area (ft^2)	Storage (*S*) (ft^2)	*S* + 1/2*O*Δ*t* (see footnote 2)
0	0	59,100	0	0
0.1	0.016	59,800	5,980	5,985
0.2	0.09	60,500	12,100	12,130
0.3	0.25	61,250	18,375	18,450
0.4	0.51	61,850	24,740	24,890
0.5	0.88	62,520	31,260	31,520
0.6	1.4	63,300	37,980	38,400
0.7	2.1	64,200	44,940	45,570
0.8	2.9	65,000	52,000	52,870
0.9	3.8	65,800	59,200	60,340
1.0	5.0	66,767	66,770	68,270
1.2	7.9	68,300	82,000	84,370
1.5	14	71,000	107,000	111,200
1.8	22	73,500	130,000	136,600
2.0	28	75,148	150,300	158,700
2.5	49	79,400	200,000	214,700
3.0	78	83,928	251,800	275,200
3.5	115	87,500	306,300	340,800

[1]Using two 90° V-notch weirs: $Q = 2(2.5H^{2.5})$.
[2]$S + 1/2O\Delta t = S + O\ (1/2\Delta t) = S + 300\ (O)$, $\Delta t = 600$ seconds.

The following suggested specifications should meet these objectives under most conditions. However, the specific pond sizes should be confirmed through continuous long-term simulations using many years of actual rainfall records for the area of interest (such modeling is possible by using WinDETPOND [available at www.WinSLAMM.com]). These guidelines therefore should be considered as a starting point and modified for specific local conditions. As an example, it may be desirable to provide less treatment than suggested by the following guidelines (Vignoles and Herremans 1996). The following guidelines were developed by Pitt (1993a and 1993b), based on literature information and on his personal experience.

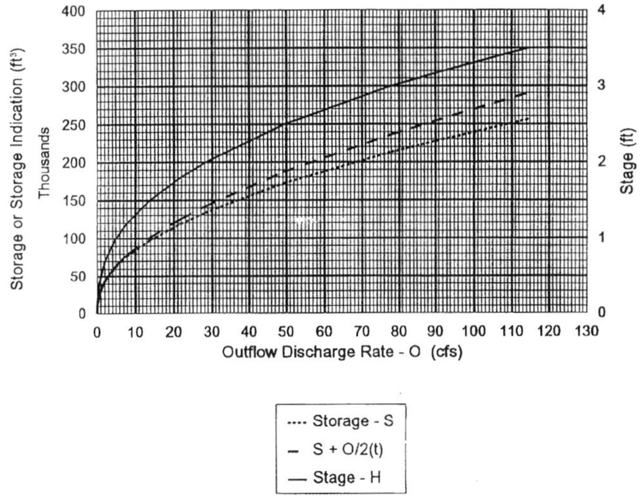

Figure 6.15. Pond-stage/storage indication curve for example problem.

TABLE 6.6. Estimates of Pond Area Based on Drainage Area Characteristics.

	Example Land Area	Pond Size Factor	Resulting Pond Area for Example
Paved area	0.6 acres	3%	0.018 acres
Undeveloped area	3.8 acres	0.6%	0.023 acres
Construction area	27.6 acres	1.5%	0.414 acres
Total	32.0 acres		0.455 acres

(1) Minimum Water Surface Area

The wet pond should have a minimum water surface area corresponding to land use and desired pollutant control. This is usually the most important aspect of the pond design that affects the pond performance. The following values were extrapolated from extensive wet detention pond monitoring, mainly from the EPA's NURP (EPA 1983) studies and other research. For construction sites, these required pond areas are 1.5% of the drainage area for approximately 90% control (arbitrary 5 μm) and 0.5% for 65% control (arbitrary 20 μm). If any undeveloped areas are in the pond drainage, the pond area would have to be increased in area by about 0.6% of those areas. Similarly, if any paved areas were in the drainage, the increase in pond area would need to be 3% of the paved area. Obviously, to be most efficient, any extra drainage areas should be kept to a minimum. Table 6.6 shows how the pond area can be estimated based on drainage area characteristics:

As will be shown in the following example, the total land area needed for the pond will be substantially larger than this value, as this area is the pond surface area during dry weather. The pond freeboard volume (for water quality control), plus the emergency spillway area, will increase the needed area dedicated for the pond.

(2) Pond Live Storage

The pond live storage (freeboard storage above the permanent pool elevation) should be equal to the runoff associated with a 1.25 inch rain from the drainage area for the land use and development type. It should be noted that this storage volume is associated with the runoff volume from a specific type of rain and not for a set runoff volume. This has the benefit of providing the same level of control for all land uses. As an example, many ordinances require capture and treatment of the first 0.5 inch, or 1 inch, of runoff for an area. Unfortunately, this has the effect of providing very uneven levels of control because of different rainfall-runoff characteristics for different land uses. As an example, a residential area may require a rain of about 1.50 inches to produce 0.5 inches of runoff. However, a commercial area, such as a strip commercial development, would only require a rain of about 0.6 inches to produce 0.5 inches of runoff. It is

TABLE 6.7. Estimated Pond Storage Volume Based on Drainage Characteristics.

	Example Land Area	Pond WQ Volume Factor[1]	Resulting Pond WQ Volume for Example
Paved area	0.6 acres	1.1 inches	0.66 acre-inches
Undeveloped area (clayey)	3.8 acres	0.3 inches	1.14 acre-inches
Construction area (clayey)	27.6 acres	0.6 inches	16.56 acre-inches
Total:	32.0 acres		18.36 acre-inches (1.53 acre-ft)

[1]If sandy soils, the pond water quality volume factors would be: paved areas: 1.1 inches (the same); undeveloped areas: 0.1 inches; and construction areas: 0.5 inches.

obvious that the residential area is providing treatment for a much more severe rain, with a correspondingly greater level of annual control, compared to the commercial area, the opposite of what should probably occur. By requiring a set level of control associated with a rain having the same recurrence interval, a more consistent effort and benefit is obtained throughout the community. About 0.3 inches of runoff would occur at construction sites for sandy soil areas and about 0.6 inches of runoff for clayey soil areas for this rain depth. Again, if other land areas are also in the drainage in addition to the construction area, the pond treatment volume would have to be increased. For any paved areas, the 1.25 inch rain would produce about 1.1 inches of runoff, and for undeveloped areas, the 1.25 inch rain would produce about 0.1 (for sandy soils) to 0.3 (for clayey soils) inches of runoff. Table 6.7 shows how the pond storage volume can be estimated based on drainage area characteristics.

Figure 6.16 is a schematic showing a cross section of the pond. The area below the invert of the lowest discharge device is the dead storage and is provided to store sediment and minimize scour of the retained particulates. At least 3 ft of "dead storage" water must be over the maximum stored sediment depth to minimize scour during large storm events. The water quality storage volume in the detention pond is the volume associated with the runoff associated with a 1.25 inch rain. The topmost layer in the detention pond is additional storage that is provided for drainage benefits. This

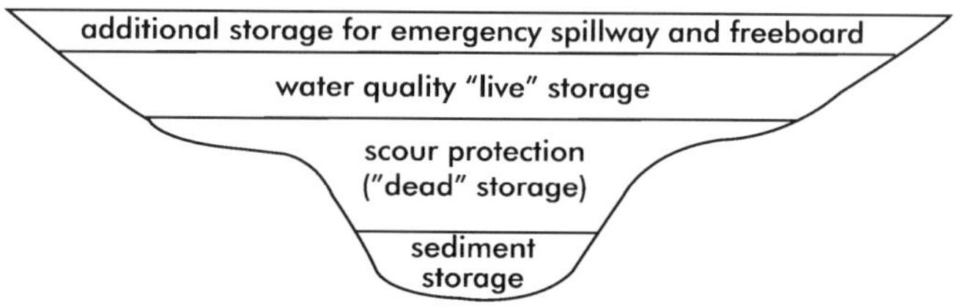

Figure 6.16. Cross-section of pond showing water quality storage portion of storage, along with other pond storage components.

storage would be provided (with the appropriate additional outlet structure) only if a basin-wide hydraulic analyses has been conducted to ensure that inappropriate interactions of the different flood hydrographs would not occur. Also, it is important to note that an emergency spillway also must be provided above the water quality storage area. Therefore, the additional storage for drainage benefits as shown in this figure would be provided to cover the range of stages of the emergency spillway, plus freeboard.

(3) Primary Water Quality Outlet Devices

The selection of the outlet devices for the wet detention pond (the primary water quality control device plus the emergency spillway) is the next step and is based on the surface area available at the maximum live storage stage. This outlet device must be selected based upon the desired pollutant control at every specific pond stage. This specification regulates the detention time periods and the "draining" period to produce consistent removals for all rains. The ratio of outlet flow rate to pond surface area for each stage value needs to be at the most 0.00013 $ft^3/sec/ft^2$ for 5 μm (about 90% annual control) and 0.002 $ft^3/sec/ft^2$ for 20 μm (about 65% annual control). In practice, the desired pond-surface-area-to-stage relationship (simply the "shape" of the hole) is compared to the minimum surface areas needed at each stage for various candidate outlet structures. As an example, Table 6.8 summarizes the minimum surface areas needed for 5 μm particle control for different stage values. Also shown are the total storage values below each

TABLE 6.8. Minimum Surface Needed for 5 μm Particle Control.

	45° V-notch weir		90° V-notch weir		24″ pipe	
Stage (ft)	Storage (ac-ft)	Surface (ac)	Storage (ac-ft)	Surface (ac)	Storage (ac-ft)	Surface (ac)
0.5	<0.01	0.032	0.02	0.08	0.07	0.28
1.0	0.05	0.18	0.15	0.44	0.39	0.98
1.5	0.22	0.5	0.56	1.2	1.1	1.8
2.0	0.60	1.0	1.5	2.5	2.1	2.4
3.0	1.6	2.8	6.2	6.8	4.5	2.4
4.0	5.9	5.8	17	14	6.9	2.4
5.0	14	10	36	25	9.3	2.4
6.0	27	16	67	39	12	2.4

elevation (assuming the noted surface areas for the shallower elevations).

Large stages above the normal wet pond depth may result in unsafe conditions for most wet detention ponds. A maximum depth of about 3 feet above the normal wet pond depth is recommended.

Tables 6.9 through 6.12 provide a quick method for selecting appropriate outfall devices for a potential pond location. These tables indicate the minimum pond surface area needed at each stage to provide a 5-μm critical control level for a variety of conventional outfall devices. Table 6.12 presents multipliers to adjust the minimum areas for other critical particle sizes. For example, in order to improve the pond performance by selecting a 2-μm critical particle size instead of 5 μm, the pond surface area would have to be increased by about 6.7 times. If the critical particle size was increased to 10 μm, then the required pond surface would be reduced by about 27% compared to the pond surface area needed for 5-μm control.

As an example, if a pond required a surface area of 3 acres at two feet above the lowest invert level, a number of outlet devices could be used to provide at least 5-μm critical control:

- All V-notch weirs from 22.5° through 90° (but not 120°)
- Only a 2 foot long rectangular weir
- All drop tubes from 8″ to 24″

Obviously, all stage levels have to be examined. The device selected must provide the desired level of control at the most critical stage (usually at the deepest depth). In most cases, the outlet device that has the largest capacity that meets the discharge requirements should be used. Under-sized discharge devices would likely cause increased flows out the emergency spillway, causing an actual decrease in sediment trapping performance.

These procedures will result in the largest storms that do not enter the emergency spillway to be treated to at least equal to the critical particle size specified. As an example, the above calculations focus on the 5 μm particle, at least, being controlled at the highest stage of the primary outfall structures in order to provide an approximate worst-case 5 μm control (90% annual control of suspended solids). The outfall device is selected to provide an outfall rate no greater than a critical value, that when divided by the pond surface area at that stage, will be no larger than the settling rate of the critical particle size. In almost all cases, the critical stage will be at the top of the primary outfall device, and all stages below that will more than meet the critical objective, and will therefore be controlling particles much smaller than the critical size specified in the objective. It may seem that the pond is therefore over-designed and that the pond is larger than needed. However, the 5 μm critical particle size is typically substantially larger than the 90th percentile particle size, and the added control provided at the lower stages in the pond is generally needed to provide this level of control on an annual basis. As indicated previously, the 90th percentile particle size is typically only 3 μm, or smaller.

(4) Emergency Spillway

An emergency spillway is always needed, even for temporary detention ponds at construction sites. Most local regulatory agencies will require an emergency spillway that

TABLE 6.9. Surface Area Requirements for 5-μm Particle Size Control for Various V-notch Weirs.

Head (ft)	Flow (cfs)	22.5° Storage (ac-ft)	Required Area (acres)	Flow (cfs)	30° Storage (ac-ft)	Required Area (acres)	Flow (cfs)	45° Storage (ac-ft)	Required Area (acres)
0.5	0.1	<0.01	0.01	0.1	<0.01	0.02	0.2	<0.01	0.03
1	0.5	0.03	0.1	0.7	0.05	0.1	1.0	0.05	0.2
1.5	1.4	0.08	0.2	1.9	0.2	0.3	2.9	0.2	0.5
2	2.8	0.1	0.5	3.8	0.3	0.7	5.9	0.6	1.0
3	7.8	0.3	1.4	11	1.6	1.8	16	1.6	2.8
4	16	1.2	2.8	22	4.4	3.8	33	5.9	5.8
5	28	3.3	4.9	38	9.6	6.6	58	14	10
6	44	7.2	7.7	60	18	10	91	27	16
	Flow (cfs)	60° Storage (ac-ft)	Required Area (acres)	Flow (cfs)	90° Storage (ac-ft)	Required Area (acres)	Flow (cfs)	120° Storage (ac-ft)	Required Area (acres)
0.5	0.3	<0.01	0.05	0.4	0.02	0.08	0.8	0.04	0.1
1	1.4	0.07	0.3	2.5	0.2	0.4	4.4	0.3	0.8
1.5	4.0	0.3	0.7	6.9	0.6	1.2	12	1.7	2.1
2	8.2	0.8	1.4	14	1.5	2.5	25	3.3	4.4
3	28	3.5	3.9	39	6.2	6.8	69	12	12
4	46	9.5	8.1	80	17	14	140	30	25
5	81	21	14	140	36	25	250	69	43
6	130	39	22	220	67	39	390	120	68

TABLE 6.10. Surface Area Requirements for 5-μm Particle Size Control for Various Rectangular Weirs.

Head (ft)	Flow (cfs)	2 ft. Storage (ac-ft)	Required Area (acres)	Flow (cfs)	5 ft. Storage (ac-ft)	Required Area (acres)	Flow (cfs)	10 ft. Storage (ac-ft)	Required Area (acres)
0.5	2.1	0.10	0.4	5.7	0.3	1.0	12	0.5	2.0
1	6	0.5	1.1	16	1.2	2.8	33	2.4	5.7
1.5	10	1.2	1.8	29	3.2	5.0	59	6.3	10
2	15	2.3	2.6	43	6.4	7.6	90	13	16
3	24	5.7	4.2	80	17	14	160	35	29
4	32	11	5.6	110	34	20	250	71	43
5	37	17	6.5	150	47	26	340	120	59
6	39	23	6.9	190	77	33	430	190	75

	Flow (cfs)	15 ft. Storage (ac-ft)	Required Area (acres)	Flow (cfs)	20 ft. Storage (ac-ft)	Required Area (acres)	Flow (cfs)	30 ft. Storage (ac-ft)	Required Area (acres)
0.5	17	0.8	3.0	23	1.0	4.1	35	1.5	6.1
1	49	3.7	8.6	66	5.1	12	99	7.3	17
1.5	90	9.9	16	120	13	21	180	20	32
2	140	20	24	190	27	32	280	40	49
3	250	54	44	340	72	59	510	110	89
4	380	110	66	510	150	89	780	220	140
5	520	190	91	710	250	120	1100	390	190
6	680	290	120	920	390	160	1400	610	250

TABLE 6.11. Surface Area Requirements for 5.μm Particle Size Control for Verticle Riser Outlets.

Head (ft)	Flow (cfs)	8″ Storage (ac-ft)	Required Area (acres)	Flow (cfs)	12″ Storage (ac-ft)	Required Area (acres)	Flow (cfs)	18″ Storage (ac-ft)	Required Area (acres)
0.5	0.5	0.02	0.09	0.9	0.04	0.2	0.6	0.07	0.3
1	0.7	0.07	0.1	2.2	0.2	0.4	4.4	0.3	0.8
1.5	0.7	0.1	0.1	2.2	0.4	0.4	6.5	0.8	1.1
2	0.7	0.2	0.1	2.2	0.6	0.4	6.5	1.4	1.1
3	0.7	0.3	0.1	2.2	0.9	0.4	6.5	2.5	1.1
4	0.7	0.4	0.1	2.2	1.3	0.4	6.5	3.6	1.1
5	0.7	0.6	0.1	2.2	1.7	0.4	6.5	4.7	1.1
6	0.7	0.7	0.1	2.2	2.1	0.4	6.5	5.8	1.1

	Flow (cfs)	24″ Storage (ac-ft)	Required Area (acres)	Flow (cfs)	30″ Storage (ac-ft)	Required Area (acres)	Flow (cfs)	36″ Storage (ac-ft)	Required Area (acres)
0.5	1.6	0.07	0.3	1.9	0.08	0.3	2.0	0.09	0.4
1	5.6	0.4	1.0	6.3	0.4	1.1	7.2	0.5	1.3
1.5	11	1.1	1.8	13	1.3	2.3	16	1.5	2.8
2	14	2.1	2.4	21	2.8	3.7	27	3.4	4.7
3	14	4.5	2.4	25	6.9	4.4	42	9.4	7.3
4	14	6.9	2.4	25	11	4.4	42	17	7.3
5	14	9.3	2.4	25	16	4.4	42	24	7.3
6	14	12	2.4	25	20	4.4	42	31	7.3

TABLE 6.12. Corrections for Needed Surface Areas for Particle Size Controls other than 5 μm.

Particle Size for Control (μm)	Typical Percentage of Particles Larger than Indicated Size	Particle Settling Rate (cm/sec)	Required Area Multiplier, Compared to 5 μm
1	100	1.5×10^{-4}	27
2	94	6×10^{-4}	6.7
5	88	4×10^{-3}	1.0
10	78	1.5×10^{-2}	0.27
20	62	6×10^{-2}	0.067
40	47	2×10^{-1}	0.02
100	28	8×10^{-1}	0.005

is capable of discharging a specific design storm, typically in the range of 25 to 100-yr events, depending on the size of the pond. The typical procedure is to use the SCS (now NRCS) (1986) version of TR-55. The graphical peak discharge method in TR-55 is commonly used to estimate the peak flow associated with the design storm, and the WinTR-55 "structure" methods are then used to estimate the emergency spillway design. This spillway design should consider the outlet device selected for water quality benefits also.

Figure 6.17 shows that for Type II and III rains, the storage volume would have to be about 55% (0.55) of the runoff volume, if the peak runoff rate is to be reduced to 10% of its influent peak flow rate.

The SCS methods can be used to size an emergency spillway. The pond is sized to provide the water quality benefits, and additional storage associated with the emergency spillway stage is V_S, read from Figure 6.17. The design storm volume that must safely be accommodated by the emergency spillway is taken as V_r. The ratio of these values can be used with this figure to estimate the peak flow attenuation that the pond will provide. The peak inflow discharge rate, q_i, can be estimated using the SCS graphical peak discharge method (or the tabular hydrograph method,

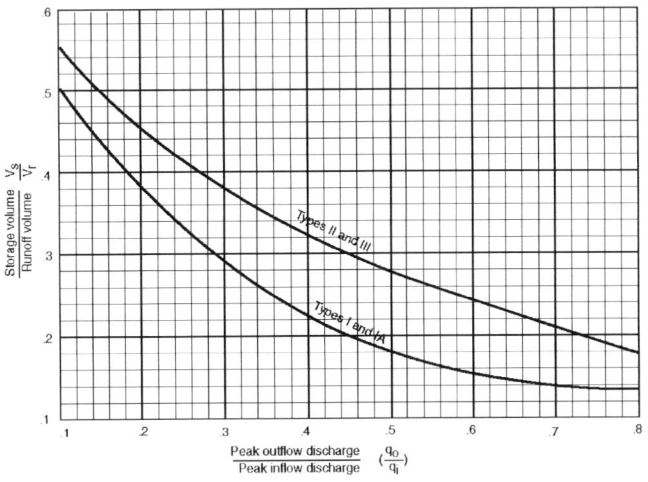

Figure 6.17. SCS TR-55 plot used to size additional freeboard needed for emergency spillway (SCS 1986).

or WinTR-55). The peak outfall discharge, q_o, is then calculated based on the measured attenuation factor.

Example: Sizing an Emergency Spillway

Given:

$V_S = 1.53$ acre-ft
$V_r = 7.5$ acre-ft
$V_S/V_r = 0.20$

Find the necessary size for the emergency spillway.

For type II or III rain categories:

$$q_o/q_i = 0.72$$

If the calculated peak discharge rate entering the pond (q_i) = 8.7 cfs, the resulting peak discharge rate leaving the pond, q_o, (through the water quality primary outlet plus the emergency spillway) is therefore: 0.72 (8.7) = 6.3 cfs. TR-55 shows how to calculate the needed emergency spillway for a specific discharge goal, considering multiple outlet structures. It helps determine the size of the spillway, plus the additional freeboard that must be added to the pond design to accommodate the emergency spillway and desired outlet flow rate.

Example: Sizing a Compound Weir Structure

The following example illustrates a compound weir structure, having a drop tube for water quality control, plus a rectangular weir for the emergency spillway. In this example, $q_o = WQ_{out} + \text{emergency spillway}_{out}$

Rain depth for the emergency spillway design (P) = 8 inches

$CN = 86$ (therefore, $Ia = 0.0366$)

Using Figure 3.17 relating P and Q through the curve number, the direct runoff (Q) = 6.2 inches $Ia/P = 0.041$

Area (A_m) = 0.021 mi^2 (13.2 acres)

T_c = 20 min (0.3 hr)

The peak unit discharge rate from the tabular hydrograph method is 498 csm/in

The peak discharge is therefore: (498 csm/in)(0.021 mi^2)(6.2 in) = 63.7 ft^3/sec

Also, the volume of runoff for this event is: V_r = [(6.2 in)(13.2 ac)]/12 in/ft = 6.82 ac-ft

As shown above, the pond surface area was determined to be about 0.4 acres at the permanent pool depth (the elevation for the water quality outlet invert). Table 6.11 confirms that a 12 inch drop tube structure would work for this pond over a wide range of stage conditions, while providing a desired worst case 5 μm particle control. The outlet flow rate for this drop tube is almost constant for heads of 1 to 6 ft (2.2 ft^3/sec). The maximum desired discharge rate for this pond (for both the water quality outlet plus the emergency spillway) is given

Vertical riser with inlet grate (MD photo) (normally the culvert discharge for the riser would be closer to the pond bottom to facilitate de-watering).

Temporary outlet made from level timber placed at correct elevation and covered with plastic to protect spillway (Auckland Regional Council).

Vertical riser having multiple outlets and wrapped with geotextile fabric (Poor design, as no cap is provided).

as 46.5 ft³/sec. The ratio of the outlet to the inlet flow rate is therefore:

$$q_o/q_i = 46.5/63.7 = 0.73$$

The ratio of the storage volume (V_S) to the runoff volume (V_R), for Type II rains (from Figure 6.17) is 0.2, for this ratio of outlet to inlet peak flow rates. The length (L_w in feet) of a rectangular weir, for a given stage (H_w in feet) and desired outflow rate (q_o in ft³/sec) can be expressed as:

$$L_w = \frac{q_o}{3.2H_w^{1.5}}$$

The desired q_o for the rectangular weir is 46.5 – 2.2 = 44.3 ft³/sec. If the maximum stage for the emergency spillway is 1 ft, then the length for the emergency spillway is:

$$L_w = \frac{q_o}{3.2H_w^{1.5}} = \frac{44\ \text{ft}^3/\text{sec}}{3.2(1\ \text{ft})^{1.5}} = 13.8\ \text{ft}$$

If the water quality outlet had a varying discharge rate for different stages (as is common), then the stage for that outlet must also be known so the actual discharge rate contribution from that outlet to the total discharge rate objective can be used in the calculation. As an example, a 45° V-notch weir would be a suitable outlet for water quality control for this pond. This weir, for a 0.4-acre pond, would provide 5 μm control up to about 1.4 feet of head, for a 0.4 acre pond (assuming the associated storage volume is adequate). At this stage, the discharge rate from the 45° V-notch weir is about 2.5 ft³/sec. With another foot of storage (as stage) for the maximum elevation of the emergency spillway (2.4 ft above the invert of the V-notch), the V-notch weir discharge rate would increase to about 10 ft³/sec. The residual discharge objective for the emergency rectangular weir would therefore now be: 46.5 – 11 = 35.5 ft³/sec, and the length for the emergency spillway would be:

$$L_w = \frac{q_o}{3.2H_w^{1.5}} = \frac{35.5\ \text{ft}^3/\text{sec}}{3.2(1\ \text{ft})^{1.5}} = 11.1\ \text{ft}$$

This method is known to be conservative with resulting over-sized emergency spillway storage volumes. A computer model should therefore be used to verify the performance of the desired pond configuration for a variety of storm conditions.

(5) Other Pond Features

Temporary construction site pond filled with sediment

The ponds must also be constructed according to specific design guidelines to insure the expected performance and adequate safety, such as those provided by the U.S. Bureau of Reclamation (1987). The guidelines need to specify such items as pond depth, side slopes, and shape.

Example Pond Design for Construction Site Sediment Control and Comparison with Modeling Results

Table 6.13 shows the conditions for an area on a construction site that requires a sediment pond. The drainage area, 53 acres, is mostly an active construction site, but some undeveloped land and paved areas also drain to the pond location. The pond therefore needs to be enlarged to accommodate the additional runoff from these areas. The table shows the drainage-area percentage needed for the

TABLE 6.13. Size of Pond for Construction Area.

Head (ft)	Area (acres)	% of Area Needed for Pond Surface	Pond Surface Area (acres)	Water Quality Volume (inches of runoff)	Pond Volume (acre-inches)
Construction area	37	1.5%	0.56	0.6	22.2
Undeveloped area	14	0.5	0.07	0.3	4.2
Paved area	2	3.0	0.06	1.1	2.2
Total	53		0.69		28.6

pond, along with the pond volume to obtain approximately 90% suspended solids reduction.

The total water quality volume ("live storage") of the pond is 28.6 acre-inches, or 2.38 acre-ft. The surface of the pond between events (during dry weather) is 0.7 acres, or about 1.3% of this drainage area. The top area of the pond during live filling and drawdown, and associated side slopes, are calculated based on various assumed pond depths, as shown in a later example.

In this example, a pond depth of 3 ft, and approximate side slopes of 12% and a top area of 0.9 acres are used. An additional 1 ft of storage to accommodate an emergency spillway is also provided, with a maximum top area needed of about 1 acre. The selection of the main discharge device is based on the water surface at the top of this water quality volume. A 12 inch vertical riser pipe, having its opening at the normal pond water surface level, seems to be a good choice, based on Table 6.11 data.

Three feet of standing water is needed above the maximum sediment depth in order to minimize scour. In addition, sacrificial sediment storage must also be provided in the pond. Using RUSLE, the total construction period sediment load to the pond can be estimated. In the following example, it is assumed that the construction period is a half year, and the following conditions apply:

$R = 350$
$LS = 4.95$ (based on typical slope lengths of 600 ft at 10% slope)
$k = 0.28$
$C = 0.25$ (assuming that 25% of the construction site area is being actively being worked, and the rest of the area is effectively protected)

The calculated unit area erosion loss for this construction period is therefore about 243 tons per acre per year. Since the construction period is one-half year and the area is 37 acres, the total sediment loss is estimated to be about 4490 tons. For a loam soil, this sediment volume is about 4600 yd^3, assuming the conventional conversion factor of tons $\times$ 1.02 = yd^3 for a loam soil. The pond area at the bottom of the 3 ft of standing water is assumed to be about 0.7 acre, requiring about 5 ft of sediment storage. Therefore, Table 6.14 lists the pond areas for each depth increment.

TABLE 6.14. Pond Areas of Each Depth Increase.

Pond Depth (ft)	Pond Area (acres)
0	0
1	0.35
2	0.50
3	0.57
4	0.63
5	0.70
6	0.77
7	0.73
8	0.90
9	0.97

This design was entered into WinDETPOND, a continuous water and sediment routing model for ponds (www.WinSLAMM.com), and evaluated. Table 6.15 shows the program results for this pond. A series of rains ranging from 0.01 to 4.0 inches was used in the evaluation. The maximum pond stage is estimated to be about 7.4 ft for the 4-inch rain, more than a half foot below the broad-crested weir emergency spillway. The peak reduction factor (the reduction of the influent peak flow rate at the outfall) is very large for the small events, as expected, and still remains about 0.5 for the largest event. Ratios in this range will help reduce erosive flows to the receiving waters. The "event flushing ratio" indicates the volume of runoff compared to the water volume in the pond before the event. Again, this value is very small for the small events and increases to greater than 1 for rains larger than about 3 inches. The last two columns indicate sedimentation performance of the pond. The flow-weighted particle size in the effluent is greater than 4 μm after 3 inches of rain. However, the expected percentage suspended solids control (assuming the

TABLE 6.15. Summarized Results from WinDETPOND to Evaluate Detention Pond at Construction Site.

Rain Number	Rain Depth (in)	Rain Duration (hrs)	Rain Intensity (in/hr)	Maximum Pond Stage (ft)	Event Inflow Volume (ac-ft)	Peak Reduction Factor (%)	Event Flushing Ratio	Flow-weighted Particle Size (Ideal)	%Part Solids Removed (Ideal)
1	0.01	3.00	0.00	5.00	0.000	1.00	0.000	0.0	100.0
2	0.05	7.00	0.01	5.00	0.002	0.99	0.001	0.0	100.0
3	0.10	8.00	0.01	5.01	0.007	0.99	0.003	0.1	99.8
4	0.25	10.00	0.02	5.07	0.052	0.99	0.022	0.1	99.5
5	0.50	12.00	0.04	5.19	0.137	0.97	0.059	0.3	98.9
6	0.75	14.00	0.05	5.30	0.230	0.94	0.099	0.5	98.2
7	1.00	14.00	0.07	5.42	0.342	0.90	0.147	0.7	96.7
8	1.50	14.00	0.11	5.64	0.610	0.85	0.262	1.2	88.5
9	2.00	14.00	0.14	5.87	0.939	0.78	0.403	1.8	80.2
10	2.50	14.00	0.18	6.26	1.528	0.67	0.656	2.9	68.1
11	3.00	14.00	0.21	6.64	2.266	0.57	0.973	4.0	57.2
12	4.00	14.00	0.29	7.37	4.014	0.50	1.724	6.5	39.1

TABLE 6.16. Performance of Temporary Sediment Pond at Construction Site (Birmingham rains).

Rain Range (inches)	Mid Point Rain (inches)	% of Annual R in Category	% Particulate Solids Removed for Pond	Weighted Total Annual Particulate Solids Removal (%)
0.01 to 0.05	0.03	0.0	100	0
0.06 to 0.10	0.08	0.1	100	0.1
0.11 to 0.25	0.18	0.7	99.8	0.7
0.26 to 0.50	0.38	3.5	99.5	3.5
0.51 to 0.75	0.63	4.8	98.9	4.7
0.76 to 1.00	0.88	8.2	98.2	8.1
1.01 to 1.50	1.26	16.1	96.7	15.6
1.51 to 2.00	1.76	15.4	88.5	13.6
2.01 to 2.50	2.26	10.9	80.2	8.7
2.51 to 3.00	2.76	7.5	68.1	5.1
3.01 to 4.00	3.5	16.3	57.2	9.3
over 4.01	5.67	16.5	39.1	6.5
4583 events	41.5 years	100.0	75.9 % annual particulate solids removal	

"low" particle size distribution—a very demanding particle size distribution that usually results in low removal estimates) remains greater than 80% for all rains less than about 2 inches. The worst case shown, for the 4-inch rain, drops down to less than 40% control.

As noted earlier in Chapters 3 and 4, most of the erosion potential is associated with the numerous moderate (greater than 1 inch) and the few large rains (up to 4 inches) that likely occur during the year. This pond will likely provide 65 to 95+% control for the moderate rains, but will drop off significantly for the largest rains. It is possible to improve the performance of the pond by changing the outlet weir to a smaller-capacity device, which would provide additional retention for the larger events. Table 6.16 illustrates how this temporary pond would affect the annual particulate solids losses from this construction site. The overall pond performance is expected to be about 75% effective, much less than the initial goal of 90% control. The performance of this pond could be improved if the design was better optimized for the larger, more erosive events. This could be done by choosing a more restrictive outlet device at higher pond stages and also by providing more storage, for example.

Example Detention Pond Shape Calculations

The following discussion presents a calculation example assuming that the wet pond surface is 1.2 acres and the runoff volume for treatment is 6.3 acre-feet

The depth associated with the wet storage volume can be estimated assuming a prismatic cross-section (simplified, compared to a conical section):

Approximately: $[1.2 + x(1.2)]y/2 = 6.3$ acre-ft.

TABLE 6.17. Top Area Multiplier.

y (depth, ft)	*x* (multiplier)	Top Area
2	4.3	4.3 (1.2 acres) = 5.2 acres
3	2.5	3.0 acres
4	1.6	1.9 acres
5	1.1	1.3 acres

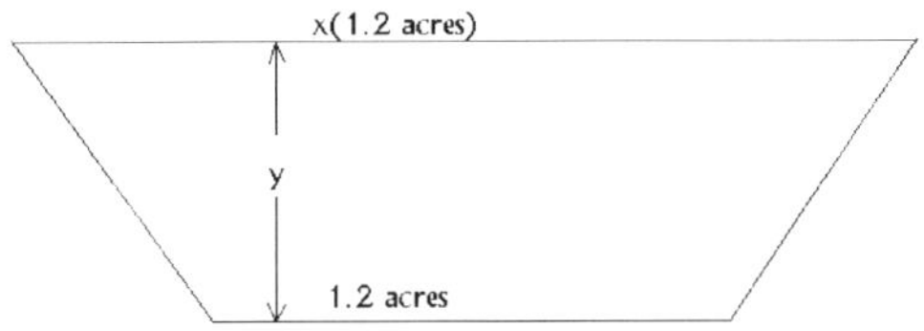

Re-arranging gives: $x = [(10.5)/y] - 1$

Table 6.17 can be used to determine the top-area multiplier, x, for various depths of the "live storage" area of the pond (the section affected by the primary water quality outlet device and located on top of the permanent pool depth, and below the invert of the emergency spillway. This includes any additional storage needed for flood control), Once x is known, the top area is simply the bottom area multiplied by x.

For this example, depths less than 2 feet are too shallow and could require very large pond top surface areas. "Live depths" greater than 5 feet may be too deep for most locations and obviously result in very steep side slopes for this example. If a circular pond is desired, Table 6.18 summarizes the calculations for the side slopes of the pond.

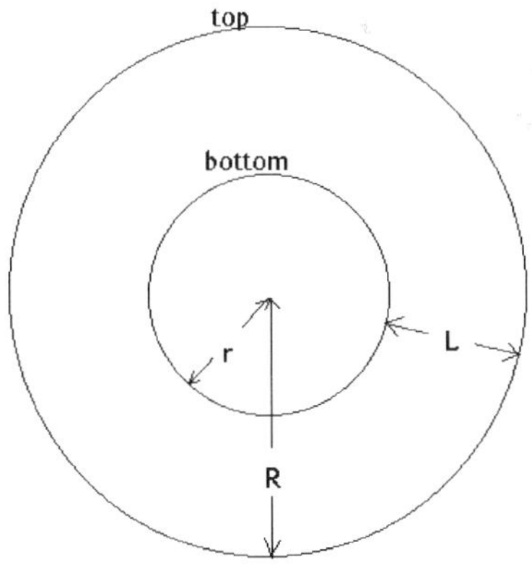

$$r = (A/\pi)^{1/2} = [1.2\text{acres}(43{,}560 \text{ ft}^2 \text{ per acre})/\pi)]^{1/2} = 130 \text{ ft}$$

TABLE 6.18. Summary of Calculations for the Side Slopes of the Pond.

Depth (ft)	Top Area (acres)	Top Radius (ft)	Slope Length (ft)	Side Slope
2	5.2	270	270 – 130 = 140	2/140 = 1.4%
3	3.0	200	200 – 130 = 70	3/70 = 4.3%
4	1.9	160	160 – 130 = 30	4/30 = 13%
5	1.3	135	135 – 130 = 5	5/5 = 100%

TABLE 6.19. Maximum Allowable Discharges.

Stage (above normal water surface, ft)	Pond Area (acres)	Maximum Allowable Discharge (cfs)
0	1.2	6.8
0.5	1.5	8.5
1	1.8	10
1.5	2.1	12
2	2.4	14
3	3.0	17 (usually most critical)

The preliminary pond cross-section is

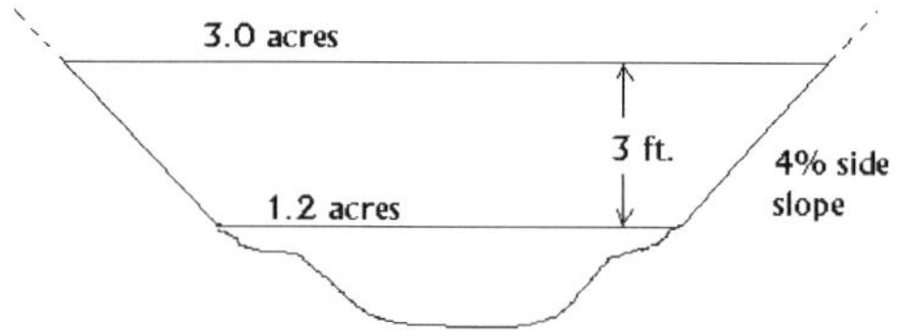

Therefore, the outfall device is selected by comparing the maximum allowable discharge rate for the surface area of the pond (surface overflow rate) at several pond depth increments. These maximum allowable discharges are compared with weir ratings (Table 6.19) to select the permissible weirs that can be used:

$$Q_{out} = vA$$

$$v = 1.3 \times 10^{-4} \text{ ft/sec for } 5\ \mu\text{m particle}$$

Hence, a single 45° V-notch weir, or two 22-1/2° V-notch weirs.

The emergency spillway (mandatory) and additional flood control storage volume (if necessary) would be selected using NRCS TR-55 (SCS 1986) procedures, as previously described.

Example Sizing of Sediment Pond at Construction Site

This example problem considers the sizing of all the major components of a sediment pond at a construction site:

- the basic pond area,
- the "live" storage volume,
- the pond side slopes, top surface area, and "dead storage" volume,
- the selection of the primary discharge device,
- the additional storage volume needed for the emergency spillway,
- the sizing of the emergency spillway, and
- the sacrificial storage volume for sediment accumulation.

TABLE 6.20. Basic Pond Area and "Lve" Storage Volume Calculations.

Site Subarea	Pond Surface Area (acres)	Pond "Live" Volume, Runoff from 1.25 Inches of Rainfall (acre-inches of runoff)
paved area (0.2 acres)	3% of 0.2 acres = 0.006 acres	1.1 inches x 0.2 acres = 0.22 ac-in
undeveloped area (1.2 acres)	0.6% of 1.2 acres = 0.007 acres	0.3 inches x 1.2 acres = 0.36 ac-in
construction area (32 acres)	1.5% of 32 acres = 0.48 acres	0.6 inches x 32 acres = 19.2 ac-in
Total:	0.49 acres	19.8 ac-in = 1.65 ac-ft

Consider the following site information:

The pond performance goal is 90% suspended-solids removal. The pond needs to safely pass the flows from the 25-yr storm. The area is characterized by clayey soils. The following are the areas associated with each land use in the drainage area:

- paved areas: 0.2 acres
- undeveloped areas: 1.2 acres
- construction area: 32 acres
- total site area: 33.4 acres

Pond side slopes and top surface area:

1. If 3 ft deep:

   Top area:

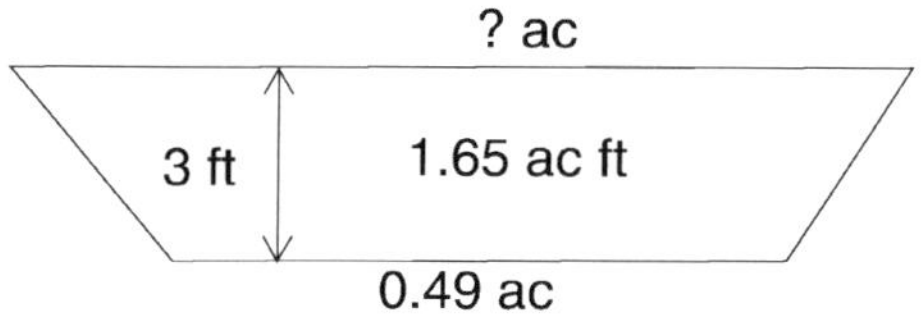

$$\frac{(0.49 \text{ acres} + X)3 \text{ ft}}{2} = 1.65 \text{ ac} - \text{ft}$$

$$X = 0.61 \text{ acres:}$$

at 0.61 acres:

$$\pi r^2 = 26{,}570 \text{ ft}^2$$

$$r = 92 \text{ ft}$$

at 0.49 acres:

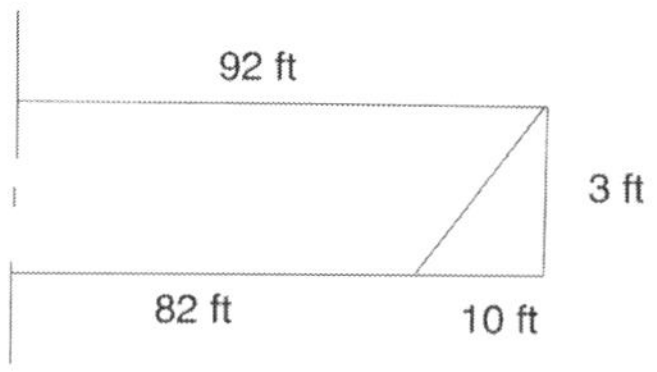

$$\pi r^2 = 21{,}340 \text{ ft}^2$$

$$r = 82 \text{ ft}$$

$$\text{side slope} = 3 \text{ ft}/(92 - 82 \text{ ft}) = 3 \text{ ft}/10 \text{ ft} = 0.3$$

$$= 30\% \text{ too steep}$$

2. If 1 ft deep:

   Top area:

$$\frac{(0.49 \text{ acres} + X)1 \text{ ft}}{2} = 1.65 \text{ ac} - \text{ft}$$

$$X = 2.81 \text{ acres}$$

at 2.81 acres:

$$\pi r^2 = 122{,}400 \text{ ft}^2$$

$$r = 197 \text{ ft}$$

at 0.49 acres, $r = 82$ ft:

side slope = 1 ft/(197 – 82 ft) = 1 ft/115 ft = 0.012
= 1.2% too shallow

3. If 2 ft deep:

   Top area:

$$\frac{(0.49 \text{ acres} + X)2 \text{ ft}}{2} = 1.65 \text{ ac} - \text{ft}$$

$$X = 1.16 \text{ acres}$$

at 1.16 acres:

$$\pi r^2 = 50{,}530 \text{ ft}^2$$

$$r = 126 \text{ ft}$$

at 0.49 acres, r = 82 ft

side slope = 2 ft/(126.82 ft) = 2 ft/44 ft = 0.045
= 4.5% suitable, but on the low side

Selection of Primary Outlet Device

At the top of the live storage volume, this pond will have provided 2 ft of stage and 1.16 acres of maximum pond area.

According to Tables 6.9 to 6.11, a 45° V-notch weir requires at least 1.0 acres of pond surface at 2 feet of stage in order to provide about 90% control of sediment. A 30° V-notch weir would require only 0.7 acres, while a 60° V-notch weir would require 1.4 acres. None of the rectangular weirs would be suitable, as the smallest 2 ft weir requires at least 2.6 acres at 2 feet of stage. The 45° weir is closest to the area available and is therefore selected for this pond. Another suitable outlet structure would be an 18″ drop tube structure which requires at least 1.1 acres.

Sacrificial Storage Volume

Calculate the sediment loss for the complete construction period for the site area draining to the pond. Chapter 4 describes how to calculate the sediment loss for different phases of the construction period and for different areas of the site. For this analysis, assume the following site conditions:

R = 350
LS = 1.28 (based on slope lengths of 300 ft at 5% slope)
k = 0.28
C = 0.24 (assuming that 5 of the 32 acres of the construction area is being actively worked with a C = 1, and the other 27 acres of the construction area is effectively protected with a C = 0.1)

The calculated unit area erosion loss for this construction period is therefore: (350)(1.28)(0.28)(0.24) = 30 tons per acre per year. Since the construction period is for one year and the active construction area is 32 acres, the total sediment loss is estimated to be about 960 tons. For a loam soil, this sediment volume is about 980 yd^3, or 0.8 acre-ft, assuming the conventional conversion factor of tons × 1.02 = yd^3 for a loam soil.

The pond water surface is approximately 0.5 acres. With a three-foot-deep dead storage depth to minimize scour, the surface area at the bottom of this 3 ft scour protection zone (and the top of the sediment storage zone), can be about 0.35 acres (about 25% underwater slope).

The sacrificial storage zone can be about 3 ft deep also, resulting in a bottom pond area of about 0.18 acre, as shown in the following calculations:

Top of sacrificial storage area is 0.35 acres,

at 0.35 acres:

$$\pi r^2 = 15{,}250 \text{ ft}^2$$

$$r = 70 \text{ ft}$$

Therefore, the area of the bottom of the sacrificial storage area needed to provide 0.8 acre-ft of storage, if 3 feet deep can be approximated by:

$$\frac{(0.35 \text{ acres} + X)3 \text{ ft}}{2} = 0.8 \text{ ac} - \text{ft}$$

$$X = 1.16 \text{ acres}$$

at 0.18 acres, $r = 50$ ft

side slope = 3 ft/(70-50 ft) = 3 ft/20 ft = 0.15 = 15%

Selection of Emergency Spillway

TR-55 can be used to estimate the peak flood flow rate that the emergency spillway must accommodate. Since these ponds are generally temporary, the design storm is usually smaller than for permanent stormwater ponds (which are commonly designed as high as controlling the 100-year event). Also, temporary ponds usually do not include an attenuated flow rate goal, like permanent ponds. These flow rate goals for permanent ponds need to be based on comprehensive basin-wide hydraulic analyses to be effective. Therefore, this example will only consider the capacity of the emergency spillway to meet the design storm flow rate. The design storm for this pond will be the 25.year event (one that has a 4% probability of occurring in any one year). The time of concentration of this small watershed was previously calculated to be 12 minutes. The watershed characteristics affecting the peak flow rate are therefore:

- Watershed area: construction area (32 acres), paved area (0.2 acres), and undeveloped area (1.2 acres) = 33.4 acres = 0.052 mi^2
- Clayey (hydrologic soil group D) soils
- Time of concentration (T_c): 12 minutes (0.2 hours).
- Since the pond is at the bottom of this watershed, there is no "travel time" through down-gradient subwatershed areas.
- Rain intensity for a "25.year" rain for the Birmingham, AL, area, with a 12 minute time of concentration (from the local IDF curve, Figure 3-4): 6.6 inches/hour (Type III rain)

Since the undeveloped area has such a comparatively low *CN* (greater than a difference of 5) from the others, and it is a very small fraction of the site, it will be ignored for these calculations. The flows from the undeveloped area will be very low and will enter the pond after the flows from the other areas. If the undeveloped area was a significant fraction of the watershed area, it should be examined as a separate subwatershed and the resulting hydrographs combined. The weighted curve number is therefore estimated to be:

$$CN_w \left(\frac{32}{32.2}\right)(94) + \left(\frac{0.2}{32.2}\right)(98) = 94$$

The *Ia* for this curve number (from Table 3-16) is 0.128 inches. The 24-hour, 25.year rain has a total rain depth (*P*) of 6.9 inches (from Table 3-3). The *Ia*/*P* ratio is therefore: 0.128/6.9 = 0.019, which is much less than 0.1. Therefore the tabular hydrograph table to be used would be Exhibit III, corresponding to a T_c of 0.2 hour. The top segment of "csm/in" (cubic feet per second per square mile of watershed per inch of direct runoff) values are therefore used, corresponding to *Ia*/*P* values of 0.1, or less. The top row is also selected as there is no travel time through downstream subwatersheds. Examining this row, the largest value is 565 csm/in, occurring at 12.3 hours. The amount of direct runoff for a site having a *CN* of 94 and a 24-hr rain depth of 6.9 inches is 6.2 inches (from Figure 3-11). The A_mQ value (area in square miles times the direct runoff in inches) for this site is: (0.052 mi^2)(6.2 inches) = 0.32 mi^2-in. This value is multiplied by the csm value to obtain the peak runoff rate for this design storm: (0.32 mi^2-in)(565 csm/in) = 182 ft^3/sec.

The first trial for an emergency spillway will be a rectangular weir, with one foot of maximum stage. At the one foot of stage for this weir plus the spillway, the 45° V-notch weir will have 3 feet of stage. The V-notch weir will discharge 16 ft^3/sec at this stage (from Table 6.9). Therefore, the rectangular weir will need to handle: 182 – 16 ft^3/sec = 166 ft^3/sec. The rectangular weir can be sized from the rectangular weir equation presented earlier:

$$L_w = \frac{q_o}{(3.2)(H_w)^{1.5}} = \frac{166 \text{ ft}^3 / \text{sec}}{(3.2)(1)^{1.5}} = 52 \text{ ft}$$

This may be large for this pond, so another alternative is to try for a rectangular weir having 2 ft of maximum stage. At this elevation (4 ft total), the 45° V-notch weir will discharge 33 ft^3/sec. Therefore, the rectangular weir will need to handle: 182 – 33 ft^3/sec = 149 ft^3/sec. The rectangular weir can be sized from the rectangular weir equation presented earlier:

$$L_w = \frac{q_o}{(3.2)(H_w)^{1.5}} = \frac{149 \text{ ft}^3 / \text{sec}}{(3.2)(1)^{1.5}} = 16 \text{ ft}$$

This is a suitable length, but does result in an additional foot of pond depth. For this example, the 52 foot long weir is selected.

Final Pond Profile and Expected Performance

This pond therefore has the following shape, and outlet structures listed in Table 6.21.

In summary, this pond has a total of 3 acre-ft of live storage, plus the needed 0.8 acre-ft for sediment storage. Table 6.22 summarizes the results of modeling the pond using WinDETPOND (www.WinSLAMM. com). Table 6.22 shows the expected pond performance for a variety of rain depths, ranging from very small rains to larger events. The maximum pond stages reflect the maximum depth of water in the pond during these events (out of the total 10 feet available). The pond has very high levels of control (using the "medium" particle size distribution) for most events.

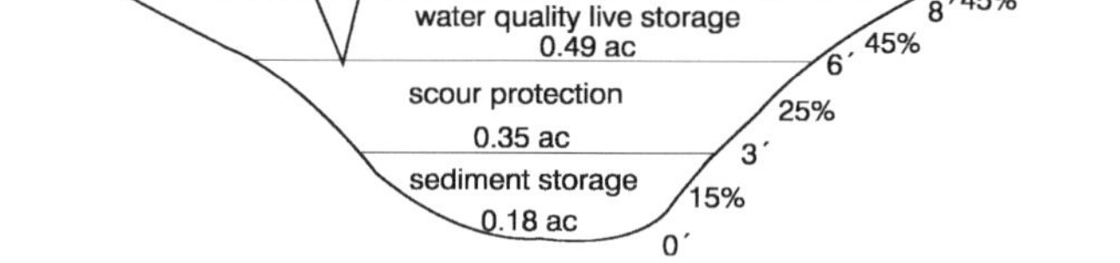

TABLE 6.21. Pond Outlet Structures.

Pond Depth (ft from bottom of pond, the datum)	Surface Area at Depth (acres)	Pond Storage Below Elevation (calculated by Detpond) (acre-ft)	Pond Slope Between This Elevation and Next Highest Noted Elevation	Notes
0	0	0	—	The pond bottom (datum) must be 0 acres for the routing calculations.
0.1	0.18	—	15%	The area close to the bottom can be the calculated/desired pond bottom area. This is the bottom of the sacrificial storage area for the sediment.
3	0.35	0.8	25%	This is the top of the sacrificial storage area for the sediment.
6	0.49	2.0	4.5%	This is the bottom of the "dead" storage area, at least 3 feet above the pond bottom (this is 6 feet above the absolute bottom, but is 3 feet above the top of the maximum sediment accumulation depth)
8	1.16	3.7	4.5%	This is the bottom (invert) of the water quality outlet structure (and live storage volume), a 45° V-notch weir.
9	1.5	5.0	4.5%	This is the top of live storage volume, and the bottom of the emergency spillway, a 52 ft long rectangular weir.
10	1.8	6.7	—	1 foot of freeboard above maximum expected water depth, the top of the pond.

The continuous simulation feature of WinDETPOND allows the user to predict the overall pond performance based on actual rain records. Table 6.23 summarizes the pond performance for a 30-year period of rain (3,346 events, ranging from 0.01 to 13.6 inches). During these 30 years, the expected maximum pond stage was slightly more than 8 ft. The emergency spillway was used a total of four times in this period. The flow-weighted particulate solids removal rate was approximately 92%. Therefore, this pond is likely over-designed for these conditions and could be somewhat reduced in area and depth.

Case Study: Example Use of Chemical-Assisted Sedimentation at Construction Sites

Larcombe (1999) of the Auckland Regional Council (ARC), New Zealand, prepared a report (*Technical Publication on Chemical Removal of Sediment from Earthworks Stormwater*) describing the use of chemical-assisted sedimentation for the control of construction site sediment. They tested both solid forms of flocculants (Magnasol Floc Blocs Allied Colloids, Australia Pty Ltd., NZ agent Chemiplas NZ Limited) and liquid chemicals at several construction areas. Test sites included areas along the extension of the northern motorway (ALPURT), and at a residential subdivision development (Greenhithe). The extensive field trials using aluminum sulfate (Alum) and polyaluminum chloride (PAC) were carried out during construction of the initial stages of the northern motorway. The ARC then developed a passive dosing system for the treatment of the construction site runoff treating the flow during passage into and through the pond. This system proved highly effective under a wide range of storm conditions. The following discussion is summarized from that report.

Conditions when Chemical Treatment may be Necessary

The requirements for sediment ponds at construction sites

TABLE 6.22. Expected Pond Performance.

Rain Depth (in)	Maximum Pond Stage (ft)	Event Inflow Volume (ac-ft)	Peak Reduction Factor (Fraction)	Event Flushing Ratio	Flow-weighted Particle Size (μm)	Particulate Solids Removed (%)
0.01	6	0	0.98	0	0	100
0.05	6	0	0.97	0	0	100
0.1	6	0.001	0.96	0	0.1	99.9
0.25	6.02	0.014	0.96	0.007	0.2	99.8
0.5	6.07	0.043	0.95	0.02	0.3	99.7
0.75	6.14	0.085	0.95	0.041	0.4	99.6
1	6.21	0.134	0.93	0.064	0.5	99.5
1.5	6.36	0.263	0.88	0.126	0.8	98.9
2	6.51	0.435	0.83	0.209	1.2	97.3
2.5	6.78	0.785	0.74	0.377	1.9	94.4
3	7.05	1.236	0.65	0.593	2.7	91.4
4	7.52	2.325	0.53	1.115	4.4	84.8

TABLE 6.23. Summary of Pond Performance for a 30-Year Period.

	Maximum Pond Stage (ft)	Event Inflow Volume (ac-ft)	Peak Reduction Factor (Fraction)	Event Flushing Ratio	Flow-weighted Particle Size (μm)	Particulate Solids Removed (%)
Maximum	8.1	23	0.99	11	6.8	100
Average	6.2	0.10	0.64	0.05	n/a	n/a
Flow-weighted Average	n/a	n/a	0.62	1.4	2.6	92
Median	6.1	0.012	0.87	0.0057	0.39	99.6
Standard Deviation	0.22	0.54	0.40	0.26	0.57	1.9
COV	0.035	5.1	0.63	5.1	1.1	0.019

are given in the Auckland Regional Council guidance (TP 90, *Erosion and Sediment Control,* 1999). The performance of ponds constructed according to these specifications is generally good, but a number of situations have been identified where chemical treatment can provide a marked improvement in sediment removal. Chemical treatment is important when a pond of the required size cannot be constructed. This may occur because of topographical constraints, difficult soil conditions, or the presence of natural habitat of high value. In some situations, the design of the pond cannot be optimized in terms of shape, depth, location of inlet and outlet, or energy attenuation of the inflow. Some soil types produce solids in the runoff that have very poor settling characteristics in a normal sediment pond. There is also a higher risk of increased erosion and sediment losses during rainstorms in areas having highly erodible soils, or having very steep or long slope lengths. Some common uses of construction sites, such as repeated machinery movement on haul roads, can result in high sediment loadings in stormwater. Finally, chemical treatment provides a means of reducing the sediment discharge to highly sensitive receiving environments.

Initial Tests

Two types of chemicals were considered for the initial bench testing and field trials, polyelectrolyte flocculants (polymer or polyacrylamide) and aluminum coagulants (aluminum sulfate (alum) and polyaluminum chloride, (PAC)).

Polyelectrolyte Flocculants

According to the ARC (Beca Carter Hollings & Ferner Ltd, undated), "anionic polyacrylamide is a negatively charged flocculant commonly used for industrial applications including raw potable water clarification, and for clarification, thickening and dewatering of wastewater and sludge. Because these polymers have a high affinity for solids, the remaining concentration in treated waters is very low in all but serious overdose situations. On the other hand, cationic polyacrylamides are positively charged and are commonly used in a number of municipal wastewater treatment plants to improve solids removal during pre-settlement. They are recognized as flocculants with greater toxicity implications for fish and other aquatic organisms than anionic or non-ionic polyelectrolytes. This is because the gills of fish are negatively charged, and the cationic polymer binds to them resulting in mechanical suffocation."

Bench testing showed that a number of polyacrylamides resulted in good removal of suspended solids from the construction site runoff water. However, they identified several difficulties hindering the use of liquid polyacrylamides at construction sites. The most serious difficulty is that liquid polyacrylamide concentrates are highly viscous and would require onsite predilution with water to achieve a suitable consistency for dosing and mixing with construction site runoff. This would require mixing equipment and storage tanks, along with electric power. In addition, the diluted polyacrylamide has a limited storage life.

Three solid polyacrylamide products (Floc Bloc), marketed by Allied Colloids, were evaluated in bench-scale tests. The products were: Percol AN1 and AN2 (both anionic polyacrylamide blends) and Percol CN1 (a cationic polyacrylamide blend). The floc blocs were 300 × 100 × 85 mm and weighed 3 kg. AN2 performed best when using runoff from sites having either clay or limestone soils. AN2, being an anonic polyacrylamides, also had a lower toxicity. Effective dose rates were between 1 and 4 mg/L of dry AN2. Higher concentrations led to reductions in flocculation and suspended sediment removal. AN2, even at excessive dosages of about 8 mg/L, did not affect pH.

Aluminum Coagulants

A major issue with aluminum coagulants is they contain large concentrations of ionic aluminum, the toxic form of aluminum. It is generally agreed that dissolved aluminum at concentrations as high as 0.050 to 0.100 mg/L and at pH values between 6.5 and 8.0 present little threat of toxicity. At lower pH, the toxicity increases due to possible mucus formations on the gills of fish. The toxic aluminum associated with the coagulant dose is very rapidly reduced by the precipitation and coagulation reactions. The insoluble precipitates (incorporating metals, nutrients, and solids) that form after aluminum coagulants are added to water are stable and denser than water. The alum floc that is formed is not toxic to benthic organisms. Most pollutants are tightly bound to the aluminum matrix with little likelihood of release from either dried or wet sludges within normal pH and redox ranges.

During the initial tests at the ALPURT site, the ARC (Beca Carter Hollings & Ferner Ltd, undated) determined that there was a need for chemical treatment of runoff from catchments having clay soils that naturally produced more acidic runoff. They decided to compare PAC with alum as a coagulant, as PAC is less acidic. Table 6.24 shows the treatment data using representative runoff samples from clay soil catchments. Samples were taken after 1 hour of settling. Longer settling times would have resulted in further reductions in suspended solids, but these tests were to compare the alkalinity and pH effects of these two alternative coagulants. These tests show that PAC has a consistently lower detrimental effect on pH reduction, and it results in higher effluent alkalinity.

Floc blocks and flume detail, initial installation (*Source:* Beca Carter Hollings & Ferner Ltd, undated).

Solid Floc Blocks

The initial tests indicated advantages to the use of the solid floc blocks, particularly on sites with difficult access; sites with only small construction areas, or sites where there was a need for short-term treatment only. They therefore followed up their initial tests with detailed field assessments to determine the best methods for using the blocs to obtain the most effective suspended solids removal under highly-variable flow conditions.

Field Trials Using Solid Floc Blocs

Preliminary field trials used an AN2 floc bloc to treat sediment-laden runoff from a construction site having limestone soils. The first trials placed the floc blocs in plastic mesh bags in plywood flumes through which the runoff from the site was directed. Those trials encountered problems with the high bedload of solids in the runoff flow that accumulated against and partially buried the floc bloc, inhibiting the dissolution of the chemical. The trial was then moved to a channel between a forebay and the settlement pond (for pre-treatment of the water to remove the large materials), and demonstrated that new floc blocs achieved good treatment for low flows (about 2 L/s) and when the suspended solids was between 10,000 to 20,000 mg/L. However, the high influent solids in the runoff continued to be a problem, and following an intense rainfall event, both the forebay and floc bloc channel were filled with sediment. As the construction site area was gradually stabilized, the quality of runoff improved. Additional tests in a new flume showed that effective treatment was achieved for new floc blocs at flows of about 2 L/s with suspended solids concentrations up to 5,000 mg/L.

The data in Table 6.25 was typical for the floc block experiments. These samples were obtained near the end of the storm event on May 12, 2000. These data show that high concentrations of suspended solids were present in the pond discharge before and after the storm. The Floc Blocks did not

TABLE 6.24. pH Data for Alum and PAC Treated Stormwater Samples.

Source	Coagulant	Al Conc. (mg/L)	pH	Alkalinity (mg/L as $CaCO_3$)	SS (mg/)L
Oteha Valley Rd SE Pond	initial test water	—	5.64	1	1504
	Alum	8	4.42	<1	71
	Alum	12	4.34	<1	71
	PAC	8	4.64	<1	107
	PAC	12	4.63	<1	85
Lonely Track Rd Gully1	initial test water	—	6.68	16	680
	Alum	8	4.64	<1	117
	Alum	12	4.54	<1	113
	PAC	8	6.03	7	81
	PAC	12	5.54	3	112
Awanohi Rd Adj. Okura Rd	initial test water	—	7.15	60	1130
	Alum	8	5.88	13	84
	Alum	12	4.85	<1	84
	PAC	8	6.71	43	229
	PAC	12	6.45	35	78

TABLE 6.25. Typical Floc Block Experiment Data.

Time	Sample Type	Flow (*L/s*)	pH	SS (mg/)l
0840	Inflow to flume	5	6.04	1,150
0850	Pond discharge	20	6.61	1,870
0900	Inflow via culvert	10	6.97	1,980
0935	Pond discharge	10	6.07	1,810
1035	Pond discharge	6	6.78	1,720

Source: Beca Carter Hollings & Ferner Ltd, undated.

Floc blocks within channel between forebay and pond (*Source:* Beca Carter Hollings & Ferner Ltd, undated).

Pond inlet channel full of sediment and buried floc blocks (Source: Beca Carter Hollings & Ferner Ltd, undated).

appear to have had any significant treatment effect during the period of peak runoff flow.

The Auckland Regional Council concluded that a constant stormwater flow through a floc bloc treatment flume is best in terms of providing the optimum chemical dose for suspended solids removal. It was difficult to set up an array of floc blocs that provided optimal dosing for highly variable flows. They concluded that for any floc bloc system, it was desirable to restrict the maximum flow to about 20 L/s. The treatment capacity of the tested floc bloc (AN2) at a limestone-soil site was about 2 L/s per bloc at 10,000 mg/L suspended solids, and about 1 L/s per bloc at 20,000 mg/L suspended solids. They concluded that floc bloc treatment has a potential for removal of suspended solids, particularly for small catchments, when flow balancing can be achieved prior to treatment, and the stormwater is of consistent quality. However, there were only moderate observed decreases in suspended solids concentrations during the floc block tests (about 50 to 75%), resulting in still very-high effluent concentrations. These limited removals were possibly due to problems associated with highly varying flows, degradation of the floc blocks, and burial of the floc blocks in sediment.

Serious cracking of the floc blocs were noted during an initial dry period of several weeks in the summer. Large pieces fell from the blocs, eventually forming a sticky mass that blocked the bottom of the bloc cages and interfered with the flow paths during subsequent periods of runoff. An intensive rain (about 30 mm of rain during 40 minutes) caused extensive site erosion and the very high sediment loads filled the forebay and treatment flume, in addition to the 60 m^3 of sediment trapped in the pond. Although the floc bloc treatment system was overwhelmed by bedload during this event, the treated pond had lower suspended solids concentrations in the discharge than the other two ponds (2,400 mg/L vs. 7,300 mg/L). During other, more-moderate events, treated pond effluent concentrations were about 500 mg/L, compared to typical effluent concentrations of about 1,000 to 2,000 mg/L from untreated ponds.

The researchers found that a construction site having saturated soils can produce runoff flows of more than 60 L/s per hectare under the intense rainfall conditions that may occur in the Auckland Region. Also, the runoff rates from construction sites can be extremely variable, making it difficult to provide an appropriate array of floc blocs that will provide optimal dosing for such variable flows. Finally, with large numbers of blocs in a single channel system, there could be some potential for overdosing in low-flow conditions.

Liquid Coagulants

Initially, the installation of a runoff-proportional dosing system was designed, which required a flow measurement weir or flume, an ultrasonic sensor and signal generating unit, and a dosing pump. Together with the cost of site preparation, chemical storage tanks and secure shelter, the cost per treatment system was estimated to be approximately $NZ12,000 (about $US9,000). Although the use of a pressure transducer for flow measurement would have reduced the cost to approximately $NZ9,000 (about $US7,000), it would have been difficult to maintain the flow measurement weir because of the large amount of eroded sediment from the construction site. An alternative system that passively provided a chemical dose proportional to rainfall intensity was developed. The rainfall-driven system had the major advantage that it did not require either a runoff flow measurement system or a dosing pump (nor electricity). This system had a total cost of approximately $NZ2,400 (about $US1,800) per installation. The following photos show an example of this system at a New Zealand construction site, including the main internal components.

The rainfall volume collected from a small roof (area proportionate to the construction site drainage area and chemical dosage desired) is used to displace the liquid chemical from a storage tank into the runoff channel before a sediment pond. This design (based on the field trails) assumes that 100% of the rainfall falling onto saturated disturbed areas and 60% of the rainfall falling onto stabilized areas, needs to be treated.

The roof runoff is drained by gravity into an elevated header tank that has a volume below an overflow equal to the detention storage of the site. A second overflow tube above the main overflow tube will cause an increased dosage rate for very high rain intensities. The overflow tubes from this elevated header tank are directed into a displacement tank that is floating in the main chemical tank. As the water flows into this floating displacement tank from the elevated header tank, the chemical is pushed out the reservoir tank and through the dosage line to the dosing location in the flow path.

Example: Volumetric Design

The following example is from the Auckland Regional Council report (Larcombe 1999), assuming a 1 ha (2.5 acre) site and using PAC. The target dosage is 8 mg/L (the actual dosage needs to be determined from bench-scale tests using actual site runoff, or runoff from a similar site). Liquid PAC obtained from Fernz Chemicals contains 10.1% Al_2O_3 by weight, equivalent to 53,500 mg/kg aluminum or 64,200 mg/L aluminum, as the density of PAC is 1.20 g/cc. Therefore, 1L of PAC would treat 8,020 L of construction site runoff at a dose rate of 8 mg aluminum per liter.

Roof Runoff Area Calculation

Each hectare of catchment area would generate about 500 m^3 of runoff per 50 mm of rainfall, assuming the soil was saturated. The volume of PAC required to treat 500 m^3 of runoff is 62.3 L at 8 mg/L. The density of PAC is 1.2 g/cc. Therefore, 74.8 L of rainwater is needed to displace 62.3 L of PAC. This would require an area of 1.5 square meters for a 50 mm rain. Table 6.26 presents the rainfall-rooftop catchment area required for different PAC dose rates (at 10.1% Al_2O_3 by weight).

TABLE 6.26. Rainfall-Rooftop Catchment Area Required for Different PAC Derived Aluminum Dose Rates.

Aluminum Dose Required (mg/L)	Roof Catchment Area per Hectare of Saturated Disturbed Ground (m^2)	Roof Catchment Area per Hectare of Stabilized Catchment (m^2)
2	0.375	0.225
4	0.75	0.45
6	1.125	0.675
8	1.5	0.90
10	1.875	1.125
12	2.25	1.35

Header Tank Size Calculation

The header tank allows initial abstraction losses on the site to be considered (provides a delayed dosage at the beginning of the rain) and continued dosing after the rain ends, but as the runoff continues. For the Auckland test sites, the header tank allows 15 mm of rainfall before dosing commences. This would require a header tank volume below the lowest overflow of 15 L per m^2 of roof rainfall catchment area. The lowest overflow consists of a 4 mm internal diameter tube, while the high rate outlet has sufficient capacity to carry the maximum predicted flow from the roof catchment during short term rainfalls of about 40 mm/hour.

Displacement Tank and Chemical Reservoir Tank Size Calculation

The displacement tank should fit neatly inside the reservoir tank when floating on the liquid chemical. A larger displacement tank and reservoir tank system will reduce the required frequency of servicing. Auckland Regional Council recommends that the minimum displacement tank capacity should be the 24-hour rainfall for a 2-year return period. In

Auckland Regional Council rainfall-driven chemical dosing system.

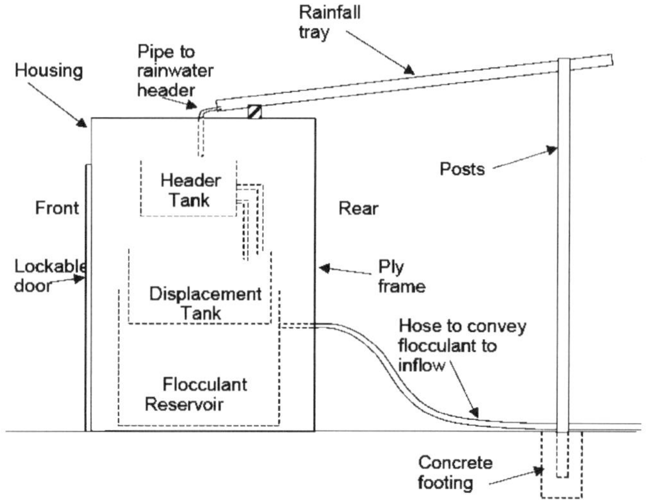

Auckland Regional Council rainfall-driven chemical dosing system, side view (Beca Carter Hollings & Ferner Ltd, undated).

their field studies, this was about 86 mm of rain. With a 1.5 m^2 roof catchment area, this would result in a volume of 129 L. Their standard design used a 400 L displacement tank inside a 550 L reservoir tank, providing dosing of up to 320 L of PAC. Their standard design called for the outlet tubing to be placed at the 400 L chemical level in the reservoir tank so it could hold the contents of two standard 200 L drums of PAC. The outlet tubing level is determined with the floating displacement tank in place to account for the slight displacement associated with the weight of the empty displacement tank.

Setup and Servicing of the Rainfall Driven Dosing System

Header Tank Setup and Maintenance

The level of the low-capacity overflow from the header tank (the vertical position of the tubing exiting the tank) is set

Auckland Regional Council rainfall-driven chemical dosing system, top plywood catchment tray (Beca Carter Hollings & Ferner Ltd, undated).

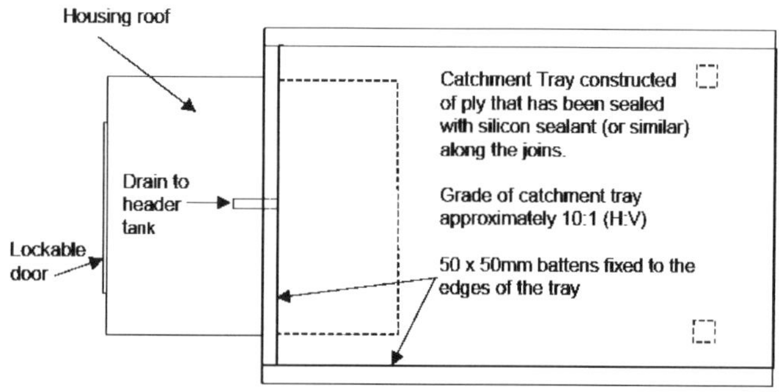

Auckland Regional Council rainfall-driven chemical dosing system, top view (Beca Carter Hollings & Ferner Ltd, undated).

to allow for initial abstractions before chemical dosing starts. In the summer, after a week or more without rain, this was found to be about 15 mm in the Auckland test areas. However, when a very intense rain of about 15 mm in 15 minutes fell on dry ground, substantial runoff occurred, and the delay in the start of dosing resulted in insufficient dosing. In wet weather, the header tank was set with no delay in dosing. During long dry periods, the header tank volume below the low capacity outlet is adjusted to provide for no dosing during the first part of the next rainfall. This is to prevent overdosing of the sediment pond which may cause

Earl Shaver of the Auckland Regional Council showing the main components of the rainfall-driven chemical dosing system.

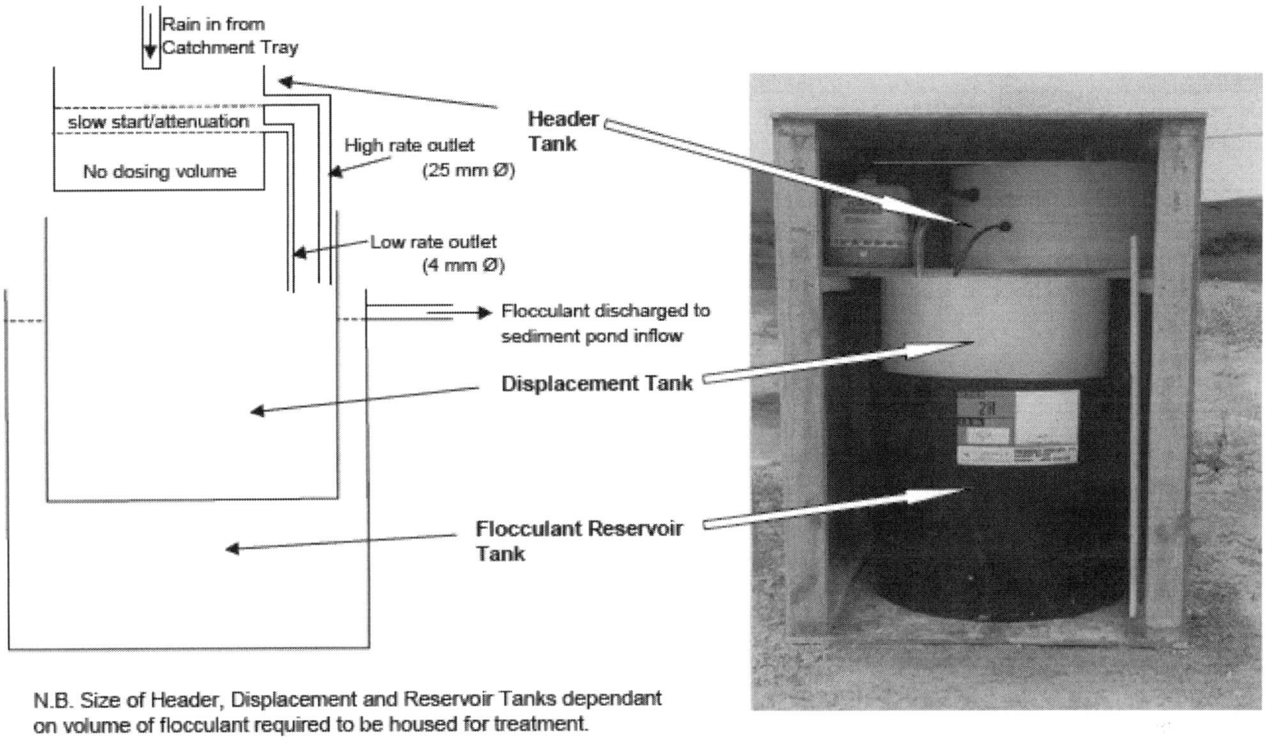

Auckland Regional Council rainfall-driven chemical dosing system showing schematic components with field installation (Beca Carter Hollings & Ferner Ltd, undated).

reduced pH levels and associated increased free aluminum concentrations, plus it also conserves PAC. After each event, the water is removed from the header tank using a siphon. It also would be possible to install a drain valve in the bottom of the header tank for easier emptying.

Displacement and Chemical Reservoir Tank Maintenance

The chemical level in the reservoir tank and the water level in the displacement tank also need to be periodically checked. If the water level is too high, or the chemical level too low, then maintenance is needed. The displacement tank may be either emptied using a siphon, or baled out by hand. The chemical reservoir can be filled using a hand operated drum pump to refill the reservoir from the 200 L delivery drum.

Monitoring and Adjustment for Changing Site Conditions

The passive chemical dosing treatment system needs to be carefully monitored during the first few runoff events to check that the system is effective, and to ensure that overdosing is not occurring. If overdosing is suspected (because the pond dead storage water is exceptionally clear), samples should be analyzed for pH and dissolved aluminum. If overdosing is occurring, reducing the size of the rainfall catchment tray can reduce the chemical dose. This can be done by placing a diagonal batten across the tray and directing some of the runoff through a waste hole.

Field Trials of Chemical-Assisted Sedimentation

Alum Additions

Initial tests indicated that alum additions (at 5.5 mg aluminum/L) worked well under a wide range of rain conditions at a site having limestone soils, including during one event having 25 mm of rain in 25 minutes. During this intense rain, the alum-treated pond had a 92% reduction in suspended solids, compared to only 10% in the same pond for a similar heavy rain during a period of no alum addition. The pH was reduced by about 0.5 pH units and the discharged dissolved aluminum concentration was about 0.1 mg/L during these tests. The pH did not undergo major reductions during bench-scale tests, even when the dosage approached 12.6 mg/L.

Polyaluminum Chloride (PAC) Additions

The runoff from test sites having clay soils had more acidic runoff than the sites that had limestone soils. At the clay sites, alum-treated runoff (after the pond) had pH values that ranged from 4.3 to 5.9, while runoff treated with PAC had pH values ranging from 5.5 to 6.7. They therefore decided that PAC was a more suitable choice, especially for clayey soil conditions. Overall, the Auckland Regional Council has data from 21 different sediment ponds that used passive PAC additions, with drainage areas ranging from 0.5 to 15 ha (1.3 to 38 acres). The overall suspended solids

TABLE 6.27. Suspended Solids Removal from PAC Treated Stormwater.

		Inflow		Outflow		
Pond	Date	Flow (L/sec)	SS (mg/L)	Flow (L/sec)	SS (mg/L)	SS Reduction
Mason's Rd	28.11.1998	3	26,300	3	144	99.4
Mason's Rd	04.12.1998	2	5,100	2	40	99.2
OVR E	13.06.1999	15	1,639	8	51	96
OVR E	04.07.1999	2	749	2	56	92
23800E	28.11.1998	8	14,800	6	966	93
23800E	22.01.1999	1	14,700	2	67	99
B1 Gully	08.04.1999	0.3	4,300	0.4	3	99.6
B1 Gully	01.05.1999	0.5	16,9000	3.0	59	99.6

Source: Beca Carter Hollings & Ferner Ltd, undated.

treatment efficiency of PAC-treated ponds has been between 90–99% for ponds having good physical designs. Lower treatment efficiencies have occurred where there have been problems with decants not operating properly, or physical problems such as multiple inflow points, high inflow energy, and poor separation of inlets and outlets. The photo on the next pages shows the typical multiple decant risers used at Auckland Regional Council sediment pond sites to allow more efficient settling of the floc.

PAC was tested for ALPURT project during the 1998/99 summer, and during the winter of 1999. A total of 21 systems were used, with contributing catchments ranging between 0.5 and 15 hectares. Table 6.27 presents representative data for PAC-dosed stormwater from sites having clay soils. The data shows that a high degree of suspended-solids reduction was achieved in the PAC dosed ponds. The influent concentrations of suspended solids for the PAC-treated ponds ranged from 750 to 26,300 mg/L (median of about 16,000 mg/), while the treated effluent ranged from 3 to 966 mg/L (median of approximately 50 to 100 mg/L). The percentage suspended solids reductions ranged from 92 to 99%, with a median of about 99%. The untreated pond had much poorer levels of treatment (about 10%).

There was considerable variation of inflow suspended solids concentrations between the different ponds sampled (Table 6.28). These large variations reflected the characteristics and condition of the construction sites. All of the treated ponds achieved good suspended solids reductions (77–98%) compared to that of untreated ponds (4–12%). The PAC dosing caused an obvious reduction in pH in all ponds, except at Lonely Culvert. It is interesting to note that the dissolved aluminum concentrations in the outflow from the untreated pond were much higher (0.29–0.31 mg/L) than in the outflows from the treated ponds (0.010–0.084 mg/L). The dissolved aluminum concentration is related to the characteristics of the suspended solids, with high concentrations of dissolved aluminum occurring in samples that also had high concentrations of very fine suspended solids. Therefore, the effluent from the untreated ponds, having high concentrations of fine sediment, also had high concentrations of dissolved aluminum. When the PAC was added at too high a concentration, the pH levels dropped to as low as 4.7, although the effluent dissolved aluminum was still low and the suspended solids concentrations were very low (as low as 10 mg/L). Typical effluent pH conditions were between 6 and 7.

The dissolved aluminum concentrations in the outflows from the treated pond samples shown in Table 6.29 were below the USEPA aquatic life chronic criterion of 0.087 mg/L (4-day average not to be exceeded; the data shown in this table are for instantaneous grab samples), and well below the acute criterion of 0.750 mg/L (1-hour average not to be exceeded). These data show very high removals of suspended solids, particularly in the ponds with

TABLE 6.28. Inflows and Discharges of PAC Treated Ponds.

		Inflow				Outflow				
Pond	Time	Flow (L/sec)	SS (mg/L)	pH	Al (mg/L)	Flow (L/sec)	SS (mg/)L	pH	Al (mg/L)	Reduction in SS (%)
1.3.8 Over	0850	12	238	8.97	0.084	9	53	6.66	0.026	77
	1135	20	253	9.97	0.077	12	55	6.79	0.068	78
Lonely	0938	40	25,830	6.83	0.052	3	266	7.62	0.072	98
	1045	15	13,310	6.62	0.093	20	214	7.02	0.018	98
21340	918	8	399	8.78	0.25	3	40	6.56	0.016	89
	1110	7	2,564	7.03	0.11	15	57	6.55	0.01	88
D5	0910	6	2,132	6.81	0.16	4	65	5.96	0.025	96
	1100	7	2,546	7.03	0.11	4	56	5.47	0.01	97
Untreated Pond	1930	12	1,571	7.88	0.22	4	1,378	7.74	0.31	12
	1100	9	1,522	8.02	0.17	4	1,459	7.83	0.29	4

Source: Beca Carter Hollings & Ferner Ltd, undated.

TABLE 6.29. PAC Dosed Sediment Retention Pond Monitoring Data, October 21, 1999.

Pond	Inlet/Outlet	Time	Flow (L/S)	pH	SS (mg/L)	Al (mg/L)	Hard (mg/)L	Reduction of SS (%)
Mason's	In	1700	3	6.44	4,704	0.02	72	
	Out	1705	3	4.44	41	1.10	49	99
OVRE	In	1720	12	8.80	23,240	0.29	65	
	Out	1725	10	9.04	272	0.07	95	98
OVRW	In	1740	8	6.86	28,845	0.02	194	
	Out	1745	10	6.89	338	0.02	85	98
2444OW	In	1750	3	—	164	0.20	58	
	Out	1745	2	4.70	15	0.34	47	90
D5	In	1815	6	7.65	770	0.03	206	
	Out	1820	5	6.15	36	0.01	159	95
2134OW	In	1825	3	10.73	128	0.31	64	
	Out	1827	4	6.84	14	0.03	81	89
Debs	In	1845	4	11.47	752	0.21	135	
	Out	1850	6	9.82	279	0.31	98	62
Lonely	In	1855	4	11.12	254	0.07	113	
	Out	1900	8	8.31	72	0.16	113	71
Untreated	Out	1835	3	8.63	712	0.06	89	

Source: Beca Carter Hollings & Ferner Ltd, undated.

high-suspended solids in the inflows. In contrast, the untreated pond had the highest concentrations of suspended solids in the outflow. The data for the Mason's Rd pond provides an example of a PAC overdose, where the pH after dosing was reduced to 4.44, and the dissolved aluminum concentration was at a high level of 1.1mg/L. The outflow data for pond 2444OW also indicates a possible PAC overdose, with a low pH of 4.70, although the dissolved aluminum was not markedly elevated.

Multi-level, perforated, floating discharges (decants) to better retain floc.

Design of Sediment Ponds with Aluminum Coagulant Treatment

Although chemical treatment using aluminum coagulants is capable of achieving effective sediment removal from stormwater (with relatively brief detention time for settlement in quiescent conditions), there are practical difficulties in achieving quiescent conditions in construction site ponds when high flows are being discharged into a small pond. The Auckland Regional Council recommends a minimum size of 1.5% (150 cubic meters per hectare) for aluminum-coagulant treated ponds. Analysis of the long-term rainfall and construction-site suspended solids data obtained during the field trials shows that more than

TABLE 6.30. Summary of Advantages of PAC Treatment of Construction Site Runoff for Normal Catchments during a Construction Season.

	Wet Sediment Pond Size (% of drainage area)		
	3%	2%	1.5%
1. Without PAC treatment:			
Total sediment discharged to receiving water (tonnes dry wt per hectare)	5.8	9.2	12.0
Efficiency of sediment removal in pond (%)	81	69	60
2. With PAC treatment:			
Total sediment discharged to receiving water (tonnes dry wt per hectare)	1.0	2.1	2.8
Efficiency of sediment removal in pond (%)	97	93	90

60% of the sediment from a construction site occurs during the two or three rainstorms per construction season which exceed 30 mm in 24 hours.

Table 6.30 shows the expected advantages of using PAC-assisted sedimentation for different sizes of wet sediment ponds in the Auckland, New Zealand, area. Chemical treatment results in a major improvement in the efficiency of sediment capture during rainstorms that exceed the hydraulic capacity of a sediment pond. This is indicated by the large improvements in sediment capture for the smaller ponds with PAC addition shown in Table 6.30.

FILTER FENCES FOR CONSTRUCTION SITE SEDIMENT CONTROL

Filter fences do not operate in the manner that their name implies. The fencing material is not acting as a filter, e.g., straining particles from the passing water. In fact, the filter fences operates by creating a small pond behind the fence, which allows runoff to slow down and pool, which allows sediment to settle in the area behind the filter fence. There are three aspects of filter fences that can be evaluated, as demonstrated in the following examples: (1) sediment capture behind the fence, (2) water flow rate reduction down slope, and (3) pressure forces on the fence from the water and resisting forces from the soil on the fence stakes. The first two aspects determine the erosion and sediment control benefits of filter fences, while the third aspect determines how filter fences may fail structurally.

Sediment Capture behind Filter Fences

Relatively few field investigations have been conducted to examine the effectiveness of filter fences, and other controls, at construction sites. Important tests have been performed by Barrett, *et al.* (1995), Horner, *et al.* (1990), Schueler and Lugbill (1990), and Smoot, *et al.* (1992). Caltrans is also currently conducting comprehensive tests of construction erosion controls and their results should become available soon. The sidebar in Chapter 1 presents a recent silt fence evaluation project conducted in Alabama.

Perhaps the most comprehensive study of filter fences was conducted by Barrett, *et al.* (1995) at Austin, TX, area highway construction sites, supplemented with controlled laboratory tests. Silt fences at six active highway construction sites were evaluated in terms of suspended solids and turbidity reduction. Two installations used non-woven fabrics, and four installations used woven fabrics. Manual grab sampling was used to obtain representative sediment samples of all size distributions during 10 rains. Uncontrolled discharges due to obvious filter fence failures (mostly undercutting flows or tears in the fabric) were excluded from sampling; only locations where the flows passed through the fabric were sampled. Samples were collected upslope of the pooled water behind the filter fence, in the pool backed up by the filter fence, and downstream of the filter fence. This sampling strategy was used to differentiate sedimentation from filtration effects, and to obtain an overall control efficiency. Because of highly variable concentrations above the pool, most of their data analysis relied on comparisons between the samples collected from the pool and the effluent from the filter fabric, reflecting filtering removal and not sedimentation.

The observed suspended solids removal rates were highly variable, ranging from –61 to 54%, with a median of 0%. Typical effluent suspended solids concentrations after the filter fence were about 500 mg/L. Similar poor results were obtained for turbidity removals (–32 to 49% range, with a median removal of 2%). As indicated by the negative removal rates, the effluent from the filter fences sometimes had greater suspended solids concentrations than were found in the pool. The removal of suspended solids due to sedimentation, however, was estimated to be about 50%, based on partial field observations. At one location where the lower portion of the fabric was clogged, a shallow upstream pool lasted for an extended period and removals of about 65% were measured.

The poor removal efficiency due to filtration was explained by comparing the particle sizes of the suspended solids and the apparent opening sizes of the fabrics (typically from 100 to 1,000 μm). Silt and clay-sized particles comprised the majority of the solids collected (68 to 100%, with a median of 96%) from the pond and below the filter fences. Any large particles present in the flowing waters were thought to have been settled in the pool before the fence. The diameters of the remaining particles passing through the fence were therefore much smaller than the openings in the fabric and were able to pass through unhindered. Earlier work by Schueler and Lugbill (1990) in Maryland substantiated the small particles observed in Texas. During settling column studies on construction site runoff, Schueler and Lugbill found that 90% of the incoming sediment was smaller than 15 μm, with the largest particles observed being only 50 μm. During their sediment pond evaluation tests, however, they did observe sediment deltas forming near the influent location, indicating that sand-sized particles were transported to the sediment ponds and represented a minor portion of the total load. These larger particles were apparently not included in the grab samples as they form part of the bed load.

Barrett, *et al.* (1995) found that filter fence installations were not designed as hydraulic structures, and frequently, failures were caused by excessive runoff. Runoff around the ends of fences, and even over-topping of the fences was observed several times during their monitoring project. However, other downstream controls were in place to

Sediment flowing under hay bale barrier.

Well-installed filter fabric fence, with bottom of fabric buried and backfilled to prevent underflow of sediment.

Hay bale barrier along edge of pavement.

Evidence of underflow erosion beneath improperly installed filter fabric fence.

Holes in filter fabric fencing.

Filter fabric fence installed close to construction area.

Large sediment load captured by filter fence, maximum load before needed maintenance (J. Voorhees photo).

Same site as photo to left showing sediment load overtopping filter fabric fence due to lack of maintenance (J. Voorhees photo).

Filter fence along edge of property line (front side to left).

Filter fence along edge of property line (back side).

Multiple rows of filter fabric fences and tree barriers to mark edge of disturbed zone.

mitigate these failures. Besides failures caused by lack of hydraulic design, they also observed deficiencies in performance that were caused by improper installation and maintenance, including:

- Inadequate filter fabric splicing
- Fence failure due to sustained over-topping
- Unrepaired holes in fabric
- Flow beneath fabric due to inadequate trenching of the bottoms of the fabric fences into the ground

Laboratory flume tests were also conducted on filter fabrics, enabling flow rates and suspended solids concentrations to be controlled at specific conditions. Austin silty clay, after passing through a 3 mm sieve, was used to make a test slurry. The median particle size in this mixture was 20 μm, and 30% was finer than 3 μm. The apparent openings in the filter fabrics tested ranged from 600 to 850 μm for 3 woven fabrics and 150 μm for the one non-woven fabric tested. During testing, the woven fabrics had median suspended solids removal rates of 68 to 87% (ranges of 46 to 97%), while the non-woven fabric had a median removal rate of 93% (range of 73 to 99%). The non-woven fabric also had the longest detention times during the tests due to its lower pass-through flow rate. In comparison, a rock berm was also tested (having the highest flow rate and therefore shortest detention time) and had a median SS removal efficiency of only 42% (36 to 49% range). The suspended-solids reductions in the testing flume were 34% without any controls in place due to sedimentation of the larger test particles while flowing over the rough bed. This high background reduction level therefore significantly reduces these reported flume test measurements with controls. The corrected rock berm removal rate was only 7%, for example, after taking into consideration the background reductions. Similar reductions would have to be made for the filter fabric test results.

An interesting observation during the flume tests was that while the detention times increased with time since the start of the tests due to partial clogging of the fabrics, the woven fabrics all had decreased detention times after being exposed to large rains. Apparently, the rains helped wash some of the caked-on mud from the fabrics. This was not observed for the non-woven fabrics where clogging was internal and more permanent. During laboratory tests on stormwater filtration, several filter types were tested by Clark and Pitt (1999). They found that all of the fabrics examined totally clogged after accumulating a layer of about 3 mm of clay. This clogging layer preferentially forms near the bottom of the fabric, usually indicating the depth of the ponding. This clogging significantly decreases the flow rates through the fabric, allowing extended detention and therefore increased sediment trapping performance.

Barrett, *et al.* (1995) concluded that the poor filtering performance of the filter fences in good condition was due to the small particles in comparison to the large fabric openings. Previously reported high filtration control efficiencies conducted during laboratory experiments were faulty due to the use of unrealistically large test particles. Median particles during field tests at construction sites indicate that almost all of the particles in the runoff are silts and clays. The relatively minor sand fractions are easily deposited during sheetflows, or in ponded areas. Sedimentation effectiveness was found to be highly dependent on the detention time in the ponded areas behind the filter fabrics. The detention time is controlled by the geometry of the upstream pond, hydraulic properties of the fabric, and maintenance of the filter fence. Holes in the fabric, under-cutting due to inadequate trenching of the fabric bottom, and overtopping or bypassing around the ends of filter fabric fences, all effectively decreased the detention time in the pond behind the fabrics and contributed to very low observed field performance of filter fabrics.

Example: Calculation of Sediment Capture Behind Filter Fence

It is possible to calculate the expected level of control for a filter fence at a specific site using the upflow velocity concept presented earlier:

$$v = \frac{Q_{out}}{A}$$

The performance of a filter fence can therefore be calculated by knowing the ratio of the discharge through the fence divided by the surface area of the ponded area. Both of these values are directly related to the depth of water detained behind the filter fence. This value can be easily calculated assuming an even slope uphill from the fence and using the manufacturer's value for unit area flow capacity. The ponded surface area increases directly with the water depth, depending on the slope. The total outfall rate also increases directly with the water depth. Therefore, the critical particles being trapped in the pond behind the filter fence is only dependent on the slope and fabric. Figure 6.17 is a plot of the particle size controlled, in μm, for different ground slopes (%) and filter fabric flow rates (ft/sec), using Stokes' law for calculating the critical particle sizes associated with the upflow velocity:

$$v = \frac{1}{18}\left[\frac{g}{\kappa}(spgr - 1)\right]d^2$$

where,

v = settling rate of particle, cm/sec
g = 981 cm/sec^2
κ = kinematic viscosity = 0.01 cm^2/sec
$spgr$ = specific gravity of particulate (often assumed to be 2.65—the specific gravity of sand)
d = particle diameter, cm

Figure 6.18 can be used to estimate the approximate

Silt captured on woven filter fabric.

Layer of silt captured against bottom edge of newly installed filter fabric fence.

Ponded area sediment accumulation and smear of silt on fence.

Bulk of sediment captured behind filter fence in ponded area.

Heavy sediment load in ponded area.

Sediment in ponded area.

Sediment in ponded area.

Sediment in ponded area.

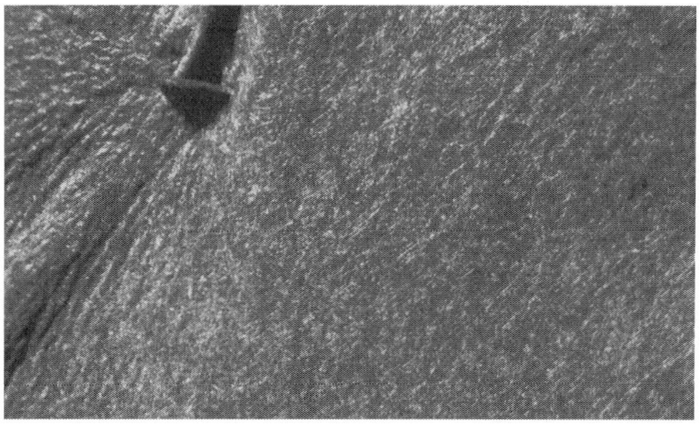
Non-woven filter fabric.

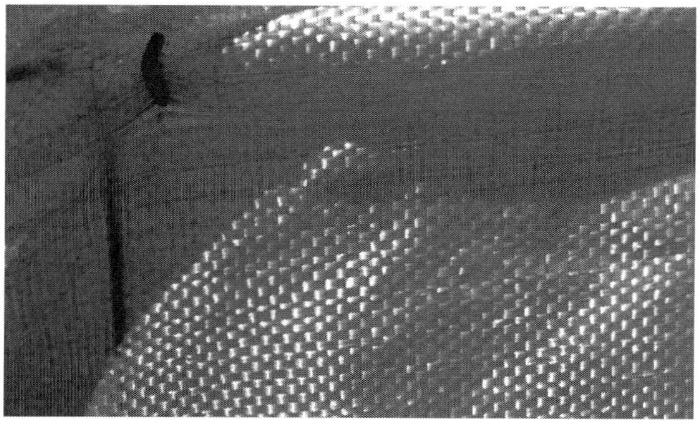
Woven filter fabric material.

Filter fabric fence on mulched slope (SCS photo).

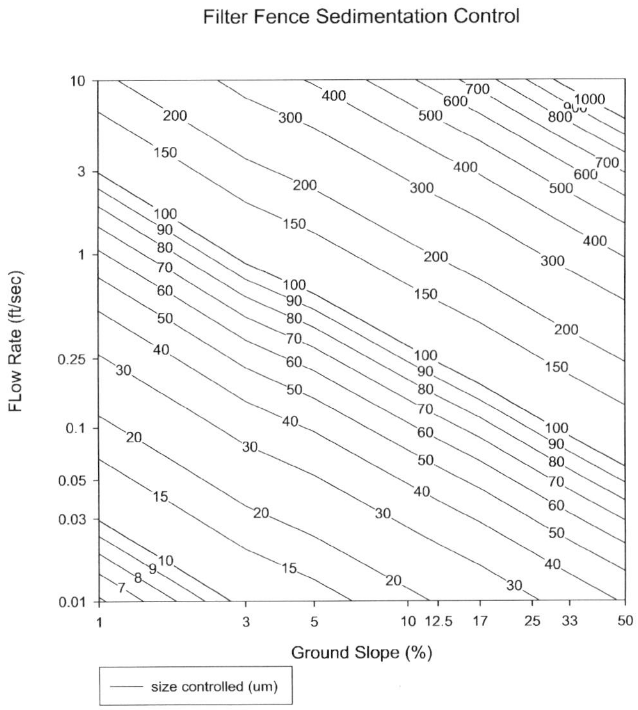

Figure 6.18. Filter fabric conditions and critical particle size controlled.

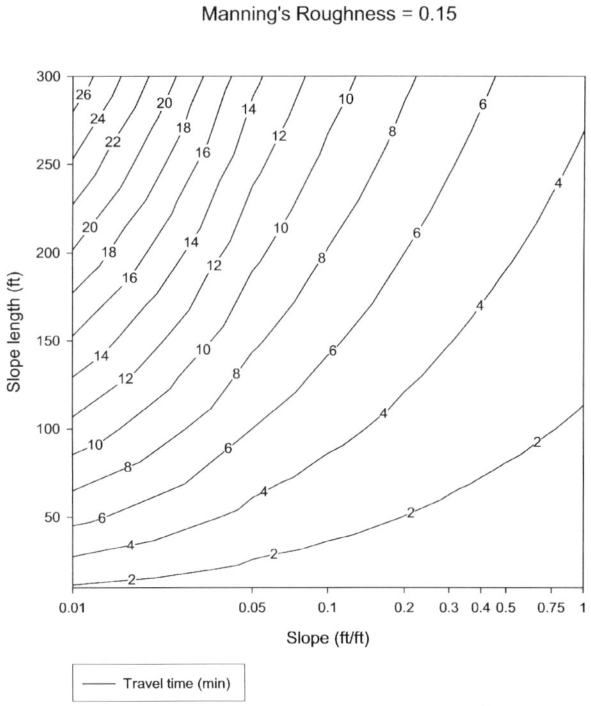

Figure 6.19. Sheetflow travel times for different slopes.

suspended solids control corresponding to the critical particle size. For example, if the calculated critical particle size is 10 μm (such as for a 2% slope and a 0.02 ft/sec filter fabric flow rate), the expected suspended-solids control would be about 25 to 45% for the size distributions likely appropriate for construction site runoff. A 5% slope and 0.25 ft/sec flow rate would result in about a 60 μm critical particle size, and the suspended-solids control would only be about 5 to 15%.

Filter Fences to Slow Water Flowing Down Critical Slopes

Filter fences intercepting sheetflows also may be used to slow the water flowing down critical slopes. The upslope length of the ponded area will be obviously protected from rain impaction and by flowing water. This length can be estimated for different water depths impounded behind a filter fence. As an example, for a 5% slope and a 1-ft water depth, the ponding would extend uphill 20 ft. In addition, some of the downslope area below the filter fence (if not installed on the toe of the slope, as generally recommended), will also have reduced flow velocities, compared to the same slope without the filter fence. WinDETPOND can be used to calculate the reduction in flow rates for flows entering the ponded area compared to the discharge water through a filter fence. Generally, non-woven filter fabrics have much lower flow rates compared to woven filter fabrics. The sheetflow calculation information in Chapter 4 also can be used to estimate the flow rates on slopes of different roughness and slopes. As an example, Figure 6.19 (a repeat of Figure 4-12) shows the sheetflow travel times for different slopes having a roughness value of 0.15, corresponding to relatively short grass. A slope of 10% that is 100 ft long would have a travel time of about 5 minutes, or a velocity of about 0.33 ft/sec. There are non-woven fabrics that have flow rates appreciably less than this value, so a filter fabric could result in critical slopes being exposed to reduced periods of high flows. Of course, using multiple filter fences along a slope could help reduce the effective speed of the flowing water, but the accumulative amount of water reaching the lowest fence may

Jamie Lyles; UA student, filter fence check dams on a construction site.

be excessive, and the silt fences would have to be closely spaced, not a very satisfactory solution.

Pressure Force on Filter Fences

The pressure equation can be used to calculate the forces acting on filter fences. The following calculation shows the resisting force needed for a 10 ft span of filter fence with 2 ft of standing water:

$$F_1 = \frac{(62.4\ \text{lb}/\text{ft}^3)(\text{Cross}-\text{sectional area of flow})(\text{Depth of flow})}{2}$$

$$F_1 = \frac{(62.4\ \text{lb}/\text{ft}^3)(20\ \text{ft}^2)(2\ \text{ft})}{2} = 1{,}248\ \text{lb}$$

The momentum equation can be used when the flow rates should be considered:

$$F_1 - F_2 = PQ(V_2 - V_1)$$

Therefore, the forces acting on a filter fence can be very large and the filter-fence stake systems must be selected to withstand this force and prevent tipping or breaking of the support posts. In addition, the resisting forces of the soil also act on the fence stake to hold it upright, which also must be considered. Wet and soft soils may need long stakes driven deeply in the ground to resist this tipping/breaking pressure.

Guidance for Filter Fence Construction

The following is excerpted from the 2003 edition of the *Alabama Handbook for Erosion Control* (USDA, 2003). As noted above, most failures of filter fences are associated with poor placement and maintenance. The following is an example of the typical guidance provided in regional construction site erosion control handbooks. These construction details are critical for proper operation of these common construction site controls.

Practice Description

A sediment barrier is a temporary structure used across a landscape to reduce the quantity of sediment that is moving farther downslope. Commonly used barriers include silt fence (a geotextile fabric which is trenched into the ground and attached to supporting posts) or hay bales trenched into the ground. Other barrier materials include sand bags, brush piles and various man-made materials that can be used in a similar manner as silt fence and hay bales. This practice applies where sheet and rill erosion occurs on small disturbed areas. Barriers intercept runoff from upslope areas to form ponds that temporarily store runoff and allow sediment to settle out of the water and stay on the construction site. Barriers can also prevent sheet erosion by decreasing the velocity of the runoff.

Planning Considerations

Sediment barriers may be used on developing sites. They should be installed on the contour so that flows will not concentrate and cause bypassing, overtopping and/or failure. The two most commonly used sediment barriers are silt fences and hay bales. Silt fences are usually preferable to hay bales because silt fences can trap a much higher percentage of suspended solids. Silt fences are the only barrier covered in the current 2003 edition of the *Alabama Handbook*. The success of silt fences depends on proper installation so as to develop maximum efficiency of sediment trapping. Silt fences should be carefully installed to meet the intended purpose.

A silt fence is specifically designed to retain sediment transported by sheetflow from disturbed areas, while allowing water to pass through the fence. Silt fences should be installed to be stable under the flows expected from the site. Silt fences should not be installed across streams, ditches, waterways, or other concentrated flow areas. Silt fences are composed of woven geotextile supported between steel or wooden posts. Silt fences are commercially available with geotextile attached to the post and can be rolled out and installed by driving the post into the ground. This type of silt fence is simple to install, but more expensive than some other installations. Silt fences must be trenched in at the bottom to prevent runoff from undermining the fence and developing rills under the fence. Locations with high runoff flows or velocities should use wire reinforcement.

Design Criteria

Silt fences are normally limited to situations in which only sheet or overland flow is expected. They normally cannot filter the volumes of water generated by channel flow. Silt fences are normally constructed of synthetic fabric (woven geotextile) and the life is expected to be the duration of most construction projects. Silt fence fabric should conform to the requirements of Table 6.31. The total drainage area behind the silt fence should not exceed 1/4 acre per 100 linear feet of silt fence for non-reinforced fence and 1/2 acre per 100 feet of wire reinforced fence. When all runoff from the drainage area is to be stored behind the fence (i.e. there is no stormwater disposal system in place) the maximum slope length behind the fence should not exceed those shown in Table 6.32.

Type A Silt Fence

Type A fence is 36″ wide (tall) with wire reinforcements. The wire reinforcement is necessary because this fabric

TABLE 6.31. Specifications for Silt Fence (USDA 2003).

Specifications	Type A	Type B	Type C
Tensile Strength (Lbs. Min.1ASTM D-4632)	Warp–260 Fill–100	Warp–260 Fill–100	Warp–260 Fill–100
Elongation (% Max.) (ASTM D-4632)	40	40	40
AOS (Apparent Opening Size) (Max. Sieve Size) (ASTM D-4751)	no. 30	no. 30	no. 30
Flow Rate (Gal/Min/Sq. Ft.) (GDT-87)	70	25	25
Ultraviolet Stability2 (ASTM D-4632 after 300 hours weathering in accordance with ASTM D-4355) With PAC treatment:	80	80	80
Bursting Strength (PSI Min.) (ASTM D-3786 Diaphragm Bursting Strength Tester)	175	175	175
Minimum Fabric Width (Inches)	36	36	22

[1]Minimum roll average of 5 specimens.
[2]Percent of required initial minimum tensile strength.
Note: 70 gal/min/ft^2 = 0.15 ft/sec and 25 gal/min/ft^2 = 0.06 ft/sec

allows almost three times the flow rate as type B silt fence. Type A silt fence should be used where runoff flows or velocities are particularly high or where slopes exceed a vertical height of 10 feet. Provide a riprap splash pad or other outlet protection device for any point where flow may overtop the sediment fence. Ensure that the maximum height of the fence at a protected, reinforced, outlet does not exceed 1 foot and that support post spacing does not exceed 4 feet.

The silt fence should be installed as shown in Figure 6.20. Materials for posts and fasteners are shown in Tables 6.33 and 6.34. Details for overlap of the silt fence and fastener placement are shown in Figure 6.23.

Type B Silt Fence

This 36″ wide (tall) filter fabric should be used on developments where the life of the project is greater than or equal to 6 months. The silt fence should be installed as shown in Figure 6.21. Materials for posts and fasteners are shown in Tables 6.33 and 6.34. Details for overlap of the silt fence and fastener placement are shown in Figure 6.23.

TABLE 6.32. Slope Limitations for Silt Fence (USDA 2003).

Land Slope (Percent)	Maximum Slope Length Above Fence (Feet)
<2	100
2 to 5	75
5 to 10	50
10 to 20*	25
>20	15

*In areas where the slope is greater than 10%, a flat area length of 10 feet between the toe of the slope to the fence should be provided.

TABLE 6.33. Post Size for Silt Fence (USDA 2003).

	Minimum Length	Type of Post	Size of Post
Type A	4′	Steel	1.3 lb/ft minimum
Type B	4′	Soft Wood Oak Steel	3″ diameter or 2 × 41.5″ × 1.5″ 1.3 lb/ft minimum
Type C	3′	Soft Wood Oak Steel	2″ diameter or 2 × 21″ × 1″ 0.75 lb/ft minimum

Type C Silt Fence

Though only 22″ wide (tall), this filter fabric allows the same unit area flow rate as Type B silt fence. Type C silt fence should be limited to use on minor projects, such as residential home sites or small commercial developments where permanent stabilization will be achieved in less than 6 months. The silt fence should be installed as shown in Figure 6.22. Materials for posts and fasteners are shown in Tables 6.33 and 6.34. Details for overlap of the silt fence and fastener placement are shown in Figure 6.23.

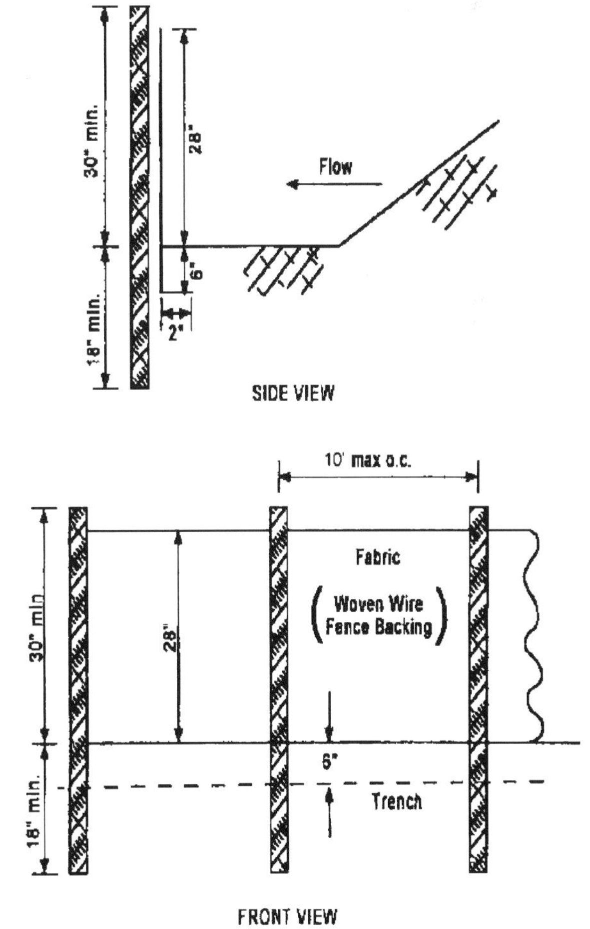

Figure 6.20. Silt Fence—Type A (1) For fabric material requirements see Table 6.31 (2) For post material requirements see Tables 6.32 and 6.33 (USDA 2003).

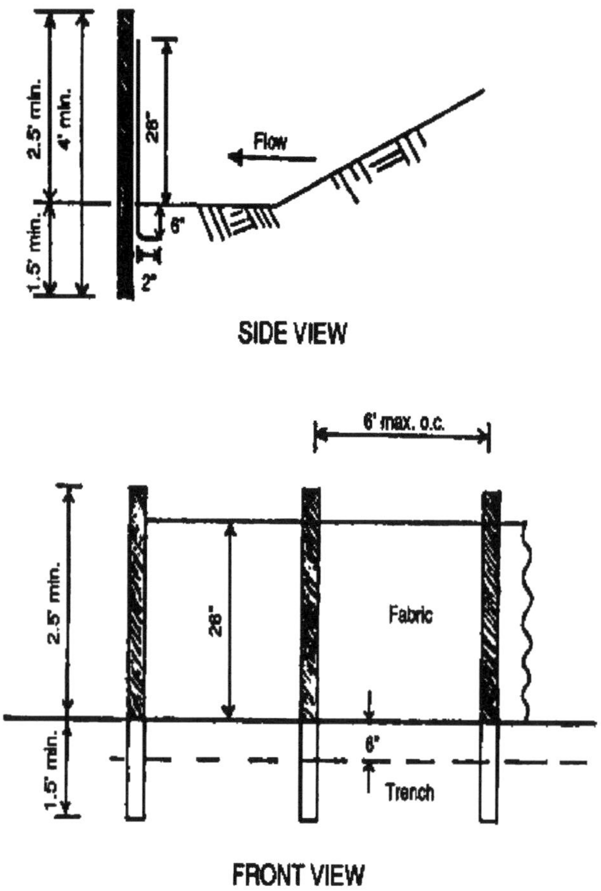

Figure 6.21. Silt Fence—Type B (1) For fabric material requirements see Table 6.33 (2) For post material requirements see Tables 6.33 and 6.34 (USDA 2003).

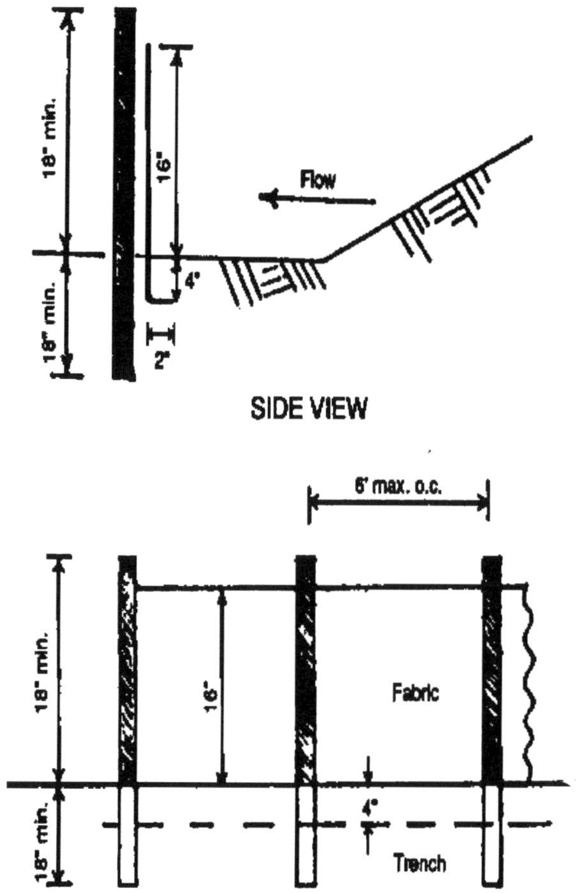

Figure 6.22. Silt Fence—Type C (1) For fabric material requirements see Table (2) For post material requirements see Tables 6.33 and 6.34 (USDA 2003).

TABLE 6.34. Wood Post Fasteners for Silt Fence (USDA 2003).

	Gauge	Crown	Legs	Staples/Post
Wire Staples	17 minimum	3/4″ wide	1/2″ long	5 minimum
		Length	Button Heads	Nail/Post
Nails	14 minimum	1″	3/4″ long	4 minimum

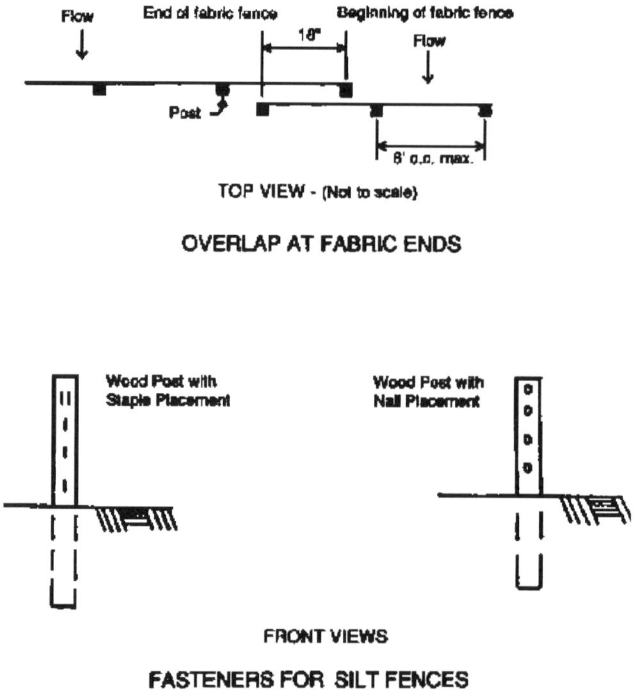

Figure 6.23. Silt Fence Installation Details (USDA 2003).

Caroline Sandel, UA student: silt fences around storage pile.

Caroline Sandel, UA student: silt fence overloaded.

Example Project Assignment on Silt Fences and Ponds

The following is excerpted from a homework assignment prepared by Mark Koopman, a student at the University of Alabama at Birmingham, as part of the Construction Site Erosion and Sediment Control Class taken during the summer of 2005. This assignment was to examine the use of silt fences and silt ponds on a construction site that has been studied during the class term.

Site Description

A current project to correct drainage problems in Birmingham's (AL) Caldwell Park, Figure 6.24, will be considered and assessed for this study. Grading operations began on July 20, 2005. The southern half of the Caldwell Park is bordered by Highland Ave. S, the northeast by Niazuma Ave. S, and the northwest by a parking lot and private property.

Caldwell Park has an area of approximately 2.5 acres and is used for various recreational and leisure activities by the community, as well as for concerts and larger events like Doo Dah Day, an annual Birmingham event sponsored by local animal rescue groups. The site has mature trees around the perimeter bounded by streets, and the sides of the park have moderately steep grassy slopes that come down to a more gradual slope in the center. There is a concrete and brick stage area at the northwestern end of the park, which is also the lowest elevation area of the site. These lower areas have had drainage difficulties after moderate to heavy rain events.

The site lies in a subdrainage of the northern slope of Red Mountain in the Five Mile Creek watershed. The USDA Soil Survey of Jefferson County, AL, indicates that the soil is in hydrological soil group B, Fullerton-Urban land complex. The top 6 inches is a dark brown cherty silt loam and the next 59 inches are a cherty silty clay loam. The soil has erosion k factors between 0.20 and 0.28 and a *T* value of 5.

Figure 6.25 shows a view of the site taken on July 22, 2005 from the south end of the park. The top soil has been moved into an unprotected pile in the middle of the park (with a shovel earth moving machine parked on top and a blue sieving apparatus to the left in this image) and the site has been initially graded. Behind the equipment and vehicles one can see a brick building, a theater that has experienced minor damage due to flooding from the park.

Site Soils

The USDA Soil Survey of Jefferson County, Alabama (map 12) indicates that the soil for the entire site is

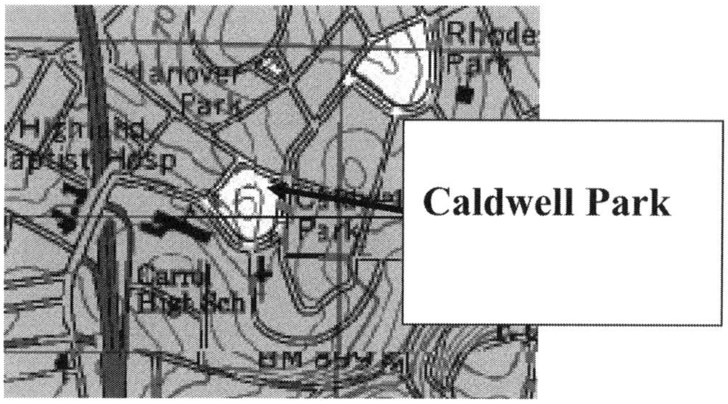

Figure 6.24. USGS Birmingham North and South quadrangle maps, spliced, with 20 ft contour intervals, showing a drainage vector slightly west of north. USGS topographical map.

hydrologic group B, Fullerton-Urban land complex, with about 6 inch depth, dark brown cherty silt loam and the next 59 inches being a cherty silty clay loam. The soils in the Fullerton Urban Land Complex are described in the soil survey as being well drained but not completely mapped and the soil has not been specifically tested. The soil has a pH between 4.5 and 5.5. The soil has an Erosion *k* Factors of 0.28 over the first six inches, 0.24 from 6 to 35 inches, a value of 0.20 from 35 to 59 inches and a *T* value of 5. The allowable shear stress for the silt loam soil is assumed to be 0.11 lb/ft^2.

Construction Plan, Excavation and Flow Paths

The preconstruction park topography is seen in Figure 6.25, with elevations given in 1 ft contour intervals. Sidewalks and utilities have been removed and top soil reserved in an unprotected pile in the active construction area, the floor of the park. The park sits in a bowl with moderately steep side slopes and a more mildly sloping floor. Side slopes are between 28 and 36% grade and the floor of the park has approximately a 2.6% grade. The active construction area is completely within the floor area of the park.

Figure 6.25. View of the site taken on July 22, 2005.

The peak run off rate for a 25 year storm was calculated to be 6.71 cfs using Win TR-55. The time of concentration, T_C, was estimated to be 2.52 minutes.

Only small changes are being made to the elevation contours of the park. These changes are significant, though, as seen if Figure 6.27 where a new sub drainage is being added to divert flow from the north eastern portion of the park to one of two new storm sewer structures. This new subdrainage is being created by raising the sidewalk elevation slightly. The blue arrows indicate the direction of water flow. A second main storm drain structure is being installed, marked 1/7, in the north western end of the park.

Phases of Construction and Soil Loss Calculations

RUSLE The Revised Universal Soil Loss Equation, was used to calculate the expected soil loss during the project. Since the project does not involve erection of a building, the time frame for construction is somewhat shorter than most construction sties, about 2.5 months. Additionally, with the small size of the park, 2.48 ac, the active construction area will occur in a single phase, followed by planting sod and various trees, shrubs and flower beds. Therefore, calculations were made for two phases, Phase I is the active construction phase and Phase II is the park in its "as finished" condition.

The Phase I analysis was divided into two zones, the first being the undisturbed steeper side slopes of the park with slopes between 28% and 36% and lengths between 10 and 50 ft, which have well established Bermuda grass in good condition. This second zone is the active construction area and is tan colored, occupying the floor of the park. This zone has a slope of 2.6% and a length of 375 ft. Phase I is shown in Figure 6.27. Phase II, with completed sidewalk construction and planting of sod, is shown in Figure 6.28.

Areas for the various zones in the two phases were calculated by converting the overlay zones into solid grey scale images in Adobe Photoshop, as seen in Figure 6.29 and converting the percentage area of the screen by pixel into sub-area values by multiplying by the total area of the site.

An annual *R* value of 350 was prorated to the relevant time frames. Phase I is estimated to run from July 18 to September 31 and Phase II was taken to be one month for a comparative analysis, but would be relevant as long as the plantings are well maintained. Length/slope or *LS* values were estimated from the contour maps and the soil erosivity factor, *k*, was taken to be 0.26 for the active construction area, an average of values for the first six inches and the next deeper layer of soil, since the excavation appears to be in the six inch range. Cover factors were taken from the text for relevant conditions. It was learned recently that the sidewalk will use

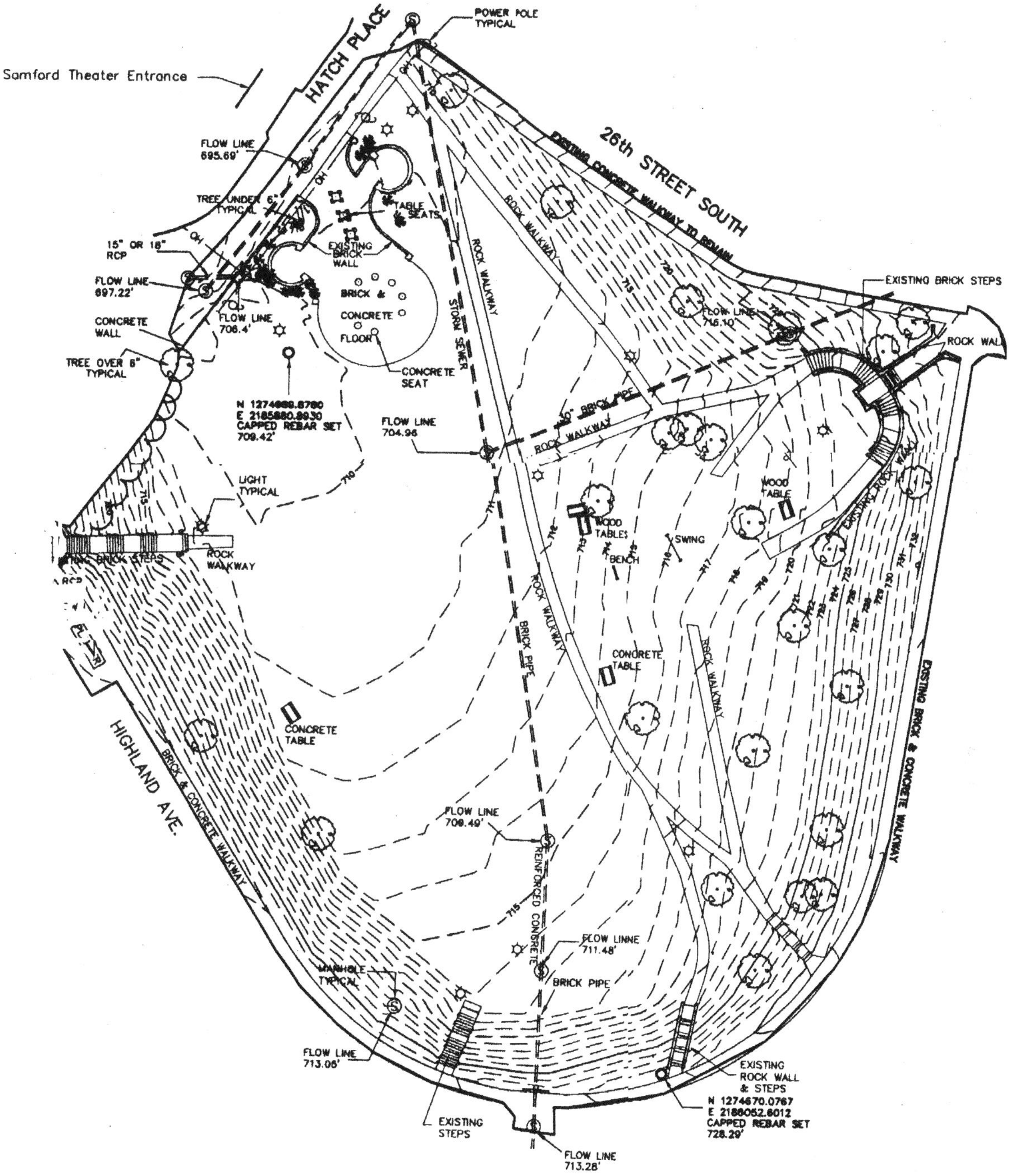

Figure 6.26. Preconstruction lay out of Caldwell Park showing storm sewer lines, contour intervals, sidewalks, paved areas and trees..

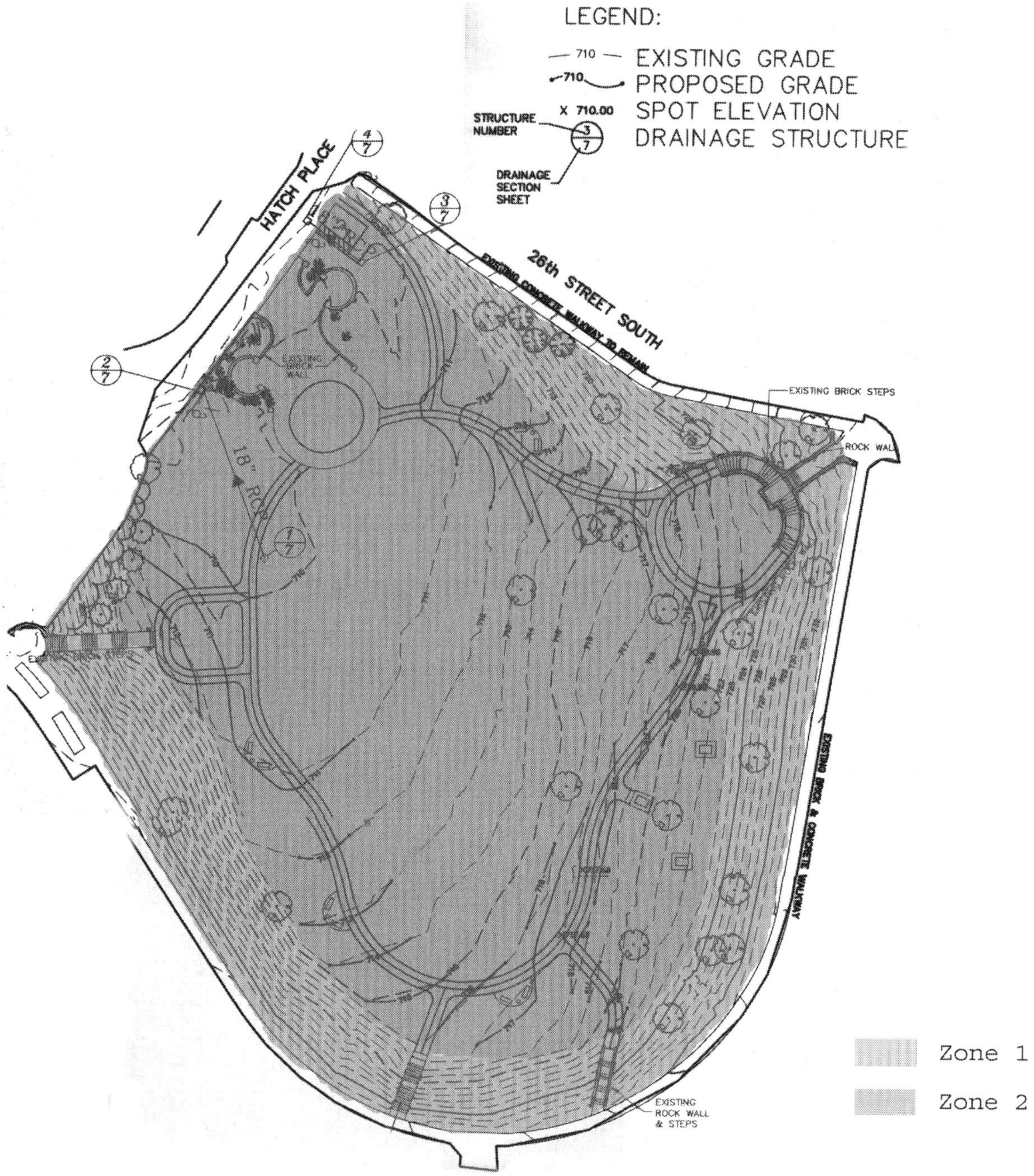

Figure 6.27. Phase I of the project showing undisturbed areas around the perimeter of the park the active construction area in the center.

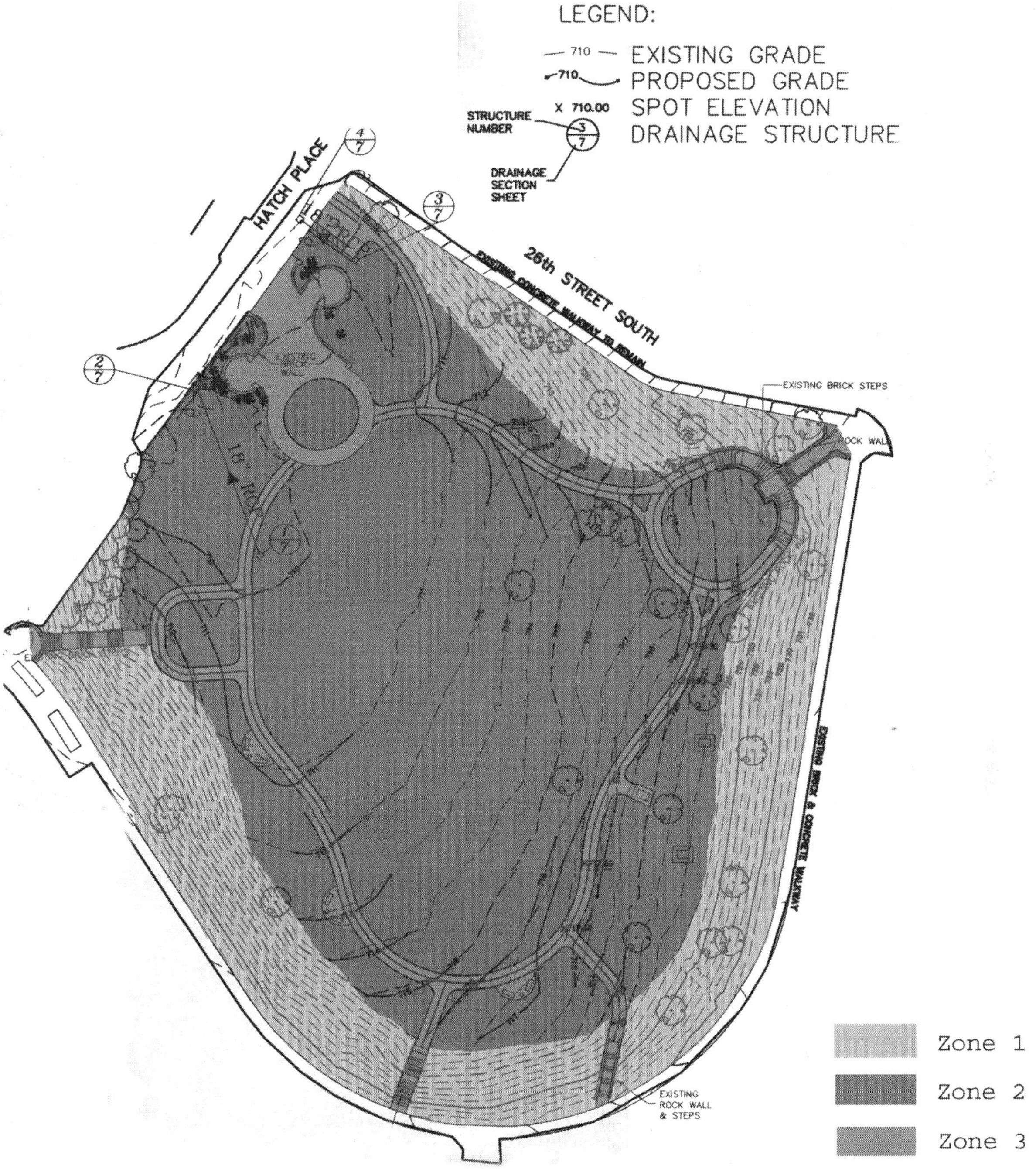

Figure 6.28. Phase II showing the complete park. The green area is undisturbed. The pink area is sod and other plantings and the blue is sidewalk and other paved or brick areas.

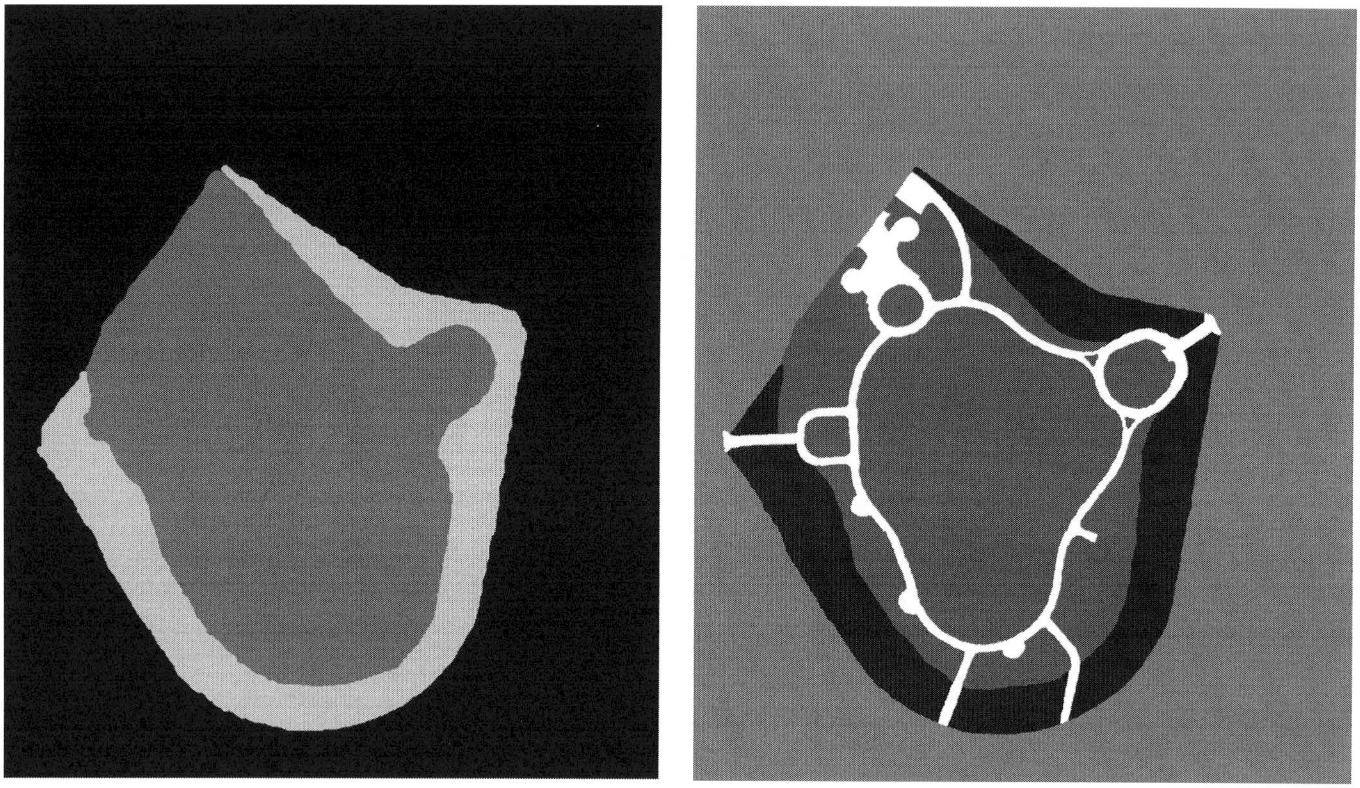

Figure 6.29. Grey scale images of Figs 2 and 3, were used to get accurate estimates of area in the various zones of each phase forRUSLE calculations.

pervious concrete in their construction. A summary of the results of the calculations are given for the two phases in Table 6.24 and 6.25, respectively. When converting the expected soil loss of Phase I, 40.6 tons, into volume with a factor of 1.02 and then dividing by the total area of the park, the construction is expected to result in roughly a 0.31 inch loss of soil.

Erosion and Sedimentation Control Plan

The goal of keeping erosion to a minimum and sediment that does erode to remain on site is facilitated by the small area involved and by the short duration of the project. Having a single phase of construction, however, limits the tools that can be appropriately applied, since access to most areas of the excavated region is continuous through the project. Silt fencing will be the primary defense against sediment being washed off site into neighboring property and into the storm sewer system. Also, a small strip of grass has been left at the lower end of the park in addition to the placement of a rock pile (rock that is being saved for re-use) that will both help trap sediment that might otherwise leave the site in heavy rain.

TABLE 6.24. Phase I RUSLE Calculations.

Site Zone	Zone Description	Land Area (acres)	R Factor for Phase Period (July 1–Sep 31)#	K Factor for Soil	LS Slope Length Factor	C Cover Factor	Calculated Soil Loss per Unit Area (tons/acre/period)	Calculated Total Acres Soil Loss (tons/period)
1	Undisturbed, grass (L = 40 ft, s = 30%)	0.73	98	0.28	2.68	0.001	0.07	0.05
2	Active construction (L = 355 ft, s = 2.6%)	1.75	98	0.26	0.91	1	23.19	40.58
Total Weight, Period		2.48						40.63 tons over 2.5 months

*K factor estimated between values for above and below 6″

#This period represents 28% of annual R (350), 0.28 × 356 = 98

TABLE 6.25. Phase II RUSLE Calculations.

Site Zone	Zone Description	Land Area (acres)	R Factor for Phase Period (July 1–Sep 31)#	K Factor for Soil	LS Slope Length Factor	C Cover Factor	Calculated Soil Loss per Unit Area (tons/acre/period)	Calculated Total Acres Soil Loss (tons/period)
1	Undisturbed, grass (L = 30 ft to 50 ft, s = 27% to 36%)	0.73	3.5	0.28	2.68	0.00	0.07	0.00
2	Park floor, sod (L = 355 ft, s = 2.6%)	1.45	3.5	0.26	0.91	0.01	23.19	0.01
3	Dewalks & paved area (L = 25 ft, s = 2%)	0.3	3.5	0.26	0.16	0.02	23.19	0.00
Total Weight, Period		2.48						0.01 tons over 1 months

*K factor estimated between values for above and below 6″
#This period represents 10% of annual R (350), 1.0 × 350 = 3.5

Figure 6.30 shows the proposed location of silt fences on the site. A Type B silt fence is recommended for the most northern silt fence, which protects the new storm drain for the

The following discussion describes the recommend placement of silt fences at this site:

Silt fence A

This silt fence protects the new storm drain structure for the newly formed small drainage in the northeastern area of the park (by raising the level of the new sidewalk area). The flow paths for the site are attached as Figure 6.31. The slope is approximately 3.3% and although the longest path length is over 100 ft, it is a very small drainage area (less than 0.3 ac) and the average path length is much shorter than 100 ft. Type B silt fence is adequate.

Silt fence B

This silt fence protects the new main storm drain structure, marked 1/7 on Figures 6.30 and 6.31. The slope is again approximately 1.5% and is over 100 ft in length. This fence would be installed immediately after the installation of the storm drain structure and would also serve to facilitate sedimentation of most of the construction site and would possibly benefit from the additional support of Type A silt fencing.

Silt fence C

This silt fence is on a roughly 0.5% grade and also exceeds a 100 ft upslope length. Again, since this fence would contribute to storage of a dry sedimentation pond, Type A would be recommended.

The slope lengths are all about 300 feet long for the main filter fences, although the slopes are generally mild. The construction area is small, so a standard silt pond is not likely practical, although a small structure may be suitable. The photo shown below, taken the morning after a small rain, shows standing water in the area where the new storm drain will be installed. This area could be modified to function as a small sediment pond.

Previously calculated values using the RUSLE for areas of undisturbed and active construction can be used to estimate the pond surface area:

Undisturbed = 0.73 ac
for undisturbed multiply by 0.6% = 0.0044 ac

Graded area = 1.75 ac
for construction multiply by 1.5% = 0.026 ac

Total = 0.032 ac

Which equates to 1,500 ft^2. This would give a small triangular pond of approximately 33 ft long and 20 ft at the outlet end. About 3 ft of standing water is also needed, above the sacrificial storage for sediment. The RUSLE calculations for the drainage gave an estimate of 40.6 tons of silt loam soil to be eroded from the site over two and a half months, which when multiplied by the 1.02 factor for volume for silt clays and clay loams, about 42 cubic yards of sediment storage would be wanted, or about 1,000 ft^3. Therefore, about 1 to 2 ft of sediment storage would also be needed.

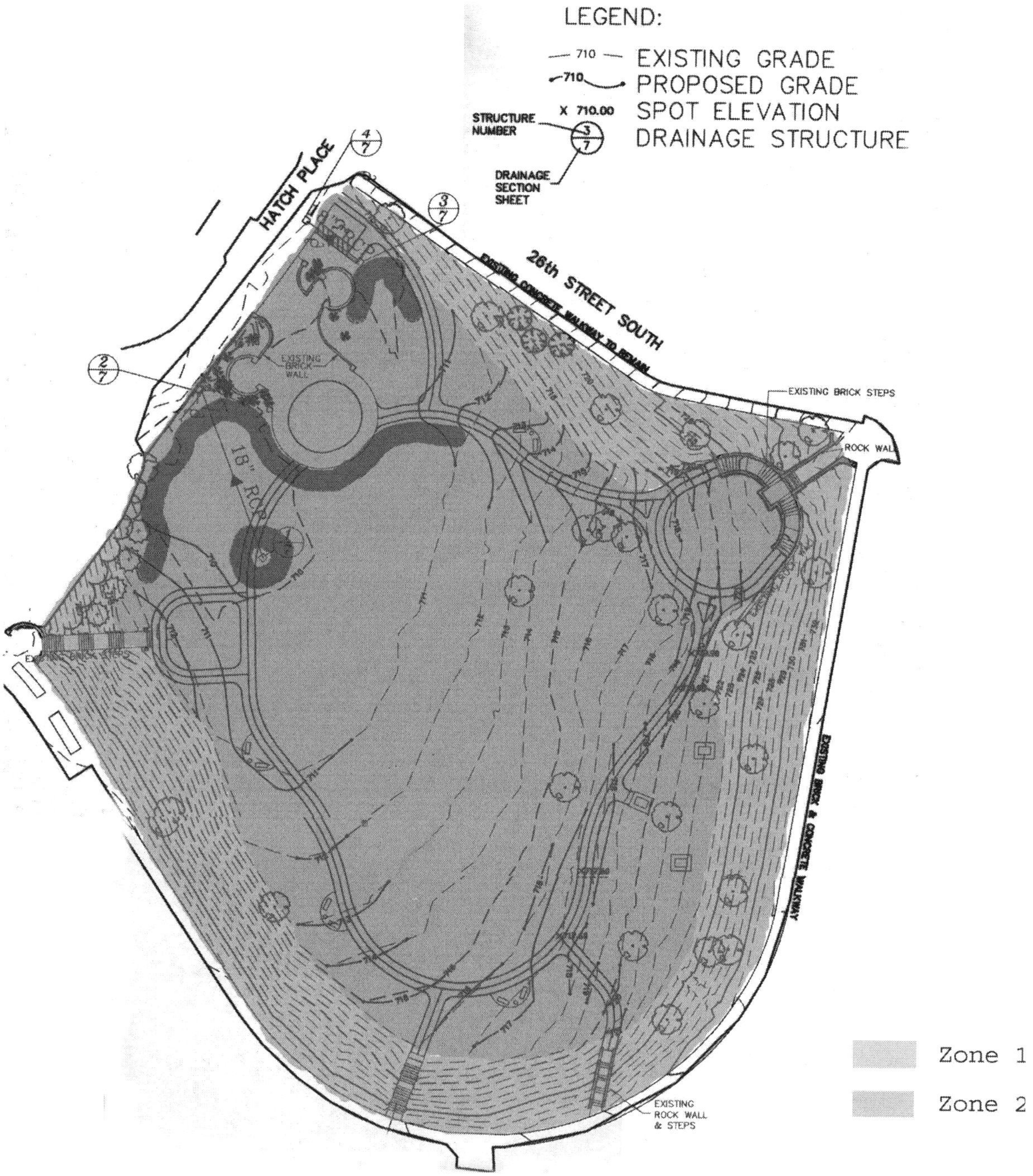

Figure 6.30. Red lines indicate the location of proposed silt fences on the site.

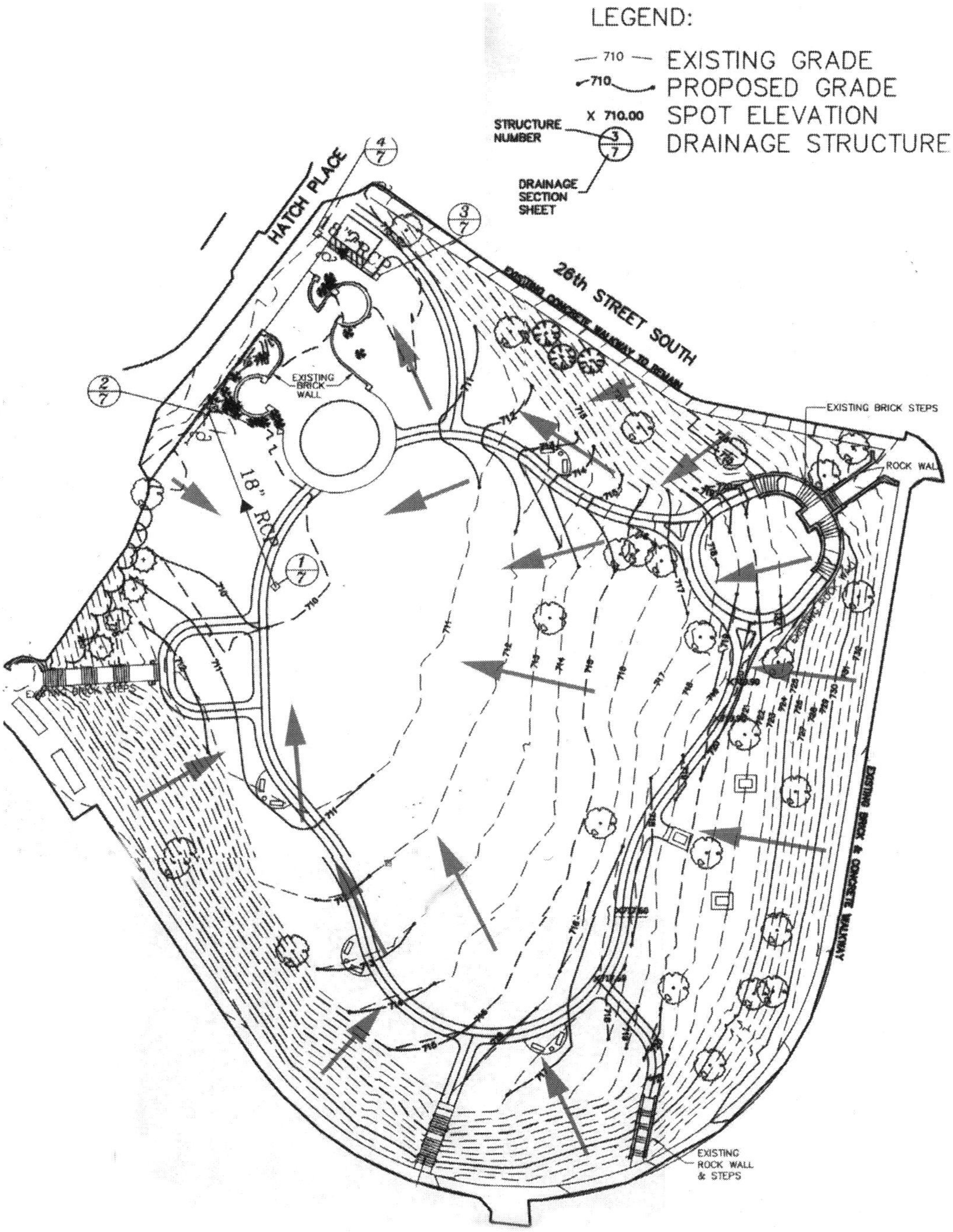

Figure 6.31. Contour map of Caldwell Park site showing drainage paths.

CONCLUSIONS

This discussion has shown that the use of relatively-simple design criteria can provide excellent water quality benefits over a wide range of storm conditions. Detention ponds are probably the most-commonly used runoff quality control devices and have substantial literature documenting their performance and problems. Wet detention ponds have been shown to be very effective, if their surface areas are sufficiently large in comparison to the drainage area and expected runoff volume. Small wet ponds and all dry ponds have been shown to be much less effective. Care must also be taken to minimize safety and environmental hazards associated with ponds.

Physical sedimentation is the main removal process occurring in wet ponds. Temporary sediment ponds at construction sites are most suitable where the area to be controlled is larger than about 10 acres (the typical upper limit for filter fencing). They have been found to be generally the most effective sediment control (after prevention).

Filter fences are suitable for much smaller areas than sediment ponds, but their maximum expected performance is less. They also act as small detention ponds by ponding water behind the fabric on the upslope side, allowing sedimentation. Common problems with filter fence installations include improper installation, placement, and maintenance. They frequently are not adequately secured along their bottom edges, allowing passage of water under the fabric. In many cases, the drainage areas also are too large.

PROBLEMS

1. Compute the settling velocity for the following particles: very coarse sand (diameter = 1.5 mm), medium sand (0.4 mm), very fine sand (0.075 mm), and clay (0.001 mm) assuming particle settling in laminar flow. Estimate the time for each particle to fall 3 feet in water.

2. The retention time in a stormwater management basin is 45 min. If the average water depth in the active zone is 4 ft, what proportion of fine sand (diameter = 0.1 mm) will settle to the sacrificial storage assuming the inflow is fully mixed?

3. A developer has designed a mixed-residential/commercial development for his property. The total acreage is 150 acres; 40 of which will be strip-commercial with paved parking and impervious roofs; 30 of which will be a townhome development (attached homes on 1/8 acre lots); and the remaining 80 acres, single-family homes on 1/2-acre lots. Prior to grading, the property is a forest with an average 5% slope. The developer will be grading the site as follows:
   - Strip commercial: slope approximately 0.2%. Slope length 2000 ft.
   - Townhome development: slope approximately 1.0%. Slope length 175 ft.
   - Single-family residential area: slope approximately 2%. Slope length 250 ft.

   The developer is planning to install a temporary erosion control pond at the lowest point in the watershed (which is where the parking lot of 25 acres is to be located). Answer the following questions about the pond the developer is planning to install:

   a. If the control is required for the 10-yr storm and assuming local rain conditions, what is the active water quality volume that is required for the pond? Assuming space is unlimited, what is the top area of a safe, well-designed pond for water quantity control? If only 5 acres is available for the pond, what changes have to be made to the design?

   b. How much sediment is anticipated to be washed off from the site assuming it is located in your current watershed and subject to your local rain conditions? The design storm for sediment loss on this site is the 25 yr storm. Assume that the contractor removes at least the top 1 ft of soil during grading and it is the underlayer that is exposed to rain events. No cover protection is put on the site during construction, e.g., all control is occurring in the pond.

   c. The development is located in a sensitive watershed, and therefore, the particle size requiring control is the 5 μm particle. For this level of control, what is the required surface area? What outlet control should be selected?

   d. How much volume is required to store sediment assuming that the development construction will last for two years in your watershed?

   e. Complete a final design for the pond assuming unlimited surface area.

4. Rework Problem 3, assuming that the construction is phased and cover practices are established for all areas after grading except for the area where the pond is located. The C value for the protective cover is assumed to be 0.1.

5. A house lot is being developed. The lot size is 80 ft wide by 125 ft long. The slope length occurs along the "long" side of the property. The developer plans to use filter fences as the primary erosion control measure. What is the maximum slope that the fence is recommended if the cover soil is clay? Loam? Silt? Sand?

REFERENCES

Akeley, R.P. "Retention ponds for control of urban stormwater quality." In *Proceedings National Conference on Urban Erosion and Sediment Control: Institutions and Technology,* EPA-905/9-80-002, Chicago, January 1980.

Auckland Regional Council. TP 90: *Erosion and Sediment Control,* Auckland, New Zealand. 1999.

Barfield, B. "Analysis of the effects of existing and alternate design criteria on the performance of sediment detention ponds." 1986 *Sediment and Stormwater Conference,* sponsored by the Maryland Water Resources Administration, Sediment and Stormwater Division, Salisbury State College, Maryland, July-August 1986.

Barrett, M.E., J.E. Kearney, J.F. Malina, R.J. Charbeneau, and G.H. Ward. *An Evaluation of the Use and Effectiveness of Temporary Sediment Controls.* Center for Research in Water Resources. Technical Report CRWR 261. The University of Texas at Austin. August 1995.

Beca Carter Hollings & Ferner Ltd. *The Use of Flocculants and Coagulants to Aid the Settlement of Suspended Sediment in Earthworks Runoff—Trials, Methodology and Design.* Report 2509784 Rev A, L1:90873. Prepared for the Auckland Regional Council. Auckland, New Zealand. Undated.

Bedner, R.E. and D.J. Fluke. *Demonstration of Debris Basin Effectiveness in Sediment Control.* Environmental Protection Agency, Industrial Environmental Research Laboratory—Cincinnati, EPA-600/7-80-154. Cincinnati, Ohio, 1980.

Bondurant, J.A., C.E. Brockway, and M.J. Brown. "Some aspects of sedimentation pond design." *Proceedings National Symposium on Urban Hydrology and Sediment Control,* University of Kentucky, Lexington, 1975.

Chambers, G.M. and C.H. Tottle. *Evaluation of Stormwater Impoundments in Winnipeg,* Report SCAT-1, Environment Canada, Ottawa, April 1980.

Clark, S. and R. Pitt. *Stormwater Treatment at Critical Areas, Vol. 3: Evaluation of Filtration Media for Stormwater Treatment.* U.S. Environmental Protection Agency, Water Supply and Water Resources Division, National Risk Management Research Laboratory. EPA/600/R-00/016, Cincinnati, Ohio. 442 pgs. October 1999.

Colston, N.V., Jr. *Characterization and Treatment of Urban Land Runoff.* EPA-670/2-74-096, U.S. Environmental Protection Agency, 1974.

Driscoll, E. D. "Detention and retention controls for urban stormwater." *Engineering Foundation Conference: Urban Runoff Quality—Impact and Quality Enhancement Technology,* Henniker, New Hampshire, edited by B. Urbonas and L. A. Roesner, published by the American Society of Civil Engineers, New York, June 1986.

Eccher. C.J. "Thoughtful design is prime factor in water safety." *Lake Line.* pp. 4–8. May 1991.

EPA (USA Environmental Protection Agency). *Final Report for the Nationwide Urban Runoff Program.* Water Planning Division, PB 84-185552, Washington, D.C., December 1983.

Gietz, R. J. *Urban Runoff Treatment in the Kennedy-Burnett Settling Pond.* For the Rideau River Stormwater Management Study, Pollution Control Division, Works Department, Regional Municipality of Ottawa-Carleton, Ottawa, Ontario, March 1983.

Grizzard, T. J., C. W. Randall, B. L. Weand, and K. L. Ellis. "Effectiveness of extended detention ponds." *Engineering Foundation Conference: Urban Runoff Quality—Impact and Quality Enhancement Technology,* Henniker, New Hampshire, edited by B. Urbonas and L. A. Roesner, published by the American Society of Civil Engineers, New York, June 1986.

Harrington, B. W. *Feasibility and Design of Wet Ponds to Achieve Water Quality Control.* Sediment and Stormwater Division, Maryland Water Resources Administration, July 1986.

Hittman Assoc. *Methods to Control Fine-Grained Sediments Resulting from Construction Activity.* U.S. Environmental Protection Agency, Pb-279 092, Washington, D.C., December 1976.

Horner, R.R., J. Guedry, and M.H. Kortenhof. *Improving the Cost Effectiveness of Highway Construction Site Erosion and Pollution Control.* Washington State Transportation Center, Washington State Dept. of Transportation, Seattle, WA. 1990.

Jones, J. E. and D. E. Jones, "Interfacing considerations in urban detention ponding." *Proceedings of the Conference on Stormwater Detention Facilities, Planning, Design, Operation, and Maintenance,* Henniker, New Hampshire, Edited by W. DeGroot, published by the American Society of Civil Engineers, New York, August 1982.

Kamedulski, G. E. and R. H. McCuen. "The effect of maintenance on storm water detention basin efficiency." *Water Resources Bulletin,* Vol. 15, No. 4, pg. 1146, August 1979.

Larcombe, M.F. *Technical Publication on Chemical Removal of Sediment from Earthworks Stormwater.* Auckland Regional Council (New Zealand). 1999.

Linsley, R. K., and J. B. Franzini. *Water Resources Engineering.* McGraw-Hill, New York, 1964.

Marcy, S. J. and J. E. Flack. "Safety considerations in urban storm drainage design." *Second International Conference on Urban Storm Drainage,* Urbana, Illinois, June 1981.

Martin, E.H. "Mixing and residence times of stormwater runoff in a detention system." Engineering Foundation Conference: *Design of Urban Runoff Quality Controls.* Potosi, Missouri. July 10–15, 1988. Edited by L. A. Roesner, B. Urbonas and M.B. Sommen. pp. 164–179. Published by the American Society of Civil Engineers, New York, June 1989.

Mason, J.M.,Jr. "On-site stormwater detention: An overview." *Public Works,* February 1982.

McCuen, R. H. *A Guide to Hydrologic Analysis Using SCS Methods.* Prentice-Hall, Englewood Cliffs, New Jersey, 1982.

Nelson, John. MSCE published thesis, Department of Civil and Environmental Engineering, University of Alabama at Birmingham. *Characterizing Erosion Processes and Sediment Yields on Construction Sites.* 1996.

Pitt, R. "Runoff controls in Wisconsin's Priority Watersheds." *Advanced Topics in Urban Runoff Research* conference. Proceedings edited by B. Urbonas and L.A. Roesner. Engineering Foundation and ASCE, New York. pp. 290–313. 1986.

Pitt, R., and J. McLean. *Toronto Area Watershed Management Strategy Study: Humber River Pilot Watershed Project.* Ontario Ministry of the Environment, Toronto, Ontario, 1986.

Pitt, R. "Detention pond design for water quality improvement." *National ASCE Hydraulic Conference.* San Francisco, California. July 1993a.

Pitt, R. "The stormwater quality detention pond model (DETPOND)." *26th Annual Water Resources Conference.* University of Minnesota. October 1993b.

Pitt, R. "Sediment Control in Alabama." *41st Annual Transportation Conference.* Montgomery, AL. February 1998.

Poertner, H.G. *Practices in Detention of Urban Stormwater Runoff.* American Public Works Association, OWRR Contract No. 14-31-0001-3722, Chicago, 1974.

Randall, C. W. "Stormwater detention ponds for water quality control." *Proceedings of the Conference on Stormwater Detention Facilities, Planning, Design, Operation, and Maintenance,* Henniker, New Hampshire, Edited by W. DeGroot, published by the American Society of Civil Engineers, New York, August 1982.

Schimmenti, F.G. "Stormwater detention basins must control more than runoff." *American City and County,* December 1980.

Schueler, T. R. *New Guidebook Review Paper No. 2, Wet Pond BMP Design.* Metropolitan Washington Council of Governments, Washington, D.C., 1986.

Schueler, T.R., and J. Lugbill. *Performance of Current Sediment Control Measures at Construction Sites.* Metropolitan Washington Council of Governments, Sediment and Stormwater Administration of the Maryland Dept. of the Environment. 1990.

SCS (now NRCS) (U.S. Soil Conservation Service). *Computer Program for Project Formulation, Hydrology.* Technical Release Number 20 (TR-20). U.S. Dept. of Agriculture, 1982.

Smith, W. G. "Water quality enhancement through stormwater detention." *Proceedings of the Conference on Stormwater Detention Facilities, Planning, Design, Operation, and Maintenance,* Henniker, New Hampshire, Edited by W. DeGroot, published by the American Society of Civil Engineers, New York, August 1982.

Smoot, J.T., T.D. Moore, J.H. Deatherage, and B.A. Tschantz. *Reducing Nonpoint Source Water Pollution by Preventing Soil Erosion and Controlling Sediment on Construction Sites.* Transportation Center of Tennessee. 1992.

Terstriep, M.L., G.M. Bender, D.C. Noel. *Nationwide Urban Runoff Project, Champaign, Illinois: Evaluation of the Effectiveness of Municipal Street Sweeping in the Control of Urban Storm Runoff Pollution.* Contract No. 1-5.39600, U.S. Environmental Protection Agency, Illinois Environmental Protection Agency, and the State Water Survey Division, University of Illinois, December 1982.

U.S. Bureau of Reclamation. *Design of Small Dams.* U.S Dept. of the Interior, A Water Resources Publication, 3rd edition, 1987.

USBR (U. S. Department of Interior—Bureau of Reclamation) *Hydraulic Design of Stilling Basins and Energy Dissipators,* Engineering monograph No. 25.

USDA. *Alabama Handbook for Erosion Control,* prepared by the Alabama Soil and Water Conservation Committee, Montgomery, AL. 2003.

Vignoles M., and L. Herremans. "Metal pollution of sediments contained in runoff water in the Toulouse city." (in French). NOVATECH 95, *2nd International Conference on Innovative Technologies in Urban Storm Drainage.* May 30–June 1, 1995. Lyon, France. pp. 611–614. Organized by Eurydice 92 and Graie. 1995.

Whipple, W. and J. V. Hunter. "Settleability of urban runoff pollution." *Journal WPCF,* Vol. 53, No. 12, pg. 1726, 1981.

Wiegand, C., T. Schueler, W. Chittenden, and D. Jellick. "Comparative costs and cost effectiveness of urban best management practices." Engineering Foundation Conference: *Urban Runoff Quality—Impact and Quality Enhancement Technology,* Henniker, New Hampshire, edited by B. Urbonas and L. A. Roesner, published by the American Society of Civil Engineers, New York, June 1986.

Construction Site Erosion Control References and Internet Sources

INTERNET SOURCES

The Internet modules at: http://unix.eng.ua.edu/~rpitt/Class/Erosioncontrol/MainEC.html contain much additional information pertaining to erosion and sediment control at construction sites. This following is a list of some of these links, as presented in the preceding chapters of this book:

Chapter 1: Introduction to Erosion and Sediment Control, problems and Regulations

The following are the main Internet links referenced in Chapter 1 and provide much additional information, especially concerning the federal programs and resources. These are likely to change with time, but current linkage addresses can usually be found by using an Internet search tool.

EPA. Office of Wastewater Management (OWM) information:
http://www.epa.gov/owm/

EPA Stormwater Program information, Final Phase II NPDES rule:
http://cfpub.epa.gov/npdes/home.cfm?program_id=6

Final Federal Register notice and supporting materials for Effluent Limits Guidelines for Erosion and Sediment control:
http://www.epa.gov/guide/construction

EPA Fact Sheet Series:
http://cfpub.epa.gov/npdes/stormwater/swfinal.cfm

EPA stormwater regulations:
http://cfpub.epa.gov/npdes/regs.cfm?program_id=6

EPA information on discharges from construction activities:
http://cfpub.epa.gov/npdes/stormwater/const.cfm

EPA National Menu of stormwater, and erosion and sediment control practices:
http://www.epa.gov/npdes/menuofbmps/menu.htm

EPA links to on-line manuals and guidance documents:
http://www.epa.gov/waterscience/guide/construction/.

State Water Pollution Control Program Grants Program
http://www.epa.gov/owm/cwfinance/pollutioncontrol.htm

Stormwater website
http://cfpub.epa.gov/npdes/home.cfm?program_id=6

Electronic Notice of Intent System
http://cfpub.epa.gov/npdes/stormwater/enoi.cfm

National Management Measures to Control Nonpoint Source Pollution from Urban Areas
http://www.epa.gov/owow/nps/urbanmm/

Smart Growth Program http://www.epa.gov/livability/
Section 319 Nonpoint Source Management Program
http://www.epa.gov/owow/nps/cwact.html

The Construction Industry Compliance Assistance Center (http://cicacenter.org/) contains information and links to a wide variety of information, including state regulatory programs and manuals for sediment and erosion controls.

State Erosion Control Handbooks Available on the Internet:

Alabama
Alabama Handbook for Erosion Control
http://swcc.state.al.us/erosion_handbook.htm

California
California Storm Water BMP Construction Handbook
http://www.swrcb.ca.gov/stormwtr/index.html

Colorado
Denver Urban Drainage Criteria Manual
http://www.udfcd.org

Delaware
Delaware Erosion and Sediment Control Handbook
http://www.dnrec.state.de.us/dnrec2000/divisions/soil/stormwater/stormwater.htm

Florida
Florida Development Manual: A Guide to Sound Land and Water Management
http://www.dep.state.fl.us/water/nonpoint/urban2.htm

Georgia
Georgia Storm Water Management Manual
http://www.atlantaregional.com/water/waterquality/stormwatertaskforce.html

Idaho
Catalog of Storm Water BMPs for Idaho Cities & Counties
http://www2.state.id.us/deq/water/stormwater_catalog/index.asp

Louisiana
State of Louisiana Nonpoint Source Pollution Management Program—Construction
http://nonpoint.deq.state.la.us/manage10.html

Maryland
Maryland Stormwater Design Manual
http://www.mde.state.md.us/environment/wma/stormwatermanual

Maryland Storm Water Design Manual, Volumes I & II
http://www.mde.state.md.us/Programs/WaterPrograms/SedimentandStormwater/stormwater_design/index.asp

Massachusetts
Erosion and Sedimentation Control Guidelines: a guide for planners, designers, and municipal officials
http://www.mass.gov/dep/brp/stormwtr/stormpub.htm

Minnesota
Protecting Water Quality in Urban Areas: A Manual
http://www.pca.state.mn.us/water/pubs/sw-bmpmanual.html

Urban Small Sites Best Management Practice Manual
http://www.metrocouncil.org/environment/watershed/bmp/manual.htm

Missouri
Protecting Water Quality: A Construction Site Water Quality Field Guide
http://www.dnr.state.mo.us/wpscd/wpcp/wpcp-guide.htm

New Hampshire
Managing Storm Water as a Valuable Resource
http://www.des.state.nh.us/dwspp/stormwater.pdf

New Jersey
Revised Manual for New Jersey: BMPs for Control of Nonpoint Source Pollution from Storm Water
http://www.state.nj.us/dep/watershedmgt/bmpmanual.htm

New York
New York State Stormwater Management Design Manual
http://www.dec.state.ny.us/website/dow/toolbox/swmanual/

New York State Standards and Specifications for Erosion and Sediment Control
http://www.dec.state.ny.us/website/dow/toolbox/escstandards/index.html

Ohio
Storm Water Program—Factsheets, Forms, & Check Lists
http://www.epa.state.oh.us/dsw/storm/

Oregon
BMPs & Storm Water Pollution Control Plan
http://www.deq.state.or.us/wq/wqpermit/wqpermit.htm

Pennsylvania
Handbook of BMPs for Developing Areas
http://www.pacd.org/products/bmp/bmp_handbook.htm

South Carolina
Sediment, Erosion, & Storm Water Management
http://www.scdhec.net/water/html/erfmain.html

Tennessee
Tennessee Erosion and Sediment Control Handbook
http://www.state.tn.us/environment/wpc/

Knoxville BMP Manual
http://www.ci.knoxville.tn.us/engineering/bmp_manual/

Texas
Texas Nonpoint Sourcebook – Interactive BMP Selector
http://www.txnpsbook.org/SiteMap.htm

Utah
UPDES Storm Water Home Page
http://www.deq.state.ut.us/EQWQ/updes/stormwater.htm

Virginia
Virginia Erosion and Sediment Control Handbook
http://www.dcr.state.va.us/sw/e&s-ftp.htm

Northern Virginia BMP Handbook: A Guide to Planning and Designing BMPs in Northern Virginia
http://www.novaregion.org/pdf/NVBMP-Handbook.pdf

Washington
Storm Water Management Manual for Western Washington
http://www.ecy.wa.gov/programs/wq/stormwater/manual.html#copies
King County Storm Water Pollution Control Manual
http://dnr.metrokc.gov/wlr/Dss/Spcm.htm

Wisconsin
Wisconsin Construction Site Best Management Practice Handbook
http://dnr.wi.gov/org/water/wm/nps/stormwater/constrforms.htm#wicon

Chapter 2: Selection of Controls and Site Planning

The following Internet Links are referenced in Chapter 2. These sites should be visited to obtain additional information. Some of the locator addresses will likely change, but the material can still likely be located using a search tool.

Alabama Department of Environmental Management (ADEM), Nonpoint Program
http://www.adem.state.al.us/Education%20Div/Nonpoint%20Program/WSNPSProgram.htm

Jefferson County Stormwater Management, Inc.
http://www.swma.com/

Alabama Soil and Water Conservation Committee
http://www.swcc.state.al.us/

Alabama Soil and Water Conservation Committee (Sediment and Erosion Control Handbook):
http://swcc.state.al.us/erosion_handbook.htm

Geological Survey of Alabama
http://www.gsa.state.al.us/

Natural Resources Conservation Service, USDA
http://www.nrcs.usda.gov/

Alabama on-line soil surveys available to download (only a few counties):
http://soils.usda.gov/survey/online_surveys/alabama/
EPA Region 4 Nonpoint Source Information
http://www.epa.gov/region4/water/nps/

EPA "Surf you Watershed" (compiled water and watershed information for your watershed)
http://www.epa.gov/surf/

USGS "Science in your Watershed" (additional water and watershed information)
http://water.usgs.gov/wsc/index.html

Microsoft TerraServer maps (maps and aerial photographs for most of U.S.)
http://terraserver.microsoft.com/

NOAA Data and Information Server (many linked environmental databases)
http://www.esdim.noaa.gov/NOAAServer/

Code of Federal Regulations, Title 40 (environmental regulations)
http://www.access.gpo.gov/cgi-bin/cfrassemble.cgi?title=199840

Chapter 3: Regional Rainfall Conditions and Site Hydrology for Construction Site Erosion Evaluations

Alabama Rainfall Atlas:
http://bama.ua.edu/~rain/

WinTR-55 computer program (new windows version, ver. 1.0.08, Jan 2005):
http://www.wcc.nrcs.usda.gov/hydro/hydro-tools-models-wintr55.html

TR-55 1986 documentation and early version of TR55 program:
http://www.wcc.nrcs.usda.gov/hydro/hydro-tools-models-tr55.html

TR-20 computer program (new windows January 2005 version):
http://www.wcc.nrcs.usda.gov/hydro/hydro-tools-models-wintr20.html

National Engineering Handbook, Part 630 HYDROLOGY
http://www.wcc.nrcs.usda.gov/hydro/hydro-techref-neh-630.html

U.S. Army Corps of Engineers, Hydrologic Management System User Guide (replacement for HEC-1) and River Analysis System User Guide for water surface profile calculations (replacement for HEC-2):
http://www.hec.usace.army.mil/

Chapter 4: Erosion Mechanisms, the Revised Universal Soil Loss Equation (RUSLE), and Vegetation Erosion Controls

The official NRCS RUSLE2 Internet site is at:

http://fargo.nserl.purdue.edu/rusle2_dataweb/Tutorial.htm.

The model can be downloaded from this site, along with supporting documents and other materials.

The Alabama Soil and Water Conservation Committee web site includes locations and contacts for local

USDA/NRCS offices where soil information can be obtained. They also recently completed an updated version to the 1993 *Alabama Handbook for Erosion Control, Sediment Control, and Stormwater Management on Construction Sites and Urban Areas*. The updated handbook (2003) is available at: http://www.swcc.state.al.us/

Sources of Commercial Seeds and Plants

Sod
- www.Gardnerturf.com
- www.usaturf.com

Hydroseeding
- www.htpa.org

Seed Suppliers
- www.sylvanative.com
- www.sroseed.com
- www.turf-seed.com
- www.seedland.com
- www.erosionseed.com
- www.seedswest.com
- www.albrightseed.com

Chapter 5: Channel and Slope Stability for Construction Site Erosion Control

U.S. Army Corps of Engineers Channel Stability Assessment Method Report: Engineering and Design – *Channel Stability Assessment for Flood Control Projects,* 1994:
http://www.usace.army.mil/inet/usace-docs/eng-manuals/em1110-2-1418/toc.htm

USDA Agricultural Handbook 667: Stability Design of Grass-Lined Open Channels, 1987:
http://www.info.usda.gov/CED/ftp/CED/AH-667.pdf

North America Green downloadable program for slope and channel protection:
http://www.nagreen.com/

ABSTRACTS FOR SELECTED REFERENCES

The following are abstracts for selected references pertaining to erosion and sediment control. Please refer to the complete publications for further information.

Allen, S.R. (1996). Evaluation and standardization of rolled erosion control products. *Geotextiles and Geomembranes.* **14**(3–4):207–221.

Abstract: The erosion control industry has grown significantly in response to continued infrastructure development and increased awareness of water quality problems. There are currently a wide variety of rolled erosion control products available, representing a broad spectrum of product construction and corresponding applications. While significant improvements in erosion control technology during the past several years have out paced associated standards and research, several important steps are underway to meet critical needs for standardization. This paper outlines the history of advancements in rolled erosion control technology. In addition, a summary is presented of the many efforts currently underway by the Erosion Control Technology Council to establish erosion control industry standards for terminology, index tests and performance criteria.

Anon. (1973) Erosion control on highway construction. National Cooperative Highway Research Program, *Synthesis of Highway Practice.* 18:52p.

Abstract: Literature review presents information on soil erosion that is available to highway planners, designers, and construction personnel.

Anon. (2000). On-site grinding and recycling of home construction debris. *BioCycle.* **41**(12):2 pp.

Abstract: Wood separated from home construction debris is reused in the form of chips for erosion control. Wallboards comprising of gypsum or calcium sulphate from the debris are capable of improving certain soils. On-site separation of the usable materials reduces a considerable amount of time and money needed for waste management. In view of increasing interest in green building, a waste management plan is suggested.

Anon. (2003). Resources for accessing erosion control information. *Public Works.* **134**(6):36–38.

Abstract: A study on the resources for accessing erosion control information is presented. The International Erosion Control Association provides education, resource information, and networking opportunities in their website www.ieca.org. State, country, and agricultural departments offer locally-based extension courses that address the specific erosion control needs of a given area. The U.S. Army Corps of Engineers Engineer Research and Development Center, Construction Engineering Research Laboratory (CERL) maintains programs with guidance and information on technology to support planning, design, execution, and management of land restoration and maintenance activities.

Barkley, T. (2004). Erosion control with recycled materials. *Public Roads.* **67**(5):12–15.

Abstract: The use of compost to control roadway soil erosion is a growing trend, thanks in large measure to fertile minds at the Texas Department of Transportation (TxD0T) and its sister agency, the Texas Commission on Environmental Quality (TCEQ). Compost has proven to be extremely effective in preventing soil runoff during and after roadway construction. It not only minimizes soil erosion but also helps prevent water contamination. And its use in the TxDOT tramsportation community also has created a dynamic market for locally produced compost. In fact, Biocycle magazine contends that TxDOT is the largest market for the material in the Nation, using more than 306,000 cubic meters (400,000 cubic yards) of compost in fiscal year 2003. The Lonestar State's award-winning program encourages the environmentally safe use of compost along the rights-of-way of federally funded highways-the type of innovation envisioned by the Intermodal Surface Transportation Efficiency Act of 1991, which recognized that transportation planning must proceed with care for both human and natural environments.

Benik, S.R.; Wilson, B.N.; Biesboer, D.D.; Hansen, B.; Stenlund, D. (2003). Performance of erosion control products on a highway embankment. *Transactions of the American Society of Agricultural Engineers.* **46**(4):1113–1119.

Abstract: Unprotected soil at construction sites often results in large rates of erosion. Five different erosion control treatments were tested on the slopes of a highway sedimentation basin to determine their impact on vegetative growth, runoff, and erosion. The treatments were a bare (no treatment) condition, a disk-anchored straw mulch, a wood-fiber blanket, a straw/coconut blanket, and a bonded-fiber matrix product (hydraulically applied). A minimum of three replicates was used for each treatment. Straw mulch was selected as the standard treatment for statistical analyses. The site was planted with native prairie seeds, and the establishment of vegetation was monitored over the growing season. Above-ground biomasses for the bare and straw-mulch treatments were statistically greater than those of the bonded-fiber matrix treatment. Statistically significant differences in above-ground biomass for the other treatments were undetected at the 10% level. Weedy grasses and forbs were the dominant plant species.. Runoff and erosion data were collected using a rotating-boom rainfall simulator for spring and fall sets of runs corresponding to little and good vegetative growth, respectively. Runoff depths were generally larger from straw-mulch and bare plots. There were no statistically significant differences in relative runoff depth between the blankets and the bonded-fiber matrix product. Under conditions with little vegetation, erosion from the straw-mulch plots was roughly one-tenth of that from the bare soil plots; erosion from the blanket and bonded-fiber matrix plots was roughly one-tenth of that from the straw-mulch plots. There were no statistically significant differences in relative sediment yield between the blankets and the bonded-fiber matrix. Erosion from bare and straw-mulch treatments was greatly reduced by vegetative growth that occurred between the spring and fall runs.

Bhatia, S.K.; Smith, J.L.; Lake, D.; Walowsky, D. (2002). A technical and economic evaluation of geosynthetic rolled erosion control products in highway drainage channels. *Geosynthetics International.* **9**(2):125–148.

Abstract: Historically, local village, town, and county highway departments have relied heavily on the use of stone fill and rock riprap to line highway drainage channels. These are often constructed without the benefit of design or the evaluation of alternatives, because they have always been done this way. In September 1999, the Munro Road reconstruction project was completed in Onondaga County, New York. As part of an erosion and sediment control demonstration project, the drainage portion of the project was redesigned using geosynthetic rolled erosion control products (RECPs) in lieu of stone fill and rock riprap on almost 1,130 m of channel. The use of the RECPs saved approximately $95,800 in construction costs and is expected to lower maintenance costs as well as provide long-term protection against erosion. The overall purpose of the project was to demonstrate to highway departments the functional and economical advantages of using RECPs for these applications. This paper summarizes the design, construction, and performance to date of seven of the ten different RECPs installed for the project.

Bjorneberg, D.L.; Lentz, R.D. (2005). Sediment pond effectiveness for removing phosphorus from PAM-treated irrigation furrows. *Applied Engineering in Agriculture.* **21**(4):589–593.

Abstract: Polyacrylamide (PAM) greatly reduces erosion on furrow-irrigated fields and sediment ponds can be constructed to remove suspended sediment from irrigation runoff. Both practices are approved for reducing phosphorus (P) loading in the Lower Boise River Pollution Trading Project in southwest Idaho, but information is not available about using both practices on the same field. The objective of this study was to measure the combined effects of PAM application and sediment ponds on sediment and P losses from a furrow-irrigated field. Small sediment ponds (5.8 m^2) with a 60-min design retention time were installed on two fields to receive runoff from PAM-treated or control furrows. Pond inflow and outflow were monitored during a total of 11 irrigations on the two fields. Three crop years of data showed that applying PAM to furrows reduced sediment and total P mass transport to the ponds 50% to 80%, which reduced the mass of sediment and total P retained in the ponds. However, PAM application did not change the percentage of sediment (86%) and total P (66%) retained. The PAM-sediment pond combination reduced average total

P loss by 86% to 98%, based on the difference between untreated inflow and PAM-treated outflow. PAM and sediment ponds had little or no effect on dissolved reactive P (DRP) concentrations. The mass of DRP retained in sediment ponds was directly related to the amount of water that infiltrated within the ponds. Applying PAM to irrigation furrows and installing sediment ponds at the end of the field can be an effective combination for reducing sediment and total phosphorus losses from furrow-irrigated fields, but these practices only reduced soluble P losses by decreasing the volume of water that ran off the fields.

Boysen, S.M. (1974). *Predicting Sediment Yield in Urban Areas.* Kentucky University, Office of Research and Engineering Services Bulletin, Jul 29–31 1974. 199–203.

Abstract: Predicting sediment loss from urban construction sites is an important aspect in the planning and enforcement phases of a sediment control program. The Universal Soil Loss Equation (USLE) is an excellent means of relating soil management, cover, and other erosion control practices to erosion in agricultural areas. The USLE, however, cannot individually be used to estimate sediment yield. A description is presented of a procedure to predict sediment yield from urban construction areas.

Boysen, S.M. (1977). Erosion and sediment control in urbanizing areas. *ASAE.* 4–77: 125–136.

Abstract: The urban sediment control program in Maryland has caused a 60 to 80 percent reduction in construction site sediment yields between 1966 and 1974. The sediment control program, based on a set of basic principles, was developed and implemented beginning in Montgomery County in 1965. Technical principles for controlling erosion and sediment were also developed. The present state law requires that all development must have a sediment control plan approved by the soil conservation district before any construction begins. The sediment control plan, developed by the owner, shows the use of erosion control practices and sediment-trapping devices to control sediment losses to an acceptable level. A manual entitled, "Standards and Specifications for Soil Erosion and Sediment Control in Developing Areas" provides guidelines for developing plans and provides criteria for the design, review, approval, installation, maintenance, and inspection of erosion and sediment control practices. Storm-water management, for offsite channel erosion control, is also required on sediment control plans in certain problem areas.

Brindle, F.A. (2003). Use of native vegetation and biostimulants for controlling soil erosion on steep terrain. *Transportation Research Record.* **I**(1819): 203–209.

Abstract: Native grass species are increasingly requested for use on revegetation and restoration projects following land disturbance. Native species can be slow to establish, a problem when the goals are to obtain quick ground cover to protect against rainfall impact and sheet and rill erosion and to promote root growth to stabilize soil on steep terrain. In addition, disturbed land is subject to invasion by weed species, creating competition with desirable plants for soil nutrients and moisture. Organic soil amendments and soil stabilizers were used on a large restoration project in northwest Oregon to provide soil conditions that would foster a healthy stand of native grasses and forbes (broadleaf plants) while holding the soil in place until plant establishment could occur. Organic soil amendments were prescribed that would improve soil structure, aid nutrient availability, and provide humic acids and bacterial activators. Organic mulch was added to provide the nutrient energy source for soil microorganisms. The amendments were mixed in a slurry and applied in a one-step hydroseeding application during the late fall of 2001. The materials were applied along with the native seed mixture with a mechanically agitated hydroseeding machine and sprayed on the surface in a pressurized spray. The results of the field evaluation were favorable for the ability of the soil to resist water erosion through the initial rainy season and the native species establishment after application.

Buchanan, J.R.; Yoder, D.C.; Denton, H.P.; Smoot, J.L. (2002). Wood chips as a soil cover for construction sites with steep slopes. *Applied Engineering in Agriculture.* **18**(6):679–683.

Also published at:

Buchanan, J.R.; Yoder, D.C.; Smoot, J.L. (2000). Controlling soil erosion on construction sites with steep slopes with wood chips. 2000 *ASAE Annual International Meeting, Technical Papers: Engineering Solutions for a New Century.* **2**:2111–2124.

Abstract: Wood chips were studied for their efficacy in controlling soil erosion on a steep construction site with disturbed soils. The purpose of the research was twofold: to determine if wood chips could be used to reduce the off-site movement of soil during construction activities, and to find an environmentally sound alternative to the landfill disposal of wood wastes generated in the urban forest. The research was conducted on field plots that received natural precipitation. Twelve erodible plots were established on an embankment with a 55% slope and an elevation change of nearly 12 m. Each plot had a width of 3 m and a horizontal slope length of 10 m. A series of flow dividers was installed at the toe of each plot to measure runoff and sediment. Four treatments were studied: large wood chips, small wood chips, a mixture of wood chip sizes, and a control with no chips. The mixture of wood chip sizes represented the size distribution that was found to occur from chippers. The wood chips were applied at a rate that covered 80% of the soil surface. The erosion rate for the small wood chip treatment was not significantly different from the zero-cover plots. The erosion rates from the large wood chip and mixture of chip sizes were not significantly different from one another, but were significantly different from the zero-cover treatment. Overall, in comparison to the zero-cover treatment, the small wood chip treatment reduced erosion by 22%, the large wood chips reduced erosion by 78%, and the mixture of chip sizes reduced erosion by 86%. The results of this project indicate that wood chips (as produced by a chipper) should be utilized as a soil cover and need not be discarded as solid waste.

Burroughs Jr., E.R.; King, J.G. (1985). Surface erosion control on roads in granitic soils. *Watershed Management in the Eighties.* 183–190.

Abstract: The impact of forest road construction on water quality and fish habitat is a serious problem for engineers, hydrologists, and foresters. Control of surface erosion is an important and expensive item in forest road construction. But, erosion control is hampered by our limited ability to estimate sediment yield from forest roads and to evaluate the effectiveness of erosion control treatments. Simulated rainfall was used to generate runoff and sediment yield from forest roads and fillslopes built in granitic soils to test the effectiveness of various surfacing materials, mulches, and barriers as erosion control treatments. An empirical equation is presented and used to estimate the relative effectiveness of gravel, oil, and bituminous surface treatments in reducing sediment yield relative to that of an unsurfaced road.

Burton, T.M.; Turner, R.R.; Harriss, R.C. (1976). Impact of highway construction on a North Florida watershed. *Water Resources Bulletin.* **12**(3):529–538.

Abstract: A 20 month study of some effects of highway construction on water quality was conducted during construction of Interstate 10 at Tallahassee, Florida. Highway construction resulted in significant increases in turbidity, suspended solids, total phosphorus, and dissolved silicon in downstream waters despite use of recommended procedures for erosion control. Highway construction did not result in significant increases in dissolved phosphorus or nitrogen.

Cabalka, D. (1996). Landfill cap erosion; Severe conditions and dramatic solutions. *Public Works.* **127**(10):32–34.

Abstract: Landfill operators risk severe damage to their landfills from stormwater runoff in both overland and channelized flow. This risk can be reduced with proactive design, construction, and maintenance practices. A variety of materials and installation methods have been developed to protect landfill caps from both water and wind erosion. The innovations available include articulating concrete block systems, turf reinforcement mats, erosion control blankets, and hydraulically-applied mulches. These alternatives need to be evaluated based on certain landfill features as they relate to erosion, including downchutes, perimeter channels, benches, cap side slopes, and the top of the cap.

Chen, C.-N.; Santomauro, F.; Fisher, J.B. (1975). Erosion control system for pipeline construction sites. *Natl Symp on Urban Hydrol and Sediment Control.* 37–49.

Abstract: The feasibility of implementing on-site erosion control measures on selected portions of the cross-county pipeline construction, proposed by the Columbia Gas Transmission Corporation, was investigated. The baseline erosion rates and the uncontrolled erosion rates from the disturbed sites, along 27 miles of the western portion and two miles of the eastern portion of the proposed pipeline route, have been estimated. Four sites representative of areas of potential severe erosion hazard were selected for detailed study. Preliminary erosion control plans for the selected sites suggest that it is impractical to limit the erosion rate from the construction sites to their baseline rates. Instead, optional criteria limiting the erosion from the construction site to either the target erosion rates suggested for the region or 1.5 times the baseline rate of the particular site is recommended. Savings in the costs of the erosion control measures can be achieved if the scheduled construction period can be shifted from the summer season which coincides with the season of high rainfall-erosion potential.

Chen, X. (2001). Reinforced turf side slope protection technology study. *Gongku Jiaotong Keji/Journal of Highway and Transportation Research and Development* (In Chinese). **18**(2):97–100.

Abstract: This paper presents the experiment of new technology of side slope protection by 3-dimensional cover plant combined with jet seeding and discusses the construction technology, its erosion resistance function and the cost-benefit analysis. The experiment shows that the technology is effective for areas with fine sandy soil, high liquid limit side slope and sandy rock soil slope, with soil erosion 20% less than that in areas with simple turf protection and that there is no conspicuous surface layer sliding even under the condition of thunderstorm. In comparison with other side slope protection measures, the cost of the reinforced turf side slope is reduced by 50%–70%, which is new alternative method for side slope protection.

Ciarla, M. (1985). Gabion weirs in water erosion control projects; Design and construction criteria. *Erosion Control: A Challenge in Our Time, Proceedings of Conference XVI.* 85–103.

Abstract: Gabion weirs have been used as an effective solution in many water erosion control projects and in soil stabilization and landslide control works. The gabion, a rectangular wire mesh basket filled with stone, is the 'basic building block'. These blocks, wired together in layers, form gabion structures that are flexible and permeable. This paper describes gabion weirs both in regard to their use in gradient control of watercourses, soil consolidation, landslide control, and check dams and in regard to the slope of their downstream face-vertical drop, stepped or sloped.

Cowherd Jr., C.; Grelinger, M.A. (1996). High-wind failure of soil moisture as a wind erosion control. *Proceedings of the Air & Waste Management Association's Annual Meeting.* 16pp.

Abstract: While windblown fugitive dust emissions can be controlled from 'inactive' sources (e.g., vacant land) through application of chemical dust suppressants or through revegetation, it is much more difficult to control windblown emissions from 'active' sites (e.g., construction projects, landfills and hazardous waste remediation sites involving contaminated soils). The objective of the study reported in this paper was to determine at what wind speeds controls for fugitive dust from exposed soils become ineffective. The specific control measure identified for testing was watering of a cover soil stockpile at a landfill prior to transfer of the soil to the active fill area. The Midwest Research Institute portable open-floored wind tunnel was used to measure the high-wind erodibility characteristics of soil surfaces as a function of soil surface moisture content. Testing was performed at a landfill site within the South Coast Air Quality Management District. Varying moisture levels of surface moisture on two test soils were achieved by blending moist soil with dry soil. This blending was accomplished by spreading and mixing loose bulk soils (moist and/or dry) on the test platform, i.e., a flat, compacted portion of the soil pile. A key conclusion from this study was that even moist soils become erodible at higher wind speeds. For example, the erodibility of a moist soil is similar to the erodibility of a dry soil if the moist soil is subjected to a wind speed that is about 15 to 20 mph higher than that impinging on the dry soil. This supports the conclusion that moisture control of wind erosion becomes ineffective at high wind speeds. Apparently, even under high soil moisture conditions, coarse particle entrainment processes result in rapid moisture depletion and collisions that shed fine particles previously bound by moisture films.

Crafton, C.S. (1987). Performance criteria for erosion and sediment control." *Erosion Control—You're Gambling Without It, Proceedings of Conference XVIII.* 39–47.

Abstract: One of the dilemmas often faced by governmental officials when establishing erosion and sediment control regulatory programs is the nature of the standards set for measuring compliance. This paper discusses alternative types of criteria, standards, and specifications, suggesting strengths and weaknesses of the various types. Examples include water quality standards for suspended solids; sediment load reductions based on a modified Uniform Soil Loss Equation with a delivery ratio; construction standards and specifications; and performance standards. The discussion will support the establishment of common sense performance criteria which describe the minimum level of control that should be achieved in a number of activities fairly common among construction projects. Such criteria are management-oriented. A list of specific criteria will be recommended, including discussion of conceivable variables.

Datta, U.; Chatterjee, P.K. (2001). Jute geotextiles for civil engineering applications. *Journal of the Institution of Engineers (India), Part TX: Textile Engineering Division.* **82**(1):6–9.

Abstract: Jute geotextiles are being increasingly employed in various civil engineering activities to facilitate construction, ensure better performance of the structure and reduce maintenance cost. The potential of jute geotextiles has caught the attention of Indian Engineering Community as well over past few years. In view of this wide spread interest, Indian Jute Industries' Research Association (IJIRA) has taken up a project under the sponsorship of Jute Manufactures Development Council (JMDC) on promotion of jute geotextiles based on successful laboratory studies and a number of field trials for erosion control of denuded slopes pertaining to environmental engineering and various other applications such as separation, futration, drainage and reinforcement in geotechnical and highway engineering. All the field experiments conducted by IJIRA in association with other R and D institutes have been monitored and their effectiveness are described brief in this paper. It has been found that open mesh jute geotextile can control slope erosion successfully by promoting growth of vegetation. In other experiments, jute geotextiles have been used successfully for protection of the river and canal banks against water action and construction of embankment on soft soil and stability of roads and railways. Jute geotextiles are also found effective for stabilization of mine spoils generated by mining activity.

Davis, C.R.; Johnson, P.A.; Miller, A.C. (2003). Selection of erosion control measures for highway construction. *World Water and Environmental Resources Congress, EWRI.* 262–271.

Abstract: Best management practices (BMPs) for erosion and sediment control are measures designed to reduce the amount of sediment leaving a construction site and prevent that sediment from entering nearby surface waters. Failure modes of these BMPs were used to develop a Failure Modes and Effects Analysis (FMEA). By conducting a FMEA, all possible failure modes and resulting consequences were determined for the selected control measures. Risk priority numbers (RPN) were used to determine the relative risk of failure for each measure. These RPNs for the different control measures could be used to aid the user in the selection of an appropriate set of BMPs for the given situation.

De Sutter, R.; Huygens, M.; Verhoeven, R. (2001). Sediment transport experiments in unsteady flows. *International Journal of Sediment Research.* **16**(1):19–35.

Abstract: By means of a test flume with semi-circular cross-section, bedload and suspended-sediment transport of non-cohesive material have been studied in transient flow. The experimental facility enables us to investigate the time evolution of friction and transport parameters. Preliminary measurements with a fixed bottom instead of a sediment bed yield a reliable assessment of flow and friction characteristics. Time sequence in unsteady flow of the relevant parameters is revealed. The influence of turbulence variation and shear stress variation on the transport is investigated. As existing transport equations are found to be in poor agreement with experimental data, a new "engineering" concept is constructed which relates friction velocity to transport.

Dallaire, G. (1976). Controlling erosion and sedimentation at construction sites. *Civil Engineering.* **47**(10):73–77.

Abstract: North Carolina is said to have one of the most progressive programs for controlling construction site erosion and sedimentation; and this article tells about some of the methods they are using. Among some of the things they suggest developers do: divert runoff water originating upstream from the construction site, so this water will not be flowing over bare-earth areas, eroding away soil; limit the area being graded on a site at any one time, so there won't be vast stretches of bare soil; limit the time any given area is laid bare; trap sediment-laden runoff in temporary or permanent basins, or filter the runoff by using silt fences, wood-chip barriers, or brush barriers.

Dillaha III, T. A.; Beasley, D.B. (1983). Distributed parameter modeling of sediment movement and particle size distributions. *Transactions of the ASAE.* 26(6):1766–1772, 1777.

Abstract: A spatially descriptive erosion submodel for estimating the particle size distribution of eroded sediment from disturbed upland watersheds and construction sites has been developed and incorporated into the ANSWERS watershed model. Model validation is accomplished using data from field plot studies. The model's usefulness as an erosion control planning tool is demonstrated on a construction site in central Indiana. The model divides the watershed into a uniform grid of square planar elements. Within each element, the model describes the processes of interception, infiltration, surface storage, surface flow, subsurface drainage, and sediment detachment, transport and deposition. The continuity equation is then used to integrate the individual elemental responses into a system response that describes the watershed as a whole. The erosion process is limited to the overland flow regime.

Eith, A.W.; Ballod, C.P.; Hart, M.F.; Case, M.E. (2005). Long-term performance of geogrid-reinforced Mechanically Stabilized Earth (MSE) berms. Geotechnical Special Publication. **130–142**:3925–3934.

Abstract: In the Fall of 2000, the first of two mechanically stabilized earth (MSB) berms was completed at the G.R.O.W.S. Landfill located in Morrisville, Pennsylvania as part of the Southwest Expansion construction. A second MSE berm was subsequently constructed as part of the Northeast Expansion and was completed during the Winter of 2001. Both of these berms have been monitored for wall movement and deflection prior to waste placement and while the adjacent landfill cells were being filled. Slope inclinometer casings were installed in the berms and monitored periodically over the past four years during the course of landfill operations. This paper will discuss the performance of the MSE berms in terms of movement monitored by the slope inclinometers.

Emerson, D.; Goldstein, M. (2004). Minnesota DOT advances compost use for erosion control. *BioCycle.* **45**(8):49–53.

Abstract: The shoreline along U.S. Highway 61 is the site of two innovative projects using compost-based erosion control systems to stabilize slopes and restore vegetation after road construction. The first is a 3-mile road reconstruction project in Grand Marais, and the second, the Silver Cliff Creek Trail in Two Harbors, involves construction of a bike and walking path along an abandoned road corridor deemed historic. These two projects set the stage for compost-based solutions for erosion and sediment control and storm water management.

Evans, M.L.; Dynes, Sheldon. (1974). *Sediment and Erosion Control Measures on Construction Sites.* Wisconsin University, College of Engineering, Engineering Experiment Station, Report, 1974. 6p.

Abstract: Various measures applied in Indiana on construction sites including dam and channel construction to reduce erosion and downstream sedimentation.

Farrar, J.A. (1993). Bureau of Reclamation experience in testing of riprap for erosion control of embankment dams. *ASTM Special Technical Publication: Symposium on Rock for Erosion Control.* **1177**: 3–14.

Abstract: The Bureau of Reclamation has accumulated

significant experience with the use of riprap for erosion control of embankment dams through its history as a major water resources design and construction agency for irrigation projects in the seventeen western states. Successful exploration, design, testing, and construction methodologies have been developed through experience. Exploration and design aspects are well documented in a series of technical manuals and design standards. Riprap quality evaluations depend heavily on expert geologic and petrographic evaluations coupled with physical properties and freeze thaw testing of rock specimens. Quality evaluation methodologies were heavily influenced by concrete technology testing resulting in the use of 75 mm (3 in) cube specimens for freeze thaw testing. Physical properties tests are performed on crushed coarse aggregate gradations. Reclamation has an available database of almost 1000 quality evaluations. Field placement, control test procedures, and riprap performance studies are reviewed.

Fifield, J.S. (1986). Erosion control measures—Are they effective? *Erosion Control: Protecting Our Future—Proceedings of Conference* XVII. 19–35.

Abstract: Various methods exist to control erosion from construction sites such as use of straw bales, construction of sediment basins, mulching, establishment of vegetation, and so forth. However, planners, hydrologists and engineers often do not have a method for evaluating the effectiveness of measures being implemented. Part of the above problem lies with the fact that all soils do not erode at the same rate, thus making an evaluation for the effectiveness of erosion control measures more difficult. To overcome such problems in the Parker, Colorado area, a Land Disturbance Hazard Map (LDHM) has been developed to assist developers, planners, hydrologists and engineers. The LDHM identifies three land zones (slight, moderate or severe) with each zone having an erosion control performance standard that must be met during development.

Flanagan, D.C.; Chaudhari, K.; Norton, L.D. (2002). Polyacrylamide soil amendment effects on runoff and sediment yield on steep slopes: Part I. Simulated rainfall conditions. *Transactions of the American Society of Agricultural Engineers.* **45**(5):1327–1337.

Abstract: Steep slopes consisting of disturbed soil are very often found in construction, landfill, and surface mining situations. Although legislation and economics dictate that vegetative cover be established on these slopes as rapidly as possible, the occurrence of large rainfall events during critical periods of vegetation establishment can frequently cause extensive soil loss. Sediment generated from erosion can impair off-site water quality, and on-site damages to the eroded region can be so extensive that expensive earthmoving, regrading, reseeding, and remulching may be necessary. We evaluated the effectiveness of two soil treatments for reducing runoff and soil loss from a silt loam topsoil placed on a constructed 32% slope. The three treatments were an untreated control, 80 kg ha^{-1} anionic polyacrylamide (PAM) applied as a liquid spray, and 80 kg ha^{-1} PAM as a liquid spray combined with a dry granular application of 5 Mg ha^{-1} of gypsum. Replicated plots were subjected to a range of rainfall intensities under a programmable rainfall simulator, and resulting runoff and sediment loss were measured. In the first event of 69 mmh^{-1} uniform rainfall applied for one hour to initially dry soil, the PAM and PAM with gypsum treatments significantly reduced runoff by almost 90% and sediment yield by 99%, compared to the control. Total runoff through a series of simulated rainfall events was reduced by 40% to 52%, and sediment loss was reduced by 83% to 91% for the plots treated with PAM and PAM plus gypsum, respectively. These results indicate that the use of PAM alone or in combination with gypsum can significantly reduce runoff and soil loss from large storm events, and may be a cost-effective approach to protect the soil during critical periods of vegetation establishment, particularly for disturbed soils on steep slopes.

Fluet Jr., J.E. (Ed.) (1987). Geotextile testing and the design engineer. ASTM Special Technical Publication: *Geotextile Testing and the Design Engineer.* 183p.

Abstract: This symposium proceedings contains 11 papers. The topics discussed are: hydraulic properties of geotextiles; geotextiles and drainage; lateral drainage design using geotextiles/geocomposites; geotextiles as filters in erosion control; geotextile tension testing methods; soil reinforcement design using geotextile/geogrids; durability testing; geotextiles construction criteria; tests for geotextile characterization/evaluation; design for geotextile applications; ASTM geotextile committee testing update.

Galagan, C.; Ziegler, C.K.; Isaji, T.; King, J. (2004). Application of a sediment transport model (BFSED) to assess the impact from bridge reconstruction in the Barrington River, RI. *Proceedings of the Eighth International Conference on Estuarine and Coastal Modeling.* 1075–1093.

Abstract: The results of an application of a sediment transport model associated with bridge reconstruction across a tide-dominated estuary were presented. A marked decrease in the speed of maximum flood currents across the bridge opening after construction of the replacement bridge. It was estimated that a maximum of 3,566.5 m^3 of sediment will be subjected to erosion as a result of new bridge configuration. The results also indicated a marked decrease in the speed of maximum flood currents across the bridge opening after construction of the replacement bridge.

Gao, P.; Liu, Z.-X.; Yu, S.-R. (2005). Soil and water loss and its prevention and control techniques on highway construction area. *Jishu Daxue Xuebao (Ziran Kexue Ban)/Journal of Liaoning Technical University (Natural Science Edition)* (In Chinese). **24**(4):543–545.

Abstract: By choosing the Pan-Hai (Panjin city to Haicheng city) highway of Liaoning province as the experimental site, the cause of formation of the soil and water loss and its happening form were analyzed systematically. The potential soil erosion amount was estimated with soil and water loss prediction models. Following engineering mechanics principle and soil and water conservation theory, the comprehensive control measures of soil erosion were put forward, which included the terrace wall projects of stabilizing the ravaged section and soil mass, the three-dimensional vegetation network fixation soil techniques, soil preparation for afforestation, vegetation restoration and utilization projects, and so on. This study can provide scientific basis for soil and water conservation and ecological environment control, and its resources use in the highway construction area.

Gharabaghi, B.; Rudra, R.P.; Whiteley, H.R.; Dickinson, W.T. (2000). Improving removal efficiency of vegetative filter strips. *2000 ASAE Annual International Meeting, Technical Papers: Engineering Solutions for a New Century.* **2**:2359–2369.

Abstract: Field experiments on vegetative filter strips (VFS) showed average sediment removal efficiency for the 2, 5, 10, and 15.m wide strips was between 80 and 95%. A power-relation existed between observed values of Einstein's dimensionless total sediment discharge, ψ, and U/w. Manning's roughness coefficient for VFS was observed to have a power-relation with Reynolds number. The observed empirical equations governing flow resistance characteristics and total sediment discharge from VFS will be adopted in a design tool (computer model) for VFS. The model will be valuable in the design of cost-effective constructed vegetative filter strips (CVFS), with enhanced removal efficiency through stem-reinforcement of vegetation with various geosynthetics products and/or through installation of drainage system to improve infiltration.

Glanville, T.D.; Persyn, R.A.; Richard, T.L.; Laflen, J.M.; Dixon, P.M. (2004). Environmental effects of applying composted organics to new highway embankments: Part 2. Water quality. *Transactions of the American Society of Agricultural Engineers.* **47**(2):471–478.

Abstract: An oversupply of composted organics, and imposition of new federal regulations governing stormwater discharges from construction sites, motivated the Iowa Department of Natural Resources (IDNR), and the Iowa Department of Transportation (Iowa DOT) to sponsor a study of the potential water quality impacts of using compost to control runoff and erosion on highway construction sites. Test areas treated with 5 and 10 cm deep blankets (unincorporated) of three types of compost (biosolids, yard waste, and bio-industrial byproducts) were constructed on a new highway embankment with a 3:1 sideslope and subjected to simulated rainfall intensity of approximately 100 mm h-1. Concentrations and total masses of N, P, K, and nine metals in runoff from compost-treated areas were compared to those in runoff from embankment areas receiving two conventional runoff and erosion control methods typically used by the Iowa DOT (light tillage and seeding of native embankment soil, or application of 15 cm of imported topsail followed by seeding). Simulations were replicated six times under both vegetated and unvegetated conditions, and the first hour of runoff was sampled to determine concentrations and total masses of soluble and adsorbed nutrient and metals. The applied composts generally contained much greater pollutant concentrations than either of the two soils used in the conventional treatments, and runoff from unvegetated plots treated with compost also contained significantly greater concentrations of soluble and adsorbed Zn, P, and K, and adsorbed Cr and Cu, than runoff from the two conventional treatments. In accordance with previously reported soil erosion research, runoff from all test plots was sampled periodically during the first hour of runoff. Due to their significantly greater infiltration capacity, however, compost-treated areas required significantly greater amounts of rainfall than conventionally treated areas to produce 1 h of runoff. In light of this significant difference in the amount of rain applied, the total mass of pollutants contained in runoff generated by equal amounts of rainfall was judged a more equitable basis for comparing the treatments. Runoff samples collected during the first 30 min of rainfall (equivalent to a 25-year return period storm at the applied intensity of 100 mm h^{-1}) were used for this purpose, and the resulting total masses of individual quantifiable soluble and adsorbed contaminants in runoff from conventionally treated areas were at least 5 and 33 times, respectively, those in runoff from compost-treated areas. Based on these results, blanket applications of compost can be used to reduce runoff and erosion from construction sites without increasing nutrients and metals in stormwater runoff.

Glanville, T.D.; Richard, T.L.; Persyn, R.A. (2003). Evaluating performance of compost blankets. *BioCycle.* **44**(5):48–54.

Abstract: The research project, which was designed to provide baseline data that would help to utilize compost for storm water and erosion control on construction projects in Iowa, is discussed. The primary objective was to compare the performance of compost-treated and conventionally-treated roadway embankments. Performance parameters have included runoff quantity, runoff quality, rill and interrill erosion, and seasonal growth of planted species and weeds.

Grismer, M.E.; Hogan, M.P. (2005). Simulated rainfall evaluation of revegetation/mulch erosion control in the Lake Tahoe basin: 2. Bare soil assessment. *Land Degradation and Development.* **16**(4):397–404.

Abstract: Slopes that have been disturbed through roadway, ski slope or other construction often produce more sediment than less disturbed sites. Reduction or elimination of sediment loading from such disturbed slopes to adjacent streams is critical in the Lake Tahoe basin. Here, use of a portable rainfall simulator (RS), described in the first paper of this series, is used to evaluate slope effects on erosion from bare volcanic and granitic soils (road cut and skirun sites) common in the basin in order to establish a basis upon which revegetation treatment comparisons can be made. Rainfall simulations (60 mm h^{-1}, approximating a 100-year, 15-minute storm) at each site included multiple replications of bare soil plots as well as some adjacent 'native', or relatively undisturbed soils below trees where available. Field measurements of time to runoff, infiltration, runoff, sediment discharge rates, and average sediment concentration were obtained. Laboratory measurements of particle-size distributions using sieve and laser counting methods indicated that the granitic soils had larger grain sizes than the volcanic soils and that road cut soils of either type also had larger grain sizes than their ski run counterparts. Particle-size-distribution-based estimates of saturated hydraulic conductivity were 5–10 times greater than RS-determined steady infiltration rates. RS-measured infiltration rates were similar, ranging from 33–50 mm h^{-1} for disturbed volcanic soils and 33–60 mm h^{-1} for disturbed granitic soils. RS-measured runoff rates and sediment yields from the bare soils were significantly correlated with plot slope with the exception of volcanic road cuts due to the narrow range of road cut slopes encountered. Sediment yields from bare granitic soils at slopes of 28 to 78 per cent ranged from [similar to] 1 to 12 g m^{-2} mm^{-1}, respectively, while from bare volcanic soils at slopes of 22 to 61 per cent they ranged from [similar to] 3 to 31 g m^{-2} mm^{-1}, respectively. Surface roughness did not correlate with runoff or erosion parameters, perhaps also as a result of a relatively narrow range of roughness values. The volcanic ski run soils and both types of road cut soils exhibited nearly an order of magnitude greater sediment yield than that from the corresponding native, relatively undisturbed, sites. Similarly, the granitic ski run soils produced nearly four-times greater sediment concentrations than the undisturbed areas. A possible goal of restoration/erosion control efforts could be recreation of 'native'-like soil conditions.

Guy, H.P. (1976). Diminution ratios for planning construction-area sediment controls. Kentucky University, Office of Research and Engineering Services, Bulletin, *Natl Symp on Urban Hydrol, Hydraul, and Sediment Control.* 91–97.

Abstract: Planning erosion-control programs to limit sediment concentration in streams where part of the drainage basin is a construction area requires knowledge of the ratio of construction-area to rural-area concentration and the relative construction area of the basin. The diminution ratio required to obtain a specific stream concentration can be computed on the basis of these construction- and rural-area concentrations determined from the universal soil loss equation. The diminution ratio is the product of the factors necessary to achieve a specific limit of average sediment concentration in the stream draining the basin. Included are the ratio of construction area to total area factor, and the conservation practice factor. A map of the eastern United States was prepared showing diminution ratio.

Hagen, S.; Salisbury, S.; Wierenga, M.; Xu, G.; Lewis, L. (2002). Soil bioengineering as an alternative for roadside management: Benefit-cost analysis case study. *Transportation Research Record.* 1794(02-2101):97–104.

Abstract: As an environmentally compatible and cost-efficient alternative for roadside management, soil bioengineering has become increasingly important and attractive. Soil bioengineering uses live plants and plant parts as building materials for engineering and ecologically sound solutions to erosion control, slope and stream bank stabilization, landscape restoration, and wildlife habitats. However, not all decision makers are aware of the specific benefits of this approach. This case study applied a benefit-cost analysis to an experimental soil bioengineering demonstration project to evaluate the cost-effectiveness of soil bioengineering as an alternative to traditional roadside management. Traditional roadside management methods (geotechnical solutions) were used as the baseline, and soil bioengineering treatments were treated as an investment alternative. Cost savings, along with other environmental benefits, were assessed and compared with construction costs. The effects of life cycle, effectiveness, and discounting were included in the analysis to ensure comparability between both treatments. The analytical results demonstrate that soil bioengineering methods, if technically feasible, could be adopted to produce equal or better economic and environmental results. The findings of the research project and the economic analysis indicate that soil bioengineering is an efficient and environmentally beneficial tool for roadside management.

Heier, T.; Starrett, S. (2003). Using SWMM to model sediment runoff from a golf course. *World Water and*

Environmental Resources Congress Proceedings. 1983–1989.

Abstract: Kansas State University has built an 18-hole championship golf course near Manhattan, Kansas, on the Little Kitten Creek watershed. The watershed, previously native grassland, underwent dramatic changes in land-uses since the beginning of the construction. A seven-year study was done to evaluate changes in surface water in terms of sediment concentrations. This paper was designed to compare PCSWMM (2002 version) water quality predictions with those measured values and predicted values from an AGNPS model. All attributes of the watershed were inputted into the program including area, slope, land-use, soil characteristics, channel characteristics, infiltration variables, precipitation values, and evaporation values. A drastic change was noticed during the construction phase compared to the operational phase. The sediment concentrations during some storms during construction were almost 10 times the concentrations observed prior to construction of the golf course. The sediment yield during the pre-construction phase was measured to be 845,000 kg/yr compared to 905,400 kg/yr predicted in AGNPS and 943,500 predicted in SWMM. During construction, the measured value was 1,574,000 kg/yr compared to 2,708,000 kg/yr predicted in AGNPS and 1,832,500 kg/yr predicted in SWMM. During operation, the concentrations returned to the level observed prior to construction. PCSWMM and AGNPS accurately predicted sediment yield increases during the construction of this golf course.

Henensal, P. (1996). La lutte contre l'erosion sur l'emprise routiere une contribution a la protection de l'environnement. Combating erosion of carriageway land: a contribution to the protection of the environment. *Bulletin de Liaison des Laboratoires des Ponts et Chaussees.* **201**:17–28 (in French).

Abstract: Erosion poses a major threat to the environment in road construction, but the risk depends to a large extent on local conditions and the organization of the working sites. This article is in three parts. The first reviews the general principles of erosion control, namely the prevention or retarding of its occurrence, and the control of sedimentation so as to alleviate its harmful effects on the environment. The second part examines five groups of techniques for reducing the risk of erosion or reducing the amount of solid matter present in water flowing away from cultivated land, roadworks or engineering structures: measures designed to modify the physical and physico-chemical characteristics of soils; surface protection; embankment modulation; control of surface water; and special protective measures in particularly affected or sensitive zones. The third part comments on the practical aspects of selecting techniques best suited to the special conditions prevailing on site.

Huang, C.-P.; Ehrlich, R.S. (2004). Erosion control and the impact of highway construction on wetland water quality: A case study. *Watershed Management, Proceedings of the Third International Conference.* **3**:117–138.

Abstract: To assess the efficacy of control on sedimentation and erosion during highway construction, the impact on water quality in adjacent wetlands was monitored. Water quality measures used were turbidity, suspended solids, conductivity, alkalinity, hardness, pH, organic carbon, dissolved oxygen, nitrate and phosphate. Heavy metal concentrations in surface waters and stream sediments were measured. Upstream measurements, used as controls, showed periodic, seasonal variations. Downstream measurements were not significantly different from upstream measurements except during an incident when erosion controls were neglected and during culert construction. Culvert construction resulted in increased sediment and turbidity from diversion and dewatering and increase in hardness, attributed to concrete leaching. Attention to sedimentation and erosion controls and seasonal scheduling of highway construction are advocated as a means of ameliorating the effects on wetlands.

Hunt, D.; Austin, D.N.; Agnew, W. (1998). Vegetation selection for rolled erosion control product. *Geosynthetics in Foundation Reinforcement and Erosion Control Systems,* Geotechnical Special Publication. **76**:130–144.

Abstract: Temporary erosion control blanket on a slope or a turf reinforcement mat for long term protection in a high channel flow are solutions when revegetating construction sites. In this regard, issues to assist engineers in the selection of these materials and in the establishment of vegetation for rolled erosion control products on construction sites are presented. Properly selected, specified and installed vegetation is paramount to successfully complimenting other best management practices to ensure that the long-term solution to slope and channel stability is not an aftersought.

Israelsen, C.E.; Clyde, C.G.; Fletcher, J.E.; Israelsen, E.K.; Haws, F.W.; Packer, P.E.; Farmer, E.E. (1980). *Erosion Control during Highway Construction.* National Cooperative Highway Research Program Report ISSN: 0077-5614. 220:36p.

Abstract: The report presents a review of the literature and describes the adaptation of the universal soil loss equation, originally developed by the Agricultural Research Service of the U. S. Department of Agriculture, for estimating the water erosion potential and the effectiveness of erosion control measures on highway construction sites. An equation for estimating wind soil loss potentials is also included.

Jin, C.-x.; Romkens, M.J.M. (2000). Modeling deposition processes in vegetative filter strips. *International Journal of Sediment Research.* **15**(1):108–120.

Abstract: Vegetative filter strips have been widely used for controlling soil erosion from agricultural land, road sides, construct sites, and other disturbed lands. The most important characteristic of a filter strip is the sediment trapping efficiency. Yet, not many studies have been conducted to quantitatively describe the trapping process and trapping efficiency. In this paper, the results of a laboratory study are presented. In the study vegetative filter strips were simulated with polypropylene broom bristles of different densities. Results show that the deposition process in filter strips is described by a time decreasing, three-parameter exponential relationship. The parameters vary with both bristle density and flume slope steepness. Bristle density and flume slope steepness are two important factors in determining the trapping efficiency. As bristle density increased from 2,500 bunches/m^2 to 10,000 bunches/m^2, trapping efficiency increased to about 45%. When slope steepness increased from 2% to 4%, trapping efficiency decreased from 50% to 5% for the low bristle density and from 90% to 50% for the high bristle density. Over 80 approx. 90% of the sediment deposited upstream from the entrance of the filter strips. Most of the sediment trapped had a particle size greater than 150 μm. As the flume slope steepness increased, deposition occurred further down into the filter strips and the sediment passing through the filter strips became larger in size. Flow rate and sediment concentration in the studied ranges (1.45 approx. 7.0×10^{-3} m^3/s for flow rates and 1.71 approx. 7.0 kg/m^3 for sediment concentrations) hardly impacted the trapping efficiency.

Jones, M.L.; Tatum, Clyde B. (1980). Sedimentation control during site development. *American Society of Civil Engineers, Journal of the Construction Division.* 106(1):17–28.

Abstract: WPPSS Nuclear Projects 3 and 5 are currently under construction in Western Washington State. Because of the difficult site topography and the stringent criteria applied to construction runoff, an extensive erosion control program was required for site development. The erosion control systems which were implemented are described, including both design criteria and configuration. Initial site development activities, along with early rainfall events and erosion problems are examined. Operational experience gained in achieving compliance with discharge criteria is then presented. Based on this experience, the writers conclude that systems to achieve compliance with permit criteria can be implemented; however the benefits do not appear to justify the large costs. In many instances, the water quality of the treated runoff exceeded that of the receiving water.

Jungerius, P.D.; Matundura, J.; Van De Ancker, J.A.M. (2002). Road construction and gully erosion in West Pokot, Kenya. *Earth Surface Processes and Landforms.* **27**(11):1237–1247.

Abstract: The study of soil erosion in Kenya is largely limited to agricultural and pastoral land. Little attention has been given to the effects of roads on soil erosion, although they cause more inconvenience than any other form of soil erosion. The object of this study is the B4, an unpaved road leading from the Marich Pass up the Kerio valley. This road runs across the footslopes of a steep scarp fault, in a climate with torrential rainfall. The footslopes consist of alluvial fans, pediments and terraces. Aerial photographs from 1963 show no erosion along the road: the 162 culverts and drifts were apparently adequate to cope with the drainage of the footslopes. Roadside gully formation is now a big problem for the road engineers. The 1 km sections of the road marked by the Ministry of Public Works have been used as the units of research, to facilitate exchange of information with the road engineers. In each section of the first 42 km of the road, the volumes of the roadside gullies were measured, along with erosion factors related to type of cross-drainage construction, stability of the roadside material, geomorphological parameters and decrease in soil cover above the road since 1963. There is a strong correlation between roadside gullies, alluvial fans and decrease of soil cover. The road engineers were guided by geomorphological principles when designing the road. They preferred drifts, although these dips in the road surface are uncomfortable to pass by car. Drifts resemble natural drainage channels more closely than culverts and cause less damage in the fields below the road. The later settlers also showed geomorphological sense by preferring the smooth surfaces and well drained fine-grained soils of the alluvial fans. The research demonstrates a common problem of road design in developing countries: however carefully the measures against erosion are designed, they become rapidly outdated because a new road attracts settlement. Deterioration of surface drainage and erosion start at unforeseeable points where people choose to settle.

Kaufman, M.M. (2000). Erosion control at construction sites: The science-policy gap. *Environmental Management.* **26**(1):89–97.

Abstract: To test the effectiveness of Michigan's soil erosion control law, 30 construction sites were evaluated in the east-central part of the state. The analytical framework lumped nine best management practices (BMPs) most closely related to the law into three categories: slope stabilization, soil stabilization, and water management. All sites were in the land clearing or foundation/framing stage of construction and were evaluated within 2 days after a rainfall event. Only four of the sites performed above the mean of the scoring scale, with the categorical scoring of BMPs indicating the worst performance for slope stabilization measures. The poor results suggest a failure to integrate scientific knowledge of erosion control with policy. A fundamental problem is the lack of basic site data on soil, topography, and hydrology, resulting in the incorrect application of BMPs, such as staging, filter fences, and berms. The current institutional framework for soil erosion control also provides disincentives to mitigate local erosion problems.

Keller, G.; Sherar, J. (2003). Low-volume roads engineering: Best management practices. *Transportation Research Record.* **1**(1819):174–181.

Abstract: The concept and application of best management practices (BMPs) for low-volume roads projects were studied. BMPs are techniques or design practices that will prevent or reduce nonpoint pollution, maintain water quality, and help produce well-built roads. A Low-Volume Roads Engineering Best Management Practices Field Guide was developed to address those key practices. Roads that are not well planned or located, not properly designed or constructed, not well drained, not well maintained, or not made with durable materials often produce negative Impacts, most of which are preventable with good engineering and road management practices. A number of key practices and design techniques can be used to prevent adverse impacts on roads. First a road must serve the needs of the user through good transportation system planning. Long-term cost-effectiveness and minimized impacts are then achieved through application of good design and maintenance practices, including a road location that avoids problematic areas such as slides or springs; positive surface drainage; adequately sized and appropriate drainage crossing structures; stable cut and fill slopes; use of erosion control measures; roadway surface stabilization; and materials source development with subsequent site reclamation.

Kuennen, T. (2005). Road science: Erosion assaults the unpaved road. *Better Roads.* **75**(2):32–44.

Abstract: Unpaved roads are destroyed mainly by water. Whether surfaced with gravel or dirt, with or without oil, or located in arid or humid climates, every element of the unpaved road's design and maintenance must be aimed at the delay of water damage, erosion, and their destructive impacts on both the riding surface and the local environment. Several approaches for achieving this objective include draining unpaved roads, constructing V-ditch, selection of good-quality surface aggregates, preventing loss of aggregates, stabilizing unpaved roads, and choosing whether or not to apply aggregate surface.

Lemly, A.D. (1982). Erosion control at construction sites on red clay soils. *Environmental Management.* 6(4):343–352.

Abstract: Five single-treatment methods used to stabilize seeded areas at urban highway constructions were tested for their ability to control erosion of red clay soils by comparisons with exposed sites and multiple treatments. Reductions in the total sediment concentration of runoff ranged from 28 to 90 percent. Larger size fractions were effectively reduced by all treatments tested regardless of slope. Established grass cover exceeded 90 percent on all plots after 60 days, but sediment release remained similar. Results indicate that current stabilization methods shift sediment composition toward a smaller particle size, causing single treatments to be minimally effective for controlling erosion of the major component of red clay soils. A multiple-treatment approach offers significantly greater control of erosion on red clay soils.

Liverman, E.; Hecklau, J.; Palmero, C. (1987). Minimization of soil erosion and siltation during construction of the Marcy-South 345kv Transmission Facilities. 1987 *Erosion Control—You're Gambling Without It, Proceedings of Conference XVIII.* 241–253.

Abstract: A Certificate of Environmental Compatibility and Public Need was issued by the New York State Public Service Commission to the New York Power Authority for construction of the Marcy-South 345kV Transmission Facilities. This certificate required that an Environmental Management and Construction Plan be prepared for this project. Among requirements set forth in this plan are guidelines for the construction of nine types of access roads, four types of permanent stream crossings and numerous erosion and drainage control features, each designed to minimize disruption of the soil and water resources found along the right-to-way. If properly designed, installed and maintained, the access roads, permanent stream crossings, and erosion and drainage control features employed on this project will minimize soil erosion and siltation.

Loew, S.R.; Haselbach, L.M.; Meadows, M.E. (2004). Life cycle analysis factors for construction phase BMPs in residential subdivisions. *Proceedings of the 2004 World Water and Environmetal Resources Congress: Critical Transitions in Water and Environmetal Resources Management.* 565–572.

Abstract: Best management practices (BMPs) are used to mitigate the impacts of development on the environment. A life cycle analysis (LCA) is a systematic method for identifying and quantifying the effects of various factors upon BMP field efficiencies. Two life-cycle factors for 2 different LCAs are studied. The first factor is the installation probability for designed and approved silt fences during the construction phase of individual lots in a residential subdivision. Ownership change of the lots during construction is also investigated to determine its effects on this factor. The second factor is the probability a permanent residential detention facility will receive uncontrolled sediment loads from individual lots during the construction phase. The study focused on 14 residential subdivisions from which 184 construction phase lots were field visited and the files of 330 lots were researched on land records. Designed and approved silt fencing was only installed at 27% of the field visited lots. Of the 330 records researched, 76% of the lots underwent a construction phase ownership change. In the 14 subdivisions, 47.4% of the permanent detention facilities were receiving uncontrolled sediment loads from 1 or more individual lots. Typical damages to the facilities included partial or full clogging of basin inlet and outlet structures, loss of reservoir volume, side-wall erosion, and vegetation damage.

Long, D.; Bender, M.J.; Sawatsky, L.F.; Anderson, P.; Metikosh, S. (1998). Sediment entrainment during construction of river pipeline crossings: Occurrence, prediction and control. *Proceedings of the International Pipeline Conference.* **2**:1045–1050.

Abstract: Sediment entrainment in rivers caused by pipeline watercourse crossing construction may represent a constraint on pipeline route selection and construction methods as designers attempt to develop a sediment and erosion control plan which meets regulatory approval without risk of costly delays. To avoid the risk of significant sediment entrainment, conventional open-cut crossing techniques may be replaced by more costly directional drilling methods. However, the concern over suspended sediment is greatest in high velocity rivers where the bed material includes a large fraction of fine sand, and in rivers with a large fraction of fine grained bed material which becomes suspended upon disturbance by construction activities. According to the current understanding of aquatic impacts due to elevated suspended sediment levels, the occurrence of suspended sediment may not be excessive at open-cut excavation in certain types of streams depending on the material consistency, fine sand content and river flow velocity. Control of sand entrainment can normally be achieved by low cost sediment control systems during construction. Methods of prediction, impact assessment, and control of sediment entrainment have been developed so that high risk crossings can be identified and impacts minimized. The application of the study findings and best management practices (BMPs) for sediment control will allow developers to choose the most appropriate crossing method while avoiding potentially adverse impacts, based on a sound understanding of river sediment transport, bed material conditions and downstream aquatic resources.

Mamo, M.; Bubenzer, G.D. (2004). Decomposition parameters for straw erosion control blankets. *Transactions of the American Society of Agricultural Engineers.* **47**(3):721–725.

Abstract: Increased environmental concerns and regulations have mandated proper planning to control erosion and sediments from construction sites. A field experiment was conducted to evaluate decomposition rates of single- and double-layer straw erosion-control blankets and to estimate their surface decomposition parameter. Single (DS1, S1) and double (DS2, S2) layer applications of two common types of straw erosion-control blankets (DS150, S150) were used as treatments in the study. The netting material for DS150 was described as having the quality of degrading more quickly than that of S150. Using randomized block design, nine single- and double-layer applications of each material type were placed within plots of a freshly tilled Plano silt loam soil (fine-silty, mixed, mesic, Typic Argiudolls). Each corner was secured in place using U-shaped wires. Every two weeks, starting at week 2, a randomly designated sample of all treatment combinations was permanently removed and analyzed for mass loss. For the 14-week study period, average percent mass loss was significantly higher for treatment DS150 (40%) than for S150 (29%) at the 0.05 level. Within each erosion-control blanket of the same netting material, mean percent mass loss for a single layer was significantly greater ($\alpha = 0.05$) than that for a double layer. Mean percent mass losses were 50%, 34%, 30%, and 24% in 14 weeks for DS1, S1, DS2, and S2, respectively. Material type (P less than or equal 0.0001), time (P less than or equal 0.0001), and their interaction (P less than or equal 0.0237) were statistically significant. The decay constant for winter wheat ($p = 0.0175$) in the RUSLE crop database was found to have adequately predicted observed decomposition for DS1 straw blanket. However, it overestimated decomposition rates for all other straw blankets.

Martin, J.S. (1985). Use of silt fences for control of sediment run-off. *Erosion Control: A Challenge in Our Time, Proceedings of Conference XVI.* 79–84.

Abstract: Construction activity can generate significant quantities of sediment even then the best erosion control practices are utilized. Traditional forms of sediment control such as hay and straw bale barriers often do not adequately control sediment produced under low flow conditions. Silt fence sediment control systems, however, have been field proven to provide a low cost, high efficiency means of retaining sediment from sheet flow construction site run-off. Silt fences are manmade vertical barriers composed of synthetic fabric and posts. This paper highlights the performance advantages of silt fences and provides easy guidelines for their installation and use.

Masters, A.; Flahive, K.A.; Mostaghimi, S.; Vaughan, D.H.; Mendez, A.; Peterie, M.; Radke, S.; Davisson, A.; Hunter, M.; Kaplan, D. (2000). A comparative investigation of the effectiveness of polyacrylamide (PAM) for erosion control in urban areas. *2000 ASAE Annual International Meeting, Technical Papers: Engineering Solutions for a New Century.* **2**:3223–3244.

Also published in:

Soupir, M.L.; Mostaghimi, S.; Masters, A.; Flahive, K.A.; Vaughan, D.H.; Mendez, A.; McClellan, P.W. (2004). Effectiveness of polyacrylamide (PAM) in improving runoff water quality from construction sites. *Journal of the American Water Resources Association.* 40(1):53–66.

Abstract: Erosion from construction sites significantly affects water quality in receiving streams. A rainfall simulator was used to evaluate the effectiveness of different methods for controlling erosion from construction sites. Erosion control methods investigated included dry and liquid applications of polyacrylamide (PAM), hydroseed, and straw mulch. Fertilizer was also applied to each plot to examine the effectiveness of the methods in reducing nutrient losses in runoff. Runoff samples were analyzed for total suspended solids, nitrate, total Kjeldahl nitrogen, ammonium, total phosphorus, and orthophosphate. A sequence of three simulated rainfall events were applied to 21 (three replications of each treatment), experimental field plots. The first sequence of rainfall events was applied shortly after application of treatments to evaluate short-term impact of treatment methods. The same sequence of rainfall events was applied a month later to evaluate long-term effectiveness of the treatments. Among all treatments investigated, straw mulch was the most effective treatment for controlling sediment and nutrients losses during both the short-term and long-term simulations. The low liquid PAM (half the recommended PAM) treatment resulted in the highest reduction in runoff, sediment-bound nitrogen, and total nitrogen concentrations and loadings. Among the PAM treatments investigated, dry PAM produced the highest reduction of sediment and sediment-bound P concentrations and yields and total phosphorous concentrations. The recommended PAM resulted in the highest reduction of total P loadings. Straw mulch and low PAM treatments were consistent in reducing sediment losses and runoff, respectively, for both short-term and long-term simulations. Dry PAM's effectiveness in reducing sediment and sediment-bound phosphorous declined for the long-term simulations. Similarly, the low PAM's effectiveness for reducing total nitrogen losses decreased substantially during the one month period between the two sets of the simulations. The study results indicate that high application rate (twice the recommended rate) of PAM could actually increase runoff and sediment losses. The low rate of liquid PAM and the dry PAM were both effective in reducing sediment and nutrient losses in runoff, however application of liquid form of PAM to construction sites is more practical and perhaps more economical than applying the PAM in the dry form.

McEnroe, B.; Treff, B.J. (1998). Temporary erosion control for highway construction in Kansas. *Proceedings of the Annual Water Resources Planning and Management Conference, Water Resources and the Urban Environment.* 722–727.

Abstract: Field observations of temporary erosion-control measures on highway construction sites revealed some common problems. Most failures result from improper location, installation, or maintenance of the erosion-control measures. A training program for field personnel could alleviate most of the misunderstandings that lead to failures. Current specifications for temporary erosion control are satisfactory, but better compliance is needed. The timely use of temporary seeding would greatly reduce soil loss.

McLain, J.L.; Anderson, S.B. (1987). Successful erosion control combines engineering and vegetation expertise at the Ridge Tahoe, Nevada. *Erosion Control—You're Gambling Without It, Proceedings of Conference XVIII.* 33–37.

Abstract: The Ridge Tahoe construction site was characterized by excessively steep slopes of decomposed granite. The major erosive forces resulted from intense summer thunder showers and spring runoff typical of the High Sierra Mountains. Initial construction at this site resulted in the removal of large expanses of native vegetation and disturbance of natural drainages leaving denuded slopes vulnerable to excessive erosion. Effective erosion control at the Ridge Tahoe was attributed to the combination and, more importantly, coordination of engineering and plant science. The erosion control plan was implemented during 1983 and 1984. Erosion control plans were phased in accordance with construction schedules. Following a two year vegetation establishment period, the site experienced a 25-year storm event. Implementation of the engineering and vegetation designs for erosion control resulted in little or no erosion of the site and confinement of any sedimentation within the property boundaries.

Middleton, L; King, M. (2003). A natural choice. *Public Roads.* **66**(5):2–5.

Abstract: Controlling erosion and reestablishing vegetation are key components in most road and highway construction or rehabilitation projects. Roadside embankments, shoulders, medians, and other nonpaved surfaces can be vulnerable to the elements, leading to excessive runoff, rutting, and damaged aesthetics. Conventional methods to prevent these conditions include hydroseeding and root reinforcement systems. In recent years, an alternative erosion control mechanism—compost—has been gaining in popularity. Several State departments of transportation have begun to experiment with compost mixtures for roadside revegetation. Recently, the Federal Highway Administration's Eastern Federal Lands Highway Division (EFLHD) had an opportunity to test the composting method. A small landslide along the Blue Ridge Parkway near Asheville, NC, took out a section of the roadside cut area, and EFLHD was charged with repairing the section and ensuring the affected area's environmental sustainability. The project was unprecedented in that not only was it the first time EFLHD had used compost in this capacity, but also never before, to EFLHD's knowledge, had compost been applied to roadside terrain this steep. The results indicate that compost shows so much potential that it should be seriously considered as a best management practice. Not only is compost as good as or better than conventional erosion control methods, but also it benefits the environment by reducing the biodegradable wastes that go to landfills.

Monlux, S. (2003). Stabilizing unpaved roads with calcium chloride. *Transportation Research Record.* **II**(1819):52–56.

Abstract: The U.S. Department of Agriculture Forest Service has stabilized unpaved road surfacing materials with relatively high concentrations of calcium chloride salt. The percentage of calcium chloride is higher than that traditionally used for dust abatement or aggregate base stabilization. Up to 2% pure salt by weight of aggregate was mixed into the top 2 in. (50 mm) of both aggregate and native road surfaces. The results were monitored for 2 to 4 years. The stabilized road surfaces resisted raveling and washboarding for several seasons and significantly reduced road blading and aggregate loss. As a result, calcium chloride stabilization may be a cost-effective treatment for roads with daily traffic volumes less than 200. Other benefits include reduced surface erosion and sedimentation; improved safety from reduced dust, raveling, and washboarding; and less frost penetration. Encouraged by these results, the Forest Service is conducting additional evaluations to determine the cost-effectiveness of surface stabilization with both magnesium chloride and calcium chloride in different environments and with different aggregate materials.

Neal, W. (1976). Specifying erosion control during construction. *Construction Specifier.* 29(1):26–30, 32.

Abstract: The author discusses what erosion does, what controls are needed, adequate specifications, and temporary control measures on erosion and sediment deposits during the construction phase.

Nyssen, J.; Poesen, J.; Moeyersons, J.; Luyten, E.; Veyret-Picot, M.; Deckers, J.; Haile, M.; Govers, G. (2002). Impact of road building on gully erosion risk: A case study from the Northern Ethiopian Highlands. *Earth Surface Processes and Landforms.* **27**(12):1267–1283.

Abstract: Although obvious in the field, the impact of road building on hydrology and gullying in Ethiopia has rarely been analyzed. This study investigates how road building in the Ethiopian Highlands affects the gully erosion risk. The road between Makalle and Adwa in the highlands of Tigray (northern Ethiopia), built in 1993–1994, caused gullying at most of the culverts and other road drains. While damage by runoff to the road itself remains limited, off-site effects are very important. Since the building of the road, nine new gullies were created immediately downslope of the studied road segment (6.5 km long) and seven other gullies at a distance between 100 and 500 m more downslope. The road induces a concentration of surface runoff, a diversion of concentrated runoff to other catchments, and an increase in catchment size, which are the main causes for gully development after road building. Topographic thresholds for gully formation are determined in terms of slope gradient of the soil surface at the gully head and catchment area. The influence of road building on both the variation of these thresholds and the modification of the drainage pattern is analyzed. The slope gradient of the soil surface at the gully heads which were induced by the road varies between 0.06 and 0.42 mm^{-1} (average 0.15 mm), whereas gully heads without influence of the road have slope gradients between 0.09 and 0.52 mm^{-1} (average 0.25 mm^{-1}). Road building disturbed the equilibrium in the study area but the lowering of topographic threshold values for gullying is not statistically significant. Increased gully erosion after road building has caused the loss of fertile soil and crop yield, a decrease of land holding size, and the creation of obstacles for tillage operations. Hence roads should be designed in a way that keeps runoff interception, concentration and deviation minimal. Techniques must be used to spread concentrated runoff in space and time and to increase its infiltration instead of directing it straight onto unprotected slopes.

Parker, S.C.; Stader, T.N.; Parker, D.G. (1995). Use of GIS to predict erosion in construction. *International Water Resources Engineering Conference—Proceedings.* **1**:839–843.

Abstract: A graphic software system is designed and implemented to allow for the analysis of erosion potential on proposed highway construction sites. The system is based on Geographic Information System technology and allows for the consideration of erosion prevention products such as straw and other mulches as well as other types of cover products designed to prevent or minimize erosion from construction practices. The use of this system will allow for effective decisions concerning erosion control before construction has begun and erosion damage has already occurred.

Persyn, R.A.; Glanville, T.D.; Richard, T.L.; Laflen, J.M.; Dixon, P.M. (2004). Environmental effects of applying composted organics to new highway embankments: Part 1. Interrill runoff and erosion. *Transactions of the American Society of Agricultural Engineers.* **47**(2):463–469.

Abstract: Construction of new highways can lead to challenges when attempting to re-establish vegetation on right-of-ways. Lack of vegetation can leave soil exposed and subject to increased runoff and soil erosion. Therefore, the Iowa Department of Transportation and the Iowa Department of Natural Resources sponsored a study to evaluate the use of composts applied as mulch blankets to decrease runoff and erosion. This article evaluates interrill runoff and erosion between three types of compost (biosolids, yard waste, and bio-industrial byproducts) and two soil conditions (existing compacted subsoil (control) and imported topsoil) on a 3:1 highway embankment. Composts were applied as 5 and 10 cm blankets on the surface of the control, and topsoil was placed on the surface of the control at a depth of 15 cm. Treatments were replicated six times over a two-year period for both bare soil and six weeks following planting of an Iowa DOT-specified cover crop. Rainfall was applied at an average intensity of 95 mm h^{-1} using a rainfall simulator, and sampling was conducted for 1 h after runoff began. All compost treatments were effective at reducing interrill erosion rates under the conditions simulated in this study. In addition, the three compost media required 30 min or longer to produce runoff, while the two conventional soils produced runoff within the first 8 min. The depth of compost application was only a factor for the runoff rate on unvegetated treatments. In this case, the 5 cm depth had a significantly greater runoff rate than the 10 cm depth. Both 5 and 10 cm compost applications had similar effects on interrill erosion rates. Although the steady-state interrill erosion rates of all three composts were 3% to 24% of the steady-state interrill erosion rates of the two soils on unvegetated treatments, and 0.1% to 30% of the steady-state interrill erosion rates of the two soils on vegetated treatments, the type of compost was also a factor in interrill erosion control. The yard waste compost was the coarsest of the three compost materials, and on unvegetated plots had a steady-state interrill erosion rate that was 17% and 33% of the steady-state interrill erosion rates of biosolids and bio-industrial compost, respectively. Interrill erodibility factors were calculated for all treatments and fell within the range of experimental rangeland values (10,000 to 2,000,000 kg sec/m^4) that are used in the Water Erosion Prediction Project.

Powell, W.; Keller, G.R.; Brunette, B. (1999). Applications for geosynthetics on forest service low-volume roads. *Transportation Research Record.* **2**(1652):113–120.

Abstract: Today's geosynthetic products have many useful, creative, and cost-effective applications for rural, low-volume roads. In the management of almost a half-million km (quarter-million mi) of low-volume roads, the U.S. Department of Agriculture, Forest Service (USFS), has developed and adopted many uses for geosynthetics. An overview is presented of many of those uses and their advantages. The USFS gained much of its experience and practice with geosynthetics while constructing a wide variety of Mechanically Stabilized Earth (MSE) retaining walls, including geotextile, timber, modular-block, and tire-faced structures, and reinforced soil slopes. More recently, the USFS has used geosynthetics for MSE bridge abutments and Deep Patch road-shoulder reinforcement. Other typical geosynthetic applications include filtration, drainage, subgrade reinforcement, and erosion control.

Pritts, J.W.; Swanson, J.E.; Collins, J.; Miles, A. (2004). National benefits of erosion and sediment control regulations. *Proceedings of the 2004 World Water and Environmental Resources Congress: Critical Transitions in Water and Environmental Resources Management.* 725–734.

Abstract: EPA is developing national effluent guidelines for discharges of stormwater from construction sites to augment the existing Phase I and II NPDES stormwater regulations. As part of this effort, EPA is required to evaluate the environmental benefits of the rulemaking. In evaluating the benefits of regulations, EPA attempts to both quantify the degree of reduction of a range of pollutants or indicators, as well as monetize the expected environmental improvements that result. Due to the national scope of this evaluation and limited data available on both the quality of discharges from construction sites and the locations of those discharges, as well as the limited range of methodologies available to quantify and monetize in-stream improvements, EPA relied on a modeled approach in lieu of using actual field-sampled data. For this evaluation, EPA estimated sediment export from construction sites using the Revised Universal Soil Loss Equation (RUSLE) and modeled sediment reductions using SEDCAD. EPA then developed a methodology for determining the geographic distribution of construction sites and modeled in-stream water quality improvements and monetized benefits of several regulatory scenarios using EPA's National Water Pollution Control Assessment Model (NWPCAM).

Reed, L.A. (1980). *Suspended-Sediment Discharge in Five Streams Near Harrisburg, Pennsylvania, Before, During, and After Highway Construction.* Geological Survey Water-Supply Paper (United States). **2072**:42p.

Abstract: Rainfall, streamflow, sediment, and turbidity data were collected as part of a study to evaluate the effects of highway construction on suspended-sediment discharges in streams. The study was also designed to evaluate the effectiveness of different erosion-control measures in reducing sediment discharge.

Reeves, P.J. (1982). After the pipeline: An ugly scar or natural scene? *ASCE Convention & Exposition,* 1982. 13p.

Abstract: Disturbance to the environment as a result of pipeline installation can be easily overlooked when solving problems of conveying fluids via pipelines from one location to another. Items which might be neglected include changes in topography, loss of vegetation, and other significant esthetic considerations. Those responsible for the selection of pipeline alignments must effectively deal with more considerations as the general public becomes more sensitive to the needs of the environment in urban, rural, and isolated areas. The potential damage and cost required to rectify erosion problems to both project and adjacent lands should be weighed against the real cost of erosion control. A fair assessment of these costs includes: an on going maintenance; riparian flood damage; riparian flood or dust caused crop damage; reduction in land values; degradation of air and water quality. An allowance should be made for maintenance costs necessary to achieve regrowth to a self-sustaining level if an investment is made to restore the pipeline right-of-way.

Renninger, F.A.; Woolief, R.S.; Bradshaw, P.J.; Laughlin, G.R.; Childers, F.A.; Nichols, F.P.; Wilkerson, J.M.; Mosley, A.S.; Breen Jr., F.L. Jr.; Canup, L.; Laws, E.P. (1974). *Georgia Highway Conference, 23rd Annual.* 121p.

Abstract: Proceedings include 11 papers that deal with geological origin and physical characteristics of aggregates, job handling of aggregates, erosion control during construction, and urban design and construction. Following is a list of titles and authors of the papers presented: Geological Origin and Physical Characteristics of Aggregates. By F. A. Renninger. Job Handling of Aggregates Used in Portland Cement Concrete Paving Products. By R. S. Woolief. Job Handling of Aggregate Used in Asphaltic Concrete Paving Mixes. By Paul J. Bradshaw. Handling of Aggregates in Base Construction. By George R. Laughlin. Aggregate Control & Record Sampling. By Fred A Childers. Utilization of Aggregates. By Frank P. Nichols, Jr. End Result Specifications. By John M. Wilkerson. Urban Design & Construction. By Albert S. Mosely. Transportation Planning & Citizens Input. By Florence L. Breen, Jr. Erosion Control During Construction. By Lewis Canup. Right of Way Enhancement. By Edward P. Laws.

Richter, B.D. (1987). Erosion control improves community water supply. *Public Works.* 118(1):80, 114.

Abstract: Urban construction activities can increase sediment loading rates as much as 40,000 times the rate that occurs from an undeveloped farm or woodland in an equivalent period of time. The Castle Pines and Castle Pines North Metropolitan districts have developed a comprehensive erosion control program for their communities, which is expected to effectively reduce erosion and sedimentation rates during construction activities by as much as 90 percent. The Metro Districts provide infrastructure construction and utility and drainage operations and maintenance services for these planned, mixed-use communities, which comprise an area of about 4,000 acres just south of Denver.

Ristic, R.; Macan, G. (1997). Impact of erosion control measures on runoff processes. *International Association of Hydrological Sciences.* 245:191–194.

Abstract: Water supplies in Serbia are based primarily on reservoir storage in protected areas. The reservoir catchments are located in hilly-mountainous regions in order to avoid the water quality problems associated with urbanization and agricultural production. One of the most important conditions for the continued effective use of such reservoirs is protection of their storage from sedimentation. Erosion problems are widespread in Serbia. 86% of the territory suffers from erosion processes of varying intensity and the total annual production of eroded material is ca. 40×10^6 m^3 year^{-1}. Erosion control is based on the construction of control structures on torrents and bio-technical works (afforestation of bare lands and restoration of degraded forests and pastures). There is currently a need for afforestation of 600 km^2 of bare land in the catchment areas of reservoirs which are currently under construction or planned. Land use change (from bare land or degraded forest to stable forest), whilst providing erosion control, also has a significant influence on runoff processes through its influence on the hydrological cycle. The impact of anti-erosive afforestation on runoff processes has been studied in the experimental catchment M-III, located on the Goc mountain in central Serbia during the period 1980–1995.

Rivas, T. (2003). Impacted erosion control selection for steep slopes with the USDA Forest Service Erosion Control Selection Guide. *Transportation Research Record.* **I**(1819):182–186.

Abstract: Low-volume forest roads are a significant source of erosion and can be challenging and expensive to control because of the steep (greater than 50% slope gradient) slopes. Most erosion control documents lack guidance on treatment selection, emphasize temporary erosion control, and do not adequately address steep slopes. To address the deficiency, the U.S. Department of Agriculture Forest Service San Dimas Technology and

Development Center created the USDA Forest Service Erosion Control Selection Guide. The primary focus of the guide is technical information to assist long-term erosion control treatment selection on steep slopes, particularly road cut and fill slopes. This guide is a synthesis of information encompassing erosion control fundamentals, erosion and sediment yield prediction models, treatment parameters, erosion control treatments, selection tips, installation tips, and additional resources. Treatments include grading, seeding, mulch, soil amendments, rolled erosion control products, soil bioengineering, and biotechnical stabilization. Significant advantages of the guide include general treatment selection guidance, inclusion of permanent treatment methods, and consideration of steep slopes. The guide is limited because of the complex nature of erosion; it does not address nontechnical issues, and it may need updating as new information becomes available. Despite these limitations, the guide is a valuable new tool for reducing erosion and sediment yield on steep slopes of low-volume roads.

Roberts, B.C. (1994). Developing erosion control plans for highway construction. Transportation *Research Record.* **1471:**38–40.

Abstract: A recommended procedure for developing erosion control plans for highway construction is presented. These procedures can be found in Best Management Practices for Erosion and Sediment Control, an FHWA manual developed through the Federal Lands Highways Coordinated Technology Implementation Program. These recommendations result in part from recent legislative requirements under the Environmental Protection Agency's National Pollutant Discharge Elimination System regulations. Erosion control plans are developed by following basic principles of erosion and sediment control. In addition, a three-phase approach based on construction stages is presented to guide the designer through the process. Finally, a brief overview of best management practices is presented.

Robinson Jr., L.R. (1974). Erosion control during pipeline construction. Kentucky University, *Office of Research and Engineering Services, Bulletin.* 175–182.

Abstract: On September 21, 1973 the Pennsylvania Environmental Quality Board adopted an Implementation Plan and Regulations Dealing with Erosion and Sedimentation Control which applies to all earthmoving activities in the Commonwealth of Pennsylvania. Progress of construction and effectiveness of erosion control structures were monitored and pictorially recorded throughout the construction season. As evidenced by photographs and turbidity measurements, it was found that in most situations very little sediment left the site of pipeline construction. Straw bales placed across small streams and draws, at critical road crossings and where bodies of water were to be protected; judicious use of pipe trenches and roadside ditches as sediment traps; utilization of the roughness and porosity of construction in the right of way as a runoff retardant and vegetation alongside the right of way as a sediment filter provided satisfactory erosion control throughout the construction season.

Robison, R. (1985). Engineering with fabric. *Civil Engineering.* 55(12):52–55.

Abstract: In 1984, use of fabrics to stabilize soil moved ahead of primary road reinforcement, in terms of yardage, for the first time. Stabilization includes dams, embankments, retaining walls, and support for parking lots, driveways and airports. Other end use categories are secondary roads, drainage erosion control, railroads. Use of the fabric reinforcement permitted construction of the embankment directly on the unstable bottom, preventing rotational and/or wedge type failure and excessive vertical displacement. The fabric provided the friction and strength necessary to prevent lateral sliding, dissipated foundation pore pressures and separated the fill from the underlying muck.

Sherman, R. (2003). Texas transportation department accelerates highway use of compost. *BioCycle.* **44**(7):24–28.

Abstract: A review on Texas transportation department accelerating highway use of compost was presented. It was stated that in the financial year 2003 the Texas Department of Transportation (TxDOT) would specify 400,000 cubic yards of compost, making the agency the largest market for compost in the nation. The compost use program was developed in part to respond to the difficulty TxDOT engineers were facing revegetating severely eroded highway construction sites.

Stevens, E.; Barfield, B.J.; Gasem, K.; Matlock, M. (2004). On and off site sediment control using silt fence. *Proceedings of the 2004 World Water and Environmental Resources Congress: Critical Transitions in Water and Environmental Resources Management.* 715–724.

Abstract: Greater public awareness and the enactment of increasingly stringent regulations have made reliable and cost-effective erosion control methods or best management practices (BMPs), particularly on construction sites, a national priority. At present, silt fence is the most widely used BMP, but unfortunately, the instances of failure are widespread. We believe that with proper assessment, following recommended installation methods, and conducting routine maintenance, better erosion control can be achieved with current silt fence technology. The current phase of our research is to investigate the current technology and develop design tools to assist users to get better results through avoiding structural failures and also provide a means of assessing performance in terms of trapping efficiency or effluent concentration. The next phase of our research will be to develop and evaluate several structural, material, and chemical enhancements to the current technology. This paper describes the progress made to date in investigation of current technology and development and application of the design tool.

Thaxton, C.S.; Calantoni, J.; McLaughlin, R.A. (2004). Hydrodynamic assessment of various types of baffles in a sediment retention pond. *Transactions of the American Society of Agricultural Engineers.* **47**(3):741–749.

Abstract: We assessed the relative improvement to the sediment trapping effectiveness of a permanent-pool sediment retention pond due to the installation of baffles composed of different materials commonly used on construction sites. A suite of experiments was performed at the Sediment and Erosion Control Research and Education Facility (SECREF) at North Carolina State University in which an acoustic Doppler velocimeter was used to record steady-state flow velocity and signal-to-noise ratio data. The data was gathered at 25 grid points at two depths within the pond for three different fixed input flow rates. The free flow maximum mean velocity in the pond, averaged over all input flow rates, was reduced by roughly 75% due to the presence of baffles composed of jute germination biotextile backed by coir fiber. Baffles made from a standard tree-protection fence, folded and tied together into three layers to reduce pore size, reduced the free flow maximum mean velocity by 65%, while baffles made of standard silt fabric fence reduced the free flow maximum mean velocity by 55%. A similar trend in the reduction of the signal-to-noise ratio along the length of the pond confirmed that the jute/coir baffles most effectively reduced the concentration of turbulent density fronts over that of the tree-protection fence or silt fence baffles, or free flow. In addition, analysis of the transverse velocity variance and vertical velocity gradients for each experiment further demonstrated that the jute/coir baffles most effectively diffused inflow momentum along the width and depth of the pond. The results of our analysis were used to calculate a sediment trapping efficiency based on Stokes settling. The minimum grain size captured would range from 30 to 42 microns with jute/coir baffles, compared to 68 to 86 microns for free flow.

Thenoux, G.; Vera, S. (2003). Evaluation of hexahydrated magnesium chloride performance as chemical stabilizer of granular road surfaces. *Transportation Research Record.* **II**(1819):44–51.

Abstract: The performance was evaluated of hexahydrated magnesium chloride ($MgCl_2 \cdot 6H_2O$) as a chemical stabilizer of granular road surfaces. $MgCl_2 \cdot 6H_2O$ is a salt with properties that are useful for chemical stabilization of granular road surfaces, such as its capacity to absorb and retain humidity from its surrounding environment, increase water surface tension, and decrease water vapor pressure. Its effect on the physical and mechanical properties of soils was evaluated in laboratory, and its functional performance and durability were evaluated in field trials. Field trials were performed in arid and semiarid regions in the Atacama Desert in northern Chile. Field results showed that roads stabilized with $MgCl_2 \cdot 6H_2O$ remain free of dust and corrugations and exhibit a considerable reduction in potholes and surface erosion. The period of effectiveness for $MgCl_2 \cdot 6H_2O$ has proven to be more than 2 years, without any type of road maintenance.

Vacher, C.A.; Loch, R.J.; Raine, S.R. (2003). Effect of polyacrylamide additions on infiltration and erosion of disturbed lands. *Australian Journal of Soil Research.* **41**(8):1509–1520.

Abstract: The removal of vegetation and disturbance of the soil surface due to a range of human activities results in the potential for soil structure degradation and sediment movement. Polyacrylamides have been used to improve infiltration and reduce erosion on agricultural lands. However, they are not commonly used as part of management and rehabilitation programs on land disturbed by construction or mining activities in Australia. A study was undertaken to investigate the potential for polyacrylamides to improve infiltration and reduce erosion of soil material from 3 Australian mine sites. The polyacrylamides were found to significantly (P less than or equal 0.05) increase total infiltration under rainfall, reduce surface hardness, and reduce sediment entrainment and erosion by both rainfall and overland flows. The effectiveness of the polyacrylamide was found to be related to clay content of the soil as well as the molecular weight and charge density of the polyacrylamide. The implications of these results for the management and rehabilitation of disturbed lands are discussed.

Vahanne, P. (1997). Urban erosion control in the Triangulo pilot study area Nacala city, Mozambique, Equatorial Africa. A multidisciplinary approach with specific emphasis on control of gully erosion by labour intensive construction work. *Publications—Technical Research Centre of Finland.* **329**:13p.

Abstract: This study is a multidisciplinary approach to erosion control in urban areas with specific emphasis on control of gully erosion by labour intensive construction work. The study was executed in cooperation with another study, 'Erosion control by vegetative measures'. Both studies were executed in the Triangulo pilot study area in Nacala, Mozambique. The aim of the study is to evaluate the impact of intrinsic and extrinsic variables on the erosion processes affecting the coastal Pleistocene sand slopes of the city. Besides analysis of the erosion processes, also dealt with are conservative methods and protective measures aimed at preventing gully initiation and development and finally at restoration of the terrain. The study includes the following parts: field inventory, rainfall analysis, estimation of discharge and estimation of soil losses. Based on the estimations, a drainage and erosion control design is presented. Execution of the work during 1990–92 is briefly described and experiences gained so far are introduced. The research methods used are time-area, rational and Cook's method for the estimation of discharge, and the methods of Elwell and Poliakov and a volumetric method for estimation of soil losses. The state of erosion in the Triangulo pilot study area is fairly severe. The main type of erosion is rill and gully erosion, the total length of erosion channels being over 5,000 m. Intrinsic variables make the system highly susceptible to disturbances in the pilot study area. The impact of extrinsic variables in accelerating the erosion process has been considerable. Calculated soil losses are about 13,000 m^3/a. Measured soil loss from the largest gullies is about 12,000 m^3. Cook's method gave the lowest values for discharge, the results of the time-area method gave approximately the same values, and the rational method slightly higher values. The drainage and erosion control plan includes cut-off, discharge and collection ditches, as well as check dams, rock protections, weirs and other gabion structures. The general plan and design values are based on the hydrological data and erosion sensitivity of the soil. The possibility to use labor intensive work methods was also a main target. Construction work was executed simultaneously with vegetative measures. Other main measures were land-use planning, resettlement of people, extension work and participation of residents. Based on experience from the Triangulo pilot study area, the erosion control method to be used in Nacala should be a combination of vegetative measures, low cost construction and some heavy construction work. The role of vegetation is crucial and a resettlement component is also needed. The term "Land husbandry" can be applied to this type of combination of urban erosion control measures.

Versteeg, J. H. (DOT, Salem, Oreg); Earley, John J. March (1982). "Erosion control at a delicate highway construction site." *Public Works,* ISSN: 0033-3840, v 113, n 3, p 82–83, Mar.

Abstract: The authors describe how stream planning of a road under construction in Oregon was protected from siltation, sedimentation and pollution during construction.

Walling, D.E. (Ed.); Probst, J.L. (Ed.). (1997). Proceedings of the 1997 5th Scientific Assembly of the International Association of Hydrological Sciences, IAHS. *IAHS Publication (International Association of Hydrological Sciences).* **245, Symp 6:** *Human Impact on Erosion and Sedimentation, Proceedings of the 1997 5th Scientific Assembly of the International Association of Hydrological Sciences.* 311p.

Abstract: The proceedings contain 29 papers from the Rabat Symposium S6 on human impact on erosion and sedimentation. Topics discussed include: soil erosion in vineyards; runoff and erosion on mountainous roads; impact of overgrazing on soil erosion; impact of deforestation in tropical areas; landslides; land use changes; sediment yields and flood protection in desert towns; sedimentation in small dams; impact of reservoir construction on river sedimentation; and impact of soil erosion control measures on runoff processes.

Watson, C.C.; Combs, P.G.; Abt, S.R.; Raphelt, N.K. (1995). Bioengineering stabilization of Harland Creek sites. *International Water Resources Engineering Conference—Proceedings.* **2**:1839–1843.

Abstract: Five sites along Harland Creek, a tributary to Black Creek in the Yazoo Basin of Mississippi, have been selected for construction of experimental bioengineered bank stabilization. Each site was selected to address different types of bank stabilization problems, and a combinations of six different bioengineering techniques will be used to stabilize these sites. Harland Creek is monitored as part of the Demonstration Erosion Control project (DEC), with stream gauging and with comparative field surveys. Although many sites have been stabilized using bioengineering, these sites will have a comprehensive monitoring and a review of conditions before and after construction.

Williams, D.T.; Austin, D.N. (1995). PC-based design of channel protection using permanent geosynthetic reinforcement mattings. *International Water Resources Engineering Conference—Proceedings.* **1**:678–682.

Abstract: EC-DESIGN is a complete erosion control design package recently published by Synthetic Industries, Construction Products Division. This program allows the user to select the most appropriate synthetic erosion control materials for either his/her construction slope or channel application. This paper describes the procedures used for the hydraulic analyses and selection of a permanent geosynthetic matting as channel lining materials featured in EC-DESIGN.

Wills, P. (1995). Use of geosynthetics in the A5 Fazeley bypass. *Highways and Transportation.* **42**(7–8):3p.

Abstract: The construction problems of the A5 Fazeley Bypass in Staffordshire, including erosion control, waste containment, subgrade reinforcement, and temporary works designs, were surmounted by the use of geosynthetics. The choice of the geosynthetic materials to be used may be influenced by several factors such as speed of construction, convenience, tighter quality control, and cost benefits. The complex highway scheme has been able to demonstrate the many simple and innovative ways that today's geosynthetic products can assist the design engineer.

Zhang, L.; Du, J.; Hu, T. (2000). Compaction and performance of loess embankments. *ASTM Special Technical Publication.* **1384**:173–184.

Abstract: This paper summarizes the practice of highway embankment compaction in the loess plateau of northwestern China, based on a field trip and the related laboratory studies. A large number of high loess embankments were built across gullies. The compaction was based on the standard Proctor method (ASTM Test Method for Laboratory Compaction Characteristics of Soil Using Standard Effort, D698-91) during 1950–1985, and the modified Proctor method (ASTM Test Method for Laboratory Compaction Characteristics of Soil Using Modified Effort, D1557-91) after 1985. The performance of these embankments is described. Stability analysis and centrifuge tests are conducted to confirm the observations and improve designs. Storm water ponds are found to be critical to both stability and settlement. For embankments compacted using the standard Proctor method, progressive failure would start with any further erosion if the slopes were steeper than 1:0.75.

Ziegler, A.D.; Sutherland, R.A., Tran, L.T. (1997). Influence of rolled erosion control systems on temporal rainsplash responseùa laboratory rainfall simulation experiment. *Land Degradation & Development.* **8**(2):139–157.

Abstract: Reduction of erosion and sediment-related pollution from urban construction sites or other degraded hillslopes often relies on the initial application of suitable rolled erosion control systems (RECS) before natural vegetation cover can be established. However, research has not clearly explained why some RECS perform better than others, or under what particular conditions one system is more suitable than another. An important link between the application of the most suitable RECS and better product design is process-based studies relating the physical properties of products to the reduction of erosion subprocesses. This study investigates time-varying reduction of rainsplash detachment and transport by 13 commonly used RECS. The results indicate that product differences in the protection they provide against splash processes vary over the duration of a rain event, and that this variation is related to individual product properties, especially surface coverage and thickness. These results should aid in the design of more effective erosion control products and in the selection of the most suitable RECS for particular hillslope applications.

REFERENCES

Akeley, R.P. "Retention ponds for control of urban stormwater quality." In *Proceedings National Conference on Urban Erosion and Sediment Control: Institutions and Technology,* EPA-905/9-80-002, Chicago, January 1980.

Alabama. *Alabama Handbook For Erosion Control, Sediment Control, And Stormwater Management On Construction Sites And Urban Areas,* produced by the Alabama Soil And Water Conservation Committee Montgomery, AL, July 1993.

Aldridge, B.N., and Garrett, J.M. *Roughness Coefficients for Stream Channels in Arizona.* U.S. Geological Survey Open-File Report, 87 p. 1973.

Allen, S.R. (TRI/Environmental, Inc, Austin, USA). "Evaluation and standardization of rolled erosion control products." *Geotextiles and Geomembranes,* v14, n3–4, Mar–Apr, 1996, Proceedings of the 1995 9th GRI Conference on Geosynthesis in Infrastructure Enhancement and Remediation, Dec 12–13 1995, Drexel, PA, USA, p 207–221 Publisher: Elsevier Science Ltd, Oxford, Engl ISSN: 0266.1144. 1996.

Anon. "Erosion control on highway construction." National Cooperative Highway Research Program, *Synthesis of Highway Practice,* n 18, 52p. 1973.

Anon. "On-site grinding and recycling of home construction debris." *BioCycle,* v41, n12, 2 pp Publisher: JG Press Inc, Emmaus, PA, USA ISSN: 0276.5055. Dec, 2000.

Arcement, Jr., G.J.and V.R. Schneider. *Guide for Selecting Manning's Roughness Coefficients for Natural Channels and Flood Plains.* United States Geological Survey Water-supply Paper 2339 and FHWA-TS-84-204. undated. (http://www.fhwa.dot.gov/bridge/wsp2339.pdf).

ASCE and WEF. *Urban Runoff Quality Management.* ASCE Manual and Report on Engineering Practice No. 87; WEF Manual of Practice No. 23. American Society of Civil Engineers, Reston, VA. Water Environment Federation, Alexandria, VA. http://www.asce.org and http://www.wef.org. 1998.

ASCE. *Sedimentation Engineering.* Edited by Vito A. Vanoni. Manuals and Reports on Engineering Practice, No. 54. 1975.

ASCE and WEF. *Design and Construction of Urban Stormwater Management Systems.* 1992.

Auckland Regional Council. *TP 90: Erosion and Sediment Control,* Auckland, New Zealand. 1999.

Barfield, B.J., R.C. Warner, and C.T. Haan. *Applied Hydrology and Sedimentology for Disturbed Areas.* Oklahoma Technical Press. Stillwater, Oklahoma. 1983.

Barfield, B. "Analysis of the effects of existing and alternate design criteria on the performance of sediment detention ponds." *1986 Sediment and Stormwater Conference,* sponsored by the Maryland Water Resources Administration, Sediment and Stormwater Division, Salisbury State College, Maryland, July–August 1986.

Barnes, H.H., Jr. *Roughness Characteristics of Natural Channels.* U.S. Geological Survey Water-Supply Paper 1849, 213 p. 1967.

Barrett, M.E., J.E. Kearney, J.F. Malina, R.J. Charbeneau, and G.H. Ward. *An Evaluation of the Use and Effectiveness of Temporary Sediment Controls.* Center for Research in Water Resources. Technical Report CRWR 261. The University of Texas at Austin. August 1995.

Beca Carter Hollings & Ferner Ltd. *The Use of Flocculants and Coagulants to Aid the Settlement of Suspended Sediment in Earthworks Runoff—Trials, Methodology and Design.* Report 2509784 Rev A , L1:90873. Prepared for the Auckland Regional Council. Auckland, New Zealand. Undated.

Bedner, R.E. and D.J. Fluke. *Demonstration of Debris Basin Effectiveness in Sediment Control.* Environmental Protection Agency, Industrial Environmental Research Laboratory—Cincinnati, EPA-600/7-80-154. Cincinnati, Ohio, 1980.

Benson, M.A., and Dalrymple, Tate. *General Field and Office Procedures for Indirect Discharge Measurements.* U.S. Geological Survey Techniques of Water-Resources Investigations book 3, chap. Al, 30 p. 1967.

BOC. *New Privately Owned Housing Units Started: Annual Data, 2001.* U.S. Bureau of the Census, Washington, DC. http://www.census.gov/const/startsan.pdf. Accessed May 23, 2002. 2001.

Bondurant, J.A., C.E. Brockway, and M.J. Brown. "Some aspects of sedimentation pond design." *Proceedings National Symposium on Urban Hydrology and Sediment Control,* University of Kentucky, Lexington, 1975.

Boysen, Stephen M. (Soil Conserv Serv, College Park, Md). *Predicting Sediment Yield in Urban Areas.* Kentucky University, Office of Research and Engineering Services, Bulletin, Jul 29–31 1974, p 199–203. Oct, 1974.

Boysen, Stephen M. (USDA, Soil Conserv Serv, College Park, Md). "Erosion and sediment control in urbanizing areas." ASAE, n 4–77, 1977, Proc of the Natl Symp on Soil Erosion and Sediment by Water, Dec 12–13 1977, Chicago, IL, USA, p 125–136 Publisher: ASAE, St. Joseph, Mich. 1977.

Brown, W. and T. Schueler. *The Economics of Stormwater BMPs in the Mid-Atlantic Region.* Prepared for: Chesapeake Research Consortium. Edgewater, MD. Center for Watershed Protection, Ellicott City, MD. 1997.

BRPC (Birmingham Regional Planning Commission). *Best Management Practices for Controlling Sediment and Erosion from Construction Activities.* Birmingham, Alabama. 1980.

Burroughs, Edward R. Jr. (USDA, Forest Service, Intermountain Forest & Range Experiment Station, Moscow, ID, USA, Station, Moscow, ID, USA); King, John G. "Surface erosion control on roads in granitic soils." *Watershed Management in the Eighties.* Proceedings of the Symposium held in conjunction with the ASCE Convention., Denver, CO, USA, p 183-190 Sponsored by: ASCE, Committee on Watershed Management, New York, NY, USA; Soc for Range Management, Denver, CO, USA; ASAE, St. Joseph, MI, USA; American Water Resources Assoc, Bethesda, MD, USA Publisher: ASCE, New York, NY, USA, 1985 ISBN: 0-87262-449-8. 1985.

Burton, Thomas M. (Mich State Univ, East Lansing);Turner, Ralph R.; Harriss, Robert C. "Impact of highway construction on a North Florida watershed." *Water Resources Bulletin,* v 12, n 3, p 529–538, Jun, 1976.

Cabalka, Dwight (American Excelsior Co, Littleton, USA). "Landfill cap erosion; Severe conditions and dramatic solutions." *Public Works,* v127, n10, p 32–34, Sep, 1996.

Chambers, G.M. and C.H. Tottle. *Evaluation of Stormwater Impoundments in Winnipeg,* Report SCAT-1, Environment Canada, Ottawa, April 1980.

Chang, H.H. *Fluvial Processes in River Engineering.* John Wiley & Sons. 1988.

Chen, Charng-Ning (Parsons, Brinckerhoff, Quade & Douglas, Inc, New York, NY); Santomauro, Frank;Fisher, J. Bruce; "Erosion control system for pipeline construction sites." *Natl Symp on Urban Hydrol and Sediment Control, Proc,* Jul 28–31 1975, Lexington, KY, USA, p 37–49. Kentucky University, Office of Research and Engineering Services, Bulletin, n 109, Nov, 1975.

Chesters, G., J. Konrad and G. Simsiman. "Menomonee River Pilot Watershed Study- Summary and Recommendations", EPA-905/4-79-029. U.S. Environmental Protection Agency, Chicago, Ill., 1979.

Chow, V.T. *Open-Channel Hydraulics.* New York, McGraw- Hill Book Co., 680 p. 1959.

Chow, V.T., Maidment, D. R, and Mays, L. W., *Applied Hydrology,* McGraw-Hill, 586 pages. 1988.

Ciarla, Massimo (Terra Aqua Inc, Reno, NV, USA). "Gabion weirs in water erosion control projects; Design and construction criteria." *Erosion Control: A Challenge in Our Time, Proceedings of Conference XVI.,* San Francisco, CA, USA, p 85–103 Sponsored by: Int Erosion Control Assoc, Freedom, CA, USA Publisher: Int Erosion Control Assoc, Freedom, CA, USA, 1985.

Clark, S. and R. Pitt. *Stormwater Treatment at Critical Areas, Vol. 3: Evaluation of Filtration Media for Stormwater Treatment.* U.S. Environmental Protection Agency, Water Supply and Water Resources Division, National Risk Management Research Laboratory. EPA/600/R-00/016, Cincinnati, Ohio. 442 pgs. October 1999.

Colston, N.V., Jr. *Characterization and Treatment of Urban Land Runoff.* EPA-670/2-74-096, U.S. Environmental Protection Agency, 1974.

Cowan, W.L.. *Estimating Hydraulic Roughness Coefficients.* Agricultural Engineering, v. 37, no. 7, p. 473–475. 1956.

Cowherd, Chatten Jr.; Grelinger, Mary Ann. "High-wind failure of soil moisture as a wind erosion control." *Proceedings of the Air & Waste Management Association's Annual Meeting,* Jun 23–28 1996, Nashville, TN, USA, 96.TP40.04, 16pp Publisher: Air & Waste Management Assoc, Pittsburgh, PA, USA, 1996.

Crafton, C. Scott (Soil & Water Conservation, Richmond, VA, USA). "Performance criteria for erosion and sediment control." *Erosion Control—You're Gambling Without It, Proceedings of Conference XVIII.,* Reno, NV, USA, p 39–47 Sponsored by: Int Erosion Control Assoc, Pinole, CA, USA Publisher: Int Erosion Control Assoc, Pinole, CA, USA, 1987.

Croke, T. "Reliable channel design using turf reinforcement mats." *Erosion Discussion.* Vol. 5, No. 1. North American Green (availabe from their web page: www.nagreen.com). June 2000.

Croke, T. "Accounting for bends in channel design." *Erosion Discussion.* Vol. 6, No. 1. North American Green (availabe from their web page: www.nagreen.com). Summer 2001.

CWP. *Design of Stormwater Filtering Systems.* Prepared for Chesapeake Research Consortium, Solomons, MD, and U.S. EPA Region 5, Chicago, IL, by Center for Watershed Protection, Ellicott City, MD. 1996.

Dallaire, Gene. "Controlling erosion and sedimentation at construction sites." *Civil Engineering* (New York), v 47, n 10, p 73–77, Oct, 1976.

Dillaha, T. A. III (Purdue Univ, Agricultural Engineering Dep, West Lafayette, Indiana, USA); Beasley, D. B. "Distributed parameter modeling of sediment movement and particle size distributions." *Transactions of the ASAE,* ISSN: 0001-2351. v 26, n 6, p 1766–1772, 1777, Nov–Dec, 1983.

Dion, T.R. *Land Development for Civil Engineers.* John Wiley & Sons, Inc. New York. 796 pgs. 2002.

Driscoll, E. D. "Detention and retention controls for urban stormwater." *Engineering Foundation Conference: Urban Runoff Quality—Impact and Quality Enhancement Technology,* Henniker, New Hampshire, edited by B. Urbonas and L. A. Roesner, published by the American Society of Civil Engineers, New York, June 1986.

Eccher. C.J. "Thoughtful design is prime factor in water safety." *Lake Line.* pp. 4–8. May 1991.

Entry, J.A., and R.E. Sojka. Polyacrylamide Application to Soil Reduces the Movement of Microorganisms in Water. In *1999 Proceedings of the International Irrigation Show,* Irrigation Association, Orlando, FL, November 9, 1999, pp. 93–99. 1999.

EPA (USA Environmental Protection Agency). *Final Report for the Nationwide Urban Runoff Program.* Water Planning Division, PB 84-185552, Washington, D.C., December 1983.

EPA. *Construction Rainfall Erosivity Waiver,* Fact Sheet 3.1. EPA 833-F-00-014. January 2001.

Evans, Max L. (US Dep Agric, Soil Conserv Serv); Dynes, Sheldon. *Sediment and Erosion Control Measures on Construction Sites.* Wisconsin University, College of Engineering, Engineering Experiment Station, Report, 1974, ASAE Pap, Annual Meeting, 67th, Stillwater, Jun

23–26 1974, Winter Meet, Dec 10–13 1974, 6p Publisher: ASAE, St. Joseph, Mich. 1974.

Farrar, Jeffrey A. (Bureau of Reclamation, Denver, USA). "Bureau of Reclamation experience in testing of riprap for erosion control of embankment dams." ASTM Special Technical Publication, n 1177, 1993, *Symposium on Rock for Erosion Control,* Jun 18 1992, Louisville, KY, USA, p 3–14 Sponsored by: ASTM Publisher: Publ by ASTM, Philadelphia, PA, USA, ISSN: 0066.0558. 1993.

Fifield, Jerald S. (HydroDynamics Inc, Parker, CO, USA). "Erosion control measures—Are they effective?" *Erosion Control: Protecting Our Future—Proceedings of Conference XVII,* International Erosion Control Association., Dallas, TX, USA, p 19–35 Sponsored by: Int Erosion Control Assoc, Freedom, CA, USA Publisher: Int Erosion Control Assoc, Freedom, CA, USA, 1986.

Fluet, Joseph E. Jr. (Ed.) (GeoServices Inc Consulting Engineers, Boynton Beach, FL, USA). "Geotextile testing and the design engineer." ASTM Special Technical Publication, 1987, *Geotextile Testing and the Design Engineer,* Los Angeles, CA, USA, 183p Sponsored by: ASTM, Committee D-35 on Geotextiles, Geomembranes, & Related Products, Philadelphia, PA, USA Publisher: ASTM, Philadelphia, PA, USA, 1987 ISSN: 0066.0558. 1987.

Fortier, S. and F.C. Scobey. "Permissible canal velocities." *Trans. ASCE,* Vol. 89, paper No. 1588, pp. 940–984. 1926.

Gietz, R. J. *Urban Runoff Treatment in the Kennedy-Burnett Settling Pond.* For the Rideau River Stormwater Management Study, Pollution Control Division, Works Department, Regional Municipality of Ottawa-Carleton, Ottawa, Ontario, March 1983.

Grizzard, T. J., C. W. Randall, B. L. Weand, and K. L. Ellis. "Effectiveness of extended detention ponds." *Engineering Foundation Conference: Urban Runoff Quality—Impact and Quality Enhancement Technology,* Henniker, New Hampshire, edited by B. Urbonas and L. A. Roesner, published by the American Society of Civil Engineers, New York, June 1986.

Guy, Harold P. (US Geol Surv, Reston, Va). "Diminution ratios for planning construction-area sediment controls." Kentucky University, Office of Research and Engineering Services, Bulletin, n 111, Dec, 1976, *Natl Symp on Urban Hydrol, Hydraul, and Sediment Control,* Proc, Jul 27–29 1976, Lexington, KY, USA, p 91–97. 1976.

Harrington, B. W. *Feasibility and Design of Wet Ponds to Achieve Water Quality Control.* Sediment and Stormwater Division, Maryland Water Resources Administration, July 1986.

HEC. *HEC-RAS User's Manual,* Version 2.0. US Army Corps of Engineers, Hydrologic Engineering Center, April 1997.

Henensal, Pierre. "La lutte contre l'erosion sur l'emprise routiere une contribution a la protection de l'environnement." "Combating erosion of carriageway land: a contribution to the protection of the environment." *Bulletin de Liaison des Laboratoires des Ponts et Chaussees,* n201, Jan–Feb, 1996, p 17–28 Publisher: Lab Central des Ponts et Chaussees, Paris, France ISSN: 0458-5860. 1996.

Hittman Assoc. *Methods to Control Fine-Grained Sediments Resulting from Construction Activity.* U.S. Environmental Protection Agency, Pb-279 092, Washington, D.C., December 1976.

Horner, R.R., J. Guedry, and M.H. Kortenhof. *Improving the Cost Effectiveness of Highway Construction Site Erosion and Pollution Control.* Washington State Transportation Center, Washington State Dept. of Transportation, Seattle, WA. 1990.

Huang, C., J.M. Bradford and J.H. Cushman. "A numerical study of raindrop impact phenomena: The elastic deformation case." *Soil Sci. Soc. Am. J.,* Vol. 47, pp. 855–861, 1983.

Hunt, Daniel (Synthetic Industries, Inc, Chattanooga, USA); Austin, Deron N.; Agnew, William. "Vegetation selection for rolled erosion control product." Geotechnical Special Publication, n 76, 1998, *Geosynthetics in Foundation Reinforcement and Erosion Control Systems,* Proceedings of the 1998 Geo-Congress, Oct 18-21 1998, Boston, MA, USA, p 130-144 Sponsored by: ASCE Publisher: ASCE, Reston, VA, USA, 1998 ISSN: 0895-0563. 1998.

Illinois. *Illinois Procedures and Standards for Urban Soil Erosion and Sedimentation Control.* Association of Illinois Soil and Water Conservation Districts, Springfield, IL 62703. 1989.

Israelsen, C. E. (Utah State Univ, Logan); Clyde, C. G.; Fletcher, J. E.; Israelsen, E. K.; Haws, F. W.; Packer, P. E.; Farmer, E. E. *Erosion Control During Highway Construction.* National Cooperative Highway Research Program Report, ISSN: 0077-5614, n 220, 36p, Apr, 1980.

Jones, J. E. and D. E. Jones, "Interfacing considerations in urban detention ponding." *Proceedings of the Conference on Stormwater Detention Facilities, Planning, Design, Operation, and Maintenance,* Henniker, New Hampshire, Edited by W. DeGroot, published by the American Society of Civil Engineers, New York, August 1982.

Jones, Michael L. (Wash Public Power Supply Syst, Elma); Tatum, Clyde B. "Sedimentation control during site development." *American Society of Civil Engineers, Journal of the Construction Division,* v 106, n 1, Mar, p 17–28, 1980.

Kamedulski, G. E. and R. H. McCuen. "The effect of maintenance on storm water detention basin efficiency." *Water Resources Bulletin,* Vol. 15, No. 4, pg. 1146, August 1979.

Kaufman, Martin M. (Univ of Michigan-Flint, Flint, USA). "Erosion control at construction sites: The science-policy gap." Springer-Verlag New York, New York, NY, USA ISSN: 0364-152X. *Environmental Management,* v26, n1, p 89–97 2000.

Kay-Shoemake, J, M. Watwood, R. Sojka, and R. Lentz. "Soil Amidase Activity in Polyacrylamide (PAM) Treated Soils and Potential Activity toward Common Amide Containing Agrochemicals". *Biology and Fertility of Soils* 31(2): 183–186. 2000.

Kinnell, P.I.A. "Rainfall intensity-kinetic energy relationships for soil loss prediction." *Soil Sci. Soc. Am. J.,* Vol. 45, pp. 153–155, 1981.

Larcombe, M.F. *Technical Publication on Chemical Removal of Sediment from Earthworks Stormwater.* Auckland Regional Council (New Zealand). 1999.

Lemly, A. Dennis (Wake For Univ, Winston-Salem, NC, USA). "Erosion control at construction sites on red clay soils." *Environmental Management,* ISSN: 0364-152X, v 6, n 4, p 343–352, Jul, 1982.

Linsley, R. K., and J. B. Franzini. *Water Resources Engineering.* McGraw-Hill, New York, 1964.

Liverman, Earl (New York Power Authority, USA); Hecklau, John; Palmero, Christina. "Minimization of soil erosion and siltation during construction of the Marcy-South 345kv Transmission Facilities." *1987 Erosion Control—You're Gambling Without It, Proceedings of Conference XVIII.* Reno, NV, USA, p 241–253 Sponsored by: Int Erosion Control Assoc, Pinole, CA, USA Publisher: Int Erosion Control Assoc, Pinole, CA, USA, 1987.

Long, Dejiang (Golder Associates Ltd, Calgary, Alberta, Can); Bender, Michael J.; Sawatsky, Les F.; Anderson, Paul; Metikosh, Serge. "Sediment entrainment during construction of river pipeline crossings: Occurrence, prediction and control." *Proceedings of the International Pipeline Conference,* IPC, v 2, 1998, Proceedings of the 1998 International Pipeline Conference, IPC. Part 2 (of 2), Jun 7–11 1998, Calgary, Can, p 1045–1050 Sponsored by: ASME Publisher: ASME, Fairfield, NJ, USA, 1998.

Madison, F., J. Arts, S. Berkowitz, E. Salmon, and B. Hagman. "Washington County Project". EPA 905/9-80-003, U.S. Environmental Protection Agency, Chicago, Ill., 1979.

Maidment, D. R. (ed.), *Handbook of Hydrology,* McGraw-Hill, 1422 pages. 1993.

Marcy, S. J. and J. E. Flack. "Safety considerations in urban storm drainage design." *Second International Conference on Urban Storm Drainage,* Urbana, Illinois, June 1981.

Martin, E.H. "Mixing and residence times of stormwater runoff in a detention system." Engineering Foundation Conference: *Design of Urban Runoff Quality Controls.* Potosi, Missouri. July 10–15, 1988. Edited by L. A. Roesner, B. Urbonas and M.B. Sommen. pp. 164–179. Published by the American Society of Civil Engineers, New York, June 1989.

Martin, J. S. (Mirafi Inc, Charlotte, NC, USA). "Use of silt fences for control of sediment run-off." *1985 Erosion Control: A Challenge in Our Time, Proceedings of Conference XVI.,* San Francisco, CA, USA, p 79–84 Sponsored by: Int Erosion Control Assoc, Freedom, CA, USA Publisher: Int Erosion Control Assoc, Freedom, CA, USA, 1985.

Mason, J.M.,Jr. "On-site stormwater detention: An overview." *Public Works,* February 1982.

McCuen, R. H. *A Guide to Hydrologic Analysis Using SCS Methods.* Prentice-Hall, Englewood Cliffs, New Jersey, 1982.

McCuen, R.H. *Hydrologic Analysis and Design,* 2nd Edition. Prentice Hall. 1998.

McEnroe, Bruce (Univ of Kansas, Lawrence, USA); Treff, Brian J. "Temporary erosion control for highway construction in Kansas." *Proceedings of the Annual Water Resources Planning and Management Conference, 1998, Water Resources and the Urban Environment,* Proceedings of the 1998 25th Annual Conference on Water Resources Planning and Management, Jun 7-10 1998, Chicago, IL, USA, p 722–727 Sponsored by: ASCE Publisher: ASCE, Reston, VA, USA, 1998.

McGee, T.J. *Water Supply and Sewerage.* McGraw-Hill, Inc., New York. 1991.

McLain, John L., Anderson, Sheila B. (Resources Concepts Inc, Carson City, NV, USA). "Successful erosion control combines engineering and vegetation expertise at the Ridge Tahoe, Nevada." 1987, *Erosion Control—You're Gambling Without It, Proceedings of Conference XVIII,* Reno, NV, USA, p 33–37 Sponsored by: Int Erosion Control Assoc, Pinole, CA, USA Publisher: Int Erosion Control Assoc, Pinole, CA, USA, 1987.

Neal, Wallace (Constr Specif Inst, Washington, DC). "Specifying erosion control during construction." *Construction Specifier,* v 29, n 1, p 26–30, 32, Jan, 1976.

Nelson, J. *Characterizing Erosion Processes and Sediment Yields on Construction Sites.* MSCE thesis. Dept. of Civil and Environmental Engineering, University of Alabama at Birmingham. 94 pgs. 1996.

Novotny, V. and G. Chesters. *Handbook of Nonpoint Pollution.* VanNostrand Reinhold, New York, 1981.

NRCS. *SITES Water Resource Site Analysis Computer Program User's Guide.* United States Department of Agriculture, Natural Resources Conservation Service. 469 pp. 2001.

NRCS. *WinTR-55 User Manual.* US Dept. of Agriculture, Natural Resources Conservation Service. Downloaded on June 23, 2002 from: http://www.wcc.nrcs.usda.gov/water/quality/common/tr55/tr55-beta.html Version dated April 23, 2002a.

NRCS. *TR-20 System: User Documentation.* United States Department of Agriculture, Natural Resources Conservation Service. 105 pp. 2002b (draft).

NRCS. National Engineering Handbook, Part 630 HYDROLOGY, downloaded June 23, 2002 at: http://www.wcc.nrcs.usda.gov/water/quality/common/neh630/4content.html

Parker, Sandra C.; Stader, Thomas N., Parker, David G.; (Univ of Arkansas, Fayetteville, USA); "Use of GIS to predict erosion in construction." *International Water Resources Engineering Conference—Proceedings,* v1, 1995, Proceedings of the 1st International Conference on Water Resources. Part 1 (of 2), Aug 14–18 1995, San Antonio, TX, USA, p 839-843 Sponsored by: ASCE Publisher: ASCE, New York, NY, USA, 1995.

Penn State. *Analysis of Flow through Porous Media as Applied to Gabion Dams Regarding the Storage and Release of Storm Water Runoff,* NAHB/NRC Designated Housing Research Center, Report No. 10, August 1992.

Pitt, R. "Runoff controls in Wisconsin's Priority Watersheds." *Advanced Topics in Urban Runoff Research* conference. Proceedings edited by B. Urbonas and L.A. Roesner. Engineering Foundation and ASCE, New York. pp. 290–313. 1986.

Pitt, R., and J. McLean. *Toronto Area Watershed Management Strategy Study: Humber River Pilot Watershed Project.* Ontario Ministry of the Environment, Toronto, Ontario, 1986.

Pitt, R. *Small Storm Urban Flow and Particulate Washoff Contributions to Outfall Discharges,* Ph.D. Dissertation, Civil and Environmental Engineering Department, University of Wisconsin, Madison, WI, November 1987.

Pitt, R. "Detention pond design for water quality improvement." *National ASCE Hydraulic Conference.* San Francisco, California. July 1993a.

Pitt, R. "The stormwater quality detention pond model (DETPOND)." *26th Annual Water Resources Conference.* University of Minnesota. October 1993b.

Pitt, R. and S.R. Durrans. *Drainage of Water from Pavement Structures.* Alabama Dept. of Transportation. 253 pgs. September 1995.

Pitt, R. "Sediment Control in Alabama." *41st Annual Transportation Conference.* Montgomery, AL. February 1998.

Pitt, R., J. Lantrip, R. Harrison, C. Henry, and D. Hue. *Infiltration through Disturbed Urban Soils and Compost-Amended Soil Effects on Runoff Quality and Quantity.* U.S. Environmental Protection Agency, Water Supply and Water Resources Division, National Risk Management Research Laboratory. EPA 600/R-00/016. Cincinnati, Ohio. 231 pgs. December 1999.

Pitt, R., M. Lilburn, S. Nix, S.R. Durrans, S. Burian, J. Voorhees, and J. Martinson *Guidance Manual for Integrated Wet Weather Flow (WWF) Collection and Treatment Systems for Newly Urbanized Areas (New WWF Systems).* U.S. Environmental Protection Agency. 612 pgs. 1999.

Pitt, R, S. Chen, and S. Clark. "Compacted urban soils effects on infiltration and bioretention stormwater control designs." *Global Solutions for Urban Drainage; 9IUCD.* CD-ROM Proceedings of the 9th International Urban Drainage Conference, edited by E.W. Strecker and W.C. Huber., Sept 8–13, 2002, Portland, OR. Sponsored by the ASCE, Reston, VA, and the International Water Association, London. 2002.

Poertner, H.G. *Practices in Detention of Urban Stormwater Runoff.* American Public Works Association, OWRR Contract No. 14-31-0001-3722, Chicago, 1974.

Ponce, V.M., *Engineering Hydrology,* Prentice Hall, 640 pages. 1989.

Powell, William (U.S. Dep of Agriculture, Portland, USA); Keller, Gordon R.; Brunette, Bruce. "Applications for geosynthetics on forest service low-volume roads." *Transportation Research Record,* v 2, n 1652, 1999, Proceedings of the 1999 7th International Conference on Low-Volume Roads, May 23–May 26 1999, Baton Rouge, LA, USA, p 113–120 Publisher: National Research Council, Washington, DC, USA ISSN: 0361-1981. 1999.

R.S. Means. *Site Work & Landscape Cost Data, 19th Edition.* R.S. Means Co., Kingston, MA. 2000.

Randall, C. W. "Stormwater detention ponds for water quality control." *Proceedings of the Conference on Stormwater Detention Facilities, Planning, Design, Operation, and Maintenance,* Henniker, New Hampshire, Edited by W. DeGroot, published by the American Society of Civil Engineers, New York, August 1982.

Raudkivi, A. J., and Tan, S. K. "Erosion of cohesive soils," *Journal of Hydraulic Research,* Vol 22, No. 4, pp 21X–233. 1984.

Ree, W.O. "Hydraulic Characteristics of Vegetation for Vegetated Waterways." *Agricultural Engineering,* April: 184–189. 1949.

Ree, W.O. and Palmer, V.J. *Flow in Channels Protected by Vegetative Linings.* Technical Bulletin No. 967, USDA: 1–22, 44–48. 1949.

Ree, W.O. *Friction Factors for Vegetation Covered, Light Slope Waterways.* Research Report No. 335, USDA: 1–34. 1960.

Reed, Lloyd A. *Suspended-Sediment Discharge in Five Streams Near Harrisburg, Pennsylvania, Before, During, and After Highway Construction.* Geological Survey Water-Supply Paper (United States), ISSN: 0083-1131, n 2072, 42p, 1980.

Reeves, Patrick J. (Boyle Engineering Corp, Fresno, CA, USA). "After the pipeline: An ugly scar or natural scene?" ASCE *Convention & Exposition,* 1982, Preprints—1982 ASCE National Convention., Las Vegas, NV, USA, 82-054, 13p Sponsored by: ASCE, New York, NY, USA Publisher: ASCE, New York, NY, USA, 1982.

Renard, K.G., Foster, G.R., Weesies, G.A., McCool, D.K, and Yoder, D.C., coordinators. *Predicting Soil Erosion by Water: A Guide to Conservation Planning with the Revised Soil Loss Equation (RUSLE).* U.S. Dept. of Agriculture, Agric. Handbook No. 703, 404 pp. 1987.

Renninger, F. A.; Woolief, R. S.; Bradshaw, Paul J.; Laughlin, George R.; Childers, Fred A.; Nichols, Frank P.; Wilkerson, John M.; Mosley, Albert S.; Breen, Florence L. Jr.; Canup, Lewis; Laws, Edward P. *Georgia Highway Conference, 23rd Annual,* Proceedings, 1974. Atlanta, GA, USA, 121p Publisher: Ga Inst of Technol, Atlanta. Engineering Sciences Data Unit, Data Items, Ga Highw Conf, 23rd Annu, Proc, Pap, Mar 4–5 1974.

Richter, Brian D. (Taranto, Stanton & Tagge Consulting Engineers, Fort Collins, CO, USA). "Erosion control improves community water supply." ISSN: 0033-3840. *Public Works,* v 118, n 1, p 80, 114, Jan, 1987.

Ristic, Ratko (Univ of Belgrade, Belgrade, Yugosl); Macan, Grigorije. "Impact of erosion control measures on runoff processes." *International Association of Hydrological Sciences,* n 245, 1997, Symp 6: Human Impact on Erosion and Sedimentation, Proceedings of the 1997 5th Scientific Assembly of the International Association of Hydrological Sciences, IAHS, Apr 23-May 3 1997, Rabat, Morocco, p 191–194 Publisher: IAHS Press, Wallingford, Engl, 1997 ISSN: 0144-7815. 1997.

Roberts, Brian C. (Water Resource Consultants, Inc, Fairfax, USA); "Developing erosion control plans for highway construction." National Research Council, Washington, DC, USA ISSN: 0361-1981.*Transportation Research Record,* n1471, p 38–40, Dec, 1994.

Robinson, Lloyd R. Jr. (Yule, Jordan & Assoc, Camp Hill, Pa). "Erosion control during pipeline construction." Kentucky University, *Office of Research and Engineering Services,* Bulletin, Jul 29–31 1974, p 175–182. Oct, 1974. Robison, Rita. "Engineering with fabric." Civil Engineering (New York), ISSN: 0009-7853, v 55, n 12, p 52–55, Dec, 1985.

Schimmenti, F.G. "Stormwater detention basins must control more than runoff." *American City and County,* December 1980.

Schueler, T. R. *New Guidebook Review Paper No. 2, Wet Pond BMP Design.* Metropolitan Washington Council of Governments, Washington, D.C., 1986.

Schueler, T.R., and J. Lugbill. *Performance of Current Sediment Control Measures at Construction Sites.* Metropolitan Washington Council of Governments, Sediment and Stormwater Administration of the Maryland Dept. of the Environment. 1990.

SCS (Soil Conservation Service). *Handbook for Channel Design for Soil and Water Conservation.* SCS-TP-61, rev. 1954.

SCS (now NRCS). *Urban Hydrology for Small Watersheds.* Technical Release 55, US Department of Agriculture, Soil Conservation Service. 91 pp. 1975.

SCS (Soil Conservation Service). *Design of Open Channels.* TR-25. 1977.

SCS. *Soil Survey of Jefferson County, Alabama.* USDA 1981—167-S/111. 1981.

SCS. *Computer Program for Project Formulation, Hydrology.* Technical Release Number 20 (TR-20). U.S. Dept. of Agriculture, 1982.

SCS. *Time of Concentration,* Hydrology Technical Note No. N4. United States Department of Agriculture, Soil Conservation Service, Northeast National Technical Center. 12 pp. 1986.

SCS. *Urban Hydrology for Small Watersheds.* Technical Release 55. U.S. Dept. of Agric., Soil Conservation Service. 156 pgs. 1986.

SEWRPC (Southeastern Wisconsin Planning Commission). *Sources of Water Pollution in Southeastern Wisconsin: 1975.* Technical Report No. 21. Waukesha, Wisconsin, 1978.

SEWRPC. *Costs of Urban Nonpoint Source Water Pollution Control Measures,* Technical Report #31. Waukesha, WI. June 1991.

Smith, W. G. "Water quality enhancement through stormwater detention." *Proceedings of the Conference on Stormwater Detention Facilities, Planning, Design, Operation, and Maintenance,* Henniker, New Hampshire, Edited by W. DeGroot, published by the American Society of Civil Engineers, New York, August 1982.

Smoot, J.T., T.D. Moore, J.H. Deatherage, and B.A. Tschantz. *Reducing Nonpoint Source Water Pollution by Preventing Soil Erosion and Controlling Sediment on Construction Sites.* Transportation Center of Tennessee. 1992.

Sprague, C.J. "Green Engineering; Design principles and applications using rolled erosion control products." *CE News On-Line,* March 1999, at: http://www.cenews.com/edenviroeng0399.html

Springer, G.S. *Erosion by Liquid Impact.* Scripta Pub. Co., John Wiley and Sons, New York, 1976.

Temple, D.M. "Flow Retardance of Submerged Grass Channel Linings." *Transactions of the ASAE:* 1300–1303. 1982.

Temple, D.M. *Erosionally Effective Soil Stress in Grass Lined Open Channels.* ASAE Paper No SWR84-102. 1984.

Terstriep, M.L., G.M. Bender, D.C. Noel. *Nationwide Urban Runoff Project, Champaign, Illinois: Evaluation of the Effectiveness of Municipal Street Sweeping in the Control of Urban Storm Runoff Pollution.* Contract No. 1-5-39600, U.S. Environmental Protection Agency, Illinois Environmental Protection Agency, and the State Water Survey Division, University of Illinois, December 1982.

Thronson, R.E. *Comparative Costs of Erosion and Sediment Control, Construction Activities.* U.S. Environmental Protection Agency. EPA430/9-73-016. Washington, D.C. 1973.

Tobiason, S., D. Jenkins, E. Molash, and S. Rush. "Polymer Use and Testing for Erosion and Sediment Control on Construction Sites: Recent experience in the Pacific Northwest". In: *Proceedings of Conference 31.* International Erosion Control Association. Palm Spring, CA, February 21–25, 2000, pp. 41-52. 2000.

U.S. Army Corps of Engineers (COE). *Engineering and Design: Channel Stability Assessment for Flood Control Projects.* EM 1110-2-1418. U.S. Army Engineer Waterways Experiment Station, Vicksburg, MS. 1994.

U.S. Army Corps of Engineers (COE). *Hydraulic Design of Flood Control Channels.* EM 1110-2-1601. U.S. Army Engineer Waterways Experiment Station, Vicksburg, MS. undated.

USBR (U.S. Bureau of Reclamation). *Design of Small Dams.* U.S Dept. of the Interior, A Water Resources Publication, 3rd edition, 1987.

USBR. *Hydraulic Design of Stilling Basins and Energy Dissipators,* Engineering monograph No. 25.

U.S. Housing Markets. *Busiest Markets, Multifamily Construction.* Meyers Real Estate Information, Inc. (http://www.housingusa.com/ushm/pmts/mulbuzz.html). Accessed October 25, 2000. 1999b.

U.S. Housing Markets. *Busiest Markets, Single-family Building.* Meyers Real Estate Information, Inc. (http://www.housingusa.com/ushm/pmts/mulbuzz.html). Accessed October 25, 2000. 1999a.

USDA, Soil Conservation Service. *National Engineering Handbook, Section 3 Sedimentation, Chapter 3 Erosion.* Undated (continuously updated).

USDA. *Handbook of Channel Design for Soil and Water Conservation.* Technical Paper TP-61, USDA: 1–34. 1954.

USDA (U.S. Department of Agriculture). *Design of Open Channels.* Technical Release No. 25, Soil Conservation Service, Washington, DC. 1977.

USDA. *Stability Design of Grass-Lined Open Channels.* Agricultural Handbook 667. 1987.

USDA. *Alabama Handbook for Erosion Control, Sediment Control, and Stormwater Management on Construction Sites and Urban Areas.* Alabama Soil and Water Conservation Committee. Montgomery, AL. July 1993.

USDA. 1997 *National Resources Inventory.* U.S. Department of Agriculture, National Resources Conservation Service, Washington, DC. (www.nrcs.usda.gov/technical/NRI/). 2000.

USDA. *Alabama Handbook for Erosion Control, Sediment Control, and Stormwater Management on Construction Sites and Urban Areas,* prepared by the Alabama Soil and Water Conservation Committee, Montgomery, AL. 2003.

USDA. *User's Guide (Technology), Revised Universal Soil Loss Equation Version 2 RUSLE2.* Agricultural Research Service, Washington, D.C. Draft, January 2003.

USEPA (U.S. Environmental Protection Agency). *Stormwater Management for Construction Activities: Developing Pollution Prevention Plans and Best Management Practices.* EPA 832-R-92-005. U.S. Environmental Protection Agency, Office of Water, Washington, DC. 1992.

USEPA. *Stormwater Management for Industrial Activities: Developing Pollution Prevention Plans and Best Management Practices.* EPA 832-R-92-006. U.S. Environmental Protection Agency, Office of Water, Washington, DC. 1992.

USEPA. *Guidance Specifying Management Measures for Sources of Nonpoint Pollution in Coastal Waters.* EPA 840-B-92-002. U.S. Environmental Protection Agency, Office of Water, Washington, DC. 1993.

USEPA. *Economic Analysis of the Final Phase II Stormwater Rule.* U.S. Environmental Protection Agency, Office of Wastewater Management. Washington, DC. 1999.

USEPA. *Development Document for Proposed Effluent Guidelines and Standards for the Construction and Development Category.* United States Environmental Protection Agency, Office of Water (4303T). EPA-821-R-02-007. Washington, DC 20460 (www.epa.gov/waterscience/guide/). June 2002.

USEPA. "Effluent guidelines for the construction and development industrial category." *Federal Register,* 40 CFR Parts 122 and 450. June 24, 2003.

USEPA. *Construction and Development Fact Sheet: Final Action—Selection of Non-Regulatory Option,* EPA 821-F-04-001; March 2004.

Vahanne, Pasi. "Urban erosion control in the Triangulo pilot study area Nacala city, Mozambique, Equatorial Africa. A multidisciplinary approach with specific emphasis on control of gully erosion by labour intensive construction work." 161, app. 13p ISSN: 1235-0621. *Publications—Technical Research Centre of Finland,* n329, 1997.

VDCHR (Virginia Dept. of Conservation and Historic Resources). *Erosion and Sediment Control Handbook.* Third Edition. Division of Soil and Water Conservation. Richmond, Virginia. 1992.

VDCR. *Erosion and Sediment Control Handbook.* Second Edition. Division of Soil and Water Conservation. Richmond, Virginia. 1980.

VDCR (Virginia Department of Conservation and Recreation). *Virginia Erosion & Sediment Control Field Manual.* Second Edition, Division of Soil and Water Conservation, Richmond, VA. 1995.

Versteeg, J. H. (DOT, Salem, Oreg); Earley, John J. "Erosion control at a delicate highway construction site." *Public Works,* ISSN: 0033-3840, v 113, n 3, p 82–83, Mar, 1982.

Vignoles M., and L. Herremans. "Metal pollution of sediments contained in runoff water in the Toulouse city." (in French). *NOVATECH 95, 2nd International Conference on Innovative Technologies in Urban Storm Drainage.* May 30–June 1, 1995. Lyon, France. pp. 611–614. Organized by Eurydice 92 and Graie. 1995.

Virginia. Erosion and Sediment Control Handbook. Second Edition. Division of Soil and Water Conservation. Virginia Dept. of Conservation and Historic Resources. Richmond, Virginia. 1980.

Walling, D.E. (Ed.); (University of Exeter, Department of Geography, Exeter, UK); Probst, J.L. (Ed.). Proceedings of the 1997 5th Scientific Assembly of the International Association of Hydrological Sciences, IAHS. IAHS Publication (International Association of Hydrological Sciences), n 245, 1997, Symp 6: *Human Impact on Erosion and Sedimentation, Proceedings of the 1997 5th Scientific Assembly of the International Association of Hydrological Sciences,* IAHS, Apr 23–May 3 1997, Rabat, Morocco, 311p Publisher: IAHS Press, Wallingford, Engl, 1997 ISSN: 0144-7815. 1997.

Watson, Chester C. (Colo. State Univ, Ft. Collins, USA); Combs, Phil G.; Abt, Steven R.; Raphelt, Nolan K. "Bioengineering stabilization of Harland Creek sites." *International Water Resources Engineering Conference—Proceedings,* v2, 1995, Proceedings of the 1st International Conference on Water Resources. Part 2 (of 2), Aug 14–18 1995, San Antonio, TX, USA, p 1839–1843 Sponsored by: ASCE Publisher: ASCE, New York, NY, USA, 1995.

Welle, P.I., Woodward, D. E., Fox Moody, H., *A Dimensionless Unit Hydrograph for the Delmarva Peninsula,* Paper No. 80-2013, ASAE 1980 Summer Meeting, 18 pp. 1980.

Whipple, W. and J. V. Hunter. "Settleability of urban runoff pollution." *Journal WPCF,* Vol. 53, No. 12, pg. 1726, 1981.

WI DNR (Wisconsin Department of Natural Resources). *Wisconsin Construction Site Best Management Practice Handbook,* revised. Madison, WI. 1994.

Wiegand, C., T. Schueler, W. Chittenden, and D. Jellick. "Comparative costs and cost effectiveness of urban best management practices." *Engineering Foundation Conference: Urban Runoff Quality—Impact and Quality Enhancement Technology,* Henniker, New Hampshire, edited by B. Urbonas and L. A. Roesner, published by the American Society of Civil Engineers, New York, June 1986.

Willett, G. Urban Erosion, in National Conference on Urban Erosion and Sediment Control; Institutions and Technology. EPA 905/9-80-002. U.S. Environmental Protection Agency, 1980.

Williams, D. T., and Julien, P. Y. "Examination of stage-discharge relationships of compound/composite channels," *Proceedings of the International Conference on Channel Flow and Catchment Runoff.* B.C. Yen, ed., University of Virginia, Charlottesville, 22–26 May 1989, pp 478–488. 1989.

Williams, David T. (WEST Consultants, Inc, Carlsbad, USA); Austin, Deron N. "PC-based design of channel protection using permanent geosynthetic reinforcement mattings." *International Water Resources Engineering Conference—Proceedings,* v1, 1995, Proceedings of the 1st International Conference on Water Resources. Part 1 (of 2), Aug 14–18 1995, San Antonio, TX, USA, p 678–682 Sponsored by: ASCE Publisher: ASCE, New York, NY, USA, 1995.

Wills, P. (Netlon Ltd). "Use of geosynthetics in the A5 Fazeley bypass." *Highways and Transportation,* v42, n7–8, July–Aug, 1995, 3p Publisher: Institution Of Highway and Transportation, London, Engl ISSN: 0265-6868. 1995.

Wischmeier, W.H. *A Rainfall Erosion Index for a Universal Soil Loss Equation.* Soil Science of America Proceedings. Vol. 23, pp.246–249. 1959.

Wischmeier, W.H. and D.D. Smith. *Rainfall Erosion Losses from Cropland East of the Rocky Mountains.* Agriculture Handbook No. 282, U.S. Dept. of Agric. Washington, D.C. 1965.

Wischmeier, W.H. and D.D. Smith. *Predicting Rainfall Erosion Losses—A Guide to Conservation Planning.* USDA. AH537. Washington, D.C. Dec 1978.

Ziegler, A.D. (Univ of Hawaii, Honolulu, USA); Sutherland, R.A., Tran, L.T. "Influence of rolled erosion control systems on temporal rainsplash response—a laboratory rainfall simulation experiment." Land Degradation & Development, ISSN: 1085-3278. v8, n2, p 139–157, Jun, 1997.

Index

About the Authors

ROBERT PITT currently is the Cudworth Professor of Urban Water Systems in the Department of Civil, Construction, and Environmental Engineering at the University of Alabama. He has been teaching for about 20 years. Prior to his teaching career, he was a senior engineer with consulting firms and government agencies for about 15 years. He also continues to consult and conduct research for businesses and governments. His current teaching and research interests include the fates and effects of hazardous materials, sediment losses and transport during land development activities, analytical methods to detect sources of contaminants in urban drainage systems, development of new analytical methods for the rapid and sensitive detection of toxicants, sources of pathogens in urban areas, modeling of urban infrastructure systems, development of stormwater control technologies, modifications of soil structure due to urbanization, and the integration of hydrology and water quality objectives in drainage design. He has published more than 100 publications, including journal articles, research reports, and several books. He received a Distinguished Service Citation from the University of Wisconsin, and has received several outstanding teacher and volunteer service awards. He is a registered Engineer and a Diplomate of the American Academy of Environmental Engineers, and a Diplomate of the American Academy of Water Resources Engineers. He has serves on numerous professional committees in the U.S. and abroad.

SHIRLEY E. CLARK currently is an Assistant Professor of Environmental Engineering at Penn State—Harrisburg (PSH). Prior to joining PSH, she was an Assistant Professor at UAB in the Department of Civil and Environmental Engineering. She also was a Post-Doctoral Research Engineer for the U.S. EPA's Urban Watershed Management Branch in Edison, New Jersey. Her bachelors' degree is in Chemical Engineering and her doctoral degree is in Environmental Health Engineering from UAB. She has worked in the environmental field since 1987 and in water resources and stormwater since 1992, including five years for an environmental engineering consulting firm. She is a registered professional engineer in the state of Alabama. She is a member of the Water Environment Federation, the American Water Resources Association and the Environment and Water Resources Institute of the American Society of Civil Engineers. She serves on various committees in WEF and EWRI. Her research interests include the interaction between environmental stressors and the health of waterbodies, and the use of structural and non-structural stormwater control technologies to improve the quality of runoff prior to its entering receiving waters. Much of her current research focuses on the removal of stormwater pollutants through natural soils and engineered filtration media.

DON W. LAKE, JR. graduated from the State University of New York at Buffalo in 1970 with a B.S. degree in Civil Engineering. He is a licensed professional engineer in New York, a Certified Professional in Erosion and Sediment Control (CPESC), and a Certified Professional in Stormwater Quality (CPSWQ). Don "retired" from the USDA-Natural Resources Conservation Service in 1995 after 27 years of service. He is the principal author of the "New York Standards and Specifications for Erosion and Sediment Control" and the "New York Contractors Erosion and Sediment Control Field Notebook". He is an adjunct Professor at Syracuse University and at The State University of New York College of Environmental Science and Forestry where he teaches a graduate engineering course on erosion control and stormwater design. He developed the Stormwater Short Course curriculum presented through Syracuse University and currently by SUNY College of Environmental Science & Forestry. He has presented over 560 training sessions in erosion and sediment control, hydrology, and stormwater management throughout the country and he is a past recipient of the International Erosion Control Association's Sustained Contributor Award. Don is currently employed by the Cortland County SWCD as Engineering Specialist to the New York State Soil and Water Conservation Committee assisting NYS-DEC with the implementation of their stormwater program.